AF167655

# Lecture Notes in Mechanical Engineering

**Lecture Notes in Mechanical Engineering (LNME)** publishes the latest developments in Mechanical Engineering—quickly, informally and with high quality. Original research or contributions reported in proceedings and post-proceedings represents the core of LNME. Volumes published in LNME embrace all aspects, subfields and new challenges of mechanical engineering.

To submit a proposal or request further information, please contact the Springer Editor of your location:

**Europe, USA, Africa:** Leontina Di Cecco at Leontina.dicecco@springer.com
**China:** Ella Zhang at ella.zhang@cn.springernature.com
**India, Rest of Asia, Australia, New Zealand:** Swati Meherishi at swati.meherishi@springer.com

Topics in the series include:

- Engineering Design
- Machinery and Machine Elements
- Mechanical Structures and Stress Analysis
- Automotive Engineering
- Engine Technology
- Aerospace Technology and Astronautics
- Nanotechnology and Microengineering
- Control, Robotics, Mechatronics
- MEMS
- Theoretical and Applied Mechanics
- Dynamical Systems, Control
- Fluid Mechanics
- Engineering Thermodynamics, Heat and Mass Transfer
- Manufacturing Engineering and Smart Manufacturing
- Precision Engineering, Instrumentation, Measurement
- Materials Engineering
- Tribology and Surface Technology

**Indexed by SCOPUS, EI Compendex, and INSPEC.**

All books published in the series are evaluated by Web of Science for the Conference Proceedings Citation Index (CPCI).

To submit a proposal for a monograph, please check our Springer Tracts in Mechanical Engineering at https://link.springer.com/bookseries/11693.

Holger Kohl · Günther Seliger · Franz Dietrich ·
Ha Thuc Vien
Editors

# Decarbonizing Value Chains

Proceedings of the 20th Global Conference on
Sustainable Manufacturing (GCSM 2024),
October 9–11, 2024
Ho Chi Minh City, Vietnam

 Springer

*Editors*
Holger Kohl
Institute of Machine Tools and Factory
Management
Technische Universität Berlin
Berlin, Germany

Franz Dietrich
Technische Universität Berlin
Berlin, Germany

Günther Seliger
Technische Universität Berlin
Berlin, Germany

Ha Thuc Vien
Vietnamese-German University
Bến Cát, Vietnam

ISSN 2195-4356    ISSN 2195-4364 (electronic)
Lecture Notes in Mechanical Engineering
ISBN 978-3-031-93890-0    ISBN 978-3-031-93891-7 (eBook)
https://doi.org/10.1007/978-3-031-93891-7

This work was supported by GPE Global Production Engineering UG.

This Springer imprint is published by the registered company Springer Nature Switzerland AG
The registered company address is: Gewerbestrasse 11, 6330 Cham, Switzerland

If disposing of this product, please recycle the paper.

# Preface

We are pleased to publish a collection of papers presented at the 20th Global Conference on Sustainable Manufacturing (GCSM), held October 9–11, 2024, in Binh Duong and Ho Chi Minh City, Vietnam. The conference is annually sponsored by the International Academy for Production Engineering (CIRP), committed to excellence in the creation of sustainable products and processes. GCSM 2024 was jointly organized by the Institute for Machine Tools and Factory Management (IWF) at the TU Berlin and Fraunhofer Institute for Production Systems and Design Technology (IPK) together with the Vietnamese-German University (VGU).

The GCSM 2024 brought together more than 140 attendees from over 25 countries providing a global forum of academics, researchers, and specialists from universities, research institutes and industry from across the globe, working on topics related to sustainable manufacturing. A unique feature of the GCSM conference series is its integration of industrial engineering perspectives, sustainable manufacturing applications in emerging and developing countries, as well as education and workforce development for advancing sustainable manufacturing. Plenary keynote speeches by experienced personalities from academics and industry, paper sessions presentations and workshops of student teams from different countries offered new insights and chances for exchange of ideas. The conference featured five keynote speakers who shared recent advances in cutting-edge research and industry practices; these prominent and internationally recognized experts elaborated how technologies in the product, process and system domains can enable sustainable manufacturing.

This volume documents almost 100 contributions presented at GCSM 2024 in 21 sessions held over three days. The proceedings are organized according to the conference program, classified into four broad categories as: Sustainable processes, sustainable manufacturing systems, sustainable manufacturing products, and crosscutting topics in sustainable manufacturing. The papers cover a variety of topics in these areas related to modelling and simulation of manufacturing processes, product design for sustainability, metrics for sustainability assessment, energy efficiency in manufacturing, strategies and business models, as well as education and workforce development for sustainable manufacturing. All papers published in these proceedings have been single-blind reviewed by experts from the international scientific committee, which are listed in the next section. These reviews resulted in an overall acceptance rate of 49,7%.

Since 2017, on the 15th GCSM in Haifa, Israel, the program is enhanced by so called student sessions. In different universities, moderated by respective teaching staff, groups of 4 to 10 students perform a respective project development to be finally presented in the GCSM. In the following chapter, papers resulting from student projects presented in two student sessions during the GCSM 2024 are presented. By presenting and discussing specific projects, we enable young researchers to face the challenges of sustainable manufacturing and share their perspectives on how to address them.

Holger Kohl
Franz Dietrich
Günther Seliger
Ha Thuc Vien

# Organization

## International Organizing Committee

| | |
|---|---|
| Holger Kohl | International Chairman |
| Franz Dietrich | Local Chairman |
| Günther Seliger | Founding Chairman |
| Valentin Eingartner | Organizational Team |
| Maxim Mintchev | Organizational Team |
| Nwaoyibo Donatus Junior | Organizational Team |

## National Organizing Committee

| | |
|---|---|
| Ha Thuc Vien | National Chairman |
| Thomas G. Aulig | National Steering Committee |
| Nguyen Quoc Hung | National Academic Advisor |
| Nguyen Hong Vi | Local Organizational Team |
| La Vinh Trung | Local Organizational Team |

## International Scientific Committee

| | |
|---|---|
| Samy Abu Salih | Braude, Karmiel |
| Feri Afrinaldi | Andalas University |
| Tülin Aktin | İstanbul Kültür Üniversitesi |
| Elita Amrina | Andalas University |
| Fazel Ansari | Technische Universität Wien |
| Afif Aqel | Birzeit University |
| Fazleena Badurdeen | University of Kentucky |
| Peter Ball | University of York |
| Thomas Bergs | Fraunhofer IPT |
| Erhan Budak | Sabancı Üniversitesi |
| Diego Cafaro | Universidad Nacional del Litoral |
| Giampaolo Campana | Università di Bologna |
| Felipe Cerdas | HaW Würzburg-Schweinfurt |
| Dannisa Chalfoun | Instituto Tecnológico de Buenos Aires |
| Jun-Ki Choi | University of Dayton |
| Pedro Filipe Cunha | Escola Superior de Tecnologia de Setubal |

| | |
|---|---|
| Roy Damgrave | University of Twente |
| Ilesanmi Daniyan | Tshwane university of Technology, Pretoria, South Africa |
| Michele Dassisti | Politecnico di Bari |
| Leopoldo De Bernardez | Instituto Tecnológico de Buenos Aires |
| Wim Dewulf | Katholieke Universiteit Leuven |
| Franz Dietrich | Technische Universität Berlin |
| Welf-Guntram Drossel | Fraunhofer IWU |
| Mohamed El Mansori | Arts et Métiers ParisTech (ENSAM) |
| Kleber F. Espôsto | Universidade de São Paulo |
| Jörg Franke | Friedrich-Alexander-Universität Erlangen-Nürnberg |
| Luigi Galantucci | Politecnico Di Bari |
| Peihua Gu | Tianjin University |
| Loice K Gudukeya | University of Johannesburg |
| Yuebin Guo | Rutgers University |
| Karl Haapala | Oregon State University |
| Steffen Ihlenfeldt | Technische Universität Dresden |
| Roland Jochem | Technische Universität Berlin |
| Min Junying | Tongji University |
| Takayuki Kataoka | Kindai University |
| Yusuf Kaynak | Marmara University |
| Karel Kellens | Katholieke Universiteit Leuven |
| Yuki Kinoshita | Kindai University |
| Thomas Knothe | Fraunhofer IPK |
| Holger Kohl | Technische Universität Berlin |
| Philip Koshy | McMaster University |
| Daniel Martinez Krahmer | Instituto Nacional de Tecnología Industrial |
| Jörg Krüger | Technische Universität Berlin |
| Asela Kulatunga | University of Peradeniya |
| Hoew Pueh Lee | National University of Singapore |
| Laura V. Lerman | Federal University of Rio Grande do Sul |
| Yongrong Li | SKF Lubrication Systems Germany GmbH |
| Barbara Linke | University of California, Davis |
| Makinde Olasumbo | Tshwane University of Technology |
| Sotiris Makris | LMS - University of Patras |
| Khaled Medini | École des Mines de Saint-Étienne |
| Laszlo Monostori | Hungarian Academy of Sciences |
| Úrsula María Montoya Rojo | Instituto Tecnológico de Buenos Aires |
| Sandra Naomi Morioka | Federal University of Paraíba |
| Anke Müller | Hochschule Hof |
| Bernd Muschard | Technische Universität Berlin |

| | |
|---|---|
| Ryosuke Nakajima | The University of Electro-Communications |
| He Ning | Nanjing University of Aeronautics and Astronautics |
| Soh Khim Ong | National University of Singapore |
| Tiaan Oosthuizen | Institute for Technology & Society |
| Emanuele Pagone | Cranfield University |
| Sina Peukert | Karlsruher Institut für Technologie |
| Julian Polte | Fraunhofer IPK |
| Franci Pušavec | University of Ljubljana |
| Paul Refalo | University of Malta |
| Konstantinos Salonitis | University of Cranfield |
| Sebastian Schlund | Technische Universität Wien |
| Matthias Schmidt | Universität Lüneburg |
| Günther Seliger | Technische Universität Berlin |
| Semih Severengiz | Hochschule Bochum |
| Vennan Sibanda | National University of Science and Technology |
| Rainer Stark | Technische Universität Berlin |
| Frank Straube | Technische Universität Berlin |
| Henning Strauß | FH Kiel |
| Nicole Stricker | Hochschule Aalen |
| John W. Sutherland | Purdue University |
| Shozo Takata | Waseda University |
| Sebastian Thiede | University of Twente |
| Olga Timoteo | Universidad Cayetano Heredia |
| Thomas Volling | Technische Universität Berlin |
| Rok Vrabič | University of Ljubljana |
| Rafi Wertheim | Technische Universität Chemnitz |
| Sudhir Yadav | Pandit Deendayal Petroleum University |
| Shuho Yamada | Toyama Prefectural University |
| Tetsuo Yamada | University of Electro-Communications |
| Hitomi Yamaguchi | University of Florida |
| Shun Yang | Karlsruhe Institute of Technology |
| Chris Yuan | Case Western University |
| Michael Zaeh | Technische Universität München |
| Zhang Weimin | Tongji University |

## Local Scientific Committee

| | |
|---|---|
| Nguyen Quoc Hung | Vietnamese German University |
| Thai Truyen Dai Chan | Vietnamese German University |
| Vo Bich Hien | Vietnamese German University |

# Contents

## Data Analytics

**Sustainability by Design**

**Product Design**

**Production of Energy Systems**

**Energy Distribution**

**Technical Processes**

## Maintenance, Repair and Lifecycle Extension

## Disassembly and Remanufacturing

## Reuse and Recycling

## Value Creation Networks and Supply Chain

**Factory Planning and Production Management**

**Student Session**

# Business Models, Policy and Compliance

# Navigating the Circular Economy: A Practical Step-by-Step Approach for Business Model Transformation

Benjamin Gellert[1], Henry Nicolai Buxmann[1]([✉]), Ronald Orth[1], and Holger Kohl[2]

[1] Fraunhofer Institute for Production Systems and Design Technology IPK, Pascalstr. 8-9, 10587 Berlin, Germany
`henry.nicolai.buxmann@ipk.fraunhofer.de`
[2] Technische Universität Berlin, Pascalstr. 8-9, 10587 Berlin, Germany

**Abstract.** The transition to a Circular Economy (CE) is paramount in fostering sustainable business practices. Recent research highlights a surge in circular inno-vations, showcasing strong technological solutions, e.g. regarding recyclability or durability, and innovative disruptive ideas for novel business models, such as sharing or repairing concepts. These concepts are often utilized by newly founded enterprises. However, existing Small and Medium Enterprises (SMEs) encounter challenges in adjusting to these new developments and are in need for support to evolve their established business model towards circularity. This paper addresses this gap by proposing a hands-on process for SMEs aiming to embrace more cir-cular practices. The process comprises several steps, starting with an assessment and analysis of the existing business model in alignment with circular economy principles, followed by an exploration of circular economy potentials within the business context. Furthermore, the steps outline specific circular business model patterns and tailored measures resulting in individual action items for the circular transformation of the enterprise' business model. Through practical implemen-tation, this approach equips enterprises with a structured process for evaluating, refining, and implementing circular business practices.

**Keywords:** circular economy · business models · business model development · transition · sustainability

## 1  Introduction

According to the Circularity Gap Report 2024 the human population has consumed over half a trillion tons of materials in the last six years, almost equaling the total consumption throughout the entire 20th century [1]. To conquer the challenges posed by this substantial overconsumption, such as increasing greenhouse gas emissions, and to stay within the planetary boundaries a shift towards a Circular Economy is unavoidable. To achieve this shift, enterprises need to integrate circularity into their current business models [2, 3].

Hereby, SMEs can play an important role. SMEs drive the productivity growth of economies, account for about 70% employment and generate about 50% to 60% value

H. Kohl et al. (Eds.): GCSM 2024, LNME, pp. 3–10, 2025.
https://doi.org/10.1007/978-3-031-93891-7_1

added in OECD member states [4, 5]. The adoption of CE principles allows SMEs to create new revenue streams, for example by utilizing recycled materials or extending the product life, while also fostering stronger connections with suppliers and customers, ultimately enhancing sustainability and supply chain efficiency [6]. Hence, SMEs can foster the transition to a global CE and limit overconsumption through the integration of circular principles into their business model.

However, most existing tools designed for the development of sustainable business models lack a specific focus on facilitating the transition to CE practices [7]. Some tools, such as those developed by Mendoza et al. (2017) and Heyes et al. (2018), do focus on CE but are primarily tailored to support corporate decision-making in the transition towards a circular economy on the product level and for service-oriented companies respectively [8, 9]. Further, the approach developed by Lewandowski (2016) focuses only on developing a theoretical framework to assess business models in the context of circularity [3].

Hence, there remains a significant gap in providing effective and practical tools or methodologies that enable SMEs to integrate circular economy principles into their business models. Therefore, this paper focuses on providing a practical step-by-step approach for SMEs to transform their whole business model within the principles of the circular economy.

## 2  Concepts

### 2.1  Circular Economy

A Circular Economy is an economic system aiming to eliminate waste and extend the use of finite resources through principles such as reuse, repair, refurbishment, or recycling [10]. It contrasts the linear economy, which predominantly relies on non-renewable fossil energies and primary raw materials in production processes, ultimately resulting in the use and disposal of materials as non-recyclable waste [11]. The shift from a linear to a circular economy needs changes at multiple levels: the macro level (cities, regions, nations, and beyond), the meso level (local ecosystems and industrial networks), and the micro level (products, companies, consumers).

For transitioning the micro level, the focus cannot be limited solely on technological solutions related to products or processes. To fully embrace sustainable and circular business practices, enterprises also need to transform their business model towards circularity.

### 2.2  Business Model Development

A business model fundamentally outlines how an enterprise operates and structures its business activities [12]. It emphasizes the creation, retention, and distribution of a value proposition, detailing the necessary business and market elements. These elements include key resources, activities, partners, and costs on the business side, as well as customer relations, customer segments, revenues, and distribution channels on the market side [13]. Typically, a business model is defined for an existing enterprise or one in development, making it unique to that organization and often impossible to transfer to other entities.

Nevertheless, there are ways how specific parts of business models can be classified as general concepts and then transferred to others. These concepts are called business model patterns. They represent general methods of how business models function and act as global design templates. Through skillful design and adaptation, they can be effectively implemented in any organization [14].

### 2.3 Circular Business Models Patterns

Business model patterns tackle in general all kinds of business fields, in context of CE the focus is on circular business model patterns (CBMPs).

CBMPs represent templates by which, after implementation, enterprises create, deliver, and capture value by decelerating, closing, or narrowing resource cycles. As a result, the transformed business model becomes more sustainable and circular [15]. According to Lacy et al. (2020) and Försterling et al. (2023) five overarching types of circular business models can be achieved through the implementation of CBMPs: Circular Inputs, Resource Recovery, Life-Cycle Extension, Product Service Systems, and Collaboration Platforms [10, 16].

Next to the positive environmental impact, CBMPs also influence business performance financially. By using these patterns, enterprises can achieve cost savings, generate new revenue streams, enhance their reputation, and reduce their dependency on resources and the volatility of resource prices [17].

## 3 Background

The development of the step-by-step process to transform the business models of SMEs towards CE took place within the publicly founded BioFusion 4.0 project. The project aimed to explore the interconnections between the principles of biological transformation and their interactions with production, services, and labor work [18].

Part of the project was the objective to apply biological transformation on business model level through the definition and application of biologically transformed business model patterns. A broad literature review was conducted, resulting in the creation of a circular business model pattern catalog that classifies over 40 CBMPs regarding their influence on business model elements as well as their benefits for the enterprise and barriers during the implementation [19]. The catalog serves as the foundation for deriving actions within the step-by-step process of circular business model development.

Given the requirement of practical application in BioFusion 4.0, an approach had to be found for applying the circular business model patterns in real business environments to enhance an SME's functioning business model with circular principles (Fig. 1).

The procedure is inspired by the integrated strategy development approach, which involves evaluating the current status for optimization, defining a future status, and identifying actions to achieve the future state [20]. It was tested with one of the project's application partners, a Berlin-based SME in the semiconductor and solar industry.

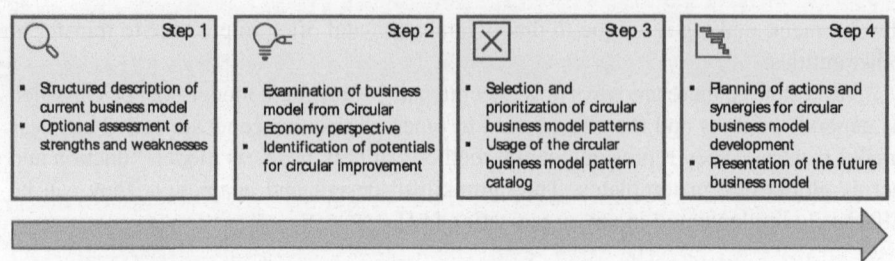

**Fig. 1.** Step-by-step approach for business model transformation

The approach follows four steps and was implemented through four workshops over a one-year period from 2023 to 2024. Each step was applied within a workshop, highlighting the close collaboration needed to achieve the transformation of a business model. Through its hands-on design the approach can be replicated with other SMEs.

## 4   Circular Business Model Development Process

In the following chapters each individual step of the approach for the circular business model development and transformation is described.

### 4.1   Assessment of Business Model

The first step is a general assessment of the business model. This analysis is particularly relevant for SMEs, as their business model is usually not redefined during ongoing operations. However, a detailed examination using the Business Model Canvas (BMC) by Osterwalder & Pigneur (2013) sharpens the focus on the company's own business model and quickly reveals areas of action for the further development of the company [13].

To support the definition of the company's business model each business model element (key resources, activities, partners, costs, customer relations, customer segments, revenues, competition) is discussed together with the representatives of the company. As dealing with the competition usually provides valuable input for an enterprise, it was taken as an element instead of distribution channels. The result is a display of the current business model (Fig. 2).

### 4.2   Examination of Business Model from Circular Economy Perspective

Once the business model has been generally defined, the focus in step two is on the CE. At this point it is helpful to create a general understanding of the topic and start with an introduction to the concepts of CE, and circular business models. Subsequently, the goal is to determine the extent to which the current business model already incorporates CE principles and where there is still potential for improvement. A questionnaire related to the BMC elements is used, which covers questions from "Could our product portfolio become more circular" to "Can costs be reduced by introducing circular economy

**Fig. 2.** Business Model Canvas (according to [13])

principles". In particular, for the element of key resource it was discussed if there are measures in place to reduce the resources used during the production process.

Now, the status quo is defined. To subsequently identify the areas within the business model that hold the greatest potential for applying circular principles, the elements are ranked. Each business model element is assessed by the potential of circular improvement and the feasibility of implementing CE principles.

The outcome is a ranking of the elements based on the assessment in the two dimensions. Elements with high potential for improvement and strong feasibility serve as starting points, for which individually suitable circular business model patterns are selected afterwards.

### 4.3 Selection and Prioritization of Circular Business Model Patterns

The selection of patterns (step 3), that match the individual business model situation of the company, is based on the information given in the CBMP catalog [19]. There, each pattern has a qualitative evaluation of its influence on the BMC elements. Patterns, which influence BMC elements positively, that have a high ranking coming from the step before, can potentially be implemented in the enterprise.

For instance, the element *Revenues* has been identified with high potential for circular improvement and good feasibility of implementing circular principles. Among the patterns clustered to positively influence *Revenues* are Take-Back-Management (potential revenues from sale of returned machines), Product Leasing (continuous revenues through leasing contract) or Upgrading (revenues through upgrading services at customer).

The selection of patterns is enriched with those that match the noted improvement suggestions. The result is a longlist of CBMPs the company can implement for the circular development of their business model.

After a detailed presentation of the patterns, a joint discussion is held between the representatives of the company to evaluate which patterns are relevant for them. This process concludes with a prioritization that establishes the implementation sequence of the CBMPs. Hereby, the patterns are clustered into priorities ranging from 1 (highest priority) to 3 (lowest priority). The priority ranking is based on the knowledge and estimations of the representatives of the company. The prioritization then forms the basis for a detailed scheduling and organizational planning of actions for the continued development of the business model.

### 4.4   Derivation of Actions for Circular Business Model Development

SMEs value the derivation of operational actions as it translates high-level business model strategies into practical, actionable steps. Therefore, in a final step, specific actions are defined based on the prioritized CBMPs. Multiple actions can result from one pattern, which are then implemented by the company. For better practicability, each action is given a detailed action plan, exemplary shown in Table 1.

**Table 1.**  Action Plan

| Action title: Introduction of machine deposit | |
| --- | --- |
| Definition | Introducing a machine deposit for customers to pay when purchasing the machine, repayment of the deposit when old machine is returned |
| Objective | Machine deposit as a component of sales contracts<br>Increased take-back rate |
| Impact | Revenues<br>Customer Relations |
| Procedure | 1. Definition of the amount of the machine deposit (e.g. € 5.000)<br>2. Clarification of legal and technical framework conditions<br>3. Customer communication about the machine deposit<br>4. Evaluation of customer responses (acceptance/rejection)<br>5. Implementation of the machine deposit in financial accounting |
| Start/End | 04/2024 – 04/2029 |

The action plan can be enriched with a visual roadmap, where all actions are mapped according to their start and end date and key milestones are defined. The primary objective is a presentation of the company's future business model, accomplished through the adoption of the prioritized CBMPs.

## 5   Conclusion and Outlook

The transition towards Circular Economy is essential to combat the significant material consumption and its environmental impacts. SMEs, in particular, play an important role in this transition [4, 5] and are in need of a practical tool to addresses the challenges faced adapting to CE [7].

This paper proposes a practical, step-by-step approach for transforming traditional business models into circular ones. The approach guides SMEs through assessing their current business model, exploring CE potentials, and implementing specific circular business model patterns tailored to their needs. By integrating assessment, exploration, and the implementation of circular business model patterns, companies are given a comprehensive pathway to evaluate, refine, and adopt circular practices. The proposed process provides a replicable framework that helps SMEs to transition from a linear economy to sustainable and circular business practices.

Future research and application should refine this approach, incorporating feedback from its practical implementation, and exploring additional circular business model patterns. Further studies could also investigate the long-term economic and environmental impacts of such transformation. By fostering collaboration between enterprises, researchers, and policymakers, the transition towards a global circular economy can be accelerated, ensuring a sustainable future for all.

# References

1. Circular Economy Foundation. Circularity Gap Report 2024. https://www.circularity-gap.world/2024. Accessed 29 May 2024
2. Asgari, A., Asgari, R.: How circular economy transforms business models in a transition towards circular ecosystem: the barriers and incentives. Sustain. Prod. Consum. **28**, 566–579 (2021)
3. Lewandowski, M.: Designing the business models for circular economy—towards the conceptual framework. Sustainability **8**, 43 (2016)
4. Owalla, B., Gherhes, C., Vorley, T., Brooks, C.: Mapping SME productivity research: a systematic review of empirical evidence and future research agenda. Small Bus. Econ. **58**, 1285–1307 (2022). https://doi.org/10.1007/s11187-021-00450-3
5. OECD. Enhancing the contributions of SMES in a global and digitalised economy. https://one.oecd.org/document/C/MIN(2017)8/en/pdf. Accessed 04 June 2024
6. Solkvint, L., Madsen, J.L.: What can companies do to adapt their business models toward a circular economy? In: Abrantes, B.F., Madsen, J.L. (eds.) Essentials on Dynamic Capabilities for a Contemporary World, pp. 87–105: Springer, Cham (2023)
7. Silvestre, W.J., Fonseca, A., Morioka, S.N.: Strategic sustainability integration: merging management tools to support business model decisions. Bus Strat Env **31**, 2052–2067 (2022). https://doi.org/10.1002/bse.3007
8. Mendoza, J.M.F., Sharmina, M., Gallego-Schmid, A., Heyes, G., Azapagic, A.: Integrating backcasting and eco-design for the circular economy: the BECE framework. J. Industr. Ecol. (J. Industr. Ecol.) **21**, 526–544 (2017). https://doi.org/10.1111/jiec.12590
9. Heyes, G., Sharmina, M., Mendoza, J.M.F., Gallego-Schmid, A., Azapagic, A.: Developing and implementing circular economy business models in service-oriented technology companies. J. Clean. Prod. **177**, 621–632 (2018)
10. Lacy, P., Long, J., Spindler, W.: The Circular Economy Handbook: Realizing the Circular Advantage, 1st edn. Palgrave Macmillan, London (2020)
11. Ellen MacArthur Foundation. Completing the picture: How the circular economy tackles climate change. https://www.ellenmacarthurfoundation.org/completing-the-picture. Accessed 30 May 2024
12. Osterwalder, A., Pigneur, Y., Tucci, C.L.: Clarifying business models: origins, present, and future of the concept. CAIS (2005). https://doi.org/10.17705/1CAIS.01601

13. Osterwalder, A., Pigneur, Y.: Business Model Generation: A Handbook for Visionaries, Game Changers, and Challengers. Wiley, New York (2013)
14. Gassmann, O, Frankenberger, K, Choudury, M.: Geschäftsmodelle entwickeln: 55 innovative Konzepte mit dem St. Galler Business Model Navigator, 2nd edn. Hanser, München (2017)
15. Lüdeke-Freund, F., Gold, S., Bocken, N.M.P.: A review and typology of circular economy business model patterns. J. Industr. Ecol. **23**, 36–61 (2019). https://doi.org/10.1111/jiec.12763
16. Försterling, G., Orth, R., Gellert, B.: Transition to a circular economy in Europe through new business models: barriers, drivers, and policy making. Sustainability **15**, 8212 (2023). https://doi.org/10.3390/su15108212
17. Salvador, R., Barros, M.V, Da Luz, L.M, Piekarski, C.M, de Francisco, A.C.: Circular business models: Current aspects that influence implementation and unaddressed subjects. J. Clean. Prod. (2020)
18. Lindow, K., Riedelsheimer, T.: Integration of Biological Principles in Industry 4.0. https://www.ipk.fraunhofer.de/content/dam/ipk/IPK_Hauptseite/dokumente/flyer/vpe-flyer-biofusion-web.pdf. Accessed 05 June 2024
19. Gellert, B., Buxmann, H.N, Orth, R.: Geschäftsmodellentwicklung für Kreislaufwirtschaft und Nachhaltigkeit. Zeitschrift für wirtschaftlichen Fabrikbetrieb **119**, 65–69 (2024). https://doi.org/10.1515/zwf-2024-1011
20. Will, M.: Integrierte Strategieentwicklung. In: Kohl, H., Mertins, K., Seidel, H. (eds.) Wissensmanagement im Mittelstand: Grundlagen - Lösungen - Praxisbeispiele, 2nd edn., pp. 87–104. Springer, Heidelberg (2016)

# Network Capabilities for the Strategic Design of Sustainable Production Networks

Alexander Schollemann[(✉)], Michael Riesener, and Seth Schmitz

Laboratory for Machine Tools and Production Engineering (WZL), RWTH Aachen University, Campus-Boulevard 30, 52074 Aachen, Germany
a.schollemann@wzl.rwth-aachen.de

**Abstract.** The increasing sustainability influences caused by internal and external stakeholders in the environment of manufacturing companies require a strategy adjustment taking into account the holistic sustainability dimensions. Within the strategic design, manufacturing in global production networks represents a decisive lever for realizing competitive advantages. At present, primarily economic advantages are exploited in terms of the capabilities developed through the network, although the globally distributed production sites can also have an impact on the environmental and social sustainability dimension. In order to address this deficit and to provide decision support for strategic network design, this paper aims to develop environmental and social network capabilities. First, the analytical development of network capabilities is realized based on sustainable production indicators (SPIs) derived from the literature in order to take into account the strategy hierarchies of global manufacturing companies. Along the allocation decisions of global production networks, design measures are thus derived for each SPI, which are consequently consolidated into environmental and social network capabilities. Finally, these are empirically tested using a Delphi study with experts from the field of global production networks in order to present a consensus on the set of environmental and social network capabilities.

**Keywords:** Global Production Networks · Network Capabilities · Sustainability

## 1 Introduction

As a consequence of globalization, value creation in global production networks has become a key competitive advantage [1]. Due to the performance frontier of a single production site [2], production in a network ensures long-term competitiveness through the realization of network capabilities [3]. The performance of the production network comprises advantages through access and cooperation in the network [4]. Therefore, a network strategy has to be developed that pursues the business goals [5]. Due to changes in the competitive environment of global manufacturing companies as a result of the influence of sustainability aspects, an adaptation of the strategic design is required [6].

Increasing stakeholder demands for environmental and social sustainability are forcing companies to adapt, leading to new production strategy objectives covering the holistic sustainability dimensions [7]. Achieving these holistic sustainability goals depends

© The Author(s) 2025
H. Kohl et al. (Eds.): GCSM 2024, LNME, pp. 11–19, 2025.
https://doi.org/10.1007/978-3-031-93891-7_2

on considering and focusing on the right levers, of which network capabilities are a key component [8]. To date, network capabilities have only been considered with an economic focus, with the emphasis on cost reduction and direct competitive pressure [5]. Consequently, the environmental and social dimension has not yet been integrated into strategic design although competitive advantages are possible [3].

To address this deficit, the aim of this paper is to develop network capabilities for the strategic design of sustainable production networks. The composition of the paper is therefore structured as follows. Section 2 introduces the concept of network capabilities and highlights the significance of sustainability in this context. Section 3 discusses the state of research. Based on these findings, the elaboration of network capabilities along the social and environmental sustainability dimensions is realized in a three-step approach in Sect. 4. Finally, the research results are summarized in Sect. 5.

## 2  Relevance of Network Capabilities and Sustainability

The network strategy of a global manufacturing company is essential for increasing competitiveness and is based on globally distributed value chains that connect different production sites with each other [1, 9]. The network strategy is defined as an overarching plan for the globally distributed production of goods with the objective of meeting the requirements of international customers [8]. In analogy to research in the field of operations management, e.g. as presented in [10], the topic of network strategy is divided into two main areas: Process of network strategy development and content of the network strategy in the form of network capabilities [11].

Existing literature emphasizes two critical relationships in the strategy hierarchy: linkage with production strategy and integration with business strategy [8, 9]. The global production network plays a key role in implementing the production strategy, which in turn contributes to achieving the objectives of the business strategy as a functional strategy [11]. This creates the need for synchronization between production and network strategy. Consequently, the network strategy is to be aligned with the differentiation factors and thus with the production targets using network capabilities [5].

The emergence of the scientific discourse on network capabilities centered on the site selection of production facilities to realize competitive advantages. According to [12], three network capabilities are developed through site selection: Access to low cost production input factors, markets and knowledge. [13] expand these network capabilities based on an empirical analysis to include access to suppliers, competitors, socio-political factors and cheap energy. In addition to these competitive advantages based on geographical site selection, [4] consider extended network capabilities, which are realized through collaboration within the network. These consist of efficiency advantages through economies of scale or scope, manufacturing mobility and the ability to learn. Network capabilities and thus competitive advantages are realized through network design mechanisms [14]. The associated allocation decisions of the network configuration are composed of the site, technology, product and volume allocation [15].

Equivalent to the existing focus on the economic dimension in determining the differentiating factors of the production strategy [16], existing network capabilities do not integrate the environmental and social sustainability dimension. However, globally distributed value creation is associated with a high number of transports and corresponding emissions [17]. On the other hand, the environmental impact of production sites is significantly influenced by location-dependent factors such as climatic conditions or access to renewable energy sources [15, 18]. To ensure competitiveness, it is therefore necessary for manufacturing companies to integrate the holistic dimensions of sustainability into their site and network design [3]. Both the identification of environmental and social network capabilities and the linking of these to sustainable production targets have not been addressed by research and are therefore been so far deficient in practice.

## 3 State of Research

In the research field of global production networks, the literature distinguishes between two streams: Approaches that focus on site selection and approaches that focus on network capabilities through network interaction [9]. Both research areas are differentiated in the following with regard to economically driven approaches and approaches in the context of the holistic sustainability dimensions.

### 3.1 Approaches from a Site Selection Perspective

In order to realize the targeted production performance, four structural decision categories exist, of which the location decision for the long-term determination of geographical value creation represents one [19]. This research field has been covered in the literature for over 50 years, with the consideration of location factors being extended to international site selection in the 1980s and 1990s [20]. Initial research on site decisions describe the hierarchy of strategic, tactical and operational criteria and also emphasize the need to integrate qualitative criteria [21]. The internationalization of value creation activities is changing these decision factors, with approaches from this field of research underlining the increasing dynamism and changing requirements of global competition in the site selection process [22]. As a result, various site selection factors exist in business practice, which contribute to increasing competitiveness, particularly from an economic perspective [20]. In the context of global production networks, various site selection factors are thus strategically combined in order to increase the performance of the network [12]. Empirical studies have identified strategic factors that determine the design of global production networks through site selection [13].

Traditional site selection strategies presented have typically prioritized economic objectives. However, over recent decades, growing concerns about environmental degradation and social sustainability have reshaped the decision-making landscape [7]. Thus, according to [18], the increasing importance of environmental considerations in the selection of manufacturing facility locations is underlined. Due to this significant influence of site selection, [23] calls for integrating manufacturing and sustainability strategy perspectives when making decisions regarding the global manufacturing footprint, whereby conflicting objectives and the contribution to competitiveness must always be taken into account.

## 3.2  Approaches from a Network Capability Perspective

Further approaches have been developed which, in addition to access to advantageous local conditions, also consider capabilities through network coordination [9]. As an initial approach [24] is considered, which conducts an empirical study to conclude that globally operating companies strive for strategic network-specific competitive advantages. In contrast, [4] adopt a practice-oriented approach to investigate the configuration and capabilities of global production networks by using knowledge-based research work to examine case studies on the basis of a conceptual model.

Further approaches examine the link between network capabilities and production and business objectives. In this context, [25] analytically derives links between the differentiation factors of the production strategy and the network capabilities, which are, however, only partially empirically validated. This qualitative derivation reflects the focus of conceptual discussions in the existing literature. To address this deficit, [26] investigate the enhancement of financial performance through the enhancement of their production networks quantitatively on the basis of survey data and interviews. Particularly in highly competitive markets, network capabilities make a significant contribution to differentiation from competitors [26].

In the context of sustainability, there are only a few approaches that address network capabilities. A methodology for assessing the environmental sustainability of production network configurations was developed by [11], however, the strategic design and alignment of network capabilities is not presented. The approach of [27] adds four additional network capabilities to the economic network capabilities described above by adding resource savings, environmental sustainability as well as internal and external social sustainability. However, a structured derivation and concretization of these environmental and social capabilities is not specified.

# 4  Elaboration of Environmental and Social Network Capabilities

The analysis of the state of research demonstrates the existing deficit in terms of environmental and social network capabilities. In addition to the structured derivation, linking the network capabilities with the strategy hierarchies of global manufacturing companies is a requirement in order to ensure the contribution of holistic network capabilities to the realization of production performance and thus competitiveness.

Therefore, a three-step approach for the elaboration of environmental and social network capabilities is applied. In the first step, environmental and social production targets are derived in order to link the network capabilities with the production strategy and at the same time derive the possible dimensions of environmental and social network capabilities. Based on this, potential environmental and social network capabilities along the allocation decisions of the network configuration are analytically derived in the second step. In the final third step, a Delphi study is carried out in order to establish an empirical consensus on the potential environmental and social network capabilities with industry experts.

## 4.1 Environmental and Social Production Targets

In the context of sustainability, indicators to measure production targets are referred to as sustainable production indicators (SPIs) [28]. Taking into account existing literature research on SPIs [11, 28–30], in which between 239 and 2112 indicators from 6 to 77 sources are consolidated, 120 indicators are analyzed with duplicates removed. In addition to differentiation along the sustainability dimensions, these are categorized in 15 targets and the description of targets are derived defined (see Table 1).

**Table 1.** Indicators consolidated from the literature research and categorized production targets along the environmental and social sustainability dimension.

Environmental sustainability

| Target | | Indicators |
|---|---|---|
| Emissions | | CO$_2$ emissions, ... |
| Energy | Energy efficiency | Energy consumption, ... |
| | Renewable energy | Rate of renewables, ... |
| Resource-conserving production | Water | Water consumption, ... |
| | Biodiversity | Land use, ... |
| | Material | Amount of virgin material, ... |
| | Waste | Hazardous waste volume, ... |
| Circularity | Recycling | Share of recycled waste, ... |
| | Remanufacturing | Quantity of remanufactured products, ... |
| | Value-adding lifespan extension | Amount of added lifespan of the product, ... |

Social sustainability

| Target | | Indicators |
|---|---|---|
| Fair business practices | Human rights | Corruption index, ... |
| | Diversity and equality | Percentage of women in management positions, ... |
| Employer attractiveness | Occupational safety | Number of occupational accidents, ... |
| | Working conditions | Absentee rate, ... |
| | Training and education | Number of training courses, ... |

## 4.2 Analytical Derivation of Environmental and Social Network Capabilities

In order to ensure consistency with existing economic network capabilities, two constituent characteristics of network capabilities were derived (see Sect. 2). Due to the required alignment of the network strategy with the production strategy, the first constituent characteristic is the required improvement of the production targets. In order to maintain the scope of global production networks, the second constitutive feature is the required development capability resulting exclusively from the allocation decisions of the network configuration. Taking into account the clustered environmental and social production targets, these are compared in the first step with the allocation decisions of the network configuration in order to analytically identify measures. The subsequent consolidation of these measures in the second step results in the derivation of potential environmental and social network capabilities, which are then examined in the final step with regard to their support for the production targets (see Fig. 1).

## 4.3 Delphi Study to Assess Environmental and Social Network Capabilities

The analytically developed network capabilities have to be discussed from a practical perspective in order to reach a consensus on the factors to be considered in the strategy process. The Delphi study is a method for determining a reliable consensus on future

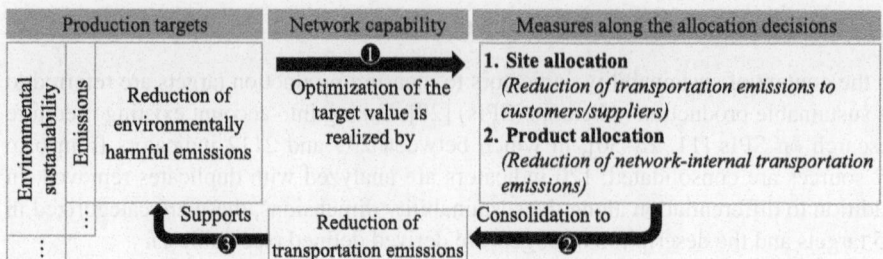

**Fig. 1.** Approach of the analytical derivation of environmental and social network capabilities.

development [31]. As part of the initial contact, 19 participants were informed about the procedure of the Delphi study, the intended objective and the benefits for the participants in a manner similar to the procedure described in [32]. The panel was composed of managers in the area of responsibility of global production networks, who were asked about environmental and social network capabilities in the first phase independently of the previous analytical derivation. After comparing the answers with the results of the analytical derivation, these were confirmed and no further potential network capabilities were added. In the course of the first evaluation round, the potential network capabilities were assessed using a nominal scale. The threshold value of more than 67% agreement is used for nominal scales to assess whether a consensus was reached [33]. As a result of the first evaluation phase, consensus was reached for seven out of nine potential network capabilities (see Table 2).

**Table 2.** Resulting environmental and social network capabilities.

| Round | Potential network capability | Number and share of agreements | Number and share of disagreements | Consensus |
|---|---|---|---|---|
| 1st round | Reduction of transportation emissions (E) | 19 (100%) | 0 (0%) | Yes |
| | Access to infrastructure for resource-conserving production (E) | 13 (68%) | 6 (32%) | Yes |
| | Environmentally optimized network allocations (E) | 16 (84%) | 3 (16%) | Yes |
| | Exploitation of economies of scope for resource-efficient production (E) | 18 (95%) | 1 (5%) | Yes |

*(continued)*

**Table 2.** (*continued*)

| Round | Potential network capability | Number and share of agreements | Number and share of disagreements | Consensus |
|---|---|---|---|---|
| | Access to countries for fair business practices (S) | 14 (74%) | 5 (26%) | Yes |
| | Safety-critical site allocation (S) | 15 (79%) | 4 (21%) | Yes |
| | Building opportunities for knowledge exchange (S) | 18 (95%) | 1 (5%) | Yes |
| | Access to renewable energy sources (E) | 12 (63%) | 7 (37%) | **No** |
| | Establishment of production sites for circular economy (E) | 11 (58%) | 8 (42%) | **No** |
| 2nd round | Access to renewable energy sources (E) | 10 (83%) | 2 (17%) | Yes |
| | Establishment of production sites for circular economy (E) | 10 (83%) | 2 (17%) | Yes |

(E): Network capability of the environmental sustainability dimension; (S): Network capability of the social sustainability dimension

All potential network capabilities for which a consensus was reached represent network capabilities in the opinion of the experts. However, no consensus was reached for the two potential network capabilities of access to renewable energy sources and the establishment of production facilities for the circular economy. The divergence of opinion on these can be attributed to the individual company situation and specifics of the business sectors. Despite the consideration of 11 different business sectors for a diverse consensus-forming process, individual opinions are therefore integrated. Consequently, a second evaluation round is required. Based on the anonymized comments of the experts, the second evaluation round was carried out by 12 remaining industry experts after seven participants dropped out. By allowing respondents to modify their responses, a consensus was reached on the remaining potential network capabilities (see Table 2), with all nine network capabilities being confirmed by the industry experts.

## 5   Conclusion and Outlook

Global production networks represent a central lever in business practice to increase the competitiveness of global manufacturing companies. In addition to the economic dimension, environmental and social network capabilities also contribute to this. The structured development of environmental and social network capabilities was carried out in order to detail this contribution to competitiveness. To ensure the strategy hierarchies, potential six environmental and three social network capabilities were developed on the basis of environmental and social production targets and then empirically tested using a Delphi study.

Further, by integrating the holistic sustainability dimensions into the strategy process of global production networks, the interrelationship between the network capabilities and differentiation factors of the production strategy have to be examined. In this way, the interaction between the sustainability dimensions should also be assessed, as the economic production targets can also be supported by the environmental and social network capabilities.

**Acknowledgements.** Funded by the Deutsche Forschungsgemeinschaft (DFG, German Research Foundation) under Germany's Excellence Strategy – EXC-2023 Internet of Production – 390621612.

# References

1. Miltenburg, J.: Manufacturing Strategy - How to Formulate and Implement a Winning Plan, 2nd edn. Productivity Press, New York (2005)
2. Schmenner, R., Swink, M.: On theory in operations management. J. Oper. Manage. **17**(1), 97–113 (1998)
3. Lanza, G., et al.: Global production networks. CIRP Ann. **68**(2), 823–841 (2019)
4. Shi, Y., Gregory, M.: International manufacturing networks-to develop global competitive capabilities. J. Oper. Manag. **16**(2–3), 195–214 (1998)
5. Friedli, T., Mundt, A., Thomas, S.: Strategic Management of Global Manufacturing Networks. Springer, Cham (2014)
6. Cagliano, R., Acur, N., Boer, H.: Patterns of change in manufacturing strategy configurations. Int. J. Oper. Prod. Manag. **25**(7), 701–718 (2005)
7. Burritt, R., Christ, K., Rammal, H., Schaltegger, S.: Multinational enterprise strategies for addressing sustainability: the need for consolidation. J. Bus. Ethics **164**(2), 389–410 (2020)
8. McGrath, M., Bequillard, R.: International manufacturing strategies and infrastructural considerations in the electronics industry. In: Ferdows, K. (ed.) Managing International Manufacturing, pp. 23–40. North Holland, Amsterdam (1989)
9. Colotla, I., Shi, Y., Gregory, M.: Operation and performance of international manufacturing networks. Int. J. Oper. Prod. Manag. **23**(10), 1184–1206 (2003)
10. Leong, G., Snyder, D., Ward, P.: Research in the process and content of manufacturing strategy. Omega **18**(2), 109–122 (1990)
11. Ferdows, K.: Mapping international factory networks. In: Ferdows, K. (ed.) Managing International Manufacturing, pp. 23–40. North Holland, Amsterdam (1989)
12. Miedler, P., Friedli, T.: Deriving a network strategy. In: Friedli, T., Lanza, G., Remling, D. (eds.) Global Manufacturing Management, pp. 87–100. Springer, Cham (2021)
13. Vereecke, A., van Dierdonck, R.: The strategic role of the plant: testing Ferdows's model. Int. J. Oper. Prod. Manag. **22**(5), 492–514 (2002)
14. Srai, J., Christodoulou, P.: Capturing Value from Global Networks. Cambridge (2014)
15. Welsing, M.: Bewertung der ökologischen Nachhaltigkeit von Produktionsnetzwerkkonfigurationen. Apprimus, Aachen (2023)

16. Schuh, G., Schmitz, S., Schlosser, T., Schollemann, A., Pfau, F.: Holistic differentiation factors for the strategic design of sustainable production networks. Procedia CIRP **120**, 446–449 (2023)
17. Citil, M.: The relationship between participation in global production networks and environmental pollution originating from the logistics industry. Doğuş Üniversitesi Dergisi **23**(2), 151–167 (2022)
18. Sihag, N., Leiden, A., Bhakar, V., Thiede, S., Sangwan, K., Herrmann, C.: The influence of manufacturing plant site selection on environmental impact of machining processes. Procedia CIRP **80**, 186–191 (2019)
19. Hayes, R.H., Wheelwright, S.C.: Restoring Our Competitive Edge. Wiley, New York (1984)
20. MacCarthy, B.L., Atthirawong, W.: Factors affecting location decisions in international operations – a Delphi study. Int. J. Oper. Prod. Manag. **23**(7), 794–818 (2003)
21. Jungthirapanich, C., Benjamin, C.: A knowledge-based decision support system for locating a manufacturing facility. IIE Trans. **27**, 789–799 (1995)
22. Badri, M.A., Davis, D.L., Davis, D.: Decision support models for the location of firms in industrial sites. Int. J. Oper. Prod. Manag. **15**(1), 50–62 (1995)
23. Chen, L., Olhager, J., Tang, O.: Manufacturing facility location and sustainability: a literature review and research agenda. Int. J. Prod. Econ. **149**(1), 154–163 (2014)
24. Bartlett, C.A., Ghoshal, S.: Managing Across Borders - The Transnational Solution, 2nd edn. Harvard Businss School Press, Boston (1998)
25. Mengel, S.: The Alignment of International Manufacturing Networks. St. Gallen (2017)
26. Flaeschner, O., Wenking, M., Netland, T., Friedli, T.: When should global manufacturers invest in production network upgrades? Int. J. Oper. Prod. Manag. **41**(1), 21–53 (2020)
27. Steier, G., Heusch, A., Voigt, J., Benfer, M., Lanza, G.: Entscheidungsfaktoren der Produktionsnetzwerkkonfiguration. WT Werkstatttechnik **113**(10), 457–462 (2023)
28. Cagno, E., Negri, M., Neri, A., Giambone, M.: One framework to rule them all. Sustain. Prod. Consum. **35**, 55–71 (2023)
29. Mengistu, A., Panizzolo, R.: Analysis of indicators used for measuring industrial sustainability. Environ. Dev. Sustain. **25**(5), 1979–2005 (2023)
30. Winroth, M., Almström, P., Andersson, C.: Sustainable production indicators at factory level. J. Manuf. Technol. Manag. **27**(6), 842–873 (2016)
31. Linstone, H., Turoff, M.: The Delphi Method - Techniques and Applications. Addison-Wesley, Reading (2002)
32. Johnson, J.: A ten-year Delphi forecast in the electronics industry. Ind. Mark. Manage. **5**(1), 45–55 (1976)
33. von der Gracht, H.: Consensus measurement in Delphi studies. Technol. Forecast. Soc. Chang. **79**(8), 1525–1536 (2012)

# Bridging the Gap: Enhancing Industry 4.0 Readiness in Low-Resource Microenterprises Through Open Source Technologies

Mohammed Omer[1]([✉]), Melina Kaiser[2], Sonja Buxbaum-Conradi[1], Manuel Moritz[1], and Tobias Redlich[1]

[1] Helmut Schmidt University, Hamburg, Germany
omerm@hsu-hh.de
[2] University of Hamburg, Hamburg, Germany

**Abstract.** Industry 4.0 (I4.0) enables companies to become more efficient and competitive, yet research often centers on larger businesses in industrialized nations. The potential of microenterprises (MEs), which absorb up to 70% of the total labor force in low and middle income countries, remains understudied. This paper investigates the readiness of MEs in low-resource settings to adopt I4.0 and explores the necessary conditions for successful adoption. Through qualitative interviews and observations conducted in manufacturing MEs in Oman, the study analyzes existing production processes, the types of machines currently used, and the workforce's skills and experiences to identify deficiencies potentially impeding the path towards digitalization and I4.0. Key findings reveal that many businesses lacking basic automated tools and having limited capital, effectively operate at Industry 2.0 levels. The research highlights the role of cost-effective open source machine tools and enhancing manual machines with open source 'Internet of Things' technologies. These solutions aim to enhance readiness and transform basic production into intelligent, interconnected systems by lowering the threshold to access. By addressing the prerequisites for I4.0 adoption, this paper contributes to advancing industrial capabilities in the global south and offers practical insights into empowering MEs through open-source technologies.

**Keywords:** Industry 4.0 · Technology Adoption · Open Source Technologies · Resource Constrained Settings · Microenterprises

## 1 Introduction

Existing research on Industry 4.0 (I4.0) readiness typically focuses on large firms or broadly on micro, small, and medium enterprises (MSMEs), often emphasizing organizational measures for adoption. Most research on MSMEs generalizes findings from small and medium enterprises to all MSMEs, overlooking the specific needs of microenterprises (MEs), which differ in terms of size, capital, and revenue. Given their limited resources and capabilities, MEs face unique challenges on shopfloor and technological levels, necessitating a more nuanced approach to I4.0 adoption. Additionally, the often

H. Kohl et al. (Eds.): GCSM 2024, LNME, pp. 20–27, 2025.
https://doi.org/10.1007/978-3-031-93891-7_3

informal nature of ME operations makes them less likely to receive government support, highlighting the need for grassroots approaches. This paper examines the conditions for adopting I4.0 technologies in migrant-run manufacturing MEs in Oman, focusing on their capabilities, literacy levels, experience, and social and infrastructure conditions. The paper particularly focuses on open-source technologies, which have recently garnered increasing importance in I4.0 development [1]. Due to recent advances in hardware and improved access to knowledge and resources, open-source machine tools (OSMT) in particular emerge as potential enablers for providing bottom-of-the-pyramid users with advanced innovations. The paper proposes a tailored strategy for I4.0 adoption, emphasizing open-source hardware, including OSMT, and open-source software.

## 2 Methodology

Methodologically, this paper is based on two rounds of a total of 15 qualitative semi-structured interviews and observations conducted among manufacturing MEs located in the Wadi al Kabir industrial cluster in Muscat, Oman (see Table 1).

**Table 1.** Key characteristics of study participants.

| | |
|---|---|
| Total number of participants | 15 microenterprises (5 owners of MEs and 10 workers of MEs) |
| Industry sectors | Manufacturing (metalwork and carpentry), small-scale production |
| Typical ME products | Predominantly low value added and simple products like railings, gates, piping, car shades, car parts and furniture |
| Number of employees | 2–5 employees |
| Geographical distribution | Wadi al Kabir industrial cluster (Muscat, Oman) |
| Demographics | Male, South Asian |

The data collection in the MEs focused on two areas of investigation, namely technological and entrepreneurial competencies, an approach loosely based on a framework to assess I4.0 readiness proposed by Naudé et al. (2019) [2]. In the two domains, interview questions were centered around the main topics laid out in Table 2.

The interviews were conducted and transcribed in Bangla, translated to English, and analyzed according to principles of qualitative content analysis [3]. The analysis aimed to identify the strengths and weaknesses of the MEs, focusing on the digitalization of production and conditions that might affect the adoption of I4.0 technologies. Based on these results, recommendations were made on how to enhance the readiness for adoption of certain I4.0 technologies in the context of resource-constrained MEs.

**Table 2.** Interview themes

| Technical Competencies | Entrepreneurial competencies |
| --- | --- |
| Types of machines used and automation levels | Skill sets and experiences of interviewees |
| Need to automate production | Educational background |
| Machine procurement sources | Overall digital and technological literacy |
| Current production practices (in-house vs. outsourced work, single production steps vs. comprehensive production from start to finish) | Familiarity with Computer Aided Design (CAD) and Computer Aided Manufacturing (CAM), and Computer Numerical Control (CNC) technologies |

## 3   Background

### 3.1   Industry 4.0

Industry 4.0 (I4.0) is an intelligent manufacturing ecosystem enabled by the augmentation of production systems with sensors and digital networks to make production smart and interconnected [4]. In essence, a pathway towards I4.0 requires the presence of the production infrastructure (hardware layer) representative of Industry 3.0 (I3.0) technologies as a prerequisite. By greatly increasing flexibility, speed, and production capacity while reducing errors, costs, and improving product quality, the benefits of I4.0 technologies for large corporations with extensive machinery and personnel are evident, leading to early research focusing on these larger entities with the necessary infrastructure, resources, and skilled workforce [4]. Recently, attention has turned to MSMEs [2, 5, 6]. However, much of the research has focused on the Global North, and findings may not be directly applicable to MSMEs in the Global South, particularly MEs, which often operate in highly resource-constrained environments [7].

### 3.2   Microenterprises

MEs contribute significantly to the GDP in developing countries and absorb the majority of the labor force, accounting for over 70 percent in many regions, including the Middle East [8, 9]. Therefore, MEs can play a decisive role in the economic development of low and lower-middle income countries, by offering labor opportunities for significant portions of the population, and they help to mitigate economic inequalities by contributing to broader economic development [5, 10]. Despite their potential, many MEs remain vulnerable, often operating within the informal economy due to the scarcity of formal job opportunities in many countries of the Global South [8]. These MEs frequently lack formal recognition, protection, and support from national governments. As a result, they face significant resource constraints, including limited financial capital, inadequate production inputs, insufficient space, and, by extension, a lack of time and knowledge. Manufacturing MEs in low and lower-middle-income countries heavily rely on manual labor due to limited resources for automation and insufficient financial capacity to invest in digital manufacturing tools, essential for Industry 4.0 adoption [11]. Additionally, much of their workforce lacks the expertise to operate these technologies, further disadvantaging them and sidelining them in economic development.

### 3.3 Open Source Technologies

In the last decade, the democratization of the internet and increased accessibility to microcontrollers and automation have given rise to open source machine tools (OSMT) such as CNC mills and laser cutters. OSMT are characterized by publicly available designs, build instructions, bills of materials, assembly guidelines, and software, allowing anyone to build, use, modify, sell, and share them freely [12]. Open source technologies can democratize access to advanced production tools, making digital fabrication affordable compared to proprietary options. This shift challenges the concentration of know-how in the Global North, fosters innovation across diverse communities, and provides affordable machine tools to marginalized individuals. OSMT promotes sustainable, inclusive production in developing countries, enabling industrial leapfrogging.

## 4  Results from Interviews and Observations

The interviews and observations conducted in several MEs in Oman revealed some variations in technological competencies, particularly in the types of machines utilized. Those MEs that focused on fabrication activities primarily relied on basic power tools such as drill presses and disc cutters, with a few possessing manual machine tools such as manual lathes and milling machines. In contrast, more machine-oriented MEs tended to have a greater number of machine tools, yet no CNC machine tools were observed. Interviewees cited prohibitive costs, insufficient workshop space and lack of knowhow among others as reasons their shops lacked CNC machine tools. Regarding needs for automation of production processes, most interviewees expressed that CNC machines might benefit their work. However, some interviewees did not see the necessity for modern CNC equipment or automation. One interviewee also stated that "CNC cannot machine certain patterns […]. An expert is always needed, the machine is only a helper", indicating both a lack of understanding of the potential of CNC capabilities and a certain risk aversion regarding new technologies. None of the interviewees were familiar with CAD, CAM, and CNC technologies, highlighting a significant lack of digital and technological literacy.

The MEs predominantly purchased their machines used, many being over 30 years in use. Most businesses have been operating for over 20 years with minimal upgrades to their equipment, indicating a lack of technological advancement. Interviewees cited "cost and shipping and taxes" as reasons. Production processes were largely conducted in-house, with businesses typically completing all production steps themselves, from start to finish. This meant that many tasks that could have been more efficiently handled with CNC technology were done manually.

These production processes were sustained by distinct entrepreneurial competencies with several prevalent skill sets and experiences among the interviewees. Most had extensive practical experience in their fields of work with a mean of ten years. Overall, a notable lack of higher education was witnessed, with some having had vocational training, but general literacy remained low. One interviewee said that "we don't have degrees or diplomas, but we have years of experience", showing confidence that their practical skills offset the lack of formal education. Apart from that, it was observed that a younger generation of microentrepreneurs is emerging, often the children of interviewees. Unlike their parents, these young entrepreneurs are more educated, often at the university level, and possess higher general and digital literacy. Their proficiency with digital devices and the internet potentially gives them an advantage in accessing and replicating open source technologies compared to prior generations.

## 5 Discussion: Open Source Technologies to Bridge the Gap

The observed MEs in Oman predominantly operate at levels equivalent to Industry 2.0 (I2.0), characterized by minimal automation. Most shops relied heavily on manual machine tools and basic power tools, with almost no automated technologies. To advance these MEs, the initial step should be to introduce basic automation at the shopfloor level, specifically through CNC-capable machine tools and robotic automation to reach levels of I3.0. These technologies could serve as essential steppingstones towards adopting I4.0 technologies. In the process of basic machine tool automation, OSMT could become a key enabler by making the blueprints for manufacturing, assembling, and commissioning of CNC machines tools freely available. Open-source hardware and software allow for seamless adaptation and integration, offering flexibility in combining diverse technologies while avoiding the limitations of proprietary systems.

As most I4.0 technologies are catered towards large companies with extensive assets, they are less relevant for small MEs with limited resources. However, with OSMT providing affordable production infrastructure, MEs can still benefit from key technologies to enhance existing production capabilities with open-source technologies. Table 3 provides guidelines for adopting six key I4.0 technologies for microenterprises, with open-source alternatives referenced, based on Naudé (2018) [13].

Other key I4.0 technologies, such as Big Data Analytics, Cybersecurity, Cloud Computing, and Digital Twins often cited in I4.0 literature for MSMEs [4] have been excluded from the recommendations. The significantly smaller scale of MEs does not justify the need for these technologies, which are more appropriate for large corporations with more employees and extensive machinery. In increasing I4.0 readiness among vulnerable MEs, it is necessary to differentiate two forms of businesses. The first type usually produces one-off custom projects which is often the case for roadside manufacturing MEs [11]. This type of ME could benefit primarily from OSMT-based shopfloor-level automation that directly increase production efficiency. Also, newer technologies like 3D printing could help expand capabilities and enable business diversification through rapid prototyping and testing of new technologies. The second type involves small-scale batch production of simple goods like utensils and agricultural tools. Such MEs, which possess more machine tools already, would benefit more from open source-based IoT implementation for machine monitoring and predictive maintenance, along with shopfloor-level automation.

Considering the inherent complexity of designing and implementing these technologies, which require advanced skills in various STEM fields over the domains of

**Table 3.** Guidelines for I4.0 adoption for microenterprises

| I4.0 technologies | Proposed open source solutions | Possible benefits for MEs |
| --- | --- | --- |
| 3D printing | Fused Deposition Modeling (FDM) 3D printing using self-built 3D printers from blueprints or kits [14, 15] | Secondary manufacturing processes, such as prototyping, creating customized jigs, tools, molds, and forms for casting in metal fabrication and carpentry |
| Industrial Internet of Things (IoT) | Open source microcontrollers such as Arduino, ESP32, or Raspberry pi to retrofit analogue or CNC machines with sensors [16] | Process automation, real-time monitoring of machine tools and manufacturing processes, predictive maintenance, energy monitoring etc |
| Robotics | Open source build-it-yourself 6-axis robot arms such as project by Annin Robotics [17] | Advanced automation of complex processes like welding for repetitive and mundane welding jobs |
| Artificial intelligence (AI) | AI-based applications [18] | Possible intelligent applications in all key I4.0 technology domains |
| Digital platforms | Cloud platforms like Wikifactory [19] | Expand market access by increasing customer base through online presence, connect MEs to each other and to larger enterprises |
| Interfacing of things | Augmented reality applications [20, 21] | Training and upskilling workers for I4.0, visualizing objects such as products, machine maintenance |

hardware, electronics, and software, implementing such solutions for the typical MEs end user would be a daunting task. Even with open source solutions and comprehensive instructional manuals, extensive prior experience, dexterity, skills, and know how are necessary. Moreover, such solutions need customization to fit the specific equipment and processes of MEs. Microentrepreneurs with higher digital, general and technological skills such as engineers, university educated users, or experienced makers will be most likely to be able to implement some of these open source solutions independently. However, for most end users with low digital literacy levels, new diffusion mechanisms will have to be explored. Furthermore, while OSMT can be produced at a fraction of the cost of proprietary machines, the building process needs a certain time and workspace which can pose difficulty for small MEs with limited human and spatial resources. The fact that most OSMT designs originate in the Global North can also affect their replicability in resource-constraint settings. One approach to meet these challenges is the collaboration between universities and local MEs to address local problems using open source hardware and software.

## 6   Conclusion

To create a more equitable society where technology justice is a reality, it is essential to make advanced technologies that enhance productivity and efficiency accessible to all. Most MEs in resource-constrained settings still operate at I2.0 levels of automation. Before they can adopt I4.0 technologies, they need affordable alternatives to build the required automation infrastructure, which is often the most expensive capital investment for firms. Well-documented OSMT projects offer viable solutions. With OSMT providing the physical infrastructure to progress to I3.0, affordable microelectronics and sensors, combined with open-source software, would make it easier to create cloud-based monitoring and automation solutions, enabling even resource-limited MEs to access I4.0 technologies.

However, despite the availability of open source hardware and software, these technologies are not easily accessible due to the high levels of digital literacy and STEM knowledge required. The diffusion of these technologies faces significant barriers because target users in resource constrained settings often have low levels of digital and general literacy. To overcome these challenges, strategies to increase replicability among users in low-resource settings as well as new business models for technology diffusion need to be developed. Studies on adoption criteria and research on ME-specific challenges to OSMT use are further necessary. These strategies are crucial for bridging the gap and ensuring that MEs can fully harness the potential of I4.0, promoting a more inclusive technological advancement.

## References

1. ten Hompel, M., et al.: Open Source als Innovationstreiber für Industrie 4.0. Munich, Germany (2022)
2. Naudé, W., Surdej, A., Cameron, M.: Ready for industry 4.0? The case of central and eastern Europe. In: Dastbaz, M., Cochrane, P. (eds.) Industry 4.0 and Engineering for a Sustainable Future, pp. 153–175. Springer, Cham (2019)
3. Kuckartz, U., Rädiker, S.: Qualitative Inhaltsanalyse: Methoden, Praxis, Computerunterstützung, 5th edn., 274 p. Beltz Juventa, Weinheim, Basel (2022)
4. Rüßmann, M., et al.: Industry 4.0: the future of productivity and growth in manufacturing industries (2015). https://www.bcg.com/publications/2015/engineered_products_proj ect_business_industry_4_future_productivity_growth_manufacturing_industries
5. Büchi, G., Cugno, M., Castagnoli, R.: Smart factory performance and Industry 4.0. Technol. Forecast. Soc. Change **150**, 119790 (2020)
6. Tortora, A.M., Maria, A., Di Valentina, P., Iannone, R., Pianese, C.: A survey study on Industry 4.0 readiness level of Italian small and medium enterprises. Procedia Comput. Sci. **180**, 744–753 (2021)
7. Seitz, H.: Microenterprises in developing countries: is there growth potential? DIW Roundup: Politik im Fokus (2017)

8.  Radic, D.: Small matters: global evidence on the contribution to employment by the self-employed, micro-enterprises and SMEs. International Labour Organization, Geneva, 48 p. (2019)

9.  Suhaili, M., Sugiharsono, S.: Role of MSME in absorbing labor and contribution to GDP. **18**(3), 301–315 (2019)

10. Achkar, S.E.: Micro and small enterprises: engines of job creation. ILOSTAT (2023). https://ilostat.ilo.org/micro-and-small-enterprises-engines-of-job-creation/. Accessed 23 Dec 2023

11. Opiyo, E., Jagtap, S., Keshwani, S.: Conceptual design in informal metalworking microenterprises of Tanzania. Sustainability **15**(2), 986 (2023)

12. Omer, M., Kaiser, M., Moritz, M., Buxbaum-Conradi, S., Redlich, T., Wulfsberg, J.P.: Democratizing manufacturing – conceptualizing the potential of open source machine tools as drivers of sustainable industrial development in resource constrained contexts, Hannover, 256 p. (2022)

13. Naudé, W.: Brilliant technologies and brave entrepreneurs: a new narrative for African manufacturing. IZA Discussion Papers (11941) (2018)

14. Bonneau, V., Yi, H.: The disruptive nature of 3D printing: offering new opportunities for verticals. Digital Transformation Monitor. European Commission (2017)

15. Petersen, E., Pearce, J.: Emergence of home manufacturing in the developed world: return on investment for open-source 3-D printers. Technologies **5**(1), 7 (2017)

16. Sri Sudha Vijay Keshav, K., Lourenço, D.M., Kumar, A.A., Plapper, P.: Retrofitting of legacy machines in the context of Industrial Internet of Things (IIoT). Procedia Comput. Sci. **200**, 62–70 (2022)

17. Annin, C.: Annin Robotics (2024). https://www.anninrobotics.com/. Accessed 23 June 2024

18. de Simone, V., Di Pasquale, V., Miranda, S.: An overview on the use of AI/ML in manufacturing MSMEs: solved issues, limits, and challenges. Procedia Comput. Sci. **217**, 1820–1829 (2023)

19. Wf, J.A.: Wikifactory | The all-in-one product development platform. Wordpress WikiFactory (2022)

20. Kühn-Kauffeldt, M., Böttcher, J.: Open source augmented reality applications for small manufacturing businesses. Augmented Reality Virtual Reality 243–251 (2020)

21. Reljić, V., Milenković, I., Dudić, S., Šulc, J., Bajči, B.: Augmented reality applications in Industry 4.0 environment. Appl. Sci. **11**(12), 5592 (2021)

# The Impact of Digitalization and Sustainability on the Resilience of Small and Medium Enterprises

Elita Amrina[✉] ⓘ, Insannul Kamil ⓘ, and Rini Syahfitri

Department of Industrial Engineering, Universitas Andalas, Padang, West Sumatra, Indonesia
elita@eng.unand.ac.id

**Abstract.** Digitalization has become a crucial focus in today's world, permeating various facets of life and business. Its benefits have notably amplified since the Covid-19 pandemic, influencing both corporate resilience and sustainability. Despite the significant role SMEs play in contributing to Gross Regional Domestic Product (GRDP), many exhibit limited understanding of how digitalization and sustainability can bolster their resilience. This study explores the effects of digitalization and sustainability on SME resilience using Partial Least Squares-Structural Equation Modeling (PLS-SEM) with SmartPLS software. The findings reveal that both digitalization and sustainability have a significant positive impact on SME resilience. Additionally, digitalization is shown to play a critical role in enhancing SME sustainability. The research aims to provide actionable insights for SMEs, guiding them in integrating digitalization and sustainability strategies to strengthen their resilience.

**Keywords:** Digitalization · Resilience · SMEs · Sustainability

## 1 Introduction

Digitalization has emerged as a significant subject in recent times. Its ubiquitous presence in various aspects of an individual's life is undoubted. The realm of digitalization encompasses various domains such as digital learning, strategy, process, and entrepreneurship [1]. Particularly in the realm of business operations, digitization stands out as a key driver that improves organizational efficiency. Moreover, digitalization also empowers businesses to innovate, customize and deliver cutting-edge products and services that can compete effectively in the market [2].

The outbreak of the Covid-19 pandemic is seen as a catalyst that drives digitalization in businesses. Post-pandemic, the benefits of digitalization are increasingly being felt, especially in terms of strengthening resilience. Through digitalization, companies can enhance their resilience by honing their capacity to deal with disruptions while maintaining service excellence [3, 4]. Resilience denotes an organization's proficiency in absorbing and responding to specific circumstances, thus enabling it to navigate transformative processes for sustainable survival [3]. This includes the ability to properly

H. Kohl et al. (Eds.): GCSM 2024, LNME, pp. 28–36, 2025.
https://doi.org/10.1007/978-3-031-93891-7_4

manage and mitigate risks, thereby minimizing the likelihood, severity, and recovery duration of disruptions [5].

The realm of digitalization is not only related to corporate resilience, but also relates to sustainability. Past research [6] has underscored the role of digitalization as an important driver in enhancing manufacturing sustainability by improving resource efficiency and operational performance. To realize this potential, leveraging digital technologies through innovative approaches is critical to achieving sustainability goals. However, many companies have yet to bridge the gap between sustainability and digitalization, with prevailing strategies often focusing on one aspect to the exclusion of the other [7]. While much research has been conducted into the impact of digitalization on large enterprises in the context of the fourth industrial revolution, little attention has been paid to assessing its impact on small and medium-sized enterprises (SMEs). Constraints such as limited resources, uncertainty around perceived benefits, and resistance to change pose significant hurdles for SMEs in the adoption of digitalization [8].

Numerous prior studies have investigated the concepts of digitization and sustainability. Brenner and Hartl utilized both qualitative and quantitative methodologies to assess the impact of digitalization on sustainability (including environmental, economic, and social aspects) [9]. Denicolai *et al.* scrutinized how digitalization indicators affect the international characteristics of Small and Medium Enterprises (SMEs), with sustainability acting as a moderating factor [8]. Meanwhile, Gao *et al.* delved into the impact of e-commerce, digital platforms, and digital strategies on the financial performance and sustainability of Micro, Small, and Medium Enterprises (MSMEs) [10]. Furthermore, there exists earlier research on the correlation between digitalization and resilience. Khalil et al. evaluated the influence of digital technology on the resilience of SMEs in developing nations amidst the pandemic. Robertson *et al.* assessed the resilience levels of SMEs that possess higher digital maturity compared to those with lower digital maturity during the Covid-19 crisis [11]. Meanwhile, Ahmić analyzed how strategic sustainable orientation affects organizational resilience, with company size serving as a moderating factor [12].

Despite the abundance of studies tackling digitalization, sustainability, and resilience, there is a dearth of research scrutinizing the interplay among these three concepts. Some studies are limited in scope, focusing solely on large corporations or organizations in general, neglecting the significance of SMEs as integral components of national and regional economies as per the National Industrial Development Master Plan (RIPIN) 2015–2035 [13]. SMEs, being vital contributors to the national economy, must thrive and possess the capability to sustain competitiveness. Furthermore, SMEs play a pivotal role in the growth of Gross Regional Domestic Product (GRDP). Hence, it is imperative to evaluate how digitalization and sustainability impact the resilience of SMEs in West Sumatra.

The main objective of this study is to examine how digitalization and sustainability impact the resilience of SMEs. The research focuses on digital literacy and digital marketing as important indicators of digitalization, while sustainability is explored across environmental, social, and economic dimensions [14].

## 2  Method

This research utilized partial least squares structural equation modeling (PLS-SEM) for inferential statistical analysis. PLS-SEM is commonly used for estimating path models with latent variables and is applicable in various fields such as management, marketing, finance, and healthcare [15]. It is particularly effective for modeling complex relationships, whether formative or reflective, without needing normal data distribution or large sample sizes [16]. The method's flexibility allows for detailed and contextual analysis, making it valuable for both research and practical applications [17]. The PLS-SEM process involves designing the inner model (structural model), designing the outer model (measurement model), evaluating the outer model, evaluating the inner model, and testing hypothesis. Data collection was conducted using a combination of direct surveys and Google Forms. The respondents were managers from SMEs involved in the manufacturing sector. They were asked to share their views on digitalization, sustainability, and resilience, rated on a Likert scale from 1 to 5 (very low to very high). A total of 105 responses were gathered and analyzed using PLS-SEM.

## 3  Results and Discussions

### 3.1  Designing the Inner Model (Structural Model)

In this stage, an inner model (structural model) is designed to present the strength of estimation between latent variables which are often called constructs. In this study, the latent variables are resilience, sustainability, and digitalization. The direction of the arrows between variables shows the hypothesis framework that is built (see Fig. 1).

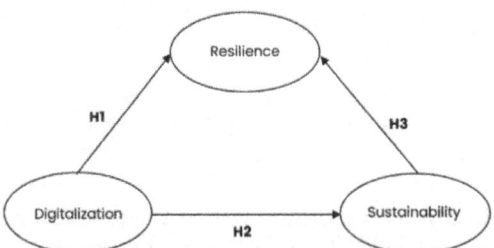

**Fig. 1.**  The inner model.

### 3.2  Designing the Outer Model (Measurement Model)

The outer model (measurement model) is then designed to show how each manifest variable, often called an indicator, relates to each latent variable, so that the manifest variable can explain the latent variable. The latent variable of resilience is unidimensional, which is a latent variable formed directly by manifest variables (four indicators). Meanwhile, the latent variables of digitalization and sustainability are multidimensional.

The latent variable of digitalization is grouped into two dimensions consisting of digital marketing (eight indicators) and digital literacy (eight indicators), while the latent variable of sustainability is grouped into three dimensions consisting of economic (seven indicators), social (nine indicators), and environmental (nine indicators) (see Table 1). The indicators are derived from relevant literatures.

### 3.3 Designing the Path Diagram

The next step is to design a path diagram to visualize the relationship between the inner and outer model paths. The path diagram forms a hierarchical component model (HCM) which is multidimensional because there are latent variables that have more than one interrelated dimension. In this study, a path diagram was designed to represent the relationship among digitalization, sustainability, and resilience.

### 3.4 Evaluating the Outer Model

Evaluation of the outer model (measurement model) aims to assess the validity and reliability of the model built.

**Validity Test.** Evaluation of the outer model with reflective indicators is carried out based on the convergent validity value and discriminant validity value of the latent construct forming indicators [15]. In this study, the validity test was carried out on indicators of the latent variables of sustainability, digitalization, and resilience.

*Convergent Validity.* It refers to the correlation between the scores of the indicators and the construct score. The assessment of convergent validity hinges on the outer loading values. For an indicator to be deemed valid, its outer loading value must exceed 0.7 [15]; otherwise, it must be eliminated and recalculated. The computation of outer loading values was carried out utilizing the PLS algorithm within the SmartPLS software. The analysis revealed that five indicators fell short of the required outer loading value. Consequently, these indicators were eliminated, and the outer loading values for the remaining indicators were recalculated. The reassessment demonstrated that all sub-indicators now possess an outer loading value surpassing 0.7. Hence, all indicators have successfully demonstrated convergent validity and can be considered valid (see Fig. 2).

*Discriminant Validity.* The examination of discriminant validity is grounded in the fundamental notion that distinct indicators (manifest variables) ought not to exhibit a substantial correlation [15]. In this study, discriminant validity was assessed by means of cross-loading values. Anticipation of the magnitude of the manifest variable is plausible provided that the cross-loading value of the measurement indicator surpasses that of the remaining measuring indicators. The results revealed that all cross-loading values satisfied the criteria for discriminant validity.

**Reliability Test.** Reliability testing seeks to demonstrate the consistency and accuracy of research instruments in assessing constructs. This study employed composite reliability (CR) values for the reliability assessment. The composite reliability should exceed 0.7 for confirmatory studies, whereas values ranging from 0.6 to 0.7 are deemed acceptable for exploratory research [15]. The findings indicate that all latent variables exhibit

**Table 1.** The variables and indicators.

| Variables | Indicators | Variables | Indicators |
|---|---|---|---|
| Resilience | Sales | Economic | Profit |
| | Profit | | Sales growth |
| | Customer satisfaction | | Material cost |
| | Delivery lead time | | Labor cost |
| Digital literacy | Information processing | | Energy cost |
| | Data collection | | Inventory cost |
| | Data evaluation | | Product quality |
| | Interaction through digital technology | Environmental | Energy consumption |
| | Collaboration through digital technology | | Energy efficiency |
| | Digital content creation | | Energy intensity |
| | Data security | | Material consumption |
| | Technical problem-solving skills | | Recycled material use |
| Digital marketing | Online advertising | | Packaging material consumption |
| | Social media capability | | Greenhouse gasses emissions |
| | Website ownership | | Wastewater discharge |
| | Ease of product discovery | | Solid waste disposal |
| | Online communication | Social | Fair salary |
| | Quality of information | | Employee turnover |
| | Improved customer service | | Employee satisfaction |
| | Online sales | | Occupational health & safety |
| | | | Employee training & development |
| | | | Lost working days |
| | | | Customer health & safety |
| | | | Customer satisfaction |
| | | | Corruption |

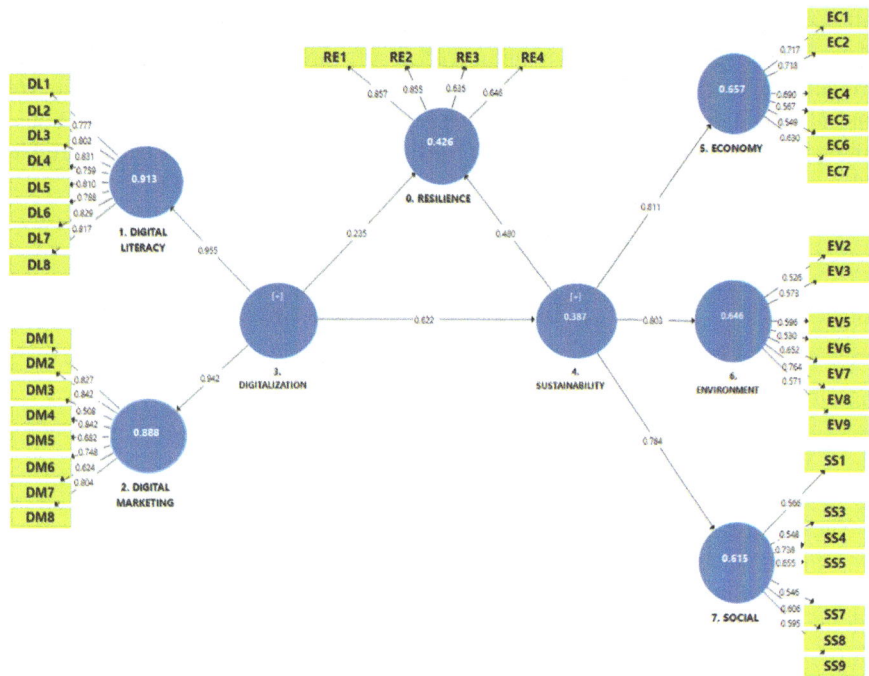

**Fig. 2.** The outer loading value.

a composite reliability value above 0.7 (see Table 2). Therefore, it is reasonable to infer that the measurement instrument's reliability requirements have been fulfilled and are deemed satisfactory.

**Table 2.** The composite reliability.

| Variable | Composite reliability |
|---|---|
| Resilience | 0.839 |
| Digital literacy | 0.935 |
| Digital marketing | 0.906 |
| Economic | 0.812 |
| Environmental | 0.799 |
| Social | 0.805 |

### 3.5   Evaluating the Inner Model

The evaluation of the inner model aims to examine the relationships among digitalization, resilience, and sustainability variables. This assessment involves calculating the path coefficients and the coefficient of determination (R2) using the bootstrapping technique with a 5% significance level in SmartPLS software. The analysis reveals that all correlation values are positive, with the strongest correlation found between digitalization and sustainability, at 0.623 (refer to Table 3). This indicates that digitalization has a significant impact on the sustainability of SMEs, more so than the other latent variables in the model.

**Table 3.** The path coefficient.

| Correlations | Original Sample | Sample Mean | Standard Deviation |
|---|---|---|---|
| Digitalization to Resilience | 0.237 | 0.224 | 0.116 |
| Digitalization to Sustainability | 0.623 | 0.631 | 0.061 |
| Sustainability to Resilience | 0.478 | 0.500 | 0.114 |

The findings reveal a $R^2$ value of 1.000 for the digitalization variable, indicating a complete explanation by resilience and sustainability variables. The resilience variable is explained by digitalization and sustainability variables at 42.2% in the weak category, with the remaining variance attributed to unexamined variables. Similarly, the sustainability variable is elucidated by digitalization and resilience factors at 38.1% in the weak category, while the residual portion is accounted for by variables not considered in this study.

### 3.6 Hypothesis Testing

Hypothesis testing is carried out using the bootstrapping method in SmartPLS. The acceptance of a hypothesis is contingent upon an alpha-value less than 0.05 or a t-statistic exceeding the t-table value (1.96) [15]. The alpha value represents the permissible threshold in research. Upon analysis, it was found that a p-value < 0.05 and a t-statistic value > 1.96 was achieved (see Table 4), leading to the acceptance of all hypotheses.

**Table 4.** The hypothesis testing.

| Hypotheses | T statistics | Alpha values | Decisions |
|---|---|---|---|
| Digitalization to Resilience | 2.044 | 0.041 | Accept |
| Digitalization to Sustainability | 10.156 | 0.000 | Accept |
| Sustainability to Resilience | 4.206 | 0.000 | Accept |

## 4  Conclusions and Recommendations

The study reveals that both digitalization and sustainability play critical roles in enhancing the resilience of Small and Medium Enterprises (SMEs) in West Sumatra. It highlights that digitalization significantly influences sustainability, with social media marketing and ease of product discovery being key factors driving digital adoption among these businesses. On the sustainability front, the focus should be on managing wastewater discharge and ensuring employee health and safety. The research also identifies sales growth and profitability as the most crucial indicators of SME resilience in the region. To strengthen SME resilience, it is essential for SMEs, government agencies, and private sector partners to work together, integrating sustainability and digitalization strategies effectively. Future research should expand the scope beyond digital literacy and digital marketing to include additional digitalization factors relevant to manufacturing companies.

**Acknowledgment.** The authors would like to thank the Department of Industrial Engineering, Faculty of Engineering, Universitas Andalas for providing publication grant 2024.

# References

1. Eller, R., Alford, P.A., Kallmünzer, A., Peters, M.: Antecedents, consequences, and challenges of small and medium-sized enterprise digitalization. J. Bus. Res. **112**, 119–127 (2020)
2. Abou-foul, M., Jose, L., Ruiz-Alba, J.L., Soares, A.: The impact of digitalization and servitization on the financial performance of a firm: an empirical. Prod. Plan. Control **32**(12), 975–989 (2020)
3. Zang, J., Long, J., Schaewen, A.M.E.V.: How does digital transformation improve organizational resilience? - findings from PLS-SEM and fsQCA. Sustainability **13**(20), 1–22 (2021)
4. Khalil, A., Abdelli, M.E.A., Mogaji, E.: Do digital technologies influence the relationship between the COVID-19 crisis and SMEs' resilience in developing countries? J. Open Innov.: Technol. Mark. Complexity **8**(2), 1–14 (2022)
5. Fitri, T.Z., Indrapriyatna, A.S., Amrina, E.: Analisis risiko pada rantai pasok kecap. J. Sains dan Teknol. **23**(2), 230–239 (2023)
6. Scholtysik, M., Reinhold, J., Koldewey, C., Dumitrescu, R.: Sustainability through the digitalization: exploring potentials and designing value co-creation architectures for product-service-systems. In: Proceedings of the Design Society, pp. 2871–2880 (2021)
7. Lichtenthaler, U.: Digitainability: the combined effects of the megatrends digitalization and sustainability. J. Innov. Manage. **9**(2), 64–80 (2021)
8. Denicolai, S., Zucchella, A., Magnani, G.: Internationalization, digitalization, and sustainability: are SMEs ready? A survey on synergies and substituting effect among growth paths. Technol. Forecast. Soc. Chang. **166**, 1–15 (2021)
9. Brenner, B., Hartl, B.: The perceived relationship between digitalization and ecological, economic, and social sustainability. J. Clean. Prod. **315**, 1–12 (2021)
10. Gao, J., Siddik, A.B., Abbas, S.K., Hamayun, M., Masukujjaman, M., Alam, S.S.: Impact of e-commerce and digital marketing adoption on the financial and sustainability performance of MSMEs during the COVID-19 pandemic: an empirical study. Sustainability **15**(2), 1–21 (2023)
11. Robertson, J., Botha, E., Walker, B., Wordsworth, R., Balzarova, M.: Fortune favours the digitally mature: the impact of digital maturity on the organisational resilience of SME retailers during COVID-19. Int. J. Retail Distrib. Manage. **50**(8/9), 1182–1204 (2022)
12. Ahmić, A.: Strategic sustainability orientation influence on organizational resilience: moderating effect of firm size. Bus. Syst. Res. **13**(1), 169–191 (2022)
13. Ministry of Industry. National Industrial Development Master Plan (2015–2035). Jakarta: Center for Public Communication, Ministry of Industry (2015)

14. Fandeli, H., Hasan, A., Amrina, E.: Model konseptual pengaruh keberlanjutan terhadap kinerja industri kecil dan menengah. J. Dampak **17**(1), 15–24 (2020)
15. Ringle, C.M., Sarstedt, M., Hair, J.F.: Partial Least Squares Structural Equation Modeling: Basic Concepts, Methodological Advances, and Applications. Springer (2023)
16. Henseler, J., Ringle, C.M., Sarstedt, M.: The use of partial least squares structural equation modeling in research: current state and future directions. Eur. J. Mark. **58**(1), 45–75 (2024)
17. Becker, J.M., Klein, K., Wetzels, M.: PLS-SEM: a comparison of two methods for analyzing data. J. Bus. Res. **140**, 229–240 (2022)

# Enhancing Sustainability – Strategic Choices from and for Small and Medium-Sized Enterprises

Dominik Saubke[(⊠)] [ID], Malin Schmitt, Pascal Krenz, and Tobias Redlich

Helmut Schmidt University/University of the Federal Armed Forces, Hamburg, Germany
dominiksaubke@hsu-hh.de

**Abstract.** The multiple crises of the recent years have shown the vulnerability of our todays dominant linear value creation process. There is a need for economic, social and ecological adjustments in the mechanisms of the production sector. In the attempt to design more sustainable value chains, increasing local production is frequently discussed. Local manufacturing at the place of need addresses some of the current problems and could make a significant contribution to the reshaping process of the manufacturing ecosystem. Small and medium-sized enterprises (SME) are considered to be the backbone of the product ion sector in many nations of the world today. Through area-wide distribution and ability to react flexibly to demands, they fulfil some of the criteria that favour a more sustainable production (proximity to the user, avoidance of transport, customised production). This article investigates *Strategic Choices* by analysing sustainability reports of manufacturing small and medium-sized enterprises using Mayring's Qualitative Content Analysis. Subsequently, these results are comparatively evaluated. The Strategic Choices identified in this process will help to reveal potentials and provide a basis for possible strategic alignments to increase sustainability in producing small and medium sized enterprises.

**Keywords:** Sustainable Manufacturing · SME · Sustainability Strategy

## 1 Introduction and Motivation

Sustainability is one of the central aspects in the design of individual corporate strategies. Since ELKINGTON coined the term "triple bottom line" framework [1], the business challenge of harmonising environmental, social and economic performance has become an essential aspect of management research and practice, often referred to as corporate sustainability [2]. The exact definition and methodological scope of corporate sustainability are still debated today. They range from a company's attempt to respond to environmental and social issues up to a set of activities that are fully integrated into a company's overall strategy and effectively contribute to the protection and improvement of the environment, social equity and economic durability [3, 4]. No doubt, increasing sustainability in companies must contribute to ensuring sustainable consumption and

© The Author(s) 2025
H. Kohl et al. (Eds.): GCSM 2024, LNME, pp. 37–45, 2025.
https://doi.org/10.1007/978-3-031-93891-7_5

production patterns. The efficient use of natural resources and significant waste reduction are essential for achieving the relevant aspects of the *17 United Nations Sustainable Development Goals* in Agenda 2030 [5]. However, the sensible utilisation of resources and waste avoidance is only a tiny part of a holistic change toward a more sustainable form of value creation. When looking at manufacturing companies in particular, KRENZ et al. currently discussed that a more distributed and local form of production close to the place of need could help reach sustainability goals [6]. Local production could make value creation more resilient, achieve ambitious climate targets, and achieve an appropriate level of production sovereignty [7]. Today, the production sector in many industrialized countries of the European Union (EU) is largely represented by small and medium-sized enterprises (SMEs) [8]. In addition to the industrialised value creation in large companies, SMEs significantly impact the manufacturing sector, with around 99.4% of all companies in Germany belonging to this category in 2018 [9]. A similar percentage distribution of companies can be seen in many high-production nations [10].

Research into applying management approaches to increase sustainability in industrial companies has already developed clear systems and is well advanced. Extensive knowledge has been gained on the implementation and impact of these approaches in industry sector [11]. In contrast, the application of these approaches in SMEs are still at an early stage, as they are more confronted with a lack of resources (personnel, qualifications, time) and, at the same time are often confronted specific challenges that require customised adaptation [12]. The following research will, therefore, examine how SME are strategically aligning themselves to act more sustainable.

## 2   Objective of Research

By systematically analysing sustainability reports, this study aims to gain new insights into the current strategic orientation of companies. The focus of the study is on analysing companies from the manufacturing sector. The study is deliberately limited to SME in order to gain specific insights into the sustainability strategies of these companies. The following research question arises in this context: *Which strategy choices can be identified to increase sustainability (social, ecological, economic) in the management strategy of manufacturing small and medium-sized enterprises (SMEs) and which strategy options result from this for other companies?* Only reports from the **Sustainability Code** [13] were used for the analysis. These reports comply with the rules of the regulation adopted within the EU, as presented later, and follow a clear structural organisation which improves traceability and systematisation. A complete year of reports is used for the methodological analysis based on *Mayring's Qualitative Content Analysis*, taking into account the restrictions, in order to provide a comprehensive and representative insight into the current research in the field of sustainability strategies. In the further course of this publication, the main definitions, the origin of the data and the methodological approach are explained in detail. This creates a clear basis for the interpretation of the results and enables the reader to fully understand the concepts used.

# 3  Current Status and Framework

## 3.1  Small and Medium Sized Enterprises in the Manufacturing Sector

In this analysis the focus is on the manufacturing sector including companies that produce, process or assemble material goods. The analysis is based on the specifications of the statistical classification of economic activities in the EU (NACE-Code) [14]. Economic Sector C "Manufacturing" was selected as the optimal branch due to its versatility and strong correspondence with production-related factors. In addition, only sustainability reports from SMEs were considered, as an untapped potential regarding ambitious climate targets is suspected. The definition of SMEs is in accordance with the criteria established by EU Commission [15]. Specifically, micro, small, and medium-sized enterprises are defined as those with fewer than 249 employees, an annual turnover below 50 million euros, or an annual balance sheet total not exceeding 43 million euros [16]. Additionally, no more than 25% of voting rights in the company should be held by a government agency or a public corporation. The manufacturing sector, synonymous with industry or the industrial sector, is characterized by machine-based production, structured division of labor, and production within designated facilities [17].

## 3.2  Overview of the Sustainability Reporting Obligation

The transparent external presentation of a company's own activities through *Sustainability Reporting (SR)* is also becoming increasingly important. The aim is for companies to disclose their sustainability performance and present it transparently by publishing sustainability reports. In Germany, holistic and sustainable production is understood as the centralisation of the aspect of sustainability in the development of the product, the manufacturing process up to the use of the product. With the introduction of the European Green Deal (EGD), further Europe-wide directives and regulations that directly affect the production sector are taking action [18]. Overview of the Sustainability Reporting Obligation. Across Europe SMEs accounted for more than 99% (24 million) of all companies in the European Union (EU) in 2022, and almost 2/3 of all people in employment are employed in these companies. Due to this importance, SMEs are either directly or immediately affected by this implementation. In Germany, SMEs can be affected by CSR reporting obligation (CSR-RUG & CSRD), EU Taxonomy Regulation (EU Regulation 2020/852) and Lieferkettengesetz (LKSG). In 2021, the EU Commission decided on far-reaching changes to CSR, which will become the Corporate Sustainability Reporting Directive (CSRD) in 2022 [19]. As a result, more companies will be required to report and, at the same time, the aim is to make sustainability-related information comparable and publicly accessible. A further expansion is planned for 2025; SMEs may be affected by the reporting obligation in the future. It is now essential to deal with the transfer and adaptation of management approaches to sustainability from industry for SMEs at an early stage.

# 4    Approach and Data Basis

## 4.1    Rating and Analysis Based on Mayring's Qualitative Content Analysis

The aim of using *Mayring's Qualitative Content Analysis* is to accumulate structured information when analysing large amounts of data (text, reports, e.g.) in order to be able to draw up new theoretical considerations [20]. As an established form of analysis in the discipline of empirical social research, the method is characterised by the use of material that is to be analysed interpretatively and meaningfully [21]. In a rule-based procedure that records the systematically analysed data in categories [22]. The method is divided into: "summarisation", "explication" and "structuring according to a coding guide". During summarisation, the content is reduced to the most essential features. In the explication, further material is used to explain the existing material. During structuring, the aspects sought are filtered and categorised into corresponding criteria. This follows a detailed, previously defined schedule (coding guide). In the coding unit, the most minor text components to be coded are defined in advance and backed up with examples. This tight structure is intended to maximise the transparency of the methodological procedure and thus ensure comprehensibility for others.

## 4.2    Reports and Companies Considered in the Analysis

A internationally recognised reporting standard for sustainability with public accessible database is the Sustainability Code [23], where used for the analysis. The structure of the reports follows a division in 20 criteria. The "Sustainability concept" includes criteria one to ten and deals with the company's general concepts on sustainability. This topic area is, in turn, divided into two subject areas. The first subject, "Strategy", covers criteria one to four and addresses the company's sustainability strategy. The second subject „Process management" including criteria five to ten, the company presents its process management regarding sustainability. Criteria 11 to 20, deal with Sustainability Aspects" divided into "Environment" (11–14) and "Society" (criteria 15 to 20). All criteria are based on performance indicators. The company has two possible bases for the performance indicators. GRI performance indicator scheme contains 29 performance indicators. European Federation of Financial Analysts Societies (EFFAS) performance indicator scheme uses 16 indicators.

There have been 1179 sustainability reports published in this database [26]. The Pandemic can be considered as a major turning point in the recent past. Reports are usually prepared by SMEs on a voluntary basis before the end of the following year, so the pandemic has affected two vintages directly and indirectly; on the one hand, the state of emergency in the companies had an indirect negative impact on the preparation of the previous report of the year 2019 and, on the other hand, the pandemic directly affected the report 2020. Both vintages were therefore not suitable for the analysis, only some reports of 2022 had been published. The focus was set on 2021, older vintages where not taken into account. The database initially contained 194 reports. After applying the filters for Criteria NACE (Economic Sector C - Manufacturing) and SME Characteristics, a dozen reports were available at the time of data access. From the selected reports, ten sustainability reports were identified as promising (Table 1).

**Table 1.** Overview of the Companies considered in the analysis

| Name | Sector | Emp | BS Total | Indicat | Duty |
|---|---|---|---|---|---|
| Comp. 1 | *vehicles and trailers* | 199 | 14.2 Mio | GRI SRS | No |
| Comp. 2 | *products of wood ex. Furniture* | 249 | 18.6 Mio | GRI SRS | No |
| Comp. 3 | *motor vehicles and trailers* | 170 | 15.6 Mio | GRI SRS | No |
| Comp. 4 | *electrical equipment* | 60 | 8.7 Mio | GRI SRS | No |
| Comp. 5 | *furniture* | 18 | t.b.d | GRI SRS | No |
| Comp. 6 | *textiles* | 8 | 1.1 Mio | GRI SRS | No |
| Comp. 7 | *glassware, ceramics, stone* | <250 | t.b.d | GRI SRS | No |
| Comp. 8 | *mechanical engineering* | 60 | 5.5 Mio | EFFAS | No |
| Comp. 9 | *mechanical engineering* | 137 | 42.8 Mio | GRI SRS | No |
| Comp. 10 | *textiles* | 24 | 1.1 Mio | GRI SRS | No |

## 5   Results

### 5.1   Identified Strategic Choices

As part of the qualitative content analysis based on Mayring, a total of ten reports were coded. 614 report pages were analysed and 699 quotations were identified, 292 of which were relevant to the analysis of strategic measures in the companies. In addition, 36 codes were developed that can be traced back to these citations. The coding was carried out by a group of four experts from the engineering discipline. The coding guide developed for the coding was iteratively refined and can be accessed permanently online (DOI: https://doi.org/10.17632/p2ffjnfwsw.1). Here is an example of a citation found in on of the sustainability reports: *"Where appropriate, the electric consume is recorded in the electricity register and is evaluated at least once a year"*. The corresponding citation was then assigned to the following code: *"Monitoring And Early Implementation Of Key Performance Indicator Systems"*. The analysis and evaluation followed a two-phase scheme. Firstly, a code structure was created by identifying various *Strategic Choices* (Codes). Secondly, a basic evaluation was derived from the mentions in the reports and the weighting added where applicable. Each *Strategic Option* was then assessed in terms of its impact on the three core elements of sustainability: environment, economic and social. The initial rating was given a higher score if an impact on two or even three of the elements could be predicted.

In this context, $n_{measures}$ is the number of measures found in the report that are connected to the coded *Strategic Choices*, $C_{Companies}$ is the number of companies in which measures of a *Strategic Choices* were implemented and $S_{Spheres}$ is the number of impacted spheres. This leads to a weighted evaluation of the three influencing variables in order to achieve an overall evaluation:

$$X_{IMPACT} = n_{Measures} * S_{Spheres} * C_{Companies} \tag{1}$$

Furthermore, the 30 highest rated codes were taken into account. Within the scoring system introduced, a maximum score of 30 points was achieved. This was set as the maximum score and, based on this, three areas were defined in which a *Strategic Choices* can occur. Thus, the *Strategic Choices* can have a marginal influence ($X > 10$), a moderate influence ($10 < X > 20$) or a high influence ($20 < X$). Secondly, the field of 30 *Strategic Choices* was divided, with the upper half being counted as the more relevant in the following and given greater emphasis in the discussion. These Strategic Choices are marked in bold on the right-hand side of the chart. The additional selection is primarily intended to limit the discussion in the context of this publication (Fig. 1).

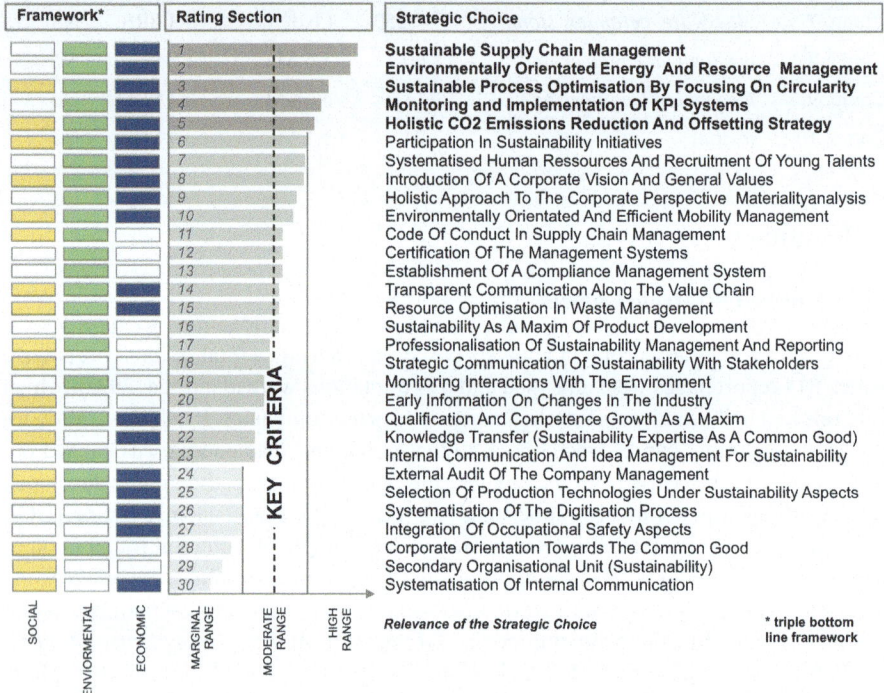

**Fig. 1.** Strategic Choices found during the analysis and rating.

## 5.2 Spread of the Strategic Choices

A *Sustainable Supply Chain Management* is recognised by most companies as relevant and as part of their own strategy. As a driver of some of the strategy decisions mentioned later, it can be seen as a key decision to introduce this and to provide appropriate resources for the area. For example, the introduction of a *Code of Conduct in Supply Chain Management* would fall into this area, and the implementation of a comprehensive monitoring and early *Implementation of Key Performance Indicator System* can be found in many reports. It can be seen as a basis for building up the necessary understanding of a company's own processes with regard to interactions with the environment and deriving optimisation measures and processes from this. The companies analysed in this report also consider the Introduction of *A Corporate Vision* and *General Values* to be an important starting point for the long-term transformation of their own company. Participation in Sustainability Initiatives is ranked on seven and was mentioned in many of the reports. Such initiatives sometimes offer free workshops and help to train your own staff. Other initiatives gather insights by sharing company data. The barrier to entry is therefore very low for all companies, it just needs to be integrated into the corporate concept, for example by integrating it into the company's value system. The analysis has shown that companies are making great efforts to fully integrate this system into their own corporate system. The systematic implementation of a *Holistic CO2 Emissions Reduction and Offsetting Strategy* is also mentioned by many different companies but slips a little further down. The analysis has shown that companies are making great efforts to fully integrate this system into their own corporate system. The *Holistic Approach to the Corporate Perspective (Materiality Analysis)*, well-known approach to analysing the company's interactions with the environment, brings together all the efforts mentioned by companies that aim to fully understand these interactions. The use of this approach was assigned a high impact by many companies. While the codes remain relatively vague and are intended to indicate *Strategic Choices*, the analysis conceals interesting actions. The concrete actions and application-orientated measures are prepared in order to provide concrete support for strategic decisions.

## 6   Critical Analysis and Outlook

The analysis has shown that there are various strategy options. Using *Mayring's Qualitative Content Analysis*, *Strategic Choices* frequently implemented in manufacturing small and medium-sized enterprises (SMEs) could be identified and systematized in sustainability reports transparently and comprehensibly. An overall assessment was derived in combination with the frequency of occurrence of a strategy option in different companies and the respective specific frequency of measures for implementing a strategy option. The results are primarily intended to help other small and medium-sized enterprises (SMEs) make their own strategy decisions more quickly by gaining insight into the strategy orientation of other manufacturing companies. As sustainability pioneers, the companies are particularly committed in this respect. Only a limited amount of company information was available for the initial research. An expansion to include more companies is necessary. In addition, there was a specific variance among the companies regarding total assets and number of employees. A finer subdivision could be used to differentiate, and further sharpening by differentiating between product groups would be interesting. In addition, a in-depth analysis should be carried out to build an ontology with strategic Options linking operational measures.

**Acknowledgements.** This research is funded by dtec.bw - Digitalization and Technology Research Center of the Bundeswehr [Project: Production Next Door]. Dtec.bw is funded by the European Union – NextGenerationEU.

# References

1. Elkington, J., John, E.: Cannibals with forks: the triple bottom line of 21st century business. Pbk. ed. Capstone, Oxford, U.K. (1999)
2. Wirtschaft und Klimaschutz, BM für. Nachhaltigkeit in der Wirt-schaft (2022). https://www.bmwk.de/Redaktion/DE/Dossier/nachhaltigkeit.html. Accessed 23 Nov 2022
3. Kurz, R., Wild, W.: Nachhaltigkeit und Unternehmen. uwf **23**, 323–328 (2015)
4. Eccles, R.G., Ioannou, I., Serafeim, G.: The impact of corporate sustainability on organizational processes and performance. Manage. Sci. **60**, 2835–2857 (2014)
5. Bundesregierung 2022. Agenda 2030: Unsere Nachhaltigkeitsziele. veröffentlicht auf der Website der Bundesregierung (2022). https://www.bundesregierung.de/breg-de/themen/nachhaltigkeitspolitik/nachhaltigkeitsziele-erklaert-232174. Accessed 23 Nov 2022
6. Krenz, P., Stoltenberg, L., Markert, J., et al.: The phenomenon of local manufacturing: an attempt at a differentiation of distributed, re-distributed and urban manufacturing. In: Bridging Smart Products and Man. Systems, pp. 1014–1022. Springer, Cham (2022)
7. Krenz, P., Saubke, D., Stoltenberg, L., et al.: Towards Smaller Value Creation Cycles: Key Factors and their Interdependencies for Local Manufacturing. Publish-Ing, Hannover (2022)
8. ifm-bonn. KMU im EU-Vergleich (2022). https://www.ifm-bonn.org/statistiken/mittelstand-im-einzelnen/kmu-im-eu-vergleich. Accessed 16 Nov 2022
9. Statistisches Bundesamt. Kleine und mittlere Unternehmen (2019). https://www.destatis.de/DE/Themen/Branchen-Unternehmen/Unternehmen/Kleine-Unternehmen-Mittlere-Unternehmen/_inhalt.html. Accessed16 Nov 2022
10. Matt, D.T., Rauch, E.: Implementation of lean production in small sized enterprises. Procedia CIRP **12**, 420–425 (2013)
11. Baumgartner, R.J.: Wertsteigerung durch Nachhaltigkeit, 2005, Montanuniversität Leoben], 1. Aufl. Sustainability Mgm. for industries, vol 1. Hampp, München, Mering (2005)
12. Mehta, K., Sharma, R.: Sustainability, Green Mgm, and Perfor. of SMEs. De Gruyter (2023)
13. Deutscher Nachhaltigkeitskodex (2022). https://www.deutscher-nachhaltigkeitskodex.de/de-DE/Home/DNK/DNK-Overview. Accessed 23 Nov 2022
14. Regulation (EC) No 1893/2006 of the European Parliament and of the Council of 20 December 2006 establishing the statistical classification of economic activities NACE No 3037/90 (2006)
15. Kommission der europäischen Gemeinschaften. Commission Recommendation of 6 May 2003, Definition of micro, small and medium-sized enterprises (C(2003) 1422) (2003)
16. IfM Bonn. KMU-Definition der EU-Kommission (2023). https://www.ifm-bonn.org/definitionen/kmu-definition-der-eu-kommission. Accessed 22 Jan 2023
17. Bundesamt, S.: Klassifikation der Wirtschaftszweige 2008 (WZ 2008) (2008)
18. Europäischer Rat. Ein europäischer Grüner Deal (2023). https://www.consilium.europa.eu/de/policies/green-deal/. Accessed 20 Oct 2023
19. Directive (EU) 2022/2464 of the European Parliament and of the Council of 14 December 2022 amending Regulation (EU) No 537/2014, Directive 2004/109/EC (2022)
20. Mayring, P.: Qualitative Inhaltsanalyse: Grundlagen und Techniken, 12, vollständig überarbeitete und, aktualisierte Beltz Pädagogik. Beltz, Weinheim (2015)
21. Baur, N., Blasius, J.: Handbuch Methoden der empirischen Sozialforschung. Springer, Wiesbaden (2014)
22. Kuckartz, U.: Qualitative Inhaltsanalyse: Methoden, Praxis, Computerunterstüt-zung, 3, überarbeitete. Grundlagentexte Methoden. Beltz Juventa, Weinheim (2016)
23. Rat für Nachhaltige Entwicklung. DNK: Mehr als ein Berichtsstandard (2023). https://www.deutscher-nachhaltigkeitskodex.de/. Accessed 07 Jan 2024

# Beyond Compliance - Towards Automated Double Materiality Analysis Under the CSRD Framework

Maximilian Nowak[1]([⊠]), Josef Baumüller[2], Ruben Hetfleisch[1], Sebastian Schlund[1,2], and Stephan Martineau[1]

[1] Fraunhofer Austria Research GmbH, Theresianumgasse 7, 1040 Vienna, Austria
maximilian.nowak@fraunhofer.at
[2] Vienna University of Technology, Karlsplatz 13, 1040 Vienna, Austria

**Abstract.** The Corporate Sustainability Reporting Directive (CSRD) has been in effect throughout the European Union since January 1, 2024, requiring companies to disclose their sustainability activities. Along with qualitative data, companies must also provide numerous quantitative key figures that were not previously required. Central to these disclosures is the double materiality assessment (DMA), which determines the relevant sustainability quantitative and qualitative data for each company. This paper explores the automation of DMA as a solution for businesses, particularly small and medium-sized enterprises (SMEs), which may lack the resources for detailed sustainability reporting. This paper explores the automation of DMA as a transformative solution for businesses, facilitating the initial and subsequent compilations of sustainability reports. It specifically seeks to identify and evaluate existing automation techniques that can effectively support this analysis, thereby answering a critical question: Which automation approaches are best suited to increase the efficiency and accuracy of a DMA methodology? Through a literature review, suitable AI-driven automation techniques are identified to enhance the efficiency and accuracy of DMA. The research proposes a new approach to help SMEs embrace sustainability reporting as a strategic opportunity, simplifying compliance while improving operations.

**Keywords:** Sustainability Reporting · Double Materiality Assessment · AI

## 1 Introduction

The introduction of the CSRD on January 1, 2024, has resulted in a significant change in the landscape of non-financial reporting within the European Union [1]. In the future, both the financial and non-financial aspects of the annual report will be subject to mandatory annual auditing by external auditors, unlike to the previously voluntary Non-Financial Reporting Directive (NFRD), where incorrect reports could result in management liability and turnover penalties [2].

The fundamental structure of the sustainability report is established through the implementation of the so-called DMA, in accordance with the European Sustainability

H. Kohl et al. (Eds.): GCSM 2024, LNME, pp. 46–53, 2025.
https://doi.org/10.1007/978-3-031-93891-7_6

Reporting Standards (ESRS) [3]. The approach offers a methodology for identifying the key sustainability activities for a company in the environmental, social, and governance dimensions along the value chain and within the company boundaries [4].

The DMA serves as the foundation for non-financial reporting, wherein the company's key areas in the environmental, social, and governance domains are identified with the involvement of internal and external stakeholders [3]. The implementation of this process (Fig. 1) for companies comprises a four-stage process: 1) Understanding the context, 2) Identification of actual and potential impacts, risks, and opportunities (IROs) and 3) assessment and determination of material IROs [5], that is particularly challenging for SMEs, especially from a financial perspective [6] and 4) the actual reporting of the outcome.

**Fig. 1.** DMA procedure [5]

It is recommended that the company's employees be supported in performing DMA by an automated solution, allowing them to focus on creative and operational activities, thus positively impacting implementation [7]. In addition to the existing guidelines for SMEs [8], it is also crucial to provide support for micro-enterprises in implementing the DMA [9]. The use of NLP-supported solutions can be beneficial in the identification of potential IROs, as demonstrated by Fischbach et al. [10]. In order to achieve this objective, BERT-based models can be employed to extract and subsequently classify keywords from a variety of reports [11], thereby enabling the definition of company-specific IROs. It is similarly crucial to ascertain and extract the specific internal sustainability KPIs of the companies in question. This can be achieved through a combination of data envelopment analysis (DEA) and a web-based information system [12].

The primary challenge is that although the specified approach delineates the framework conditions for implementing the DMA, it permits a considerable degree of flexibility, which may result in disparate outcomes in certain instances [13]. This compromises the comparability and reproducibility of the method [14]. SMEs, either directly affected by the directives or as part of large corporations' supply chains, often find it challenging to identify all key areas [15]. For those lacking training or experience in the field of CSRD or DMA, this can result in confusion. The concurrent introduction of other regulations like the EU-Taxonomy, the Carbon Border Adjustment Mechanism, or the Corporate Sustainability Due Diligence Directive presents additional challenges for SMEs, with around 50,000 companies needing to perform or update their DMA for the first time in 2026 under the CSRD reporting process [16].

To achieve the goal of a simplified and standardized DMA, this paper employs a narrative literature review (NLR) to demonstrate that the automation of the DMA is both possible and necessary. Existing literature shows that solutions for systematic data collection already exist (e.g., Nayak et al. [17] Yu et al. [18]), although these typically refer to existing reports or alternative requirements such as GRI.

A proposed approach should demonstrate how potential automation solutions can facilitate the DMA process. The objective is to establish a standard that will enable future DMA to be comparable and reduce the workload for SMEs.

This leads to the research question of this paper: *"How can DMA be efficiently implemented with the help of ML-based automation approaches?"*.

The purpose of this paper is to identify technological approaches in the context of DMA and highlight their potential for DMA. We focus on steps 1 and 2 of the process, depicted in Fig. 1, which appear as the most promising starting points for automation.

The rest of the paper is structured as follows: Sect. 2 presents the NLR that investigates the research gap. Section 3 offers the insight's outlook on implementing DMA with automation techniques and Sect. 4 summarizes findings, outlines limitations, and identifies avenues for further research.

## 2   State-of-the-Art Literature Review on DMA Automation

An NLR was conducted with the objective of establishing the basis for the approach presented in Sect. 3. In accordance with a NLR, the subject area to be researched was subjected to examination regarding the requirements of planning, implementation and reporting of the results [19]. The research highlighted the significant challenges SMEs face when implementing DMA for the first time, leading to the decision to prioritize automation of the process.

The NLR commenced with a targeted search determined through a preliminary analysis of key concepts from relevant literature, focusing on critical themes such as sustainability reporting, automation, and double materiality within the context of CSRD requirements. 11 search strings ('NLP in sustainability reporting', 'Automation in reporting', 'Double materiality and CSRD', 'Sustainability reporting automation CSRD', 'DMA data mining', 'ML sustainability reporting', 'ML sustainability reporting CSRD/ESRS', 'NLP corporate sustainability reporting', 'double materiality assessment/analysis') were applied across the databases Scopus, IEEE, SpringerLink, and JSTOR, designed to capture the relevant literature. Following the definition of the strings, the synthesis of the results commenced. Figure 2 depicts the entire process that culminates in the final publications, which were subsequently analyzed in the paper. In the end, 46 articles were deemed eligible for the final analysis, with a detailed review of publications on non-financial reporting, automation, and the transferability of methods to DMA.

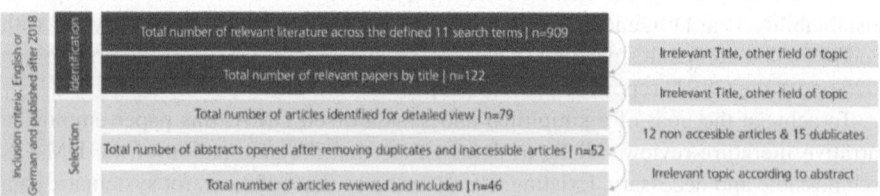

**Fig. 2.**  Synthesis throughout the NLR

The analysis revealed a clear picture of the literature in terms of its delimitation. Publications with a more in-depth technical description of the ML methods used had little or no reference to the starting points in the context of the new CSRD reporting guideline [10, 11].

The fifth string, addressing CSRD and DMA concepts, resulted in five publications focusing on DMA methodology and its organizational, legal, and accounting challenges. These publications, from law, accounting, and business journals, lack technological approaches but highlight challenges that automation in ML can address. Materiality assessments (MA) face issues such as significant variation and lack of standardization in definitions and reporting, inadequate stakeholder engagement, complex implementation processes, and management discretion leading to biased or incomplete disclosures [20–22]. By automating these aspects, companies can increase the speed of data analysis, ensure consistency in materiality identification, and enhance the overall effectiveness of sustainability reporting [23].

In the domain of technological solutions, a plethora of support options exists, including T-BERT, FinBERT, and RoBERTa-Large, for the ESG (Environmental, Social and Governance) domain [11]. However, these approaches are predominantly employed in the context of analyzing existing sustainability reports [24, 25].

In their study, Dumitru et al. propose a method for advancing ERP systems through automation that can support sustainability reporting. This method emphasizes the integration of Robotic Process Automation (RPA) and Intelligent Process Automation (IPA) [26]. This approach is particularly promising for the consolidated collection of data required under DMA.

Kassem & Trenz uses a sustainability assessment model for SMEs to automate the process, including the collection and selection of KPIs, the development of a comprehensive assessment model for economic, environmental, social and governance dimensions, and the implementation of a web-based platform to automate the assessment. This model will help SMEs to efficiently prepare sustainability reports and to improve their sustainability practices [12].

The study by Kulkarni et al. emphasizes how ML can optimize processes, reduce costs, and improve governance in Micro, Small, and Medium Enterprises (MSMEs) by providing effective decision-making tools for environmental, social, and governance factors [27]. In addition, ML can improve the accuracy, reliability, and timeliness of sustainability reporting by automating data collection, analysis, and interpretation, leading to improved decision-making and sustainability performance [28].

No paper offering a comprehensive technical ML-based solution for DMA was found. However, several papers discuss the importance of DMA in sustainability reporting [29], the emergence of the double materiality perspective in reporting requirements [21, 30], and the DMA process in different companies, including stakeholder engagement and formalized processes [20].

In summary, while the NLR did not uncover a fully developed technical solution for automating DMA, it provided substantial evidence supporting the use of NLP, ML, and RPA as foundational technologies. These insights guide the conceptual model in Sect. 3, emphasizing automation's role in improving consistency, comparability, and scalability for SMEs' DMA processes.

## 3   A 4-step Approach Towards DMA Automation

The DMA is comprised of the four steps depicted in Fig. 1. The objective of this paper is to identify potential automation solutions, especially for the steps 1 and 2. These steps offer the greatest potential for automation, as they require extensive data integration, continuous updating, complex analyses, efficient stakeholder involvement, and a reduction in manual processes.

### 3.1   Step 1: Understanding the Context

In the first step of the DMA, AI methods play a crucial role in systematically evaluating a company's activities, business relationships, and stakeholders across the entire value chain. The business model canvas is a well-established methodology for identifying pertinent areas along the value chain and pinpointing crucial points of intervention in the context of sustainability.

NLP techniques, particularly BERT-based models like FinBERT, are employed to extract Environmental, Social, and Governance (ESG)-relevant keywords from both internal and external reports. These keywords are then analyzed and prioritized based on criteria such as frequency and relevance within sustainability contexts.

Technologically, these NLP models are integrated into the company's data analysis workflow. This integration automates the continuous scanning of internal and external documents, feeding identified key topics directly into the materiality assessment. As a result, companies can adopt a more dynamic and data-driven approach to identifying material issues. Additionally, RPA tools, such as UiPath, are utilized to automatically extract data from disparate systems, including Enterprise Resource Planning (ERP) and Customer Relationship Management (CRM) platforms. This ensures that critical metrics, such as energy consumption or $CO_2$ emissions, are always current and accurate, further enhancing the robustness of the materiality analysis.

In practice, companies implement these technologies to monitor ESG-relevant terms continuously and identify emerging trends and risks that may not be immediately apparent from internal data alone. By using RPA tools, they can streamline the collection of vital data, which feeds directly into the DMA process. This technological integration not only increases efficiency but also improves the accuracy of the evaluation of the company's fields of action.

### 3.2   Step 2 – Identification of Actual and Potential IROs

The second step in the DMA process involves the identification of actual and potential IROs, which necessitates the integration of both internal and external data sources along with stakeholder perspectives. Here, ML algorithms are pivotal for analyzing large volumes of structured and unstructured data. These algorithms are applied to internal data sources, such as operational KPIs and risk reports, to identify patterns and anomalies that signal significant IROs.

Simultaneously, NLP and sentiment analysis tools process external data, including market studies, regulatory changes, and social media, to detect emerging trends and risks. Stakeholder opinions are also evaluated and categorized based on their relevance and urgency, ensuring that the stakeholder perspectives are continuously integrated into the IRO identification process.

For seamless implementation, NLP and ML models are embedded into existing data management systems like ERP and CRM platforms, enabling automated and ongoing analysis of both internal and external data sources. This integration allows companies to proactively identify material issues, ensuring that identified risks and opportunities are not only data-driven but also aligned with relevant stakeholder needs. The automated nature of this process enhances consistency and comparability of DMA results, particularly benefiting SMEs by optimizing data collection and analysis workflows.

### 3.3 Comprehensive View of the Process

The initial two steps of the DMA process exhibit considerable overlap, particularly with regard to their utilization of NLP technologies for data analysis and the identification of pertinent topics. Both steps place significant emphasis on the integration of stakeholder feedback and continuous data analysis, which are regarded as core components. However, the two steps serve distinct purposes. While Step 1 focuses on the systematic identification and prioritization of action fields along the company's value chain, Step 2 is specifically concerned with identifying and integrating risks and opportunities that impact the company's sustainability efforts.

The primary distinction between the two steps lies in the focus of each respective step. Step 1 is primarily concerned with mapping out critical areas of intervention, whereas Step 2 delves more deeply into the specific risks and opportunities that could influence these areas. Additionally, the methods used in Step 2 are more dynamic and responsive, continually integrating external and internal data to reflect evolving regulatory and market conditions.

## 4 Conclusion

The automation of DMA with ML offers SMEs a transformative opportunity to enhance sustainability reporting by streamlining the process and ensuring greater accuracy, consistency, and efficiency. AI-driven approaches, such as NLP and RPA, provide significant potential for simplifying the identification, assessment, and reporting of material sustainability factors. By integrating these technologies into the DMA process, SMEs can shift sustainability reporting from a regulatory burden to a strategic advantage. Automated processes enable more efficient resource allocation by optimizing both data collection and analysis workflows, thereby not only improving accuracy but also significantly reducing the complexity of managing DMA requirements.

Limitations: A key limitation of this study lies in the narrow selection of search terms during the NLR, which may have led to a restricted view of alternative methods. Additionally, the automation solutions proposed are context-dependent and may not be universally applicable, necessitating careful consideration when applying the findings.

Future research should aim to refine these models, explore new algorithms for handling sustainability data, and validate the proposed approaches through case studies across diverse industries, which is part of an ongoing research project. Ethical considerations such as transparency and fairness in AI algorithms must also be addressed to prevent biases. Ultimately, the scalability and adaptability of these solutions across different company sizes and sectors remain critical for their widespread adoption.

**Acknowledgements.** The authors would like to acknowledge the financial support provided by the Austrian Research Agency (FFG) for the project "SustainTool" (project No. 48897807).

# References

1. Amt für Veröffentlichungen der Europäischen Union, L-2985 Luxemburg: Delegierte Verordnung (EU) 2023/2772 der Kommission vom 31. Juli 2023 zur Ergänzung der Richtlinie 2013/34/EU des Europäischen Parlaments und des Rates durch Standards für die Nachhaltigkeitsberichterstattung (2023)
2. Europäisches Parlament und Rat der Europäischen Union: Richtlinie (EU) 2022/2464 zur Änderung der Verordnung (EU) Nr. 537/2014 und der Richtlinien 2004/109/EG, 2006/43/EG und 2013/34/EU Amtsblatt der Europäischen Union, L 322, pp. 15–25 (2022)
3. European Comission: ANNEX to the Commission Delegated Regulation (EU) .../... supplementing Directive 2013/34/EU of the European Parliament and of the Council as regards sustainability reporting standards. ANNEX 1 (2023)
4. Baumüller, J., Schönauer, K.: Die neue Wesentlichkeit in der europäischen Nachhaltigkeitsberichterstattung. Darstellung und Diskussion der Wesentlichkeitsanalyse gem. ESRS (Teil 1) (2023)
5. EFRAG: IG 1 Materiality Assessment (2024)
6. EFRAG: Cover Letter on the Cost-benefit analysis of the First Set of draft ESRS (2022)
7. Ferreira, N., Potgieter, I., Coetzee, M. (eds.): Agile Coping in the Digital Workplace: Emerging Issues for Research and Practice. Springer (2021)
8. EFRAG: ESRS LSME ED (2024)
9. EFRAG: VSME ED January 2024 (2024)
10. Fischbach, J., et al.: Automatic ESG assessment of companies by mining and evaluating media coverage data: NLP approach and tool. In: 2023 IEEE International Conference on Big Data (BigData). IEEE (2023)
11. Gupta, A., Chadha, A., Tewari, V.: A natural language processing model on BERT and YAKE technique for keyword extraction on sustainability reports. IEEE (2024)
12. Kassem, E., Trenz, O.: Automated sustainability assessment system for small and medium enterprises reporting (2020)
13. Nielsen, C.: ESG reporting and metrics: from double materiality to key performance indicators. Sustainability (2023)
14. DIN German Institute for Standardization: The German Standardization Strategy (2009)
15. EFRAG: EFRAG IG 2 Value Chain (2024)
16. European Parliament: Sustainable economy: Parliament adopts new reporting rules for multinationals [Press release] (2022)
17. Nayak, R., Balasubramaniam, T., Kutty, S., Banduthilaka, S., Peterson, E.: A semi-automatic data extraction system for heterogeneous data sources: a case study from cotton industry (2021)
18. Yu, Y., Scheidegger, S., Elliott, J., Löfgren, Å.: climateBUG: a data-driven framework for analyzing bank reporting through a climate lens. Expert Syst. Appl. (2024)

19. Tranfield, D., Denyer, D., Smart, P.: Towards a methodology for developing evidence-informed management knowledge by means of systematic review. Br. J. Manag. 207–222 (2003)
20. Miettinen, M.: Are Materiality determination practices evolving in the wake of increasing legislation on sustainability reporting? Findings from EUpharmaceutical companies' reports. Int. J. Law Manage. **66**(3) (2024)
21. Dragomir, V., Chersan, I.: Double materiality disclosure as an emerging practice: the assessment process, impacts, risks, and opportunities. Account. Eur. (2024)
22. Lee, J., Serafin, A.M., Courteau, C.: Corporate disclosure, ESG and green fintech in the energy industry. J. World Energy Law Bus. (2023)
23. De Cristofaro, T., Gulluscio, C.: In search of double materiality in non-financial reports: first empirical evidence. Sustainability (2023)
24. Niveditha, R., Naveen, K., Parimi, M.R., Raam, A., Babu, S.: Develop CSR themes using text-mining and topic modelling techniques. In: 2020 IEEE International Conference on Cloud Computing in Emerging Markets (CCEM). 2020 IEEE International Conference on Cloud Computing in Emerging Markets (CCEM). IEEE (2020)
25. Yim, T.Y., Zhang, Y., Tan, W., Lam, T.-W., Yiu, S.M.: Meticulously analyzing ESG disclosure: a data-driven approach. In: 2023 IEEE International Conference on Big Data (BigData), Sorrento, Italy, 15–18 December 2023, pp. 2884–2889. IEEE (2023)
26. Dumitru, V.F., Ionescu, B.-Ş., Rîndaşu, S.-M., Barna, L.-E., Crîjman, A.: Implications for sustainability accounting & reporting in the context of the automation-driven evolution of ERP systems. Electron. (Switz.) (2023)
27. Kulkarni, A., Joseph, S., Patil, K.: Role of artificial intelligence in sustainability reporting by leveraging ESG theory into action. In: 2023 International Conference on Advancement in Computation & Computer Technologies (InCACCT). IEEE (2023)
28. Karbekova, A.B., Makhkamova, S.G., Inkova, N.A., Pakhomova, O.K.: Automation based on datasets and AI of corporate accounting and ssustainability reporting in quality management in industry 4.0. PES (2023)
29. Correa-Mejía, D.A., Correa-García, J.A., García-Benau, M.A.: Analysis of double materiality in early adopters. Are companies walking the talk? Sustain. Account. Manage. Policy J. (2024)
30. Fiandrino, S., Tonelli, A., Devalle, A.: Sustainability materiality research: a systematic lit. review of methods, theories and academic themes. QRAM (2022)

# Automating EU Taxonomy Reporting: Can Generative AI Facilitate Corporate Sustainability Reporting?

Ruben Hetfleisch[1]([✉]), Maximilian Nowak[1], and Fazel Ansari[1,2]

[1] Center for Sustainable Production and Logistics, Fraunhofer Austria Research GmbH,
Theresianumgasse 7, 1040 Vienna, Austria
Ruben.Hetfleisch@fraunhofer.at
[2] Research Group of Production and Maintenance Management, TU Wien, Theresianumgasse
27, 1040 Vienna, Austria

**Abstract.** To achieve global sustainability goals, enterprises are obliged to declare sustainability of their economic activities as part of EU Taxonomy reporting. Due to a lack of capacity and expertise, SMEs are unable to adequately fulfil this reporting requirement. A significant degree of automation is required to enable SMEs to efficiently comply with reporting. Recent advances in generative AI entails potentials to overcome several technical challenges inter alia i) heterogeneity of data, ii) necessary semantic intelligence, iii) usability requirements, and iv) assuring quality and reproducibility of automatically generated reports. This paper proposes a novel AI framework for automating sustainability reporting, thus assisting sustainability managers. It uses a hybrid approach incorporating knowledge graphs (KG) and large language models (LLM). Firstly, the EU taxonomy is converted into a Taxonomy-KG that specifies required KPIs and identifies target sources for data collection, thus retrieving necessary information for feeding KPIs. Subsequently, the Taxonomy KG guides an LLM to extract relevant information from corporate data, thereby enabling the EU Taxonomy reporting. By feeding the KPIs, a sustainability manager chatbot automatically generates an EU taxonomy report.

**Keywords:** Sustainable Development Goals · Corporate Sustainability · Reporting · SMEs · Generative AI · Knowledge Graph

## 1 Corporate Sustainability Reporting: Complexities and Challenges

To improve corporate sustainability, various global and corporate strategies and regulations have been developed among others UN's Sustainable Development Goals (17 SDGs), the European Green Deal, Corporate Sustainability Reporting Directive (CSRD), European Sustainability Reporting Standards (ESRS), and Net-Zero 2050. The lack of transparency regarding corporate sustainability makes it extremely challenging to identify, plan, monitor and control sustainability improvements in industrial context.

Focusing on transparency, the EU has implemented the so called EU Taxonomy, as part of the CSRD and ESRS, which affects the entire European industrial landscape, e.g.

H. Kohl et al. (Eds.): GCSM 2024, LNME, pp. 54–62, 2025.
https://doi.org/10.1007/978-3-031-93891-7_7

more than 46,000 enterprises in various sizes in Austria by 2026 [1]. It establishes criteria for corporate sustainability reporting, and involves enterprises disclosing their turnover, capital expenditures (CapEx) and operating expenses (OpEx), i.e. as either sustainable or non-sustainable. This is based on rigorous criteria related to six environmental objectives namely 1) climate change mitigation and 2) adaptation, 3) sustainable use and protection of water and marine resources, 4) transition to a circular economy, 5) pollution prevention and control, 6) protection and restoration of biodiversity and ecosystems, cf Fig. 1.

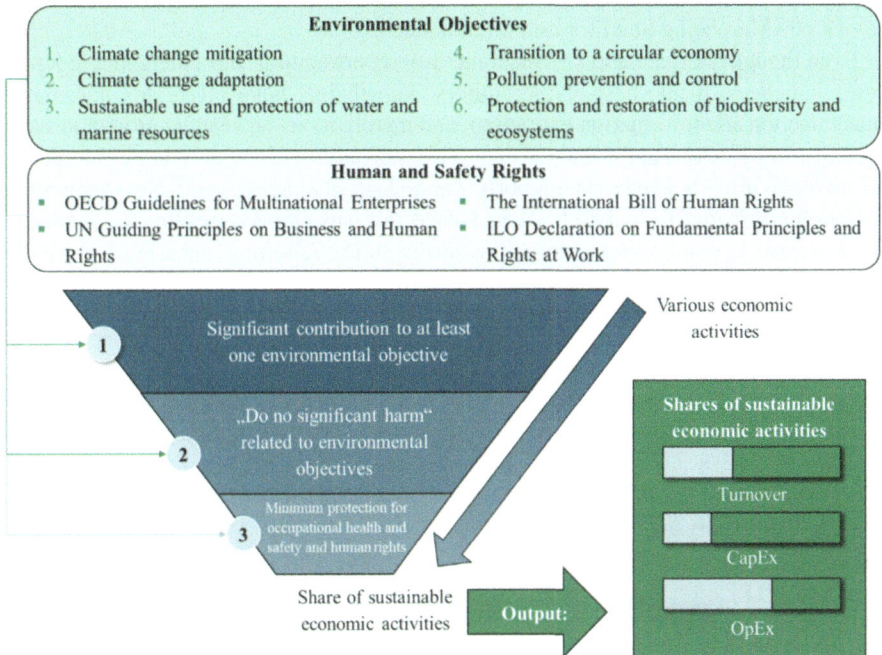

**Fig. 1.** Process of declaring sustainability of economic activities according to EU Taxonomy

After identifying taxonomy eligible economic activities, determining and declaring the alignment of these is constructed in a three step process (see Fig. 1), where it is assessed if i) an economic activity makes a significant contribution to one environmental objective, ii) it causes harm any to other objectives, and iii) it respects basic human and security rights. In practice, assessing these criteria according to the aforementioned objectives is a highly complex process, which requires extensive expert knowledge [2], and indeed demands high transparency and accessibility to copious amounts of diverse corporate data. This raises major challenge and requires considerable investments on resources, infrastructure and organizational restructuring.

In particular, Small and Medium-Sized Enterprises (SMEs) as the backbone of European industrial landscape confront the following challenges: 1) **Lack of human resources and capacity:** SMEs often struggle to hire the necessary one or two full-time equivalents dedicated to sustainability reporting, which limits their innovation and cooperation capacities, unlike large enterprises [3], 2) **Lack of expertise:** SMEs struggle with the complexity of the EU Taxonomy, which makes it challenging to effectively and efficiently map existing, diverse data sources to the reporting criteria [4], 3) **Lack of automation:** SMEs shall process an overwhelming amount of data manually. This involves repetitive overhead communication with various corporate departments to locate the right data, thus non-automated sustainability reporting causes significant problem in terms of effective time management and (labor) productivity [4], and 4) **Lack of reproducibility:** SMEs are primarily concerned with the reproducibility and trustworthiness of specific KPIs and sustainability reports. A lack of traceability of the results increases the risk of SMEs being fined for non-compliance [5].

Even though a large market of sustainability reporting tools has emerged to counter these challenges, to the best of the authors' knowledge, none offers any automated data collection and information extraction, and merely focus on visualization of already manually preprocessed and collected data. This indicated also that even large enterprises are unable to provide appropriate reports, due to lack of concise, traceable, trustworthy and reproducible metrics and KPIs [6–8]. Hence, this paper aims at tackling the challenge and introduces a solution for automating sustainability reporting, ensuring tractability and reproducibility of results. In particular, it proposes an AI framework that automates the manual reporting using hybrid methods with knowledge graphs (KG) and large language models (LLM).

## 2  Automating Sustainability Reporting: Technological Challenges

As illustrated in Fig. 2, sustainability manager (SM) deals with several tasks that can be partially and/or fully automated.

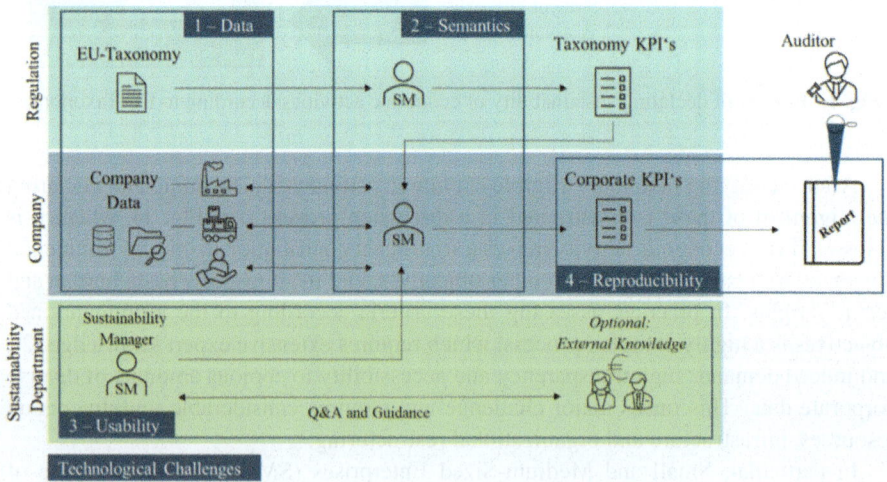

**Fig. 2.** Roles of the sustainability manager (SM) and resulting technological challenges

SM activities are divided in three levels: i) Sustainability Department level, ii) Company (corporate) level, and iii) Regulatory level. In Sustainability Department level, SM initiates and manages sustainability reporting according to EU Taxonomy. It is worth mentioning that large enterprises often utilize external knowledge for support and advice, while for SMEs this is mostly financially not feasible.

At the regulatory level, SM is mainly concerned with understanding the text documents of the EU Taxonomy. Core activities include identifying economic activities and their taxonomy eligibility as well as understanding and identifying the necessary KPIs (i.e. taxonomy KPIs) that must be reported to declare an activity sustainably. The knowledge of the necessary taxonomy KPIs is introduced at the level of company data. In company level, the core activities of SM are to collect the necessary data from various departments and stakeholders across and external to the company. This data must then be analyzed to identify and ensure factors like quality, relevance, and trustworthiness, and then processed and aggregated into specific KPIs/metrics. Ultimately, these metrics shall feed and provide required taxonomy KPIs. The majority of SM efforts is dedicated to communication with various departments and stakeholders. The proper interpretation of the data is also not trivial and requires domain expertise and depends on experiential knowledge. Once the necessary corporate KPIs are available, the sustainability report can be prepared and approved by the auditor.

Considering the above discussion, automating the sustainability reporting poses various challenges due to the complexity of the endeavor and the diversity of human activities. Four main technological challenges can be summarized as:

- **Technological challenge 1 – Data:** Management of growing amount of multimodal and heterogeneous data from different company sources and IT systems, and transferring them in a targeted manner to meet the EU taxonomy requirements. This involves technical challenges like interoperability across IT landscape.
- **Technological challenge 2 – Semantic Intelligence:** Unstructured corporate data must be automatically transformed into corporate KPI, which match and answer the required Taxonomy-KPI. Therefore, a semantically correct linkage between the unstructured corporate data, and the necessary Taxonomy-KPI is required.
- **Technological challenge 3 – Usability:** To provide appropriate support and guidance for SM, while dealing with EU taxonomy reporting, system acceptance and usability should be taken into account.
- **Technological challenge 4 – Reproducibility:** As taxonomy reporting is complex, determining correct, secure, and reliable information and metrics is a challenge. This renders explainability, traceability, and especially reproducibility of the results a main focus.

Following this line of research, the following sections discuss the implementation of AI technologies for automating sustainability reporting. First a brief state-of-the-art on AI technologies is provided (cf. Sect. 3). Further Section 4 introduces the AI framework as an artefact for automating sustainability reporting. Section 5 concludes the discussion and identifies the pathways for future research.

## 3   AI for Sustainability Reporting: Technological Possibilities

Natural Language Processing (NLP) is a subfield of AI that concentrates on processing textual data and currently undertakes significant research progress, especially in the thematic areas of generative AI and foundation models mostly represented by Large Language Models (LLM), and also semantic technology, i.e. integration of KGs.

KGs represent semantic data models that describe entities, their attributes, and the relationships between them within a graph-based structure. KGs are central to orchestrating knowledge structuring and organization, enhancing the precision of semantic search capabilities, and to supporting data-centric decision-making processes [9]. While research focuses on the construction of KG, the enrichment of data quality and the advancement of semantic interoperability are demanded [10].

LLMs (e.g. GPT-4, Mistral, Luminous, LLama, etc.) are advanced AI systems capable of understanding and generating human-like text, which renders them as very usable and easily accessible (cf. Technological Challenge 3 above). They excel in various tasks, including writing, summarizing, translating, and even coding, by learning from vast datasets of human and programming language. However, LLMs face significant challenges, inter alia they may produce hallucinated or factually incorrect content [11], which poses questions about the trustworthiness and responsibility of their outputs [12]. To increase trustworthiness and reproducibility of LLM outputs, the combination of KG (providing prior domain knowledge) and LLMs is proposed [13], i.e. fine-tuning and advancing LLMs by means of KG. Hereby KG-enhanced LLM-Inference is an appropriate approach, where information from a KG is used during the inference of an LLM, to guide the LLM to output the right information and avoid hallucinations. This can be used to enhance understanding, reasoning [14] and semantic search [13]. Especially in reporting, transparency is crucial in this integration, as stakeholders need to understand how the system arrives at its conclusions [15], which addresses the challenge of reproducibility (cf. Sect. 3). This is where KGs play a vital role, as their structured nature allows for better tracing of information sources and the reasoning paths taken by the LLM. This shows the potential for sustainability reporting and gives access to heterogenous data, which is seen as a key challenge (cf. Sect. 3). Since the construction of KG is time-consuming, LLMs can also be used for automated KG construction. Here, LLMs automatically recognize entities and relationships from unstructured text. This automated process could reduce the time needed for construction and the risk of errors drastically, while simplifying further updates of KG in the future. Hillebrand et al. discuss the potential of this approach in sustainability reporting, and underline the semantic capabilities of LLM [16], cf. Challenge of semantic linkage (cf. Sect. 3). However most scientific works focus on analyzing reports, instead of report generation [17–19]. In this context, assuring the correctness of the KG by an expert seems inevitable [20].

## 4   Generative AI Framework for Automating Sustainability Reporting

Considering the discussion in Sect. 3, Fig. 3 shows how various AI technologies are incorporated in a novel framework to automate repetitive SM activities (cf. Fig. 2). Here, generative AI models are divided into the three core functionalities "Automated KG Construction", "KG-enhanced LLM Inference" and "Human-AI Interaction".

**Automated KG Construction** uses a language model to automatically create a taxonomy KG from the EU taxonomy. The EU taxonomy and related appendices (regulations, directives, annexes) are incorporated as input, making comprehensive mapping of the EU taxonomy possible. By recognizing entities and relationships, a structured mapping is provided in which concrete and specific KPIs are filtered out of the EU Taxonomy and made retrievable by storing them with all dependencies in a KG. In the second step, **KG-enhanced LLM inference** is used with a language model that extracts the necessary corporate KPIs from the company data, under guidance of the taxonomy KG. The language model first needs access to all relevant company data. By accessing the previously created taxonomy KG, the language model knows which corporate KPIs it needs to extract. As it is not possible to ensure that all the necessary KPIs are already available in the company data, the core challenge is that the language model must have semantic knowledge of how the necessary corporate KPIs could be processed out of raw data.

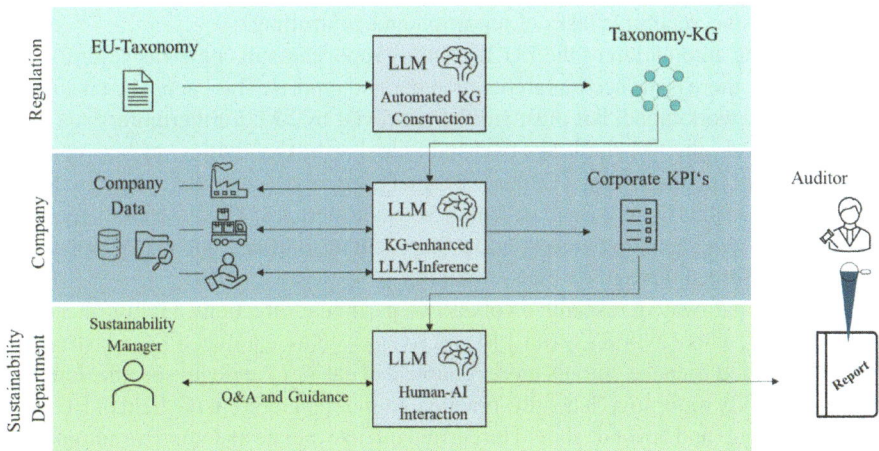

**Fig. 3.** Generative AI-Framework to automate sustainability reporting

The third core functionality focuses on **Human-AI interaction**. A chatbot-like system guides the SM through the reporting process, answering questions, integrating the previously extracted corporate KPI and transferring them into a complete report. The system must identify relevant economic activities for the company and integrate the knowledge and responsibility of the human in an appropriate and effective way. Furthermore, the system shall recognize, when it does not have certain data and instruct the human to collect specific data. When developing such chatbot-like systems, a high level of user acceptance needs to be prioritized, to make sure that it does not frustrate or mislead the SM.

As sustainability reporting is a complex and multifaceted endeavor, the interoperability of the individual core systems needs to be ensured and tested extensively in prototype environments to ensure the confidentiality and reproducibility of the outcoming reports. In addition to the positive effect of efficiency and high-quality sustainability reporting, the sustainability of the framework itself shall also be considered. Here, care should be taken not to use larger models than necessary and, if possible, to use foundation models that only need to be fine-tuned and not be trained from scratch.

## 5 Conclusion and Outlook

In the era of twin transformation, i.e. green and digital, there is an urgent need for automating sustainability reporting. This involves organizational, economic and technological challenges for all types of enterprises, especially SMEs. Using recent advances in generative AI, i.e. integrating LLM and KG, a framework is established that facilitates sustainability reporting. With the realization of such an AI framework, it is assumed that the effort of SMEs for reporting can be significantly reduced, and the quality of the reports significantly increased. Furthermore, the realization of such a framework can improve sustainability across industrial landscape, as it increases transparency. Enterprises can, therefore, focus on operational sustainability projects instead of time and resource intensive, repetitive tasks of reporting and controlling.

Notably, the automation of the EU Taxonomy reporting still confronts several limitations. Firstly, there is the accountability of generated reports. This cannot be taken over by an AI framework alone, but must still be approved by SM. Furthermore, even with an AI framework, the SM shall take over individual critical activities, such as controlling accuracy of the taxonomy KG, processing of particularly complex corporate KPIs, and collection of data in the case of decentralized data storages. This essentially requires involves identifying handling measures, development of solutions, and management and should be reinforced through on the job training.

The future pathway of research involves two particular directions. Firstly, on a technological level, where hybrid use of LLM and KG as well as automated KG construction shall be advanced to make outputs more reliable and easier to apply to specific domains like sustainability reporting. Secondly, practical implementations in the field of EU Taxonomy reporting (and beyond) should be carried out to evaluate and improve automation efficiency and effectiveness of the AI framework. Finally, the realization and evaluation of the framework, as a software demonstrator, is planned in two maturity levels, i.e. simulated and near-to-industry environments, and then in industrial environments.

**Acknowledgements.** The authors would like to acknowledge the financial support provided by the Austrian Research Agency (FFG) for the project "AI Enabled Sustainability Jurisdiction Demonstrator" (project No. 915229).

# References

1. Bundesministerium für Digitalisierung und Wirtschaftsstandort. KMU im Fokus 2022 (2022)
2. Europäische Kommission (ed) European Green Deal
3. EFRAG - European Financial Reporting Advisory Group. Draft European Susainability Reporting Standard (2022)
4. Giacomelli, A.: EU sustainability taxonomy for non-financial undertakings: summary reporting criteria and extension to SMEs (2022)
5. Bannier, C.: Nachhaltigkeitsberichterstattung - Aktuelle Herausforderungen und Chancen für Großunternehmen und Mittelständler. Springer, Wiesbaden (2023)
6. Arnold, J.L., Cauthorn, T., Eckert, J., et al.: Let's talk numbers: EU taxonomy reporting by German companies: what can we learn from the first EU taxonomy reporting season? (2023)
7. van der Heijden, T.: First EU taxonomy reports under the microscope: comparing reporting practice and regulatory requirements (2022)
8. Arvidsson, S., Dumay, J.: Corporate ESG reporting quantity, quality and performance: where to now for environmental policy and practice? (2022)
9. Chen, X., Jia, S., Xiang, Y.: A review: Knowledge reasoning over knowledge graph. Expert Systems with Applications (2020)
10. Peng, C., Xia, F., Naseriparsa, M., et al.: Knowledge Graphs: opportunities and challenges. Artif. Intell. Rev. 1–32 (2023)
11. Xu, Z., Jain, S., Kankanhalli, M.: Hallucination is inevitable: an innate limitation of large language models (2024)
12. Xu, M., Yin, W., Cai, D., et al.: A survey of resource-efficient LLM and multimodal foundation models (2024)
13. Pan, S., Luo, L., Wang, Y., et al.: Unifying large language models and knowledge graphs: a roadmap (2023)
14. Hu, L., Liu, Z., Zhao, Z., et al.: A survey of knowledge enhanced pre-trained language models. IEEE Trans. Knowl. Data Eng. 1–19 (2023)
15. Rajabi, E., Etminani, K.: Knowledge-graph-based explainable AI: a systematic review. J. Inf. Sci. (2022)
16. Hillebrand, L., Pielka, M., Leonhard, D., et al.: sustain.AI: a recommender system to analyze sustainability reports, pp. 412–416 (2023)
17. Bronzini, M., Nicolini, C., Lepri, B., et al.: Glitter or gold? Deriving structured insights from sustainability reports via large language models (2023)
18. Angin, M., Taşdemir, B., Yılmaz, C.A., et al.: A RoBERTa approach for automated processing of sustainability reports (2022)
19. Ni, J., Bingler, J., Colesanti-Senni, C., et al.: Paradigm shift in sustainability disclosure analysis: empowering stakeholders with chatreport (2023)
20. Huaman, E., Kärle, E., Fensel, D.: Knowledge graph validation (2020)

# Sustainability Evaluation

# Evaluation and Impact Analysis of Corporate Climate Protection Measures in Small and Medium-Sized Enterprises

Felix Budde[(✉)] and Benjamin Gellert

Fraunhofer-Institute for Production Systems and Design Technology IPK, Pascalstraße 8-9, 10587 Berlin, Germany
felix.budde@ipk.fraunhofer.de

**Abstract.** Corporate carbon reduction measures are crucial in mitigating climate change, particularly within small and medium-sized enterprises (SMEs). This paper presents a methodological approach for assessing the impact of climate protection interventions and corporate measures. The approach is applied on the example of measures initiated through a three-year publicly-funded applied research project in Germany. In this example, the impact is measured in two impact chains: (1) Specific support for SMEs in selecting and implementing climate protection measures through workshops and seminars, (2) Broad information campaign to inform enterprises about climate strategies and motivate climate protection measures. Through a comprehensive analysis of data collected from the SMEs participating in the project, the research reveals substantial impacts resulting from the implementation of climate protection measures during the project. The measures are categorized into distinct solution clusters and evaluated based on their effects on reducing the corporate carbon footprint (CCF). This paper offers valuable perspectives on the effectiveness of different approaches and their potential implications for business sustainability and climate protection strategies.

**Keywords:** climate protection management · climate protection measures · project impact · SMEs

## 1 Introduction

Despite the global goal set in the Paris Agreement of limiting global warming to 1.5 °C, current efforts are projected to result nearly in a 3 °C rise [1]. This alarming gap underscores the need for intensified action, as highlighted in the latest UN Climate Conference of the Parties (COP) 28, which called for the establishment of funds for the damage and losses caused by the climate crisis and set ambitious targets for expanding renewable energy and enhancing energy efficiency [2].

SMEs play a crucial role in this context. In 2022, about 24.3 million SMEs are active in the EU-27 and these SMEs accounted for 99.8% of all enterprises in the non-financial business sector [3]. SMEs are responsible for approximately 60% of all greenhouse gas emissions produced by the business sector in the EU [4]. Considering the global

© The Author(s) 2025
H. Kohl et al. (Eds.): GCSM 2024, LNME, pp. 65–72, 2025.
https://doi.org/10.1007/978-3-031-93891-7_8

responsibility to mitigate climate change and the rising regulatory requirements, effective climate management has become critically important for SMEs.

Climate management can be defined as the "identification, recording, active reduction and avoidance of relevant emission sources and emissions at the site as well as from upstream and downstream activities along the value chain [...]" [5]. The basic structure of corporate climate management can therefore be divided into four steps, where step 2 and 3 form the climate strategy [6]:

(1) assessment of the carbon footprint,
(2) development of corporate climate protection goals,
(3) implementation of measures to reduce the carbon footprint and
(4) reporting and communication internally and externally [5].

## 2   Scope of Paper

Every successful management system requires a process that allows to review the impact of implemented measures and evaluate their success, quality, and efficiency [7]. In case of the German publicly-funded advanced research project "KliMaWirtschaft - Nationwide Climate Protection Management for the Economy", which focuses on a company's environmental impact on the climate as part of the corporate sustainability performance, a method for the evaluation of corporate climate protection measures was needed. Over 250 German SMEs were supported in implementing a corporate climate management system over a total project timespan of three years to proactively engage in climate protection measures and thus reducing greenhouse gas emissions to minimize their CCF and environmental impact. The research project had a set target for the total GHG reduction of 333,520 t CO2e. Subsequently, a methodological approach for the evaluation of the project interventions and the total quantitative impact on greenhouse gas emissions in the industry had to be derived.

This paper shows a method to evaluate the success of climate protection measures regarding their GHG emission reduction using an "impact chain"-oriented approach. Furthermore, it gives insight in the analysis results of successful measures for climate protection. This approach enables SME to quantify implemented climate protection measures and allows for impact-based decision making in their prioritization.

## 3   Methodology and Conceptual Approach

In the research project, the impact was measured through two distinct impact chains (ICs), each involving different types of interventions and target groups. Impact Chain 1 (IC 1) focused on providing tailored support to SMEs by implementing a structured intervention model that involved workshops and seminars. This model was designed to equip SMEs with a scientifically grounded decision-support framework, helping them to identify, select, and implement climate protection measures effectively. Impact Chain 2 (IC 2) involved a broad information dissemination strategy. This chain utilized a variety of communication channels including blog posts, online events, and a climate protection toolbox [8] to inform enterprises about climate strategies and stimulate climate protection measures. These two impact chains offer reproducible models for evaluating climate action capabilities among SMEs.

## 3.1 Definition of Impact Chains

The basic systematics of an impact chain is shown in Fig. 1. An intervention triggers a $CO_2e$ reducing measure, which leads to a yearly reduction and has a cumulative impact over its duration. The calculation of the individual impact chains is based on the guidance document for determining the greenhouse gas reduction of the German National Climate Initiative (NKI). Intervention refers to those work steps or actions in a project that are aimed at triggering GHG reductions in the target group. GHG-reducing measures are the changes ultimately made by the actors in the target group [9].

However, important data is usually missing for the calculation, i.e. there are data gaps (variables x, y, z in Fig. 1). These gaps are specific to the two impact chains of the project example and will be closed in the following section (see Fig. 2).

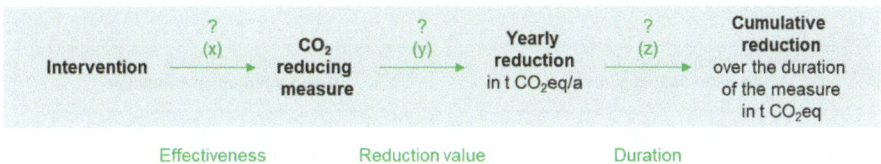

**Fig. 1.** Systematics of impact chains to determine cumulative carbon emissions reductions [9]

## 3.2 Specific Description of the Impact Chains in the Project

### Impact Chain 1: Specific SME Support
Impact Chain 1 examines the outcomes of targeted SME support (workshops and accompanying assistance) with the strategic implementation of corporate climate management. The support consisted of detailed information and tools for the assessment of the CCF, the development of climate protection goals as well as the derivation and implementation of measures to reduce the CCF.

Regarding the evaluated climate protection measures, the project focuses on a very broad range. The evaluated measures are divided into behavioral and technical measures in various impact areas in accordance with the NKI guidance document for determining the greenhouse gas reduction [9] to provide a better overview.

During the project, the GHG-reducing measures of the participating companies were systematically documented during the workshops. The following information was recorded: Description of the reduction measure, assignment of the effect to the emission categories of the GHG protocol and the annual reduction effect (in t $CO_2e$ /a). As the amount of GHG reducing measures were documented directly, no effectiveness estimations (gap x) were needed. The calculation of the annual reduction effect (gap y) of the individual measures was carried out independently by the companies. The estimation of the reduction effect duration (gap z) was based on the expected time a measure is carried out in the companies. Technical measures are expected to yield a more permanent effect (8-20 years) than behavioral measures (2 years).

### Impact Chain 2: Broad Information Campaign
The second impact chain relates to the information campaign carried out during the

project. The interventions of the broad information campaign are carried out via a climate protection toolbox, the information events, online courses on climate protection
management and as well as the information provided on the project website in blog posts
and social media activity.

The interventions led to a number of companies that have taken part in information
events (1,937) as well as online courses (156) as well as users of the climate protection
toolbox (1,712). With a conservative estimate of the effectiveness (x) of 8% (climate
protection toolbox) or 2% (information events) [9] and 42% (online courses, estimated
effectiveness derived from documented number of implemented GHG reducing measures
in relation to participating companies in the workshops) and 1% (information on project
website), this results in 279 measures. With an average reduction value (y) of 61 t $CO_2$e
of the measures implemented in IC 1 and a duration z of 4 years the cumulative GHG
reduction of IC 2 can be estimated (Fig. 2).

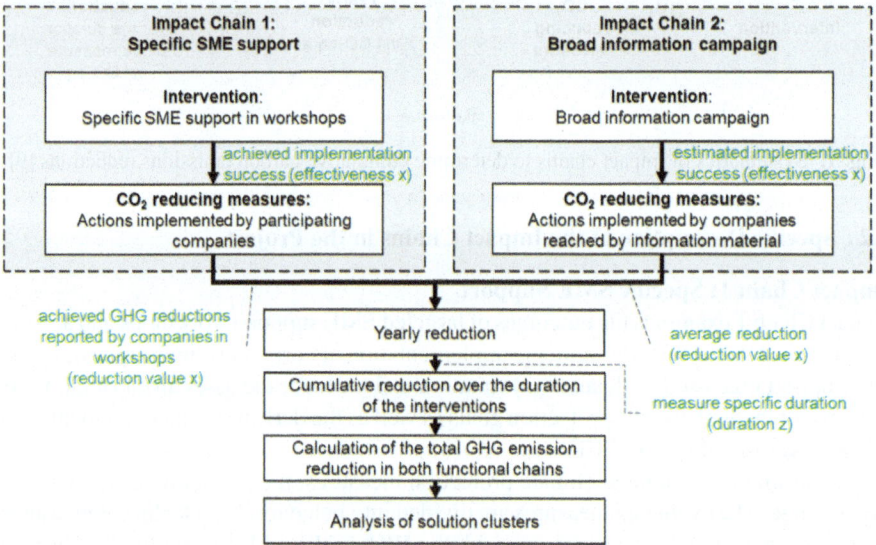

**Fig. 2.** Methodological approach to measure the impact in the two impact chains

## 4   Results and Interpretation

Through a comprehensive analysis of data collected from the SMEs participating in
the project, the research reveals substantial impacts resulting from the implementation
of climate protection measures during the project. The measures are categorized into
distinct solution clusters and evaluated based on their effects on reducing the CCF.
This offers valuable perspectives on the effectiveness of different approaches and their
potential implications for corporate sustainability and climate protection strategies.

Additionally, the method for evaluating the impact of climate protection measures,
combined with broad information campaigns, enables SMEs to quantify their strategic

initiatives for reducing GHG emissions. By integrating Impact Chain 2, SMEs can also assess the effectiveness of measures targeting behavioral change, as opposed to relying solely on strict policies or technological improvements.

## 4.1 Analysis of the Project Impact

Subject to the final evaluation after completion of the project, the interventions of IC 1 of the project resulted in an annual saving of 134,723 t $CO_2$e and thus a cumulative GHG reduction effect over the specific measure durations of 482,515 t $CO_2$e. The interventions of results of IC 2 result in an annual saving of 18,952 t $CO_2$e and thus in an additional cumulative GHG reduction effect of 75,808 t $CO_2$e. Overall, this results in a provisional project-specific reduction effect of 558,324 t $CO_2$e.

When interpreting the values, it should be noted that only those corporate measures were considered where specific statements on the reduction effect were reported. It is assumed that the participating companies have implemented further measures, which will only be evaluated after the end of the project and therefore cannot be documented as part of the project. The reduction effect of the project could therefore increase over time. On the contrary, it is also possible that the measures recorded were not necessarily initiated by the project but were already being implemented before. In addition, companies potentially reported specifying measures coming from other funded projects, leading to double counting across publicly-funded initiatives.

In comparison with the reduction potential estimated at project start (IC 1: 133,440 t $CO_2$e and IC 2: 200,080 t $CO_2$e and a total GHG reduction of 333,520 t $CO_2$e), a shift of the impact from IC 2 to IC 1 can be observed. Overall, it is apparent that the measures in IC 1 have achieved a greater reduction effect than originally predicted, while the measures in Impact Chain 2 lag the expected reduction effect. This is partially due to the late publishing of the climate protection toolbox as well as increased reductions achieved by single participating companies. As further project activities such as the broad information campaign and online events will take place until the project is completed, the results above should be regarded as provisional. However, the project impact is already greater than initially expected.

## 4.2 Evaluation and Impact Analysis of Solution Clusters from IC 1

For the evaluation and impact analysis of IC 1 the reported climate protection measures from the participating companies were classified in distinct solution clusters, see Table 1.

The clusters with the most reported measures concern electricity and heat procurement (C3), mostly in terms of green power purchase agreements, as well as mobile equipment (C2), particularly transformation to e-mobility. While C3 shows high average $CO_2$e reduction per measure, C2 measures only achieve relatively modest reductions. As a result, while the implementation of measures in C2 were often reported and can be considered as quick wins, each measure individually does not significantly impact the company's overall emissions.

In contrast, measures from the behavioral cluster that address upstream resource consumption and transportation (C7) and energy efficiency (C6) show greater impact. Measures of cluster C7 typically target scope 3 emissions associated with purchased

**Table 1.** Climate protection clusters in Impact Chain 1

| Cluster | Type of measures | Description | Measures documented | Average GHG reduction per year in t $CO_2e/a$ | Duration | Cumulative total reduction in t $CO_2e/a$ |
|---|---|---|---|---|---|---|
| C1 | technical | Stationary systems | 4 | 86.51 | 8 | 2,768.29 |
| C2 | technical | Mobile equipment, in particular company cars and fleet management | 24 | 153.23 | 8 | 29,420.38 |
| C3 | behavioral | Electricity and heat procurement | 27 | 2,055.67 | 2 | 111,006.09 |
| C4 | behavioral | Energy efficiency (buildings) | 6 | 31.73 | 2 | 380.76 |
| C5 | technical | Energy efficiency (buildings) | 19 | 171.33 | 20 | 65,106.46 |
| C6 | technical | Energy efficiency (process technologies) | 8 | 2,573.21 | 8 | 164,685.67 |
| C7 | behavioral | (Upstream) resource consumption/transportation | 13 | 3,302.94 | 2 | 85,876.54 |
| C8 | behavioral | Employee mobility | 6 | 340.30 | 2 | 4,083.65 |
| C9 | behavioral | Downstream logistics, use and value assessment | 7 | 110.31 | 2 | 1,544.28 |
| C10 | technical | Electricity and heat procurement | 17 | 66.84 | 8 | 9,090.20 |
| C11 | behavioral | Mobile equipment, in particular company cars and fleet management | 5 | 855.36 | 8 | 2.768,29 |

goods and services, a category in which manufacturing companies report particularly high emissions in their CCF. For instance, replacing input materials with high emissions with those that have a lower product carbon footprint, making it a substantial lever for emission reduction.

Mostly, the companies opted for quick wins in energy efficiency. Even though the overwhelming majority (92%) of emissions disclosed by European companies in 2022 were in Scope 3, with the use of sold products (57%) and purchased goods and services (17%) cited as companies' key hotspots [10], the highest cumulative total reduction is observed in clusters (C6, C3, C5) reducing emissions in Scope 1 and 2. C7 still shows high emission reductions.

# 5  Summary and Outlook

This paper presents a methodological approach for the impact evaluation of climate protection interventions and corporate measures to reduce the CCF. The approach was derived within a publicly-funded research project, where it was used to evaluate corporate climate protection measures implemented by over 250 SMEs in Germany. Specifically, two kinds of interventions were analyzed:

(1) Strategic implementation of climate protection measures through targeted support in workshops and seminars. This involved the development of a procedural workshop concept and assistance tools designed to help SMEs with identifying, selecting, and implementing climate protection measures.

(2) Information dissemination through broad-reaching campaign, including blog posts, online events, and a climate protection toolbox. The aim was to create an environment promoting climate action by increasing awareness and motivation externally, aiming to engage a broad spectrum of SMEs throughout the project phase, while also enhancing internal support within SMEs for the strategic implementation of corporate climate management practices.

The proposed approach demonstrated significant potential for GHG reduction when corporate climate protection measures are strategically implemented, achieving a cumulative reduction of 482,515 t $CO_2$e across all reporting enterprises. Additionally, interventions such as a comprehensive information campaign, including online events, the climate protection toolbox, and social media activities, contributed to a substantial reduction of 75,808 t $CO_2$e.

These findings suggest that SMEs aiming to transition to environmentally sustainable practices should prioritize the implementation of effective climate protection measures. Simultaneously, these efforts should be supported by internal information campaigns to further drive the adoption of GHG-reducing measures. Finally, by adopting this approach, SMEs can use a straightforward method for quantifying the impact of climate protection interventions and systematically prepare it for future decision-making.

A critical next step is the broader application of the method in corporate contexts, enabling enterprises to systematically assess the long-term impact of their climate protection measures. Enterprises must adopt a continuous improvement approach, incorporating the duration of measures into their impact calculations. Only interventions that are sustained over time can contribute meaningfully to long-term emission reductions, supporting the achievement of climate neutrality by 2050 or earlier.

**Acknowledgement.** The joint project between the Fraunhofer Institute for Production Systems and Design Technology IPK and the Bundesverband Der Mittelstand. BVMW is funded by the German Federal Ministry of Economics and Climate Protection (BMWK) as part of the funding call for innovative climate protection projects of the National Climate Protection Initiative (NKI) under the code 67KF0166B.

# References

1. United Nations Environment Programme. Emissions Gap Report: Broken Record – Temperatures hit new highs, yet world fails to cut emissions (again) (2023)
2. UNFCCC. COP 28: What Was Achieved and What Happens Next? (2023)
3. Di Bella, L., Katsinis, A., Lagüera-González, J., et al.: Annual report on European SMEs 2022/2023: SME performance review 2022/2023. EUR, vol. 31618. Publications Office of the European Union, Luxembourg (2023)
4. European Commission. Annual Report on European SMEs 2021/2022: SMEs and environmental sustainability: SME Performance Review 2021/2022, Luxembourg (2022)
5. Kube, M., Rhiemeier, J.-M., Stern, F., et al.: Unternehmerisches Klimamanagement entlang der Wertschöpfungskette - eine Sammlung guter Praxis (2016)
6. Budde, F., Gellert, B., Orth, R., et al.: Klimamanagement zur Verbesserung der CO2-Bilanz in der Produktion. Zeitschrift wirtschaftlichen Fabrikbetrieb **118**, 40–44 (2023). https://doi.org/10.1515/zwf-2023-1017
7. Naslund, D., Norrman, A.: A performance measurement system for change initiatives. BPMJ **25**, 1647–1672 (2019). https://doi.org/10.1108/BPMJ-11-2017-0309
8. Budde, F., Kohl, H.: Corporate climate protection measures to improve the carbon footprint in production: a systematic development of a toolbox with climate protection actions for the reduction of greenhouse gas emissions in small and medium-sized enterprises (2023)
9. Tews, K., Schumacher, K., Eisenmann, L., et al.: Arbeitshilfe zur Ermittlung der Treibhausgasminderung (2020)
10. CDP; Capgemini Invent. From stroll to sprint: A race against time for corporate decarbonization (2023)

# Use of the Consumption Performance Sustainability Index as a Decisional Tool at a Preliminary Stage of Project Development

Leopoldo De Bernardez[1], Giampaolo Campana[2(✉)], and Sebastian Mur[1]

[1] Buenos Aires Institute of Technology (ITBA), Iguazú 341,
C1437 Ciudad Autónoma de Buenos Aires, Argentina
[2] University of Bologna, Viale del Risorgimento 2, 40136 Bologna, Italy
giampaolo.campana@unibo.it

**Abstract.** The choice of materials and manufacturing processes based on the design of industrial products depends on several factors related to common practice, availability of materials and machines and reliability of the production processes. Furthermore, industrial products must achieve a zero-defect policy and be safe and sustainable by design. This work implements the Consumption Performance Sustainability Index and extends its use to compare two Additive Manufacturing processes that transform the same polymers into a reference geometry. The index considers the part design and production parameters related to the material transformation due to the manufacturing processes. In particular, the product's mechanical performance, materials consumption, and energy used for manufacturing. It is here developed to evaluate, at an early stage, the product design by using the production equipment's available technical data, manufacturing times and material consumption - estimated through specific software - and other data from the scientific literature. The index is proposed to assess the sustainability of products and find the best manufacturing alternative as a complementary tool to standard approaches based on life cycle sustainability assessment. The procedure includes optimising the geometry using topology or generative design, assessing applied stresses and production times, and evaluating the performance of the transformed material based on part orientation to optimise the manufacturing process.

**Keywords:** Product Design · Polymers · Performance Index · Sustainability · Additive Manufacturing

## 1 Introduction

During the design stage of a part, the initial step is to identify the requirements it must meet. If the part is subjected to mechanical stresses, the design must ensure optimal performance by selecting the appropriate design and material and defining the necessary manufacturing processes. From a sustainability perspective, environmental impact assessments are typically conducted after the product's manufacturing, often using the

H. Kohl et al. (Eds.): GCSM 2024, LNME, pp. 73–81, 2025.
https://doi.org/10.1007/978-3-031-93891-7_9

Life Cycle Sustainability Assessment (LCSA) [1], a comprehensive and labour-intensive analysis. Identifying the most suitable combination of product geometry, material and manufacturing processes during the design phase would be far more beneficial if various alternatives, including sustainability considerations, are compared to determine the optimal options.

This work aims to strike the optimal balance between the mechanical performance of the part and the consumption of materials, energy, and time required for its production, focusing on additive manufacturing (AM) technologies [2]. Specifically, it compares two distinct AM processes: Fused Deposition Modelling (FDM) and Selective Laser Sintering (SLS). Additive manufacturing, due to its versatility, allows for the optimisation of part design through generative design (GD) and topology optimisation (TO) procedures [3, 4]. This means that once the dimensional constraints of the part are defined, specialised software can be used to optimise the final design by imposing objectives and minimising material usage. This study compares two different AM methods, reflecting a common decision-making scenario for designers. The primary goal is to develop a design methodology that enables the selection of the most appropriate manufacturing method from a sustainability perspective by utilising an index also accounting for the mechanical performance of the part, as different technologies can introduce anisotropy in the material, leading to variations in mechanical properties depending on the print orientation, which could result in service failures [5]. Furthermore, the proposed index must be simple to apply and contain elements not typically included in the LCA or other indices under development [6, 7]. A description of other indices under development with similar purposes was already investigated in [6] and is not reported here for brevity.

To this end, a procedure has been developed to estimate the sustainability of a given manufacturing process based on a modified version of the previously defined Consumption Performance Sustainability (CPSI) Index, which was applied to parts manufactured using the FDM process with varying infill patterns and densities [6, 7]. To the authors' knowledge, no other index includes all the factors considered here.

Section 2 presents a reference part used as a case study. The production machines are introduced there, and the manufacturing parameters, which can be predicted using open-source software specific to fused deposition modelling (FDM) and selective laser sintering (SLS) technologies, are evaluated. Section 3 focuses on evaluating the Consumption Performance Sustainability Index, combining data from existing scientific literature and data calculated using open-source software.

## 2 Proposed Approach, Methods and Tools

### 2.1 The Proposed Approach in a Digital Environment

This work aimed to develop a tool that makes it possible to evaluate different manufacturing methods in a digital environment without carrying out prior laboratory tests. The result could later be verified by materialising the design. For a first estimate, data must be collected to calculate the CPSI.

The process begins by defining the requirements the part must meet, from which the design is made. To manufacture a part that meets the target performance, the mechanical requirements or stresses that it has to withstand must be considered. The piece's dimensions, the connection points with the rest of the structure and any other condition that represents a geometric restriction must also be considered. The appropriate software can then be used to optimise the part's design for production through additive manufacturing [3]. In the case of AM processes, generative design or topology optimisation can be used as deemed appropriate. The objective of generative design is to minimise the amount of material used.

After obtaining the optimised 3D design of the part, specific software corresponding to each manufacturing equipment that is to be compared is used to obtain the G-code file that will be used for printing. The software typically calculates and provides information on the amount of material to be used, the support material required, and the printing time. Additionally, the energy required for each manufacturing process can be estimated using literature data, which may include information from the equipment manufacturer.

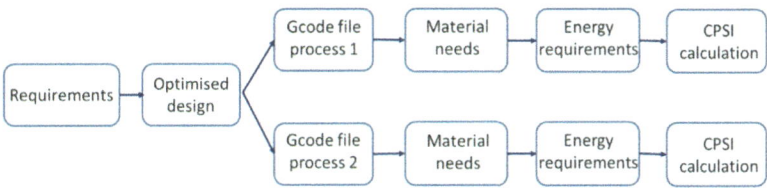

**Fig. 1.** Steps to follow to calculate the CPSI for two different additive manufacturing processes.

The process to be followed is represented in Fig. 1. "Requirements" refers to mechanical and geometric requirements, such as dimensions and applied stresses, and geometric constraints, such as part shape and position of connection points. "Optimised design" is the result obtained from the utilised software, considering the requirements and restrictions and the material that has been selected for the part. From this point, the flowchart is divided into branches, one for each AM process. "G-code file" refers to the output of the slicer software that each printing equipment needs to manufacture the part. "Material needs" refers to the material used to build the final part, the necessary support material, and all the wasted material from the process. "Energy requirements" include the energy needed to produce the part, the energy required to make the semi-finished products used in the manufacturing process, and the energy used to recycle the material if applicable. Material and energy requirements depend on the chosen manufacturing process.

The CPSI for a given design, material and manufacturing process is defined in Eq. 1–8, starting from the previous definition already published in [6, 7]. This index is calculated through a simple formulation that includes the most relevant aspects for the sustainability of the manufacturing process: energy and materials consumption. Furthermore, it also includes the mechanical properties, allowing the designer to consider the expected resistance of the manufactured part [8]. The general formulation is reported in Eq. 1. It is a multiplication of terms that are described in Eq. 2–8. $C_i$ are coefficients, $0 < C_i \leq 1$, that are introduced to account for the local relevance of each analysed factor.

$$CPSI = \prod_i I_i = I_{prop}^Y \times I_{design} \times I_{mat} \times I_{time} \times I_{s.e} \tag{1}$$

$$I_1 = I_{prop}^Y = C_1 \times \frac{UTS_m}{\rho_m} \tag{2}$$

$I_{prop}^Y$ represents material properties (Eq. 2); $UTS_m$ and $\rho_m$ are the feedstock material's ultimate tensile strength and density, respectively.

$$I_2 = I_{design} = C_2 \times \left( \frac{UTS_s}{UTS_m} \times \frac{\rho_m}{\rho_s} \times \frac{\sigma_{eq}}{\sigma_{eq,lim}} \right) \tag{3}$$

$I_{design}$ accounts for the increase or reduction of the mechanical properties observed in the final part, and the part density $\rho_s$ compared with the feedstock material. $\sigma_{eq}/\sigma_{eq,lim}$ represents the maximum value of the equivalent stress over the limit equivalent stress and describes the material exploitation (e.g. evaluated by von Mises). UTS values can be evaluated experimentally for the different manufacturing technologies and printing orientations. The material consumption index $I_3 = I_{mat}$ accounts for any quantity of required material for the production process that is not a portion of the final piece (Eq. 4):

$$I_3 = I_{mat} = C_3 \times \frac{m_s}{m_t} \tag{4}$$

where $m_s$ and $m_t$ are the mass of the part and the total mass involved in the production process, respectively; $I_{mat}$ represents the mass of the material wasted during the manufacturing process, $m_m$, that will not be part of the final piece (the support materials in the case of AM). The total mass can be calculated by the sum $m_t = m_s + m_m$. This factor always diminishes the index value. The material consumption index is dimensionless, as is the design index.

The machine utilisation index is calculated as in Eq. 5:

$$I_4 = I_{time} = C_4 \times \frac{t_a}{t_p} \tag{5}$$

where $t_p$ is the average cycle time for the complete process (conditioning, printing, unloading), and $t_a$ is the average time available for production. As for the previous factors, the machine utilisation index is dimensionless; in this case, it is less than 1. The last multiplication factor represents the energy consumption index (Eq. 6):

$$I_5 = I_{s.e} = C_5 \times \frac{m_s}{E_t} \tag{6}$$

where $E_t$ is the total consumed energy that can be calculated as $E_t = E_m + E_f + E_{np}$, with $E_m$ the energy used to produce the semi-finished material from material feedstock, $E_f$ the energy consumed for the component production, and $E_{np}$ the energy consumed during non-productive hours due to specific requirements of production machines; the unit of this index is $s^2/m^2$ if the International Standard of Units is used.

Replacing (2) to (6) in (1) and considering for simplicity $C_i = 1$ for all $i$, CPSI results in a dimensionless number (Eq. 7):

$$CPSI = \frac{UTS_m}{\rho_m} \times \frac{UTS_s}{UTS_m} \times \frac{\rho_m}{\rho_s} \times \frac{\sigma_{eq}}{\sigma_{eq,lim}} \times \frac{m_s}{m_t} \times \frac{t_a}{t_p} \times \frac{m_s}{E_t} \tag{7}$$

CPSI will be higher if the achieved specific modulus or UTS and the machine time utilisation are higher and if the design optimises the mechanical performance. It will be lower if the material or the energy consumption is higher. Comparing indices corresponding to two or more different manufacturing processes is possible. So, the process that results in a product with a better relationship between performance and consumption of materials and energy can be identified.

## 2.2   The Case Study

The proposed methodology can be used to compare different processes, for example, additive and subtractive manufacturing, ensuring that the material and energy consumption data include the production of the corresponding semi-finished products if these were different for each process being compared. The data required for calculating CPSI can, in many cases, be obtained from the literature. Besides, material consumption and manufacturing time can be estimated from specific software; energy data can be estimated from literature or direct measurements of the equipment.

In this paper, Fused Deposition Modelling (FDM) and Selective Laser Sintering (SLS) processes were considered for product manufacturing using polyamide (Nylon). A benchmark geometry, represented in Fig. 2, was chosen as a case study. The optimisation criteria by the generative design were used to achieve the maximum stiffness of the part since a thermoplastic material was considered (Nylon).

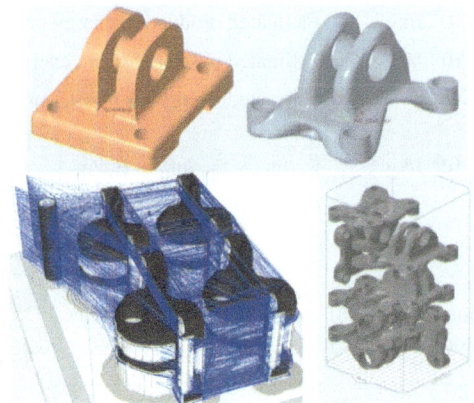

**Fig. 2.** Benchmark part (upper left) and its optimisation by generative design (upper right). Their possible arrangement for manufacturing purposes (slicer virtual environments: FDM, lower left and SLS, lower right).

Table 1 and Table 2 collate the estimated values to calculate the CPSI. It is supposed to manufacture four pieces. Available time is considered equal to one day. This value is compared with the production time, which, in the case of the SLS process, is the sum of the printing and cooling time.

The mathematical relationship between the part's mass and the powder's total mass in SLS is calculated, considering that 5% of the remaining powder is wasted due to the thermal cycle and cannot be reused.

In the case of the FDM process, the layer thickness was set to 0.2 mm; supports were generated by considering the standard geometry and a typically used polymer compatible with Nylon (White Breakaway).

**Table 1.** Factors to estimate the CPSI (per piece for four pieces, SLS technology. Printing machine: Fuse 1 by FormLabs. Slicer: PreForm v. 3.37.2 by FormLabs).

| Factor | Value & units | Description and references |
|---|---|---|
| $TS_p/\rho_p$ | **48/1.18** MPa/kg/m$^3$ | Part tensile strength/material density [5] |
| $t_p = t_p^{1} + t_p^{2}$ | 21765 s | Estimated production time per piece for 4 pieces [5] |
| $t_p^{1}/t_p^{2}$ | 12060 s/9705 s | Estimated printing/cooling time per piece for 4 pieces |
| $t_A/t_p$ | **3.97** | Available time (1 day)/production time |
| $m_p$ | 0.162 kg | Part mass (porosity 2%) [6] |
| $m_T$ | 0.585 kg | Estimated total powder mass |
| $m_p/m_T$ | **0.88** | Estimated material efficiency (5% wasted material) |
| $P_p/P_c$ | 1725 W | Installed power for printing and cleaning |
| $P_b$ | 690 W | Installed power for blasting |
| $E_p/m_p$ | $2.32 * 10^8$ J/kg | Estimated production energy per part mass [7] |
| $E_{pp}/m_p$ | $3.83 * 10^6$ J/kg | Estimated post-processing energy per part mass |
| $E_w/m_p$ | $1.0\ 10^8$ J/kg | Estimated energy for powder production per part mass [10] |
| $E_T/m_p$ | **$3.36 * 10^8$** J/kg | $E_p/m_p + E_{pp}/m_p + E_w/m_p$ |
| CPSI | **0.426** | $(TS_p/\rho_p)*(t_A/t_p)*(m_p/m_T)*(m_p/E_T)$ |

**Table 2.** Factors to estimate the CPSI (per piece for four pieces, FDM technology. Printing Machine: Ultimaker Factor 4. Slicer: Cura v. 5.7.2 by Ultimaker).

| Factor | Value & units | Description and references |
|---|---|---|
| $TS_p/\rho_p$ | **60/1.17** MPa/kg/m$^3$ | Part tensile strength/material density [5] |
| $t_p$ | 62010 s | Estimated production time per piece for 4 pieces |
| $t_A/t_p$ | **1.39** | Available time (1 day)/production time |
| $m_p/m_S$ | 0.160 kg/0.54 kg | Estimated part mass/support mass |

*(continued)*

**Table 2.** (*continued*)

| Factor | Value & units | Description and references |
|--------|---------------|---------------------------|
| $m_T$ | 0.214 kg | Estimated total part mass |
| $m_p/m_T$ | **0.75** | Estimated material efficiency (5% wasted material) |
| $P_p$ | 500 W | Installed power for printing |
| $E_p/m_p$ | $1.94 * 10^8$ J/kg | Estimated production energy per part mass |
| $E_w/m_p$ | $1.00 * 10^8$ J/kg | Estimated energy for filament production per part mass |
| $E_T/m_p$ | $\mathbf{2.94 * 10^8}$ **J/kg** | $E_p/m_p + E_w/m_p$ |
| CPSI | **0.181** | $\mathbf{(TS_p/\rho_p)*(t_A/t_p)*(m_p/m_T)*(m_p/E_T)}$ |

## 3   Results and Discussion

Due to the anisotropy of the mechanical properties in the material printed by FDM [9], where the bond between the deposited layers is weak and can cause layer delamination, an orientation of the part during printing was sought that would prevent the stresses during use were perpendicular to the joint between layers. Additionally, a linear fill pattern but with a raster angle of ±45° and a fill density of 100% was used to reduce the anisotropy within each layer of deposited material and improve the strength of the material.

The calculation of the CPSI was carried out for the two chosen manufacturing processes, FDM and SLS and for the same material, PA12, since it is possible to use it for both manufacturing processes. The variables' values must be known to calculate the index. To compare 3D printing technologies of similar complexity, two pieces of machines were selected: Ultimaker Factor 4 for FDM and Formlabs Fuse 1 for SLS.

If for simplicity $\sigma_{eq}/\sigma_{eq,lim} = 1$, Eq. 7 is reduced to:

$$CPSI = \frac{UTS_s}{\rho_s} \times \frac{m_s}{m_t} \times \frac{t_a}{t_p} \times \frac{m_s}{E_t} \tag{8}$$

The values used for the CPSI calculation are shown in Tables 1 and 2.

From Tables 1 and 2, it can be seen that the CPSI is higher when using SLS technology than when using FDM. The result does not mean SLS technology is generally more sustainable than FDM, but is limited to this case study. In other cases, with different geometries, materials or numbers of pieces to produce, the result could be different. However, it is interesting to understand the reasons behind the difference in the index. Figure 3 shows a spider or radar chart of the factors or partial indices and the CPSI for both printing technologies under investigation. It can be seen from the figure that the main factors that generate the difference are the energy used per unit of deposited/sintered mass and the relationship between the available time and the printing time.

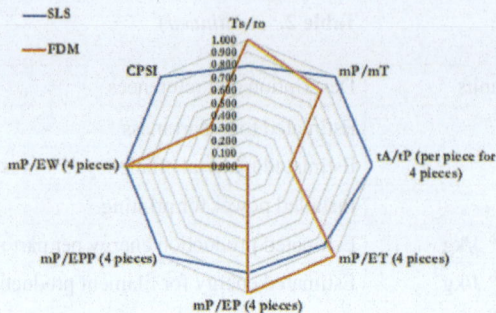

**Fig. 3.** Spider chart of the CPSI and the factors used to calculate it for SLS and FDM.

## 4   Conclusions

This work presents a decisional tool that allows comparing, from the design stage, two or more technologies or pieces of equipment that can be used to produce a part that must meet specific mechanical requirements. The proposed Consumption Performance Sustainability Index can be calculated in the early stages of the design process using data available in the scientific literature, specific data from the equipment used to manufacture the part, and estimating manufacturing times and material consumption using specific software for each piece of equipment. Recognising why one manufacturing process has a higher index than another would also allow for modifying some aspects of the design to improve the overall performance of the design.

## References

1. Ribeiro, I., et al.: Framework for life cycle sustainability assessment of additive manufacturing. Sustainability **12**(3), 929 (2020)
2. Prabhakar, M.M., Saravanan, A.K., Lenin, A.H., Mayandi, K., Ramalingam, P.S.: A short review on 3D printing methods, process parameters and materials. Mater. Today: Proc. **45**, 6108–6114 (2021)
3. Prathyusha, A.L.R., Babu, G.R.: A review on additive manufacturing and topology optimization process for weight reduction studies in various industrial applications. Mater. Today: Proc. **62**, 109–117 (2022)
4. Vlah, D., Žavbi, R., Vukašinović, N.: Evaluation of topology optimization and generative design tools as support for conceptual design. In: Proceedings of the Design Society: DESIGN conference, vol. 1, pp. 451–460. Cambridge University Press (2020
5. Zohdi, N., Yang, R.: Material anisotropy in additively manufactured polymers and polymer composites: a review. Polymers **13**(19), 3368 (2021)
6. De Bernardez, L., Campana, G., Mele, M., Mur, S.: Towards a comparative index assessing mechanical performance, material consumption and energy requirements for additive manufactured parts. In: Lecture Notes in Mechanical Engineering, pp. 302–310 (2023)
7. De Bernardez, L., Campana, G., Mele, M., Sanguineti, J., Sandre, C., Mur, S.: Effects of infill patterns on part performances and energy consumption in acrylonitrile butadiene styrene fused filament fabrication via industrial grade machine. Progr. Addit. Manuf. **8**, 117–129 (2023)

8. Terekhina, S., Tarasova, T., Egorov, S., Guillaumat, L., Hattali, M.L.: On the difference in material structure and fatigue properties of polyamide specimens produced by fused filament fabrication and selective laser sintering. Int. J. Adv. Manuf. Technol. **111**, 93–107 (2020)

9. Telenko, C., Conner Seepersad, C.: A comparison of the energy efficiency of selective laser sintering and injection molding of nylon parts. Rapid Prototyp. J. **18**(6), 472–481 (2012)

10. Su, D., Yang, J., Liu, S., Ren, L., Qin, S.: Preparation of polyamide 12 powder for additive manufacturing applications via thermally induced phase separation. e-Polym. **22**(1), 553–565 (2022)

# Sustainability Assessment and Optimization in Construction Site: A Simulation-Based Approach

Alexander Schlosser[1]([⊠]) [iD], Peter Schuderer[2], and Jörg Franke[1]

[1] Institute for Factory Automation and Production Systems, Friedrich-Alexander-Universität Erlangen-Nürnberg, Erlangen, Germany
alexander.schlosser@faps.fau.de
[2] Technische Hochschule Ingolstadt, Ingolstadt, Germany

**Abstract.** The construction industry is currently facing significant challenges. In order to address these challenges, the REMUS simulation model library for the construction industry is being developed. To this end, the physical modules are divided into stationary and mobile modules, as well as information objects. To create the simulation model and conduct the simulation experiment, a requirements cluster with the most important parameters of construction sites is created. The elements of sustainability—environmental, economy, and social aspects— are employed to assess the simulation results and to optimize the model. To this end, corresponding KPIs, methods, and procedures are delineated, which are documented during the various simulation experiments and evaluated subsequently. The equipment and environment exert an influence on the "economy". This is reflected in the costs associated with the model components and their operation. The area of "environmental" is represented by the consumption of input materials. Alternative consumption and recovery concepts are implemented and compared here. The "social" aspect is represented by the human-machine collaboration. As part of the simulation experiments, the recorded variables are continuously adapted and refined. This process enables the simulation to improve the sustainability of the construction site environment.

**Keywords:** Simulation · Construction Side · Sustainability

## 1 Introduction

The construction industry, a major economic sector in Germany, accounted for 6% of gross value added in 2023. Dominated by small and medium-sized enterprises (SMEs), with only 13.1% employing more than 200 people in 2022, the industry faces significant sustainability challenges [6, 12]. In particular, the European Financial Reporting Advisory Group (EFRAG) has issued a draft Voluntary Sustainability Reporting Standard (VSME Standard) for SMEs, and the German Sustainable Building Council (DGNB) offers certification for sustainable construction sites [3, 8].

H. Kohl et al. (Eds.): GCSM 2024, LNME, pp. 82–90, 2025.
https://doi.org/10.1007/978-3-031-93891-7_10

Sustainably certified sites must consider aspects such as environmental, economical, socio-cultural and functional aspects, technology and processes. As SMEs are particularly affected to cost and efficiency pressures, maintaining competitiveness remains crucial. German construction SMEs have yet to establish substantial links with automation providers - there remains a gap in comprehensive digitization and networking on construction sites [11]. However, experience from other sectors highlights the potential benefits of digitalization and automation processes for SMEs, with several studies emphasizing the role of digitalization in increasing productivity in the construction sector [5, 10].

The REMUS simulation model provides components for automated construction sites, facilitating the rapid creation of models that reflect typical conditions and sustainability aspects. This allows simulation experiments to be conducted before and during construction phases to optimize supply chains.

## 2 Construction Site Simulation – Status and Challenges

Project RoMuLuS enhances simulation-based material provisioning for autonomous construction, addressing gaps in industry-specific modeling. The REMUS simulation model database is being developed in this project.

### 2.1 Simulation Model Library REMUS

The REMUS simulation model library, currently under development, aims to represent construction sites in logistics and supply chain simulations. REMUS houses multiple model elements to simulate the operational level of a construction site (see Fig. 1).

Both modules and objects represent the operational level, with modules grouped into physical and logical categories, and objects grouped into physical and informational classes. Within REMUS, physical modules are further categorized into stationary and mobile subsets.

The simulation model focuses primarily on the automated masonry robots and their system components to map the effects, impacts, and potential responses to disruptions. In addition, Key Performance Indicators (KPIs) will be implemented to evaluate and improve the sustainability of the site. Associated parameters will be defined, recorded and analyzed, while customizable global variables will provide any additional or customized statistical parameters [9].

### 2.2 Framework Conditions

Simulation components in REMUS are employed to generate models for construction sites. Before creating a simulation model and carrying out an experiment, a requirement cluster is formulated, highlighting the key construction site parameters. These parameters ought to be selectively variable in the simulation experiments. If applicable or transferable from parallel fields, the implementation utilizes pertinent standards or guidelines. Information objects leveraged in this endeavor encompass work plans, environment, weather, energy, and maintenance (see Fig. 2).

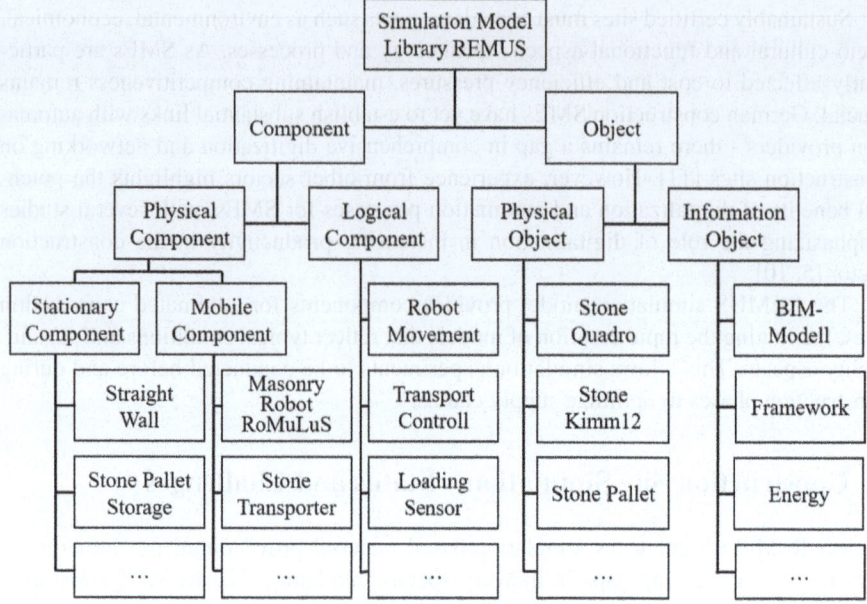

**Fig. 1.** Simulation model bank REMUS

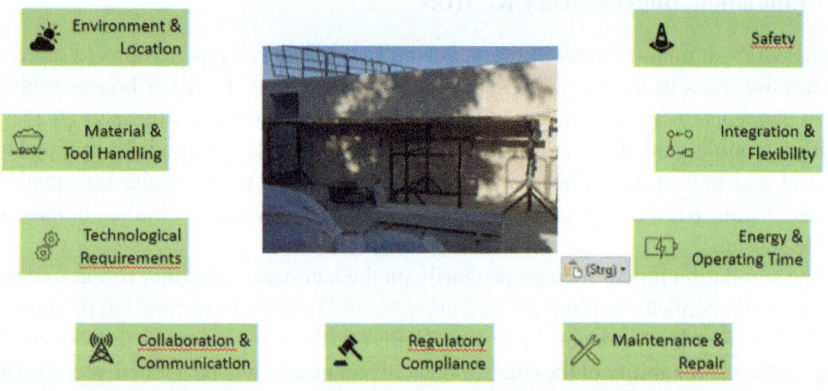

**Fig. 2.** Actual considered key construction site parameters

The REMUS simulation model considers several factors, including environmental conditions, technological requirements, safety, integration and flexibility, energy consumption, material and tool handling, maintenance, regulatory compliance, communication, and cost effectiveness. Environmental and logistical aspects include weather resilience, simulated disruptive events from varying conditions, and the ability to self-adapt to volatile environments through sensor-based perception. From a safety perspective, safe protocols are established based on workplace standards to prevent accidents, requiring the integration of features such as emergency shutdown options and protection against unauthorized data access.

Based on the concept of integration and flexibility, REMUS aims to transfer existing construction processes uniformly across projects and promotes standardized interface design. It also supports various data transfer formats. In terms of energy efficiency, the model includes provisions for autonomous power sources and extended operational capability. Transition times between operating states are measured. Material and tool handling prioritizes ergonomic designs to minimize physical stress, taking into account the impact of uneven terrain on travel speed and stoppage frequency. Wear and tear from environmental/working conditions necessitates preventive maintenance and service, which is built into REMUS with variable intervals and durations. Regulatory compliance requires adherence to safety laws and standards, and ensures that automated equipment complies with site regulations.

Communication support between stakeholders is robustly designed to protect against any systemic failure, with fail-safes for wireless communication. Finally, cost-effectiveness is addressed through careful monitoring of expenses and profitability, allowing for a prospective evaluation of the profitability of the overall system and its individual components. The energy and operational time are monitored through the logging of resource utilization, material, and energy consumption.

In essence, REMUS provides a time-sensitive, customizable, cost-effective, and safety-focused simulation environment for construction sites, considering factors ranging from environmental resilience to communications robustness.

### 2.3 Implementation of the Dimensions of Sustainability

The content of sustainability reporting is specified by European standards, the European Sustainability Reporting Standards (ESRS) [4]. Various models distinguish between the ecological, economical and social dimensions of sustainability [7]. The widely used sustainability triangle combines these into a whole [7].

The consideration and evaluation of these dimensions in simulation studies is based on the framework conditions. Selected parameters are recorded, displayed and evaluated during the simulation experiments. These parameters must be typical and meaningful for the respective sustainability dimension. They must be representable, recordable and analyzable in the simulation model. Quantitative parameters are preferable to qualitative ones. In numerous publications, various economic indicators are assigned to the dimensions of sustainability [1]. These were compared with the framework conditions. For the first simulation experiments, KPIs are selected that allow the simulation model to be optimized in terms of process development and sustainability. These KPIs can therefore be assigned to different sustainability dimensions and different framework conditions (see Fig. 3).

*Resource Utilization, Material* and *Energy* are selected for the **Environmental Dimension** [1]. *Resource Utilization* is recorded and evaluated for each component, robot and transport device as part of the statistical data. *Material* is determined by the quantity and type of building blocks used and the mortar required. They are assigned to the conditions "Environment and Location". Energy consumption can also be mapped and displayed in the simulation model. This correlates with the "Energy and Operating Time".

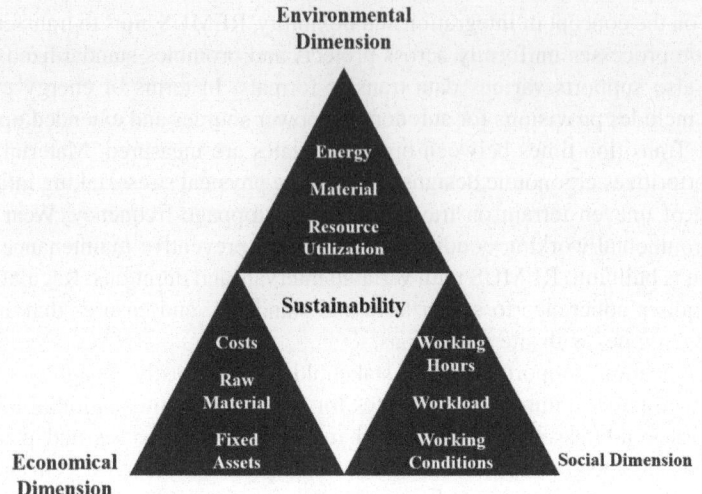

**Fig. 3.** Dimensions of sustainability with selected KPIs

The KPIs *Working Hours, Workload* and *Working Conditions* are recorded for the **Social Dimension** [1]. *Working Hours* are logged and differentiated in the static data of the respective simulation components. Shift calendars can be used to differentiate shift times by calendar and specify them for different simulation components. The Working Conditions can be mapped in a first step via the disruptions to the processes caused by weather conditions and break times. These are refined and supplemented during the simulation experiments. The *Working Conditions* and *Working Hours* are assigned to "Regulatory Compliance and Safety". The *Workload* is defined via user-defined attributes for individual components. Statistics tables, diagrams and frequency histograms are available for evaluation purposes. The *Workload* maps aspects of "Collaboration".

The **Economical Dimension** is evaluated using the KPIs *Fixed Assets, Raw Material*, and *Costs* [2]. The *Fixed Assets* are the building blocks used in the current simulation model for machines, robots and transportation devices. There are numerous key figures offered by the simulation system that can be recorded and evaluated for each device. In addition to availability, mean time to repair (MTTR), different times, such as processing times and dwell times, this also includes data on electrical power consumption and costs, investment costs, depreciation periods and operating costs. In this way, evaluations, analyses and optimization can be carried out for each module regarding necessary fixed assets, based on the occupancy times/working time shares. "Costs and Profitability" are recorded and prepared for optimization. The *Raw Material* used is recorded via the sources and sinks of the model. The number of inputs and outputs and the occupancy times and types are recorded here. In addition, any user-defined variables can be generated and evaluated. The raw material is also assigned to "Environment and Location". *Costs* are determined from the combination of disruptions or maintenance and operating costs and can subsequently be optimized. The "Costs and Profitability" areas are assigned here.

The framework conditions are modified in various simulation experiments. The effects of the changes are evaluated and targeted optimizations are carried out. The framework conditions can thus be integrated into the sustainability reports (CSRD and ESRS) via KPIs. It is possible to add to the KPIs considered and this will be done continuously during future simulation studies. The optimization thus considers and evaluates both process-related and sustainability criteria.

## 2.4  Simulation Experiments

The simulation experiment represents the construction of a masonry wall of pre-specified height, with unique conditions progressively incorporated into the model for subsequent experiments. Modifications to the wall provisions for windows and doors are addressed, requiring custom cut bricks that could affect the loading of brick pallets. Adherence to Just-In-Sequence (JIS) during removal is critical. The simulation experiment represents the construction of a wall with a given height, with special conditions gradually being added to the model for subsequent experiments. Changes to the wall specifications for windows and doors are addressed, requiring specially cut bricks, which could affect the load on the brick pallets. Compliance with just-in-sequence (JIS) in the finishing process is crucial. The process flow includes the supply chain from delivery to the construction of the wall and is represented by corresponding components in the simulation model (see Fig. 4).

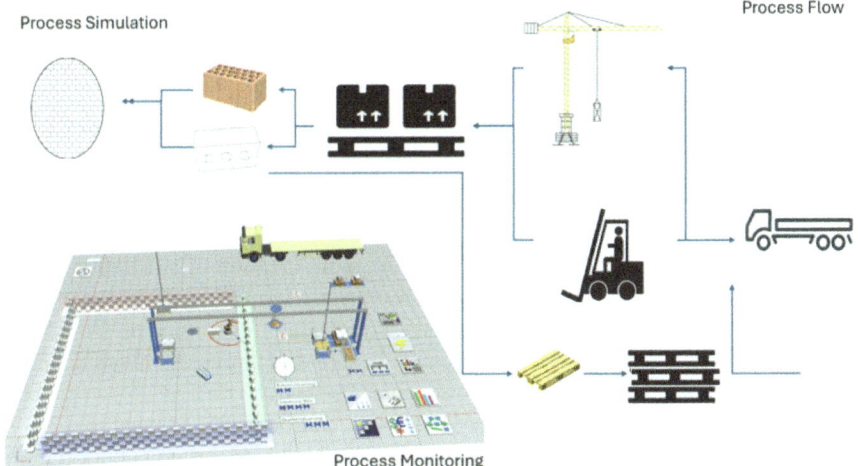

**Fig. 4.** Model of the construction site with process Monitoring using PlantSimulation [9]

The various dimensions of sustainability are considered for evaluation and optimization. The simulation runs demonstrate the basic functionality of the construction site environment. In the first experiments, different supply concepts and their effects on the buffer spaces and the supply of the construction robots are analyzed. The delivery frequency correlates with the buffer space required to ensure the supply of goods to the construction site. In addition, the KPIs for monitoring sustainability are recorded and evaluated.

From an **Economical Dimension**, costs and systems are directly linked. They can be changed by adding additional equipment or equipment with additional functions. For the current application, one bricklaying robot, one AGV and one truck have proven to be sufficient. If the wall gets bigger, at least more AGVs will be needed. The amount of raw material required depends primarily on the number of bricklaying robots used. Above a certain output, the AGVs become a bottleneck and need to be supplemented by others.

In the **Environmental Dimension**, the energy for each system component and the resource use are evaluated. Low resource use is advantageous in terms of energy requirements, as the systems are operated in an optimal load range. The material used, sand-lime bricks, correlates as expected with the resources used and their performance.

Within the **Social Dimension**, different working time models are mapped. The coordination of these models increases output as there are no waiting times due to missing parts, deliveries or machine operation. The change in workload shifts the performance of the entire system until bottlenecks are reached in various components. The working conditions lead to unplanned interruptions of the processes and a reduction of the overall performance.

The simulation experiments can be used to model the processes on the construction site. Using the environmental conditions and sustainability dimensions, specific aspects of the site environment are mapped. KPIs are used to evaluate and assess the different aspects, also regarding interdependencies. With the help of the REMUS simulation database, the stored environmental conditions and the considered sustainability dimensions as well as key aspects of a future-oriented construction site can be mapped and optimized. The simulation model is subject to continuous adaptation, refinement, and expansion.

## 3   Results

The REMUS Simulation Model Bank allows the rapid creation of simulation models for construction sites. It allows the direct insertion of significant model components that can be easily parameterized to reflect the construction environment. Component replacement is also possible at a rapid pace. This approach provides valuable insights that can be applied to real-world scenarios at an early stage.

Simulation components, such as the wall robot, can be easily adapted to different requirement clusters and successively integrated. The bank model is crucial for the simulation of construction project processes. It includes various scenarios and parameters such as weather, resource availability, and labor. By mapping the dimensions of sustainability, the site environment can be optimized with respect to important sustainability criteria.

Virtual testing and optimization reduce the risk and cost of errors during the construction phase. Different scenarios can be quickly tested to understand the impact of changes. Overall, the simulation model bank improves the efficiency, quality and profitability of construction projects, while increasing the success and satisfaction of all parties involved. This enables process-based, sustainable optimization of the construction site environment.

# References

1. Díaz Caselles, L.M., Guevara, J.: Sustainability performance in on-site construction processes: a systematic literature review. Sustainability **16**(3), 1047 (2024). https://doi.org/10.3390/su1 6031047
2. Donhauser, T.: Ressourcenorientierte Auftragsregelung in einer hybriden Produktion mittels betriebsbegleitender Simulation. Dissertation, Friedrich-Alexander-Universität Erlangen-Nürnberg (2019). https://www.researchgate.net/publication/344199391_Ressourcenorientie rte_Auftragsregelung_in_einer_hybriden_Produktion_mittels_betriebsbegleitender_Simu lation
3. EFRAG: Voluntary ESRS for Non-Listed Small- and Medium-Sized Enterprises (VSME ESRS). VSME ED January 2024 (2024). www.efrag.org. Accessed 24 May 2024
4. EU: Delegierte Verordnung - EU - 2023/2772 - DE - ESRS. DELEGIERTE VERORD-NUNG (EU) 2023/2772 DER KOMMISSION vom 31. Juli 2023 zur Ergänzung der Richtlinie 2013/34/EU des Europäischen Parlaments und des Rates durch Standards für die Nachhaltigkeitsberichterstattung (2023). https://eur-lex.europa.eu/legal-content/DE/TXT/? uri=CELEX%3A32023R2772&qid=1716457536043. Accessed 23 May 2024
5. Infra-Bau: Digitalisierung der Baubranche (2023). https://www.infra-bau.com/studie-zur-dig italisierung-im-bauwesen-als-ebook/. Accessed 3 Apr 2024
6. KfW: Förderprodukte für Ihren Neubau (2024). https://www.kfw.de/inlandsfoerderung/Pri vatpersonen/Neubau/F%C3%B6rderprodukte/F%C3%B6rderprodukte-PB-Neubau.html. Accessed 3 Apr 2024
7. Kropp, A.: Grundlagen der Nachhaltigen Entwicklung. Springer, Wiesbaden (2019)
8. Neue DGNB Zertifizierung für nachhaltige Baustellen: Erste Auszeichnungen vergeben I DGNB (2024). https://www.dgnb.de/de/dgnb-richtig-nutzen/newsroom/presse/artikel/neue-dgnb-zertifizierung-fuer-nachhaltige-baustellen-erste-auszeichnungen-vergeben/. Accessed 16 May 2024
9. Plant Simulation: SimBauLog (2022). https://plant-simulation.de/bausteine/simbaulog/. Accessed 24 March 2024
10. PricewaterhouseCoopers: Die Bauindustrie in Krisenzeiten: Fortschritte bei ESG, Stillstand bei der Digitalisierung. Eine PwC-Studie zum Umgang der Baubranche mit den aktuellen Herausforderungen (2024). https://www.pwc.de/de/managementberatung/capital-projects-and-infrastructure/bauindustrie-unter-druck.html#studie. Accessed 3 April 2024
11. Schrage, T., Schuderer, P., Barth, M., Franke, J.: Entwicklung und realisierung einer modellbibliothek für ein entscheidungsunterstützungssystem in der kalksandsteinproduktion. 20. ASIM Fachtagung Simulation in Produktion und Logistik 2023, p. 61 (2023). https://doi.org/ 10.22032/DBT.57787
12. Statista: Bauhauptgewerbe in Deutschland. Statistik-Report zum Bauhauptgewerbe in Deutschland (2022). https://de.statista.com/statistik/studie/id/13186/dokument/bauhauptg ewerbe-in-deutschland/. Accessed 21 Mar 2024

# Assessment of Product Carbon Footprint Reduction Potential Using Lightweight Rotor Components for Electric Traction Motors

Nicolaus Klein[1]([✉]) [iD], Leon Franken[1], Markus Heim[1], Florian Kößler[1], Benjamin Dönges[2], and Jürgen Fleischer[1] [iD]

[1] Karlsruhe Institute of Technology, 76131 Karlsruhe, Germany
nicolaus.klein@kit.edu
[2] Muhr und Bender KG, 57586 Weitefeld, Germany

**Abstract.** The mobility sector is currently undergoing one of the most profound transformation processes in its history. With ambitious climate targets on the horizon, there is a pronounced shift towards electric traction drives. However, in comparison to drivetrains with internal combustion engines, electric drivetrains and batteries have a larger carbon footprint during manufacturing. Thus, product innovations become important to reduce the carbon emissions associated with the traction motor. Among these innovations are lightweight components such as rotor shafts, balancing disks, and magnetic fixations. This study compares three novel components with spring behavior designed for lightweight rotors in terms of their carbon footprint, contrasting them with state-of-the-art electric traction rotors. Given the substantial carbon footprint resulting from the use of rare earth magnets in rotors, efforts are directed towards highest value preservation. With regard to the potential of lightweight components, carbon emissions can be reduced during the electric motor operation. This paper discusses a life cycle assessment of electric traction rotors using lightweight components. A hotspot analysis shows that the environmental impact of the components can be significantly reduced by avoiding the use of chrome steel and by recycling of permanent magnet materials.

**Keywords:** Permanent Magnet Synchronous Machine · Life Cycle Assessment · Electric Mobility

## 1 Introduction

### 1.1 Motivation

Comparative life cycle assessments oftentimes come to the same conclusion when tackling different drivetrain solutions. Regardless of methodology or assumptions, battery electric vehicles (BEVs) enter the use phase with higher emission numbers as compared to internal combustion engine vehicles (ICEVs) [2, 9]. While research in the development of ICEVs is focused on reducing driving emissions, there is a significant need for cleaner and more circular production of BEVs [19]. Without a doubt, the battery significantly

© The Author(s) 2025
H. Kohl et al. (Eds.): GCSM 2024, LNME, pp. 91–98, 2025.
https://doi.org/10.1007/978-3-031-93891-7_11

contributes to the lifecycle emissions of BEVs, accounting for 40 - 60%. However, this study focuses on the environmental impact of electric traction motors, which are essential for every electric drivetrain concept and can be optimized for specific applications [16]. These motors comprise the rotor shaft, a lamination stack, balancing disk, and permanent magnets, which pose substantial environmental risks due to their material composition. This issue is exacerbated by the limited availability and resource constraints of these critical materials [13]. One of the most critical resources for electric traction motors, rare earth elements (REE), are still not commonly recycled or reused, with a recycling rate below 10% in the EU [15]. Given that material costs are estimated to account for 73% of the cost of permanent magnet synchronous machine (PMSMs), a circular economy presents significant potential due to these scarce resources [16]. This study emphasizes the potential of three key components of PMSMs to reduce emissions.

## 2  Innovation Rotor

The key difference in emissions can be achieved during the production of these electric drivetrain components. Hence, this study compares the cradle-to-gate (C2G) emissions of a state-of the-art rotor and an innovative rotor concept using spring-loaded rotor components (see Fig. 1). These spring-loaded components are designed to facilitate weight reduction and require alternative manufacturing processes, compared to the manufacturing of state-of-the-art rotors.

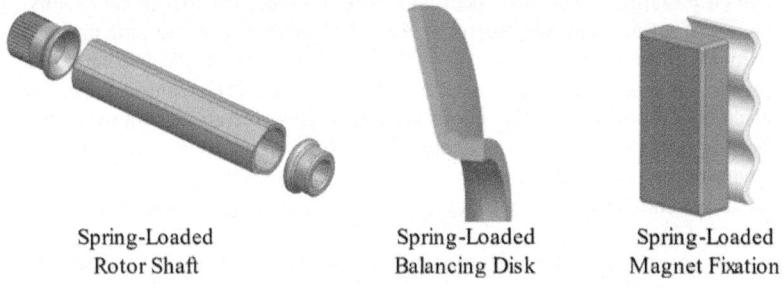

Spring-Loaded          Spring-Loaded          Spring-Loaded
Rotor Shaft            Balancing Disk         Magnet Fixation

**Fig. 1.**  Innovative rotor components.

### 2.1  Spring-Loaded Rotor Shaft

The three-part structure of the rotor shaft is a novel design feature. It enables continuously thin wall thickness and a larger inner diameter of the lamination stack package, which contributes to weight savings. The core of the rotor shaft consists of a drawn polygonal tube, which exhibits radial spring action. This design enables a joining process of the rotor shaft and lamination stack at room temperature, while also ensuring the transmission of a consistently high torque when the lamination stack expands under temperature and rotational speed. At the same time, the polygonal tube exhibits low sensitivity to deviations in the inner diameter of the lamination stack, allowing for larger manufacturing tolerances in this regard. Furthermore, the three-part construction of the rotor shaft provides ample design flexibility in selecting the outer diameter, without significantly affecting the volume to be machined [17].

## 2.2 Spring-Loaded Balancing Disk

Traditional balancing disks often require additional tension rods to create axial preload on the lamination stack. However, the introduction of spring-loaded balancing disks eliminates the need for these rods. These thin-walled disks, through their resilient behavior, serve the dual purpose of balancing the rotor and reducing weight. A non-ferromagnetic material lacking chromium and nickel, offering high cost efficiency, is utilized. The plate spring design allows for axial pre-tensioning of the lamination stack as well as a force-locking connection with the rotor shaft [6, 17].

## 2.3 Spring-Loaded Magnet Fixations

At the end of its lifecycle, a waved spring strip enables the disassembly of the rare-earth magnets. These magnets can be reused or separated from the remaining materials and recycled to a high standard. The force-locking, detachable fixation within the magnet pocket of the lamination stack needs no modification of the typically cuboid-shaped magnet pockets. The gap previously required for material-locking fixation in the magnetization direction can be filled by the spring strip. It thus not only serves the function of fixation but also facilitates the conduction of magnetic flux and provides thermal bonding of the magnets to the lamination stack. This allows for an increase in efficiency of 0.1% across different operating points of a traction drive compared to transfer molding [7, 17].

## 2.4 Manufacturing Innovations

On the production side, balancing through mass application or redistribution aims to reduce the amount of material needed for a conventionally subtractive balancing process. The previously thermally assisted joining process of the shaft-hub connection can now be reliably conducted at room temperature due to the resilient behavior of the rotor shaft. The resilient magnet fixation also does not require any thermal curing processes. Unlike conventional rotor shafts, the polygonal tube does not require additional machining in the press fit area, thus reducing manufacturing times. As the radial preload between the shaft and balancing disk occurs only at the end of the joining process, joining at room temperature is possible here as well [17].

# 3 Comparison of Rotor Components

## 3.1 Cradle-to-Gate Comparison

Both the innovative rotor as well as the reference rotor have been modeled in openLCA software utilizing the ecoinvent database and primary data. The reference rotor has a global warming potential (GWP) of 265.9 kg $CO_2$ eq in the C2G assessment, while the innovative rotor reduces this by 4.38% to 254.2 kg $CO_2$ eq without magnet recycling.

For the innovative rotor, 98.5% of the emissions are caused by the rotor components, and only 1.5% are attributable to the joining processes. The permanent magnets are responsible for 54% of the emissions, followed by the lamination stack with 25%, the rotor shaft with 16.5%, and the balancing disks with 3%. The balancing disk reduces emissions compared to the reference rotor by 8.6 kg $CO_2$ eq, while the rotor shaft achieves a reduction of 2.5 kg $CO_2$ eq (Table 1).

**Table 1.** Environmental impacts of rotor production.

| Impact | Innovation | Reference | Unit |
|---|---|---|---|
| Acidification | 1.042 | 1.101 | kg $SO_2$ eq |
| GWP$_{100}$ | 254.247 | 265.896 | kg $CO_2$ eq |
| Ecotoxicity: freshwater (FAETP) | 645.385 | 686.201 | kg 1,4-DCB eq |
| Ecotoxicity: terrestrial (TETP) | 1.350 | 11.042 | kg 1,4-DCB eq |
| Eutrophication | 0.498 | 0.520 | kg PO4 eq |
| Human toxicity | 484.859 | 764.037 | kg 1,4-DCB eq |

All impacts are calculated according to the CML v4.8 2016 assessment method. The greatest differences can be found in the terrestrial ecotoxicity (89.7% reduction) and the human toxicity (36.5% reduction). This can be attributed to the avoided use of chrome steel in the balancing disks compared to the reference (Fig. 2).

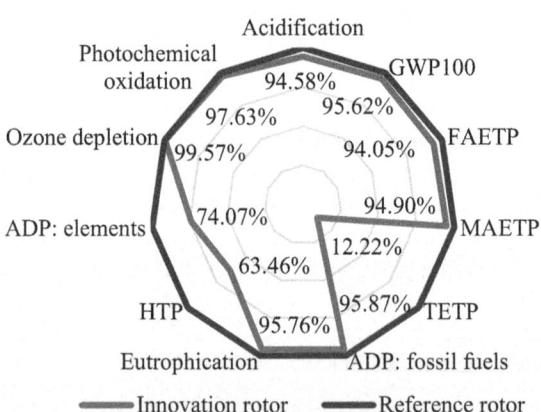

**Fig. 2.** Environmental impacts of rotor production processes.

As seen above, terrestrial ecotoxicity (TETP) is immensely improved. TETP studies how environmental pollutants affect land organisms and their environment. It involves three elements: a pollutant source, receptors, and exposure pathways and is especially highlighted when working with various metals [3, 10]. A holistic view of the different impact categories is very important to avoid shifting emissions between different categories [4, 14]. It is important to clarify that, at this point, improvements in environmental impacts are primarily attributed to the different production processes and materials used. In the following sections, the potential of lightweight design for reducing production emissions as well as the immense recycling potential for permanent magnets, are highlighted.

### 3.2 Lightweight Design Improvements

To reach the design goal of a lightweight rotor, the innovative rotor components enable a weight reduction of 0.25 kg in the lamination stack by removing additional tension rods and about 1 kg in the rotor shaft. This decreases the amount of raw materials and energy required for production resulting in lower emissions of 14.32 kg $CO_2$ eq without affecting performance during the use phase. Although the weight reduction compared to the reference rotor is notable, the approximately 1 kg difference is a small contribution to the energy consumption of a two-ton electric vehicle.

### 3.3 Magnet Recycling

Environmental improvements for BEVs are constrained by high emissions during raw material extraction and the limited application of recycling possibilities [1]. Over 90% of total energy consumption for permanent magnet production is needed for mining and refining of REEs with about 50% of the material being lost in the process [15]. Preparation processes of REEs have to be repeated several times using high amounts of chemicals and energy to gain the required purity [12]. Figure 3 shows the distribution for permanent magnet production, as modelled in openLCA. The annual consumption of REE magnets is expected to increase, making recycling crucial to avoid shortages and supply risks and enable a transition away from fossil fuels. The mining of REEs is highly centralized, with China holding a monopoly position at 83% leading to high import-dependency of 100% for the EU and price volatility [15, 18].

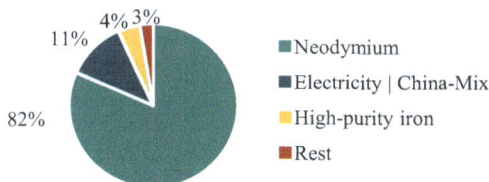

**Fig. 3.** Emissions in permanent magnet production according to the manually modelled openLCA system.

Three different routes can be used for magnet recycling: Functional recycling, elemental recycling, and direct reuse, each utilizing different amounts of primary material. While the former focuses on extracting the permanent magnet alloys or the REEs, the focus of this paper is on the direct reuse enabled by the enhanced magnet fixation for the innovative rotor. This provides a range of possible improvements through the recycling of magnets. Lixandru et al. prove the concept of magnet-to-magnet recycling with up to 100% of magnetic properties possible to be restored [11]. Accardo et al. estimate a 33% GWP reduction potential in (NdDy)FeB magnet production by using a hydrogen decrepitation recycling process with 60% recycled material [1]. This would correlate to 45,3 kg $CO_2$ eq reduction or 17,8% of the overall GWP of the innovative rotor. Zakotnik et al. adopt a magnet-to-magnet recycling which incorporates only 1.9 wt % of Nd-Pr hydride additive and enables energy savings of around 95% [18]. This recycling route removes the costly mining and purifying of the REEs and replaces them with demagnetization and hydrogen decrepitation. Adopting such a procedure for the innovate rotor would reduce the GWP by 124.9 kg $CO_2$ eq or 49.1%. Jin et al. also explore the possibilities of magnet-to-magnet recycling, reaching a potential GWP improvement of 80% [8] (Fig. 4).

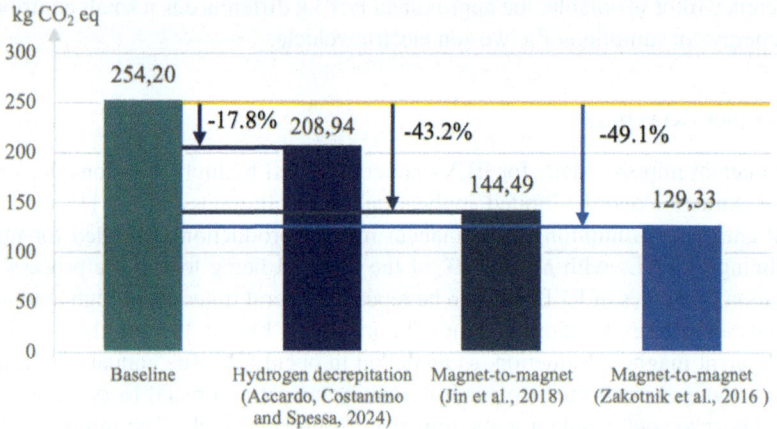

**Fig. 4.** GWP comparison of rotor production with different magnet recycling scenarios.

## 4   Conclusion and Outlook

In this paper, three major changes to the design of a rotor for an electric traction motor have been highlighted regarding their environmental benefits compared to a reference rotor. The product has been modelled in openLCA software to explore hotspots and ways of improvement. Results show, that through different manufacturing routes, lightweight design and enabled magnet recycling, the lifecycle emissions of the innovative rotor can be significantly reduced. The potential in GWP reduction lies especially in the possibility to reuse the highly critical permanent magnets. Figure 5 illustrates, how the individual improvements effect the GWP of the rotor.

As the demand for electric traction motors is projected to increase, the potential environmental benefits of widespread implementation of these innovative concepts, particularly magnet recycling is highlighted. With a projected growth of up to 47 million

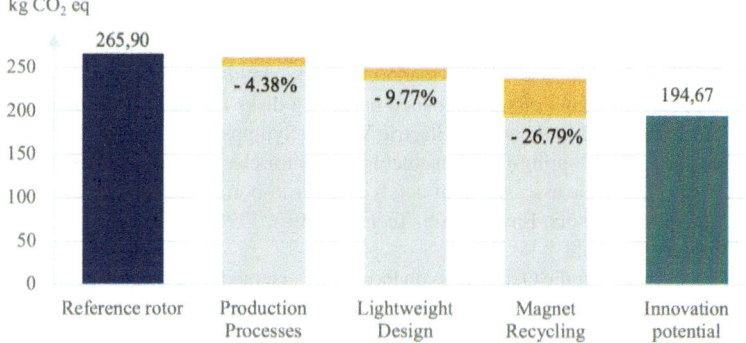

**Fig. 5.** GWP reduction potential via different innovations.

motors by 2035 [5], the potential savings can be significant, as illustrated in Fig. 6. The spring-based magnet fixations offer a first stepstone in a more circular economy but to capitalize on these opportunities, it is crucial that magnet-to-magnet recycling is further improved.

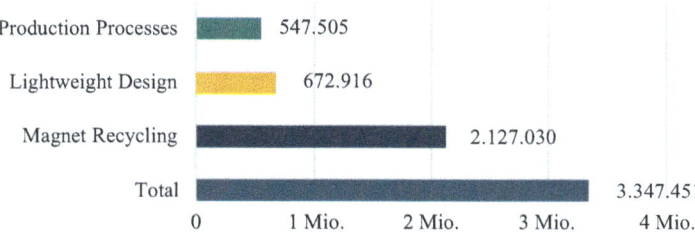

**Fig. 6.** Scenario analysis for $CO_2$ reduction potential by 2035 [t $CO_2$ eq].

**Acknowledgement.** The authors gratefully acknowledge financial funding from the German Federal Ministry of Economic Affairs and Climate Action and organizational support by the Project Management Jülich (grant no. 03LB3041D).

# References

1. Accardo, A., et al.: LCA of recycled (NdDy)FeB permanent magnets through hydrogen decrepitation. Energies. **17**(4), 908 (2024). https://doi.org/10.3390/en17040908
2. Bothe, D., Steinfort, T.: Cradle-to-grave life-cycle assessment in the mobility sector (2020)
3. Fairbrother, A., Hope, B.: Terrestrial Ecotoxicity, 2nd Chapter NA. Encyclopedia of Toxicology, pp. 138–142. Elsevier (2005)
4. Giolito, F., et al.: Evaluation of the environmental benefit of an eco-design strategy on the life cycle assessment of a permanent magnet synchronous high-speed electric motor. Transp. Res. Procedia. **70**, 241–248 (2023). https://doi.org/10.1016/j.trpro.2023.11.025

5. Hammer, H.: Automobil Industrie: Milliardenmarkt E-Achsen: Die wichtigsten Zuliefererer 2022 (2022). https://www.automobil-industrie.vogel.de/milliardenmarkt-e-achsen-die-wichtigsten-zulieferer-a-9f6451697051aad50d692ac53a6a9d50/

6. Heim, M., et al.: Potential of spring-loaded balancing disks for electric traction motors. In: EVS37 Symposium on International Electric Vehicle Symposium and Exhibition (2024)

7. Heim, M., et al.: Wave spring-based magnet fixations for electric traction motors (2024)

8. Jin, H., et al.: Life cycle assessment of neodymium-iron-boron magnet-to-magnet recycling for electric vehicle motors. Environ. Sci. Technol. **52**(6), 3796–3802 (2018). https://doi.org/10.1021/acs.est.7b05442

9. Koch, T., et al.: VDI-Studie Ökobilanz von Pkws mit verschiedenen Antriebssystemen (2020)

10. Larsen, H.F.: LCA of wastewater treatment. In: Hauschild, M.Z., et al. (eds.) Life Cycle Assessment, pp. 861–886. Springer, Cham (2018). https://doi.org/10.1007/978-3-319-56475-3_34

11. Lixandru, A., et al.: A systematic study of HDDR processing conditions for the recycling of end-of-life Nd-Fe-B magnets. J. Alloys Compd. **724**, 51–61 (2017). https://doi.org/10.1016/j.jallcom.2017.06.319

12. Nordelöf, A., et al.: A scalable life cycle inventory of an electrical automotive traction machine (2016)

13. Nordelöf, A., et al.: Life cycle assessment of permanent magnet electric traction motors. Transp. Res. Part Transp. Environ. **67**, 263–274 (2019). https://doi.org/10.1016/j.trd.2018.11.004

14. Rosenbaum, R.K., et al.: Life cycle impact assessment. In: Hauschild, M.Z., et al. (eds.) Life Cycle Assessment, pp. 167–270 Springer, Cham (2018). https://doi.org/10.1007/978-3-319-56475-3_10

15. Schönfeld, M., et al.: Recycling of rare earth permanent magnets for advanced electric drives - Overcoming the criticality and supply risk (2018)

16. Stanek, R., et al.: Wertschöpfungspotenziale von E-Motoren für den Automobilbereich in Baden-Württemberg. Presented at the Cluster Elektromobilität Süd-West c/o (2021)

17. Wößner, W., et al.: Federnde Rotorkomponenten für elektrische Antriebe. Z. Für Wirtsch. Fabr. **117**(10), 667–672 (2022). https://doi.org/10.1515/zwf-2022-1103

18. Zakotnik, M., et al.: Analysis of energy usage in Nd–Fe–B magnet to magnet recycling. Environ. Technol. Innov. **5**, 117–126 (2016). https://doi.org/10.1016/j.eti.2016.01.002

19. Zhang, X., et al.: Carbon emission analysis of electrical machines. In: ICEMS 2021: 2021 24th International Conference on Electrical Machines and Systems, 31 October–3 November 2021, HICO, Gyeongiu Korea, pp. 1678–1683. IEEE, Piscataway (2021)

# A Convergent Systems Approach for Sustainable Manufacturing

Asela K. Kulatunga[1]($\boxtimes$) (iD), Fazleena Badurdeen[2], and I. S. Jawahir[2]

[1] University of Exeter, Exeter, UK
`a.k.kulatunga@exeter.ac.uk`
[2] University of Kentucky, Lexington, USA
`{badurdeen,i.s.jawahir}@uky.edu`

**Abstract.** Sustainable Manufacturing is applied at product, process, and systems levels towards achieving sustainable consumption and production. Many present-day problems related to sustainable consumption and production are known to be dynamic, adaptive, and complex in nature and fall into the category of wicked problems. In general, only the events and consequences resulting from such problems are apparent on the surface. These are, however, akin to the tip of the iceberg whereas the underlying patterns or trends, system structures, and drivers, as well as the predominant social paradigms and mental models that are often root causes of such wicked problems, are invisible and untraceable. Understanding these complex interdependencies and solving complex problems cannot be achieved using conventional techniques or a singular domain of expertise. Challenges related to sustainable consumption and production are such problems that require contributions from beyond transdisciplinary boundaries to assess the potential underlying causes and for the implementation of robust solutions. To address this need, this research presents a four-phase framework based on the convergent approach to solving complex problems. An in-depth examination of the complex problem is presented in first phase, elaborating the underlying principles, tools, and techniques that could be used concurrently to identify the transboundary team to solve the problem identified and its operation is presented in phase two. The third phase presents the solutions generation process and the fourth presents the final solution selection and implementation by considering nexus of sustainable consumption and production, the product design perspective.

**Keywords:** Convergence Approach · Sustainable Manufacturing · Sustainable Consumption and Production · System Thinking · product design · Team Science

## 1 Introduction

### 1.1 Necessity for Sustainable Consumption and Production

The planet's ecosystem has deteriorated in the post industrialization era due to anthropogenic activity to such an extent that we are currently beginning to witness global warming and many other socio-environmental issues. After the industrial revolution, the

© The Author(s) 2025
H. Kohl et al. (Eds.): GCSM 2024, LNME, pp. 99–107, 2025.
https://doi.org/10.1007/978-3-031-93891-7_12

demand for natural resources increased as a result of the need to simultaneously meet the needs of the agricultural, industrial, and service sectors of society. These trends only prioritized the economic prosperity at the expense of social inequality and environmental deterioration. The UN has developed a more focused development strategy called the Sustainable Development Goals (SDGs). Despite efforts to reorient global development through the Bruntland Commission's [1] outcomes and the sustainable development concept, there was no discernible improvement from 1987 to 2015. Through the 12th SDG, Responsible Consumption and Production (RCP), the UN has acknowledged that to achieve sustainability, both production and consumption components must be concurrently aligned. The notion of sustainable consumption and production, also known as RCP or SCP, extends beyond the level of intervention that was implemented in the field of sustainable manufacturing following the introduction of the Sustainable Development concept in 1987 [1]. Though sustainable manufacturing is used at the product, process, and system levels [2], SCP has created new opportunities to see sustainability from a different angle. One such angle is systems thinking, which has the potential to better capture the complexity of the issue than earlier approaches.

## 1.2   Complex Problems and Systems Thinking

The socioeconomic and environmental factors surrounding problems related to sustainable development have changed over time, making them more complex than they were when solutions were first proposed. Moreover, the resulting difficulties are often caused by several unseen factors that are generally impossible to identify using traditional methods. These unanticipated problems are brought about by the interdependencies among the causes, which result in unsustainable behaviors that are evident in society. Adaptive problems are complicated due to unanticipated concerns like interdependencies, the variabilities of the interdependencies and other factors, which is why they are called wicked problems [3]. Managing complex adaptive systems can be especially difficult when addressing "wicked" or unsolvable issues that are not amenable to conventional problem-solving techniques. Even if wicked problems are unpredictable and have extended timelines, it is vital to better understand these unique systems and identify methods to better understand and manage system performance to progress towards to goals.

Since many of the earlier attempts to address such problems were unsuccessful, it is necessary to view the problem or think about it differently to recognize its complexity and adaptive character. This could be achieved through systems thinking based problem formulation. Systems thinking is a way of looking at the world that enhances our ability to make decisions by highlighting connections and interactions, using a collection of techniques and resources [4].

One way ahead is to conceive of systems thinking as a collection of techniques and tools as well as a way of looking at the world that enhances our ability to make decisions by highlighting linkages and interactions [5]. The secret to systems thinking is to ask more questions to improve understanding and look for ways to push system function in desirable directions rather than focusing on prediction or control. Seeing that there are usually multiple coordinated acts that lead a system towards a desired state or outcome rather than a single solution to an issue is made easier with the aid of systems thinking.

To ensure that the strategies developed are viable for comprehending complex problems and identifying workable solutions, those involved in solving complex problems must collaborate across disciplines and with community stakeholders and partners at the local, state, national, and international levels when addressing sustainability problems and developing solutions [6]. Traditional academic training places a strong emphasis on the accomplishments of independent scholars, the advancement of individual disciplines, and the production of new knowledge. Hence, academics with traditional training frequently lack the knowledge and expertise required to tackle wicked problems [7].

### 1.3 Convergent Approaches

Convergence is strategy for reaching a shared goal in a system by adapting, innovating, and integrating diverse forms of knowledge [10]. While convergence science has largely been used in biology, engineering, mathematics, and computational sciences. There are several attempts to overcome the gaps of forming research teams to solve complex problems stated by [6, 7]. A convergence science framework extending beyond transdisciplinary team formation has been proposed to address food system sustainability due to uncertainties of climate change [8]. However, it has not considered the complexities due to interdependence of causes for the main problem and the concept was limited to agri-food sector sustainability. A road map for value proposition in convergence research presented by [9] was limited to identifying needs and proposing approach, benefits, and completion aspects. It did not present any specific methodologies to identify the complexities of the problem, establishing, team to solve the problem and how to attain a robust solution. Principles of convergence in nature and society and their applications was presented by [10]. Proposing the used the convergence and divergent cycles along with the innovation spiral to solve complex problems. The significance of identifying the problem and finding solutions through the convergence phase to then propel the innovation spiral in the divergence phase for generating multiple solutions is emphasized. However, the different tools and techniques on how to identify the complexities, development of transdisciplinary teams, how to maintain the integrity of the teams and achieve a final robust solution are not presented. The framework, developed primarily from nano-scale research point of view.

This paper attempts to bridge these gaps in existing studies focusing on the convergence approaches [8–10] by proposing a systems thinking-based, integrated and innovative methodology to identify the complexity of sustainable consumption and production problems and to find a robust solution.

## 2   Methodology

The proposed Framework consists of four phases.

**Phase I:** In this phase, identification of inherent complexities of the events of unsustainable practices followed by underlying patterns and structures of the events/activities leading to unsustainable practices are investigated. This is used to identify the system structures and drivers which leads to the predominant social paradigms (mental models/world views) that cause the resultant outcomes to emerge.

**Phase II:** Once the complexity of the problem is identified in phase I, setting up a team to refine the complex problem and identify solutions will be done in phase II. During this phase, it is paramount to identify the tipping points where interventions could initiate the changes to adapt the adapt the world views/the mental models for succeeding with the solutions. This must be done through the designing and development of appropriate artifacts depending on the context of the complex problem which requires cross disciplinary team plus the representation from all the stakeholders involved in the complex problem. Therefore, identifying the wider stakeholder representation needs and teaming mechanism must be established. Furthermore, mechanism for seamless operation of the team consisting of multiple disciplines and different stakeholders must be established. In addition, proper leadership and robust framework must be established about the individual's roles and responsibilities towards converging towards focus direction.

**Phase III:** The established team will examine the previously identified complex issue and, after consulting with other team members, redefine its scope. It is anticipated that, a larger range of professional and stakeholder perspectives about the problem will be solicited with numerous rounds of discussions and refinements until the team is pleased with the depth and breadth of the work. This repetitive process is vital to capture some of the unforeseen concerns of the complex problem. The primary goal of the cross-functional team is to create artifacts that should influence the stakeholders' mental models to reconsider unsustainable practices, leading them to act or make decisions that will promote sustainable practices. The team should be equipped with data and information where required and establish links whenever required with policy makers and civil society organizations too. To ensure long-term success, it's crucial to build a supportive environment and secure funding for ongoing research. Team members should have confidence in the leadership, which must demonstrate strong qualities to maintain trust. Even if leadership changes, the solution generation process should continue. The team will eventually transition to the divergent mode of phase III, generating (search the solution space) solutions and identifying robust solutions that could be implemented within a reasonable timeframe.

**Phase IV:** This phase will drive towards implementing the best solution with the support of all the stakeholders including policy makers. It is expected that the robust solution will enable developing and implementing appropriate artifacts which will act drivers/levers for the system to react differently/change the world view on unsustainable practices towards transitioning to more a sustainable mindset and decision making.

The activities at each phase of the proposed method are given in Fig. 1. The framework consists of two vertically oriented bicone (created by joining two congruent, right, circular cones at their bases), where diameters of the common surface and respective heights of each individual cones of each bicone can be different as indicated in the Fig. 1. As indicated in the Fig. 1, these cones are denoted as C1, C2, C3. C4 and radii of the common surface and heights are denoted as r1, h1 and r2, h2, respectively. The heights represent the timeline taken (length) to complete respective phases and radius indicates the breath of the problem. The driving force of each phase is represented by the spiral moving in a divergent (C1 &C3) and convergent (C2 & C3) manner.

| Phase | Gaps identified from Literature | Tools & Techniques |
|---|---|---|
| I/C1 | Identify the complexities | Adaptation of Systems thinking, use iceberg analogy, Adapt Causal loop & stock & flow diagrams. Use divergence spiral to capture all the complexities |
| II/C2 | Establishment and operate transdisciplinary research team. Refine and define problem boundaries, and identify comprehensive stakeholder representation | Use Team Science to form and operate the team, adapt RACI Matrix for project Management, Adapt EDI concept to build the trust among team members, use citizen science concept to get active stakeholder participation. Use convergence spiral to form comprehensive and cohesive transdisciplinary team |
| III/C3 | Generate comprehensive solutions with contributions from transdisciplinary team | Integration of research leadership, data sharing & interoperability, long term existence of the team & linking policy, decision makers with researchers. Adapting Divergence spiral towards generating multiple solutions |
| IV/C4 | Selecting and implementing the most appropriate and robust solution out of the alternatives | Integration of research leadership, data sharing & interoperability, long term existence of the team & linking policy, decision makers with researchers. Adapting convergence spiral towards finalize and implement robust solution |

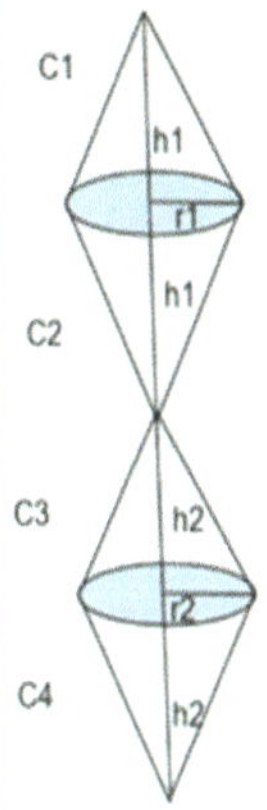

**Fig. 1.** The phases of the proposed Framework and the replication of dual bicone analogy

## 3 Results – Proposed Framework

This section presents on how dual dicone framework could be used effectively to solve the nexus of the SCP by taken into account the sustainable product design along with designing of artifacts to transform the society towards environmentally conscious society. The phase I, facilitates understanding the complexities of the problems and underlying patterns and structure leading to identify the mental models. In this case, nexus of SCP, that is the interaction between customers and designers taken into account. Conventionally, marketing teams act as the interface between the consumers and the designers of the products. Market research provides designers with information related to consumers whereas advertisements, for example, convey information in the opposite direction. Appropriate artifacts are essential to change the mental models of consumers transforming them make environmental conscious purchasing and consumption decisions. These artifacts have to be made in parallel to the latter part of the eco designs or eco innovations so that marketing and design teams will jointly developing them. Artifacts such as eco labels, EPDs (environmental Product Declarations), informational flyers, advertisements and media promotions etc. can be used to influence consumer mental models and transform their purchasing and consumption behavior society knowledgeable of on environmental issues and anthropogenies contribution towards it.

Figure 2 illustrates the comprehensive process of identifying the complexity of SCP issues. It highlights the gap between product designers and consumers, as well as the lack of commitment or conviction among marketing professionals to utilize persuasive artifacts. This often results due to insufficient interaction or active involvement during the product design phase. The proposed bi-cone dual DCDC model, particularly in Phase II, points out the approach to artifact design and create development teams, extending beyond the conventional core design team and external parties typically involved. This is achieved by establishing transdisciplinary development teams that include essential

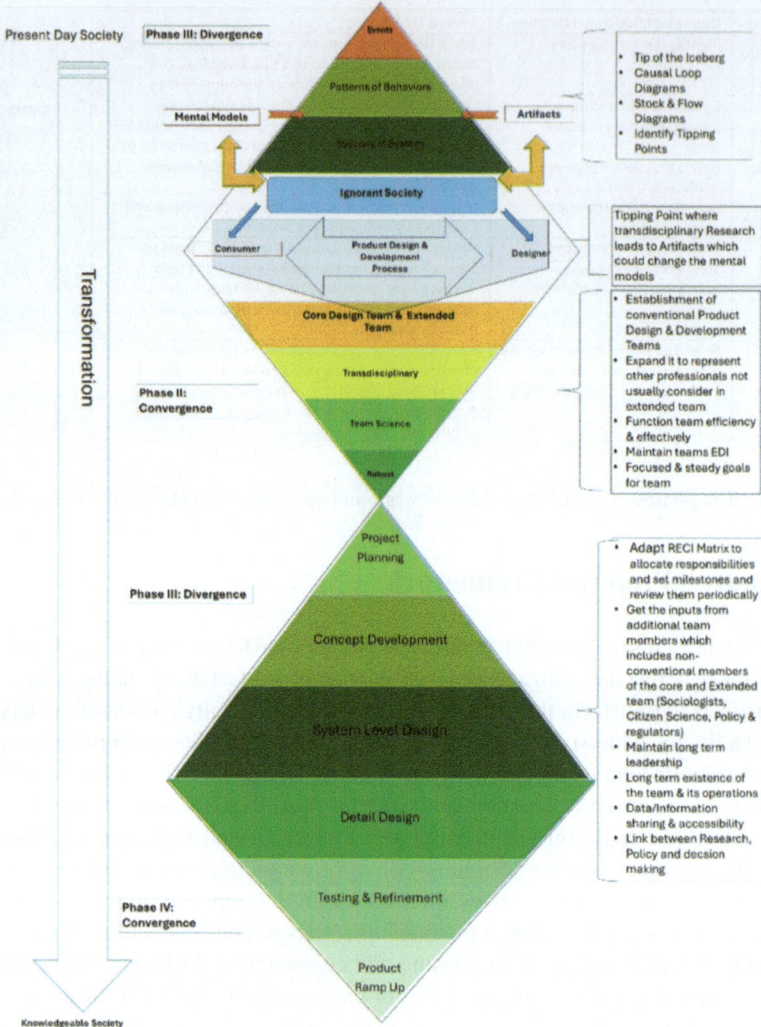

**Fig. 2.** Proposed framework of dual bicone Framework adapted to promote SCP through Sustainable product and artifacts design

core team members, an external stakeholder, representatives from policy and regulatory bodies (where applicable), and civil society representatives through citizen science initiatives. In the proposed dual bicone framework, phase II presented how to form the products and artifacts design and development teams going beyond conventional core product design team and external parties used in the conventional setting. This is achieved by establishing beyond transdisciplinary development teams which includes requited core team members, extended team plus reparations from policy and regulator representations where applicable and civil society representations through citizen sciences. To facilitate teamwork and for project management team science [12] (forming, norming, storming, performing, adjoining/transforming) concept while maintaining Equity, Diversity, and Inclusiveness (EDI) to build the trust among team members has been established along with RECI (Responsible, Accountable, Consulted, and Informed) Matrix developed by James R. Robinson [13] respectively.

How the products life cycle environment is considered and its link with SCP concept, and interventions to be made at each of those stages by the different members of the establish team is shown in Fig. 3 the potential artifacts developments at different stages and how those contribute towards SCP as well.

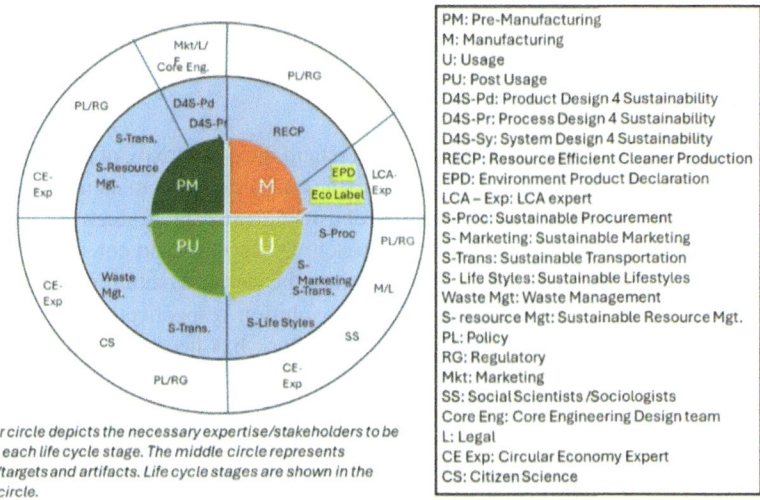

Note: Outer circle depicts the necessary expertise/stakeholders to be engaged at each life cycle stage. The middle circle represents objectives/targets and artifacts. Life cycle stages are shown in the innermost circle.

PM: Pre-Manufacturing
M: Manufacturing
U: Usage
PU: Post Usage
D4S-Pd: Product Design 4 Sustainability
D4S-Pr: Process Design 4 Sustainability
D4S-Sy: System Design 4 Sustainability
RECP: Resource Efficient Cleaner Production
EPD: Environment Product Declaration
LCA – Exp: LCA expert
S-Proc: Sustainable Procurement
S- Marketing: Sustainable Marketing
S-Trans: Sustainable Transportation
S- Life Styles: Sustainable Lifestyles
Waste Mgt: Waste Management
S- resource Mgt: Sustainable Resource Mgt.
PL: Policy
RG: Regulatory
Mkt: Marketing
SS: Social Scientists /Sociologists
Core Eng: Core Engineering Design team
L: Legal
CE Exp: Circular Economy Expert
CS: Citizen Science

**Fig. 3.** How the Transdisciplinary team contribute towards designing of sustainable products and respective artifacts by considering the overall product Life cycle thinking

The phase III and phase IV follow the standard product design and development process where core team deals with the core product functionalities and specifications develop the concepts, system level design and detail designs before testing and refinement in consultation where necessary with other members of the team shown in Fig. 3. Since all the parties co-existing in a design environment it is very much productive towards design and developing products, process, supply chain along with the necessary artifacts by the team in consultation with marketing and social science experts, meantime policy and regulatory representatives also can update any pertaining policies and regulations or any amendments to accommodate core requirements of transforming the mental models of the consumers towards environmentally conscious and responsible citizens. Civil society representatives or citizen science could also share their views to both product and artifact design on and social scientists from theoretical perspectives on social and behavioral aspects. The rest of the parties conventionally involved in design and development such as legal, finance etc. will play their usual role and could assist if any implications to be clarified in artifacts developed in addition to commenting on the product designs.

In both Phase III and Phase IV, strict adherence is maintained regarding innovation and research leadership, accessibility and sharing of data and information, the long-term continuity of the team and its operations, and the ongoing connection between the research team and policymakers or decision-makers. These elements are preserved until the entire process concludes with the successful implementation of robust solutions. With the implementation of the developed robust solution, it is expected that society will be transform from ignorance to knowledgeable society which has the awareness of the risk factors from environmental problems, human heath vs planetary health, it is expected that implementing the developed solutions will transform society to be more knowledgeable of the risks associated with environmental problems and impacts on human health.

# 4 Conclusion

This research proposed a framework to understand the complexity of the problems, form transboundary teams to develop solutions to the complex problem identified and to facilitate the team to develop and implement robust solution. This framework uses multiple tools and techniques as an integrated approach. Divergence and convergence are taken place in each bicone to facilitate on expanding the problem/solutions and establishing transboundary team and generating a robust solution respectively in first and second bicone of the framework. The proposed framework has been used to demonstrate on how to promote SCP through environmentally friendly product design through concurrent design of sustainable product and respective artifacts to transform the consumers. With this framework, the nexus of sustainable manufacturing can be influenced through the early decision making of the product design since designers will actively be engaged with the complex problem and transdisciplinary team and their final outcomes. In future, it is expected to test this framework through agent-based simulation modelling and to validate with selected supply chains/product categories.

# References

1. World Commission on Environment and Development. Our Common Future, p. 27. Oxford University Press, Oxford (1987). ISBN 019282080X
2. Huang, A., Badurdeen, F.: Metrics-based approach to evaluate sustainable manufacturing performance at the production line and plant levels. J. Clean. Prod. **192**, 462–476 (2018). https://doi.org/10.1016/j.jclepro.2018.04.234. ISSN 0959-6526
3. Head, B.W., Alford, J.: Wicked problems: implications for public policy and management. Adm. Soc. **47**(6), 711–739 (2015). https://doi.org/10.1177/0095399713481601
4. Anderson, V., Johnson, L.: Systems Thinking Basics: From Concepts to Causal Loops. Pegasus Comm. Inc., Waltham (1997)
5. Paxton, A., Frost, L.J.: Using systems thinking to train future leaders in global health. Glob. Public Health **13**(9), 1287–1295 (2018)
6. Bieluch, K., McGreavy, B., Silka, L., Strong, A., Hart, D.: Empowering sustainability leaders: variations on a learning-by-doing theme. Dev. Change Agents 123 (2019). https://scholars.unh.edu/nh_epscor/123

7. Kreuter, M.W., De Rosa, C., Howze, E.H., Baldwin, G.T.: Understanding wicked problems: a key to advancing environmental health promotion. Health Educ. Behav. **31**(4), 441–454 (2004). https://doi.org/10.1177/1090198104265597
8. Sixt, G.N., et al.: A new convergent science framework for food system sustainability in an uncertain climate. Food Energy Secur. **11**, e423 (2022). https://doi.org/10.1002/fes3.423
9. Westerhoff, P., Wutich, A., Carlson, C.: Value propositions provide a roadmap for convergent research on environmental topics. Environ. Sci. Technol. **55**(20), 13579–13582 (2021). https://doi.org/10.1021/acs.est.1c05013
10. Roco, M.C.: Principles of convergence in nature and society and their application: from nanoscale, digits, and logic steps to global progress. J. Nanopart Res. **22**(11), 321 (2020). https://doi.org/10.1007/s11051-020-05032-0. Epub 2020 Oct 22. PMID: 33106748; PMCID: PMC7577848
11. Betley, E., Sterling, E.J., Akabas, S., Paxton, A., Frost, L.: Lessons in conservation, vol. 11, no. 1, pp. 9–25 (2021)
12. Stokols, D., Misra, S., Moser, R.P., Hall, K.L., Taylor, B.K.: The ecology of team science. Am. J. Prev. Med. **35**(2), S96–S115 (2008)
13. www.lerningloop.io. https://learningloop.io/glossary/raci-matrix. Accessed 01 June 2024

# Life Cycle Assessment (LCA) of a Prototypical Hairpin Stator

Achim Kampker, Henrik Born, Michael Nankemann$^{(\boxtimes)}$, Sebastian Hartmann, and Tim Franitza

Chair of Production Engineering of E-Mobility Components, RWTH Aachen University, Aachen, Germany
M.Nankemann@pem.rwth-aachen.de

**Abstract.** Current forecasts indicate a significant increase in demand for electric traction drives, including the corresponding impact on the environment. Modern motor designs and sustainability assessments thereof focus on efficiency in use, among other reasons motivated by the major environmental impact during this phase caused indirectly by carbon intensive energy production. However, due to the use of limited resources and energy intensive manufacturing processes, the production phase is increasingly critical for a sustainable life cycle of electric motors. In the prototyping stage, a major part of future product development, the direction for the sustainability of a product throughout its life cycle is being set. Therefore, the prototyping stage can already be used to gather useful insights into the environmental impact of a future product and its production system. In this research, a life cycle analysis of a prototypical production of one of the main components of the electric drive, the stator, is performed. The process chain of stator production of the research project ScaleUp E-Drive is investigated in two parts: First, a Life Cycle Inventory is conducted by direct data collection supplemented by the data sets from the commercial database ecoInvent v3.91 as well as publicly available data from other research. Subsequently, a Life Cycle Impact Assessment (LCA) is performed with the target of a comprehensive and accurate evaluation of the input and output flows of the prototypical production. The assessment is focused on the impact categories global warming potential (GWP), terrestrial acidification, human toxicity and freshwater ecotoxicity.

**Keywords:** Life cycle assessment · hairpin stator · e-motor prototyping

## 1 Introduction

Current forecasts predict a massive increase in demand for electrically powered vehicles. The primary background to this is the massive potential of electrification of the mobility sector in terms of $CO_2$ reduction and consequently slowing down climate change. While electrically powered vehicles offer an advantage over conventional vehicles, especially in the use phase, up to 80% more $CO_2$ is emitted during the production phase [1]. Optimization of the production processes to reduce emissions is therefore essential. There are already many approaches and publications on this subject, particularly in the

© The Author(s) 2025
H. Kohl et al. (Eds.): GCSM 2024, LNME, pp. 108–116, 2025.
https://doi.org/10.1007/978-3-031-93891-7_13

battery [2–4]. The electric motor, on the other hand, is hardly considered at present, even though it is also a central component of the electric powertrain and is used in all vehicles with electrified drive train designs.

In addition, the production capacities needed to manufacture the forecast number of units are not yet available and are still being built up. However, in this early phase of prototyping, small and pre-series production, the fundamental framework conditions are already established which define the subsequent sustainability at the product and production system level. Conversely, valuable knowledge can therefore already be gathered in this early phase to optimize sustainability at the product and production system level of later series production. This study is carried out within the research project ScaleUp E-Drive [5]. Its objective is to perform a life cycle assessment (LCA) based on a reference process chain for a variant-flexible stator production in the context of the HaPiPro2 [6] project of the PEM at RWTH Aachen University in order to identify the greatest challenges and potentials within the hairpin stator production chain with regard to sustainable component production for the electric drive in the prototypical context and to derive possible implications for series production. Furthermore, this work is intended to highlight overriding challenges regarding the sustainability assessment of production processes of electric motors to create a basis on which future needs for action can be derived.

## 2 System Boundary Conditions and Target Definition

The LCA is carried out according to the methodological framework of DIN-EN-ISO 14040 and consists of four successive, iterative process steps [7].

The investigation is done as part of the ScaleUp E-Drive research project, which utilizes a variant-flexible pilot line for hairpin stator production, that has been set up within the research project HaPiPro$^2$.

### 2.1 Objective and Functional Unit

The focus of this LCA is the prototypical production phase of a hairpin stator using the HaPiPro2 demonstration production line as an example. Accordingly, the aim is to quantify the influences of individual processes and material inputs of the hairpin stator on ecological sustainability aspects in the prototypical environment and thus to be able to draw conclusions about subsequent series production. To this end, the production of 50 HaPiPro2 stators is analyzed by means of a "cradle-to-gate" approach. One of these produced stators is the functional unit of this LCA.

### 2.2 Objects of Study and Life Cycle Scope

The object of consideration is a prototype stator and the prototype production line of the HaPiPro2 research project. The stator under consideration consists of a laminated core of electrical steel, insulated flat copper wire, slot insulation paper, epoxy resin, phase cable and the connection assembly. Basic information on the dimensions and weight of the stator is given in the table below. No performance data of the stator is available since the stator is manufactured exclusively for research of the production processes (Table 1).

*Framework of a life cycle assessment*

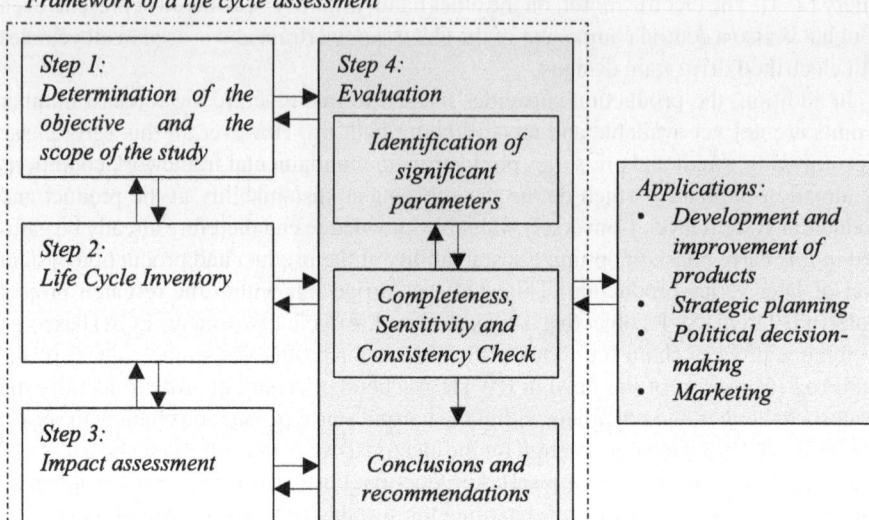

**Fig. 1.** Structure of a life cycle assessment according to DIN EN ISO 14040 [8]

**Table 1.** Properties of the reference object HaPiPro$^2$-stator

| Product properties | Value |
| --- | --- |
| Outer Stator Diameter | 190 mm |
| Inner Stator Diameter | 126,6 mm |
| Length electrical steel core | 160 mm |
| Number of hairpins | 144 pins |
| Number of slots | 48 slots |
| Total mass | 19,36 kg |

## 2.3  System Boundaries

The use of the prototype is limited to the investigation of process parameters during production. No further use or recycling takes place, so that the life cycle of the HaPiPro2 stator is considered complete at the end of the production process. The prototypical production line is located in Aachen, Germany. For this reason, within the framework of the research project, suppliers from the German economic area are primarily used. Based on this, the emission data was selected using the database data available for Germany. Emission data were not collected from the suppliers involved due to the effort involved and the accessibility of such data. Transport routes were excluded from the analysis due to the intended general applicability of the results.

## 3   Life Cycle Inventory

The process chain of the prototypical production line is divided into a total of eight stations, which are schematically sketched in Fig. 2.

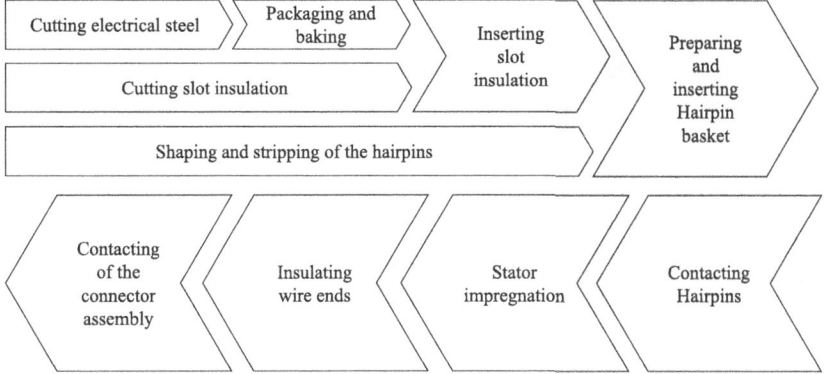

**Fig. 2.** HaPiPro$^2$ prototype production process chain [9]

The energy consumption of the equipment was recorded using current sensors during active operation. To determine the emissions caused by primary energy for each process, the German electricity mix in 2022 was used [10] and modeled using the corresponding data sets for the individual types of energy generation from ecoinvent. 3.9.1. Other external media flows are neglected in this study.

The production system incorporates the externally purchased raw materials coated electrical sheet, enameled copper wire, epoxy resin and insulation paper. The processes and media flows for their production are modeled using external literature and data sources. The starting point here is the data stored in the commercial database ecoInvent, current version 3.91. So-called system processes are used, which consider the entire material production according to the "cradle-to-gate" principle. The allocation of the data is carried out according to the cutoff system model. The overall process flow is documented in the Fig. 3. Here, the quantities of electrical energy used per process are listed with each of the stations. The process steps included within the facility of the research project are marked as such.

The 0,3 mm thick electrical steel, which is purchased in sheet form and already coated with baking varnish, is cut into the stator geometry via a laser cutting machine. The relatively high amount of waste is directly taken into account in the model, since low utilization of the steel strip is a key cost factor in stator prototyping. The cut sheets are clamped in a fixture and baked in an oven. The grooves of the laminations are lined with an n-aramid insulation paper. This paper is approximated by melamine formaldehyde resin in the ecoinvent database due to its membership in the polyamide group. In the absence of alternatives, this simplification had to be applied in this work also. The paper is obtained in rolls, shaped using embossing rollers, then folded and inserted by hand into the sheet package.

The prototype stator has a hairpin winding. The flat wire with dimensions 4 × 2 mm is provided with a 100 µm thick PAI coating. Since no data are available for the PAI coating material, this material is approximated using PE-HD. This assumption as well as the energy input for the coating of copper wire are based on literature data [11]. Each hairpin stator has 144 individual hairpins, each of which is 520 mm long. To produce these pins, the wire is first straightened and then bent into shape and cut using a CNC bending machine. In preparation for contacting, the PAI varnish is removed from the wire ends using a laser stripping process.

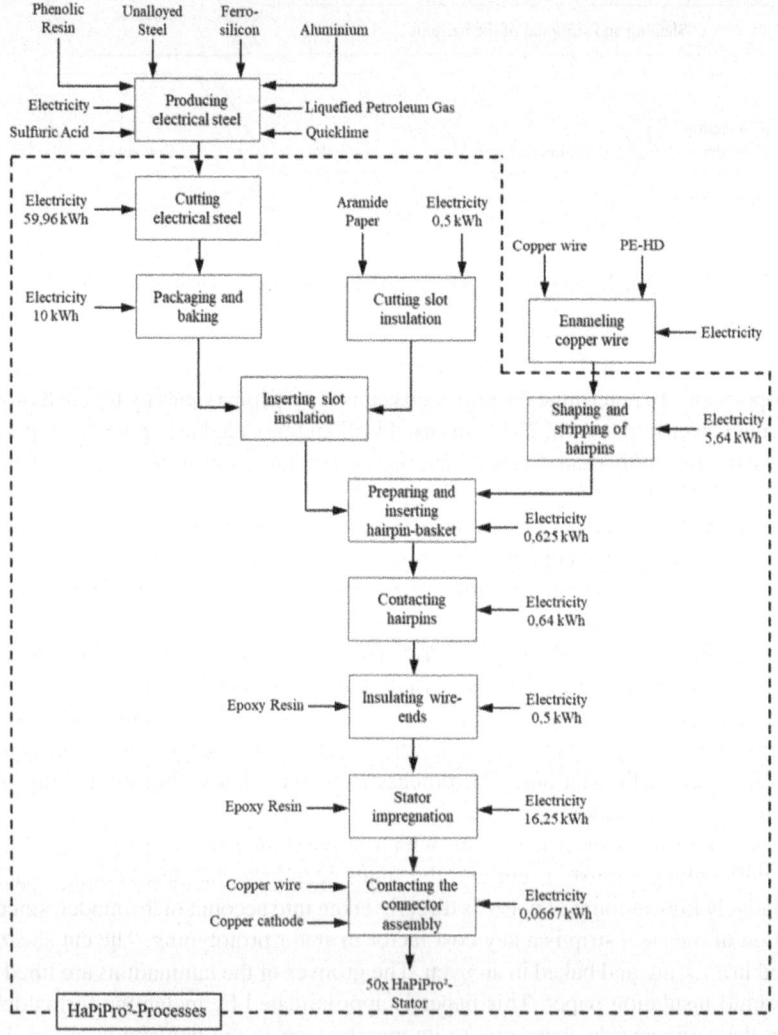

**Fig. 3.** Process flow chart including energy use per stator

The stripped hairpins are then assembled into a hairpin basket and inserted into the laminated stack. In preparation for the welding process, the hairpin ends are separated and twisted. The twisting tool is driven by a hand crank. The hairpin ends are then welded using a solid-state laser with a wavelength of 1070 nm. After this process, the stator is moved to a joint station for impregnation and powder insulation. The stator is preheated in an oven before the winding is impregnated with a liquid epoxy resin. For this purpose, a resin is dripped into the winding head of the rotating, preheated stator. After application, the resin is cured by reheating in the oven. Finally, the copper ends are coated with an epoxy resin powder. In this study, the same data is used for the two different epoxy material variants. In the last step, a phase cable and a connection assembly are welded on. The exact material quantities per stator are listed in Table 2. The quantities given are considered as "input quantities" into the system.

**Table 2.** Total material input of the stator

| Material | Amount |
| --- | --- |
| Electrical steel | 46,816 kg |
| Slot insulation paper | 0,05 kg |
| Copper wire (winding) | 5,04 kg |
| Copper wire insulation | 0,12 kg |
| Epoxy resin (impregnation) | 0,08 kg |
| Epoxy resin (Secondary insulation) | 0,10 kg |
| Copper wire (phase cable) | 0,21 kg |
| Copper cathode (Connection assembly) | 0,02 kg |

## 4   Life Cycle Impact Assessment

The environmental impacts are modeled using the open-source platform Open LCA, combined with the information from the ecoinvent 3.91 database as well as additional research. The impact categories used for this LCA are a selection of the most frequently used categories in the life cycle assessment of electrically powered vehicles [12]. The calculation is performed according to the ReCiPe 2016 1.03 midpoint (H) calculation method. The environmental impact of these impact categories in relation to the reference value of a prototypical stator is shown in Table 3.

In the following figures, the results in the four investigated midpoint categories are shown across the top four process elements contributing to the respective category. Firstly, the high impact of copper is visible in categories considering toxicity and acidification, with values 10 to 100 times higher than for the remaining materials and processes. Secondly, the electrical steel and its processing shows to be a leading contributor to the overall Global Warming Potential of the prototypical stator. This is mainly due to energy intensive processing during the manufacturing of the steel strip (Fig. 4).

**Table 3.** Results per stator

| Material | Amount |
|---|---|
| Climate Change GWP100 | 199.7 kg CO2-Eq |
| Human toxicity – non-carcinogenic | 3,443.5 kg 1.4-DCB-Eq |
| Ecotoxicity: freshwater | 176.7 kg 1.4-DCB-Eq |
| Acidification: Terrestrial | 3.587 kg SO2-Eq |

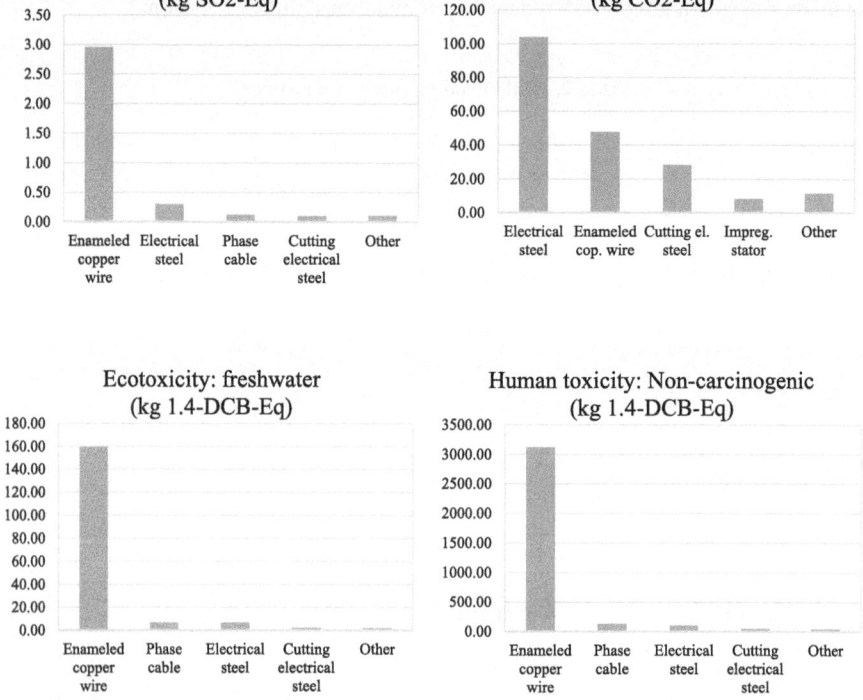

**Fig. 4.** Midpoint category impacts calculated per stator according to ReCiPe 2016

## 5  Discussion

The aim of the study was to quantify the ecological impact of a hairpin stator in the prototypical environment. The results of the investigation show that the enameled copper wire and the coated electrical sheet contribute heavily to the environmental impact of a hairpin stator. The influence of the enameled copper wire is particularly prominent in this analysis. A further challenge is that the results are not directly transferable due to technical differences between prototype construction and series production. This becomes particularly clear when taking a closer look at the sheet package production. Since individual sheets are used here and the inner part of each stator sheet cannot be used, large amounts of waste are produced. In series production, it is often possible to use this material to manufacture the rotor stack, increasing material utilization significantly. Additionally, the investigation has shown that existing data sets can only be applied to hairpin stator production with numerous simplifying assumptions. One particularly significant example is the lack of data concerning PAI, one of the most common wire insulation materials used in hairpin stator production, which had to be modelled using PE-HD. This lack of data is a major barrier for the use of LCA in early stages of product development of stators. Since large parts of the production system are still to be defined, gathering accurate and case specific data is often not feasible. Since the user must rely on general data, the value of the analysis is heavily impacted by poor data quality. This study helps to spotlight the most significant gaps regarding hairpin stator production in current data. In future work, it will be necessary to build up a specific database to allow users to gather more accurate results quickly.

## References

1. Kämper, C., Helms, H., Biemann, K.: Wie klimafreundlich sind Elektroautos? Institut für Energie- und Umweltforschung Heidelberg (2020)
2. Chen, Q., et al.: Investigating carbon foot-print and carbon reduction potential using a cradle-to-cradle LCA approach on lithium-ion batteries for electric vehicles in China. J. Clean. Prod. **369** (2022)
3. Du, S., Gao, F., Nie, Z., Liu, Y., Sun, B., Gong, X.: Comparison of electric vehicle lithium-ion battery recycling allocation methods environmental. Sci. Technol. (2022)
4. Manurkar, E.F., Ahmad, A., Tariq, M.N., Wu, F., Chen, R., Li, L.: Life cycle assessment of lithium-ion batteries: a critical review; resources. Conserv. Recycl. **180** (2022)
5. https://www.hub-edrive.de/. Accessed 10 Apr 2024
6. https://www.pem.rwth-aachen.de/cms/pem/forschung/projekte/~lenop/hapipro2/. Accessed 10 Apr 2024
7. Deutsches Institut für Normung DIN EN ISO 14040:2021-02. Umweltmanagement-Ökobilanz-Grundsätze und Rahmenbedingungen. Beuth, Berlin (2021)
8. Blazejczak, J., Edler, D.: Nachhaltigkeitskriterien aus ökologischer, ökonomischer und sozialer Perspektive: ein interdisziplinärer Ansatz. Vierteljahrshefte zur Wirtschaftsforschung. **73**(1), 10–30 (2004). https://doi.org/10.3790/vjh.73.1.10. ISSN 1861-1559
9. PEM RWTH Aachen homepage. https://www.pem.rwth-aachen.de/cms/PEM/For-schung/Projekte/~lenop/HaPiPro2/. Accessed 21 July 2023
10. ISE Fraunhofer homepage. https://www.ise.fraunhofer.de/de/presse-und-medien/presse-informationen/2023/nettostromerzeugung-in-deutschland-2022-wind-und-photovoltaik-haben-deutlich-zugelegt.html. Accessed 21 July 2023
11. Nordelöf, A., Grunditz, E., Tillman, A.M., et al.: A scalable life cycle inventory of an electrical automotive traction machine - part I: design and composition. Int. J. Life Cycle Assess. **23**, 55–69 (2018). https://doi.org/10.1007/s11367-017-1308-9
12. Iganova, I., Rödl, A., Bach, V., Kaltschmitt, M., Finkbeiner, M.: A review of life cycle assessment studies of electric vehicles with a focus on resource use. Resources **9** (2020). https://doi.org/10.3390/resources9030032

# Data Analytics

# Simplified Data Acquisition for Product Carbon Footprints in the Automotive Industry and Their Accuracy

Kai Rüdele[1]([⊠]), Lukas Nagel[2], and Matthias Wolf[1]

[1] Institute of Innovation and Industrial Management (IIM), Graz University of Technology, Kopernikusgasse 24, 8010 Graz, Austria
kai.ruedele@tugraz.at
[2] Institute for Production Management, Technology and Machine Tools (PTW), Technical University of Darmstadt, Otto-Berndt-Street 2, 64287 Darmstadt, Germany

**Abstract.** Manufacturing processes significantly impact greenhouse gas (GHG) emissions, but can also contribute to climate change mitigation through sustainable practices and product stewardship. Life cycle assessments (LCAs) and product carbon footprints (PCFs) are crucial tools for assessing the environmental impact of products. However, the data acquisition process for conducting such analyses can be complex and time-consuming. This article delves into the comparative analysis of different PCF studies for a vehicle component. Our detailed PCF study is mainly based on measured data and used as the reference value for accuracy. The other three PCFs were derived from available data sources, such as technical drawings or the International Material Data System (IMDS). Our research reveals that the determined value of the detailed study is significantly higher than the results that only take material input into account, but 18% lower than the PCF, which is based on rough assumptions and secondary data. This indicates a relationship between the depth of information available and the accuracy of outcomes. However, less accurate methods may be suitable for initial estimations, especially when resources are lacking or production has not yet begun.

**Keyword:** carbon footprint · automotive · life cycle inventory · ex-ante LCA

## 1 Introduction

The product manufacturing stage is one of the main contributors to greenhouse gas (GHG) emissions [1], making the decarbonization of industry a crucial aspect of mitigating climate change [2]. Within the automotive sector, vehicle manufacturers have implemented GHG reduction measures into their business practices [3] and related suppliers must demonstrate strong environmental commitment [4], often by documenting, or justifying environmentally relevant processes [5].

To manage and improve the environmental impact of its product portfolio, it is useful for a company to use life cycle assessments (LCA) to evaluate the magnitude and significance of a product's impact on the environmental dimension of interest such as

© The Author(s) 2025
H. Kohl et al. (Eds.): GCSM 2024, LNME, pp. 119–127, 2025.
https://doi.org/10.1007/978-3-031-93891-7_14

generation of air pollutants, water use, or energy consumption. For this, companies must compile their product's in- and outputs throughout their life cycles (life cycle inventory analysis, LCI) [6]. Also, product carbon footprint (PCF) studies follow this systematic procedure but exclusively measure the amount of GHGs caused by a specific good.

The unavailability of LCI data is a common hurdle encountered in LCA and often overcome by using secondary data in combination with a sensitivity analysis. Occasionally, LCAs or PCFs are required even before the product's market entry. Such prospective or 'ex-ante' assessments are challenging since some aspects such as used technologies/processes or applicable policies are not predictable with certainty. At the same time, it is recommended to not assume factors of previous assessments as fixed [7], or to transfer data from the lab/pilot directly to industrial series production [8].

In this article, we use an industrial case study from the automotive industry to evaluate the accuracy and applicability of early available product and production data for PCFs and compare them with 'ex-post' primary data results.

## 2   State-of-the-Art

LCI data can be collected through measurements, calculations, or estimations, of the inputs and outputs of a process [9]. Data based on direct on-site measurements, also known as primary data, is considered the most specific, precise, and thus most reliable [10]. To save LCA practitioners' time and effort, existing data from databases, publications, default values from national inventories, and validated estimates can be used [9, 10]. The appropriateness of secondary data's quality and fitness for purpose depends on objective, credibility, and feasibility, as well as cost and capacity constraints [11].

Moreover, non-automated exchange of data, and lacking integration of IT tools hinder the compilation of relevant information and quick decisions [12]. Therefore, research has already focused on both the simplification of data acquisition and the application of LCAs even before production ramp-up.

### 2.1   Previous Research on Simplified/Streamlined and Screening LCAs

Simplified or streamlined LCAs (the vast majority of publications use both terms synonymously, [13, 14]) aim to get acceptable results with reduced data collection efforts. This is mainly achieved by using generic data as a proxy and/or exclusion of presumably irrelevant aspects [15]. Compared to a full scale LCA based on detailed LCI, simplified LCAs are created more efficiently and faster. Also, automatically generated LCIs combining product information with environmental data is seen as a simplification practice [15, 16]. However, such software solutions reach their limits as soon as product complexity increases [16].

Occasionally, screening LCAs are distinguished from simplified LCAs [17], as the former are more comprehensive in coverage to identify environmental hot spots (e.g., processes where the most emissions occur) and only superficial in detail [18]. Simplified as well as screening LCAs are suitable in the product development process when there is just limited information about the final product and its characteristics [19].

## 2.2 Definitions and Concepts for Ex-Ante LCAs

Since screening LCAs use early available data and information to draw preliminary conclusions about the studied system, they are seen as ex-ante LCAs [17]. Ex-ante LCA concepts are conducted before a product or technology is commercially deployed at scale and insights into the system are still lacking [8, 10]. Using LCAs in an ex-ante manner introduces a systemic, but still less structured approach than conventional LCA [20], and provides implications for potential avoidable environmental impacts [10]. To do so, early available basic information (e.g., theoretical prior knowledge) about the product, the techno-economic context and scale on which it will be manufactured are sufficient [8]. Especially the combination of ex-ante LCAs with a range of exploratory scenarios can support decisions in early development stages and further refined later on [20–22]. Thus, ex-ante LCAs can guide developers towards the improved environmental performance of the product [21], and facilitate promising innovations early on [22]. Usually, ex-ante LCAs deal with emerging technologies [10, 20, 21].

## 2.3 LCAs Based on Engineering Documentation

Due to the prospective nature of ex-ante LCAs, only limited data is available and significant degrees of uncertainty are present [22]. In the case of industrial goods, primary this limited data most probably includes technical drawings, computer-aided design (CAD) models, or bill of materials (BOM). However, this data is often scattered among different actors in the supply chain, differs in format, and was originally designated for purposes other than an LCA [23]. BOM-based methods to quantify GHG emissions of the production of goods [24], as well as concepts for integrating (existing) CAD-models into LCAs [25, 26], are subjects of current research.

## 2.4 Existing Studies Using IMDS for Inventory Data Acquisition

The International Material Data System (IMDS) is the automotive industry's database to record, maintain, analyze, and archive the detailed chemical composition of all (sub)components used in a vehicle. This system was jointly developed by leading automobile manufacturers and became the global standard. Today, the web-based repository is used by 46 name-brand manufacturers (representing more than 70 brands) and more than 120,000 suppliers [27]. IMDS traces hazardous and controlled substances to meet (declaration) obligations by standards, laws, and regulations, but also supports recyclability and recoverability of materials in a vehicle and addresses the disposal of substances of concern.

Numerous scientific works used IMDS as a data source for LCAs and PCF studies [23, 28–31]. Some of them explored how using IMDS data at different levels of completeness affects LCA results and investigated trade-offs between workload and accuracy [23, 28]. For the analyzed components and scenarios, it was found that a reasonable outcome can be obtained even with reduced effort and thus lower level of detail. Especially, the impact category of climate change was satisfactorily quantified, making IMDS-based approaches suitable for PCF studies [23, 28].

# 3  Approaches for Inventory Data Acquisition

To compare different LCI modeling scenarios and related impact assessments, we take the subframe from [32] as a case study. We used the well-established ISO standards [6, 9] which are also recommended when an LCA is applied in an ex-ante manner [8, 10]. Assumptions and methodological choices made in our study are described hereinafter.

## 3.1  Objective, Scope Definition, and Applied Methods

The functional unit is one unit of the manufactured subframe. Similarly to the approach used in [23] and [28], it is a cradle-to-gate study ending with the assembly of the component by the supplier, so neither in-vehicle assembly, nor use stage and end-of-life emissions were assigned to it. Results were generated in the software GaBi (Version 10.6.1) using the regionally specific datasets from the Sphera database. For simplicity, we limited our assessment to the impact category of climate change (GWP100 according to CML 2001–2016 impact assessment method).

While study object and system boundaries remain the same, the LCIs and subsequently the PCF studies are based on different types of data sources. A total of four scenarios were assessed. Figure 1 illustrates the characteristics of these.

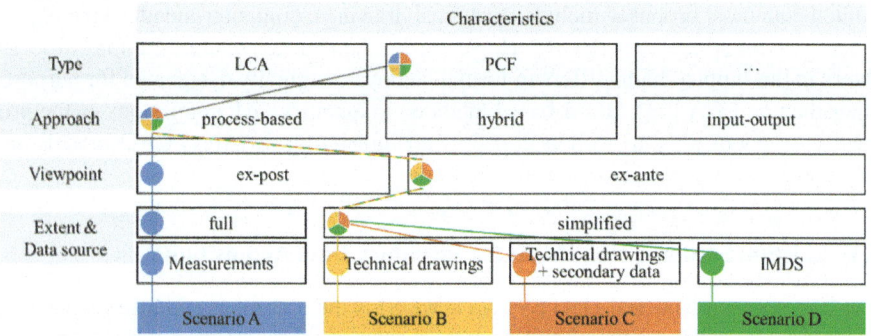

**Fig. 1.** Characteristics of LCI modeling scenarios

## 3.2  Scenario A: LCI Based on Measurements

Like [28] the scenario with the highest level of detail requiring the most effort is used as a benchmark for evaluating the accuracy of all the other scenarios. However, our reference scenario is based on primary data that is mainly derived from measurements. Important data points include material inputs, energy consumption of different process steps, detailed information on waste generation, transportation, utilities (e.g., lubricants), and auxiliaries, as well as primary data from sub-suppliers.

This scenario considers the existing production of all components, including raw material extraction, material transformation, and assembly (ex-post). Information about electricity generation and consumption, transport, and waste was well documented or retrieved from site-specific data. Allocation problems were solved by partitioning.

### 3.3   Scenario B: LCI Based on Technical Drawings

The second scenario only considers material input of the final product from technical drawings. We used weight and material information taken from documents for all parts. Such information is usually available long before the start of production and can be used for initial assessments of the product's environmental impact. This scenario assumes waste- and transport-free production and is associated with a very low workload for data gathering.

### 3.4   Scenario C: LCI Based on Technical Drawings and Secondary Data

In order to overcome the expected shortcomings of scenario B, we created another scenario where we added losses and rough assumptions regarding production and transportation efforts. Available common knowledge and experience as well as secondary data about processes (e.g., casting and machining), were used to quantify energy consumption and scrap. Although CAD models and BOMs might be created before the supplier selection, placeholder values for the transportation from sub-suppliers to the assembly site can be estimated to some extent. In our case, the average distance to eligible plants was taken into account.

### 3.5   Scenario D: LCI Based on IMDS

Via IMDS, the composition of components is declared by the suppliers in great detail. In addition to the breakdown by unique materials and substances for every single part, standardization and mandatory disclosure are also advantages [29, 30]. However, IMDS only provides information on materials in the finished product without information on manufacturing and assembly processes (such as material losses and utilities) and cannot be used in the early vehicle development stages [23, 27, 29].

We discovered an inconsistency in the IMDS similar to [31]. This inaccuracy was reflected in the total weight of the subframe, as it deviated by almost 2% from the value of the other scenarios. Additionally, the very detailed breakdown in IMDS reveals a higher proportion of low-carbon materials than indicated in the technical drawings.

## 4   Results of the Study and Discussion

### 4.1   Impact Assessment and Interpretation of All Scenarios

To further compare the LCI data, a life cycle impact assessment was carried out. Figure 2 shows the composition and amount of the PCFs (measured in $CO_2$ equivalents) for all four scenarios under study.

Scenarios B and D underestimate the actual PCF by 22% and 33% respectively. It is evident that the emissions omitted from processing, generated waste, and transportation are significant. Scenario C with its estimated values results in a PCF that is 19% above the reference scenario. Especially the production of the die-cast side parts is much more efficient and thus lower in emissions in reality than in the generic dataset. Although

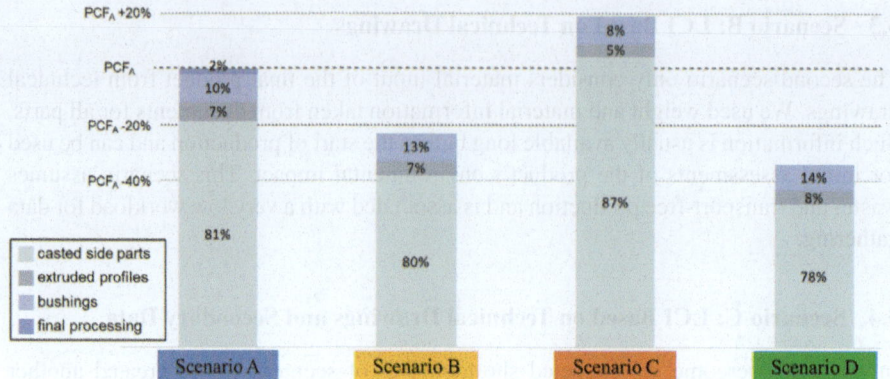

**Fig. 2.** Resulting PCFs for the scenarios (B to D) compared to the reference scenario (A)

modeled in Scenario C, the proportion of final processing (i.e. joining of individual parts and machining) by the supplier is well below one percent.

It can also be seen that the relative composition of the emission sources is very close to reality. The most extreme case - the contribution of the side parts (Scenario A vs. C) - showed a deviation of six percentage points; this can be explained by the state-of-the-art manufacturing process (A) that produces considerably fewer emissions than the average efficient casting process from an older reference period, as reflected by the used secondary data (C). Particularly for the hydro bushings, the absolute values determined were very close to each other.

Whether the accuracy of scenarios B to D is satisfactory (e.g., for forecasts when awarding contracts) certainly depends on the individual case. In the present case study, a corridor spanned by scenarios B and C, or the derived average value, almost exactly matches the values from the real scenario A. Nevertheless, data availability is still the most important criterion to decide on the level of detail to which a study can be conducted actually [17].

## 4.2 Effort-Accuracy Ratio of the Scenarios

The reference scenario is characterized by the highest workload, but also the highest degree of completeness and representativeness. In our case study, data collection took several weeks, as it was the first PCF study for the involved parties, and comparable LCIs are expected to be compiled more quickly in the future. Nevertheless, due to the volume of data involved, the reference scenario is characterized by a feasibility limit [28]. Reproducibility is also limited due to changing scrap rates and capacity utilization, or measurement errors and inaccuracies.

Compiling and evaluating the technical drawings took less than a workday while the export of data from IMDS was finished within minutes. In both cases, the most work was associated with the modeling in GaBi. The main difference in terms of workload between scenarios B and C was primarily in research work, which however took less than two person days.

It is important to emphasize that the stated workloads of the different modeling options are only rough guidelines (esp. when applied to other products), but offer an indication of the relative differences. As in other cases [28, 33], saving effort by omitting information has led to an underestimation of GHG emissions. It can be expected that the time required to get additional data initially shows a good cost-benefit ratio, but results will later approach the real value asymptotically. Literature indicates that trade-offs between sophistication and accuracy and effort spent are influenced by the quality of databases [34]. Here, especially Scenario C still has potential for improvement.

## 5  Conclusion

The automotive industry must question itself regarding its ecological impact. PCF studies allow fast identification of main contributors and thus enable organizations to make informed decisions to reduce their carbon intensity. However, the amount of accessible data is commonly limited in early design phases. To enhance transparency and accountability, while keeping efforts low, already existing information can be used.

Our research using the case of a chassis component confirms that detailed data collection through measurements leads to the most accurate PCF, while workload-reduced methods using technical drawings or IMDS may still be suitable for initial estimates. Our results show that the average of the results of two simplified approaches yields a realistic PCF and could be achieved with less work than the reference study. Undoubtedly, further studies and sensitivity analyses are needed to confirm our findings.

Future research could focus on other products, system boundaries, (industry-specific) input sources, and/or impact categories. Such additional case studies may also validate our findings. This article may also contribute to designing a comprehensive automation of (parametrized) LCI modeling.

## References

1. Crippa, M., Guizzardi, D., Pagani, F., et al.: GHG emissions of all world countries. Publications Office of the European Union, Luxembourg (2023). https://doi.org/10.2760/953332
2. Material Economics: Industrial Transformation 2050: Pathways to net-zero emissions from EU heavy industry. https://materialeconomics.com/publications/industrial-transformation-2050. Accessed June 2024
3. Poligkeit, J., Fugger, T., Herrmann, C.: Decarbonization in the automotive sector: a holistic status quo analysis of original equipment manufacturer strategies and carbon management activities. Sustainability **15**, 15753 (2023). https://doi.org/10.3390/su152215753
4. Wissuwa, F., Durach, C.: Turning German automotive supply chains into sponsors for sustainability. Prod. Plann. Control **34**(2), 159–172 (2023). https://doi.org/10.1080/09537287.2021.1893405
5. Finkbeiner, M., Krinke, S., Oschmann, D., et al.: Data collection format for life cycle assessment of the German association of the automotive industry (VDA). Int. J. Life Cycle Assess. **8**, 379–381 (2003). https://doi.org/10.1007/BF02978511
6. ISO 14044:2006. Environmental management – Life cycle assessment – Requirements and guidelines

7. Rajagopal, D., Zilberman, D.: Environmental lifecycle assessment for policy decision-making and analysis. In: Lifecycle Carbon Footprint of Biofuels Workshop (2008)
8. Tecchio, P., Freni, P., De Benedetti, B., et al.: Ex-ante life cycle assessment approach developed for a case study on bio-based polybutylene succinate. J. Clean. Prod. **112**, 316–325 (2016). https://doi.org/10.1016/j.jclepro.2015.07.090
9. ISO 14067:2018. Greenhouse gases – Carbon footprint of products – Requirements and guidelines for quantification
10. van der Giesen, C., Cucurachi, S., Guinée, J., et al.: A critical view on the current application of LCA for new technologies and recommendations for improved practice. J. Clean. Prod. **259**, 120904 (2020). https://doi.org/10.1016/j.jclepro.2020.120904
11. Pandey, D., Agrawal, M., Pandey, J.: Carbon footprint: current methods of estimation. Environ. Monit. Assess. **178**, 135–160 (2011). https://doi.org/10.1007/s10661-010-1678-y
12. Saling, P.: Eco-efficiency assessment. In: Special Types of Life Cycle Assessment. LCA Compendium, pp. 115–178. Springer, Dordrecht (2016). https://doi.org/10.1007/978-94-017-7610-3_4
13. Nakamura, S., Nansai, K.: Input-output and hybrid LCA. In: Special Types of Life Cycle Assessment. LCA Compendium, pp. 219–291. Springer, Dordrecht (2016). https://doi.org/10.1007/978-94-017-7610-3_6
14. Yang, Y., Heijungs, R., Brandão, M.: Hybrid life cycle assessment (LCA) does not necessarily yield more accurate results than process-based LCA. J. Clean. Prod. **150**, 237–242 (2017). https://doi.org/10.1016/j.jclepro.2017.03.006
15. Beemsterboer, S., Baumann, H., Wallbaum, H.: Ways to get work done: a review and systematisation of simplification practices in the LCA literature. Int. J. Life Cycle Assess. **25**, 2154–2168 (2020). https://doi.org/10.1007/s11367-020-01821-w
16. Haun, P., Müller, P., Traverso, M.: Improving automated life cycle assessment with life cycle inventory model constructs. J. Clean. Prod. **370**, 133452 (2022). https://doi.org/10.1016/j.jclepro.2022.133452
17. Klöpffer, W., Grahl, B.: Life cycle assessment (LCA): a guide to best practice. Wiley-VCH, Weinheim (2014). https://doi.org/10.1002/9783527655625
18. Jensen, A., Elkington, J., Christiansen, K., et al.: Life cycle assessment (LCA): a guide to approaches, experiences and information sources. Environ. Issues Ser. **6** (1998)
19. Arena, M., Azzone, G., Conte, A.: A streamlined LCA framework to support early decision making in vehicle development. J. Clean. Prod. **41**, 105–113 (2013). https://doi.org/10.1016/j.jclepro.2012.09.031
20. Villares, M., Işıldar, A., van der Giesen, C., et al.: Does ex ante application enhance the usefulness of LCA? A case study on an emerging technology for metal recovery from e-waste. Int. J. Life Cycle Assess. **22**, 1618–1633 (2017). https://doi.org/10.1007/s11367-017-1270-6
21. Cucurachi, S., van der Giesen, C., Guinée, J.: Ex-ante LCA of emerging technologies. Procedia CIRP **69**, 463–468 (2018). https://doi.org/10.1016/j.procir.2017.11.005
22. Delpierre, M., Quist, J., Mertens, J., et al.: Assessing the environmental impacts of wind-based hydrogen production in the Netherlands using ex-ante LCA and scenarios analysis. J. Clean. Prod. **299**, 126866 (2021). https://doi.org/10.1016/j.jclepro.2021.126866
23. De Oliveira, F., Nordelöf, A., Sandén, B., et al.: Exploring automotive supplier data in life cycle assessment–precision versus workload. Transp. Res. Part D: Transp. Environ. **105**, 103247 (2022). https://doi.org/10.1016/j.trd.2022.103247
24. Zhou, G., Zhou, C., Lu, Q., et al.: Feature-based carbon emission quantitation strategy for the part machining process. Int. J. Comput. Integr. Manuf. **31**(4–5), 406–425 (2018). https://doi.org/10.1080/0951192X.2017.1328561

25. Ben Slama, H., Gaha, R., Benamara, A.: Proposal of new eco-manufacturing feature interaction-based methodology in CAD phase. Int. J. Adv. Manuf. Technol. **106**, 1057–1068 (2020). https://doi.org/10.1007/s00170-019-04483-7

26. Merschak, S., Hehenberger, P.: Ecodesign methods for mechatronic systems: a literature review and classification. In: 20th International Conference on Research and Education in Mechatronics, pp. 1–8 (2019). https://doi.org/10.1109/REM.2019.8744105

27. DXC International Material Data System (IMDS). https://public.mdsytem.com/docments/10906/633283/DXC_IMDS_Making_Manufacturers_Greener_A4.pdf/3251a31d-b1d0-215b-ddac-eea47bc575dc?t=1631695498082. Accessed 13 June 2024

28. Accardo, A., Dotelli, G., Spessa, E.: Impact of different LCI modelling scenarios on the LCA results, a case study for the automotive sector. SAE Technical Paper 2023-01-0884 (2023). https://doi.org/10.4271/2023-01-0884

29. Yu, M., Kim, Y.: Development of environmental assessment system of vehicle. In: Lecture Notes in Electrical Engineering, pp. 1151–1160. Springer, Berlin (2013). https://doi.org/10.1007/978-3-642-33738-3_19

30. Koffler, C., Krinke, S., Schebek, L., et al.: Volkswagen slimLCI: a procedure for streamlined inventory modelling within life cycle assessment of vehicles. Int. J. Veh. Des. **46**(2), 172–188 (2008). https://doi.org/10.1504/IJVD.2008.017181

31. Teng, C.: Implementing simplified LCA software in heavy-duty vehicle design: an evaluation study of LCA data quality for supporting sustainable design decisions (2020). https://urn.kb.se/resolve?urn=urn:nbn:se:kth:diva-273973

32. Rüdele, K., Wolf, M.: Identification and reduction of product carbon footprints: case studies from the Austrian automotive supplier industry. Sustainability **15**(20), 14911 (2023). https://doi.org/10.3390/su152014911

33. Parvatker, A., Eckelman, M.: Comparative evaluation of chemical life cycle inventory generation methods and implications for life cycle assessment results. ACS Sustain. Chem. Eng. **1**, 350–367 (2019). https://doi.org/10.1021/acssuschemeng.8b03656

34. Ciroth, A., Burhan, S.: Life cycle inventory data and databases. In: Life Cycle Inventory Analysis. LCA Compendium, pp. 123–147. Springer, Cham (2021). https://doi.org/10.1007/978-3-030-62270-1_6

# The Lean and Green Imperative
# of Manufacturing Data

John Patsavellas[(✉)], Yousef Haddad, and Konstantinos Salonitis

Sustainable Manufacturing Systems Centre, Cranfield University, Bedford MK43 0AL, UK
John.Patsavellas@cranfield.ac.uk

**Abstract.** This study introduces a stochastic model-based framework for the prediction and measurement of the environmental impact of manufacturing systems' digitalization. Utilising a Monte Carlo simulation experimental framework, this paper forecasts the $CO_2$e emissions from the entire lifecycle of manufacturing data over long-term time horizon under different scenarios. The analysis proceeds to estimate the maximum, average, and minimum potential $CO_2$e emissions, under different growth models. Findings reveal that, with the current exponential growth of data that exceeds data centres' efficiency improvements and carbon intensity decay rates, the environmental footprint associated with the entire lifecycle of data can have a potential adverse impact on the realisation of net-zero goals. The proposed approach provides a viable pathway for manufacturing enterprises aiming to align their data management practices with environmental sustainability and operational efficiency.

**Keywords:** Digitalization · Industry 4.0 · Monte Carlo Simulation · Sustainable Data

## 1 Introduction

The advent of Industry 4.0, with digitization being a founding cornerstone [1], has resulted in an exponential growth of data generation. This increase in the volume of data, resulting from the propagation of advanced information and communication technology (ICT), has far exceeded the growth in population and industrial activity [2]. Indeed, it was estimated in 2019 that 90% of all data back then was generated within the preceding two years [3]. This exponential growth is, however, accompanied by a significant growth in the energy required to generate, transmit, and store these data in data centres [4]. Indeed, the International Energy Agency estimates that 1–1.3% of the worldwide electricity consumption comes from data centres, and 0.6% of total greenhouse gas (GHG) emissions comes from these data centres [5].

In general, carbon intensity, which is the amount of $CO_2$e per unit energy, of energy mixes worldwide has significantly decreased over the last two decades. For example, China has managed to decrease the carbon intensity of its energy mix by 56% over a 15 year period from 2005 to 2020 [6]. In addition, energy efficiency of data centres has also been improving consistently. For example, the authors in [7] report that an increase

H. Kohl et al. (Eds.): GCSM 2024, LNME, pp. 128–136, 2025.
https://doi.org/10.1007/978-3-031-93891-7_15

an excess of fivefold of computing power in data centres has resulted in a mere increase of 6% in energy consumption over an 8 year period from 2010 to 2018. However, this increase might still be insufficient to deal with the exponential growth of data. These improvements in energy efficiency and the carbon intensity of such energy pose the risk of the rebound effect where improvements in efficiency and carbon intensity might be counterpart and overcompensated by an unproportional increase in data usage [8]. The degree of the risk of the rebound effect with regards to the growth of data consumption and its impact on the environment is, however, still unclear. For example, the authors in [6] found that the rebound effect is not significant with regards to the ICT industry, which is tightly related with data (i.e., the use phase of ICT devices), while in another study, the authors in [9] found that digitalization can have a strong rebound effect with regards to energy consumption.

In this paper, a stochastic model-based framework for the prediction of digitalization-related growth in data over its entire lifecycle, along with its environmental footprint, in the manufacturing industry is developed. The framework considers two growth models under a Monte Carlo simulation experimental framework; unbounded (exponential) growth, and bounded (logistic) growth. The rest of the paper is organised as follows: the next section presents the overarching model-based framework, Sect. 3 presents a numerical example, and finally Sect. 4 presents and discusses the concluding remarks.

## 2   Model-Based Framework

In this section, the overarching approach for the data and $CO_2e$ prediction is presented and discussed. The determination of the growth of data and their subsequent environmental footprint are determined under Monte Carlo simulations to reveal possible future pathways of the carbon footprint of manufacturing digitalization. Monte Carlo simulation is a stochastic modelling approach that involves random sampling of an experiment's stochastic parameters from probability distribution over several independent replications [10].

The first step is to estimate the amount of data generated by the manufacturing system. This can be done through analysis of historical data and identification of data growth. Although analysis of historical data by itself is insufficient to determine data growth, it can still shed some light on potential growth pathways. This is because data growth depends also on the emergence of new technologies (e.g., sensors, new machine models, advanced software tools etc.…). Therefore, the growth-parameters identification should be conducted carefully by analysing different factors such as the prospects of updating new machines, acquiring software tools etc.… Next, other key parameters such as carbon intensity's decay rate are determined. It is important to point out here that carbon intensity can vary greatly, depending on the data storage location (i.e., local vs data centres). This is because, as mentioned earlier, data centres are highly efficient with regards to energy consumption.

The next step is to calculate the $CO_2e$ emissions over the chosen period. Any growth model can be chosen, however, as discussed earlier, the size of data has demonstrated exponential growth traits. Nevertheless, long-term exponential growth is highly unlikely, even impossible. Indeed, in their seminal work *The Limits to Growth* [11], the authors discussed how limitations in resources will inevitably inhibit exponential growth and result in a slowdown in growth. Therefore, exponential growth often gradually slows down after a certain period of time elapses, resulting in a sigmoid-shaped growth, most recognizable of which is the logistic curve [12].

After the data growth model has been determined, energy consumption of data lifecycle stages has to be determined next. These stages are: data generation, data transmission, and data storage. These can be calculated as shown in Eq. (1) below:

$$E_{total} = E_{generation} + E_{transmission} + (E_{storage} \times t) \tag{1}$$

where $E_{total}$ is the total energy over the entire lifecycle of each unit of data (e.g., per bit), $E_{generation}$ is the energy required to generate each unit of data, $E_{transmission}$ is the energy required to transmit each unit of data to the storage location, $E_{storage}$ is the energy required to store each unit of data for a period of $t$ units of time. It should be noted here that the energy required to store data typically constitutes most of the data's lifecycle energy consumption, particularly with regard to cooling requirements [4].

In order to reflect the trend of the ever-decreasing carbon intensity of most grids, carbon intensity is assumed to exponentially decay. Therefore, the amount of CO2e emissions at any year for the logistic and exponential growth models, respectively, is calculated as follows

$$CO_2e_{t+1} = E_t CI_0 (1 - s_t)^t \left[ D_t + r_t D_t \left( 1 - \frac{D_t}{K} \right) \right], \quad t \in \{0, 1, \ldots, n-1\} \tag{2}$$

$$CO_2e_{total} = (E_t \times CI_0) \times a \prod_{t=1}^{n} ((1 + r_t)(1 - s_t)) \tag{3}$$

where $CO_2e_{t+1}$ is the amount of $CO_2e$ emissions at the end of year $t$ (note that $CO_2e_0 = CI_0 D_0$), $K$ is the carrying capacity (i.e., the maximum amount of data that can be realistically sustained), $D_0$ is the *current* data generation during time period $t = 0$, $r$ is the growth rate, $s$ is the carbon intensity's decay rate, $E_t$ is the total energy consumption per unit of the data's entire lifecycle, and $t$ is the time period.

## 3   Numerical Example

In order to test and validate the framework, and to demonstrate its applicability, a numerical example constituting various representative scenarios is presented and discussed in this section. The numerical example consists of 18 scenarios that depict different manufacturing company sizes/ grid's carbon intensity configurations, and growth models as presented in Table 1 below.

The modelled scenarios consist of three manufacturing systems sizes (small, medium, and large) and three geographical locations, representative of different levels of grids' carbon intensities. It should be noted here that one of the primary assumptions of this study is that the amount of data generated by a manufacturing enterprise is proportionally related to its size, which is not always necessarily valid. This assumption follows the logical view that bigger manufacturing firms have more machines, more processes, more people, hence more data associated with the higher number of resources.

**Table 1.** The examined scenarios

| Scenario | $D_0$ (GB/day) | $CI_0$ (gCO$_2$e/ kWh) | Growth model (Unbounded - bounded) | Carrying capacity (GB) |
|---|---|---|---|---|
| $S_1$ | 0.5 | 258 | U | N/A |
| $S_2$ | 0.5 | 258 | B | $10^4$ |
| $S_3$ | 0.5 | 369 | U | N/A |
| $S_4$ | 0.5 | 369 | B | $10^4$ |
| $S_5$ | 0.5 | 582 | U | N/A |
| $S_6$ | 0.5 | 582 | B | 104 |
| $S_7$ | 5 | 258 | U | N/A |
| $S_8$ | 5 | 258 | B | $10^5$ |
| $S_9$ | 5 | 369 | U | N/A |
| $S_{10}$ | 5 | 369 | B | $10^5$ |
| $S_{11}$ | 5 | 582 | U | N/A |
| $S_{12}$ | 5 | 582 | B | $10^5$ |
| $S_{13}$ | 50 | 258 | U | N/A |
| $S_{14}$ | 50 | 258 | B | $10^6$ |
| $S_{15}$ | 50 | 369 | U | N/A |
| $S_{16}$ | 50 | 369 | B | $10^6$ |
| $S_{17}$ | 50 | 582 | U | N/A |
| $S_{18}$ | 50 | 582 | B | $10^6$ |

In addition, big manufacturing facilities are more likely to adopt advanced technologies and wide variety of ICT infrastructure and software tools. It should be acknowledged, however, that there might be exceptions to this rule, especially with the falling cost of data generation means, primarily sensors [3].

The model's global parameters (i.e., those that are not scenario-specific), were chosen to reflect the current, and the forecasted future state of data growth and carbon intensity decay as displayed in Table 2 below.

The data storage period is assumed to be 25 years since it is unlikely that companies would want to store *all* data for more than that period, especially since most equipment and software tools have a lifetime of less than 25 years. As for the carbon intensity figures, they were chosen to reflect regional discrepancies. These carbon intensity values correspond with average grids' carbon intensity in the European Union [13] and the USA and China [14]. These carbon intensity values represent a wide range of carbon intensities in areas with active manufacturing industries.

**Table 2.** The model's global parameters

| Parameter | Value | Unit |
|---|---|---|
| Simulation period | 25 | Year |
| Growth rate (μ, σ) | (0.2, 0.04) | - |
| Decay rate U[$\lambda_{min}, \lambda_{max}$] | [0.01–0.03] | - |
| $E_{total}$ | 1 | kWh/GB/year |
| Number of independent replications | 1000 | Replication |

## 3.1   Results and Discussion

When the model was run for a total of 1000 independent replications, using random number seeds for each replication, the following results, displayed in Fig. 1 below, were obtained.

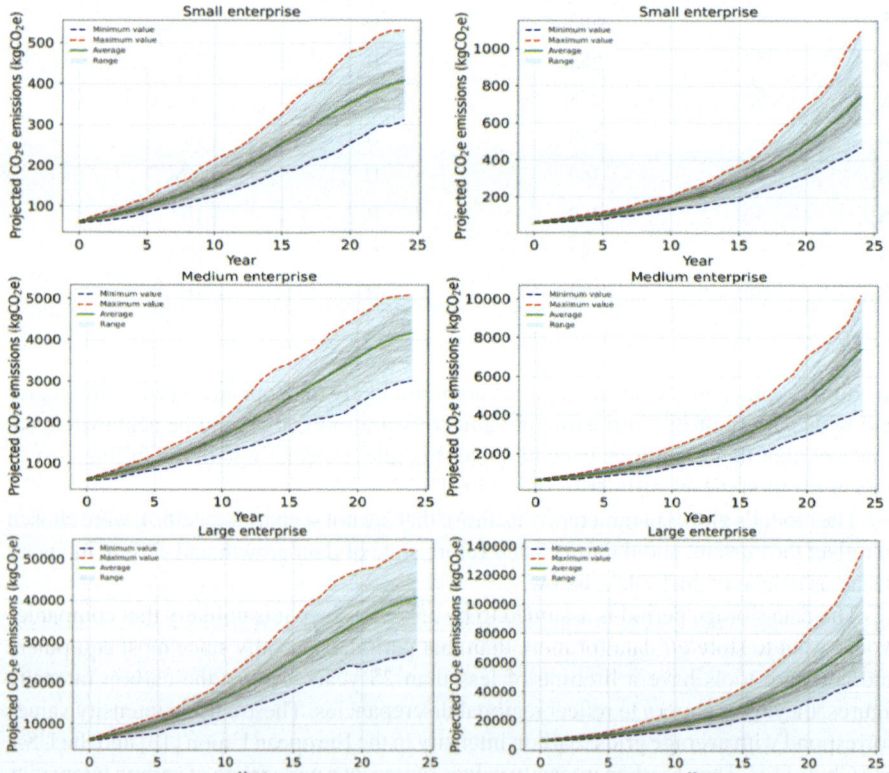

**Fig. 1.** Total $CO_2e$ emissions for the small, medium and large manufacturing enterprises for the logistic (left) and exponential (right) growth models when carbon intensity is medium

It could be noticed from Fig. 1 that although the carbon intensity of the energy mix decreases exponentially, and although relatively low carrying capacities were assumed, substantial amount of $CO_2e$ emissions were emitted from each of the presumed enterprises. Another important remark, is that the uncertainty with regards to data generation growth, and the decay rate of carbon intensity, can have a significant impact on the prediction of total $CO_2e$ emissions, differing by as much as 50%.

In order to further examine the behaviour of $CO_2e$ emissions and volumes of data at individual years, Fig. 2 below displays a heatmap, at logarithmic scale, for $CO_2e$ emissions over the years.

Another important insight from the results is that although carbon intensity of the grid is assumed to decrease exponentially (and its decrease indeed resembles exponential decay), this did not have a determining impact on the resulting amounts of $CO_2e$ emissions. This might be attributed to data generation growth rate exceeding the growth rate of the renewables' share in most energy grids.

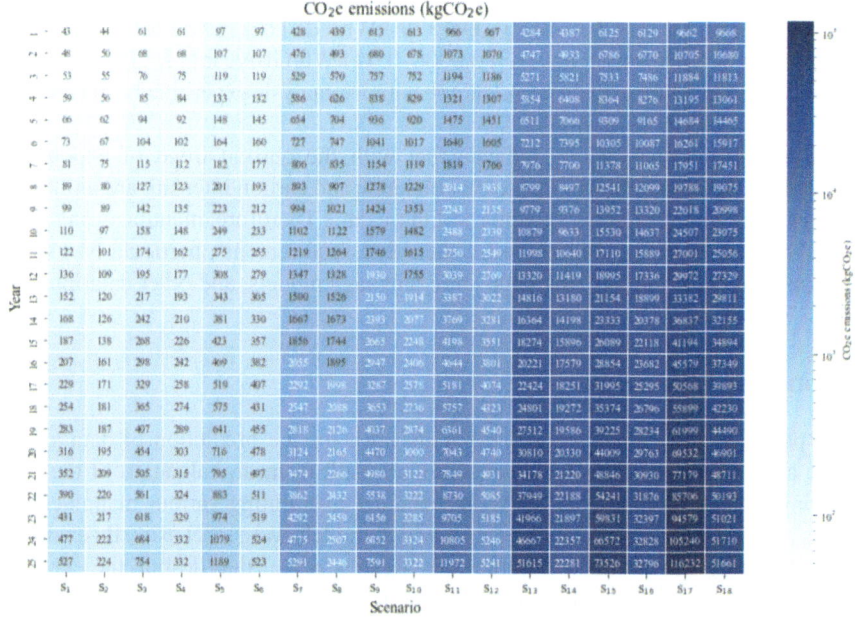

**Fig. 2.** Heatmap (logarithmic scale) displaying total amount of $CO_2e$ emissions for all scenarios during all years

Figure 2 shows that scenarios 13 to 18 generate the largest amount of $CO_2e$ emissions. This is an intuitive outcome since these scenarios are associated with large enterprises. However, it is important to emphasize here that the limit to growth does not have a substantial impact on mitigating the environmental impact of data. To explain more, although the scenarios where growth followed logistic distribution had significantly lower data volumes, and subsequently $CO_2e$ emissions, these projected emissions still constitute a significant burden on the environment. Considering the thousands of large manufacturing enterprises worldwide, the total amount of $CO_2e$ emissions from data alone will be several million metric tons, which can have an adverse effect on the prospects of achieving net-zero emissions by 2050. Indeed, this result is in line with [9] where the authors found that ICT (which is the cornerstone of digitalization) can have a strong rebound effect, hence rendering efficiency and carbon intensity improvements insufficient to match the growth in data volumes.

Since there is a lack in studies that model data growth over the long-term, direct comparison of the results with previous studies is difficult. However, in order to validate the results obtained from the numerical example, input data were validated by two subject matter experts [15], who are senior managers at small manufacturing enterprises in the Middle East and the UK. The rest of the input data were mostly obtained from official sources (e.g., the IEA and the EEA, amongst others). Finally, the mathematical formulations are well-established and widely used in population dynamics modelling, therefore, they have high validity. One point, however, that requires further exploration is the amount of energy required per GB per year. There is very little agreement about this in the literature. For example, estimations of the energy required for data transmission has been reported to vary between 0.004–135 kWh/GB [16], while estimations with regards to data storage varied between 0.0006–0.04 kWh/GB/year [17]. Therefore, it is recommended for future studies to conduct sensitivity analysis experiments, in particular with respect to energy consumption.

The aforementioned results indicate that with the current trend of exponential increase of data, improvements in energy efficiency and carbon intensity are unlikely to compensate for the environmental impact of these data. Therefore, manufacturing enterprises should carefully how to prioritize their data requirements, particularly with the propagation of cloud storage, big data, large language models, and other data-intensive tools.

## 4 Conclusion

In this paper, a stochastic model-based framework for the prediction of data growth and the associated environmental footprint that result from digitalization of the manufacturing industry is developed. Several scenarios constituting various manufacturing systems' sizes, geographical locations and growth models were stochastically examined under a Monte Carlo simulation framework. Results reveal that despite the exponentially decreasing carbon intensity, and the limitations posed on data growth (the carrying capacity), the environmental footprint of data over the long-term is significantly high. Another interesting insight from the results is that although uncertainty was significantly high in the long-term, the lower bound of emissions within all scenarios for the medium and large manufacturing enterprises was significantly high.

This research can be extended in a number of directions. First, more scenarios can be examined, particularly with regards to long-term energy-efficiency. Another possible direction is to carry out sensitivity analysis experiments on key parameters, particularly regarding energy consumption and carrying capacity for the logistic growth model. Finally, it is worthwhile to investigate the viability of optimization models that can determine the optimal amounts of data that manufacturing companies can sustain over the long-term with minimal environmental footprint.

# References

1. Ghobakhloo, M.: Industry 4.0, digitization, and opportunities for sustainability. J. Clean. Prod. **252**, 119869 (2020)
2. Belkhir, L., Elmeligi, A.: Assessing ICT global emissions footprint: trends to 2040 & recommendations. J. Clean. Prod. **177**, 448–463 (2018)
3. Patsavellas, J., Salonitis, K.: The carbon footprint of manufacturing digitalization: critical literature review and future research agenda. Procedia CIRP **81**(March), 1354–1359 (2019)
4. Zhang, Q., Clements-croome, D.: Innovative approaches for deep decarbonization of data centers and building space heating networks: modeling and comparison of novel waste heat recovery systems for liquid cooling systems. Appl. Energy **357**(December 2023) (2024)
5. Data Centres and Data Transmission Networks. International Energy Agency (2023)
6. Sun, H., Kim, G.: Technology in society the composite impact of ICT industry on lowering carbon intensity: from the perspective of regional heterogeneity. Technol. Soc. **66**(March) (2021)
7. Farfan, J., Lohrmann, A.: Gone with the clouds: estimating the electricity and water footprint of digital data services in Europe. Energy Convers. Manag. **290**(June), 117225 (2023)
8. Pohl, J., Hilty, L.M., Finkbeiner, M.: How LCA contributes to the environmental assessment of higher order effects of ICT application: a review of different approaches. J. Clean. Prod. **219**, 698–712 (2019)
9. Lange, S., Pohl, J., Santarius, T.: Digitalization and energy consumption. Does ICT reduce energy demand? Ecol. Econ. **176**(June), 106760 (2020)
10. Raychaudhuri, S.: Introduction to Monte Carlo simulation. In: Proceedings of the 2008 Winter Simulation Conference, pp. 91–100. IEEE (2008)
11. Meadows, D.H., Meadows, D.L., Randers, J., Behrens, W.W.: The Limits to Growth. Universe Books, New York (1972)
12. Tsoularis, A., Wallace, J.: Analysis of logistic growth models. Math. Biosci. **179**, 21–55 (2002)
13. Greenhouse gas emission intensity of electricity generation in Europe. European Environment Agency (2024). https://www.eea.europa.eu/en/analysis/indicators/greenhouse-gas-emission-intensity-of-1?activeAccordion=546a7c35-9188-4d23-94ee-005d97c26f2b
14. Statistical Review of World Energy. Ember; Energy Institute (2024). https://ourworldindata.org/grapher/carbon-intensity-electricity
15. Law, A.M.: How to build valid and credible simulation models. In: Proceedings - Winter Simulation Conference, pp. 24–33. IEEE (2019)
16. Aslan, J., Mayers, K., Koomey, J.G., France, C.: Electricity intensity of internet data transmission untangling the estimates. J. Ind. Ecol. **22**(4), 785–798 (2018)
17. Al Kez, D., Foley, A.M., Laverty, D., Del Rio, D.F., Sovacool, B.: Exploring the sustainability challenges facing digitalization and internet data centers. J. Clean. Prod. **371**(August), 133633 (2022)

# Chances for Decreasing the Carbon Footprint in Hospitals by Means of AI-Based Recognition of Surgical Instruments

Anna-Maria Paust[1(✉)], Ümit Ejder[2], Jan Lehr[1], Marian Schlüter[1], Clemens Briese[1], Ole Kroeger[1], and Jörg Krüger[3]

[1] Fraunhofer IPK, Pascalstraße 8-9, 10587 Berlin, Germany
annapaust@googlemail.com
[2] Charité Facility Management GmbH (CFM), Charitéplatz 1, 10117 Berlin, Germany
[3] Technische Universität Berlin, Straße des 17. Juni 135, 10623 Berlin, Germany

**Abstract.** Central Sterile Services Departments (CSSDs) play a pivotal role in the reprocessing of surgical instruments significantly impacting the healthcare sector's environmental footprint. This study evaluates the resource and energy consumption of CSSD processes, revealing that the sterilisation and reprocessing of surgical instruments are major contributors to greenhouse gas emissions. By integrating Cir.Log, an AI-based camera system, into the CSSD packing process, unnecessary sterilisation cycles can be reduced, resulting in annual savings of 726.64 kg $CO_2$-eq, lowering the Global Warming Potential of CSSDs. This research underlines the importance of improving CSSD efficiency to reduce the healthcare sector's environmental impact.

**Keywords:** life cycle assessment · reprocessing surgical instruments · AI-based recognition · resource and energy consumption · Central Sterile Services Department · CSSD

## 1 Introduction

The healthcare sector is responsible for 4.4% of the total global emissions, with three-quarters of these emissions occurring as indirect (Scope 3) emissions along the value chain [1]. Medical devices and medicines account for almost half of Scope 3 emissions [2], which indicates an urgent need for action in this sector. In this context Central Sterile Services Departments (CSSDs) play a fundamental role in the reprocessing of surgical instruments and cannot be overlooked as a contributing impact factor for the healthcare sector. Nevertheless, research has demonstrated that the sterilisation processes are accountable for a considerable expenditure of resources and energy [3], which amplifies the Scope 3 emissions of hospitals. It is therefore imperative that improvements in the efficiency of CSSDs have to be made in order to reduce the environmental impact in healthcare.

H. Kohl et al. (Eds.): GCSM 2024, LNME, pp. 137–145, 2025.
https://doi.org/10.1007/978-3-031-93891-7_16

Furthermore, there is a significant deficit of qualified personnel in numerous sectors within Germany. In addition to medical professionals, clinics and hospitals require a multitude of other service personnel to ensure the functioning of their processes. This is particularly evident in the CSSD. At the Charité - Universitätsmedizin Berlin approximately 14 million surgical instruments are reprocessed annually. The manual processing of this volume of instruments is dependent on the expertise of qualified personnel. Within a limited time frame 160 different instruments have to be packed up, while many of them are confusingly similar. This can result in missing or mixed up instruments which can lead to complications during surgery.

To meet these challenges a smart camera system, Cir.Log, is developed by Fraunhofer IPK in collaboration with Charité CFM Facility Management GmbH (CFM, 100% subsidiary of Charité Berlin) to assists during the packing process and increases efficiency and patient safety. Moreover, considerable energy and resource savings can be attained, which result in a reduction of the Global Warming Potential (GWP) of the CSSD and a concomitant reduction in the environmental impact on the healthcare sector.

## 2   Methodology

Cir.Log uses markerless and AI-based image processing techniques to recognise surgical instruments reliably. The objective is to ensure that the packing process in the CSSD is error-free. This encompasses the errors "instrument missing" and "wrong instrument packed". Further insights into the state of the art of the AI can be obtained from the paper by J. Lehr et al. [4]. Each of these errors significantly increases the risk of opening a new tray with up to 160 surgical instruments. The reprocessing of this tray can be avoided if the error rate is reduced to zero. In order to ascertain the potential for environmental reduction through Cir.Log, the following procedure was employed:

Two Life Cycle Assessments (LCAs) are performed for Cir.Log and for the processing cycle of a basic instrument tray in the CSSD of the Charité. The OpenLCA software from GreenDelta was employed in conjunction with primary data from Cir.Log and the CFM, the Ecoinvent 3.5 dataset, and secondary literature sources. Supplementary data on emission factors of metals, plastics and other materials from the German Environment Agency (UBA) was utilised to ascertain the impact from the value chains. Cir.Log was analysed from cradle to grave, while the boundaries of the CSSD were constrained to the environmental impact of the supply chain of medical devices in the tray and their process cycle in the department. For both systems the GWP was determined using the IPCC 2013 GWP 20a impact category and the cumulative energy demand (CED) was calculated. Additionally, water and steam consumption and wastewater were assessed for the CSSD. In this study, the number of errors and reprocessings of a basic tray from the previous year is used.

## 3   System Boundaries

To narrow down the Cir.Log system for the LCA, it was divided into four parts: hardware, software, use phase and end-of-life. The category software included data acquisition, AI training and software development, which were adapted to the use case of a single tray and a processing factor of 2.25% per year. The processing factor results from the fact that the basic tray accounts for 2.25% of all processed trays in the CFM for last year. The end-of-life phase is merely assumed as an undefined type of recycling. As Cir.Log is still in the development stage and manually assembled from purchased single parts, production is not considered. Therefore emission factors are utilized for hardware, software, hardware development and use phase. Simultaneously, $CO_2$-equivalents ($CO_2$-eq) are employed, for instance for the computers, screens and other devices, such as those used in the hardware e.g. industrial cameras, in accordance with the data provided by manufacturers or literature sources. This data is adjusted for the application scenario, which consists of the lifetime of the individual devices taken from the manufacturer's specifications, as well as the parameters of the one tray. To calculate the energy consumption with the CED impact analysis, the performance in each life cycle phase is taken from the aforementioned devices and calculated from the data collected from Cir.Log. These were also reduced to the use case, thereby enabling the evaluation from the viewpoint of the basic instrument set.

The CSSD system boundaries start with the emissions from the value chain of the medical devices. For this purpose emission factors from metals and plastics, such as aluminum, stainless steel and hard metals have been used to determine the GWP from the supply chain. As the central sterilisation facility is located within the hospital premises, the logistics associated with the transportation of surgical instruments and other materials from and to the operating rooms are disregarded. The resource consumption has only been analysed for the CSSD itself, in which water, energy and steam were considered. The process cycle which the utility set undergoes comprises of a washing process, with the washer and disinfector DS1000 from Steelco, a sterilisation process with the Selectomat PL from MMM and the packing process of the medical devices (Fig. 1).

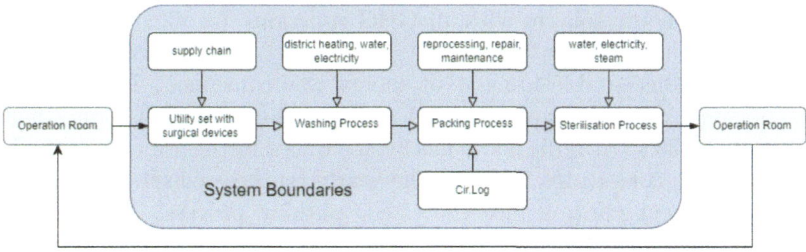

**Fig. 1.** CSSD process and system boundaries

The washing process lasts for one hour and comprises a prewash with cold water, a washing cycle with a water temperature of 55 °C, a disinfection at 69 °C and a drying process with hot air at a temperature of 91 °C. Once the set has cooled down the instruments are packed manually into the tray, by using a given packing list. Subsequently, the instruments are conveyed to the sterilisation machine, where they are vacuum sealed and subjected to steam sterilisation at 134 °C followed by an additional drying cycle. During these processes, district heating, electricity, water and steam produced by an in-house boiler are used. All data on the duration of the cycles, the water and steam volumes and the temperatures were provided by CFM.

In the event of an error a new tray has to be opened. As there is currently still a risk of these errors, trays are prepared and made available twice. The consequence is that the machines must be utilised twice, resulting in a corresponding doubling of resource consumption. This encompasses energy, water, and steam, among other resources.

To evaluate the given use case, Cir.Log was theoretically implemented into the packing process. Afterwards, the error rate provided by the CFM was employed as an indicator for the potential annually energy, resources and emission reduction analysis.

## 4   Results of the Life Cycle Assessments

### 4.1   LCA of Cir.Log

The analysis revealed that Cir.Log produces 73.19 kg $CO_2$-eq throughout its lifecycle and a given lifetime of seven years. The GWP is primarily attributable to indirect Scope 3 emissions such as hardware components, which include the camera system, SBC system and the case. These account for up to 40% of the $CO_2$-eq, reflecting the substantial quantities of metals and electronics utilized. The software development subsequently contributes the majority of emissions followed by the hardware development, largely due to the computers and monitors employed. While the energy consumption is often the primary factor contributing to emissions of AI-based products [5], this is not the case here, as the Fraunhofer IPK is exclusively powered by eco electricity, which is derived from solar and wind energy sources without GHG emissions [6].

Despite the utilisation of explicitly green energy for the development and AI training, a considerable amount of energy was consumed. The cumulative energy demand was calculated to be 1,197.90 MJ-eq, which is equivalent to approximately 333 kWh. This figure has been reduced to the use-case of the basic utility set, which constitutes 2.25% of the yearly reprocessed sets at Charité. In order to implement Cir.Log into the CSSD packing process, the values were divided by the 1,414 reprocessings of a basic tray per year, thereby scaling the usage to one cycle.

## 4.2   LCA of the CSSD

Table 1 presents the results of the analysis which was conducted based on the resource and energy consumption in each process, as well as the resulting GWP of 13.84 kg $CO_2$-eq.

**Table 1.** Resources, energy consumption and GWP of the CSSD

| Processes | Units | |
| --- | --- | --- |
| **Basic Surgery Utility Set** | kg | kg $CO_2$-eq |
| surgical instruments | 6.81 | 7.02 |
| reusable tray | 1.95 | 1.11 |
| other components | 0.05 | 0.21 |
| Sum | 8.81 | 8.35 |
| **Washing and Disinfecting** | | |
| *Resources* | l | kg $CO_2$-eq |
| water | 203 | 0.07 |
| MediClean and MediKlar | 0.309 | 0.19 |
| Subtotal | 203.31 | 0.27 |
| *Energy* | kWh | kg $CO_2$-eq |
| electricity | 15.34 | 0 |
| district heating | 16.66 | 4.9 |
| Sum | 32 | 5.2 |
| **Packing** | Wh | kg $CO_2$-eq |
| LED lamp | 6 | 0.1 |
| monitor | 6.67 | 0.11 |
| Sum | 12.67 | 0.21 |
| **Sterilisation** | | |
| *Resources* | $m^3$ | kg $CO_2$-eq |
| water | 0.2 | 0.07 |
| demin. water | 0.013 | - |
| compressed air | 0.2 | 0.028 |
| steam | 14.48 | 0.013 |
| Sum | | 0.11 |
| *Energy* | kWh | kg $CO_2$-eq |
| electricity | 44.10 | 0 |
| **Sum GWP** | | 13.84 |

As demonstrated in Table 1, the greatest contributor to the GWP is the value chain, with surgical instruments accounting for the majority of the emission. Given that there are over 100 instruments in this set, it can be estimated that approximately 60 g $CO_2$-eq per surgical instrument are emitted. District heating is the next highest source of GWP, with 294 $\frac{g}{kWh}$ $CO_2$-eq [7].

The total energy demand for one cycle was calculated to be 274.77 MJ-eq, which equates to 76.33 kWh. Although the electricity is considered to be GHG neutral, the warm water, which is essential for the washing process, is derived from district heating, is primarily sourced from natural gas and coal. The steam required for sterilisation is generated by an in-house steam boiler with electrical connection and does not contribute to additional GHG emissions, apart from the direct emissions produced by the water vapour itself.

### 4.3   Reduction Analysis of Cir.Log Implemented in the Packing Process

As previously stated, the consequence of an error is a doubling of resource and energy consumption, due to the repetitive nature of the cycle. This results in an increase in the GWP to up to 19.45 kg $CO_2$-eq per reprocessing cycle, which is accompanied by a twofold increase in vapor, water, and energy consumption, while the emissions from the supply chain of surgical instruments remain unchanged. It is common for errors to result in the CSSD being unable to charge for the preparation of the tray. However, only the cost of error is included and not the expenditure of time and personnel resources involved in the reprocessing.

The implementation of Cir.Log into the CSSD packing process will serve to avoid these factors. To demonstrate the extent of the savings achievable through the AI-driven automation, the one-time reduction from a single error scenario is multiplied by 132 errors in 2023. Furthermore, the costs (German factors) associated with each consumable are included in the calculation of the annual cost savings. These are presented in Table 2.

**Table 2.** Annual resource, energy and cost savings due to the usage of Cir.Log

| Consumables | Error | One-time reduction | Savings p.a. | Saved costs p.a. |
|---|---|---|---|---|
| Water | $0.832\,m^3$ | $0.41\,m^3$ | $54.91\,m^3$ | 93.02 € |
| District heating | 33.31 kWh | 16.66 kWh | 2,198.48 kWh | 222.18 € |
| Electricity | 119.33 kWh | 59.66 kWh | 7,875.26 kWh | 1,937.31 € |
| Waste water | $0.807\,m^3$ | $0.403\,m^3$ | $53.24\,m^3$ | 114.73 € |
| Costs per error | 35 € | 35 € | 4,620 € | 4,620 € |
| Sum costs per year | | | | 6,987.24 € |

As the steam is produced on-site and no additional steam is required, the costs correlated with this have been incorporated into the electricity costs. The results for the total savings of vapour quantify to $1,911.36\,m^3$ per year. In total an energy demand of a value exceeding 10 MWh was economised.

Emissions of $5.5\,kg$ $CO_2$-eq can be saved per cycle which represents approximately 30% per tray. Consequently, Cir.Log would attain an overall reduction of $726.64\,kg$ $CO_2$-eq per year.

## 5    Discussion

With regard to the AI-based camera system, it can be stated that the lifecycle emissions are rather insignificant. This is due to the fact that in the context of a single utility set, the initial GWP is additionally reduced to less than one percent per cycle. As the instruments in this set will be used in other sets and numerous hospitals in Germany, the assumption is no +t entirely accurate. It would be reasonable to posit that the emissions and energy demand would decrease in line with an increase in Cir.Log usage. However, as the system boundaries were set for one CSSD, the calculation was conducted accordingly.

The GWP of the process cycle of the CSSD can be validated through the study of Rizan et al. [8] in which it is stated that reprocessing surgical instruments generates approximately $15.31\,kg$ of $CO_2$-eq per cycle which is not dissimilar to the calculated results of $13.84\,kg$. Given that the majority of emissions are attributable to the value chain of instruments, an analysis of the material wear out resulting from the implementation of unnecessary sterilisation and washing cycles would be beneficial. It seems reasonable to conclude that, given the processing volume at Charité, the lifespan of the surgical instruments could potentially decrease. The extension of their service life by the AI-based packing control would result in a further reduction of Scope 3 emissions and could provide the basis for a new research topic.

This study has demonstrated that the automation of the packing process in CSSDs is not merely in means of alleviating the burden on employees, it is also an essential tool for reducing water, steam, and energy consumption. It should be noted that for the purposes of this research, the yearly error rate was taken for all types of utility sets, while only the cycle of the basic set was calculated. This may result in discrepancies in the values based on the number of instruments in the sets. It is also important to state that the carbon reductions are specific to the CSSD of Charité and cannot be directly applied to other hospitals. If the German electricity mix is used instead of green energy, a yearly reduction of up to $3,459.84\,kg$ $CO_2$-eq can be expected.

## 6    Conclusion and Outlook

This study presents an in-depth analysis of the lifecycle emissions and energy consumption associated with the CSSD and the integration of the AI-based camera system Cir.Log into the packing process. The results show that Cir.Log reduces the environmental impact of reprocessing surgical instruments by decreasing unnecessary sterilisation cycles. The lifecycle $CO_2$-eq emissions for Cir.Log were calculated to be $73.19\,kg$ over seven years, with the hardware components contributing the most to the GWP.

In the context of the CSSD, the greatest contributor to GWP was found to be the value chain of medical instruments and considerable emissions originated from the district heating required for the washing process. When Cir.Log is implemented, it mitigates the repetitive resource and energy consumption inherent in the reprocessing cycle caused by reclamations, leading to a reduction of up to 726.64 kg $CO_2$-eq per year. This highlights the system's potential for several environmental benefits by reducing the energy and material demands associated with medical instrument reprocessing. The implementation of Cir.Log in CSSDs across Germany could contribute to the overall prevention of Scope 2 and Scope 3 emissions in the healthcare sector by eliminating the error rate for missing instruments. If only 20% of hospitals in Germany adopted Cir.Log, a potential reduction of between 275.40 and 1,311.29 tons of $CO_2$-eq (dependent on the electricity mix) could be achieved.

In conclusion, the deployment of Cir.Log in CSSDs not only enhances operational efficiency but also has a positive environmental impact by reducing the GWP. With the commercialisation of the product, this has implications for the healthcare sector. Additionally, the study underscores the importance of considering the longevity and wear off of surgical instruments. By extending the lifespan of these instruments through reduced processing volumes, Cir.Log could further lower Scope 3 emissions. Another aspect is that the implementation of the AI-based system for the detection of defects and contamination, as part of an ongoing initiative, has the potential to facilitate the resolution of remaining errors within the CSSD. This enhancement could therefore cover all errors and achieve even higher annual savings.

# References

1. Karliner, J., Slotterback, S., Boyd, R., Ashby, B., Steele, K., Wang, J.: Health Care's Climate Footprint, How the Health Sector contributes to the Global Climate Crisis and Oppertunities for Action. Health Care Without Harm (2019). https://noharm-global.org/sites/default/files/documents-files/5961/HealthCaresClimateFootprint_092319.pdf
2. Franke, B., et al.: Greenhouse gas balancing using the example of Heidelberg University Hospital. Institute for Energy and Environmental Research Heidelberg (2022). https://www.klinikum.uni-heidelberg.de/fileadmin/KliOL/KliOL-ifeu_-_Treibhausgasbilanzierung_am_Beispiel_des_Universitaetsklinikums_Heidelberg_-_klik_green_Netzwerktreffen_-_21_Nov_2022.pdf
3. Pelzeter, A., et al.: Carbon footprint of services – findings from German non-medical hospital processes. In: Conference: EuroFM Research Symposium (2023). https://doi.org/10.5281/zenodo.10051131
4. Lehr, J., Kelterborn, K., Briese, C., et al.: Image-based recognition of surgical instruments by means of convolutional neural networks. Int. J. CARS 18, 2043–2049 (2023). https://doi.org/10.1007/s11548-023-02885-3

5. Rohde, F., et al.: Sustainability challenges of artificial intelligence and policy implications. Ecol. Econ. J. **36**(O1), 36–40 (2021). https://doi.org/10.14512/OEWO360136
6. Berlin Public Utilities: electricity mix 2022. https://berlinerstadtwerke.de/strommix/. Accessed 2024
7. Senate Department for Finances. Sustainability Report on the Berlin state companies (2022). http://www.berlin.de/sen/finanzen
8. Rizan, C., et al.: Minimising carbon and financial costs of steam sterilisation and packaging of reusable surgical instruments. Br. J. Surg. **109**(2), 200–210 (2022). https://doi.org/10.1093/bjs/znab406

# EcoSentry: A Cost-Effective IoT System for Efficient Real-Time Forest Monitoring

Hung Nguyen Trung, Quang Khanh Le, Tieu My Lam, Kim Duy Vu, and Le The Dung$^{(\boxtimes)}$

FPT University, Ho Chi Minh Campus, Ho Chi Minh City, Vietnam
{hungntse182394,khanhlqse182420,myltse180542,
duyvkse182407}@fpt.edu.vn, dunglt96@fe.edu.vn

**Abstract.** Nowadays, climate change is a severe issue that is attracting global attention. Effective forest monitoring is vital among various methods to cope with climate change. Unfortunately, current approaches to forest monitoring are costly and sophisticated, making them difficult to apply widely or affordable for developing countries. In this paper, we propose a cost-effective Internet of Things (IoT) system called EcoSentry, which is low-cost and straightforward but still provides reasonably accurate, real-time data for efficient forest monitoring. The proposed system is comprised of two main components: a data-collecting hardware and a data-processing software. The first component includes multiple sensing nodes managed by a single gateway. Each node can collect various data such as temperature, soil moisture, air humidity, and rain level, then transfer them to the gateway via a 2.4 GHz wireless connection. Meanwhile, the second component utilizes Firebase, Angular, and Google Maps platforms for a web interface and Java, Google OAuth, and Google Maps platforms for an Android application on mobile devices. Experimental results show that the proposed system can efficiently support multi-point real-time monitoring and automatically trigger alerts to warn relevant authorities in emergent cases. Furthermore, the proposed system is scalable and can easily integrate with other systems.

**Keywords:** IoT system · real-time forest monitoring · cost-effective · simple and efficient

## 1 Introduction

Forests play a critical role in mitigating climate change by absorbing carbon dioxide. However, deforestation rates continue at an alarming pace, further compounded by intense wildfire activity. A recent report reveals that wildfires in Canada alone released a staggering 480 million tonnes of carbon emissions in 2023, nearly five times the average for the past 20 years [1].

Forest monitoring is the systematic recording of a sequence of measurements related to environmental variables within a designated forest area to yield reliable insights into the area's condition and changes over time. Traditional approaches

© The Author(s) 2025
H. Kohl et al. (Eds.): GCSM 2024, LNME, pp. 146–154, 2025.
https://doi.org/10.1007/978-3-031-93891-7_17

often rely on expensive and complex technologies such as drone, laser scanning, satellite imagery [2], inhibiting their widespread adoption, particularly in developing countries. For example, drones can be used for monitoring [3]. The cost of each drone can be up to $2000 and require special hardware for charging; scaling this out to more giant forests means the price can go up significantly. Additionally, its maintenance requires specialized knowledge, making it unsuitable for developing countries. Another popular method can be detection using image and infrared sensors [4]. However, this relies on complex and proprietary algorithms, which make it hard to audit its efficiency and reliability. Currently, some systems are also aiming to solve this problem. A camera system based on the Raspberry Pi was introduced in [5] to monitor forest canopy development. This method was cheap and straightforward to assemble, but the performance was comparable to that of commercial solutions. An Arduino can also be used along with an AMG8833 IR camera to detect possible fire [6], which is then transmitted with LoRaWAN connectivity. This method also used simple off-the-shelf components, making it cheap and simple to deploy at a large scale.

Based on these findings and a commitment to advancing accessible solutions, we introduce EcoSentry, an innovative Internet of Things (IoT) system designed to address this challenge. EcoSentry is a cost-effective, open-source solution that utilizes easily accessible technologies for real-time forest health monitoring. These valuable features of EcoSentry enable them to make well-informed decisions regarding forest health and resource management.

## 2  System Design

The proposed EcoSentry forest monitoring system consists of two main components: data collecting hardware and data processing software, as illustrated in Fig. 1. Details of the data-collecting hardware and data-processing software are presented in the following subsections.

### 2.1  Hardware and Data Collection

For real-time forest monitoring solutions, the data collecting hardware consists of several primary hardware elements such as Arduino UNO microcontroller board, RF 2.4GHz wireless module, and ESP32 Wi-Fi integrated system-on-chip (SoC). The hardware is cheap and straightforward to buy and assemble, thanks to standard microcontrollers and a simple-to-use platform.

**Core Components.** Each node is built around an Arduino UNO microcontroller board and is powered by an external battery module. Each node includes a capacitive soil moisture sensor, a DHT22 temperature-humidity sensor, a rainfall sensor, an MQ9 CO sensor, and a Sharp GP2Y1014AU0F dust sensor. All collected data is sent to a gateway via an NRF24L01+ wireless module. Each gateway is made up of two simple components, i.e., the ESP32-S WiFi-enabled SoC and the NRF24L01+ wireless module for receiving data from each node.

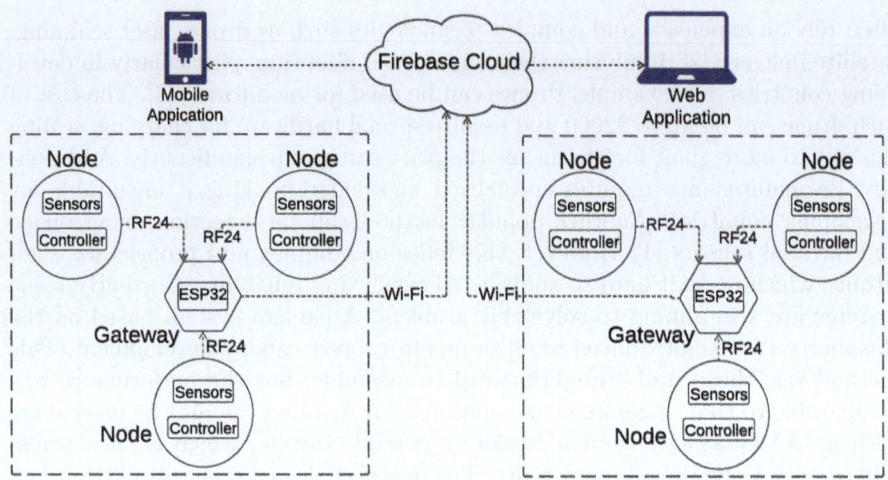

**Fig. 1.** Data collection procedure in EcoSentry system.

**Data Collection Mechanism.** The data collection mechanism of the EcoSentry system relies heavily on sensors at each node and is designed to efficiently acquire environmental parameters essential for monitoring various aspects of the target environment. It has the following processes: ecological data collection, node-to-gateway data transfer, and data upload.

## 2.2    Software Development

The software suite integrates cutting-edge technologies: a firmware system for collecting, processing and transmitting data to Firebase, a web application powered by Angular, along with an Android mobile application that serves as a means to monitor forest data and alerts users in case of critical events, such as wildfire or landslides, and a data management system utilizing Firebase Firestore for real-time updates. These features are described as follows.

**Low-Power Firmware for Edge Hardware.** The firmware for every node is built on the Arduino platform [7] because of its rapid development speed and inherent simplicity. Based on this platform, the firmware employs RadioLib [8], a robust library for wireless communication, to ensure reliable data transmission between nodes and gateways. The firmware for the ESP32 gateway embraces a streamlined approach similar to that of the node firmware, promoting simplicity and efficiency across the monitoring system.

**Web Application with Interactive Data Visualization.** The web application of the EcoSentry system serves as the primary user interface for forest managers to monitor forest health in real time. The features of the web application are as follows: (i) to utilize Google Maps JavaScript API to visually

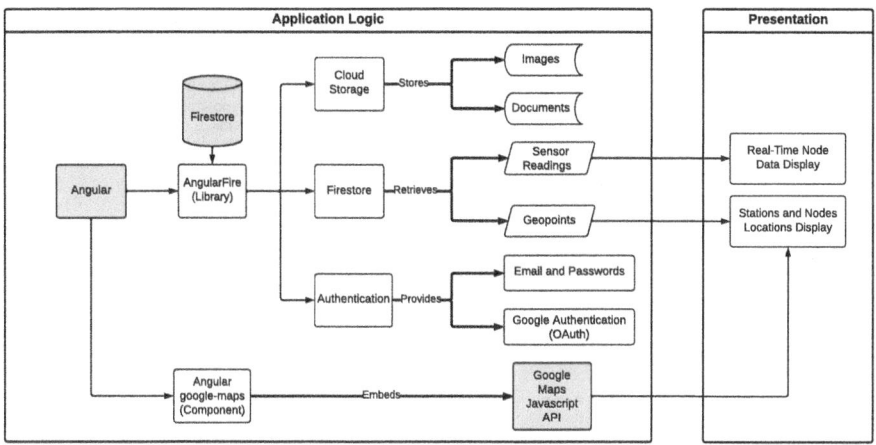

**Fig. 2.** The architecture of the web application of EcoSentry system.

represent the geographical locations of all stations in the forest, (ii) to use the Angular framework and Firebase to provide a user-friendly experience for data visualization and analysis, and (iii) to adopt the model-view-controller (MVC) architecture for efficient data management, as presented in Fig. 2.

**Mobile Application for Remote Monitoring.** The mobile application of the EcoSentry system utilizes a colour-coded threshold, real-time station statistics, and push notifications. The application endeavours to provide extensive assistance in monitoring forest activities and promptly addressing incidents.

**Data Management with Firebase.** Firestore and Google's NoSQL cloud database play vital roles in forest monitoring by offering real-time data management and historical data storage. The whole NoSQL structure of the Firestore database is illustrated in Fig. 3. This data structure ensures that every environmental parameter in the monitored location is recorded and readily accessible for analysis and interpretation.

## 3    Forest Health Monitoring and Real-Time Analysis

Forest health is vital for a balanced ecosystem. This section delves into EcoSentry's fire detection algorithms and real-time data acquisition mechanisms.

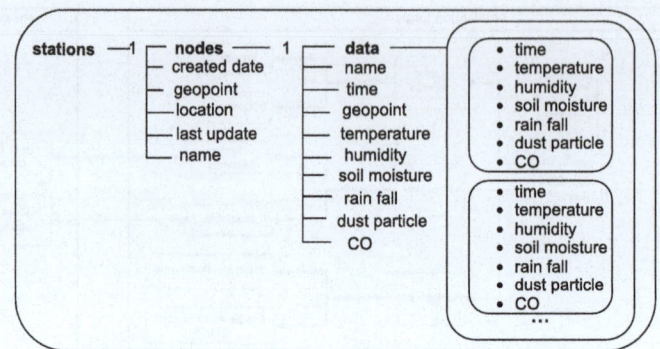

**Fig. 3.** NoSQL structure of Firestore database.

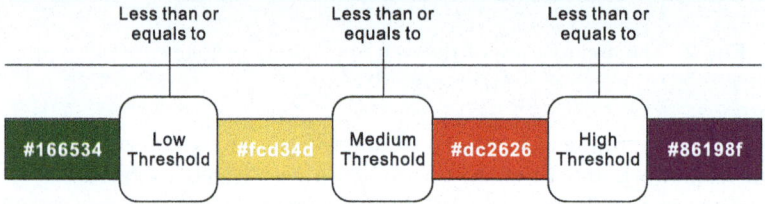

**Fig. 4.** Color-based thresholds of various features in EcoSentry system.

## 3.1 Fire Detection Algorithm

**Color-Coded Threshold.** As a feature of the EcoSentry system, each station node detects independent critical situations when the sensing data exceed thresholds listed in Table 1. These thresholds are monitored and adjusted through many empirical experiments. Furthermore, a colour-based approach shown in Fig. 4 is implemented to improve user awareness of each feature's status. The colour codes for safe, warning, danger, and urgent levels are also provided.

**Table 1.** Predefined threshold of each feature.

| Feature | Unit | Low | Medium | High | Feature | Unit | Low | Medium | High |
|---|---|---|---|---|---|---|---|---|---|
| Temperature | °C | 15 | 25 | 50 | CO | PPM | 70 | 150 | 200 |
| Air humidity | % | 10 | 70 | 90 | Rain fall | % intensity | 1 | 30 | 70 |
| Soil moisture | % | 10 | 60 | 90 | Dust particle | μg/m³ | 15 | 50 | 150 |

**Normalization.** Normalizing the values of each feature is crucial to simplify their range, ensuring they all fall within the range of (0, 1). The min-max normalization is adopted in the EcoSentry system because it is an optimized and efficient multi-purpose normalization technique [10], i.e.,

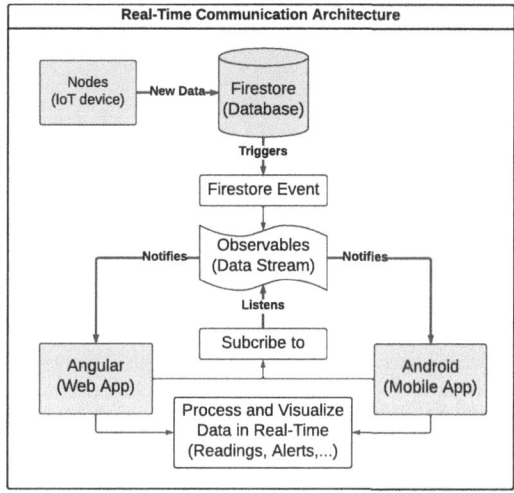

**Fig. 5.** The real-time data flow of EcoSentry system.

$$x_i' = a + \frac{(x_i - \min(X)) \cdot (b - a)}{\max(X) - \min(X)}. \tag{1}$$

From (1), we can see that the result is scaled. Specifically, we have $x_1' = 0 + \frac{(3-3) \cdot (1-0)}{19-3} = 0$, $x_2' = 0 + \frac{(7-3) \cdot (1-0)}{19-3} \approx 0.2$, $x_3' = 0 + \frac{(11-3) \cdot (1-0)}{19-3} \approx 0.4$, $x_4' = 0 + \frac{(15-3) \cdot (1-0)}{19-3} \approx 0.6$, and $x_5' = 0 + \frac{(19-3) \cdot (1-0)}{19-3} = 1$. In brief, the set of scaled values is $X' = \{0, 0.2, 0.4, 0.6, 1\}$.

**Weighted Average.** Following the above normalization process, the fire detection algorithm calculates the weighted average value of all features based on multi-criteria decision analysis (MCDA) [11], i.e.,

$$\mathscr{D} = \sum_{i=1}^{n} \eta_i \times \mathrm{w}_i. \tag{2}$$

where $\mathscr{D}$ is the danger score, $\eta_i$ is the normalized $i$th feature, and $w_i$ is the weight of the $i$th feature.

In the proposed EcoSentry system, the weights of the features listed in Table 1 are 0.4, 0.15, 0.25, 0.1. 0.05 and 0.05, respectively.

## 3.2    Real-Time Data Acquisition and Processing

As shown in Fig. 5, the proposed EcoSentry system handles real-time communication to ensure timely access to forest health monitoring. Sensing data are transmitted from nodes to Firebase Firestore. Whenever new sensing data is updated, the Firestore will trigger an event. The Observables then notify the

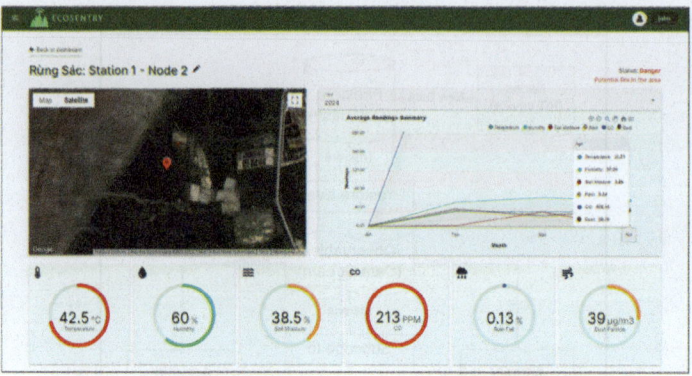

**Fig. 6.** Web application interface of EcoSentry system.

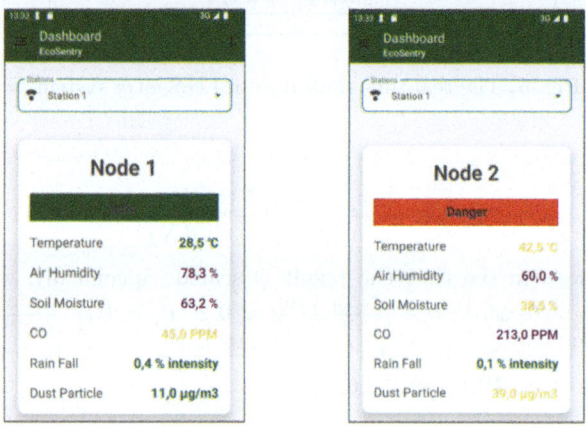

**Fig. 7.** The Android interface of EcoSentry system.

subscribed apps about the data changes, prompt them to fetch the latest information from the data stream, and require the views to be re-rendered [9, p. 105]. This mechanism ensures that the web interface and mobile app consistently reflect the most recent forest health conditions and alert the forest managers if anomalies occur.

## 4  Results and Discussion

To evaluate the performance of our proposed EcoSentry forest monitoring system, we deployed a prototype in Sac forest, Can Gio, Ho Chi Minh City, Vietnam. This prototype of the EcoSentry system comprises sensor devices, a Firebase Firestore backend, and an Angular web application frontend. Throughout Jan. 2024, sensor data was continuously collected, providing real forest health monitoring. It is worth noting that the low-cost nature of the sensor devices introduced limitations. Specifically, while the system functioned as intended, CO and dust readings might exhibit inaccuracies due to sensor limitations. Further testing and calibration with higher-grade sensors are needed to achieve more precise data.

Figure 6 demonstrates the web application interface, which displays a page with detailed node information and a chart to provide user access to specific sensor data points and trends. Figure 7 depicts the Android mobile application interface where data collected at each node is constantly updated in real-time. The results on the web application interface and the Android mobile application interface are well-matched. These results allow forest managers to visualize and analyze the forest conditions. It is noted that the EcoSentry system can also send alerts to relevant authorities when warning, danger, and urgent levels appear.

## 5   Conclusion

In this study, we have successfully investigated the development and implementation of EcoSentry, a cost-effective IoT system for forest monitoring. Overall, the proposed EcoSentry system significantly advances IoT technology in forest monitoring. When including additional sensor types (e.g., wind speed, thermal imaging) alongside improvements in sensor quality, the system can provide a more accurate environmental picture. Furthermore, machine learning algorithms can be used for anomaly detection, and even predictions can significantly enhance forest management practices. Future work could facilitate collaborative forestry management (CFM) [12] by following authorized personnel from different departments or organizations to share sensor data and insights through EcoSentry. This collaborative approach could expand the system's reach and overall impact.

## References

1. Copernicus A. M. S.: 2023: A year of intense global wildfire activity. Copernicus Atmosphere Monitoring Service (2023). https://atmosphere.copernicus.eu/2023-year-intense-global-wildfire-activity
2. Fassnacht, F.E., et al.: Remote sensing in forestry: current challenges, considerations and directions. Forestry Int. J. Forest Res. **97**(1), 11–37 (2024)
3. Paneque-Gálvez, J., et al.: Small drones for community-based forest monitoring: an assessment of their feasibility and potential in tropical areas. Forests **5**(6), 1481–1507 (2014)
4. Matthews, S., et al.: Field evaluation of two image-based wildland fire detection systems. Fire Saf. J. **47**, 54–61 (2012)
5. Wilkinson, M., et al.: A Raspberry Pi-based camera system and image processing procedure for low cost and long-term monitoring of forest canopy dynamics. Methods Ecol. Evol. **12**(7), 1316–1322 (2021)
6. Apriani, Y., et al.: Design and implementation of LoRa-based forest fire monitoring system. J. Robot. Control (JRC) **3**(3), 236–243 (2022)
7. Arduino: Arduino platform (2024). https://www.arduino.cc
8. Gromes, J.: Radiolib library (2018). https://github.com/jgromes/RadioLib

9. Wilken, J.: Angular in Action. Simon and Schuster (2018)
10. Patro, S.G.O.P.A.L., Sahu, K.K.: Normalization: a preprocessing stage (2015). arXiv preprint arXiv:1503.06462
11. Ishizaka, A., Nemery, P.: Multi-criteria Decision Analysis: Methods and Software. Wiley, Hoboken (2013)
12. Petheram, R.J., et al.: Collaborative forest management: a review. Aust. Forestry **67**(2), 137–146 (2004)

# OPC UA Information Model
# for Energy-Flexible Aqueous Parts
# Cleaning Machines

Lina Kramer[1(✉)], Daniel Fuhrländer-Völker[1], Magnus von Elling[1],
Sebastian Karnapp[1], Maximilian Moser[2], and Matthias Weigold[1]

[1] Institute of Production Management, Technology and Machine Tools (PTW),
Technical University of Darmstadt, Otto-Berndt-Str. 2, 64287 Darmstadt, Germany
L.Kramer@PTW.TU-Darmstadt.de
[2] Verband Deutscher Maschinen- und Anlagenbau, 60528 Frankfurt am Main,
Germany
http://www.ptw.tu-darmstadt.de

**Abstract.** To reduce carbon emissions, renewable energies are being
integrated into power grids worldwide, but their inherent volatility leads
to overproduction or underproduction of electricity. Besides using costly
electrical storage, supply and demand can be synchronized utilizing
energy flexibility of industrial consumers. Industrial aqueous parts cleaning machines have significant potential for energy-flexible operation. VDI
5207 suggests various energy flexibility measures, such as utilizing the
inherent storage capacities of rinsing tanks. This paper introduces an
energy synchronization platform and an OPC UA interface for operating
machines in an energy-flexible manner. It combines data on composition
and electricity prices to generate flexibility measures. A central OPC
UA server communicates with the market platform and distributes these
measures to machine-specific servers, automating the machine's operation and documenting energy consumption and $CO_2$ emissions. The
concept is tested on the BvL YukonDAD cleaning machine at the ETA
factory of TU Darmstadt.

**Keywords:** Cyber-physical production system · sustainability ·
production machine · inherent energy storage

| **APCM** aqueous parts cleaning machine | **EPEX** European Power Exchange |
|---|---|
| **CPPS** cyber-physical production system | **ID** identity document |
| **CS** Companion Specifications | **IM** information model |
| **DM** data model | **JSON** JavaScript Object Notation |
| **DR** demand response | **OPC UA** Open Platform Communications |
| **EFDM** Energy Flexibility Data Model | Unified Architecture |
| **ENTSO-E** European Network of Transmission | **XML** Extensible Markup Language |
| System Operators for Electricity | |

## 1 Introduction

Purchasing electricity has become expensive for industrial companies. Average
annual electricity spot market prices in Germany increased from 28.20 €/MWh

© The Author(s) 2025
H. Kohl et al. (Eds.): GCSM 2024, LNME, pp. 155–163, 2025.
https://doi.org/10.1007/978-3-031-93891-7_18

in 2016 to 92,29 €/MWh in 2023 [1]. Between 2011 and 2022, total renewable electricity generation in the European Union almost doubled from 1971.68 TWh to 3118.06 TWh [2]. The transition to renewable energy requires an adaptation of the electricity system. In the future, consumers must adapt to fluctuating renewable electricity generation through the integration of energy storage and the use of demand response (DR), thereby reducing energy costs [3].

A review of the literature revealed a significant body of research articles that addressed the topic of energy-flexible or energy efficient scheduling of single machines. We analysed the DR potential of industrial aqueous parts cleaning machine (APCMs) in previous research [4]. Additionally, we developed an automation data model (DM) for the communication between DR services and the machine automation in cyber-physical production system (CPPSs) [5].

In [6], we implemented a CPPS based on the aforementioned automation DM, which included a simulation model of an APCM and a simple rule-based DR service to control the tank heater of the machine.

An Energy Flexibility Data Model (EFDM) was developed to describe the potential capabilities of an energy-flexible system to deviate its performance from the reference operation with the help of key figures and technical parameters [7]. General descriptions of the EFDM are provided in [8], with more details in [9].

For this purpose, of operating machines in an energy-flexible manner, the Open Platform Communications Unified Architecture (OPC UA) information model (IM) is extended by the EFDM, considering the procedure for creating OPC UA Companion Specifications (CS) for an industrial APCM. A literature review identified a gap in CPPS implementations focusing on single machine scheduling with energy-related objectives. To address this gap, we present a scheduling model integrated into a CPPS with an energy-cost objective.

## 2   Method for OPC UA Data Modeling

We develop a generic and extensible IM which enables simple data exchange from field devices to the connected world level. The EFDM includes three classes: *Flexible Load*, *Flexible Storage*, and *Dependencies*, covering all potential variations in energy output compared to the reference case.

The OPC UA standardises DMs, known as OPC UA CSs, developed by working groups of employees from various manufacturers. Their aim is standardised data exchange between systems from different manufacturers [10]. We applied the following approach in cooperation with the VDMA [11]. The first step is to involve all interested parties and identify the need for an OPC UA CS. There is currently no OPC UA CS for cleaning machines. The second step is to standardise terms, functions and properties. Firstly, the use cases are defined, from which a communication plan is derived. The information that is represented in the form of the parameter list is also defined there. The object hierarchy ensures the development of a generalised architecture. Finally, the content is designed, implemented in OPC UA and published. It is then used in industry and implemented by the manufactures.

1. **Definition of Scope and Use Cases**
   The use case and scope are defined to limit the development framework.
2. **Creating a Communication Plan**
   The communication plan examines which systems, machines, and IT components communicate and how this is mapped in the CS.
3. **Creation of a Parameter List**
   The use cases and the communication plan result in a list of parameters that are to be mapped by the specification.
4. **Creation of an Object Hierarchy**
   OPC UA information models are hierarchical, so components and data points, including parameters and properties, are arranged hierarchically before modeling in OPC UA.
5. **Modelling of the information model**
   Modelling of the OPC UA interface with a modelling tool such as UaModeler. The result of this step is a type hierarchy from which objects can be derived for modelling a specific system/machine. Finally, the IM is saved and exported as a set of nodes in Extensible Markup Language (XML) format for further use.

External factors, such as electricity mix data and generation forecasts, flow via ENTSO-E to the central OPC UA server. This platform is operated by the transmission system operators. This server derives energy flexibility measures from the electricity market data using EFDM measures. The central server exchanges energy management data with the proprietary OPC UA server of the APCM, which controls and monitors the machine. A placeholder allows control of other machines using EFDM measure objects. To map the information DM onto a machine, it must be executed directly on the controller or connected through a mapping table to a proprietary server.

We develop the IM for the YukonDAD APCM from the manufacturer BvL OBERFLÄCHENTECHNIK in the ETA facility at TU Darmstadt. The system has a total rated output of 20.7kVA and is equipped with two closed treatment chambers and two 500litre detergent tanks heated by a 60kW electric tank heater [12].

## 3 Application of the OPC UA Data Model

The definitions of scope and use cases are set out below. This IM aims to enable energy flexibility of machines and systems via an OPC UA interface. The potential analysis [13] has shown that the energy inherently storing and interrupting the process from VDI 5207 is particularly suitable for cleaning machines. The machine must be able to control variables via an OPC UA server. To map machines from different industries, a generic IM makes the following use cases possible:

- Identification of the machine
- Emissions accounting
- Balancing of consumption

- KPI calculation for energy management
  - With standardised EFDM object
  - List of implemented energy flexibilisation measures and load profiles

A communication plan exists in both the presence and absence of a mapping table. As there is already an OPC UA server on our machine controller, which is not standardised, only the communication plan with mapping table is discussed below. To standardise the information contained in this server, the client part of the standardised OPC UA server connects to the proprietary server and synchronises the server nodes with each other. To achieve this, it must be known which node in the proprietary IM corresponds to which node in the standardised IM. The central OPC UA server manages the connection to ENTSO-E. This server sends energy flexibilisation measures to one or more OPC UA servers assigned to individual machines. The machines' OPC UA servers also read electricity market data, including the current electricity price and the emission factor of the electricity, from the central server to enable cost and emission balancing. We created this structure with security and scalability in consideration (Fig. 1).

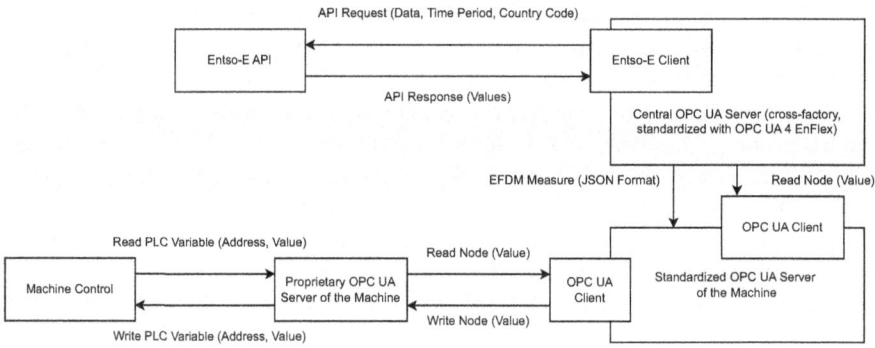

**Fig. 1.** The communication plan considers both internal and horizontal or vertical communication

A parameter list now results from the use cases and the communication plan. As the energy flexibility measures define the central part of this work, only this section of the parameter list is shown below.

- Metadata

  - ID of the flexibilisation measure

  - Timestamp of creation

- Flexibilisation measures
  - ID of the measure
  - ID of the associated flexible load
  - Load change profile
    - Services
    - Units
    - Timestamp
  - Reward
    - Unit
    - Value

Depending on these parameters and the flexibility space of the factory in which the central OPC UA server runs, the server generates an EFDM measure object. The specific values that the parameters of the JavaScript Object Notation (JSON) file can assume are mapped via the flexibility space, which is also available in JSON format. This file defines the sources from which energy is available, the maximum amount of energy that can be utilised at certain times, and the costs associated with energy use or storage. The measures must therefore comply with the limits of the flexibility space. The central server maps and processes energy market data and distributes energy flexibility measures as EFDM objects to machine-specific servers. It contains folders for energy market data and energy flexibility measures.

Figure 2 shows the `EnFlexCentralServer`'s model. The EFDM measures are stored in the `EFDMMeasures` folder, supporting standard operations such as reading, writing, and deleting JSON files. The `EnergyMarketData` folder aggregates data from the market platform (e.g., ENTSO-E), including the current $CO_2$ emission factor, the bidding zone, and generated electrical energy. Subfolders include `DayAheadData` for day-ahead information and `EnergyProductionTypes` for energy source details.

The `DayAheadData` folder aggregates intervals describing the trading of electricity for the following day. Each timeframe includes its start and end times, predicted electricity price, total electricity generation, and generation from wind and solar, allowing calculation of the renewable energy share. For the electricity price values, we use the day ahead spot prices of the EPEX electricity exchange. We use these data points to calculate the share of these weather-dependent, renewable energies in the total electricity mix, which is also represented by a variable in the timeframe.

To aggregate information on the individual energy sources, at least one source must be contained in the folder `EnergyProductionTypes`, but as many energy sources should be created as are available in the electricity mix of the underlying bidding zone. An energy source is always mapped via its currently generated energy, `CurrentEnergy`, and its $CO_2$ emission factor, `CarbonEmissionFactor`. In addition, a third variable `PercentOfTotalEnergy`, which is calculated from its `CurrentEnergy` variable and the `CurrentTotalEnergyGeneration` variable from the `EnergyMarketData` folder, is assigned to the energy source.

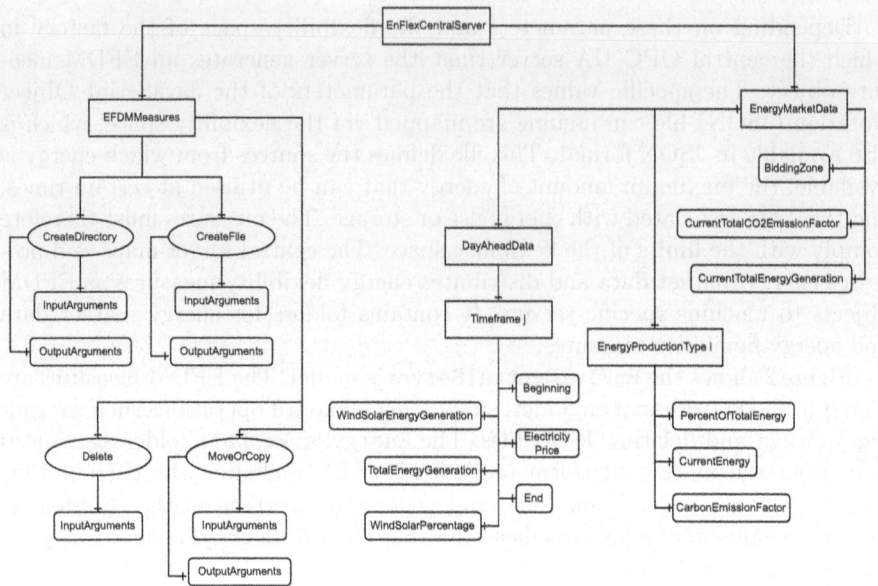

**Fig. 2.** EnFlexCentralServer shown in OPC UA notation

We then modelled the object types, which are published in Git. Data types, including descriptions, dependency information, and modeling rules, must be created in OPC UA to model the described hierarchy and obtain standardized objects consistently. Due to space limitations, we only present the EnFlexMeasure type here.

An energy flexibilisation measure is mapped in the EnFlexMeasureType with parameters according to the EFDM standard. This includes strings for the assigned flexible load and the measure's identity document (ID), as well as a State string indicating the measure's status as *planned*, *completed* or *error*. The reward is represented by an object of RewardType. All load changes associated with the measure are collected in the LoadChanges folder, consisting of LoadChangeType objects.

## 4 Implementation

Finally, the developed OPC UA interface is implemented as a prototype on the BvL APCM at ETA-Fabrik to validate the results. A central OPC UA server, which connects to the ENTSO-E database, is programmed for this purpose. In addition, an OPC UA server assigned to the APCM, which maps the flexible loads of the machine, is programmed and connected to the central server. The electricity mix data is retrieved via the ENTSO-E API. The data series from the DE-LU bidding zone on electricity generation, electricity prices and electricity generation per energy source are imported from this platform.

The Python libraries used, along with their respective versions, are documented in the Git repository. In this subparagraph, we provide an overview of the most important ones.

- **Asyncio**: A basic library for programming I/O-bound network functions with support for asynchronous loops [14].
- **Asyncua**: asyncio-based implementation of the OPC UA protocol. Supports both low-level and high-level interfaces and provides functions to implement basic functionalities with few lines of code [15].
- **ENTSO-E**: provides methods for querying various types of data related to electricity transmission, as well as high-level functions for connecting to the ENTSO-E Web API, eliminating the need to request it with http-get requests [16].

The central server interfaces with the ENTSO-E database to retrieve electricity market data via the ENTSO-E API. This data is processed and stored in an OPC UA IM. In addition, the server generates EFDM measures in JSON format based on this market data, which are then sent to the APCM to implement energy flexibilisation. Before programming the OPC UA server functions, the IM is exported from UaModeler in XML format for integration into the Python script.

In the following paragraph, we describe the generation of EFDM measures. To enable energy-flexible operation of the APCM, the central server generates EFMD measures using ENTSO-E data. Three functions are implemented for this purpose: `gen_efdm`, `efdm_from_WS_percentage`, and `efdm_from_el_price`. The `gen_efdm` function requires inputs such as the index of a flexible load in the dictionary, switch times, load levels, and a JSON blueprint document `efdm_measure_blueprint`. The functions `efdm_from_WS_percentage` as well as `efdm_from_el_price` compute load profile values based on wind and solar shares or electricity prices from day-ahead data. These values are then used by `gen_efdm` to generate EFDM measures. Once the energy-flexible central server and machine-specific servers communicate, they exchange EFDM measures and electricity market data. This facilitates energy-flexible operation of the APCM, focusing on scenarios like electricity cost savings or $CO_2$ emission reductions. The model was tested and validated in initial two-shift trials and compared to the reference operation of the machine without the model, showing the following savings. Compared to the reference operation of the machine, costs were reduced by 10.08% and $CO_2$ emission by 14.06% in the early shift. In the late shift, costs were reduced by 36.82% and $CO_2$ emission by 9.46%.

# 5   Summary, Conclusion and Outlook

This paper proposes a modular and consistent IM for industrial CPPS in terms of automation, simulation and communication, which is intended to enable demand-response algorithms. The model was applied to a continuous APCM. Furthermore, the object-oriented design ensures the potential for applications outside the field of energy flexibility, as the number of processes optimised by intelligent algorithms is expected to increase. The application of machine learning strategies based on simulation, communication and automation is of great importance for the development of control strategies, as they usually require a simulation-based training phase. The use of a standardised IM is of significant advantage in this context.

Moreover, the IM enables the construction of data science applications in the context of CPPS by virtue of the consistency of the data structures, which are explicitly specified as variable types throughout. Future work will aim to extend the applicability of the IM by transferring it to other research areas. Furthermore, we will validate it in the application of optimised operation. This may involve the generation of energy efficiency indicators and the allocation of emissions over the production cycle. To this end, the IM will be applied to various use cases inside the ETA research factory, including a transfer of the model to other production machines. As our IM has been demonstrated to be of utility, we intend to leverage its principal characteristics to establish a direct coupling between the real systems and the corresponding simulation models.

**Acknowledgements and Appendix.** The authors gratefully acknowledge the financial support of the Kopernikus-Project "SynErgie" (Grant Number 03SFK3A0-3) by the Federal Ministry of Education and Research of Germany (BMBF) and thank the Projektträger Jülich (PtJ) for the project supervision.

The software code for the presented work is available as an open-source project on GitHub: https://github.com/Jtopschall/GCSM2024-OPC-UA-for-cleaning-machines. git.

# References

1. Burger, B.: Annual spot market prices in Germany (2024). https://www.energy-charts.info/?l=de&c=DE
2. Ritchie, H., Rosado, P., Roser, M.: Energy. Our World in Data (2024). https://ourworldindata.org/energy
3. Walther, J., et al.: A methodology for the classification and characterisation of industrial demand-side integration measures. Energies **15**(3) (2022)
4. Fuhrländer-Völker, D., Magin, J., Weigold, M.: Automation architecture for harnessing the demand response potential of aqueous parts cleaning machines. Prod. Eng. **17**(6), 785–803 (2023)
5. Fuhrländer-Völker, D., Borst, F., Theisinger, L., Ranzau, H., Weigold, M.: Modular data model for energy-flexible cyber-physical production systems. In: 55th CIRP Conference on Manufacturing Systems, vol. 107, pp. 215–220 (2022). https://www.sciencedirect.com/science/article/pii/S2212827122002529
6. Grosch, B., Fuhrländer-Völker, D., Stock, J., Weigold, M.: Cyber-physical production system for energy-flexible control of production machines. Procedia CIRP **55** (2022)
7. Buhl, H.U., Duda, S., Schott, P.: Das energieflexibilitätsdatenmodell der energiesynchronisationsplattform (2021)

8. Schott, P., Sedlmeir, J., Strobel, N., Weber, T., Fridgen, G., Abele, E.: A generic data model for describing flexibility in power markets. Energies **12**(10), 1893 (2019). https://www.mdpi.com/1996-1073/12/10/1893

9. Reinhart, G., Bank, L., Brugger, M., et al.: Konzept der energiesynchronisationsplattform. diskussionspapier v3 (2020)

10. OPC Foundation: Ua companion specifications (2024). https://opcfoundation.org/about/opc-technologies/opc-ua/ua-companion-specifications/

11. Lindner, M., et al.: Industrie 4.0 - Interoperabilität durch OPC UA mit Companion Specifications : Mehrwerte für Stakeholder des Maschinen- und Anlagenbaus. VDMA e.V., Frankfurt am Main (2023)

12. LoTuS: Lotus - leistungsoptimierte trocknung und sauberkeit: Energetische trockenprozessoptimierung und -vernetzung für eine energieeffiziente reinigungsanlage (2023). https://www.enargus.de/pub/bscw.cgi/?op=enargus.eps2&q=lotus&m=2&v=10&s=8&y=1&id=1323683

13. Fuhrländer-Völker, D.: Automation architecture for demand response on aqueous parts cleaning machines. Ph.D. thesis, TU Darmstadt (2023)

14. Python Software Foundation: asyncio—asynchronous I/O (2024). https://docs.python.org/3/library/asyncio.html

15. OPC UA Contribution: OPC UA asyncio (2024). https://github.com/FreeOpcUa/opcua-asyncio

16. Pecinovsky, J., Boerman, F.: entsoe-py (2024). https://github.com/EnergieID/entsoe-py

# Predicting Selling Status of Second-Hand Products on EC Platform Based on Morphological Analysis and Classification

Hiromasa Ijuin(✉) and Aya Ishigaki

Tokyo University of Science, Noda, Chiba 278-8510, Japan
ijuin@rs.tus.ac.jp

**Abstract.** Currently, the depletion of resources has become a serious problem worldwide. In order to promote sustainable development, people are encouraged to minimize the waste from their products, while companies should produce remanufactured products such as second-hand laptops and smartphones. On the Electric Commerce (EC) platform, people engage in trading the second-hand products among themselves and with businesses. The expansion of the second-hand product market on EC platforms contributes to reducing the amount of product waste. On EC platforms, individuals or groups acting as online retailers list second-hand products. However, these second-hand products often fail to accurately reflect online shopper demand. This study proposes a prediction model for the selling status of second-hand products on the EC platform based on morphological analysis and classification results. First, morphological analysis is performed on online shopper comments for second-hand products to reveal frequently used noun words. Second, the features of these frequently used words are calculated using principal component analysis, and online shoppers are classified into clusters. Finally, a prediction model for the status of second-hand products is proposed based on the cluster information.

**Keywords:** Principal Component Analysis (PCA) · Random Forest (RF) · Term Frequency-Inverse Document Frequency (TF-IDF)

## 1 Introduction

### 1.1 The Second-Hand Market on EC Platforms

For the development of society, product manufacturing and waste generation have grown significantly [1]. To manufacture these products, it has been reported that many resources are mined and consumed. For example, The International Resource Panel annual global announced that annual material consumption increased by an average of more than 2.3% per year [2]. Consequently, depletion of resources has become a serious problem worldwide. In order to promote sustainable development and reduce material consumption, people should be encouraged to minimize the waste [3] from their products, while companies should produce remanufactured products such as second-hand laptops and

© The Author(s) 2025
H. Kohl et al. (Eds.): GCSM 2024, LNME, pp. 164–172, 2025.
https://doi.org/10.1007/978-3-031-93891-7_19

smartphones. Second-hand products are often traded in the second-hand product market on Electronic Commerce (EC) platforms. On EC platforms, individuals or groups acting as online retailers list second-hand products. The second-hand product markets on EC platforms have grown in Japan [4], and the expansion of this market contributes to reducing the amount of product waste. However, these second-hand products often fail to accurately reflect online shopper demand because the quality and condition of second-hand products vary based on their usage. For example, Dou et al. [5] introduced a conflict case related to the quality of secondhand products. They suggested that retailers with information advantages are often reluctant to disclose the true quality of used products to customers. Consequently, consumers may avoid purchasing secondhand products from a retailer unless product quality can be verified through transactions [5].

Machine learning enables the identification of trends and online shopper demand from their comment data [6]. For example, to analyze this data, morphological analysis is conducted to find elements that share the same form and meaning [7]. Morphological analysis automatically determines the part of speech for each word in the text [8]. Additionally, Principal Component Analysis (PCA) can extract a few meaningful factors by reducing noise [9]. PCA identifies new variables that are linear combinations of the original features. These components are orthogonal to each other and are ordered to maximize the variance explained [10]. Furthermore, Random Forest (RF) is used for general-purpose classification predictions [11]. RF is an effective algorithm for non-linear classification tasks, especially when dealing with imbalanced large datasets [12].

Therefore, the research objective of this study is to combine these machine learning methods to uncover the online shopper demand and predict the selling status classification of second-hand products based on the identified online shopper demand.

## 1.2 Literature Review

With respect to the second-hand market on EC platforms, Liu and Wan [13] analyze the influence factors of the user's emotional inclination through the methods of text feature analysis such as the latent Dirichlet allocation, the topic model and the sentiment analysis. However, they did not refer to user demand based on classification of cluster. Regarding classification, the Guiot and Roux [14] proposed the model to summarize motivations for second-hand shopping by non-hierarchical cluster analysis. Their proposed model offers a validated measurement tool for assessing second-hand shoppers' motivation and manifests the critical attitude as distancing and avoidance behaviors toward the classic market system. However, they did not consider applying their cluster information for other methods such as prediction or regression.

## 1.3 Purpose of this Study

This study proposes a prediction model for the selling status of second-hand products on EC platforms based on morphological analysis and classification results. First, morphological analysis is performed on online shopper comments for the second-hand products to reveal frequently used noun words. Second, the features of these frequently used words are calculated using principal component analysis, and online shoppers are classified into clusters. Finally, a prediction model for the status of second-hand products is proposed based on the cluster information.

## 2  Procedure

Figure 1 shows the procedures of this study. Step 1 conducts morphological analysis to find the frequently used noun words in the online shoppers' comment data. Additionally, to evaluate the frequency of each noun word, *TF-IDF* (Term Frequency-Inverse Document Frequency) is calculated for the word importance indicator. *TF-IDF* is a numerical statistic that represents the relevance of key words to some specific text [15]. To clarify the online shopper demand model, the features of those frequently used noun words are calculated by PCA in Step 2. Furthermore, based on the PCA results, online shoppers on the EC platform are classified. Step 3 predicts the selling status of second-hand laptops on the EC platform using the RF model and $K$-Fold cross-validation. The RF model can be biased due to the splitting data used for training and testing [16]. To mitigate this bias, $K$-Fold cross-validation is conducted, which involves repeated splitting of the data and model training $K$ times [16].

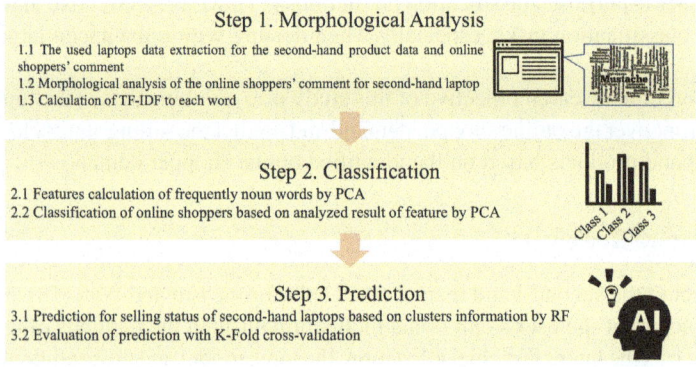

**Fig. 1.**  Procedures of proposed prediction model in this study.

## 3  Method and Data

### 3.1  Methods for Morphological Analysis, Classification, and Prediction

Table 1 shows the methods for morphological analysis, classification, and prediction.

Step 1 applies morphological analysis to the online shoppers' comment data to find the frequently used noun words with the GiNZA library. Based on the result of the frequently used words, Step 2 classifies the online shoppers to clarify their demand model on the EC platform. After classification, the sale status for each second-hand laptop is predicted by using the extracted variables from the online shopper classification in Step 3.

The explanation of each method is described as below.

**Table 1.** Summary of process and method for each step.

| Step | Process | Method | Output |
|------|---------|--------|--------|
| 1 | Morphological analysis | GiNZA | Frequently used noun words |
| 2 | Classification of online shopper | PCA | Clusters of online shoppers |
| 3 | Prediction | RF | Result of selling status |

**GiNZA.** GiNZA is a Japanese natural language processing library and released in 2019 [17]. GiNZA has features such as high-speed and highly accurate Japanese analysis processing, dependency structure analysis, named entity extraction, and other advanced natural language processing technologies that can be used on an internationalized framework [17, 18]. This study uses GiNZA to split words and extract noun words from online shopper comments.

**TF-IDF.** To apply scikit-learn API, is calculated as shown in Eq. (1).

$$TF - IDF(t, d) = TF(t, d) \times (IDF(t, d) + 1) \tag{1}$$

Equations (2) and (3) are presents the formulations for *TF* (Term Frequency) and *IDF* (Inverse Document Frequency). *TF* is a measure method that how many times a term is present in a text [15]. *IDF* assigns lower weight to frequent words and assigns greater weight for the words that are infrequent [15].

$$TF(t, d) = \frac{n_{t,d}}{n_d} \tag{2}$$

$$IDF(t, d) = log \frac{1 + \sum_d n_d}{1 + tf(t, d)} \tag{3}$$

After morphological analysis by GiNZA, this study applies the *TF-IDF* to online shoppers' comment data to identify the frequently used noun words.

**Principal Component Analysis (PCA).** The PCA is a technique of data analysis for building linear multivariate models of complex data [9, 10]. The PCA models the statistically significant variation int the data as well as the random measurement error [9]. Additionally, the PCA model enables to eliminate the principal components associated with noise. This study proposes the online shopper demand model by PCA based on the frequently used noun words.

**Random Forest (RF).** The RF is a machine learning method that combines several diction trees and aggregates their predictions [11, 12]. The algorithm of RF is non-parametric regression estimation and the goal is to predict the square integrable random response by estimating the regression function [11]. This study predicts the selling status of second-hand laptop as the "sold out" or "on sale". Furthermore, this study conducted *K*-Fold cross-validation to estimate the bias of training error [16].

## 3.2  Data

To classify of online shopper demand and predict the selling status, this study uses two types data from one EC platform as below.

**Online Shoppers' Comment Data.** Online shoppers' comment data refers to raw online shopper comments about second-hand laptops. On the EC platform, some online shoppers negotiate discounts on selling prices and delivery methods through their comments. By analyzing the online shoppers' comment data, this study models the online shopper classification.

**Second-Hand Product Data.** Second-hand product data refer to the selling history of second-hand laptops. The second-hand product data include information such as the selling price, second-hand product descriptions by individual online sellers, and the selling status of "sold out", which means the second-hand laptop is sold out, or "on sale", which means the second-hand laptop keeps to sell.

**Step 1.1 The second-hand laptops data extraction for the second-hand product data and online shoppers' comment.** From the second-hand product data, this study illustrates the frequency distribution of the selling price of the second-hand laptops. The selling price of the second-hand laptop is ranged from 10,000 Yen to 200,000. Additionally, over 80% of the selling prices are lower than 70,000 Yen.

# 4  Results

## 4.1  Result of Morphological Analysis

**Step 1.2 Morphological analysis of 7the online shoppers' comment for second-hand laptop.** To identify the online shopper demand, morphological analysis is applied to online shoppers' comment data and *TF-IDF* is calculated for the noun word. However, some noun words such as "Laptop", "Computer", and "Purchase" are excluded because these words are always included and cannot be considered characteristic words for online shoppers' comment data.

**Step 1.3 Calculation of *TF-IDF* to each word.** Figure 2 presents the top 10 words based on *TF-IDF* from online shoppers' comment data for second-hand laptops. The word "Request" has the highest *TF-IDF* among all nouns and is used by online shoppers for negotiations, such as requesting discounts on the price and the delivery method. Additionally, it is found that some noun words for using in negotiate situation have appeared. From the result of the top 10 most frequent words, each the online shoppers' comment data was represented as a binary feature vector. This vector indicates the presence or absence of each of the top 10 most frequent words. Then, the feature vectors are classified into two categories using PCA in Step 2.

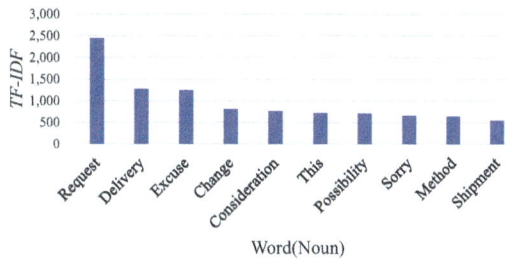

**Fig. 2.** *TF-IDF* of comment data for the second-hand laptops.

## 4.2   Result of Classification by PCA

**Step 2.1 Features calculation of frequently noun words by PCA.** From the results of Step 1, it was found that the top 10 frequently used words in the online shoppers' comment data were identified. In Step 2, PCA is applied to those top 10 frequently words to classify online shoppers. Additionally, since online shoppers tend to specify the delivery method such as "Delivery option 1", "Delivery option 2", and "Unspecified", the PCA results are color-coded for each delivery method, as shown in Fig. 3.

When the value of PCA1 is higher, delivery methods of "Delivery option 1" and "Delivery option 2" are selected in the online shoppers' comment data. It appears that the online shoppers who have less experience using EC platform, i.e., beginners, do not select the delivery methods. On the other hand, the online shoppers with a lot of experience using EC platform, i.e., experts, do select the delivery methods. Therefore, this study considers PCA1 as the parameter for the frequency of EC platform usage, where higher PCA1 values indicate more experienced online shoppers.

When the value of PCA2 is higher, the delivery methods of "Delivery option 1" and "Delivery option 2" are not specified in the online shoppers' comment data. It appears that the online shoppers who plan to buy a second-hand laptop with low specifications and a lower selling price do not care about the delivery methods. Conversely, the online shoppers who plan to buy a second-hand laptop with high specifications and a higher selling price do care about the delivery methods. Therefore, this study considers PCA2 as the sensitivity to the selling price of second-hand laptop, where higher PCA2 values indicate that online shoppers prefer to buy lower-priced second-hand laptop.

**Step 2.2 Classification of online shoppers based on analyzed result of feature by PCA.** From results of PCA analysis, this study classifies two clusters for online shoppers as below.

**Cluster 1: Beginner of EC platform and prefer cheap second-hand products.** The online shoppers in Cluster 1 have less experience with the EC platform and look for cheap second-hand laptops. It is assumed that they prioritize low cost when choosing a second-hand laptop on the EC platform.

**Cluster 2: Expert of EC platform and prefer good condition products.** Online shoppers in Cluster 2 have a lot of experience with EC platforms and prefer second-hand laptops in good condition. It is assumed that they have specific requests for the delivery method. PCA analysis of the online shopper's comments on second-hand products

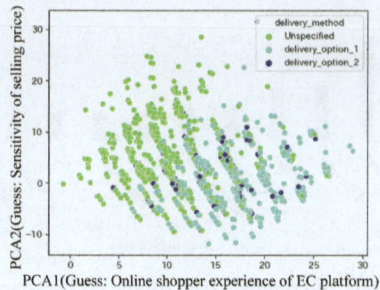

**Fig. 3.** Features plot of frequently noun words with delivery method.

revealed 2 clusters based on the words used. This indicates that analyzing the online shopper's comments can effectively uncover the characteristics of remanufactured and second-hand products that shoppers are seeking.

### 4.3   Result of Prediction by RF

**Step 3.1 Prediction for selling status of second-hand laptops based on clusters information by RF.** From Step 2, the online shoppers are classified into 2 clusters based on the features centered around the selling price and the delivery method. Therefore, this study extracts this information from the second-hand product data and uses it to predict the selling status using the RF model. In addition, the value of $k$ for $K$-fold cross-validation is set to 5 ($|k| = 5$).

**Step 3.2 Evaluation of prediction with K-Fold cross-validation**

**Table 2.** Result of accuracy of selling status.

| | $K$-Fold | | | | | Average |
|---|---|---|---|---|---|---|
| | $k = 1$ | $k = 2$ | $k = 3$ | $k = 4$ | $k = 5$ | |
| Accuracy Rate [%] | 70% | 72% | 71% | 77% | 74% | 73% |

Table 2 displayed of the accuracy of selling status using the RF model with $k$-fold ($|k| = 5$) cross-validation. From result of it, the average accuracy rate is 73%. In addition, Fig. 5 illustrates the decision tree sample for predicting selling status using RF. It is observed from Fig. 4 that cheaper second-hand laptop tends to be the status of "sold out", and when the second-hand laptops are delivered with "delivery option 2", they are more likely to be the selling status are "on sale".

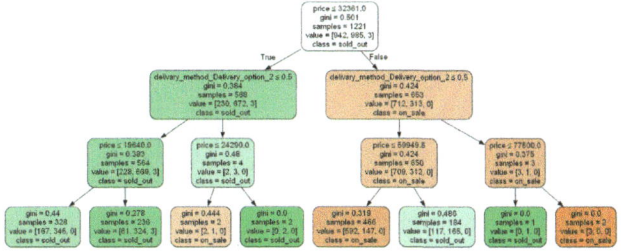

**Fig. 4.** Decision tree sample of the selling status by RF with k-fold ($k = 5$) cross-validation.

## 5 Conclusion

This study proposes a prediction model for the selling status of second-hand products on EC platforms based on morphological analysis and classification results. By combines machine learning method, this study classifies online shoppers based on their comment and predict the selling status. The experiments in this study demonstrate a usable finding that selecting explanatory variables for prediction based on morphological analysis and PCA is more effective than using a trial-and-error approach. Future study should apply this procedure for other types of products.

**Acknowledgements.** This work was supported by Japan Society for Promotion of Science (JSPS), KAKENHI, Grant-in-Aid for Research Activity Start-up, 24K23151, from 2024 to 2025.

## References

1. Yang, H., Ma, M., Thompson, J.R., Flower, R.J.: Waste management, informal recycling, environmental pollution and public health. J. Epidemiol. Community Health **72**(3), 237–243 (2018)
2. United Nations Environment Programme: Global Resources Outlook 2024: Bend the trend: Pathways to a liveable planet as resource use spikes. https://www.resourcepanel.org/sites/def ault/files/documents/document/media/gro24_full_report_1mar_final_for_web.pdf
3. Dong, K., Lv, W.Y.: Cost optimization model of second-hand product for extended non-renewing replacement-repair warranty. Open Cybern. Systemics J. **10**(1), 101–111 (2016)
4. Ministry of Economy, Trade and Industry, Market research report on electronic commerce for 2021, May 2021. https://www.meti.go.jp/press/2022/08/20220812005/20220812005-h.pdf. (in Japanese)
5. Dou, X., Zenglu, L., Chun, L.: Secondhand product quality disclosure strategy of the retailer under different supply chain structures. Manag. Decis. Econ. **43**(7), 2982–2999 (2022)
6. Jordan, M.I., Mitchell, T.M.: Machine learning: trends, perspective, and prospects. Science **349**(6245), 255–260 (2015)
7. Aronoff, M., Fudeman, K.: What is Morphology? Wiley, Hoboken (2022)
8. Mengliyev, B., Shahabitdinova, S., Khamroeva, S., Gulyamova, S., Botirova, A.: The morphological analysis and synthesis of word forms in the linguistic analyzer. J. Lang. Linguist. Stud. **17**(1), 558–564 (2021)
9. Gemperline, P.J.: Principal component analysis. In: Gempeline, P.J. (eds.) Practical Guide to Chemometrics, pp. 69–103. CRC Press, New York (2006)

10. Jolliffe, I.T., Jorge, C.: Principal component analysis: a review and recent developments. Philos. Trans. R. Soc. A: Math. Phys. Eng. Sci. **374**(2065) (2016)
11. Biau, G., Scornet, E.: A random forest guided tour. Test Official J. Spanish Soc. Stat. Oper. Res. **25**, 197–227 (2016)
12. Paul, A., Mukherjee, D.P., Das, P., Gangopadhyay, A., Chintha, A.R., Kundu, S.: Improved random forest for classification. IEEE Trans. Image Process. **27**(8), 4012–4024 (2018)
13. Liu, Y., Wan, Y.: Consumer satisfaction with the online dispute resolution on a second-hand goods-trading platform. Sustainability **15**(4), 3182 (2023)
14. Guiot, D., Roux, D.: A second-hand shoppers' motivation scale: antecedents, consequences, and implications for retailers. J. Retail. **86**(4), 383–399 (2010)
15. Qaiser, S., Ali, R.: Text mining: use of TF-IDF to examine the relevance of words to documents. Int. J. Comput. Appl. **181**(1), 25–29 (2018)
16. Fushiki, T.: Estimation of Prediction Error by Using K-Fold Cross-Validation. Springer, Cham (2011)
17. Sakahira, F., Hiroi, U.: Creating a disaster chain diagram from Japanese newspaper articles using mechanical methods. J. Adv. Comput. Intell. Intell. Inform. **25**(3) (2021)
18. Bonnell, J., Ogihara, M.: Rule-based adornment of modern historical Japanese corpora using accurate universal dependencies. DHQ: Digit. Hum. Q. **16**(4), (2022)

# Sustainability by Design

# Designing a Greener Future - Publishing Platforms for Open Source Hardware

Dominik Saubke[1]([⊠]) [iD], Paul Ihle[2], Pascal Krenz[1], and Tobias Redlich[1]

[1] Helmut Schmidt University/University of the Federal Armed Forces, Hamburg, Germany
dominiksaubke@hsu-hh.de
[2] Hamburg University of Applied Sciences, 20099 Hamburg, Germany

**Abstract.** In a time change shaken by crises, the currently dominant value creation concepts are being increasingly questioned and the focus is more and more on sustainability, longevity and individuality. Various initiatives are promoting the idea of producing more locally close to the point of need in an attempt to realise a circular economy (see FabCity Initiative). Global collaboration and local manufacturing have become the maxim of these new economic concepts. Under the Open Design paradigm, object-based artifacts are created that are accessible, modifiable and reproducible for everyone. The outcome is known as Open Source Hardware. In the last decade, the number of Open Source Hardware has risen and platforms that bundle many thousands of designs are growing steadily. As a result, product designs are gaining more and more visibility and arousing people's interest. But where can one find Open Source Hardware and how does the business behind this work? In this study, 48 web-based sources for open source hardware were identified and analysed in order to classify them and identify patterns.

**Keywords:** Open Source Hardware · Open Design · Openness · Open Innovation

## 1 Introduction and Motivation

All industries need to actively participate in generating sustainable value creation chains to meet the aspirations of future generations [1]. The concept of the circular economy is expected to play a central role here [2]. Driven by sustainability, longevity, and individuality, new approaches must already be pursued in product development [3]. The idea of making product development open and its documentation freely accessible has been around for many years and offers innovative solutions [4]. The Open Source Hardware (OSH) movement is pursuing ambitious goals that promise benefits for society and the economy [5]. The aim is to promote reparability and the free exchange of information, enabling broader and more accessible innovations. OSH reduces redundancies and facilitates collaboration and the sharing of resources, leading to efficiency gains. It creates a more inclusive and diversified innovation environment, and can help develop sustainable and robust technological solutions accessible to a broad community [6]. This is supported in latest research on shaping knowledge transfer processes and methods during product development on digital platforms [7]. However, this inclusivity can lead to conflicts,

© The Author(s) 2025
H. Kohl et al. (Eds.): GCSM 2024, LNME, pp. 175–183, 2025.
https://doi.org/10.1007/978-3-031-93891-7_20

such as liability issues in the event of quality problems and dealing with unauthorised reproductions. Furthermore, *Open Business Models* need to be developed that strengthen the competitiveness of domestic value creation structures. This is further emphasized by Acatech's Circular Economy Initiative calling for new sustainable business models in their latest circular economy strategy proposal [8]. The integration of soft OSH concepts in industry could initially contribute to increasing the value of products. The importance of OSH is evident in the fact that the topic is already included in the draft of the German National Circular Economy Strategy [9].

## 2   Clarifying the Objective - Open Source Hardware

If an object-based artifact is developed in line with the Open Design paradigm, this can be referred to as OSH. The term OSH originates from open source software development, which goes back to an initiative in the early 1990s [10] and led to the founding of the Open Source Initiative (OSI) in 1998 [11]. However, a widely accepted attempt to define the concept was published in 2011: the Open Source Hardware Definition 1.0 [10]. Today, current research refers to DIN SPEC 3105 [6] as an approach to describing OSH [8]. According to these standards, certain information (e.g. name, author, functionality, technical documentation and more) have to be provided and a valid licence has to be chosen. A debate around OSH is that it relies on authors being aware of the concept to apply it. Since many projects might be published without this awareness, for this reason, OSH is referred to as Open Hardware (OH) in this research [12]. An study on OH form 2017 examined 132 object-related artifacts, today there are already several platforms that publish far more then 1000 projects alone [13]. In this dynamic development, the first platforms for publishing OH from the early 2000s have evolved so that today multi-part assemblies can be published there and platforms tend to have their own structures for product development processes (e.g. Github). In the following a study is carried out, based on a cross-linked literature- and multi-layer review, which focuses exclusively on digital sources for publishing OH. Following the investigation, following research question has to be answered to gain a better understanding of this digital sources of OH: *How can open hardware sources be systematically categorized, and is it possible to identify patterns through a comparative analysis?* To effectively address this, the platform will be assessed based on various criteria and reviewed by experts. Certain groups of object-based artifacts such as toys, decorative objects and art objects are not included in this search.

## 3   Methodical Approach

Many OH projects are created as part of academic and scientific programmes or in a related environment, those projects be easily identified by a literature review since there is often the obligation to share and publish research results. A consortium of German research institutes has just announced the *Open Source Hardware InnOvation Plattform* (oSHOP) for sharing hardware research projects. However, there is no universal system for indexing OH without limitations for example the licensing system from OSH Foundation only covers projects that have been officially published in compliance with their obligations. This means that there are only limited and highly dispersed collections of OH projects, therefore a structured approach is required to identify all sources. Figure 1 shows the search procedure. The introduced *Cross-linked Literature- and Multi-Layer Review* is a two-stage approach combing a classic literature review with web research using different technology layers. To this end, the participating scientists first defined the search strings. Due to the problems described above, the terms *Open Hardware* OR *Open Source Hardware* OR *OSH* were selected as search terms. The same search query was conducted in two scientific databases to identify the literature: Web of Science (WOS) and Scopus. The search options chosen were 'topic' for WOS and 'article title, abstract, keywords' for Scopus. As far as possible, both databases were searched in the same way. The sources found were screened for OH publications; sources, names, and publishers were listed. Following the identification of projects, a rigorous validation process was undertaken to confirm their continued existence. The projects themselves were not the sole focus, but rather the platforms hosting them. The data found was then also transferred to the *Multi-Layer Review*. As a final step in this process, all web presences found were analyzed for references. These could be other platforms with a different focus or other project pages. The *Multi-Layer Review* generally involves different search entities (generative AI, search engine, and knowledge databases), with each entity describing one layer of the search. The aim is to systematically search the internet at different levels using identical search terms. The last check for new publications was carried out in the first calendar quarter 2024. Subsequently, the platforms were evaluated by experts with engineering background specialised in Open Hardware research, who were already part of the search process. The evaluation of the platforms followed two dimensions of Open Business Models in the context of digital platforms derived from the Data. The first is *Business Accessibility* and the second is *Business Commercialisation*.

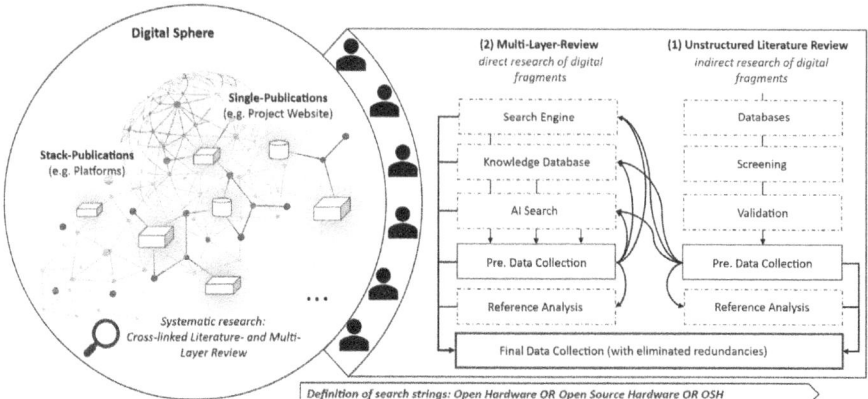

**Fig. 1.** Procedure for the Cross-linked Literature Review and Multi-Layer Analysis.

# 4   An In-Depth Analysis of Open Hardware Sources

A total of 49 platforms were identified during the search. To compare them and recognize patterns, evaluation parameters were defined for the initial description. The following parameters were determined: *Number, Name of the platform and year of foundation, webaccesses per Open Hardware, Business Accessibility (BA), Business Commercialisation (BC), Available Technologies (AT) and Focus Rate (fr).* The share of the latter was then calculated from the number of publications and the number of OH projects; this percentage value is described as the focus rate (fr). The results of the search are shown in Table 1.

**Table 1.**  Overview of the Platform identified during the Review

| Platform Data | wA/OH | BA | BC | AT | fr |
|---|---|---|---|---|---|
| **P1** - Adafruit [2005] | >1k | High | High | mix | >95% |
| **P2** - Airtable [2012] | <0,1k | Top | Medium | mix | <0,5% |
| **P3** - Appropedia [2006] | <0,1k | High | Low | high-tech | >95% |
| **P4** - Arduino [2006] | >1k | Inter | High | high-tech | >95% |
| **P5** - Bauanleitung.org [2009] | <1k | Medium | Medium | mix | >95% |
| **P6** - Cgtrader [2011] | <1k | High | Top | high-tech | >95% |
| **P7** - CircuitMaker [1988] | <0,1k | Top | Medium | mix | <0,5% |
| **P8** - Cults [2014] | <0,1k | High | Top | mix | 10% |
| **P9** - DIY Drones [2007] | >50k | Inter | Top | low-tech | <0,5% |
| **P10** - Electroschematics [2008] | <1k | Medium | Medium | low-tech | >95% |
| **P11** - FarmHack [2015] | <0,1k | Top | Medium | high-tech | >95% |
| **P12** - Fritzing [2016] | <0,1k | High | Medium | high-tech | 17,5% |
| **P13** - Gasblender [2021] | nv | Top | Medium | mix | >95% |
| **P14** - GitHub [2008] | <0,1k | Top | Medium | high-tech | >95% |
| **P15** - GitLab [2011] | <0,1k | Top | High | mix | >95% |
| **P16** - GrabCAD [2009] | <0,1k | Top | High | mix | <0,5% |
| **P17** - Hackaday.io [2004] | <0,1k | Top | Medium | mix | <0,5% |
| **P18** - Hackster [2013] | <0,1k | Top | High | mix | 9% |
| **P19** - Hardware-x [2017] | >5k | Inter | Low | mix | 57% |
| **P20** - HeimwerkerTipps [2008] | <1k | Low | Medium | mix | 12% |
| **P21** - Instructables [2005] | <0,1k | Top | High | high-tech | >95% |
| **P22** - MyMiniFactory [2013] | <1k | Inter | Top | low-tech | >95% |
| **P23** - Oho.wiki [2020] | <0,1k | High | Low | mix | 72% |

*(continued)*

**Table 1.** (*continued*)

| Platform Data | wA/OH | BA | BC | AT | fr |
|---|---|---|---|---|---|
| **P24** - Open Design [2017] | <0,1k | High | Low | low-tech | 31% |
| **P25** - Open Electronics [2011] | <0,1k | High | Top | high | >95% |
| **P26** - Open Hardware [2024] | <0,1k | High | Top | low-tech | >95% |
| **P27** - Open Motors [2013] | <1k | Low | Top | high-tech | >95% |
| **P28** - OS Ecology [2003] | nv | High | Medium | high-tech | >95% |
| **P29** - OS Eco.Germany [2018] | <1k | Top | Low | mix | - |
| **P30** - OpenBuilds [2012] | <1k | Top | High | mix | >95% |
| **P31** - Openfabpdx [2013] | <1k | Low | High | mix | 88% |
| **P32** - OpenStructures [2012] | <0,1k | Inter | Inter | low-tech | >95% |
| **P33** - OSHWA [2012] | >1k | Top | Low | low-tech | 12% |
| **P34** - Ottodiy [2016] | >1k | Inter | Top | mix | 89,5% |
| **P35** - Pinshape [2013] | <0,1k | High | High | mix | >95% |
| **P36** - Pinterest [2008] | <0,1k | Top | High | high-tech | >95% |
| **P37** - Preciousplastic [2013] | >10k | Top | High | lowtech | <0,5% |
| **P38** - Printables [2022] | nv | High | High | mix | <0,5% |
| **P39** - PrusaPrinters [2012] | >> | Inter | Top | high-tech | >95% |
| **P40** - Public Lab [2010] | >1k | Top | Medium | low-tech | - |
| **P41** - Raspberry Pi [2009] | >100k | High | Top | high-tech | >95% |
| **P42** - RepRap [2004] | >1k | High | Low | mix | >95% |
| **P43** - Selbermachen.de [1974] | >1k | Low | Medium | high-tech | >95% |
| **P44** - Sensorica [2011] | <1k | High | Low | high-tech | >95% |
| **P45** - SparkFun [2003] | nv | High | Top | lowtech | 59% |
| **P46** - Thingiverse [2008] | <0,1k | Top | High | high-tech | >95% |
| **P47** - WikiHouse [2011] | >10k | Medium | Inter | high-tech | - |
| Bottom (B): 0%–16% | | | Intermediate (I): 51%–67% | | |
| Low (L): 17%–33% | | | High (H): 68%–84% | | |
| Medium (M): 34%–50% | | | Top (VH): 85%–100% | | |

Entering various search parameters was often a complex task, as not all sources have sophisticated filter or search functions. In one case, 10% of all projects were viewed to ensure statistically reliable statements about the total number of projects. Furthermore, the operators of all identified sources have been contacted and asked to confirm those results. The sources where investigated according to the available object-related artifacts: low-tech includes furniture, consumer goods, and various other less advanced items, while high-tech includes robotics, automotive technology, machine tools, and medical technology. Estimating the correct number of OH on some websites can be challenging. For instance, the source *P40 Printables* offers a variety of project data, but most entries are spare parts, toys, or purely illustrative objects. Nonetheless, there are numerous multi-part assemblies with additional instructions and background information using links to data sharing platforms. Another example is *P25 Oho Wiki*, which is supported by several initiatives and academic institutions. Its explicit goal is to create a comprehensive database of OH projects with special sustainable approaches. Another source, *P47 Sparkfun*, unfortunately lacks suitable filtering methods or systems to retrieve data from the source and to differentiate between OH projects and non-commercial products.

The research indicates a mature mix of low-tech, mixed-tech, and high-tech sources, highlighting that the original concept from the software sector initially spilled over into related hardware technologies like microcontrollers and computer technologies. Figure 2 illustrates the collected data, supplemented by site access numbers. The diagrams in the lower section display various combinations of collected or calculated parameters, with specific labels indicated on the axes. The timeline shows the establishment dates of the sources, and the peak period is easily identifiable. The pie charts illustrate the technology mix and the proportion of hardware focusing sources. Most of the investigated sources emerged in the late 2000s and many of them are still active and regularly updated by their operators. The next step was to determine the number of projects and OH projects, which proved quite challenging. There is a high expectation for numerous sources with highest focus rates (fr), meaning these sources are specifically designed for publishing OH.

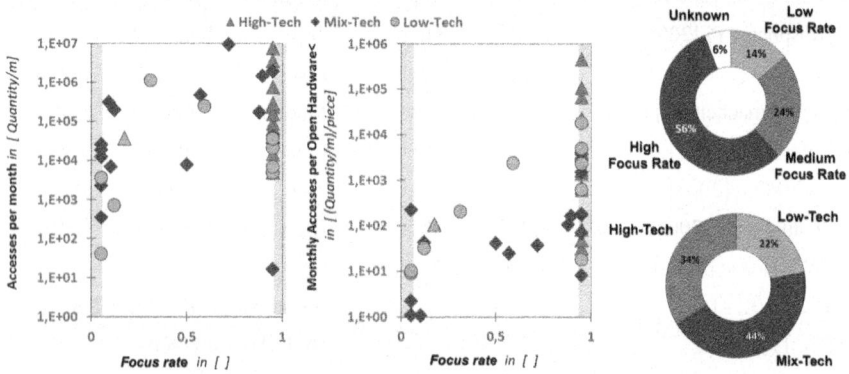

**Fig. 2.** Research data and typification based on various analysis approaches.

The focus rate (fr) is not detailed below 0.05 or above 0.95, leading to data clusters. The latter will be referred to as *Single Publications*. These typically consist of project pages where individual OH and their derivatives are published or sources that are used in a massively different way than originally intended. Generally, sources of OH seems to be divided into two different classes, which can be understood as *Single Publication* and *Stack Publication*. These *Stack Publications* are also to be understood as sources behind which there is usually a platform technology (User account-depended upload). These can be further typified into *Full Type with fr < 95%*, *Half Type* with *5% < fr < 95%* and *Loose Type with fr > 5%*. Looking at the list of sources, it is evident that *P16 GitHub*, for example, has a very low focus rate. This is due to the fact that mainly software projects are hosted there, with hardware only playing a subordinate role. Other platforms have even lower adoption values, with the lowest values <0.01% for *P37 Pinshape* and *P38 Pinterest*, which can probably be explained as a kind of misapplication. Notably, the cluster of pure OH sources with high access numbers suggests that specialization appears to be a success factor (left diagram). It is also interesting to note the prevalence of high-tech-focused OH platforms. A closer look at the middle diagram reveals that successful platforms exist across all areas, regardless of the number of published projects. Combining the number of OH projects with access numbers yields a quotient describing access per OH, shows that sources with a high-tech focus are notably more successful in the sense of higher access rate.

# 5   An Attempt at Categorising Sources

While the study has already revealed an initial differentiation of the sources of OH into the classes *Single Publication Form* and *Stack Publication Form*, this is being broadened to include an evaluation of how to categorize platforms. The data indicates that there are different principles in the layout and operation of platforms. Two categorisation dimensions for the categorisation of platforms were derived from the observations. The concept of open business deals as a base to define the target dimensions of *Business Accessibility (BA)* and *Business Commercialisation (BC)*. *BA* is used to evaluate aspects of openness in the positioning of the platform. The second categorisation dimension, BC, is intended to evaluate the approaches to commercialisation implemented on the platform. The overall evaluation is based on two different evaluation approaches. On the one hand, an assessment was made from the data collected based on platform characteristics; this also includes, for example, usage of dummy variables from the elicitation such as general openness towards submission or whether commercialisation approaches are currently implemented. Furthermore, the experts responsible for the research conducted an assessment for each target dimension. For BA, the ability for the user to participate in OH development for the user was assessed; this includes the free possibility to participate, the possibility to fork projects and the possibility to fully access projects. For BC, the degree of commercialisation for the operator, for other companies and for the user was assessed. The evaluation factors were weighted with a higher values on the obtained data. Finally, an attempt was made to categorise the platform by introducing four descriptive categories of business orientation in the diagram. Most sources receive a minimum rating here, as the systematic publication already provides kind of basic introduction to the market. The standardised evaluations were plotted against each other in Fig. 3 with the respective percentage value. Sources with a relatively closed system for OH but a comprehensive, fully integrated business system fall into the PROFIT EXCAVATOR area; they mostly offer a single OH system with a strong customer focus. Sources with a more open opportunity for participation are referred to below as NEW VALUE CREATORS and GREEN FRONTIERS; the latter have no integrated commercialisation strategies. Both groups to the right of the doted line should be investigated further. In the New Value

Creator group in particular, the four sources P37, P41, P9 and P39 have high accesses per OH and are particularly suitable for a business model analysis.

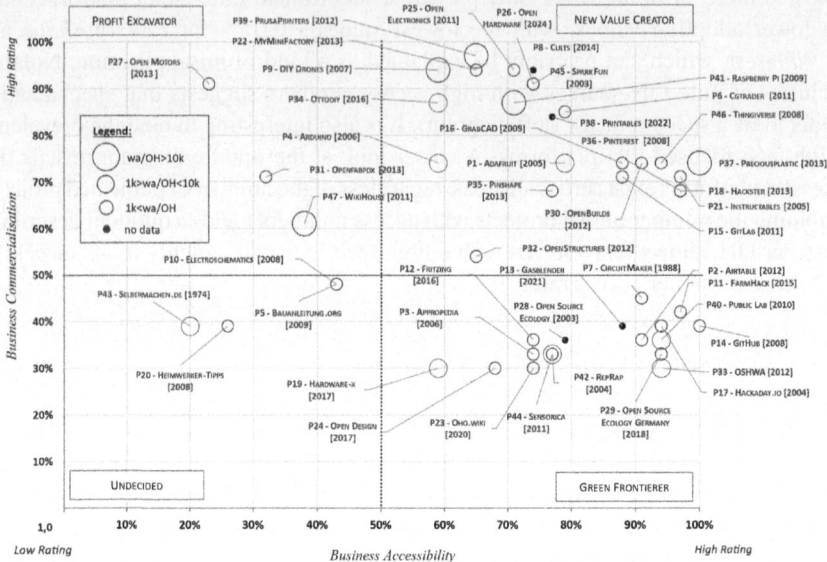

**Fig. 3.** Platform Map - Rating and categories of business orientation based on the analysis.

## 6  Critical Analysis and Outlook

The analysis shows that OH is increasingly developing into a broader movement. In the first part, an analysis of sources for publishing OH was presented. The sources often have different technology mixes, showing that high-tech publications generate more accesses. In addition, classes were introduced to differentiate between sources of OH; here a basic distinction could be made between *Single* and *Stack Publication Form.* Furthermore four descriptive categories of business orientation of the sources where introduced. Moreover, the proportion of OH projects suggests that many platforms were not originally designed for collaborative product development of physical artifacts and are often used inappropriately. The approach points out some challenges in data collection, such as the lack of confirmation of all data by platform operators and the lack of filtering functions to identify all OH projects on a source. Aggregator websites such as OHO Wiki lead to problems of data redundancy. Another problem is the use of search strings, suggesting that an unknown number of OH exists. Future research should focus on assessing the quality of OH projects answering the question if successful platforms correlate with high-quality OH projects. In addition, the business models should be further investigated.

# References

1. Larsson, M.: Circular BM: Developing a Sustainable Future. Palgrave Macmillan (2018)
2. Mazur-Wierzbicka, E.: Towards circular economy—a comparative analysis of the countries of the European union. Resources **10**, 49 (2021)
3. Lieder, M., Asif, F.M.A., Rashid, A., et al.: Towards circular economy implementation in manufacturing systems using a multi-method simulation approach to link design and business strategy. Int. J. Adv. Manuf. Technol. **93**, 1953–1970 (2017)
4. Balka, K.: Open Source Product Development. Gabler, Wiesbaden (2011)
5. (2022) Open design manifesto. https://opendesignmanifesto.org/. Accessed 01 Mar 2023
6. DIN e.V. DIN SPEC 3105-1:2020-07. Open Source Hardware_-Teil_1: Anforderungen an die technische Dokumentation
7. Saubke, D., Krenz, P., Redlich, T.: Product Development through Co-Creation Communities - General Measures for a Distributed and Agile Planning Preparation in Cross-Company Production. Hannover: Publishing (2023)
8. Kadner, S., Kobus, J., Hansen, E.G., et al.: Circular Economy Roadmap for Germany. acatech - Deutsche Akademie der Technikwissenschaften (2021)
9. Bundesumweltministeriums. Entwurf einer Nationalen Kreislaufwirtschaftsstra-tegie (NKWS) (2024). https://www.bmuv.de/DL3288. Accessed 31 Aug 2024
10. Antoniou, R., Bonvoisin, J., Hsing, P.-Y., et al.: Defining success in open source hard-ware development projects: a survey of practitioners. Des. Sci. **8** (2022)
11. Open Source Initiative. History of the OSI (2006). https://opensource.org/history/. Accessed 14 Apr 2023
12. Mies, R., Bonvoisin, J., Stark, R.: Development of OSH in online communities: investigating requirements for groupware. In: Proceedings of the Design Society: DESIGN Conference, vol. 1, pp. 997–1006 (2020)
13. Bonvoisin, J., Mies, R., Boujut, J.-F., et al.: What is the "source" of open source hardware? J. Open Hardw. **1** (2017)

# Wood-Based Components to Reduce the Carbon Footprint in the Body-in-White Production

Jannis Heise[1]([⊠]), Stefan Böhm[1], Nils Ratsch[1], Martin Kahlmeyer[1], Andreas Winkel[1], Moira Burnett[2], and Dirk Berthold[2]

[1] Department for Cutting and Joining Manufacturing Processes, University of Kassel, Kurt-Wolters-Str. 3, 34125 Kassel, Germany
j.heise@uni-kassel.de

[2] Technology for Wood and Natural Fiber-Based Materials, Fraunhofer Institute for Wood Research, Riedenkamp 3, 38018 Braunschweig, Germany

**Abstract.** Private transportation will continue to be car-based in the next one to two decades, especially in the global North. Even if vehicles with electric or hydrogen-based drives have an even smaller carbon footprint, the carbon footprint of vehicle body-in-white has not yet been optimized. The publication describes how a large number of projects have been carried out over the last 10 years to integrate wood-based components, which even have a negative carbon footprint, into body-in-white construction. First, the properties of possible wood-based components were characterized and the structure of the components was optimized to meet vehicle requirements. The process chain in vehicle body-in-white construction was then examined and the extent to which wood-based components can be integrated here was investigated. The current status of work on wood-based car body components with advantages and disadvantages is presented and it is shown which challenges still have to be overcome to be able to produce vehicle bodies in a significantly more environmentally friendly way in the future.

**Keywords:** automotive · body in white · cathodic dip coating · hybrid wood composites · adhesive · production

## 1 Introduction

Recently, the potential of wood and hybrid wood composites has been rediscovered, leading to several publicly funded research projects involving automotive Original Equipment Manufacturers (OEMs). Notable projects include HAMMER and For(s)tschritt (funded by BMBF/BMEL, Germany) and WoodC.A.R (funded by FFG, Austria). These initiatives aim to exploit and combine the unique properties of wood (e. g. high specific strength, good fatigue behavior [1]) to develop innovative hybrid wood composites for use in car body shells. Through these projects, it has been demonstrated that it is possible to tailor specific characteristics, such as bending strength and impact resistance, to meet the demands of automotive applications. However, significant steps are still required to fully qualify these new wood composites for widespread use in vehicle structures, addressing challenges related to durability, performance, and integration into existing manufacturing processes [2–5].

© The Author(s) 2025
H. Kohl et al. (Eds.): GCSM 2024, LNME, pp. 184–191, 2025.
https://doi.org/10.1007/978-3-031-93891-7_21

The painting process in the automotive industry is a multi-stage process that combines various pre-treatment methods (e.g. cleaning, passivating, phosphating) with electrochemical dipping baths and robot-guided spraying to apply several layers of paint [6]. However, while the spray coating of wood-based materials is a common process in furniture manufacturing, construction, and automotive interiors, the effects of the aqueous dip coating baths and the following rapid oven drying/curing on wood-based components (e.g. swelling, shrinkage) in automotive body shell construction, have not been researched. Electric dip coating uses electric current to deposit paint particles on the car body shell which is usually designed as a cathode (CDC/CDP). The automotive industry uses the dip coating (DC) process for the first layers of coating to achieve complete corrosion protection and to ensure a uniform coating quality of metal surfaces. This well-established process is crucial for the longevity and appearance of vehicle parts. During DC the substrate is immersed in a water-based dip coating. The car body undergoes several meticulous steps: it is first alkaline degreased (pH > 10–11), then rinsed with demineralized water, immersed in an activation solution, treated with a phosphoric acid solution, and finally rinsed again with demineralized water. During the DC bath, the body is exposed to a mixture of pigments, binder, water, and acid. After these steps, the car body is subjected to high temperatures ranging from 150 °C to 180 °C, which results in the curing of the coating. This step ensures excellent corrosion protection and imparts additional chemical and mechanical properties to the coated surface [7, 8].

The DC process is a standard and essential procedure for any composite material used in the bodywork sector, ensuring that all components meet stringent quality and durability standards. This process must also theoretically apply to hybrid wood-based materials used in car body shells to ensure they meet the same rigorous standards as traditional materials. The integration of wood-based components into the DC process presents unique challenges but also offers the potential for significant environmental benefits, aligning with the automotive industry's goals of reducing the carbon footprint and enhancing sustainability. By overcoming these challenges, wood and hybrid wood composites could revolutionize vehicle manufacturing, leading to more environmentally friendly and sustainable automotive solutions in the future.

## 2   Materials and Experimental Set-Up

### 2.1   Wood-Metal-Hybrid

This paper aims to demonstrate the fundamental capability of integrating wood-based material systems into the body-in-white without significant adverse effects on mechanical performance. For this purpose, 5-layer beech veneer (1.5 mm) plywood was partly combined with an outer layer of aluminum sheets (AlMg3/EN AW-5754, 1.0 mm). In the first step, the wood-wood and wood-aluminum bonds were prepared with the same adhesive, using 2 one-component polyurethane (PUR) adhesives and a one-component emulsion polymer (EPI) system. Furthermore, the bonding of wood and aluminum using a two-component PUR (2C-PUR) including an activator and black primer was also considered.

The plywood consisted of 5 veneer sheets in a locked layer orientation (0°/90°) to ensure a quasi-isotropic behavior of the main material. The grammage of the four adhesive layers was 200 g/m$^2$ and the veneer boards were produced by hot pressing at 120 °C for 8 min at a pressure of 1 N/mm$^2$. The aluminum-wood bonds were subsequently realized by pressing for 5 min at 80 °C and 1 N/mm$^2$.

## 2.2  Dip Coating Process

This work subdivides into the DC pre-treatment process (DCP), the DC main process (DCM), and a combination of these two (DCC). For all baths, the test specimens are clamped in such a way that floating or sinking is avoided and extensive exposure to the corresponding liquid is ensured. The CDP process consists of a dip degreasing in alkaline detergent (pH≈10) for 10 min at 60 °C, rinsing with potable water, dip phosphating for 3 min at 50 °C and then double rinsing with potable and demineralized water. Oil-free compressed air is used for superficial drying. The actual dip coating in the following DCM process takes place for 5 min at room temperature (RT) and is followed by ultrafiltration (production of rinsing liquid through filtration) for 10 min at RT. Finally, the unclamped test specimens are immediately oven-dried with circulating air for 25 min at 200 °C.

## 2.3  Material Characterization

A universal testing machine was used to carry out 3- and 4-point bending tests following DIN EN 310 and DIN 53293, respectively, and tensile tests following DIN 52377. The tensile strength ($R_m$) is calculated using the formula:

$$R_m = \frac{F_{max}}{a * b_2} \tag{1}$$

The bending strength ($f_m$) was determined according to the following formula:

$$f_m = \frac{3 * F_{max} * l_1}{2b * t^2} \tag{2}$$

Dynamic differential scanning calorimetry tests (DSC) were performed under isothermal conditions at a temperature of 20 °C.

# 3  Results and Discussion

## 3.1  Influence of the CDP Process Fluids and Temperatures on Beech Veneer Plywood

Test specimens for 3-point bending according to DIN EN 310 with the adapted dimensions 150*24*7.5 mm$^3$ (length/width/height) and tensile test specimens according to DIN 52377 were used. The two external plywood layers of the timber composite have a fiber orientation longitudinal to the load direction. As described in Sect. 2.2, destructive tests were carried out on CDP-, CDM-, and CDC-treated, as well as untreated reference specimens, while the CD variants were tested after re-conditioning.

The following Fig. 1 (l) shows the tensile strengths for the different test series. The gray samples represent the unconditional reference bonds with the PUR-A, the red samples represent the results after the single considered DCP, and the blue samples describe samples that were only loaded with the DCM. Finally, the green samples represent those samples that have undergone the combined DCC.

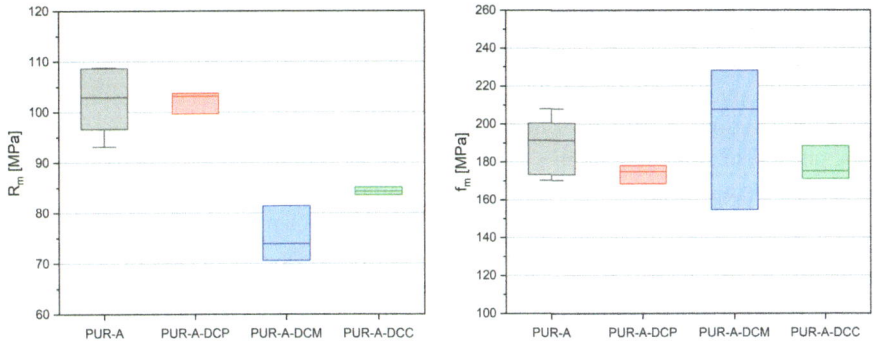

**Fig. 1.** Results of the tensile tests (l) and 3-point bending tests (r) for beech veneer plywood after varying treatment processes.

Concerning the pure tensile load, no change in the strength after DCP (102.23 MPa) can be observed in relation to the arithmetic mean value of the reference (102.17 MPa). Instead, a reduction in the standard deviation can be observed ($-70\%$). Significant losses with accompanying value scattering on the other hand can be attested in the case of a stand-alone treatment with the DCM. At only 75.35 MPa, the values are around 26.3% below those of the reference. The combined DCC load in turn results again in a higher tensile strength value (84.42 MPa).

Figure 1 (r) depicts the results of the 3-point bending experiments with the color scheme described above. In the case of bending strength, which represents a more complex load situation with tensile and compressive areas compared to pure tensile loading, no significant influence of the CD process sections was observed. At 189.11 MPa (Reference), 173.84 MPa (DCP), 196.85 MPa (DCM), and 178.29 MPa (DCC), the values are roughly the same. An increased standard deviation can be observed only in the case of DCM.

### 3.2 Improvement of the CDP Suitability Through Isocyanate Impregnation for Wood-Metal-Compounds (WMC)

As described in 2.2, CDP includes expansion due to moisture and temperature changes and is therefore a challenge for beech wood which is generally sensitive to shrinkage and swelling [9]. In the tests carried out on laminated beech plywood, no significant influence on the mechanical performance could be observed for DCP and DCC, especially regarding $f_m$ when reconditioned. However, vehicle structures are multi-material systems that also involve joining operations between dissimilar materials. The different thermal expansion coefficients ($\alpha$) of the joining partners can lead to critical stress states during temperature changes in operation, the so-called $\Delta\alpha$ problem. While the typical temperature range for the use of vehicles is $-20$ to $+80$ °C, a temperature increase of 180 °C occurs in automotive body shell construction in the oven lines as described above. For this reason, experiments were carried out with a wood-aluminum composite, as the two substrates clearly possess different $\alpha$ values: $\alpha_{Al} \approx 23, 1*10^{-6}K^{-1}$; $\alpha_{plywood} \approx 3, 6 - 4, 7*10^{-6}K^{-1}$ [10]. Furthermore, the selection of adhesives was expanded to include another PUR and an EPI adhesive (as mentioned in Sect. 2.1). As a reduction in tensile strength was particularly observed with DCM-treated samples under tensile load in the previous tests, a combination of these treatments resulted in a less pronounced reduction in tensile strength, a modified DCP+ process is used in the following observations. DCP+ combines immersion at a controlled temperature (55 °C) in a zinc phosphate emulsion for an extended period of 30 min and drying of the samples in a circulating air oven at 200 °C for another 30 min.

Figure 2 (l) shows the tensile strength values of specimen prepared with the adhesives EPI/PUR-A/PUR-B. The gray bars describe the reference specimens without DCP+ treatment, the DCP+ specimens after reconditioning are shown in red. Results regarding specimens that were impregnated with an isocyanate finish in a 10-min vacuum infusion process (i) are labelled blue. This variant was also subjected to the DCP+ process, also shown in the chart (green).

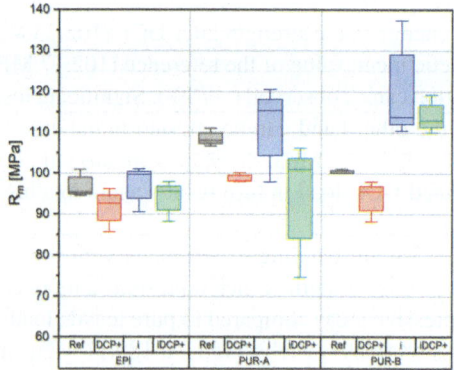

**Fig. 2.** Results of the tensile tests for WMC after adapted DC-process (l) and failure in the wooden substrate (r).

All the tensile test specimens failed in the wood substrate (Fig. 2, r). For non-impregnated samples, the CDP+ treatment had a negative influence on the tensile strength. Concerning the choice of adhesive, it was found that EPI is less suitable for joining the plywood-metal composite than the PUR adhesives under consideration. In some samples, delamination between wood and metal can be observed even before testing. Impregnation of the EPI-bonded test specimens showed no significant improvement in the general strength or CDP suitability of the bond. The situation is different with PUR composite bonds: The impregnation leads to an increase in stiffness, whereby higher strength values than in the reference can be achieved, at the same time the standard deviation increases drastically. After the DCP+ run, there is no decrease in the average strength of PUR-B; in particular, all sample values are above the original reference. The effect is less pronounced for PUR-A but is noticeable for some samples. The wide standard deviation is mainly due to partial pre-failure between aluminum and wood substrate.

Using the same selection of sample series, the bending behavior of the WMC (4-point bending configuration, wood on top) was also analyzed as shown in Fig. 3. By applying the force via two pressure points, a constant bending moment is realized in between these points. The color coding follows the logic in Fig. 2.

Similar to the behavior of bare wood specimens in the 3-point bending test, no statistically relevant negative effect of DCP+ treatment can be noted for any configuration in the 4-point bending of WMC. The non-impregnated PUR-A and PUR-B joined test specimens, however, show a slightly increased force absorption, while the isocyanate-impregnated EPI samples even withstand 11.3% higher forces after the CDP+ run on arithmetic mean before bond failure. In general, impregnation of the EPI samples allows the composite to be strengthened almost to the level of the two PUR adhesives with regard to this load situation; after DCP+ treatment, the values are even slightly better than those of PUR-A.

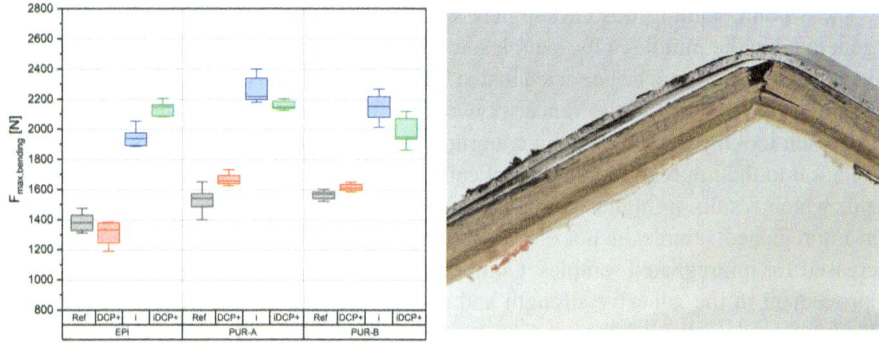

**Fig. 3.** Results of the 4-point bending tests for WMC after the adapted DC process.

### 3.3  Adaption of the Adhesive Selection to Counter the $\Delta\alpha$-Challenge in the Area of the Aluminum-Wood Connection

As already described for the WMC tensile tests, delamination between wood and aluminum in certain samples occurs even after minimal loads or no loading at all. In view of the joining process at RT, 30 min immersion bathing at 55 °C, and subsequent oven drying at 200 °C, including reconditioning to RT, the adhesives used for wood-based materials do not seem to be suitable. A flexible, structural two-component PUR adhesive is therefore selected which can better handle complex stress situations (e.g. $\Delta\alpha$ problem) in the seam. To ensure that the production of the veneer plywood and bonding of the aluminum sheet does not significantly extend the overall process, the PUR-B system is designed for a short curing time, as shown in the isothermal DSC measurement at RT in Fig. 4 (l). The adapted WPC was tested in a 3-point bending test as shown below. In Fig. 4 (r), the color coding is as follows - gray: reference without DCP+, red: DCP+, blue: impregnated, green: impregnated after DCP+.

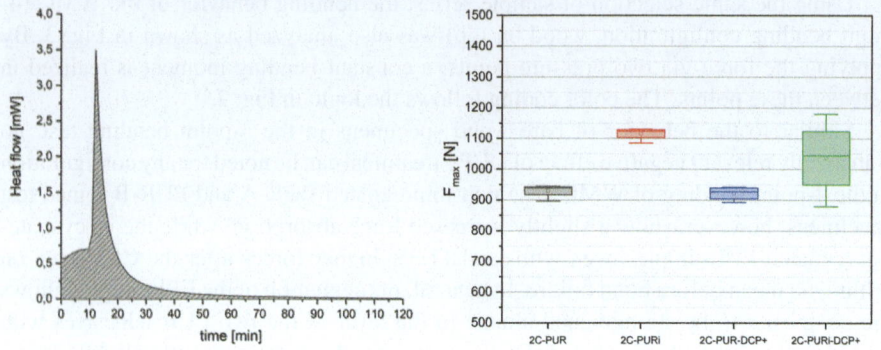

**Fig. 4.** Curing time at 20 °C and results of the 3-point bending of the adapted WMC.

The 3-point bending was chosen here so that the largest possible bending moment is generated in the middle of the sample and the behavior of the adhesive bond between wood and aluminum can be better evaluated. In this first analysis, the bond was produced with the same pressure as the veneer plywood before. Thus, the adhesive joint is very thin. When looking at the results, a congruent picture to the previous tests can be seen in relation to the impregnation: the application of treatment leads to a stiffening of the bond, which results in higher strength values. An influence of the DCP+ process on non-impregnated samples is not observed, while the standard deviation after DCP+ is increased for impregnated samples. Compared to the previous composite, a significant improvement in the adhesive strength and adhesion of the 2C-PUR on the aluminum substrate is evident. While some of the previous wood-metal bonds showed delamination before testing, no early bond failure was observed here. The samples allow a high elongation before the shear forces cause a failure in the 2C-PUR adhesive joint. A mixed fracture image with adhesive and cohesive components can be seen here.

## 4   Summary and Outlook

Wood-based materials offer great potential for technological applications and have an unique ecological potential when grown sustainably. When using wood or wood-based materials in the car body, these components have to undergo the manufacturing process without being damaged. This includes in particular the painting process which comprises several temperature-controlled dipping baths and drying sections. In this work, it was shown that the mechanical performance of wood or WMC decrease due to DC if no additional protection is provided. By impregnating the WMC with isocyanate-containing resin, the stiffness of the composite and its suitability for CDC are improved. By exclusively impregnating layers close to the surface in a minimally invasive process, the lightweight and sustainability factor of the wood structure remains unaffected.

Furthermore, residual stresses initiated by the production process can also be equalized with an adapted bond between the wood and metal sheet. In future work, the connection between wood and metal needs to be analyzed in more detail, in particular the influence of differing layer thicknesses on the mechanical and elastic behavior. Vehicles are exposed to cyclical loads during use and highly dynamic loads in the event of a crash. The influence of such load cases on the WMC must be investigated in further studies.

## References

1. Henning, F., Weidemann, K., Bader, B.: Hybride Werkstoffverbunde. In: Handbuch Leichtbau, pp. 413–428 (2011)
2. Kohl, D., Link, P., Böhm, S.: Wood as a technical material for structural vehicle components. Procedia CIRP **40**, 557–561 (2016)
3. Kohl, D., Long, T.H.N., Böhm, S.: Wood-based multi-material systems for technical applications –compatibility of wood from emerging and developing countries. Procedia Manuf. **8**, 611–618 (2017). https://doi.org/10.1016/j.promfg.2017.02.078
4. Käse, D.B., et al.: Potential for use of veneer-based multi-material systems in vehicle structures. Key Eng. Mater. **809**, 633–638 (2019). https://doi.org/10.4028/www.scientific.net/KEM.809.633
5. Domitner, J., Silvayeh, Z., Predan, J., Graf, E., Krenke, R., Gubeljak, N.: Mechanical performance and failure behavior of screw-bonded joints of aluminum sheets and cross-laminated birch veneer plates. Eng. Failure Anal. **146**, 107074 (2023)
6. Porsche Leipzig GmbH. https://www.porsche-leipzig.com/produktion/lackiererei. Accessed 02 July 2024
7. Brock, T., Groteklaes, M., Mischke, P., Strehmel, B.: Lehrbuch der Lacktechnologie, 5th edn. Vincentz Network, Hannover (2017)
8. Bayerische Motoren Werke AG: Process specification 98028 (2012)
9. Bodig, J., Jayne, B.: Mechanics of Wood and Wood Composites. Reprint Edition, Krieger Publishing, Florida (1993)
10. Sonderegger, W., Niemz, P.: Investigation of swelling and thermal expansion of fibreboard, particleboard and plywood. Eur. J. Wood Wood Products **64**(1), 11–20 (2006)

# A Service-Based Approach for Predicting the Carbon Footprint in the Supply Chain of Plastic Injection-Moulded Parts During Product Development

Lukas Nagel[1]([⊠]) [iD], Rainer Gerstbauer[2] [iD], Levon Harutyunyan[2] [iD], Thomas Trautner[2] [iD], Friedrich Bleicher[2] [iD], and Matthias Weigold[1] [iD]

[1] Institute for Production Management, Technology and Machine Tools (PTW), Technical University of Darmstadt, Otto-Berndt-Str. 2, 64287 Darmstadt, Germany
L.Nagel@PTW.TU-Darmstadt.de
[2] Institute of Production Engineering and Photonic Technologies (IFT), TU Wien, Getreidemarkt 9, 1060 Vienna, Austria

**Abstract.** Driven by the European Union's commitment to achieving climate neutrality by 2050 and reducing EU emissions by 55% relative to 1990 levels by 2030, a critical assessment of manufacturing processes for their environmental impacts is necessary. During the strategic planning stage of product development and manufacturing, limited data is available to predict the environmental impacts across supply chains. However, decisions made at this phase significantly influence these impacts. This paper presents a data-provision method that aims to support decision-making in product development with quantifiable metrics of ecological sustainability. This is achieved through predicting carbon footprints, while leveraging the European data infrastructure Gaia-X, which enables secure service offerings and data sharing across the supply chain. The prediction service is implemented as a web application that facilitates the input and variation of hypothetical production scenarios. In a specific use-case, the method is applied to the supply chain of an injection-moulded cup. This concept offers a vision for carbon-optimized product development, enabling significant carbon avoidance in later production stages and showcasing the influence of different production parameters on the supply-chain carbon footprint.

**Keywords:** $CO_2$ emission · sustainable development · data spaces

## 1 Motivation

The discussion on sustainability strategies is closely linked to the concept of planetary boundaries, defining ecological limits within which human activities do not endanger Earth's stability. Nonetheless, six out of the nine boundaries have already been crossed [1]. Policymakers and societal actors strive to slow down climate change by adhering to these boundaries. In this context, the development of effective $CO_2$ emission reduction

H. Kohl et al. (Eds.): GCSM 2024, LNME, pp. 192–199, 2025.
https://doi.org/10.1007/978-3-031-93891-7_22

strategies is becoming increasingly important. The European Union (EU) has committed to achieving climate neutrality by 2050 with the EU Green Deal and to reducing EU emissions by 55% by 2030 compared to 1990 levels [2]. These ambitious goals necessitate a comprehensive assessment of the environmental impacts of industrial processes, as they can significantly contribute to achieving these targets. In Germany in the year 2023, 24% [3] of total domestic emissions, amounting to 144 million tons of $CO_2$ equivalents, are attributed to the industrial sector.

Design and manufacturing are essential elements in the engineering process of production. The design stage can affect for up to 80% of a product's environmental impact [4]. However, this phase is characterized by limited availability and quality of data to predict future environmental impacts. Consequently, understanding and managing emissions becomes particularly challenging. The highest emissions for most companies in the manufacturing industry are found in Scope 3 (indirect emissions), surpassing emissions in Scope 1 (direct emissions) and Scope 2 (indirect emissions) [5]. At the same time, this area presents great complexity in emission calculations, posing challenges for companies in their accounting processes [6].

The following sections outline future possibilities for data-driven sustainable $CO_2$ prediction methods in production. First, the challenges of data transfer for $CO_2$ accounting are highlighted. Then, a possible approach for the structured development of corresponding prediction service within the Gaia-X framework will be presented. Finally, the method will be applied to an injection-moulded component.

## 2 State of the Art

An increasing number of regulatory measures have been introduced in recent years to enforce climate protection agreements and reduce industrial greenhouse gas emissions. The Corporate Sustainability Reporting Directive is a prominent example, compelling companies to produce transparent sustainability reports with detailed information on their environmental [7]. These regulations incentivize companies to adopt sustainable technologies voluntarily. The carbon footprint (CF) is the most widely used metric for measuring environmental impacts. However, determining the CF along entire supply chains is complex due to varying data availability and quality, and the multitude of standards such as the Greenhouse Gas Protocol, DIN EN ISO 140XX, and BSI PAS 2050. Existing standards are interrelated to varying degrees, and their specification during practical application presents problems that lead to inconsistent results [8].

New approaches are needed for precise and consistent CF predictions before a product is manufactured. These approaches must be flexible enough to cover the diversity of production processes and supply chains while ensuring the protection of sensitive data. Currently, comprehensive prediction tools are lacking, representing a significant research gap in the field of sustainability strategies. Existing tools for holistic Life Cycle Assessments (LCA), such as GaBi, Umberto and SimaPro typically require an expert level of usability, relying on own databases or external data sources to provide LCA outcomes. However, they are often characterized by high cost, limited availability, and complexity, making them time-consuming and challenging to navigate. In addition, CAD-LCA integrated tools like SOLIDWORKS Sustainability, Autodesk Sustainability, and Siemens NX Sustainability offer user-friendly interfaces, with features such as energy impact assessment, material recommendations, and in some cases providing real-time sustainability feedback but still rely on secondary data from various databases. Data confidentiality remains a significant hurdle, as companies often consider detailed information about production processes and supply chains competitively sensitive [9]. This leads to limited transparency and hinders the effective data transfer necessary for accurate CF analysis.

The Gaia-X initiative addresses these issues by providing a framework for secure and trusted data sharing, emphasizing data sovereignty and transparency. Gaia-X data spaces facilitate interoperability and secure data exchange, thereby enhancing CF predictions accuracy and fostering collaboration to reduce greenhouse gas emissions. [10].

## 3    Methodology

This section addresses the use case "$CO_2$ Footprint in Product Development" within the context of the EuProGigant project [11]. The presented results stem from interdisciplinary project work involving domain experts in polymer processing, data scientists, and software developers.

Hoffmann et al. [12] have already elucidated the underlying business models and provided a hypothetical outlook for the subsequent content. Figure 1 illustrates the three-stage approach. At the most advanced stage, stage 3, the objective is to generate a $CO_2$ optimized product design based on a comprehensive set of requirements. This involves integrating multiple factors to achieve the most environmentally efficient design possible. In stage 2, the value proposition is limited to compare different types of manufacturing processes, enabling a broader evaluation of potential environmental impacts. In stage 1, the service examines a predefined part design and calculates the CF for various production scenarios. This provides a foundation for optimizations within a specific material group, machine, mould design, and process. This approach is implemented in the CF prediction service described in this paper.

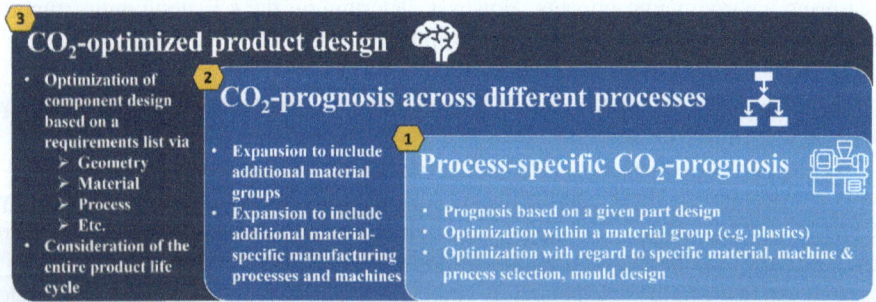

**Fig. 1.** Development stages of the business model "$CO_2$ footprint in production creation" according to [12]

To achieve this, a CF prediction service for the injection moulding industry was developed, integrating data collection, multiple calculation logics, and a user-friendly web application. To establish the prediction service, the current state of the process chain in the injection moulding domain was first mapped in detail to understand the information flows between all stakeholders within the ecosystem, including suppliers, manufacturers, and customers. By comprehensively mapping these information flows based on current standards and norms to provide added value for future regulatory processes, critical data points and interactions necessary for accurate CF predictions were identified. This approach ensures that the predictions are not only accurate but also timely, providing valuable insights during the initial stages of product development. By integrating these elements into a cohesive system, the authors created an early-stage tool that supports sustainability efforts in the injection moulding industry. To validate this methodology, a case study was conducted on an injection-moulded product. This validation underscores the potential of the prediction service to drive significant environmental improvements in the injection moulding industry.

## 4 Application and Discussion

### 4.1 Prediction Service as a Method

The prediction service envisions three possible levels of precision. In the first level, users receive a semi-automated estimate of the emission load for the corresponding emission areas in the value chain. In the subsequent levels, the degree of data availability and communication effort increases significantly, thereby also substantially extending the calculation time. At this point, the paper only considers the first level of precision, which focuses on the minimal amount of data necessary for a sufficient CF prediction, aimed at identifying areas with high reduction potential at an early stage.

To achieve this, users must initially provide the data outlined below, which can be entered through the interface of the web application for a potential production scenario. Based on this data, the data transfer and query among the involved entities in the Manufacturing Data Space are initiated. The web application receives individual CFs and energy consumption calculations, then computes the overall CF for the specified scenario (see Fig. 2).

To predict the CF of the injection mould, the tool manufacturing company Haidlmair developed a calculation logic that estimates the mould's weight within a product category using user-provided parameters, such as product dimensions, the number of cavities in the tool, and the wall thickness of the product. Based on the estimated weight of the mould, the CF can be predicted by utilizing initial CF data for the raw material supplied by steel supplier voestalpine. The injection moulding machine manufacturer Arburg provides data for selected machines based on user-supplied information, including the product carbon footprint of the machines and the predicted energy consumption per hour, encompassing necessary auxiliary units during the injection moulding process. Arburg supplies energy consumption values based on the specific injection moulding machine and the type of plastic granule used, which are utilized in the prediction logic. Additionally, Arburg offers a logic within the web application that provides users with direct feedback, identifying suitable injection moulding machines for producing the entered production scenario. This is achieved by analyzing machine capabilities and comparing them with the product volume, the number of cavities in the mould, and the required clamping force. For both Haidlmair's and Arburg's prediction logic, specific process data tailored to the user's production scenario is required. This data is provided through a mould filling simulation conducted by Simcon. Using the CAD model and material data from the plastic granule supplied by Covestro, the simulation software can provide relevant data such as injection pressure and melting temperature specific to the production scenario. Finally, the user receives the CF of the injection moulded product for the specific production scenario as a result within the web application. Users can add individual $CO_2$ data from their own company to obtain a comprehensive $CO_2$ analysis, thereby establishing a foundation for sustainability reporting. This enables users to modify various parameters or adjust the CAD model to generate comparative predictions, allowing them to select the optimal production scenario. The scalability of the approach is dependent on the standardization of data interfaces and formats to ensure seamless integration with various manufacturing processes and machines.

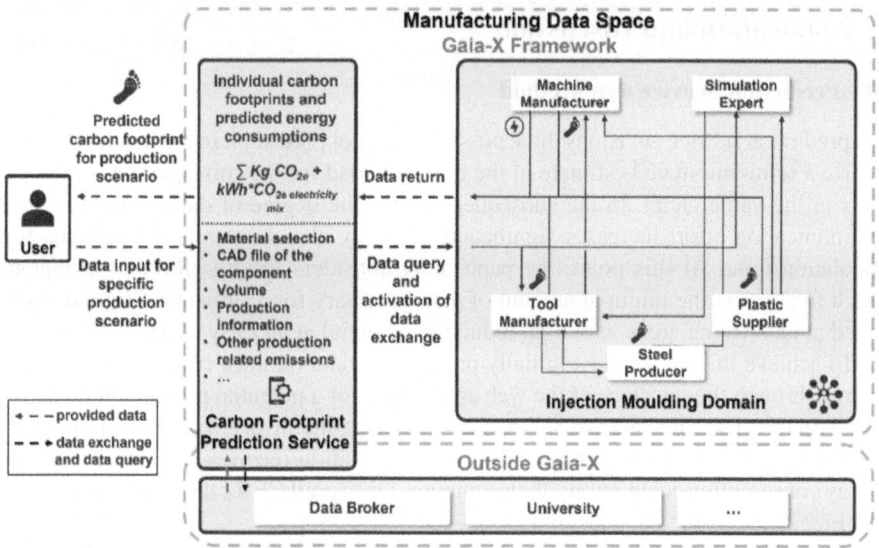

**Fig. 2.** Overview and data exchange logic of the prediction service within the Manufacturing Data Space and Gaia-X Framework.

Challenges during implementation arise from the complexity within the injection moulding domain. Predicting the cradle-to-gate CF of an injection moulded product requires numerous data points from various companies along the supply chain. Consequently, companies must have the appropriate infrastructure to collect, store, and exchange relevant data. A Manufacturing Data Space facilitates the sharing of services and data among all entities within the injection moulding domain. Through interfaces, data is made available via cloud solutions to transmit the defined data between companies and the developed web application. The realization is achieved through the EuProGigant-Portal, specifically provided by A1 Digital using the Exoscale cloud infrastructure. This approach allows other companies to access and utilize the data upon completion of a data usage agreement, while the data remains within the IT system environment of the respective owners until then. When data is made granular and selectively accessible in compliance with data protection, security regulations, and other applicable laws, it enables orchestrated data exchange that can drive additional value creation. This approach opens potential synergies and shared benefits for various stakeholders within the ecosystem. The Manufacturing Data Space is based on the Gaia-X framework, which not only ensures secure and compliant data sharing but also facilitates future integration with other data spaces using Gaia-X standards. [10].

Beyond the Gaia-X framework, the prediction service can communicate with additional entities, such as data brokers for obtaining $CO_2$ factors from public databases or universities that can provide or access research data. This extended communication capability enhances the robustness and utility of the prediction service, allowing for more comprehensive and accurate environmental impact assessments.

### 4.2 Calculation Example

To align with industry needs and to validate the method, a case study was developed. A company, in this case study, Arburg, aims to produce a specific number of injection-moulded products within a certain timeframe. For this case study, an injection-moulded cup was chosen due to its widespread presence in the plastics industry, making it ideal for evaluating the CF along the supply chain. Employees from the product development department, who represent the target user group, initially determine various production parameters and provide the CAD model of the product. These inputs are entered into the interface of the web application. The backend initiates the data transfer within the injection moulding domain to predict the CF for the specific production scenario. As mentioned, different standards dictate what to include in the analysis. To ensure consistency, the Pathfinder Framework, based on the Greenhouse Gas Protocol, was chosen [13]. This framework provides guidelines for determining carbon emissions, ensuring that all companies along the supply chain follow a unified method for calculating and reporting emissions. This promotes transparency and accurate comparisons across different entities.

Considering a production scenario of 100,000 cups made from a blend of Polycarbonate (PC) and Acrylonitrile-butadiene-styrene (ABS) granule, with CF data sourced from the ecoinvent v3.7.1 database. An Arburg 370 A 600 - 170 Comfort injection moulding machine was selected as suitable for the cups through Arburg's calculation logic. The machine is powered by electricity with an emission factor of 350 g $CO_2$e/kWh and equipped with a one-cavity injection moulding tool. The composition of the cradle-to-gate CF of one cup is displayed in Fig. 3.

The cradle-to-gate CF analysis reveals that the primary contributors to total carbon emissions are the plastic granule, accounting for 88.36%, and the energy consumed during the injection moulding process, accounting for 9.04%. In contrast, the CF of the injection moulding machine, the injection mould and the transport of the injection mould and the granule to Arburg contribute only 0.29%, 1.86% and 0,45%. Initial reduction measures could include the use of alternative plastic granules, such as those with a higher amount of recycled material, or energy efficiency measures like heat recovery during the injection moulding process.

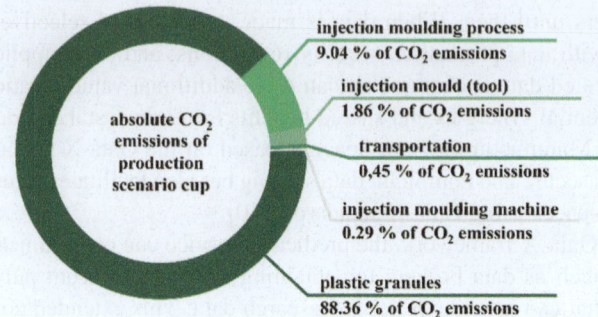

injection moulding process
9.04 % of CO₂ emissions

injection mould (tool)
1.86 % of CO₂ emissions

absolute CO₂
emissions of
production
scenario cup

transportation
0,45 % of CO₂ emissions

injection moulding machine
0.29 % of CO₂ emissions

plastic granules
88.36 % of CO₂ emissions

**Fig. 3.** Calculation of the cradle to gate carbon footprint for the production scenario of a cup

Additionally, carbon emissions from the factory at Arburg, which do not directly arise from the production process of the plastic products (e.g. emissions resulting from the heating of the factory), were not considered in the conducted case study. However, in the web application, there is the possibility for users to enter this information for their specific production scenario to be included in the final CF of the plastic product.

## 5  Conclusion and Final Remarks

This paper presents a methodology for predicting the CF of products during the early stages of development, particularly focusing on the injection moulding industry. By integrating data collection, sophisticated calculation logics, and a user-friendly web application, this approach allows stakeholders to make informed decisions aimed at minimizing environmental impacts. Simultaneously, the developed prediction service aligns with the EU's ambitious climate goals by offering a practical tool for industries to proactively manage and reduce their CFs. SMEs may face initial challenges with CF prediction services due to setup costs, but these can be offset by long-term $CO_2$ savings and resource efficiency. The approach helps suppliers enhance market positioning through sustainability compliance and offers potential revenue from data contributions. Global standards within Gaia-X ensure easier implementation, benefiting even non-European SMEs. The use of primary data from supply chain partners, combined with scenario comparison, provides a robust foundation for optimizing production processes for ecological sustainability. The case study on an injection-moulded cup reveals that the primary sources of carbon emissions are the plastic granules and energy consumption during the injection moulding process. By leveraging the Gaia-X framework with a Manufacturing Data Space for secure data sharing, this approach also addresses the critical issue of data sensitivity, ensuring that companies can collaborate effectively without compromising proprietary information. Furthermore, a feedback system planned for stage 2 will ensure the quality of decisions by providing validation and adjustments. Future research and development can expand this methodology to other manufacturing processes and refine the prediction algorithms to enhance accuracy and applicability across diverse industrial contexts. Furthermore, the project aims to achieve the next stages of the underlying business models.

*The presented work is supported by the Austrian Research Promotion Agency and the German Federal Ministry for Economic Affairs and Energy.*

# References

1. Richardson, J., et al.: Earth beyond six of nine planetary boundaries. Sci. Adv. **9**(37) (2023)
2. European Commission. https://climate.ec.europa.eu/eu-action/climate-strategies-targets/2050-long-term-strategy_en. Accessed 20 June 2024
3. Statista Homepage, Distribution of $CO_2$ emissions in Germany 2023, by source. https://www.statista.com/statistics/1385874/co2-emissions-source-germany/. Accessed 20 June 2024
4. Fuchs, S., et al.: Product sustainability: back to the drawing board. https://www.mckinsey.com/capabilities/operations/our-insights/product-sustainability-back-to-the-drawing-board. Accessed 20 June 2024
5. Schmidt, M., Nill, M., Scholz, J.: The relevance of the supply chain for the climate footprint of companies. Chem. Ing. Tec. **93**(11), 1692–1706 (2021)
6. Seyfried, S., Nagel, L., Weigold, M.: Empirical investigation of climate neutrality strategies of companies in industrial production. In: Kohl, H., Seliger, G., Dietrich, F. (eds.) Manufacturing Driving Circular Economy, GCSM 2022, pp. 999–1007. Springer, Cham (2023)
7. European Commission. https://finance.ec.europa.eu/capital-markets-union-and-financial-markets/company-reporting-and-auditing/company-reporting/corporate-sustainability-reporting_en. Accessed 20 June 2024
8. Weyand, A., et al.: Analysis of uncertainty factors in part-specific greenhouse gas accounting. Sustainability, MDPI **15**(24) (2023)
9. Lohmöller, J., et al.: On the need for strong sovereignty in data ecosystems. In: Deco'22@VLDB, pp. 51–63 (2022)
10. Gast, F., et al.: Automatic publication of data to data- and service ecosystems from the shopfloor. In: Procedia CIRP CMS 2024, Portugal. Elsevier (2024)
11. Weber, M., Weigold, M., Koch, T.: The European production giganet: towards a green and digital manufacturing ecosystem. In: Smart and Networked Manufacturing. Conference Proceedings. Wiener Produktionstechnik Kongress, Wien, pp. 95–100. New Academic Press, Wien (2022)
12. Hoffmann, F., et al.: Development of data-based business models to incentivise sustainability in industrial production. In: Proceedings of the Conference on Production Systems and Logistics CPSL, Hannover, pp. 199–209 (2023)
13. WBCSD. https://www.wbcsd.org/resources/pathfinder-framework-version-2-0/. Accessed 20 June 2024

# Decision Support for Lightweight Design and Design for Circularity: A Trade-Off Analysis

Kristian König$^{(\boxtimes)}$ and Michael Vielhaber

Institute of Product Engineering, Saarland University, Campus E2 9, 66123 Saarbrücken, Germany
kristian.koenig@uni-saarland.de

**Abstract.** The need for resource efficiency in sustainable development and environmental responsibility emphasizes the importance of lightweight design and design for circularity in product development. However, the complexity and multidisciplinary nature of decision-making in these areas pose a major challenge. Therefore, the present work proposes a novel approach for trade-off analysis to facilitate decision-making in conceptual design. Illustrated by a case study on the development of a semi-mobile handling system, five fundamental design strategies for decision-making are identified and discussed. This enables a nuanced understanding of resource optimization in the context of complex design considerations and environmental requirements.

**Keywords:** Lightweight Design · Design for Circularity · Trade-Off Analysis · Ecodesign

## 1 Introduction

In response to mounting environmental challenges and resource constraints, the design paradigms 'lightweight design' and 'design for circularity' have emerged as key strategies for resource conservation. Lightweight design aims to reduce the weight of structures, enhancing material efficiency and energy savings [1]. Conversely, the circular economy (operationalized in design for circularity) seeks to extend product lifetimes as well as to narrow and close product and material loops [2]. Integrating these sustainability approaches presents complex decision-making challenges, requiring careful balancing of the benefits of weight reduction and circularity. Effective decision support systems are essential but often face obstacles, such as accurately assessing trade-offs between weight reduction and recyclability, lacking comprehensive frameworks for holistic environmental impact evaluation [3]. Focusing on the nano level of circular economy [4] for comparing products and components, in this paper, the following research question is addressed: 'how can trade-offs between lightweight design and circular economy principles be determined to optimize a product's environmental performance?'.

© The Author(s) 2025
H. Kohl et al. (Eds.): GCSM 2024, LNME, pp. 200–207, 2025.
https://doi.org/10.1007/978-3-031-93891-7_23

## 2 Literature Review

### 2.1 Evaluating Lightweight Design for Environmental Sustainability

Lightweight design plays a pivotal role in enhancing environmental sustainability, particularly in dynamic products such as automobiles, aerospace, trains, or ships [1, 5]. The primary focus of lightweight design lies in reducing energy consumption throughout the dynamic usage phase centering on its potential to mitigate $CO_2$ emissions [6]. However, other life cycle issues such as recyclability, repairability, or the technical lifespan of a product may also play a significant role in comprehensive evaluations [7]. Considering these aspects, lightweight design is not only seen as a means to reduce environmental impact but also as a measure to enhance functionality. Therefore, sustainability assessments should consider not only $CO_2$ reduction potential but also technical robustness and the extent to which functional requirements are met in product development in order to more effectively satisfy customers. A holistic approach thus ensures that lightweight solutions contribute to environmental targets while maintaining or improving product performance, all within acceptable cost limits [8].

### 2.2 Methodologies for Assessing the Circularity of Products

The fundamental principles of circular economy for products and components are rooted in extending product lifespans as well as closing and narrowing product and material loops [2]. Consequently, assessment methodologies for circularity aim to systematically evaluate these principles either qualitatively or quantitatively. As this publication focuses on the nano level of circular economy, an overview of potentially relevant circularity indicators at this level can be found in the articles by De Oliveira et al. [9] and Patil and Ramakrishna [4]. To detail two specific examples, the 'material circularity indicator' [10] focuses on restoring material flows and assesses the degree of linearity or closed-loop nature of material flows within a product, leveraging life cycle assessment data for a straightforward determination. Conversely, the 'resource duration indicator' [11] (also known as the 'longevity indicator') evaluates not only the circularity of a resource but also its duration of active utilization. In essence, circularity indicators necessitate a departure from traditional sustainability metrics associated with the linear economy paradigm.

### 2.3 Decision Making Between Lightweight Design and Design for Circularity

Historically, lightweight design has been predominantly discussed independently of other design for environment strategies. However, within frameworks of circular economy development, lightweight design is often identified as a partial solution to minimize resource consumption [12], as exemplified in the 'resource pressure' method proposed by Desing et al. [13]. Specific methodologies for evaluating circular economy design strategies against each other are sporadic and suffer from methodological deficiencies. A common approach to decision-making and evaluation involves considering the so called 'Ashby Maps' proposed by Michael Ashby [14], comparing performance indicators of lightweighting (e.g., specific strength or stiffness) against environmental impacts (e.g.,

CO2 emissions of primary material production), partly relevant for product circularity. For instance, Allwood et al. [15] assessed the technical lightweight performance (e.g., specific strength or stiffness) against the energy input of primary material production. Additionally, Ferro and Bonollo [16] examined the lightweight potential concerning raw material criticalities using 'criticality indicators', centering on product mass while neglecting environmental impacts throughout the entire product life cycle. Addressing the entire product life cycle represents the gap targeted in the 'functional life cycle energy analysis' by König et al. [3], which, however, exhibits weaknesses in identifying global optimization potentials and trade-offs due to the underlying mechanism in their bar plot [17]. In this context, the subsequently proposed methodology aims to complement these existing approaches addressing the research question.

## 3   Trade-Off Analysis for Supporting Decision Making in Product Development

### 3.1   Assumptions and Distinctions

In this publication, we consider the implementation of a circular economy and lightweight design as two essential strategies for enhancing resource conservation. Therefore, our methodology primarily targets products with dynamic usage phases, which can environmentally benefit from weight reduction. In contrast, realizing a circular economy may be relevant for all types of physical products. Thus, the proposed methodology may also be applied to non-dynamic products, although the identification of optimization potential may be limited. Given this background, we assume that dynamic products predominantly offer lightweighting potentials to minimize environmental impacts during their usage phase (mid-of-life, MoL). In contrast, the potential of implementing a circular economy lies in reducing environmental impacts at the beginning-of-life (BoL, e.g., by consuming secondary materials) as well as at the end-of-life (or end-of-use, EoL, e.g., by implementing a resource value retention option).

Our methodology is intended to be applied at the functional level of product development. Therefore, measurable impacts of physical components are methodically allocated to technical functions, for example, through a pairwise scheme, as these represent the functional unit at the smallest level in terms of performance comparison [3]. This approach evaluates the value of a product and allows questioning the necessity of each function. It also offers the greatest optimization potential, as entirely new solutions may subsequently be found using systematic solution-finding methods [18].

### 3.2   Quantifying and Assessing Trade-Offs

The fundamental objective of this research lies in determining the trade-offs between lightweight design and design for circularity. Therefore, the respective positive environmental benefits (or amount of reduced negative impacts) of both design approaches have to be evaluated. In general, a fundamental principle is that both design approaches can be implemented either in conflict, synergistically or with subordinate influence on each other, as discussed by König et al. [3]. Therefore, to determine potential trade-offs, it is essential to select indicators that enable a decision-making process effectively.

In this publication, we select energy consumption across the three main life cycle stages (BoL, MoL and EoL) as the indicator and measurement criterion for quantifying the environmental effects of both lightweight design and design for circularity, as previously done in an earlier work [3]. We attribute the sum of energy consumptions from the BoL and EoL (regained energy at the EoL stage is credited negatively in the calculation) stages to the circularity of a product (respectively technical functionality). Conversely, for assessing the benefit of lightweight design over the product's usage, we select the energy consumption during the MoL as the indicator.

As energy consumption represent life cycle data and not an impact assessment indicator, it is fundamentally possible to choose indicators other than energy. The rationale for using energy consumption lies in its nature as a physically measurable quantity, which, as part of the life cycle inventory, has many useful implications for a product's environmental impacts (e.g., $CO_2$ emissions) or costs (e.g., less energy consumption may mean lower costs). However, it is not a direct measure of a product's environmental impact and circularity. Probably, it is also more challenging to determine in supply chains compared to costs (direct exchange medium) and the $CO_2$ footprint in future (within the framework of the 'corporate sustainability reporting directive'). Fundamentally, the choice of an indicator is ultimately up to the user of the methodology. The approach to determining trade-offs remains unaffected.

Conversely, particularly the circularity indicators discussed in Sect. 2.2 could be a meaningful choice for assessing the circularity of a product and thus its environmental sustainability. In that case, a different indicator would need to be chosen to evaluate the environmental benefit of implementing lightweight design. Here, potential $CO_2$ savings during the usage phase of a lightweight product design could be considered. However, such an approach quickly becomes enormously challenging, as nearly infinite implications and potential impact shifting to other dimensions of sustainability (social and economic aspects) would need to be considered (and an evaluation of the indicators against each other may get necessary: 'is material circularity more important than $CO_2$ footprint reduction?'). The resulting increase in complexity would then not be proportionate to the added value of the methodology, which is why we refrain from this for the time being. Simplicity and low complexity with goal-oriented insights are symbolic goals for the early phase of product development. The concept of trade-off identification does not fundamentally exclude such considerations and also includes a possible cost analysis in future iterations of methodological adjustments.

### 3.3 Determining Strategies for Reducing Resource Consumption

Considering our assumptions and the selection of indicators, a diagram can be constructed as illustrated in Fig. 1. In our case, we have indexed the energy consumption of the usage phase on the x-axis, representing the environmental potential of lightweight design. On the y-axis, the sum of the energy expenditure of the BoL and EoL stages is plotted as the indicator of the potential for implementing a circular economy. Each technical function is represented as a single point in the diagram, with its total energy consumption calculated by summing the energy expenditures of all three life cycle stages.

There are four possible directions in which a function can shift in terms of energy expenditure in the diagram. Increasing energy consumption without exploiting any

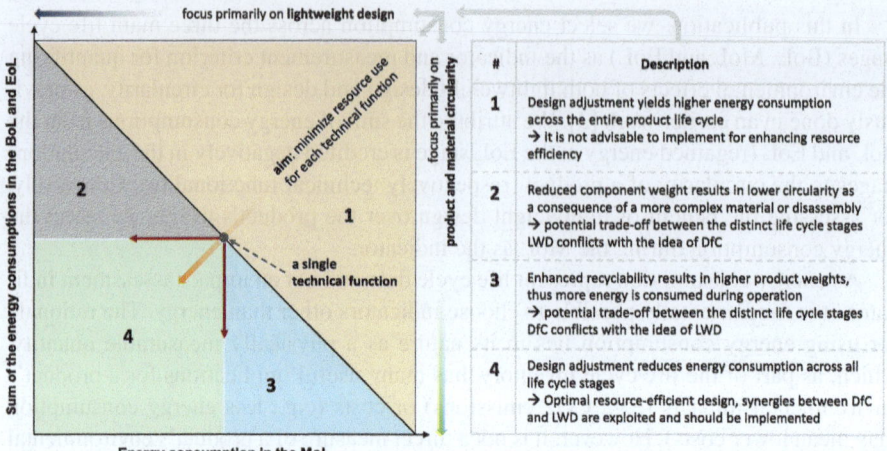

**Fig. 1.** Concept for trade-off analysis methodology to support decision making between lightweight design and design for circularity.

potential benefits should generally be avoided (option 1) as it may lead to higher environmental impacts. The primary development goal of engineering should be to reduce energy expenditures by optimizing the product considering the surrounding product system across all life cycle stages (options 2–4). Thereby, each function may enhance a potential of implementing a circular economy (option 2) or a lightweighting measure (option 3), with the other strategy might be negatively impacted. The aim should be to ensure that the benefit of implementing the environmentally superior solution in one strategy does not result in a greater negative impact in the other. Ideally, lightweight design and design for circularity should be synergistically implemented (option 4).

### 3.4    Identifying Trade-Off as Local and Global Optimizations

From a global perspective, the most environmentally friendly product is the one that is not needed, as it requires no resources. Such a function would ideally combine lightweight design and circular economy principles (option 4) and would be positioned at the origin of the diagram. Thus, to find an absolute global minimum, it is crucial to first determine whether a function is truly necessary or if it can be dematerialized to minimize resource use, invoking the concept of sufficiency.

Many products inherently have negative environmental impacts. These impacts do not necessarily need to be completely offset, as long as operations remain within the Earth's biophysical limits [19]. Therefore, for essential functions that ensure customer satisfaction and are of high importance to the product, global minima cannot be at the origin. Such functions will always use resources (e.g., material and energy) throughout the product life cycle and may return them to the Earth's ecosystem. Depending on the function's characteristics, their impacts are more pronounced on either the BoL/EoL axis or the MoL axis. The goal is to find an acceptable level of resource consumption and identify the lowest impact while ensuring functionality along options 2–4. Local minima can also be identified within iterations: an existing solution is initially optimized slightly in one strategy; rethinking the functionality entirely and searching for an altered solution principle may harness greater potential.

One way to move beyond the depicted quadrant, which only illustrates negative environmental impacts, and instead, achieve positive effects across all life cycle stages may be through the application of the 'positive impact product engineering' methodology proposed by Mörsdorf and Vielhaber [20]. Their approach suggests compensatory measures in each impact category (in our example: energy) to mitigate impacts.

## 4 An Application Example: Optimization Potentials of a Semi-mobile Handling System

### 4.1 Description of the Use Case

To ensure that the preceding considerations are not purely theoretical, we discuss the potential identification in a semi-mobile handling system as an illustrative application example. This use case has been previously explained in earlier works [3, 21] and is revisited here. The system involves a portal robot mounted on a carriage that travels along a gantry (approximately 25 m in length), transporting tools from a tool storage to operating machines with a cycle time of about one minute per tool change.

### 4.2 Results of the Use Case Implementation

For Fig. 2, the proposed methodology was applied to the functional structure of the described system and the associated energy expenditures previously determined in an earlier study [3]. The results depict five possible implementation strategies along with illustrative descriptions. As an example of optimizations, material selection was considered, whereby different materials (recycled steel, aluminum, carbon fiber-reinforced plastics) were proposed for improving energy consumption for various technical functions compared to a baseline scenario (new primary steel for the respective components). The analysis indicates that, from an environmental perspective and when balancing the entire life cycle, the implementation of sophisticated lightweighting measures is preferable to other circular economy strategies for highly dynamic functions or components (e.g., 'generate holding force'). In contrast, the opposite is true for static components: closed-loop systems or materials (e.g., using recycled steel for the function 'place robot on duct') drastically improve a product's environmental performance, while lightweight measures seem elaborate and might only be minimally beneficial. For semi-mobile functions (e.g., 'move tool'), one-dimensional optimizations may cancel each other out, resulting in local trade-offs that need to be evaluated in detail to achieve minimal environmental impacts while maintaining functionality.

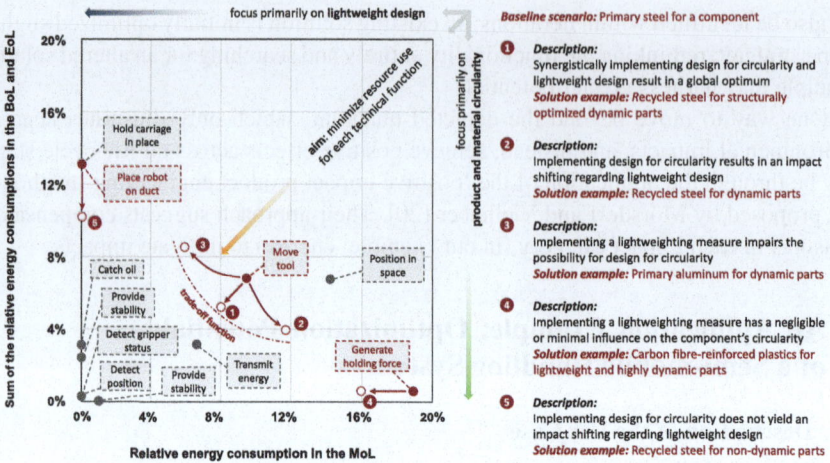

**Fig. 2.** Results of the implementation of the trade-off analysis on the use case example of the resource-efficient design of a semi-mobile handling system.

## 5   Conclusion and Outlook

Driven by the global imperative to reduce resource consumption, this study was motivated by the exploration of two fundamental strategies: implementing lightweight design and design for circularity. Therefore, the research question focuses on developing a methodology to identify trade-offs between these strategies. Based on theoretical foundations, a proposed methodology was presented and validated at the functional level of product development, specifically through material selection for technical functions of a semi-mobile handling system. This validation process yielded five distinct optimization strategies, which were examined and discussed. Future work may involve the usage of other indicators as well as exploring the methodology's applicability to physical components to enhance comprehension and application in practical contexts.

## References

1. Herrmann, C., Dewulf, W., Hauschild, M., Kaluza, A., Kara, S., Skerlos, S.: Life cycle engineering of lightweight structures. CIRP Ann. **67**, 651–672 (2018)
2. Bocken, N.M.P., De Pauw, I., Bakker, C., Van Der Grinten, B.: Product design and business model strategies for a circular economy. J. Ind. Prod. Eng. **33**, 308–320 (2016)
3. König, K., Mathieu, J., Vielhaber, M.: Resource conservation by means of lightweight design and design for circularity-a concept for decision making in the early phase of product development. Resour. Conserv. Recycl. **201**, 107331 (2024)
4. Patil, R.A., Ramakrishna, S. (eds.): Circularity Assessment: Macro to Nano: Accountability Towards Sustainability. Springer, Singapore (2023)
5. König, K., Vielhaber, M.: Incorporating lightweight design and design for sustainability in product development - a meta-synthesis based on a systematic literature review. In: Proceedings of the NordDesign (2024, in press)
6. Kaspar, J., et al.: SyProLei - A systematic product development process to exploit lightweight potentials while considering costs and CO2 emissions. Procedia CIRP **109**, 520–525 (2022)

7. König, K., Vielhaber, M.: Analysis of the interrelationships between lightweight design and design for sustainability. Procedia CIRP **122**, 324–329 (2024)
8. Klein, B.: Leichtbau-Konstruktion. Springer Fachmedien Wiesbaden, Wiesbaden (2013)
9. De Oliveira, C.T., Dantas, T.E.T., Soares, S.R.: Nano and micro level circular economy indicators: assisting decision-makers in circularity assessments. Sustain. Prod. Consumption **26**, 455–468 (2021)
10. Ellen Mac Arthur Foundation: Circularity indicators: An approach to measuring circularity (2015)
11. Franklin-Johnson, E., Figge, F., Canning, L.: Resource duration as a managerial indicator for Circular Economy performance. J. Clean. Prod. **133**, 589–598 (2016)
12. International Organization for Standardization: ISO/DIS 59004:2023 (E) Circular Economy - Terminology, Principles and Guidance for Implementation (2023)
13. Desing, H., Braun, G., Hischier, R.: Resource pressure – a circular design method. Resour. Conserv. Recycl. **164**, 105179 (2021)
14. Ashby, M.F.: Materials Selection in Mechanical Design. Butterworth-Heinemann, Amsterdam (2005)
15. Allwood, J.M., Ashby, M.F., Gutowski, T.G., Worrell, E.: Material efficiency: a white paper. Resour. Conserv. Recycl. **55**, 362–381 (2011)
16. Ferro, P., Bonollo, F.: Lightweight design versus raw materials criticalities. Sustain. Mater. Technol. **35**, e00543 (2023)
17. König, K., Mathieu, J., Vielhaber, M.: A sustainability-driven comparison of methods for the identification of lightweight design potentials in product generation engineering. Procedia CIRP **122**, 145–150 (2024)
18. König, K., Zeidler, S., Walter, R., Friedmann, M., Fleischer, J., Vielhaber, M.: Lightweight creativity methods for idea generation and evaluation in the conceptual phase of lightweight and sustainable design. Procedia CIRP **119**, 1170–1175 (2023)
19. Steffen, W., et al.: Planetary boundaries: guiding human development on a changing planet. Science **347**, 1259855 (2015)
20. Mörsdorf, S., Vielhaber, M.: Positive impact product engineering (PIPE) model - the way to net-positive sustainable products. Procedia CIRP **116**, 474–479 (2023)
21. Scholz, J., Kaspar, J., König, K., Friedmann, M., Vielhaber, M., Fleischer, J.: Lightweight design of a gripping system using a holistic systematic development process - a case study. Procedia CIRP **118**, 187–192 (2023)

# Context-Based Methodology for a Holistic Development of Circular Products and Business Models

Niels Demke[1]([✉]), Simon Mörsdorf[2], Dominik Neumann[2], Fabian Rusch[1],
Wilke Willems[1], Jonas Mohnke[2], Frank Mantwill[1], and Michael Vielhaber[2]

[1] Helmut-Schmidt-University, 22043 Hamburg, Germany
niels.demke@hsu-hh.de
[2] Saarland University, 66123 Saarbrücken, Germany

**Abstract.** Developing circular products and business models requires a holistic perspective on the product life cycle and its context. This contribution introduces a methodology consisting of five phases, categorized into three perspectives, linked by analysis and synthesis activities. This new approach embraces the adaptability of the dynamic product context through the creation of new material and information cycles. Contextual elements, which influence product development yet are not inherent parts of the product, are systematically considered. Moreover, it identifies starting points for the development of new circular product and business model scenarios, which can be also used to systematically discover problem shifts in the impact areas of sustainability. Circularity and sustainability assessments of the product portfolio, the business model, and the context serve as the methodological foundation for deriving requirements in product development. For this purpose, the scenario technique is further developed and integrated into evaluation methods. By assessing technical feasibility in the preliminary concept, requirements and technical constraints are integrated into cycle planning processes. This ensures that holistic circularity considerations are initially addressed. Thereby, the gained context sensitivity provides manufacturing companies with decision-relevant information, allowing them to leverage their potential from an integrated view of products and material/information cycles.

**Keywords:** Circular design · Context Engineering · methodology · product planning and design · scenario and evaluation methods

## 1  Introduction

The circular economy (CE) intends to transform material, emission, and energy industries into circular and sustainable systems [1]. As a result, CE markets are increasingly featured alongside consumers in the development of new products and business models (BM) [2]. The inclusion of material and information cycles in product development (PD) means that the context of a product to be developed is expanded to include the variability of existing and planned cycles. Consequently, the context can no longer be assumed to

H. Kohl et al. (Eds.): GCSM 2024, LNME, pp. 208–216, 2025.
https://doi.org/10.1007/978-3-031-93891-7_24

be static [3]. The System Context is defined as a "part of the system environment that is relevant for the definition as well as the understanding of the requirements of a system to be developed" [3]. Accordingly, the context for the example of a battery electric vehicle (BEV) encompasses, among other elements, material cycles of rare earths and charging infrastructures. The requirements and effects of the context must be continuously evaluated during the development of product ideas and BMs, particularly concerning the technical feasibility of cycles. At the same time, CE strategies, enable the product to be used over several life cycles [1]. In this way, savings can be achieved in terms of raw material consumption, energy consumption, pollutant emissions and costs. To leverage this potential in the sense of a CE, issues affecting several product generations must be addressed, leading to challenges in capturing requirements. To develop more circular products a better understanding of the interrelationships between context, product and BM development is needed. For this purpose, the aim of this research is to review the state of the art, subdivided into context considerations, design strategies, and circularity analysis approaches. Based on the need for actions, a conceptual support methodology for holistically aligning the dynamic context of BM and PD with its product life cycle (PLC) is introduced.

## 2  State of the Art

### 2.1  Context Consideration in Product and Business Development

The context analysis of PD processes constitutes a pivotal source of situational awareness about requirements for the product, which are not obvious and only arise through chained dependencies, often remaining unknown to stakeholders. A holistic, goal-oriented, and iterative analysis of the context is often not standard practice, but the following approaches do exist. To provide targeted support to product developers, the "SASTPD" approach [4] pursues the goal of holistically considering the system and context boundaries of raw materials up to the end of life and uses iterative assessments with Life Cycle Assessments (LCA) for this purpose. Fernandes et al. [5] consider the context analysis with reference to the product type. Relevant context variables of different product categories are collected and categorized on an empirical basis by Gericke et al. [6]. The model-based MESSIAH approach [7] analyzes functions and processes in the early phase of PD regarding their environmental and social value as well as circularity. GIGA-MAPS [8], offer a way to capture contextual information on a visual map to gain knowledge through visual enrichment. An approach was introduced by Langer and Lindemann [9] to evaluate the effects of dynamic changes in PD on the context, considering numerous criteria including reaction delay, sensitivity, response characteristics, and criticality.

### 2.2  Circular Design Strategies

Transitioning from a linear to a CE necessitates simultaneous evolution of BM and PD strategies [10–12]. Therefore, the successful adoption of a CE mandates close collaboration among businesses, governments, academia, and society [13]. Additionally, it requires innovation and advancements in supply chains, product and service designs, and emerging technologies for circular business models (CBMs) [14]. Due to a lack of standardized classification for circular PD, the approach of Bocken et al. [10] "Closing, Slowing, and Narrowing the Loop" subdivides methods to direct their focus on specific areas of the PLC. Based on the ISO 59004 framework [15], approaches can operate at various levels, including the system, product, component, or material level and contribute to one or more of the "Ten Retention Options", depicted by Reike et al. [16]. CBMs constitute a pivotal element in bolstering the CE [13], however, prevalent CBMs also often lack standardization [17]. Consequently, companies implementing circularity principles must innovate their BM [18]. The implementation of the CE affects numerous components of a company's BM [19], necessitating substantial change. To facilitate this transition, Geissdoerfer [17] outlines four CBM strategies: Cycling, Extending, Intensifying, and Dematerializing. Although the ISO 59010 [20] defines the CE as an economic system that inherently contributes to sustainable development, sustainability is often not initially considered in CBMs [21] or considerations are merely restricted to recycling practices and use-oriented product service systems [22].

### 2.3 Circularity Analysis and Evaluation

CE is conceptualized across Macro, Meso, Micro, and Nano levels, which provide a framework for implementing CE strategies at different scales [23, 24]. According to de Oliveira et al. [23], each level necessitates different methods to assess circularity adequately. For instance, to evaluate circularity at the product level, various indicators and methodological approaches exist, but none has yet prevailed as the standard. Building upon the Material Circularity Indicator (MCI) [25], the EMA initiated the Circulytics project, wherein the MCI has been further refined and tested in real-life environments with companies [26]. The ISO 59020 [27] establishes circularity assessment guidelines to analyze circularity at all levels in three steps: system boundaries, measuring circularity, and reporting the results. Luthin et al. [28] propose Circular Life Cycle Sustainability Assessment (C-LCSA), merging CE indicators with LCSA methods to evaluate circularity across PLC and addressing trade-offs between circularity and environmental, economic, and social impacts. So far, only a few methods exist for evaluating CBMs. An approach for developing a set of indicators that link CE principles, CBM, and the pillars of sustainability is proposed by Rossi et al. [29].

### 2.4 Need for Action

Building on the main findings in the state of the art, this chapter will derive the requirements for a new methodology. Context analysis approaches [9] enable the classification of external context factors. The inclusion of material and information cycles necessitates the formalization and continuous assessment of the context throughout the PLC. However, semantic models for capturing context [8] don't consider differentiation of known/unknown and relevant/irrelevant information, which is crucial for future cycle scenarios. Approaches observing BM innovation focus on individual stages, neglecting the ongoing activities needed to align a company's capabilities with the dynamic changes mandated by a CE [30]. Most publications on CBMs assume a single economic context, which biases the proposed BM [31]. Considering a variety of context scenarios would promote the implementation and evaluation of CBMs. The majority of existing methods and tools are not suitable for some CE-specific challenges [19], and none covers all phases of CBM innovation. Sassenelli et al. [32] also emphasizes the need for a methodology for conducting a CE performance evaluation of CBMs to provide practitioners with an assessment tool that quantifies the benefits of CE. Additionally, there is a lack of decision-making procedures for influencing product concepts at an early stage of PD [33]. The necessary feedback from technology to circular markets via BM does not take place sufficiently, meaning that products are not assessed to context compliance and circularity. Given the mentioned deficits, there is a need for methodological support to systematize the continuous consideration of a dynamic context, provide improved understanding, and better align context with the development of circular products and BMs. This contribution proposes a methodology designed to support the development of more circular products from the early stages of development throughout their entire PLC. Additionally, the approach should bidirectionally consider the interrelationships between context, PD, and BM development to assess and optimize each other's influence for greater circularity. To ensure this comprehensive consideration, various analysis and synthesis methods need to be included (Sect. 2), supporting circularity across all levels, from material to global scale.

## 3   Context-Based Methodology

The context-based methodology, displayed in Fig. 1, comprises the context and the common PLC with its corresponding BM, systematically interconnected through various synthesis and analysis activities, forming circular considerations across five stages, which are continuously exemplified with an example of a BEV development.

The **context** consists of four internal viewpoints, collectively forming the holistic context model. Firstly, the impact models simulate the reality and formulate objectives concerning sustainability parameters. Transformation models, such as the approach proposed by Rusch et al. [34, 35], are necessary to translate the target values of the impact models into addressable values for PD. These models should also enable product developers to identify problem shifts in an early phase. Processing methods such as GIGA-Maps [8] or Model-based Systems Engineering [3] are employed to capture and process the context, facilitating traceability and the accumulation of relational knowledge. These methods may be also supported by an engineering graph, as proposed by Schweizer et al. [36]. Lastly, the evaluation models, e.g. Langer & Lindemann [9], evaluate the context and validity of information under the influence of factors such as frequency of change, criticality of effect, precision, quantity, actuality, and permissible deviations. The subdivision into these four viewpoints effectively supports the partial integration of various model approaches. Through the interaction of these models, it consistently enhances the situational awareness of product developers. Changes in both product and context can be systematically analyzed and synthesized with reference to circular considerations. This process enables the derivation of requirements for the product and the BM from the continuously gained knowledge to enhance circularity. The **product life cycle** consists of five life cycle phases: planning, design, production, usage and end of life, known from

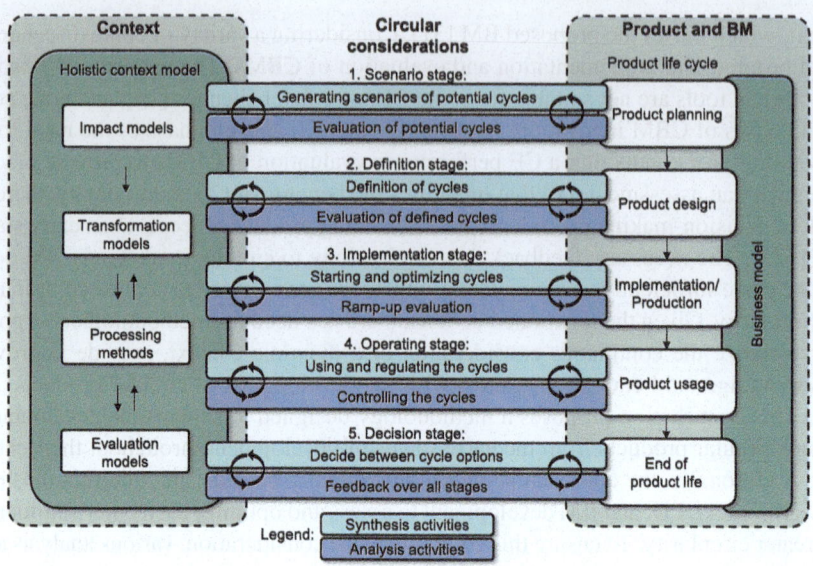

**Fig. 1.** Context-based methodology for a holistic development of circular products and business models

various development processes [37]. Circular strategies can also lead to an extension of the product usage and place greater emphasis on the end-of-life phases in the early phases of BM and PD. The understanding of a dynamic context as a result of the implementation of circular considerations leads to the need for phase-specific valuation methods. It should be noted that, at this point, the sequence of the steps described may vary and be iterative depending on the task. **Circular considerations:** To develop products that have a higher level of circularity, continuous alignment with the context, and vice versa, must occur throughout the entire PLC. This alignment is maintained for each phase of the PLC across the five stages, as depicted in the center of Fig. 1. Each stage is subdivided into synthesis and analysis activities that iteratively interact with each other. With the help of analysis methods, the interrelationships can be understood better, while synthesis methods help to generate more circular solutions. The analysis throughout the PLC is carried out with the ISO 59020 [27] or C-LCSA [28], as they can assess a product beginning at the early stages of development. Starting from the realization stage, the MCI [25] can also be used for the assessment on material level. In the first stage, the *scenario stage*, various scenarios for potential cycles based on the product planning by using scenario techniques [38] are generated and projected into the context model. Through appropriate assessment methods (Sect. 2.3), the potential cycles are evaluated within the future context, and conclusions for further product planning are drawn. For example, political scenarios or future BEV drive technologies could be developed, in which the circularity of the used energy storage systems and the availability of resources are considered. This stage is iterated until the most relevant and promising material and information cycles for the considered products have been identified. These cycles result in requirements for the product and BM. In the second stage, the *definition stage*, subsequent cycles are progressively clarified and defined, leveraging CBM design and circular product design strategies as outlined in Sect. 2.2. As granularity increases, these cycles can be more precisely harmonized and synchronized with the dynamic context through the more intensive application of circularity analysis and evaluation techniques to avoid problem shifts. For example, cycles could be defined that consider the reuse of individual BEV components. The third stage, the *implementation stage*, follows the product design and focuses on implementing and optimizing the planned and developed cycles. During the Ramp-up, physical interactions between the PLC and the context occur through the utilization of existing and the establishment of new material and information cycles. These interactions should be monitored from the outset for analysis purposes using appropriate tools. For instance, make or buy decisions, contracts and procurement processes must be designed to establish the cycles. The fourth stage, the *operating stage*, continues these activities to effectively control all relevant cycles throughout the use-phase and continuously assess them within the dynamic context and, if necessary, cycles can be adjusted through appropriate CBM design or circular PD strategies. In this manner, materials that have become critical due to dynamic changes in the context can be substituted with suitable alternatives. Namely, changing prices and demand of recyclates can influence the further production and recycling of BEV components. In the fifth and final stage, the *decision stage*, decisions are made about the cycles that do not behave as planned at the end of the PLC. In addition, to dissolve or transfer cycles into other cycles whereby closing cycles should be prioritized. The closing of material and information cycles is realized using a feedback loop with the help of the control mechanisms established in the previous stages, such as tools for material identification or maintenance in databases. This enables materials to be integrated into new product generations or existing material cycles within the context, while information about circularity practices can be documented and lessons learned can be reflected. As an example, it can happen that less parts as planned can be reused due to a deviating BEV utilization.

## 4 Discussion

Given the stated requirements, the methodology provides an overview of the interrelationships between context and PLC phases by visualizing the possible methodological connections. The user is provided with several appropriate synthesis and analysis methods to guide the development of more circular products within the circular considerations. The combination of these can cover the different levels of circularity and bidirectionally consider the interrelations. The continuously updated holistic context model allows the developer to observe the dynamic context. Through these measures, the alignment of the context to the BM and PD, resulting in the potential to develop more circular products, is being elevated. Nevertheless, it should be noted that the introduced methodology, and its non-linear application can be very complex, even for relatively simple products. These shortcomings are mainly due to the size of the context and the number of interactions with it. Furthermore, quality, architecture, actuality, documentation, and comprehensiveness of necessary data utilized in this iteration of the methodology need to be further elaborated. The methodology provides the framework for further research into digital assistance systems at the interface between circular markets and companies, so that, for example, the ability to forecast the availability of materials can be supported. Similarly, the intersections between the different models and stages need to be further defined. In this sense, the choice of methods and the application of each method can be adapted depending on the specific use case to ensure compatibility. Beyond that, the PLC and the BM must further be integrated to contribute to a holistic development model. Lastly, methodological limitations arise from a further consideration of a longer PLC or multiple product generations.

## 5   Conclusion and Outlook

This contribution highlighted in the state of the art that existing methods lack sufficient holistic and temporal formalization for context processing, preventing the evaluation of context validity. Additionally, existing tools do not adequately address the phases of the CE, and BMs are not sufficiently adapted to dynamics and scenarios. Furthermore, it has been derived from the state of the art that the interrelationships between context, BM and PD are unclear. Therefore, this paper presents a novel methodology that systematically connects the dynamic context and the development of circular products and BMs through circular considerations. The foundations for circular assessment methods are derived from product, business, and context scenarios, while various assessment methods are also proposed for the specific phases of the PLC. Due to the high level of abstraction and the scope of the methodology, the focus was placed on the basic idea and priorities for further research were identified. Based on an application example, the methodology will be detailed and applied in further contributions within real-life environments. The realization of closed-loop strategies leads to an extension of the lifetime of a product and its components. Thus, the interoperability between individual product generations and the development of a single product becomes crucial, so that further research based on the proposed methodology will follow.

## References

1. WiGeP & WGP: Update-Factory für ein industrielles Produkt-Update (2022)
2. Zink, T., Geyer, R.: Circular economy rebound. JIE (2017)
3. Rupp, C., Pohl, K.: Requirements Engineering Fundamentals, 2nd edn. Rockynook (2015)
4. Luthe, T., Kägi, T., Reger, J.: A systems approach to sustainable technical product design. JIE (2013)
5. Fernandes, P., Canciglieri Júnior, O., Sant'Anna, Â.: Method for integrated product development oriented to sustainability. Clean. Tech. Environ. Policy (2017)
6. Gericke, K., Meißner, M., Paetzold, K.: Understanding the context of product development. In: ICED (2013)
7. Halstenberg, F., Lindow, K., Stark, R.: Leveraging circular economy through a methodology for smart service systems engineering. Sustainability (2019)
8. Kvam, E.: Giga-mapping and knowledge of boundary object theory in sustainable product development. In: EPDE (2021)
9. Langer S., Lindemann, U.: Managing cycles in development processes - analysis and classification of external context factors. ICED (2009)

10. Bocken, N., de Pauw, I., Bakker, C., van der Grinten, B.: Product design and business model strategies for a circular economy. JIPE (2016)
11. den Hollander, M., Bakker, C., Hultink, E.: Product design in a circular economy: development of a typology of key concepts and terms. JIE (2017)
12. Moreno, M., De los Rios, C., Rowe, Z., Charnley, F.: A conceptual framework for circular design. Sustainability (2016)
13. Fraccascia, L., Giannoccaro, I., Agarwal, A., Hansen, E.: Business models for the circular economy: empirical advances and future directions. Bus. Strat. Environ. (2021)
14. Rashid, A., Asif, F., Krajnik, P., Nicolescu, C.: Resource conservative manufacturing. JCP (2013)
15. ISO 59004: Vocabulary, principles and guidance for implementation. ISO (2024)
16. Reike, D., Vermeulen, W., Witjes, S.: The circular economy: new or Refurbished as CE 3.0? Resour. Conserv. Recycl. (2018)
17. Geissdoerfer, M., Pieroni, M., Pigosso, D., Soufani, K.: Circular business models: a review. JCP (2020)
18. Zucchella, A., Urban, S.: Circular Entrepreneurship. Springer, Cham (2019)
19. Mentink, B.: Circular business model innovation. ESB (2014)
20. ISO 59010: Guidance on the transition of business models and value networks. ISO (2024)
21. Nußholz, J.: Circular business models. Sustainability (2017)
22. Rosa, P., Sassanelli, C., Terzi, S.: Circular business models versus circular benefits: an assessment in the waste from electrical and electronic equipments sector. JCP (2019)
23. de Oliveira, C.T., Dantas, T.E.T., Soares, S.R.: Nano and micro level circular economy indicators: assisting decision-makers in circularity assessments. SPC (2021)
24. Kirchherr, J., Reike, D., Hekkert, M.: Conceptualizing the circular economy. RCR (2017)
25. Granta Design and Ellen MacArthur Foundation. https://ellenmacarthurfoundation.org/mat erial-circularity-indicator. Accessed 08 Feb 2024
26. Ellen MacArthur Foundation: Circulytics® Method introduction. Ellen MacArthur Foundation (2022)
27. ISO 59020: Measuring and assessing circularity performance. ISO (2024)
28. Traverso, L.M., Crawford, R.H.: Circular life cycle sustainability assessment. JIE (2023)
29. Rossi, E., Bertassini, A., Ferreira, C., Amaral, W., Ometto, A.: Circular economy indicators for organizations considering sustainability and business models. JCP (2020)
30. Pieroni, M.P.P., McAloone, T.C., Pigosso, D.C.A.: Business model innovation for circular economy and sustainability: a review of approaches. JCP (2019)
31. Lüdeke-Freund, F., Gold, S., Bocken, N.M.P.: A review and typology of circular economy business model patterns. JIE (2018)
32. Sassanelli, C., Rosa, P., Rocca, R., Terzi, S.: Circular economy performance assessment methods: a systematic literature review. JCP (2019)
33. Kamp Albæk, J., Shahbazi, S., McAloone, T., Pigosso, D.: Circularity evaluation of alternative concepts during early product design and development. Sustainability (2020)
34. Rusch, F., Willems, W., Demke, N., Manwill, F.: Context-based derivation of holistic sustainability requirements in the early phase of product development. CIRP (2024)
35. Rusch, F., Demke, N., Willems, W., Manwill, F.: Linking product development's and society's view on sustainability to enhance the contextual derivation and validation of requirements. INCOM (2024)
36. Schweitzer, G., Mörsdorf, S., Bitzer, M., Vielhaber, M.: Detection of cause-effect relationships in life cycle sustainability assessment based on an engineering graph. Proc. Des. Soc. (2022)
37. VDI 2221 - Blatt 1, Beuth (2019)
38. Gausemeier, J., Plass, C.: Zukunftsorientierte Unternehmensgestaltung. Hanser (2014)

216    N. Demke et al.

# Introduction of Closed Loop Engineering: Product as a Resource!

Diana Völz$^{(\boxtimes)}$ and Maria Heckel

Frankfurt University of Applied Sciences, Nibelungenplatz 1,
60318 Frankfurt, Germany
voelz@fb2.fra-uas.de

**Abstract.** Sustainability in traditional engineering education is often considered as a cross-cutting issue or is reduced purely to ecological and/or economic aspects in embedded courses. However, nowadays it is assumed that engineers act in the interests of sustainability. At Frankfurt University of Applied Sciences, a lecture course "Sustainable Product Development" is firmly anchored as a module. Here, a holistic view of the product life cycle emphasizes the product's end-of-life to enable an extended circular economy. The lecture course is project-orientated with both, practical elements and individual research accompanied by lectures. In addition, ongoing Life Cycle Assessment (LCA) methods to determine life-phase-oriented optimization are taught. The product's "end-of-life" is a phase that is not sufficiently represented in LCA and lately not as much considered as the core of optimization as it could be. Thus, students should critically examine "resources and materials" in the sense of "closed loop", recycling processes, and their economic efficiency, as well as the challenges of current legislative changes (e.g., the act "Right to Repair"). The entire process is carried out in small teams for their special product. The course aims to provide students with a clear view of LCAs, their differences as well as the various framework conditions that usually should be leveraged against each other, to develop more sustainable products. The course outcome is a detailed proposal for product optimization.

**Keywords:** Sustainable Product Development · Engineering Education · End-of-Life Phase

## 1 Introduction and Motivation: Sustainability Teaching in Product Development

The UNESCO Program "Education for Sustainable Development: Learn for our Planet. Act for Sustainability" (ESD 2030)[1] focuses on transforming education systems worldwide to promote sustainability. This program is part of the global framework of the Agenda 2030 and aims to support the Sustainable Development

---

[1] ESD: Education for Sustainable Development — UNESCO, latest access 2014/06/12.

H. Kohl et al. (Eds.): GCSM 2024, LNME, pp. 217–225, 2025.
https://doi.org/10.1007/978-3-031-93891-7_25

Goals through education. The need for stronger integration of sustainability in engineering curricula is not new. In [1] international studies are highlighted and strategic guidelines are recommended. Recent political initiatives indicate effectiveness: a systematic literature review shows that sustainability is increasingly incorporated into engineering education worldwide [2]. Efforts in Australia also show an increase in the teaching of sustainability in engineering, as outlined by a text mining review in [3]. Internationally, many courses dealing with sustainable product development were found and of course the implementation of the Sustainable Development Goals is also an issue in Germany. The BilRess network[2] created an interactive map on teaching "sustainability" in Germany [4]. About one hundred programs and courses on resource conservation were found, but only a few in product development. Ecological requirements are often taken into account and LCA methods are taught, but the optimization cycle and an in-depth examination of materials, their recycling, and the development of new concepts are missing in mechanical engineering education. An interesting academic program "Sustainable Design - Ecodesign" at the Ecosign Academy for Design, for example, addresses the comprehensive challenges of product design: providing maximum benefit to people while minimizing the environmental impact [5].

The Frankfurt University of Applied Sciences does not have a specific degree program focused on sustainability, but within the Bachelor's of Engineering program, the course "Sustainable Product Development" is an integral part. In this lecture course, LCA methods are taught in the context of the design of sustainable products, primarily for electrical devices. The lecture aims to shift from linear thinking patterns in product development to a circular approach for increased resource conservation in the future. Currently, the product lifecycle is often taught as a linear process with the phases: production, distribution, use, and end-of-life. Then a new product passes through these process steps, maybe using recycled material. While this makes ecomomic sense from a business perspective, it is counterproductive in terms of conserving resources. Therefore, this lecture course is designed to introduce students to closed loop thinking, an approach in which materials and products are shared, leased, repaired, refurbished, and recycled for as long as possible. Recycled materials or components are reintegrated into the cycle as secondary resources (see Fig. 1). The goal is to keep a product and its materials in circulation for as long as possible.

## 2    Didactic Concept of Sustainable Product Development

To implement the concept of "product as a resource" (see Fig. 1), product designers need to have a deep understanding of end-of-life material flows, recyclability of materials, and recycling methods including material recovery rates. Currently, the left part of Fig. 1 is not adequately integrated into engineering product design education. The proof of products' recyclability and/or re-use is not required in current regulations so far, although guidelines exist (e.g., [6]). With the introduction of the Repair Act, expected in Germany in 2025, the longevity

---

[2] An association of stakeholders of resource conservation.

of products is already being promoted. Therefore, it is reasonable to assume that the concept of "products as resources" will become increasingly important in the future, as rising energy prices and diminishing resources necessitate a rethink.

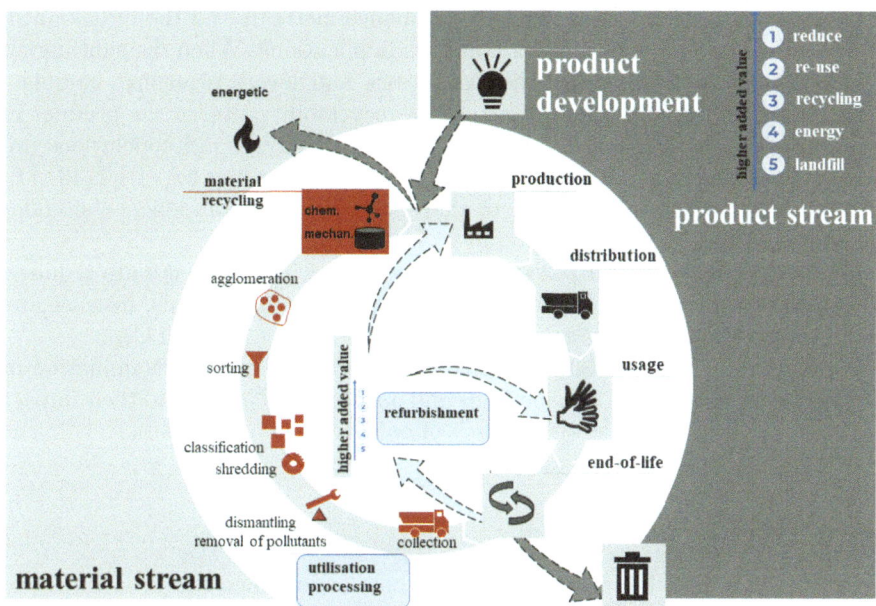

**Fig. 1.** "Product as a resource": considering material and product streams

To bridge this gap, future product development engineers will need to focus on creating sustainable products. To achieve these educational objectives, the course will use the following didactic methods:

1. **Project-based learning:** Students will engage in projects that require them to design products with end-of-life considerations, emphasizing recyclability and material recovery. The practical exercises will help students to understand the complexities of material flows and recycling methods to be more sensitized to the Circular Economy.
2. **Reflection:** Students are required to perform three different (short) balancing methods (Ecolizer 2.0 [7], Carbon Footprint [8], and Cumulative Energy Expenditure [9]) and reflect the outcomes based on the provided background during the lectures. As a result, students gain a holistic view of the product's life cycle.

Those students wishing to participate in the course are required to have a solid basic education in engineering design. By integrating the mentioned methods into the curriculum, we aim to equip future engineers with the skills and the knowledge necessary to develop sustainable products that minimize environmental impact.

## 2.1   Lecture Course Organization

The course is structured in two phases and an accompanying reflection:

- **Phase 1:** Students should become aware of the discrepancies between the outcomes of different LCAs and short-balance methods and the interconnection between design optimization and its implications. When dismantling an electrical household device, the weaknesses and levers regarding modularity, durability, repairability, disassembly, recyclability, etc. of the product in accordance with [10] should be detected. The analysis and conclusions are then evaluated, and improvements are defined. According to VDI 2243 [11], the product must be analyzed to determine whether components/modules can be re-used or whether material recycling is necessary.
- **Phase 2:** The technical redesign process begins. The students are required to undertake a critical examination of design recommendations, because any improvement on one side may have a negative impact on the LCA.
- **Reflection:** An important part of this course is the critical examination of the content, including LCA methods and optimization levers and their impact on the results, as well as the efficacy of optimization approaches.

## 2.2   Lecture Course Content

In the case of reuse, the target system may be a maintenance, refurbishment, or re-work of the product. Therefore, a certain amount of creativity is required, as many companies have already demonstrated (see Fig. 1, loops for products on the right side). Several ongoing practical examples will be presented and discussed during the lecture:

- **Closed-loop or mining-free concepts:** robots that dismantle their own old electric devices into their components and separate materials for re-use in new products for a closed loop in a company.
- **Long lifespan through high-quality products:** focus on repairability, maintenance, and upgrading potential, e.g., a consequent module concept for do-it-yourself repair like modular smartphones.
- **Resource-efficient usage:** optimization of energy, water, and material consumption for production or in use.
- **Considering the next life:** design for disassembly and usage in new contexts, e.g. car batteries used for energy storage ("Circular Economy"/"Zero Waste").
- **Optimized packaging:** by reduced material usage or innovations, e.g., mycelium packaging.
- **Innovative material concepts:** e.g. using biobased polymers for disposals/phytomining and reduction of material employment.
- **Use of recycled materials/substitutes** and avoidance of hazardous substances in design as re-use of material through special concepts like deposit systems.

– **Implementation of ideas for design optimizations**, e.g. weight reduction, realizing homogeneous material fractions for better recycling, etc.

When the level of added value cannot be maintained through re-use, material recycling is considered. This requires specific knowledge of material flows and recycling routes to support recycling through product design. Therefore, different materials are discussed, including their hazard potential and recyclability, as well as the possibilities of sorting, separating, classifying, and agglomerating are addressed (see Fig. 1, left side). Component-based processing, such as for printed circuit boards or batteries, is also part of the lectures.

## 3    Example: Electrical Product "Air Fryer"

This section presents an example of the student's work [12] on an air fryer that typically uses hot air to cook foods (WEEE according to [13]). The objective is to illustrate the process of reflection rather than the solution itself. Figure 2 (see phase 1, left side) lists the important points of the approach and assumptions.

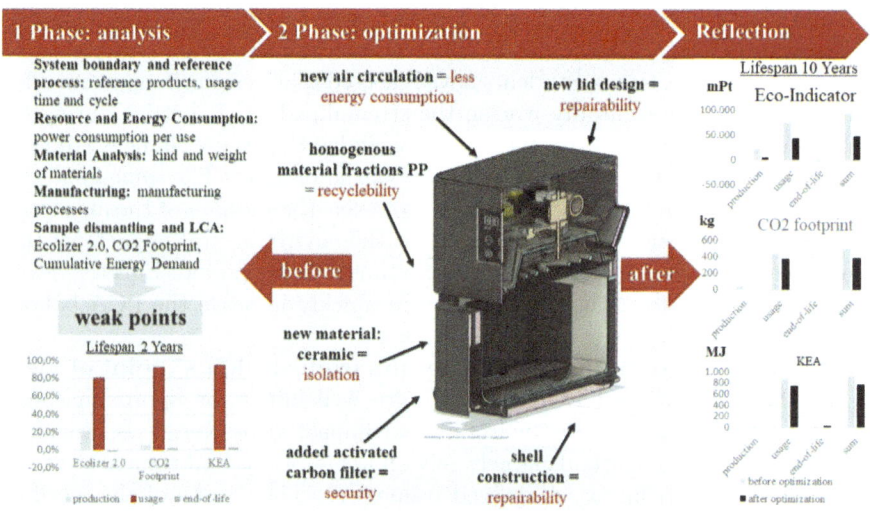

**Fig. 2.** Project Overview "Air fryer" based on [12]

– **Identified weak points.** The greatest environmental impact of the air fryer is the preparation of meals (meat dishes). A design feature of concern is the venting system at the back of the device. The energy input is used to heat the air to the required temperature. During the cooking process, the hot air becomes saturated with moisture from the food. This saturated air is then expelled, leading to high energy consumption as fresh air must be reheated upon entering the system. Re-using the moist air could reduce the required energy. The design of the air fryer complicates repair and disassembly due to the large number of components. In general, repairs are not anticipated in the business model, and spare parts are not available.

- **Identified levers for optimization.**
  - Reduce energy consumption by at least 20%.
  - Increase homogenous material fractions/reduce diversity of materials.
  - Design for durability: ensure a lifespan of at least 10 years before first repair.
  - Protect components from environmental impacts.
  - Enable repairability through the provision of spare parts.
- **Enhancements.** A larger cooking chamber ensures more effective use without increasing the air fryer's overall footprint. Modifications to the internal air circulation system lead to reduced waste heat and increase energy efficiency. A newly implemented activated carbon filter at the exhaust removes moisture and odors from the expelled air. Another design enhancement is the heat element splash guard, which aims to increase the heating element's durability, prevent smoke formation, and improve safety. For further design improvements see Fig. 2 (phase 2). In addition, standardizing materials and using materials with less environmental impact are used. At least new business concepts are suggested.
- **Students' reflections.** Students have reflected that design changes and additional components initially worsen the streamlined LCA metrics during the production phase. However, they have concluded that these changes are justified over time by the increased durability of the product. The same applies to the measures to improve repairability. However, the lifespan of the air fryer is significantly extended, which means that the environmental impact per unit of time decreases when calculated over a much longer period. Effective levers such as transport, or the energy mix can quickly improve the LCA balance on paper.
- **Reliability of streamlined LCA results from students' point of view.** The three streamlined LCA methods are well-suited for approximating a product's lifecycle and its environmental impact. However, it is crucial to note that all three methods largely rely on assumptions and comparative values, operating within a self-defined framework. Thus, none of the methods can be considered entirely fact-based. Nevertheless, comparing the methods and their results provides a general understanding of the product's lifecycle weaknesses and areas for improvement.
- **Weaknesses of the Streamlined LCA Studies from students' point of view.** All three streamlined LCA methods have notable weaknesses, particularly in capturing the production accurately. Detailed information on production sites, transport routes, materials, and manufacturing processes of individual components (e.g., heating element) is often missing but would have a high impact on LCA results. The end-of-life phase is also inadequately represented in LCA, as life cycle, disassembly, and reuse concepts are not

integrated. Additionally, social aspects, such as working conditions, are not taken into account in the LCA methods used.

## 4   Course Evaluation

The effectiveness of the course "Sustainable Product Development" was evaluated through a survey conducted among the 15 participating students. Overall, the course was perceived as relevant to their future professional careers by the students (Fig. 3).

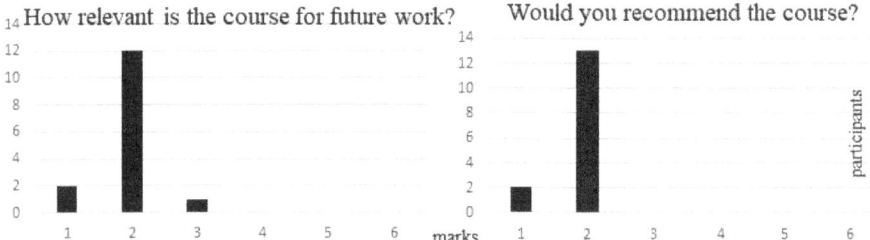

**Fig. 3.** Course evaluation, 15 participants (1: very relevant/recommendable to 6: absolutely not)

Students highlighted the importance of engaging critically with LCA models and the process of revising a product with sustainability aspects in mind as key technical skills acquired. They reported that the course had heightened their awareness of environmental and sustainability issues, prompting them to scrutinize corporate practices and advertising claims. The hands-on experience of revising a real product was particularly valued by the students. A significant challenge in this course is that although students are equipped with specialized knowledge, they are entering an entirely novel domain with the product parts. This means that all practical details (like materials and manufacturing processes) have to be worked out independently for each part. Consultation hours, close supervision and discussions with experts supported the students in their project work. Although the course already includes a broader view of sustainability, it is notable that a special focus is placed on the end-of-life phase in addition to LCA. There is a need to further enhance the course's focus on sustainability by incorporating more economic and social aspects. From a lecturer's perspective, students can be more encouraged to develop new business models viable for the future.

# 5   Conclusion

The "Sustainable Product Development" course aims to sensitize future product developers to the importance of considering the entire life cycle when developing products. Students are taught to critically scrutinize the results of LCA and to understand the impact of boundary conditions. The course focuses on the end-of-life phase, as this is becoming increasingly important in the context of legislation and measures. Product designers are often unaware of the recycling processes of products, but they do play a significant role in determining. In the future, products must be seen as a valuable resource whose life cycle needs to be extended to maintain a high level of added value. Therefore, product development should be more than just designing for the context of use. Rather, innovative ideas for longer or further life cycles should be considered, such as service concepts, innovative ideas for re-use, etc. (Closed Loop Engineering). The evaluation of the course results by the students proves that the course content ensures an adequate foundation for future professional life. Thus, this approach represents a significant step towards the fulfillment of the ESD 2030 objectives. To further optimize the course content, additional assessment tools should be introduced. It is also advisable to prepare students for new regulations, like the creation of the Digital Product Passport. In addition, further business models for products should be integrated into the lecture course. In sum, it is crucial to incorporate sustainability in product development for future engineers.

# References

1. Desha, C., Hargroves, K., Smith, M., et al.: The importance of sustainability in engineering education - a toolkit of information and teaching material. In: Engineering Training & Learning Conference (2007)
2. Thürer, M., Tomašević, I., Stevenson, M., Qu, T., Huisingh, D.: A systematic review of the literature on integrating sustainability into engineering curricula. J. Clean. Prod. **181**, 608–617 (2018)
3. Arefin, M.A., Nabi, M.N., Sadeque, S., Gudimetla, P.: Incorporating sustainability in engineering curriculum: a study of the Australian universities. Int. J. Sustain. High. Educ. **22**(3), 576–598 (2021)
4. BilressNetwork: Netzwerk.bilress.de/map. Accessed 13 June 2024
5. Ecosign.     www.ecosign.net/allgemein/oekologie-design/nachhaltiges-design-unsere-definition-von-ecodesign/. Accessed 13 June 2024
6. DIN EN 45554:2020, General methods for the assessment of the ability to repair, re-use and upgrade energy-related products. Beuth Verlag GmbH, Berlin
7. Ecolizer: EcolizerEN_1180.pdf. Accessed 28 May 2024
8. DIN EN ISO 14024:2018, Environmental labels and declarations - Type I environmental labelling - Principles and procedures. Beuth Verlag GmbH, Berlin
9. VDI 4600:2012, Cumulative energy demand (KEA) - Terms, definitions, methods of calculation, Beuth Verlag GmbH
10. DIN ISO Guide 64:2020 Guide for addressing environmental issues in product standard, Deutsches Institut für Normung
11. VDI 2243:2002 Recycling-oriented product development. Beuth Verlag GmbH

12. Lindner, A., et al.: Air fryer – product documentation as part of the lecture course "Sustainability Product Development", Frankfurt University of Applied Sciences (2024)
13. VDI 2343:2014 Recycling of electrical and electronical equipment, 1-7. Beuth Verlag GmbH

# Product Design

# Effect of the Core on Mechanical Characteristics of the Crash Box

DucHieu Le[1], MinhKet Tran[2], VanTrang Dam[3], MyTien TranThi[4],
PhucThien Nguyen[5], and TrongNhan Tran[6(✉)]

[1] Faculty of Transportation Engineering, School of Mechanical and Automotive Engineering,
Hanoi University of Industry, Hanoi, Vietnam
[2] Faculty of Automotive Engineering, School of Technology, Van Lang University, Ho Chi
Minh City, Vietnam
[3] Faculty of Mechanical Engineering, District Twelve Technical Economic College, Ho Chi
Minh City, Vietnam
[4] Faculty of Engineering and Technology, Dong Sai Gon College, Ho Chi Minh City, Vietnam
[5] Van Hien University, 665–667–669 Dien Bien Phu St, Ward 1, Dist. 3, Ho Chi Minh City,
Vietnam
[6] Faculty of Mechanical Engineering, Industrial University of Ho Chi Minh, Ho Chi Minh City,
Vietnam
trantrongnhan@iuh.edu.vn

**Abstract.** This study examines the influence of a lotus root-inspired core structure on the mechanical properties of energy-absorbing structures in automobiles.
The research employs a simulation method to analyze the energy absorption and
compression resistance of the structure while varying the core wall thickness and
core size. Compared to the baseline 7030601.8 tube, the polycaprolactone (PCL)
and specific energy absorption (SEA) of the 7050652.2 tube exhibited a 15.3%
increase and a 2.3% decrease, respectively. The results demonstrate that maintaining a constant bottom size and altering the top size of the core is an effective design
approach to enhance the mechanical properties of the energy-absorbing structure.
A larger top size corresponds to a greater maximum force, thereby improving the
compression resistance of the structure.

**Keywords:** Thin-walled tube · Bionic design · Crashworthiness

## 1 Introduction

With the global development of the automobile manufacturing industry, energy saving, environmental protection, and safety have become the three main topics of today's
automobile industry. Consequently, the safety index has emerged as one of the most
important indicators of an automobile. The growing interest in automobile crash safety
has led to extensive studies on the structural performance of thin-walled metal tubes.
These columns are used in the design and production of energy absorbers for vehicles
[1–3]. The design of new energy absorption devices is an effective method to enhance the
collision safety of automobiles. This is because energy absorption devices significantly
influence an automobile's ability to withstand collisions [4].

© The Author(s) 2025
H. Kohl et al. (Eds.): GCSM 2024, LNME, pp. 229–237, 2025.
https://doi.org/10.1007/978-3-031-93891-7_26

To enhance the mechanical characteristics of energy dissipation devices, such as resistance to deformation and capacity for energy absorption, extensive research has been conducted on the design of thin-walled tubes. This is because the energy absorbers of automobiles are often made of thin-walled structures. Research results show that the thin-walled structure can effectively absorb the automobile's kinetic energy during a collision [5]. Some researchers have investigated the energy absorption capabilities of tubes featuring varied cross-sectional shapes, such as square [6] and rectangular [7] sections. Additionally, several modified structures were also studied. The proper use of thin-walled structures can also improve crashworthiness performance [8].

In addition, foam material is also considered one of the components that can enhance the energy absorption capability of energy-absorbing devices. Foam materials can be made from metals [9] or polymer [10]. The research results from the aforementioned studies show that the use of foam material can increase the energy absorption of tubes, proving to be very beneficial in automobile design work.

Despite the significant improvements in the energy absorption ability of impact devices through previous research, the mechanical properties of these devices remain limited due to constraints in the layout or connection of their components. To tackle this concern and explore the impact of the structural core on both the energy absorption capability and the crush resistance of the energy-absorbing structure, this research suggests a core configuration inspired by the morphology of a lotus root. Following this, the crashworthiness performance of the composite structure comprising the outer tube and inner core is analyzed.

## 2  Sample and Methodology

### 2.1  Sample

The structure of the sample, as presented in Fig. 1, is proposed to consider the influence of core parameters on the mechanical properties of the structure. The structure of the sample comprises an outer tube and an inner core. In the design of energy absorbing structures, predicting the deformation process is crucial. With nested structures, deformation control can be applied to the outer tube, inner tube, or both tubes. However, this study focuses on the deformation control of the inner tube. The lotus root core has a special shape consisting of rhizomes connected by nodes, which inspired the proposed structure of the inner tube. This proposed structure is intended to support the deformation process of the energy absorbing structure. To evaluate the influence of the core on the mechanical properties of the structure, two design cases are proposed: i) Case 1: The bottom size is fixed while the top size varies, and ii) Case 2: The top size is fixed while the bottom size varies. The thickness of the core ranges from 1.8 to 2.2 mm. Consequently, the main parameters of the structure are presented in Table 1.

### 2.2  Methodology

The energy-absorbing structure plays a crucial role in absorbing impact energy during a collision. To study the mechanical properties of the energy absorption structure within an impact-resistant structure, a simulation study, as shown in Fig. 2, was conducted. The test specimen, made of steel, has its mechanical properties detailed in Fig. 2. The arrangement depicted in Fig. 2 demonstrates that one end of the test specimen is fixed to a stationary wall, while the other end is subjected to the impact of an object with a mass of 500 kg moving at a speed of 15.6 m/s, in accordance with the Euro NCAP axial impact standards. The simulation was performed using the LSdyna.

**Table 1.** Tube parameter.

| Sample | Mass (kg) | Outer tube (mm) | | Core (mm) | | | | Length (mm) |
|---|---|---|---|---|---|---|---|---|
| | | a × b | Thickness | Top size | Bottom size | Height | Thickness | |
| 7030601.8 | 1.367 | 70 × 70 | 2 | 30 × 30 | 60 × 60 | 40 | 1.8 | 200 |
| 7040602 | 1.466 | | | 40 × 40 | 60 × 60 | | 2 | |
| 7050602.2 | 1.579 | | | 50 × 50 | 60 × 60 | | 2.2 | |
| 7050551.8 | 1.419 | | | 50 × 50 | 55 × 55 | | 1.8 | |
| 7050602 | 1.515 | | | 50 × 50 | 60 × 60 | | 2 | |
| 7050652.2 | 1.621 | | | 50 × 50 | 65 × 65 | | 2.2 | |

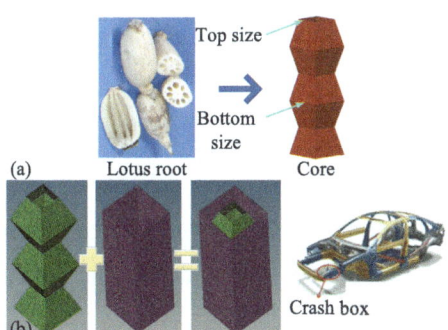

**Fig. 1.** a) Bio-inspired core and b) Sample.

# 3   Crashworthiness Indexes

To evaluate the mechanical properties of energy absorption devices, the following parameters: Energy absorption (kJ), Specific energy absorption (kJ/kg), Maximum force (kN), and Average force (kN) should be defined:

(i) Energy absorption (kJ) is the energy absorbed by the structure through its plastic deformation. It is calculated as follows:

$$Energy\ absorption = \int_0^d F(x)dx,  \qquad (1)$$

(ii) Specific Absorption Energy (kJ/kg) is the energy absorption of the structure relative to its mass (m). Therefore, it can be calculated using the formula:

$$Specific\ energy\ absorption = \frac{Energy\ absorption}{m},  \qquad (2)$$

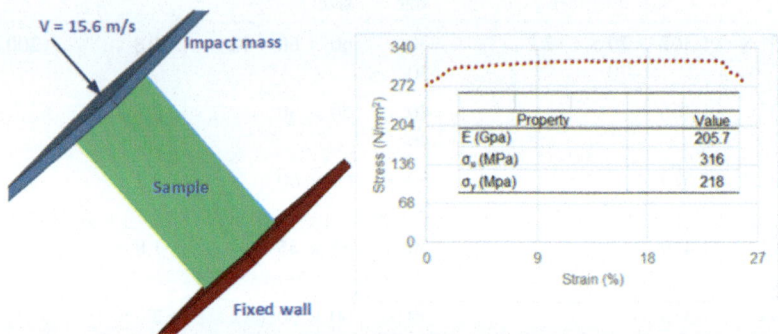

**Fig. 2.** Impact schema and mechanical properties of steel.

(iii) Maximum force (kN) refers to the highest impact force experienced during the initial stage of crushing.
(iv) Average force (kN) is calculated by dividing the energy absorption of the thin-walled pipe by the displacement (d) as follows:

$$Average\ Force = \frac{Energy\ absorption}{d},  \qquad (3)$$

## 4  Discussion

The deformation process of all considered samples is depicted in Figs. 3 and 4. For the tested samples, where the bottom size of the core is 60 × 60 mm and the top size of the core varies from 30 × 30 mm to 50 × 50 mm (samples 7030601.8, 7040602, and 7050602.2), the deformation of the sample occurs from the upper end to the lower end in a progressively gradual manner (as shown in Fig. 3). However, due to the variation in core size, these samples also exhibit slight differences in deformation (Fig. 3). With the variation in the size of the core, the interaction between the core and the outer tube also differs. This difference is clearly demonstrated in the deformation of the core of sample 7030601.8 compared to the two samples 7040602 and 7050602.2. The deformation of the core of sample 7030601.8 occurred at the lower constriction area, while in the other two samples, the deformation occurred at the upper constriction area.

For the experimental samples where the top size of the core is 50 × 50 mm and the bottom size varies from 55 × 55 mm to 60 × 60 mm (samples 7050551.8, 7050602, and 7050652.2), the deformation process of all three samples exhibits progressive deformation (Fig. 4). However, the deformation of two samples, 7030601.8 and 7040602, occurs from the upper end to the lower end, while the deformation of sample 7050652.2 starts from the middle of the sample. Thus, as the size of the core changes, the deformation of these samples also differs (Fig. 4). The deformation of the core of sample 7050602 in Fig. 4(b) is more chaotic than the cores of the other two samples in Fig. 4(a) and 4(c).

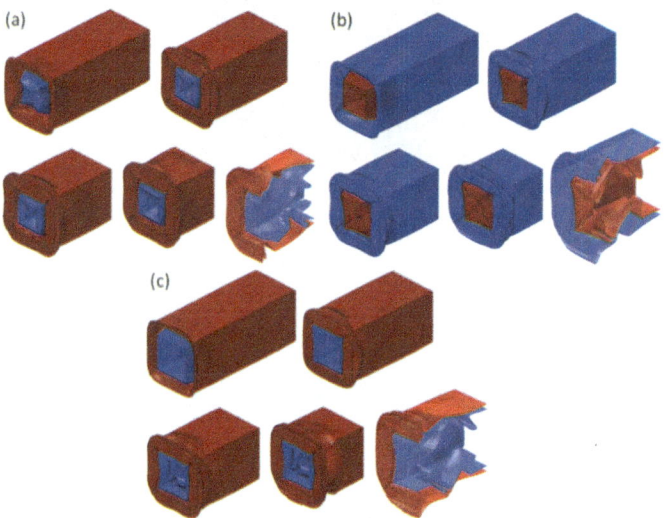

**Fig. 3.** Deformation modes: a) 7030601.8, b) 7040602 and c) 7050602.2.

The force response of all samples is depicted in Figs. 5 and 6. The force response of three samples, 7030601.8, 7040602, and 7050602.2, demonstrates similar behavior (as shown in Fig. 5). Initially, the impact force increases to an initial value and then to a maximum value. Following this, the impact force fluctuates around the average value until the impact process concludes. The evolution depicted in Fig. 5(b) indicates that the average forces of the two samples, 7050602.2 and 7030601.8, are the largest and smallest, respectively. Consequently, the energy absorptions of the two samples, 7050602.2 and 7030601.8, are also the largest and smallest, respectively.

Regarding the force response of the three samples 7050551.8, 7050602, and 7050652.2 depicted in Fig. 6, the impact force of these specimens is similar to the impact force of the three samples 7030601.8, 7040602, and 7050602.2 in the beginning. However, they diverge at the later stages of the impact process. The impact forces of the three samples 7050551.8, 7050602, and 7050652.2 fluctuate around their average values with a small difference. Figure 6(b) shows that the average forces of two samples, 7050551.8 and 7050602, are approximately the same and larger than the average force value of sample 7050652.2. Consequently, the absorbed energy of the two samples, 7050551.8 and 7050602, is greater than the absorbed energy of sample 7050652.2

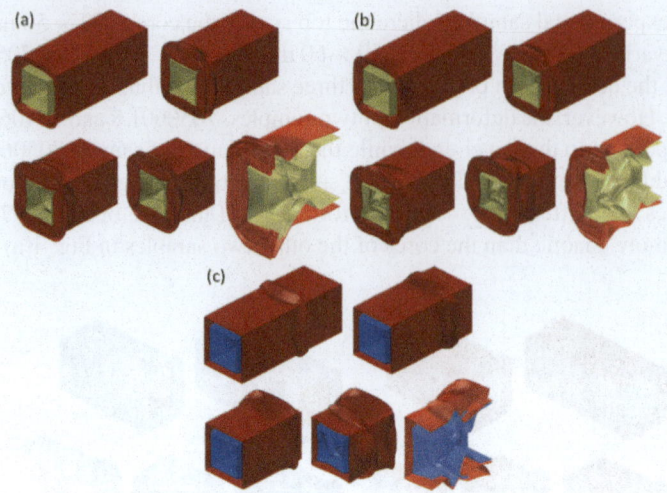

**Fig. 4.** Deformation modes: a) 7050551.8, b) 7050602 and c) 7050652.2.

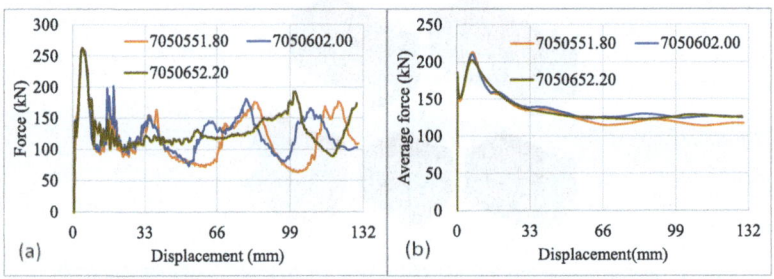

**Fig. 5.** a) Displacement-force curve and b) Displacement-average force curve of 7030601.8, b) 7040602 and c) 7050602.2 samples.

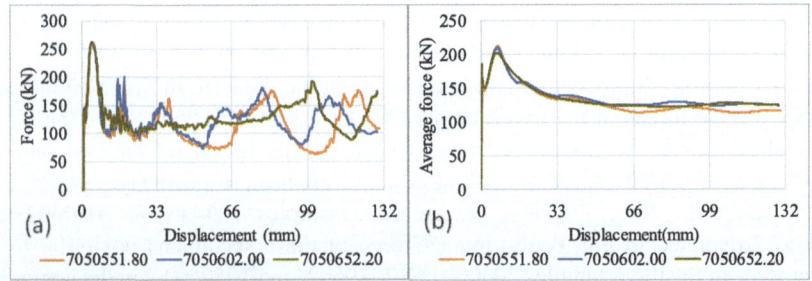

**Fig. 6.** a) Displacement-force curve and b) Displacement-average force curve of 7050551.8, b) 7050602 and c) 7050652.2 samples.

To understand the influence of the core on the mechanical properties of the energy-absorbing component, it is necessary to consider and compare the peak force, average force, and energy absorption indices. The parameters of Case I for the three samples 7030601.8, 7040602, and 7050602.2 are shown in Fig. 7. The figure illustrates that the maximum force, average force, and specific energy absorption gradually increase with the increase in both parameters: the size of the upper base and the thickness of the core. The difference between the maximum and minimum values of all three indices, including maximum force, average force, and specific energy absorption, for the two samples 7030601.8 and 7050602.2 is 18.86%, 21.26%, and 7.33%, respectively. Thus, although the mass of the three pieces gradually increased from sample 7030601.8 to sample 7050602.2, the indicators evaluating the ability to resist deformation and absorb energy all increased. This demonstrates that the positive effect of specimen type 1 is to keep the size of the lower base unchanged while changing the size of the upper base.

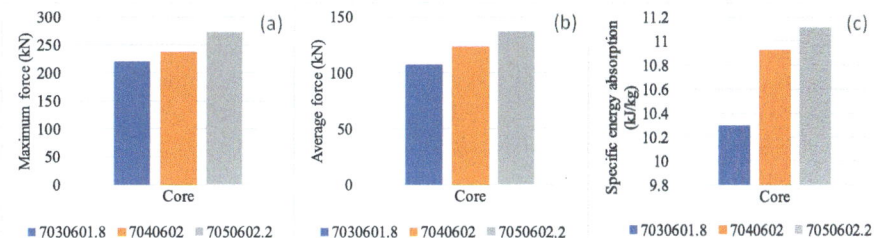

**Fig. 7.**  a) Key parameters of 7030601.8, b) 7040602 and c) 7050602.2 samples.

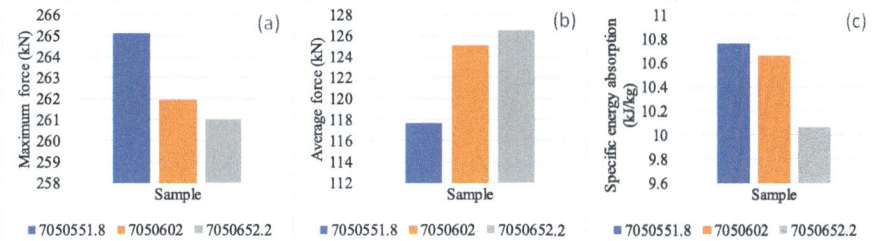

**Fig. 8.**  a) Key parameters of 7050551.8, b) 7050602 and c) 7050652.2 samples.

The comparison of the peak force, average force, and specific energy absorption indices of Case 2 for the three samples 7050551.8, 7050602, and 7050652.2 is shown in Fig. 8. Similar to the three samples in Case 1, the mass of all three samples in Case 2 gradually increases in order from sample 7050551.8 to sample 7050652.2. However, they differ when comparing indicators that evaluate the mechanical properties of energy-absorbing structures. The peak force and energy absorption of all three samples gradually decrease. The difference between the maximum peak force of sample 7050551.8 and that of sample 7050652.2 is 1.57%, and the difference between the maximum effective absorbed energy of sample 7050551.8 and that of sample 7050652.2 is 6.91%. Although the ratio of 6.91% also represents the difference between the largest and smallest average force between samples 7050551.8 and 7050652.2, the average force of sample 7050652.2

is larger than the average force of sample 7050551.8. Thus, the changes in the indices of the samples in Case 2 are different from the samples in Case One. Comparing the two results of Fig. 7 and Fig. 8 shows that the structure of core type 1 is more effective than the structure of core type 2.

## 5   Conclusion

The core structure is designed by mimicking the shape of a lotus root. The influence of the core on the mechanical properties of the structure is then studied in the context of impact loading. Research results show that, compared to maintaining the top size and changing the size of the lower one in Case 2, keeping the bottom size unchanged and altering the top size of the core in Case 1 results in a better design that improves the mechanical properties of the energy-absorbing structure. Compared with 7030601.8 tube, the PCL and SEA of 7050652.2 one increased and decreased by 15.3% and 2.3%, respectively. In both cases, increasing the thickness of the core increases the maximum force. However, in Case 1, the specific energy absorption increases, while the effective energy absorption in Case 2 decreases. This decrease is not desirable in the design of energy absorption devices.

**Acknowledgement.** Van Lang University is acknowledged.

## References

1. Lu, G., Yu, T.: Energy Absorption of Structures and Materials. Woodhead Publishing (2003)
2. Baroutaji, A., Arjunan, A., Niknejad, A., Tran, T., Olabi, A.-G.: Application of cellular material in crashworthiness applications: an overview. In: Reference Module in Materials Science and Materials Engineering. Elsevier (2019)
3. Ha, N.S., Lu, G.: A review of recent research on bio-inspired structures and materials for energy absorption applications. Compos. B Eng. **181**, 107496 (2020)
4. Wang, S., Zhang, M., Pei, W., Yu, F., Jiang, Y.: Energy-absorbing mechanism and crashworthiness performance of thin-walled tubes diagonally filled with rib-reinforced foam blocks under axial crushing. Compos. Struct. **299**, 116149 (2022)
5. Tran, T., Le, D., Baroutaji, A.: Theoretical and numerical crush analysis of multi-stage nested aluminium alloy tubular structures under axial impact loading. Eng. Struct. **182**, 39–50 (2019)
6. Abdullahi, H.S., Gao, S.: A novel multi-cell square tubal structure based on Voronoi tessellation for enhanced crashworthiness. Thin-Walled Struct. **150**, 106690 (2020)
7. Gupta, N.K., Khullar, A.: Collapse of square and rectangular tubes in transverse loading. Arch. Appl. Mech. **63**(7), 479–490 (1993)
8. Duan, L., Sun, G., Cui, J., Chen, T., Cheng, A., Li, G.: Crashworthiness design of vehicle structure with tailor rolled blank. Struct. Multidisc. Optim. **53**(2), 321–338 (2016)
9. Reyes, A., Hopperstad, O.S., Langseth, M.: Aluminum foam-filled extrusions subjected to oblique loading: experimental and numerical study. Int. J. Solids Struct. **41**(5), 1645–1675 (2004)
10. Tarlochan, F., Ramesh, S.: Composite sandwich structures with nested inserts for energy absorption application. Compos. Struct. **94**(3), 904–916 (2012)

# Design and Initial Experimental Testing of a Haptic Steering-by-Wire System with Integrated Magneto-Rheological Brake

Nguyen Duy Hung[1], Do Xuan Phu[1], Nguyen Hoang Vinh Khang[1],
Hoang Long Vuong[2], Bui Quoc Duy[2], Truong Thuy Duy[1],
and Nguyen Quoc Hung[2(✉)]

[1] Faculty of Engineering, Vietnamese-German University (VGU), Bến Cát, Vietnam
[2] Faculty of Mechanical Engineering, Industrial University of Ho Chi Minh City, Ho Chi Minh City, Vietnam
nguyenquochung@iuh.edu.vn

**Abstract.** This paper presents the integration and experimental testing of a magneto-rheological (MR) brake in a steering-by-wire (SbW) system to provide haptic feedback. The research focuses on incorporating the MR brake into the SbW architecture to generate realistic steering sensations for drivers. Experimental tests were conducted to evaluate the performance of the integrated system. Results demonstrate the system's ability to produce variable steering resistance, simulating road conditions and enhancing the driving experience. This study contributes to the advancement of SbW technologies by addressing the challenge of providing tactile feedback in electric vehicles, potentially improving driver control and safety.

**Keywords:** Haptic steering-by-wire · magnetorheological brake (MR brake) · driver feedback · experimental evaluation

## 1 Introduction

Steering-by-wire (SbW) systems have revolutionized automotive technology, replacing mechanical linkages with electronic controls [1]. These systems enhance vehicle performance, design flexibility, and integration of advanced driver assistance features [2]. However, a significant challenge in SbW systems is providing realistic haptic feedback, crucial for maintaining control and evaluating road conditions [3, 4].

This study proposes integrating a magneto-rheological (MR) brake into SbW systems to address the haptic feedback challenge, particularly in personal electric vehicles.

MR brakes offer variable, controllable resistance in response to electrical or magnetic stimuli, making them suitable for replicating tactile sensations in SbW systems. By achieving this objective, this research contributes to advancing SbW technologies in electric vehicles, enhancing driving experience, safety, and control.

H. Kohl et al. (Eds.): GCSM 2024, LNME, pp. 238–245, 2025.
https://doi.org/10.1007/978-3-031-93891-7_27

## 2    System Design

### 2.1    The SbW System with MR Brake-Based Haptic Feedback

Figure 1 compares three steering architectures: conventional, standard SbW, and SbW with MR brake-based haptic feedback.

The traditional setup directly connects the steering wheel to road wheels, providing natural feedback but limiting design flexibility. Standard SbW systems replace mechanical linkages with sensors, actuators, and an electronic control unit (ECU), offering design freedom but lacking natural feedback. The proposed system enhances the standard SbW architecture by incorporating an MR brake controlled by the ECU. This integration combines SbW benefits with realistic feedback, simulating tactile sensations of conventional steering systems through variable resistance at the steering wheel.

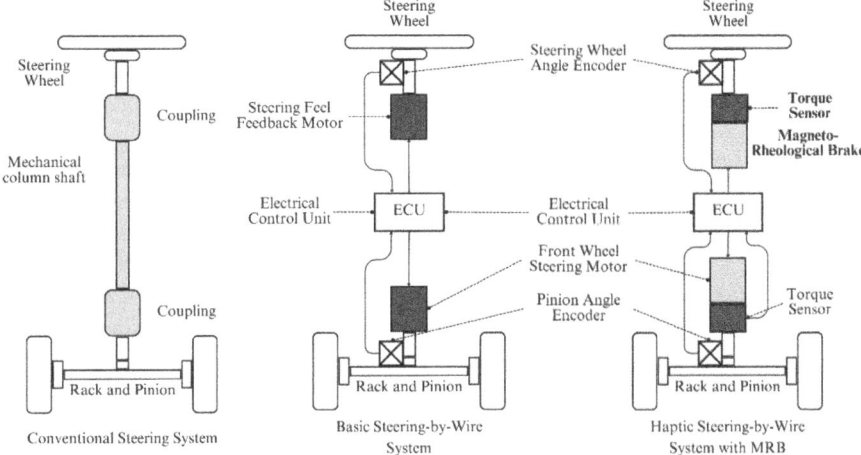

**Fig. 1.** Comparative Overview of Three Steering System Architectures

### 2.2    Steering Control System Design

Figure 2 illustrates the schematic diagram of our proposed Steering-by-Wire (SbW) system, focusing on the control path from the steering encoder to the Front Wheel Steering motor output.

As shown in Fig. 2, the steering wheel angle ($\delta_{sw}$) is measured by the steering encoder, while the feedback torque ($\tau_{fm}$) is generated by the MR brake. The relationship between these two parameters is crucial for providing an authentic steering feel to the driver. The feedback torque is controlled based on various factors including:

- Steering wheel angle ($\delta_{sw}$)
- Vehicle speed (v)
- Road conditions
- Desired steering characteristics

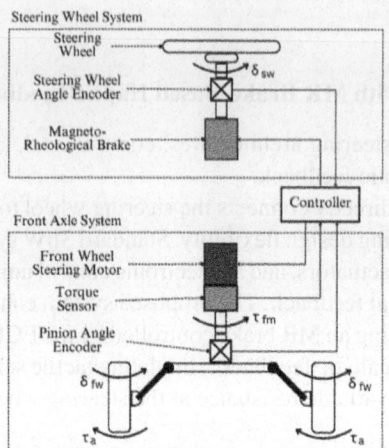

**Fig. 2.** Schematic diagram of the SbW system with MR brake and Front Wheel Steering motor

The schematic diagram of our proposed SbW system specifically focuses on the control path from the steering encoder to the Front Wheel Steering motor output. In this diagram, the steering wheel angle ($\delta_{sw}$) is measured by the steering encoder, while the feedback torque ($\tau_{fm}$) is generated by the MR brake. The interaction between these two parameters is crucial for delivering an authentic steering experience to the driver. To achieve this, the feedback torque is controlled based on several factors, including the steering wheel angle ($\delta_{sw}$), vehicle speed ($v$), road conditions, and desired steering characteristics. The relationship between these factors can be described by the following general expression:

$$\tau_{fm} = f(\delta_{sw}, v, road\_condition, steering\_profile) \tag{1}$$

In which, *steering_profile* is a predefined set of characteristics that determine the overall steering feel.

The relationship between the steering wheel angle ($\delta_{sw}$) and the front wheel angle ($\delta_{fw}$) plays a crucial role in the design of the steering system. In conventional vehicles, this relationship is determined by the steering ratio, which typically ranges from 12:1 to 20:1 for passenger cars. For instance, a steering ratio of 16:1 means that for every 16 degrees of steering wheel rotation, the front wheels turn by 1 degree. In our proposed SbW system, we have the flexibility to incorporate a variable steering ratio that can adapt to different driving conditions. During low speeds, such as during parking maneuvers, a lower ratio (e.g., 12:1) can be utilized to reduce the number of steering wheel turns required. Conversely, at higher speeds, a higher ratio (e.g., 20:1) can be implemented to enhance stability and reduce steering sensitivity. This variable ratio can be mathematically expressed as:

$$\delta_{fw} = \delta_{sw}/R(v) \tag{2}$$

The variable steering ratio, denoted as $R(v)$, represents the steering ratio as a function of the vehicle speed $v$. This function is carefully designed to ensure optimal steering feel and control in different driving scenarios. It aims to enhance maneuverability at low speeds and promote stability at high speeds. The adaptive nature of the steering ratio is a significant advantage of our proposed SbW system, providing an enhanced driving experience and improved safety when compared to conventional fixed-ratio steering systems.

The core of our design focuses on controlling the Front Wheel Steering motor based on the input from the steering encoder. This process involves:

Steering Angle Mapping: The steering wheel angle ($\delta_{sw}$) is mapped to the desired front wheel angle ($\delta_{fw}$) using a non-linear function that may vary with vehicle speed [11]:

The core of our design focuses on controlling the Front Wheel Steering motor based on the input from the steering encoder. This process involves:

*Steering Angle Mapping:* The steering wheel angle ($\delta_{sw}$) is mapped to the desired front wheel angle ($\delta_{fw}$) using a non-linear function that may vary with vehicle speed [11]. This mapping is essential to translate the driver's steering input into the appropriate front wheel angle required for vehicle control.

$$\delta_{fw} = h(\delta_{sw}, v) \tag{3}$$

*Motor Position Control:* To accurately position the Front Wheel Steering motor based on the desired front wheel angle, a closed-loop control system is implemented. The control law can be expressed as [12]:

$$\theta_m = g(\delta_{fw}, v) \tag{4}$$

where $\theta_m$ is the motor angular position.

Control Strategy: The control system employs a combination of feedforward and feedback control to ensure quick response and accurate positioning [13]:

$$u = K_p\left(\theta_{m\_desired} - \theta_{m\_actual}\right) + K_d\left(d\theta_{m\_desired}/dt - d\theta_{m\_actual}/dt\right) + u_{ff}\left(\theta_{m\_desired}\right) \tag{5}$$

where u is the control input to the motor, $K_p$ and $K_d$ are proportional and derivative gains respectively, and $u_{ff}$ is a feedforward term based on the system model.

## 2.3  Experiment Design

The steering control system design integrates the hardware components described above to create a robust and efficient system. Figure 3 illustrates the functional block diagram of the open-loop haptic moment control system for the MR brake. The system operates as follows: The PC communicates with the PLC, sending control commands. The PLC then performs two parallel tasks: generating pulse signals to control the stepping motor, which simulates a driver's input on the steering wheel, and producing analog signals to control the programmable power supply. The power supply provides a fixed voltage and variable current to the MR brake, with the current determined by the following equation.

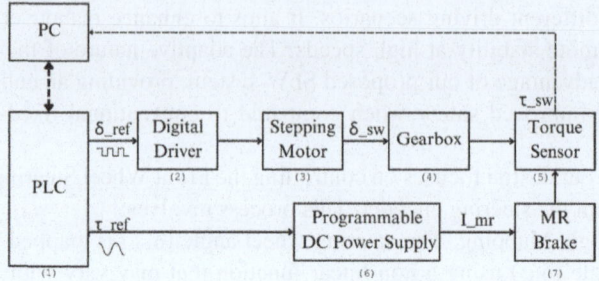

**Fig. 3.** Functional block diagram of the open-loop haptic moment control system for MR brake

$$I_{mr} = \alpha * \delta_{sw} \qquad (6)$$

where $\delta_{sw}$ is the steering wheel angle. The MR brake generates a braking torque based on this applied current.

Finally, the torque sensor measures the actual torque produced by the MR brake and sends this data back to the PC for analysis. It is important to note that a stepping motor is used in place of actual driver input. This substitution allows for precise and stable setting values, which are crucial for the accuracy of the experiment. The primary objective of this experiment is to determine the appropriate value of $\alpha$ for the characteristic relationship between current (I) and steering wheel angle ($\delta_{sw}$). This relationship is fundamental to the effective control of the MR brake in response to steering input. This open-loop control system enables the investigation of the relationship between steering wheel angle, applied current, and resulting torque in the MR brake system, providing valuable insights for the development of haptic feedback in steering applications.

Based on the block diagram in Fig. 3, the experimental setup is arranged as shown in Fig. 4 below.

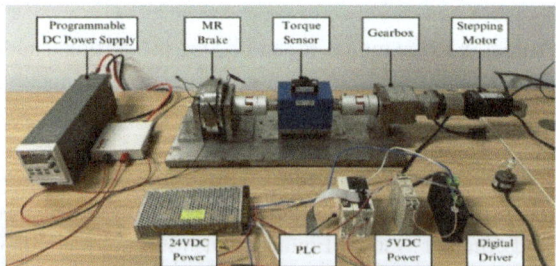

**Fig. 4.** Experimental setup for the open-loop haptic moment control system

After obtaining the survey results from the Experiment 1 abovementioned, we can proceed to construct experiment 2 as shown in Fig. 5, addressing the basic components of the SbW system in Fig. 2.

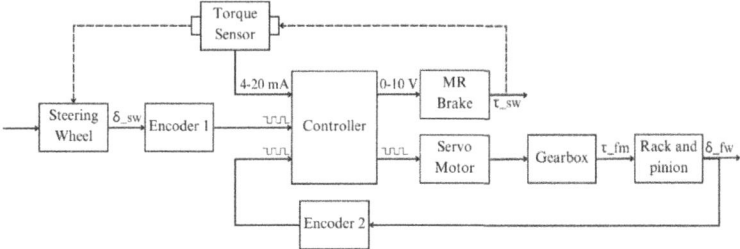

**Fig. 5.** Block diagram of the steering control system

In contrast to Experiment 1, this investigation incorporates a direct application of the MR Brake to the steering wheel. This configuration necessitates the installation of a torque sensor between the MR brake and the steering wheel to accurately measure the applied torque. The control mechanism for the MR brake involves the controller generating an analog output signal in the form of voltage (0-10VDC) to a programmable power supply. This power supply, in turn, produces the control current for the brake, mirroring the methodology employed in Experiment 1. A second encoder (Encoder 2) is integrated into the system to monitor the actual steering angle of the front wheels ($\delta_{fw}$). This measurement allows for a comparative analysis against the desired steering angle ($\delta_{ref}$), enabling precise control and error correction in the steering system.

Our experiment focuses on two main aspects of the steering system: Position Control and Haptic Feedback Simulation.

– *Position Control*

The primary focus is on controlling the position of the main branch from $\delta_{sw}$ (steering wheel angle) to $\tau_{fm}$ (feedback motor torque). This involves accurately translating the steering wheel's rotation into the corresponding position of the feedback motor.

– *Haptic Feedback Simulation*

We implement a linear equation to simulate the relationship between $\tau_{fm}$ (feedback motor torque) and $\tau_{sw}$ (steering wheel torque). This equation is used to adjust the current intensity supplied to the MRB, which provides the haptic feedback to the driver.

The experiment procedure is as follows:

- The steering wheel is rotated, and the angle ($\delta_{sw}$) is measured.
- The position control system translates this rotation into the corresponding feedback motor position.
- The linear equation calculates the required $\tau_{sw}$ based on the $\tau_{fm}$.
- The system adjusts the current supplied to the MRB accordingly.
- The torque sensor measures the actual torque applied to the steering wheel.

From the final results collected by the torque sensor, we aim to derive two critical relationships:

+ $\delta_{sw}$ vs. $I$ (current): This graph will show how the steering wheel angle relates to the current supplied to the MRB.

+ $\delta_{sw}$ vs. $\tau_{sw}$: This graph will illustrate the relationship between the steering wheel angle and the torque experienced at the steering wheel.

These relationships will provide valuable insights into the system's performance and the effectiveness of our haptic feedback simulation. They will also serve as a foundation for future improvements and refinements to the steering control system.

## 3   Results and Discussions

A crucial aspect of this study's results comes from the torque sensor used in the experiments. Although the current implementation represents an open-loop control of the haptic feedback value, future studies can build upon this foundation to create a more comprehensive model. Such a model could incorporate a tilting steering wheel operating on a road. In this scenario, Eq. (1) would be expanded to incorporate all relevant variables, forming a closed-loop control system for the haptic feedback simulation (Fig. 6).

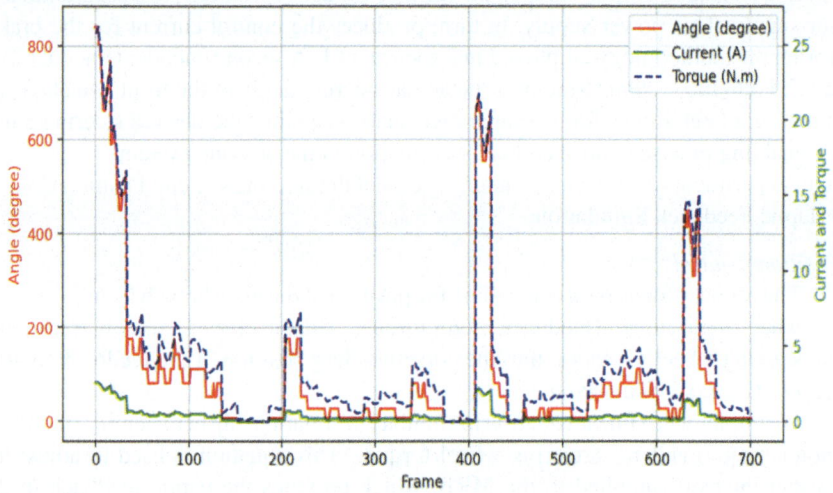

**Fig. 6.**  Current and Torque Response

The presented figure illustrates the relationship between the current supplied to the MRF brake and the torque generated, as measured by the torque sensor. It is evident from the data that there is a linear correlation between the angle of rotation and the torque values. This finding highlights the effectiveness of MRF brakes in providing precise control in Steering by Wire systems, demonstrating their potential for enhanced performance and reliability in dynamic applications.

## 4   Conclusion

This study developed a haptic feedback steering control system for automotive applications, demonstrating real-time force feedback and stability. While the open-loop system effectively simulates steering forces, improvements are needed to reduce vibrations and overcorrection. Future research should focus on a closed-loop system with real road conditions. This work advances automotive control systems, supporting the need for accurate steering in evolving autonomous vehicles, and provides valuable insights for future developments.

**Acknowledgement.** This work was supported by research fund from 2024 VGU-level project under the grand no. DTCS2024-002.

# References

1. Wang, R., Wang, J.: Fault-tolerant control with active fault diagnosis for four-wheel independently driven electric ground vehicles. IEEE Trans. Veh. Technol. **61**(9), 3799–3808 (2012)
2. Yao, Y.: Vehicle steer-by-wire system control. In: SAE Technical Paper 2006-01-1175. SAE International, Warrendale (2006)
3. Abbink, D.A., Mulder, M., Boer, E.R.: Haptic shared control: smoothly shifting control authority? Cogn. Technol. Work **14**(1), 19–28 (2012)
4. Cetin, A.E., Adli, M.A., Barkana, D.E., Kucuk, H.: Implementation and development of an adaptive steering-wheel force feedback for a steer-by-wire system. IEEE Trans. Veh. Technol. **61**(5), 1800–1814 (2012)
5. Choi, S.B., Han, Y.M.: Magnetorheological Fluid Technology: Applications in Vehicle Systems. CRC Press, Boca Raton (2013)
6. Jolly, M.R., Bender, J.W., Carlson, J.D.: Properties and applications of commercial magnetorheological fluids. J. Intell. Mater. Syst. Struct. **10**(1), 5–13 (1999)
7. Nguyen, Q.H., Choi, S.B.: Optimal design of MR shock absorber and application to vehicle suspension. Smart Mater. Struct. **18**(3), 035012 (2009)
8. Atabay, O., Ozkol, I.: Application of a magnetorheological brake system to a vehicle steering mechanism. Proc. Inst. Mech. Eng. Part D: J. Automobile Eng. **228**(7), 846–859 (2014)
9. Kim, J.H., Jeon, D.: Haptic feedback using a magnetorheological brake for a steer-by-wire system. Mechatronics **59**, 134–148 (2019)
10. Do, X.P., Choi, S.B.: High-performance torque control of a magneto-rheological brake for haptic systems. Smart Mater. Struct. **29**(3), 035013 (2020)
11. Wong, J.Y.: Theory of Ground Vehicles. Wiley, Hoboken (2008)
12. Rajamani, R.: Vehicle Dynamics and Control. Springer, Cham (2011)
13. Åström, K.J., Murray, R.M.: Feedback Systems: An Introduction for Scientists and Engineers. Princeton University Press (2010)

# Effect of Cross-Section on Anti-crushing and Energy Absorption Features of Single and Nested Square Tubes

TrongNhan Tran[1], PhucThien Nguyen[2], NhatTan Nguyen[3], and DucHieu Le[4(✉)]

[1] Faculty of Mechanical Engineering, Industrial University of Ho Chi Minh City, Ho Chi Minh City, Vietnam
[2] Van Hien University, 665–667–669 Dien Bien Phu St, Ward 1, Dist 3, Ho Chi Minh City, Vietnam
[3] Vietnam – Japan Center, Hanoi University of Industry, Hanoi, Vietnam
[4] Faculty of Transportation Engineering, School of Mechanical and Automotive Engineering, Hanoi University of Industry, Hanoi, Vietnam
hieuld@haui.edu.vn

**Abstract.** This study investigates the anti-crushing and energy absorption performance of single and nested tube structures used as crushing absorbers. The tube sections analyzed had dimensions of $60 \times 60$ mm$^2$, $70 \times 70$ mm$^2$, and $90 \times 90$ mm$^2$. Compared to the S1 tube, the S3 tube's Peak Force (PF) was 86.6% larger, while its Specific Energy Absorption (SEA) was 29.4% smaller. Additionally, the NT sample exhibited 86.6% higher PF, 85.6% higher Mean Force (MF), and 85.6% higher Energy Absorption (EA) compared to the S1 sample. The study found a direct correlation between section size and mass: as these increase, PF increases while SEA decreases. The Crushing Force Ratio (CFR) emerged as a critical parameter in the selection and design of energy-absorbing tubes. Among the samples tested, the NT nested configuration, which demonstrated superior crushing resistance and more efficient space utilization than single-tube samples, is well-suited for energy-absorbing devices.

**Keywords:** Thin-walled tube · Anti-crushing · Energy absorption · Crashworthiness

## 1 Introduction

As efficient energy-absorbing structures, thin-walled metal tubes are widely used in various engineering fields to absorb energy under compressive conditions. The outstanding feature of these sections is that they are always closed, especially when subjected to axial loads. Among the various deformation patterns of the thin-walled tube, progressive deformation from one end to the other with successive folds has attracted significant attention from researchers.

In the last century, Wierzbicki and Abramowicz [1] first implemented a theoretical method for the axial compression of square or multi-angle prismatic tubes. From the

© The Author(s) 2025
H. Kohl et al. (Eds.): GCSM 2024, LNME, pp. 246–253, 2025.
https://doi.org/10.1007/978-3-031-93891-7_28

1990s onwards, thin-walled tubes have been widely used in the automotive industry. The force response of hollow tubes [2], foam-filled tubes [3] are often chosen as research topics. In recent years, multi-cell tubes have been found to be much more impact-resistant than single tubes. Many studies on the axial crushing behavior of multicellular tubules have been performed by Chen and Wierzbicki [4] or Kim [5].

Regarding the circular or square sections, Mamalis et al. [6] studied the behavior of telescopic structures with circular cross-sections made of two different materials. They built a theoretical model to predict the average crushing force of this structure. Hsu and Jones [7] conducted static and dynamic tests on tubes made of aluminum alloy, mild steel, and stainless steel. They observed a significant difference in the energy absorption capacity of these tubes. Alavi Nia and Hamedani [8] performed static and dynamic tests on tubes of different cross-sections and showed that the tube cross-section plays an important role in influencing the energy absorption capacity of thin-walled tubes.

To improve the impact performance of thin-walled tubes, researchers have used aluminum powder or honeycomb materials as fillers. Seitzberger et al. [9] or Rajendran et al. [10] studied the energy absorption efficiency of circular tubes filled with foam and showed that thin-walled tubes filled with foam can absorb energy through permanent deformation. They also demonstrated that the filler can change the deformation pattern of the tube. Hou et al. [11] demonstrated that foam-filled tubes could absorb more energy than hollow tubes, and they used multi-objective optimization to support the design process. Yin et al. [12] used a multi-objective optimization algorithm to study the energy absorption characteristics of honeycomb polygonal tubes. Acar et al. [13] used a multi-objective optimization method to determine the maximum crushing force efficiency (CFE) and specific energy absorption (SEA) of frusta structures.

Thus, studies on the influence of the cross-section on the bearing capacity and energy absorption of the structure are limited. In this study, the effect of cross-sectional size, as well as their mass, on the bearing capacity and energy absorption efficiency of single and nested tubes is examined. Based on experimental results, a comparison of indicators affecting energy absorption and bearing capacity is also made.

## 2 Sample and Experiment

### 2.1 Sample

This study investigates the impact of cross-sectional size on the bearing capacity and energy absorption of thin-walled tubes, which have square section. The tubes, with a wall thickness of 1.3 mm, were cut from commercially available material and made of mild steel CT3, which has mechanical properties including an elastic modulus of 205.7 GPa, a yield stress of 218 MPa, an ultimate tensile stress of 316 MPa, and a Poisson's ratio of 0.3.

The research examined both single and nested tube configurations, all with a height of 200 mm. The single tubes were labeled S1, S2, and S3, with cross-sections of $60 \times 60$ mm$^2$, $70 \times 70$ mm$^2$, and $90 \times 90$ mm$^2$, respectively. The nested tube (NT) combined $60 \times 60$ mm$^2$ and $90 \times 90$ mm$^2$ tubes. Figure 1(a) illustrates the samples, with their detailed parameters listed in Table 1.

## 2.2 Experiment

The experimental process was conducted using a 500 kN New LUDA (China) machine, as shown in Fig. 1(b). The test specimen was placed on the lower platen, with the upper platen fixed. The lower platen moved upward at a speed of 7 mm/min. The machine consisted of a compression unit and a computer. All tests were performed under ambient conditions. Specimens were subjected to axial compression until they reached a length of 165 mm, a point at which densification was observed in all samples. A data acquisition system precisely recorded load-displacement data throughout the testing process, while a camera captured deformation.

**Table 1.** Tube parameter.

| Sample | Dimension | | | |
|--------|-----------|----------|----------------|-------------|
| | Width (mm) | High (mm) | Thickness (mm) | Length (mm) |
| S1 | 60 | 60 | 1.3 | 200 |
| S2 | 70 | 70 | | |
| S3 | 90 | 90 | | |
| NT | 60/90 | 60/90 | | |

**Fig. 1.** (a) Samples and (b) Universal testing machine.

## 3 Main Indexes

When subjected to axial compression, thin-walled tubes undergo plastic deformation and exhibit a typical force response, as depicted in Fig. 2. The deformation process consists of three stages: (1) the elastic stage, (2) the plastic stage, and (3) the densification stage. During this compression, the compressive force increases to a peak value, then falls and begins to fluctuate around the mean value. To compare the bearing capacity and energy absorption of thin-walled tubes subjected to axial compression, important indices such as Absorbed Energy (EA), Specific Absorbed Energy (SEA), Peak Force (PF), Mean Force (MF), and Crushing Force Ratio (CFR) should be defined:

(i) Absorbed energy (EA) is determined by integrating the area of the compressive force with respect to displacement in Fig. 3 as follows.

$$EA = \int_0^d F(x)dx, \tag{1}$$

(ii) Specific Absorption Energy (SEA) is considered the more important criterion for measuring the energy absorption capacity of a tube, defined as the ratio of absorbed energy to mass m of the tube.

$$SEA = EA/m, \tag{2}$$

(iii) Peak force (PF) is the maximum value of compression force (see Fig. 3).
(iv) Mean force (MF) is defined as below.

$$MF = EA/d, \tag{3}$$

(v) Crushing force ratio (CFR) is calculated as the ratio of MF to PF as follows.

$$CFR = MF/PF, \tag{4}$$

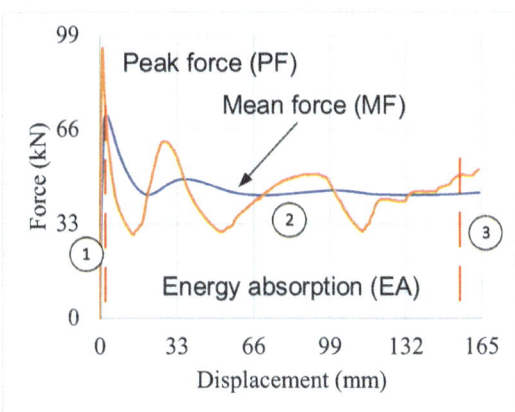

**Fig. 2.** Force response.

## 4   Results and Comparison

Figures 3, 4, 5 and 6 present the deformation history and displacement-force graphs for the four specimens. Figures 3(a), 4(a), 5(a) and 6(a) specifically illustrate the deformation mechanisms of specimens S1, S2, S3, and NT, respectively. The illustrations show that all four specimens experience progressive deformation, with their characteristics conforming to the extensional deformation mode. This deformation mode is especially advantageous for energy-dissipating structures due to its effectiveness in managing energy dissipation during crushing events.

The deformation history of the S1 sample is similar to that of the S2 and NT samples, but they differ from the S3 one. All three samples deform from top to bottom, while the deformation of the S3 one occurs from the lower end to the upper end. The number of lobes for the S1 and S2 samples is three, while the S3 sample has two lobes. The number of lobes of the NT one is one and a half, which is smaller than those of the three tubes mentioned above.

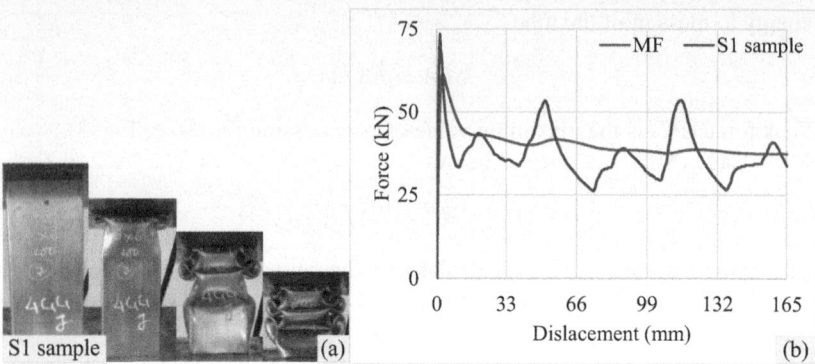

**Fig. 3.** a) Deformation and b) Displacement-force curve of S1 sample.

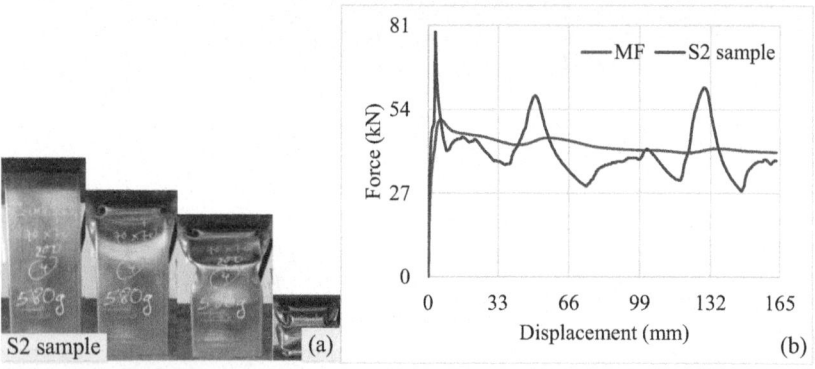

**Fig. 4.** a) Deformation and b) Displacement-force curve of S2 sample.

The wavelengths of the lobes for the S1 and S2 samples are approximately the same. However, for the S3 and NT samples, the wavelengths of the lobes differ from those of the S1 and S2 ones. The wavelengths of the folds of the S3 and NT samples are, respectively, larger and smaller than those of the other two samples. The differences and similarities in the height of the lobes can be explained by their cross-sections. Thus, the larger the tube is, the greater the wavelength of the fold. Additionally, the fold's wavelength of the nested tube is smaller than that of the single tube.

The displacement-force diagrams for all four tubes are depicted in Figs. 3(b), 4(b), 5(b) and 6(b). In these figures, the formation of lobes is associated with fluctuations in the compressive force. When compared to the S1 and S2 samples, the compressive force fluctuations of the S3 and NT ones are more gradual, which contributes to reducing the impact load on the interior cabin structure. Furthermore, these figures reveal that the load level of the NT sample is the highest, while the load level of the S1 sample is the lowest.

**Table 2.** Main indexes.

| Tube | m (kg) | PF (kN) | MF (kN) | CFR | EA (kJ) | SEA (kJ/kg) |
|------|--------|---------|---------|-----|---------|-------------|
| S1 | 0.444 | 73.54 | 37.60 | 0.51 | 6.05 | 13.63 |
| S2 | 0.580 | 79.06 | 40.28 | 0.57 | 6.48 | 11.17 |
| S3 | 0.710 | 94.48 | 43.69 | 0.46 | 7.03 | 9.91 |
| NT | 1.168 | 137.29 | 69.81 | 0.51 | 11.23 | 9.61 |

From the displacement-force diagrams in Figs. 3(b), 4(b), 5(b) and 6(b), key indices such as mass (m), Peak Force (PF), Mean Force (MF), Crushing Force Ratio (CFR), Energy Absorption (EA), and Specific Energy Absorption (SEA) were determined and are listed in Table 2. These values help to clarify the influence of tube cross-sections on bearing capacity and energy absorption.

The results in Table 2 show that as the cross-sectional size increases, PF, MF, and EA also increase, while CFR and SEA decrease. PF and MF for the samples rise as their masses increase. However, the CFR of the NT sample increases slightly and is equivalent to that of the S1 sample. Among the samples, S3 has the smallest CFR, while S2 has the largest.

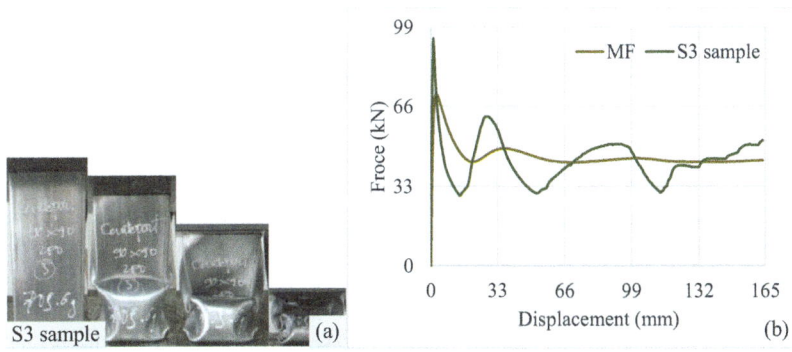

**Fig. 5.** a) Deformation and b) Displacement-force curve of S3 sample.

The key indices including PF, MF, CFR, and EA of the S2 sample increase by 7.5%, 7.1%, 11.7%, and 7.1%, respectively, compared to those of the S1 sample; however, its SEA is 17.9% less than that of the S1 sample. Compared with the PF, MF, and EA of the S1 sample, the PF, MF, and EA of the S3 sample increase by 28.4%, 16.2%, and 16.2%, respectively. The CFR and SEA of the S3 sample decrease by 9.8% and 27.3%, respectively. Similar to the S3 sample, the PF, MF, and EA of the NT sample are, respectively, 86.6%, 85.6%, and 85.6%, larger than those of the S1 sample. However, its CFR and SEA decrease by 0.6% and 29.4%, respectively. The decreases in CFR and SEA of the last three samples compared to the first sample are due to their larger PF and mass.

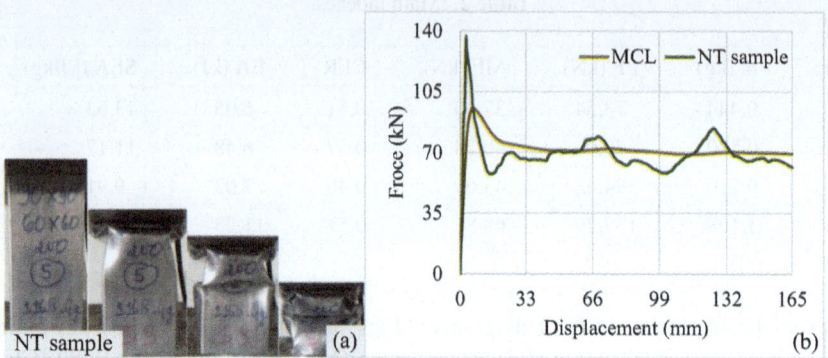

**Fig. 6.** a) Deformation and b) Displacement-force curve of NT sample.

As such, they show that the heavier the tube is, the smaller the SEA is, and the larger the tube is, the greater the compressive force required to trigger the tube deformation. The S2 sample has the largest CFR value due to large MF and small PF, which reduces the impact of compression on the structure attached to this energy-absorbing tube. Although the NT sample's SEA is 2.97% smaller than that of the S3 one, it has the highest crushing resistance, and the space it occupies is equal to the space occupied by the S3 one.

## 5   Conclusion

This paper examines the influence of tube cross-section on bearing capacity and energy absorption. The study focuses on square cross-sections. Experimental results show that both the cross-section and mass of the tube significantly affect its bearing capacity and energy absorption efficiency. Larger and heavier tubes exhibit greater peak force but lower energy absorption efficiency. The CFR is a crucial factor in the selection and design of energy absorbers, as a high CFR is typically desired. The PF of the S3 sample is 86.6% greater, while its SEA is 29.4% lower compared to the S1 sample. Additionally, the NT sample showed increases of 86.6% in PF, 85.6% in MF, and 85.6% in EA compared to the S1 sample. In terms of crashworthiness design, the NT nested sample offers the greatest crushing resistance and more efficient use of space compared to single-tube samples.

## References

1. Wierzbicki, T., Abramowicz, W.: On the crushing mechanics of thin-walled structures. J. Appl. Mech. **50**(4a), 727–734 (1983)

2. Langseth, M., Hopperstad, O.S.: Static and dynamic axial crushing of square thin-walled aluminium extrusions. Int. J. Impact Eng 18(7–8), 949–968 (1996)
3. Santosa, S.P., Wierzbicki, T., Hanssen, A.G., Langseth, M.: Experimental and numerical studies of foam-filled sections. Int. J. Impact Eng 24(5), 509–534 (2000)
4. Chen, W., Wierzbicki, T.: Relative merits of single-cell, multi-cell and foam-filled thin-walled structures in energy absorption. Thin-Walled Struct. 39(4), 287–306 (2001)
5. Kim, H.-S.: New extruded multi-cell aluminum profile for maximum crash energy absorption and weight efficiency. Thin-Walled Struct. 40(4), 311–327 (2002)
6. Mamalis, A.G., Manolakos, D.E., Demosthenous, G.A., Johnson, W.: Axial plastic collapse of thin bi-material tubes as energy dissipating systems. Int. J. Impact Eng 11(2), 185–196 (1991)
7. Hsu, S.S., Jones, N.: Quasi-static and dynamic axial crushing of thin-walled circular stainless steel, mild steel and aluminium alloy tubes. Int. J. Crashworthiness 9(2), 195–217 (2004)
8. Alavi Nia, A., Haddad Hamedani, J.: Comparative analysis of energy absorption and deformations of thin walled tubes with various section geometries. Thin-Walled Struct. 48(12), 946–954 (2010)
9. Seitzberger, M., Rammerstorfer, F.G., Gradinger, R., Degischer, H.P., Blaimschein, M., Walch, C.: Experimental studies on the quasi-static axial crushing of steel columns filled with aluminium foam. Int. J. Solids Struct. 37(30), 4125–4147 (2000)
10. Rajendran, R., Prem Sai, K., Chandrasekar, B., Gokhale, A., Basu, S.: Impact energy absorption of aluminium foam fitted AISI 304L stainless steel tube. Mater. Des. 30(5), 1777–1784 (2009)
11. Hou, S., Li, Q., Long, S., Yang, X., Li, W.: Crashworthiness design for foam filled thin-wall structures. Mater. Des. 30(6), 2024–2032 (2009)
12. Yin, H., Wen, G., Hou, S., Chen, K.: Crushing analysis and multiobjective crashworthiness optimization of honeycomb-filled single and bitubular polygonal tubes. Mater. Des. 32(8–9), 4449–4460 (2011)
13. Acar, E., Guler, M.A., Gerçeker, B., Cerit, M.E., Bayram, B.: Multi-objective crashworthiness optimization of tapered thin-walled tubes with axisymmetric indentations. Thin-Walled Struct. 49(1), 94–105 (2011)

# Safety and Sustainability in Medical Devices

Maria Heckel[1]([⊠]) [iD], Simon Barrans[2] [iD], and Diana Völz[1] [iD]

[1] Frankfurt University of Applied Sciences, 60318 Frankfurt am Main, Germany
maria.heckel@fb2.fra-uas.de
[2] University Huddersfield, Queensgate, Huddersfield HD1 3DH, UK

**Abstract.** The European Union (EU) is promoting the integration of biobased plastics into industrial processes to support the circular economy framework and reduce greenhouse gas emissions. Biobased plastics have proven to be a low-carbon solution in various industries. This paper examines the challenges and opportunities of transitioning to circular bioplastics for medical devices. The medical sector presents unique challenges, including degradability, leaching, sterilization, biocompatibility, recycling, and incineration of bioplastics. In addition, inconsistencies in LCA methodologies complicate the evaluation process. A network analysis highlights the interconnectedness of these factors, with leaching, degradation, and incineration behaviors emerging as primary influencers. This review proposes a framework for transitioning to a circular plastics economy in medical devices, emphasizing the need to reduce overall plastics demand, explore sustainable end-of-life options, conduct standardized LCAs, and develop suitable bioplastic compounds. Future research directions include developing durable, low-toxicity bioplastics, advancing chemical recycling methods, and expanding comprehensive LCA studies to ensure the safe and sustainable integration of bioplastics into medical devices.

**Keywords:** Bioplastics · Medical Devices · Life Cycle Assessment

## 1 Introduction

Medical devices are employed for a variety of purposes, including diagnosis, prevention, monitoring, and treatment of human diseases. The versatility and cost-effectiveness of plastics have contributed to the efficiency of modern medical devices. The popularity of disposables (single-use devices) increases in response to concerns about the potential for infection and cross-contamination. Consequently, the healthcare sector generates a considerable quantity of waste, with 20–25% being plastics [1], which contributes to environmental issues. It has been estimated that up to 5% of global greenhouse gas emissions can be attributed to the healthcare sector [2]. As a result, the environmental impact of plastics has prompted regulatory initiatives at the European Union level, including the call for at least 20% of the carbon content of chemical and polymeric products to be from renewable sources by 2030 and the use of biobased materials in instances where environmental advantages can be created [3, 4]. The utilization of biobased plastics is

© The Author(s) 2025
H. Kohl et al. (Eds.): GCSM 2024, LNME, pp. 254–261, 2025.
https://doi.org/10.1007/978-3-031-93891-7_29

regarded as a potential strategy in line with the principles of the circular economy (CE), with the objective of reducing resource consumption, waste generation, and environmental impacts [5]. However, integrating them in medical devices necessitates a careful consideration of regulatory compliance and environmental performance to ensure patient safety and product efficacy over the life cycle of the devices.

This paper introduces a framework for transitioning to circular medical plastics, addressing key challenges identified through a network analysis using Gephi 0.10, an open-source software for network analysis and visualization. The shortcomings of the current LCA methodologies for bioplastics in medical devices are reviewed and suggestions for potential adaptations that could enhance their applicability are developed. The objective of this paper is to facilitate further investigation into the potential of circular plastics in the field of medical technology, which is still in its infancy. Despite the regulatory challenges faced by the medical devices industry, the insights gained from this analysis allow the development of a framework for the use of circular medical polymers with improved environmental performance.

## 2   Potentials and Limitations of Bioplastics

Bioplastics are formulations of biopolymers such as polylactic acid (PLA), which, like fossil bioplastics, are mixed with additives and fillers for industrial application. They can be designed to be biodegradable under controlled conditions or non-biodegradable. For the purposes of this paper, we define bioplastics as polymeric materials made from renewable feedstock. Bioplastics may have a chemically novel structure, having no direct fossil-based equivalents, such as PLA. In contrast, bio-based polyethylene (bio-PE) is chemically identical to fossil-based polyethylene (PE), but is manufactured from bio-based raw materials and designated as a "drop-in" material [6]. To illustrate, PLA is a substitute for polystyrene (PS), while bio-based polyethylene terephthalate (bio-PET) is a substitute for PET. They can be processed using conventional processing technologies, such as injection molding and sheet extrusion, as evidenced by the existence of industrially available products like biobased and even biodegradable bags and other products. Nevertheless, the suitability of biobased plastics for all applications remains uncertain, as they degrade more rapidly than petrochemical plastics when exposed to light, high temperatures, and increased humidity. This also results in alterations to their mechanical properties and durability. To address these challenges and facilitate the transition to a CE based on biobased raw materials, biobased plastics must be developed that are tailored to every single specific application and, in particular, allow the materials to be returned to the material cycle [5], for example as recycled raw materials.

As with all other sectors, the medical industry is facing increasing pressure to reduce its environmental impact [3]. This is particularly relevant regarding the use of disposable products containing plastics [7]. These include infusion bags, endoscopes, syringes, etc. As a result, the substitution of fossil-based plastics with bio-based plastics is now also being researched in medical technology. Bio-based plastics are a promising alternative, potentially reducing greenhouse gas emissions by up to 45% over their lifecycle and aligning with the European Green Deal's focus on the circular economy and climate goals [6]. However, research results also indicate the potential for "burden shifting" where changes in land use and the acidification and eutrophication of soils represent a significant concern. This is largely attributed to the use of fertilizers and pesticides in conventional agriculture for the production of plant-based feedstock. Nevertheless, extensive agriculture, for instance, can mitigate the potential for the burden-shifting [8].

# 3 Life Cycle Assessment of Bioplastics

LCA is a quantitative methodology employed to evaluate the environmental impacts of products throughout their entire life cycle, from the extraction of raw materials to their final disposal. This process is commonly referred to as "cradle-to-grave" analysis. According to the ISO 14040 standard and the Product Environmental Footprint (PEF) methodology proposed by the European Union, which was developed to facilitate the comparison of assessments and prescribe the impact categories that should be employed [9], an LCA is divided into four phases:

1. Goal and Scope Definition: This involves defining the purpose of the LCA, the system boundaries, the level of detail required and the functional unit to be assessed.
2. Life-cycle inventory (LCI): The LCI phase collects and quantifies data on all inputs (materials, energy) and outputs (emissions, waste) throughout the product life cycle, from raw material extraction to disposal.
3. Life-cycle impact assessment (LCIA): The LCIA phase involves the analysis of the LCI data in order to assess potential environmental impacts, categorizing them into impact categories such as climate change or land use. The assessment can be concluded at this point, or it can be further aggregated into broader areas of protection, such as human health, ecosystem quality and natural resources [10].
4. Interpretation and recommendations: The results are interpreted to identify key environmental issues, assess the robustness of the results, and provide recommendations for improving the environmental performance of the product.

## 3.1 Challenges in LCA of Bioplastics

LCA studies that incorporate multiple impacts beyond just climate change, such as acidification, eutrophication, and land use, have demonstrated that bioplastics can be disadvantageous in certain aspects at their current development stage [8]. Although the methodology for conducting LCAs for products is standardized according to ISO 14040 and the EU advocates the Product Environmental Footprint methodology for the assessment of environmental impacts, the freedom of choice regarding the employed methodology leads to inconsistencies and reduced relevance in LCA outcomes of biobased plastics [11]. In the light of the current state of knowledge, it is not possible to recommend either fossil-based or biobased plastics on the basis of their environmental impact.

## 3.2 Opportunities in LCA of Bioplastics

Today, LCAs are typically comparable only in terms of the climate change impact category, as this is the most consistently covered in LCA studies [11, 12]. Nevertheless, the existing literature indicates that by incorporating only six key impact categories (see

Table 1), namely climate change, land use, ozone depletion, acidification, eutrophication and ecotoxicity, it is possible to successfully address up to 92,3% of the variabilities observed in LCA studies [12]. As previously stated, the most specifically relevant impact categories for evaluating bioplastics in LCA are land use, acidification, and eutrophication [9]. The aggregation of these findings permits the implementation of a comprehensive yet efficient approach to LCAs, as illustrated in Table 1. The application of the six impact categories ensures the coverage of all three damage level categories of the LCIA phase (human health, ecosystem quality, and natural resources [10]) with the use of the lowest number of impact categories possible.

**Table 1.** Integrating Six Key Categories in LCAs for Comprehensive Coverage.

| Key Impact categories | Relevance | Damage categories (areas of protection) | | |
|---|---|---|---|---|
| | | Human health | Ecosystem quality | Natural resources |
| Climate change | ●○ | ● | ● | ○ |
| Land use | ●● | ○ | ● | ● |
| Ozone depletion | ●○ | ● | ● | ○ |
| Acidification | ●● | ○ | ● | ● |
| Eutrophication | ●● | ○ | ● | ○ |
| Ecotoxicity | ●○ | ○ | ● | ○ |

# 4 Challenges in Medical Device Plastics

To mitigate the risk of infection by viruses or spore-forming bacteria, disposable items are frequently employed in the medical device industry [13]. While bio-based plastics are already utilized in various industries, their integration into medical devices presents interrelated challenges due to specific material properties and their impact on ecosystems and human health. Based on a literature review, six key challenges were identified: degradability, leaching, sterilization, biocompatibility, recycling, and incineration.

Bioplastics, composed of a polymer base and various additives, are often sensitive to environmental conditions. In medical devices, these materials often interact with liquids such as infusion solutions, medications, blood and other body fluids. This interaction **accelerates degradation**, as the molecular chains of the biopolymers can break down on contact with these fluids. Therefore, the degradability of bioplastics poses a challenge for their medical application, especially regarding their durability and biocompatibility, which may necessitate early disposal and replacement. The Medical Device Regulation (MDR, 2017, Annex II) for the European Union stipulates that any material utilized in a medical device must not compromise its clinical function or the safety and health of patients or users. **Biocompatibility**, defined as "the ability of a material to perform with an appropriate host response in a specific application" [14], is crucial for biopolymers employed as material substitutes, regardless of the device's invasiveness or application duration. The human tissue response depends on the material's chemical composition and the leaching of additives. Zimmermann et al. have demonstrated that although bioplastics are known to be biocompatible, they contain thousands of chemicals that could induce toxicity in humans [15] through leaching. The **leaching of these additives**, which are not covalently bound to the material, poses significant health risks and impacts the safety and efficacy of medical devices.

In addition to degradation processes, other factors such as the original composition, the quality of the components, and the effects of sterilization, which is required for the safety of medical devices, can also influence the biocompatibility of a material. **Sterilization processes can alter the chemical structure** of bioplastics, potentially releasing harmful substances and compromising the integrity of the material [16]. Such issues can result in a reduction in the lifespan of the product, necessitating earlier disposal or impeding effective recycling, which in turn affects the results of life-cycle assessments. In order for the transition to bioplastics in the circular economy to be successful, it is essential to determine a safe and contamination-free end-of-life process with the lowest environmental footprint. Currently, the **EU prohibits material recycling** for contaminated plastics from medical settings, classifying them as hazardous waste and mandating thermal recycling (incineration) [17]. In order to maintain the value of bioplastics without compromising material quality, chemical recycling could be employed. Nevertheless, this necessitates an assessment of the chemical composition and its impact on the recyclability of the material in question. Incineration has been identified as a significant risk to human health and ecosystems. The practice of **thermal recycling (incineration) for contaminated plastics** from medical settings has the potential to release hazardous substances into the environment. Evaluating the chemical composition of bioplastics is crucial to developing new materials that reduce environmental impacts compared to fossil-based plastics.

## 5  Developing a Framework for Circular Medical Plastics

Research has outlined a strategy aimed at achieving a circular plastics industry through four interconnected targets: reducing plastics demand by 50%, achieving a 95% recycling rate, incorporating renewable feedstocks such as biopolymers, and decarbonizing the industry [5, 18]. These goals provide a framework for addressing specific challenges in the adoption of circular practices within the medical device industry.

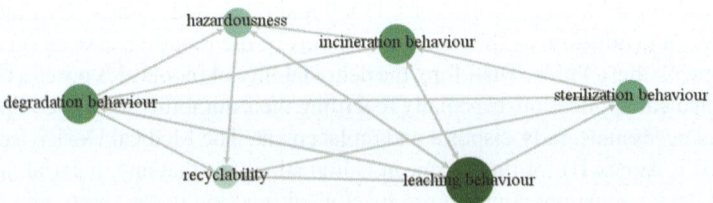

**Fig. 1.** Network Graph of Influencing Factors

To understand the cause-and-effect relationships among the six factors influencing the transition to circular medical plastics, a network analysis using Gephi 0.10 was conducted, visually represented in Fig. 1. Nodes in the network represent these factors, with darker green nodes indicating greater influence. Of central importance are the leaching behavior, degradation behavior, and incineration behavior as primary influencers within the network. Leaching behavior was identified as a critical factor due to its direct impact on biocompatibility and potential toxicity. The behavior of chemicals leaching from bioplastics can affect both the safety of medical devices and the environment. Degradation behavior is critical because it affects the lifetime of medical devices and the environmental degradation of bioplastics. Rapid degradation can lead to more frequent replacement and potentially adverse environmental effects. Incineration of bioplastics poses a significant risk to human health and the environment. The importance of this factor is underscored by the emission of hazardous substances and the challenges in achieving safe disposal methods. Recyclability and biocompatibility do not influence other factors.

The network analysis of the seven critical factors influencing the adoption of bioplastics in medical devices reveals the complexity of the issue, highlighting the interconnectivity and impact of these factors. With these challenges and influencing factors in mind, the framework of Vidal et al. (2024) has been adapted to facilitate the transition to a circular plastics industry for medical devices. The framework consists of four key pillars (see Fig. 2) that need to be addressed simultaneously due to their interrelated nature:

- To adopt CE principles regarding medical devices, it is necessary to reduce the overall demand for plastics by **focusing on essential uses**.
- To address the challenges of incinerating disposable medical devices, explore **sustainable end-of-life options** like chemical recycling and controlled degradation with carbon capture to reintroduce materials into value chains.
- Conduct accurate life cycle assessments of bioplastics in medical devices using **standardized LCA** methodologies.
- Ensure **bioplastics' suitability** for medical devices by developing compounds that enhance performance while meeting safety, efficacy, and regulatory standards. Use reliable biocompatibility testing methodologies.

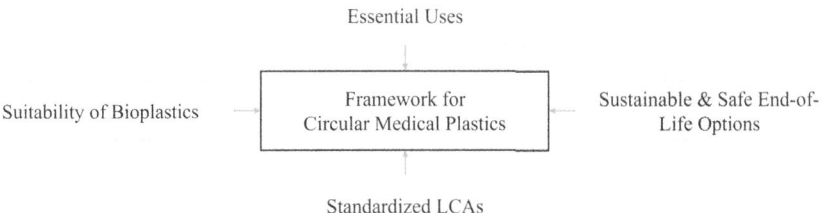

**Fig. 2.** Framework for Transitioning to Circular Plastic Industry in Medical Devices

To effectively implement the four pillars and guide future work, the framework must be quantifiable. First, to reduce plastic demand, research should focus on identifying key areas of high plastic use in medical devices through sources like hospital and treatment statistics. This will help target efforts where they can be most impactful. Medical devices that can be sterilized and reused should be excluded from further analysis, narrowing the focus to single-use devices. Understanding the plastic volumes required for these non-reusable devices is crucial for sustainability assessments through comprehensive LCAs, which should explore alternative end-of-life scenarios beyond incineration to find the most sustainable options.

Standardized LCA approaches should be applied, utilizing the PEF framework's impact categories, especially those relevant to bioplastics from agricultural feedstock. For biocompatibility testing, future decision-making should prioritize material tests focusing on leaching behavior, degradation, and incineration to ensure the safe integration of bioplastics into medical devices. By emphasizing these quantifiable aspects, the framework will better guide sustainable practices in the medical device industry.

## 6 Conclusion and Future Outlook

Integrating bioplastics into medical devices offers sustainability but comes with challenges. A network analysis identified key factors affecting bioplastics' environmental performance. Leaching and degradation significantly influence biocompatibility, toxicity, and device lifespan, while incineration poses substantial health and environmental risks. To address these challenges, a framework for a circular economy in the medical plastics sector is proposed. This framework encompasses the following elements: reducing the demand for plastics, developing sustainable end-of-life options, applying standardized LCA methods, and making informed decisions in the development of new bioplastics with regard to current and future regulatory requirements in terms of sustainability and safety. Future efforts should focus on developing durable, low-toxicity bioplastics, advancing chemical recycling methods, and expanding life cycle assessments. This approach can reduce the environmental impact of medical plastics and improve patient safety.

## References

1. Lee, B.-K., Ellenbecker, M.J., Moure-Eraso, R.: Analyses of the recycling potential of medical plastic wastes. Waste Manag. **22**, 461–470 (2002)
2. Lenzen, M., et al.: The environmental footprint of health care: a global assessment. Lancet. Planet. Health **4**, e271–e279 (2020)
3. European Commission: Communication from the Commission to the European Parliament and the Council: Sustainable Carbon Cycles. COM(2021) 800 final. Brussels, Belgium (2021)
4. European Commission: Communication from the Commission to the European Parliament, the Council, the European Economic and Social Committee and the Committee of the Regions. A New Circular Economy Action Plan for a Cleaner and More Competitive Europe. COM(2020) 98 final (2020)
5. Vidal, F., et al.: Designing a circular carbon and plastics economy for a sustainable future. Nature **626**, 45–57 (2024)

6. Zuiderveen, E.A.R., et al.: The potential of emerging bio-based products to reduce environmental impacts. Nat. Commun. **14**, 8521 (2023)
7. Ivanović, T., Meisel, H.-J., Som, C., Nowack, B.: Material flow analysis of single-use plastics in healthcare: a case study of a surgical hospital in Germany. Resour. Conserv. Recycl. **185**, 106425 (2022)
8. Weiss, M., et al.: A review of the environmental impacts of biobased materials. J. Ind. Ecol. **16** (2012)
9. Walker, S., Rothman, R.: Life cycle assessment of bio-based and fossil-based plastic: a review. J. Clean. Prod. **261**, 121158 (2020)
10. Verones, F., et al.: LCIA framework and cross-cutting issues guidance within the UNEP-SETAC Life Cycle Initiative. J. Clean. Prod. **161**, 957–967 (2017)
11. Hottle, T.A., Bilec, M.M., Landis, A.E.: Sustainability assessments of bio-based polymers. Polym. Degrad. Stab. **98**, 1898–1907 (2013)
12. Steinmann, Z.J.N., Schipper, A.M., Hauck, M., Huijbregts, M.A.J.: How many environmental impact indicators are needed in the evaluation of product life cycles? Environ. Sci. Technol. **50**, 3913–3919 (2016)
13. Hospodková, P., Rogalewicz, V., Králíčková, M.: Gynecological speculums in the context of the circular economy. Economies **11**, 70 (2023)
14. Williams, D.F., Zhang, X. (eds.): Definitions of biomaterials for the Twenty-First Century. Proceedings of a Consensus Conference held in Chengdu, People's Republic of China, June 11th and 12th 2018, organized under the auspices of the International Union of Societies for Biomaterials Science & Engineering: hosted and supported by Sichuan University, Chengdu and the Chinese Society for Biomaterials, China. Materials Today, vol. Elsevier, Amsterdam, Netherlands (2019)
15. Zimmermann, L., Dombrowski, A., Völker, C., Wagner, M.: Are bioplastics and plant-based materials safer than conventional plastics? in vitro toxicity and chemical composition. Environ. Int. **145**, 106066 (2020)
16. Bernard, M., Jubeli, E., Pungente, M.D., Yagoubi, N.: Biocompatibility of polymer-based biomaterials and medical devices - regulations, in vitro screening and risk-management. Biomater. Sci. **6**, 2025–2053 (2018)
17. Kane, G.M., Bakker, C.A., Balkenende, A.R.: Towards design strategies for circular medical products. Resour. Conserv. Recycl. **135**, 38–47 (2018)
18. Bachmann, M., et al.: Towards circular plastics within planetary boundaries. Nat Sustain. **6**, 599–610 (2023)

# Guidelines for Conceptualization and Design of Human - Centered Empathetic Intelligent Machines in Production Systems Within the Industry 5.0 Paradigm

Domenico Monopoli[1]([✉]), Pietro Miciaccia[1], Concetta Semeraro[2],
and Michele Dassisti[1]

[1] Department of Mechanical, Mathematics and Management (DMMM), Polytechnic University
of Bari, Bari, Italy
d.monopoli2@phd.poliba.it

[2] Department of Industrial and Management Engineering, University of Sharjah, Sharjah,
United Arab Emirates

**Abstract.** In recent years, the uncontrolled proliferation of new technologies has brought numerous benefits to both industry and society, improving productivity and the quality of production processes. However, an increasingly common sentiment is that developing innovative technologies, if not properly regulated, could hinder sustainable progress. The emergence of Industry 5.0 seeks to address these issues. The widespread integration of intelligent technologies, such as AI, into production systems has resulted in unforeseen adverse effects, such as declining human skills, social isolation, increasing social inequality, and ethical and privacy concerns. These challenges can be addressed by developing empathetic and collaborative intelligent systems and innovative human-machine interaction modes. These intelligent and empathic collaborative machines must be able to detect the operator's physical, cognitive, and emotional state, predict the operator's intentions, and behave accordingly. This manuscript aims to establish a framework within the Industry 5.0 paradigm to serve as a guideline for conceptualizing and designing intelligent, empathic, collaborative technologies in production systems capable of empowering operators while respecting their innate qualities as thinking, emotional, ethical, and social beings.

**Keywords:** Human-Centered Manufacturing · Collaborative Intelligent machines · Empathetic Technologies

## 1 Introduction

In production systems, adopting new technologies has freed human beings from performing repetitive tasks, improved worker safety, and reduced physical fatigue by improving the ergonomics of workstations. Technological development has thus undoubtedly brought many benefits to society at large. Despite this, there is a common feeling in the scientific community that, if not properly developed, technologies can lead to unsustainable development. Dieste et al. [1] in their work identify some unexpected effects of Industry 4.0 technologies on society, including negative impacts on jobs, loss of worker privacy, loss of worker autonomy, unhealthy work-life balance due to constant connectivity, safety issues related to the coexistence of autonomous humans and machines at workstations. In response to these difficulties, the European Union has changed its policy and produced a manifesto document announcing the launch of Industry to ensure Sustainable, Human-centric, and Resilient development of European industry [2]. In this vision, machines would work alongside operators, improving production efficiency, flexibility of production lines, and operator well-being. Although the topic is relatively new, some authors have tried to rationalize the concepts behind developing these new forms of collaboration between humans and machines. As pointed out in [3], to achieve these goals, the human being and the machine must achieve the "symbiotic understanding" allowing the system to achieve optimal performance in action and decision-making. In [3], a three-component framework is proposed to rationalize the symbiotic collaboration between man and machine. Given a goal, the decomposition level allows one to define the various steps to achieve it, the decision level permits one to understand the best alternative to reach the goal, and in the action plan, the human and the machine collaborate and implement the developed plan. A similar concept is identified in [4], where they state that the machine must possess Know How (KH) and Know How To Cooperate (KHTC) to achieve maximum efficiency and flexibility in production. In the framework they developed, KH and KHTC are fundamental to building the behavioral models of the two entities needed to make decisions that maximize objectives at the strategic, tactical, and operational levels. Although these works represent a first step towards improving human-machine collaboration, they do not consider the human being as a whole, as they do not consider cognitive and emotional well-being. Furthermore, they do not guide professionals who want to implement these collaborative human-machine systems. This work is proposed as an initial effort to conceptualize intelligent and empathic collaborative systems in productive environments that can promote human beings' physical, cognitive, and emotional well-being with a reference-based approach. It is essential to specify that in this work, we refer to empathy as a communicative process in which we understand and respond to the feelings and emotions perceived by others [5]. In this work, specifically in Sect. 2, the challenges that hinder deep human-machine collaboration are collected; in Sect. 3, these challenges are declined into system requirements, and in Sect. 4, the requirements are grouped into the Sense, Understand, Predict, Act framework, and some examples of technical implementation in real systems are shown. Finally, future research directions are identified in the conclusions.

## 2    Challenges of Human-Machine Collaboration

The dominant view concerns the development of work cells consisting of two entities, man and machine, and within this, each entity can express its strengths to the full. The first difficulty in human-machine collaboration lies in the communication between the two entities [6]: in exchanging actions, intentions, and information. Intending to improve human-machine collaboration, several authors have reported on the difficulties encountered in achieving the goal. These have been collected in Table 1. The changes required of human beings are mainly concerned with training and education. Knowledge and understanding of the machine are the primary enablers for predicting machine behavior [3]. Indeed, if the machine behaves as expected, it is readily accepted by the human [7]. Similarly, the machine will have to understand the human being through training through the creation of behavioral models, characteristics, and preferences of the human being to understand its limitations, predict its behavior, and act accordingly [8]. Starting from the challenges that hinder the realization of fully collaborative human-centered systems, as summarised in Table 1, the functional requirements in Table 2, which are that the machine must be collaborative, empathetic, and intelligent, were derived.

**Table 1.** Human-Machine Collaboration Challenges.

| Type | N | Challenge | Description /Rationale | Reference |
|------|---|-----------|------------------------|-----------|
| Social Challenges | 1 | Technology acceptance and trust | Man perceives the machine as practical, easy to use, and reliable | [5, 7, 9–16] |
| | 2 | Change of team Dynamics | The man accepts that his role within the production may change | [10, 11, 14, 15, 17, 18] |
| | 3 | Ethics and Privacy | The man acknowledges that the machine will use his data for good and will not be passed on to third parties | [1, 12–16] |
| | 4 | Lifelong learning | Man accepts that training will be continuous | [7–10, 13–15, 18] |
| Technical Challenges | 5 | Human Centric, personalized and adaptive | The machine adapts to man's physical, cognitive, and emotional conditions and allows the customization that operators prefer | [7, 8, 10–12, 14–18] |
| | 6 | Transparency and Explainability | The machine acts transparently, showing its intentions | [7–9, 13–16] |

(*continued*)

**Table 1.** (*continued*)

| Type | N | Challenge | Description /Rationale | Reference |
|------|---|-----------|------------------------|-----------|
| | 7 | Performances Measures | The performances to be monitored are not only those related to the production process but also those concerning the well-being of the operator and the quality of collaboration | [5, 7, 11, 14, 15] |
| | 8 | Interaction modes | The machine allows new and different modes of interaction depending on the task, role performed, and characteristics and preferences of the human being | [7, 8, 10, 13–16] |

## 3  Functional Framework for Conceptualization and Design of Intelligent, Collaborative, and Empathic Machine

As can easily be seen from Table 2, the functional requirements contain keywords reflecting the main functions that the main subsystems constituting the collaborative, intelligent empathic machine will have to perform. In particular, the machine will consist of the sensing, understanding, prediction, and action subsystems.

The findings summarized in Table 2 confirm what has been found in the literature regarding conceptualizing enhanced human-machine collaboration. It also adds some components to the frameworks presented by [3] and [4]. To achieve "Symbiotic understanding" and the "know-how to cooperate," it is necessary for the machine to gather the information required to be able to construct the behavioral models of the operator. This will also require the framework to contain the sensing subsystem. Having collected the relevant information about intentions, preferences, skills, and physical, cognitive, and emotional states, this information is used to model the operator in the understanding subsystem. It is important to note that the models constructed by the machine will generally be valid only for the operator from whom the data were collected because of the high subjectivity of the information collected. In this way, the machine will be able to understand whether, for example, the worker is fatigued rather than bored and whether or not it is necessary to change the work mode. These evaluations will be made in the predicting subsystem where the machine will assess whether any change in the mode of operation is in line with the production goals set. Finally, thanks to the acting subsystem the identified action is performed. It is essential to notice that the part relating to sensing,

**Table 2.** System Requirements for collaborative empathic and intelligent machines.

| N | System Requirement | Ch. Add. | Ref. |
|---|---|---|---|
| 1 | Humans shall be trained and educated on machine Limits | 1,2,4, | [12, 13, 17] |
| 2 | Humans shall be trained and educated on machine use and Behavior | 1,2,3,4 | [8, 12, 13, 17] |
| 3 | Humans shall be educated on Privacy concerns and cybersecurity issues | 3 | [14, 17] |
| 4 | The machine should adapt itself to Human's skills | 1,2,4 | [6, 19, 20] |
| 5 | The machine should adapt itself to Human's preferences and behavior | 1,3 | [9, 14, 19, 21] |
| 6 | The machine shall be able to train the user | 4 | [8] |
| 7 | The machine shall sense Human Physical, Cognitive, Emotional load | 5 | [4, 5, 22, 23] |
| 8 | The machine shall understand Human Physical, Cognitive, and Emotional Limits | 5,6 | [4, 20, 22–24] |
| 9 | The machine shall understand Human intentions and Behavior | 5,6,8 | [4, 6, 8, 9, 21, 24] |
| 10 | The machine shall understand Human Skills, Expertise and Working preferences | 5,6,8 | [6, 8, 9, 20] |
| 11 | The machine shall communicate with the human | 5,6,8 | [4, 6, 9, 24] |
| 12 | The machine shall evaluate human well-being performances | 7,8 | [5, 21, 22, 25, 35] |
| 13 | The machine shall evaluate the collaboration quality performances | 7,8 | [9, 25, 26] |
| 14 | The machine shall evaluate process performances | 7,8 | [4, 24, 25] |
| 15 | The machine shall behave according to the Human Physical, Cognitive, and Emotional load | 8 | [9, 20–22] |
| 16 | The machine shall behave according to Human Behavior | 8 | [8, 9, 21, 22] |
| 17 | The machine shall behave according to the Skills, Expertise and Working preferences | 8 | [7, 8, 18, 20] |
| 18 | Machines shall have different working modes according to human and process performances | 8 | [6, 9, 18, 20] |
| 19 | The machine shall have different interaction roles | 8 | [18–20] |

Understanding, and Predicting to complete the task is beyond the scope of this work. The complete framework is shown in Fig. 1.

### 3.1 Sensing Subsystem

This subsystem must capture the states of the operator, namely, human workers' physical, emotional, and cognitive condition. This subsystem is responsible for collecting signals and information from the operator. According to requirements 7 and 8, the system must take up the physical, cognitive, and emotional load. Generally, the types of physical, cognitive, and emotional load assessment fall into three categories: subjective, those derived from observations, and those by direct measurement [27]. Regarding physical load, the characteristics to be measured allow us to make kinetic and kinematic analyses of movements. For example, Lorenzini et al. [28] Seventeen inertial measurement units were used to capture body movements, and a force plate was used to measure loads. Tirupachuri et al. [29] perform online physical load assessments using accelerometers and innovative sensor shoes. An additional example is addressed in [26]. Regarding cognitive load, on the other hand, numerous studies report the possibility of intercepting this dimension through the measurement of physiological parameters and behavioral observations. The most measured physiological parameters are EDA, EEG, and HRV, while behavioral measures include those related to eye tracking, gesture analysis, and voice analysis. Planke et al. 2021 [21] use a system capable of receiving EEG, cardiac activity analysis, eye movement analysis, and voice analysis to assess the operator's cognitive state. Jo et al. 2022 [23] use sensors for EGG analysis, cardiac analysis, and electrodermal activity as physiological measurements, eye analysis, facial expressions, and mouse control to perform behavioral assessments. Finally, Lagomasino et al. [11] use two subsystems for cognitive analysis: a kinematic tracking system for gesture analysis to assess body language, evaluating, for example, hyperactivity or self-touching, and attention tracking by head positioning in addition to measurements of physiological parameters such as EGG, EDA, and HRV. Emotional load is another factor to consider in improving human-machine collaboration [30]. Methods of measuring and assessing emotional load include those based on measures of physiological parameters, e.g., those derived from EGG measurements. In [30, 31], they use measurements derived from 12 sensors placed on the operator's head to measure emotional responses in different scenarios.

### 3.2 Understanding Subsystem

According to [3], for man and machine to achieve the highest collaboration standards, each entity in the system must understand the other and predict its behavior before taking action. Once the operator's physical, cognitive, and emotional signals have been received, the machine must possess the patterns to compare them to modify its behavior. The machine, therefore, must contain the static and dynamic models representative of the operator with whom it is working [22, 25, 32]. Some examples of static models of the operator concern the biographical characteristics, skills possessed, personality, and physical dimensions. As for the dynamic ones, we have the dynamic and kinematic characteristics, physical, cognitive, and emotional fatigue [20, 33]. For example, in [29], they use a kinematic and dynamic geometric model of the human body consisting of rigid bodies connected by joints to estimate the load and optimize the robot's actions. In [26], on the other hand, the geometric model is constructed by the vision system based on the mutual positioning of the markers worn by the human being. In [21], an Adaptive Neuro-Fuzzy Inference System, calibrated offline, allowed the data to be clustered into four different states characterizing the operator. Subsequently, the model could accurately predict the operator's state in real time. An innovative approach is adopted by Lago-marsino et al. [11] where, in addition to signals derived from physiological parameters, behavioral manifestations such as hyperactivity or self-touching are used to construct the operator's behavioral and cognitive model. Regarding emotional models, it is first worth noting that there are very few application cases in the literature. An example is described by Ardito et al. [31]. Machine learning, deep learning, and neural networks are the most commonly used technologies to develop these models.

## 3.3 Predicting Subsystem

This subsystem uses the models created in the previous subsystem to formulate an operational decision before action in the following subsystem. To make a decision, the system must know the objectives, the task to be performed, and the human being [34]. Objectives can be related to productivity performance, quality, operator welfare, and environmental and economic sustainability [20, 24]. Lemoine et al. [4] strategic, tactical, and operational sub-blocks. In the strategic one, objectives are defined, possible alternatives are formulated, and the best alternative (tactical and operational plan) is selected through prediction. According to [9], the decision is made considering several dimensions: the capabilities, availability, welfare, preferences of the worker, and the requirements of the process. Simulations, optimizations, and predictive algorithms are mainly used to realize this sub-system.

## 3.4 Acting/Interacting Subsystem

At this point, the system must act to achieve the task objective. Depending on the task, the action can occur in the physical or information world. Thus, we realize a variation in the physical or informational world, producing, for example, information. Both actions occur at the interface. An application case in which the machine understands the user's intentions and behaves accordingly is addressed in [35]. The authors use an assisted milling tool that allows inexpert users to make artifacts from a block of Balsa Foam. The tool contains the geometric model of the artifact to be made and corrects the spindle rotation speed when the physical model approaches the actual model. In addition, the machine is equipped with force sensors that allow it to understand the user's intentions and override the machine's recommendations. For example, [25] shows the design of an intelligent wearable robot to support physical assembly operations. The robot placed on the shoulders of the worker contains the models of the task and the operator. The robot should be able to minimize worker fatigue by reallocating task steps and respecting process constraints. Another fascinating work by Eyam et al. [36] involved using a cobot to complete assembly operations. The system uses EGG signals to classify the operator's emotional state. It adjusts the robot's speed of movement and the delay time of operations to bring the emotional state back within acceptable limits. Depending on the task, the suitable technologies may change. Augmented reality or I/O interfaces can provide information to the operator, while cobots, exoskeletons, or robotic arms can perform physical tasks.

Human Entity                                                    Machine Entity

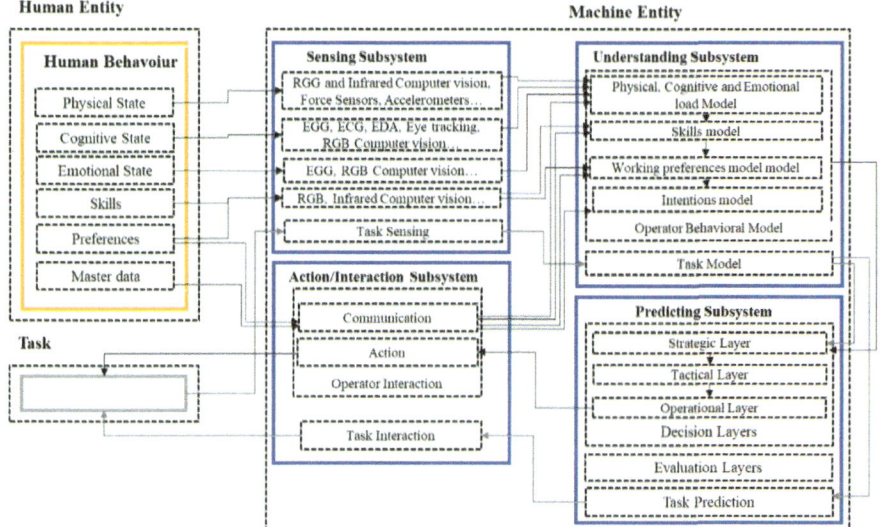

**Fig. 1.** Sensing, Understanding, Predicting, Acting Framework

## 3.5 Use Case Scenario

An example is used in this section to consolidate the use of the framework. Let us take the case of the flexible production of a plate to be drilled at a particular axial angle and a certain depth. Each hole in every processed plate must have a different angle and depth. At the initial use, the machine will not have any information about the operator; therefore, it will need to collect data about the operator performing the task, using the sensing subsystem and, with the data, build models within the understanding subsystem. Let us assume for simplicity's sake that the training part of the machine, i.e., the creation of the understanding subsystem models, has already been carried out. While working, the operator holds the drill, supported by the robotic arm, and approaches the drilling site. The machine perceives that the operator is fit via the sensing and understanding subsystem, sets targets for maximum productivity, and sends information about the movement speeds to the predicting subsystem. Strategically, the machine knows that the sequence of holes and the angle and depth of each hole must be defined. In this case, the human will decide which hole to drill first, and the machine will provide the right axial angle and depth, then check that the working parameters are correct, thus executing the task by the action subsystem. As the task progresses, the operator may become physically, cognitively, and emotionally fatigued. The machine will be able to understand the state of the operator and modify the goals for the recovery of the optimal condition of the worker and thus his behavior accordingly, e.g., by reducing the forward speed, overriding decisions that appear to be unwise.

# 4   Conclusion

Industry 5.0 sets the course for sustainable, human-centered, and resilient development. One of the fundamental building blocks is the large-scale adoption of collaborative technologies to support workers. However, numerous challenges are identified in Sect. 2, which must be addressed to achieve this goal. These challenges can be overcome by adopting empathic and intelligent collaborative machines. In this work, from the challenges, the system requirements in Sect. 3 that these machines must fulfill to achieve the objectives were derived. In Sect. 4, examples of the implementation of technical solutions are shown, following the framework, sense, Understand, Predict, and act derived from the identified system requirements. Although these solutions are promising, few real-life application cases in the literature address the identified challenges. A further experimental effort in this direction would, therefore, be desirable. The framework will be enriched with an in-depth literature search and experimentally validated in the following work.

# References

1. Dieste, M., Orzes, G., Culot, G., Sartor, M., Nassimbeni, G.: The 'dark side' of Industry 4.0: how can technology be made more sustainable? Int. J. Oper. Prod. Manag. **44**, 900–950 (2023). https://doi.org/10.1108/IJOPM-11-2022-0754
2. Industry 5.0 Towards a sustainable, human-centric and resilient European industry
3. Inga, J., et al.: Human-machine symbiosis: a multivariate perspective for physically coupled human-machine systems. Int. J. Hum. Comput. Stud. **170**, 102926 (2023)
4. Pacaux-Lemoine, M.P., Trentesaux, D., Zambrano Rey, G., Millot, P.: Designing intelligent manufacturing systems through Human-Machine Cooperation principles: a human-centered approach. Comput. Ind. Eng. **111**, 581–595 (2017). https://doi.org/10.1016/j.cie.2017.05.014
5. Janssen, J.H.: A three-component framework for empathic technologies to augment human interaction. J. Multimodal User Interfac. **6**(3–4), 143–161 (2012)
6. Abbass, H., Petraki, E., Hussein, A., McCall, F., Elsawah, S.: A model of symbiomemesis: machine education and communication as pillars for human-autonomy symbiosis. Phil. Trans. Royal Soc. A: Math. Phys. Eng. Sci. **379**(2207), 20200364 (2021). https://doi.org/10.1098/rsta.2020.0364
7. Stephanidis, C., et al.: Seven HCI Grand Challenge. Taylor and Francis Inc. (2019)
8. Gweon, H., Fan, J., Kim, B.: Socially intelligent machines that learn from humans and help humans learn. Phil. Trans. Roy. Soc. A: Math. Phys. Eng. Sci. **381**(2251), 20220048 (2023). https://doi.org/10.1098/rsta.2022.0048
9. Yuqian, L., et al.: Outlook on human-centric manufacturing towards Industry 5.0. J. Manuf. Syst. **62**, 612–627 (2022). https://doi.org/10.1016/j.jmsy.2022.02.001
10. Xiong, W., Fan, H., Ma, L., Wang, C.: Challenges of human—machine collaboration in risky decision-making. Front. Eng. Manag. **9**(1), 89–103 (2022). https://doi.org/10.1007/s42524-021-0182-0
11. Lagomarsino, M., Lorenzini, M., Balatti, P., De Momi, E., Ajoudani, A.: Pick the right co-worker: online assessment of cognitive ergonomics in human-robot collaborative assembly. IEEE Trans. Cogn. Dev. Syst. **15**, 1928–1937 (2022). https://doi.org/10.1109/TCDS.2022.3182811
12. Adel, A.: Future of industry 5.0 in society: human-centric solutions, challenges and prospective research areas. J. Cloud Comput. **11**(1), 40 (2022). https://doi.org/10.1186/s13677-022-00314-5

13. Bechinie, C., Zafari, S., Kroeninger, L., Puthenkalam, J., Tscheligi, M.: Toward human-centered intelligent assistance system in manufacturing: challenges and potentials for operator 5.0. Procedia Comput. Sci. **232**, 1584–1596 (2024). https://doi.org/10.1016/j.procs.2024.01.156

14. Garibay, O.O., et al.: Six human-centered artificial intelligence grand challenges. Int. J. Hum. Comput. Interact. **39**(3), 391–437 (2023). https://doi.org/10.1080/10447318.2022.2153320

15. Pizoń, J., Witczak, M., Gola, A., Świć, A.: Challenges of human-centered manufacturing in the aspect of industry 5.0 assumptions. IFAC-PapersOnLine **56**(2), 156–161 (2023). https://doi.org/10.1016/j.ifacol.2023.10.1562

16. Xu, W., Dainoff, M.J., Ge, L., Gao, Z.: Transitioning to human interaction with AI systems: new challenges and opportunities for HCI professionals to enable human-centered AI. Int. J. Hum. Comput. Interact. **39**(3), 494–518 (2023). https://doi.org/10.1080/10447318.2022.2041900

17. Li, L.: Reskilling and upskilling the future-ready workforce for industry 4.0 and beyond. Inf. Syst. Front. **26**(5), 1697–1712 (2022). https://doi.org/10.1007/s10796-022-10308-y

18. Siemon, D.: Elaborating team roles for artificial intelligence-based teammates in human-AI collaboration. Group Decis. Negot. (2022). https://doi.org/10.1007/s10726-022-09792-z

19. Kaasinen, E., Anttila, A.-H., Heikkilä, P., Laarni, J., Koskinen, H., Väätänen, A.: Smooth and resilient human–machine teamwork as an industry 5.0 design challenge. Sustainability **14**(5), 2773 (2022). https://doi.org/10.3390/su14052773

20. Liu, X., et al.: Human-centric collaborative assembly system for large-scale space deployable mechanism driven by Digital Twins and wearable AR devices. J. Manuf. Syst. **65**, 720–742 (2022)

21. Planke, L.J., Gardi, A., Sabatini, R., Kistan, T., Ezer, N.: Online multimodal inference of mental workload for cognitive human machine systems. Computers **10**(6), 81 (2021). https://doi.org/10.3390/computers10060081

22. Buerkle, A., Matharu, H., Al-Yacoub, A., Lohse, N., Bamber, T., Ferreira, P.: An adaptive human sensor framework for human–robot collaboration. Int. J. Adv. Manuf. Technol. **119**(1–2), 1233–1248 (2022)

23. Jo, W., Wang, R., Cha, G.E., Sun, S., Senthilkumaran, R.K., Foti, D., Min, B.C.: MOCAS: a multimodal dataset for objective cognitive workload assessment on simultaneous tasks. IEEE Trans. Affect. Comput. **16**, 116–132 (2024)

24. Pacaux-Lemoine, M.P., Berdal, Q., Guérin, C., Rauffet, P., Chauvin, C., Trentesaux, D.: Designing human–system cooperation in industry 4.0 with cognitive work analysis: a first evaluation. Cogn. Technol. Work **24**(1), 93–111 (2022)

25. Marvel, J.A., Bagchi, S., Zimmerman, M., Antonishek, B.: Towards effective interface designs for collaborative HRI in manufacturing: metrics and measures. ACM Trans. Hum.-Robot Interact. **9**(4), 1–55 (2020). https://doi.org/10.1145/3385009

26. Manghisi, V.M., Uva, A.E., Fiorentino, M., Bevilacqua, V., Trotta, G.F., Monno, G.: Real time RULA assessment using Kinect v2 sensor. Appl. Ergon. **65**, 481–491 (2017)

27. Lorenzini, M., Lagomarsino, M., Fortini, L., Gholami, S., Ajoudani, A.: Ergonomic human-robot collaboration in industry: a review. Front. Rob. AI **9**, 813907 (2023). https://doi.org/10.3389/frobt.2022.813907

28. Lorenzini, M., Kim, W., De Momi, E., Ajoudani, A.: An online method to detect and locate an external load on the human body with applications in ergonomics assessment. Sensors (Switzerland) **20**(16), 1–18 (2020). https://doi.org/10.3390/s20164471

29. Tirupachuri, Y., et al.: Online non-collocated estimation of payload and articular stress for real-time human ergonomy assessment. IEEE Access **9**, 123260–123279 (2021)

30. Neerincx, M.A., Harbers, M., Lim, D., van der Tas, V.: Automatic feedback on cognitive load and emotional state of traffic controllers. In: Harris, D. (ed.) Engineering Psychology and

Cognitive Ergonomics: 11th International Conference, EPCE 2014, Held as Part of HCI International 2014, Heraklion, Crete, Greece, June 22-27, 2014. Proceedings, pp. 42–49. Springer International Publishing, Cham (2014). https://doi.org/10.1007/978-3-319-07515-0_5

31. Ardito, C., et al.: Brain Computer Interface: Deep Learning Approach to Predict Human Emotion Recognition. https://www.emotiv.com/product/emotiv-insight-5-channel-mobile-
32. Wang, B., et al.: Human Digital Twin in the context of Industry 5.0. Elsevier Ltd. (2024)
33. Wang, L., et al.: Symbiotic human-robot collaborative assembly. CIRP Ann. **68**(2), 701–726 (2019). https://doi.org/10.1016/j.cirp.2019.05.002
34. Millot, P., Pacaux-Lemoine, M.-P.: A common work space for a mutual enrichment of human-machine cooperation and team-situation awareness. IFAC Proc. Vol. **46**(15), 387–394 (2013). https://doi.org/10.3182/20130811-5-US-2037.00061
35. Zoran, A., Shilkrot, R., Paradiso, J.: Human-computer interaction for hybrid carving. In: UIST 2013 - Proceedings of the 26th Annual ACM Symposium on User Interface Software and Technology (2013)
36. Toichoa Eyam, A., Mohammed, W.M., Martinez Lastra, J.L.: Emotion-driven analysis and control of human-robot interactions in collaborative applications. Sensors **21**(14), 4626 (2021)

# Study on Vibration of Electric Vehicles Manufactured in Vietnam on Road Class C According to ISO 8608:2016

Le Minh[1(✉)], Cao Hung Phi[1], and Trinh Minh Hoang[2]

[1] Faculty of Automotive Engineering, Vinh Long University of Technology, 73 Nguyen Hue Stress, Ward 1, Vinh Long City, Vinh Long Province, Vietnam
{minhl,caohungphi}@vlute.edu.vn

[2] Hanoi University of Science and Technology, No. 1 Dai Co Viet, Hanoi Capital, Hai Ba Trung District, Vietnam
hoang.trinhminh@hust.edu.vn

**Abstract.** This study explores how the damping coefficient affects vibration in small electric cars. By developing a mathematical model and simulating different damping coefficients, the authors found that this coefficient is crucial for controlling vehicle vibration, leading to a smoother and safer driving experience. Unlike traditional gasoline cars, small electric vehicles require different damping coefficients due to their distinct weight and structure. The study offers specific recommendations for optimizing damping coefficients in current electric car models, aiming to enhance user comfort and safety. This research is vital for improving small electric vehicles in Vietnam and supporting the shift to green transportation. Detailed findings are available in the full research paper.

**Keywords:** Damping coefficients · small car suspension design · Electric vehicle vibration · Vehicle Dynamic · Green Car

## 1 Introduction

The transition to zero-emission vehicles is a global trend aimed at mitigating the negative impacts of traffic emissions on the environment and human health [9]. In Vietnam, Decision No. 867/QD-TTg dated July 22, 2022, has set the target of achieving net zero greenhouse gas emissions by 2050 [9]. In this context, small electric vehicles are seen as a feasible solution to meet personal mobility needs while minimizing environmental impact.

However, specific challenges for small EVs in Vietnam make it essential to consider local road conditions, particularly with the prevalence of Category C roads as defined by ISO 8608:2016. Vietnam's road infrastructure often includes rough, uneven surfaces with a high density of mixed traffic, which creates unique demands on vehicle performance and durability [3] (Fig. 1).

These conditions impact vehicle operation, with suspension systems playing a critical role in maintaining stability, reducing shocks, and ensuring passenger comfort [4]. Small

H. Kohl et al. (Eds.): GCSM 2024, LNME, pp. 273–281, 2025.
https://doi.org/10.1007/978-3-031-93891-7_31

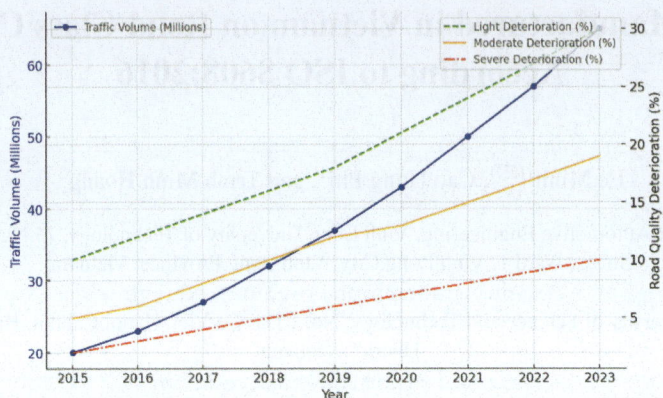

**Fig. 1.** The relationship between the number of vehicles and the level of road deterioration in Vietnam from 2015 to 2023

EV suspension systems typically incorporate components like Macpherson dampers and springs, designed to reduce vibrations and manage road impacts [5]. The development of electric vehicles in Vietnam has shown considerable progress, promising benefits such as reduced air pollution, energy savings, and lower operating costs [5]. This growth aligns with Vietnam's commitment to fostering a sustainable future and supporting the domestic electric vehicle industry, ensuring that EV technology not only meets global standards but also addresses local operational demands [6].

Electric vehicles with adjustable suspension systems allow customization based on road conditions and driving style, which can enhance performance and comfort for users. This study evaluates how vehicle vibrations influence comfort and stability, with a focus on the impact of factors like speed and road conditions, especially on Category C roads as per ISO 8608:2016. Reducing vehicle vibrations not only improves passenger comfort and protects cargo but also extends vehicle lifespan, aligning with safety and durability standards. Simulations using a 7 DOF model will be applied to assess the damping coefficient's role in reducing vibration on rough road surfaces.

## 2   Dynamic Models of a Small Electric Car

The spatial dynamics model of tiny automobiles is used to determine the research model for vibrations in small cars as they are driving on the road. The author of this paper creates a dynamic model using the structural decomposition of many-body systems approach.

The vehicle body is connected to the axles through the suspension system, characterized by stiff-ness parameters C and damping coefficient K.

- The axle is connected to the surface via elastic wheels with CL hardness
- Ignore the tire's resistance component and the impact of wind resistance
- Wheels are always in contact with the road surface

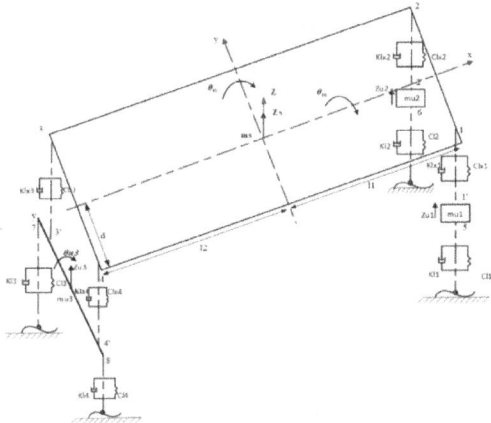

**Fig. 2.** Dynamics diagram of Vinfast VF e34 [8]

Thus, the structural model will have 7 generalized coordinates (7 DOF) (Fig. 2).

Using the method of structural separation and the system of Newton-Eulerian differential equations, the author has obtained a system of 7 equations as follows:

– The equation of motion of the body in the Z direction of the car:

$$m_s \ddot{Z}_s = -K_{Lx1}\left(2\dot{Z}_s + 2\dot{\theta}_{sy}l_1 - \dot{Z}_{u1} - \dot{Z}_{u2}\right) - C_{Lx1}\left(2Z_s + 2\theta_{sy}l_1 - Z_{u1} - Z_{u2}\right) \\ -K_{Lx2}\left(2\dot{Z}_s - 2\dot{\theta}_{sy}l_2 - 2\dot{Z}_{u3}\right) - C_{Lx2}\left(2Z_s - 2\theta_{sy}l_2 - 2Z_{u3}\right) = 0 \tag{1}$$

– Equation of vertical swing $\ddot{\theta}_{sx}$ of the car:

$$J_{sx}\ddot{\theta}_{sx} = -K_{Lx1}d\left(2\dot{\theta}_{sx}d - \dot{Z}_{u1} + \dot{Z}_{u2}\right) - C_{Lx1}d\left(2\theta_{sx}d - Z_{u1} + Z_{u2}\right) - K_{Lx2}d\left(2\dot{\theta}_{sx}d\right) \\ -C_{Lx2}d\left(2\theta_{sx}d\right) = 0 \tag{2}$$

– Equation of horizontal sway $\ddot{\theta}_{sy}$ of the suspended mass:

$$J_{sy}\ddot{\theta}_{sy} = -K_{Lx1}l_1\left(2\dot{Z}_s + 2\dot{\theta}_{sy}l_1 - \dot{Z}_{u1} - \dot{Z}_{u2}\right) - C_{Lx1}l_1\left(2Z_s + 2\theta_{sy}l_1 - Z_{u1} - Z_{u2}\right) \\ +K_{Lx2}l_2\left(2\dot{Z}_s - 2\dot{\theta}_{sy}l_2 - 2\dot{Z}_{u3}\right) + C_{Lx2}l_2\left(2Z_s - 2\theta_{sy}l_2 - Z_{u3}\right) = 0 \tag{3}$$

– The system of equations for lateral dynamics of the right front axle:

$$m_{u1}\ddot{Z}_{u1} = -K_{Lx1}\left(\dot{Z}_{u1} - \dot{Z}_s - \dot{\theta}_{sy}l_1 - \dot{\theta}_{sx}d\right) - C_{Lx1}\left(Z_{u1} - Z_s - \theta_{sy}l_1 - \theta_{sx}d\right) \\ -K_{L1}\left(\dot{Z}_{u1} - \dot{h}_{11}\right) - C_{L1}(Z_{u1} - h_{11}) \tag{4}$$

– The system of equations for the lateral dynamics of the left front axle:

$$m_{u2}\ddot{Z}_{u2} = -K_{Lx1}\left(\dot{Z}_{u2} - \dot{Z}_s - \dot{\theta}_{sy}l_1 + \dot{\theta}_{sx}d\right) - C_{Lx1}\left(Z_{u2} - Z_s - \theta_{sy}l_1 + \theta_{sx}d\right) \\ -K_{L1}\left(\dot{Z}_{u2} - \dot{h}_{12}\right) - C_{L1}(Z_{u2} - h_{12}) \tag{5}$$

– The system of equations for the horizontal dynamics of the rear axle of the suspension system depends:

$$m_{u3}\ddot{Z}_{u3} = -2K_{Lx2}\left(\dot{Z}_{u3} - \dot{Z}_s + \dot{\theta}_{sy}l_2\right) - 2C_{Lx2}\left(Z_{u3} - Z_s + \theta_{sy}l_2+\right) \\ -K_{L2}\left(\dot{Z}_{u3} - \dot{h}_{21} - \dot{h}_{22}\right) + C_{L2}(Z_{u3} - h_{21}) \tag{6}$$

– The equation for the horizontal swing angle of the unsuspended mass follows:

$$J_{u3}\ddot{\theta}_{u3} = -2K_{Lx2}d^2(\ddot{\theta}_{u3} - \dot{\theta}_{sx}) - 2C_{Lx2}d^2(\theta_{u3} - \theta_{sx}) - K_{L2}B(2\dot{\theta}_{u3}B - \dot{h}_{21} + \dot{h}_{22}) \\ -C_{L2}B(2\theta_{u3}B - h_{21} + h_{22}) \tag{7}$$

ISO Class C Road surfaces were selected for simulation to evaluate ride comfort using the RMS acceleration [m/s$^2$] index, proportional to gravitational acceleration (9.81 m/s$^2$). The formula:

$$\sigma\ddot{z}_t = [\frac{1}{T}\int_0^T \ddot{z}_t(t)dt]^{1/2} \tag{8}$$

In experiments lasting 8 h at oscillation frequencies considered sensitive to humans (from 4 to 8 Hz), the RMS values affecting humans will create the following feelings (Figs. 3, 4 and Table 1):

– Comfortable: 0.1 m/s$^2$.
– Causes fatigue: 0.315 m/s$^2$.
– Causes negative effects on health: 0.63 m/s$^2$.

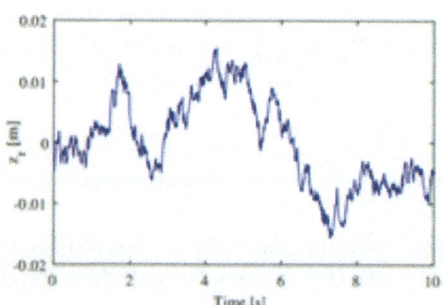

**Fig. 3.** ISO 8608 Road type C [5]

## 3   Simulation Results

### 3.1   When the Damping Coefficient is 410 Ns/m

The displacement graph shows that the initial oscillation has a large amplitude of about 0.03 m, which gradually decreases over time. The damping system works effectively, progressively reducing the oscillation amplitude, though the oscillation continues with a

**Fig. 4.** Damping experimental Tester equipment in VLUTE

**Table 1.** Specifications of Vinfast VF e34 in simulation [6]

| No | Name | Symbol | Unit | Value |
|----|------|--------|------|-------|
| 1 | Suspended mass | $m$ | kg | 1490 |
| 2 | The mass moment of inertia is suspended about the x-axis | $J_{sx}$ | $kgm^2$ | 1279.1 |
| 3 | The mass moment of inertia is suspended around the y trục axis | $J_{sy}$ | $Kgm^2$ | 5117.6 |
| 4 | Mass without suspension right front axle | $m_{u1}$ | kg | 100 |
| 5 | Mass without suspension left front axle | $m_{u2}$ | kg | 100 |
| 6 | Mass not hanging left rear axle | $m_{u3}$ | kg | 115 |
| 7 | Mass without suspension right rear axle | $m_{u4}$ | kg | 115 |
| 8 | Front elastomer stiffness | $C_{Lx1}$ | N/m | 16000 |
| 9 | Rear elastomer stiffness | $C_{Lx2}$ | N/m | 16000 |
| 10 | Stiffness of the front tire | $C_{L1}$ | N/m | 160000 |
| 11 | Rear tire stiffness | $C_{L2}$ | N/m | 160000 |
| 12 | Front damping coefficient | $K_{Lx1}$ | Ns/m | [410;1910] |
| 13 | Rear damping coefficient | $K_{Lx2}$ | Ns/m | [410;1910] |
| 14 | Tire drag coefficient | $K_{L1}$ | Ns/m | 14000 |
| 16 | Distance between left and right wheel link points | $D$ | m | 1,860 |
| 17 | Distance from center of mass to be suspended to center of preload | $l_1$ | m | 1.118 |
| 18 | Distance from center of mass to be suspended to center of rear load | $l_2$ | m | 1.493 |

smaller amplitude after around 12 s and does not completely cease. The system reaches relative stability after about 8 s, with the oscillation amplitude decreasing to a range between -0.01 m and 0.02 m) [5, 6].

The acceleration graph shows an initial peak of nearly 3 m/s$^2$, which gradually decreases over time. The oscillation amplitude reduces with each cycle, stabilizing around 8 s, with values oscillating between $-1$ m/s$^2$ and 1 m/s$^2$, indicating effective damping but with continued low-intensity oscillations [5, 6].

The body rotation angle graph shows an initial small oscillation with an amplitude near 0.001 rad. Over time, the oscillation amplitude gradually decreases. After approximately 8 s, the oscillation stabilizes with values fluctuating between $-0.0005$ rad and 0.0005 rad, indicating effective damping and very low-intensity oscillations remaining [5, 6] (Fig. 5).

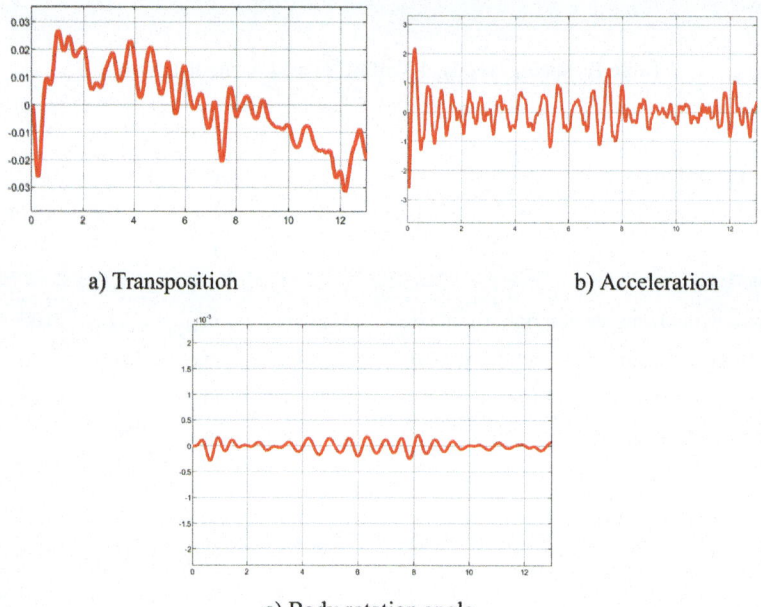

a) Transposition                         b) Acceleration

c) Body rotation angle

**Fig. 5.**  Survey results when running at 60 km/h with damping coefficient of 410 Ns/m

## 3.2   When the Damping Coefficient is 1910 Ns/m

The displacement graph shows an initial large amplitude of 0.06 m, which gradually decreases over time. The oscillations persist but diminish steadily, indicating effective damping. After about 10 s, the amplitude reduces significantly but the oscillations have not completely ceased. [5, 6].

The acceleration graph shows an initial high amplitude of around 6 m/s$^2$, which gradually decreases over time. The oscillations continue but with a diminishing amplitude, indicating effective damping. By around 10 s, the amplitude has significantly reduced, though minor oscillations are still present. [5, 6].

The body rotation angle graph shows an initial oscillation with a small amplitude of about 0.0005 rad (Fig. 6).

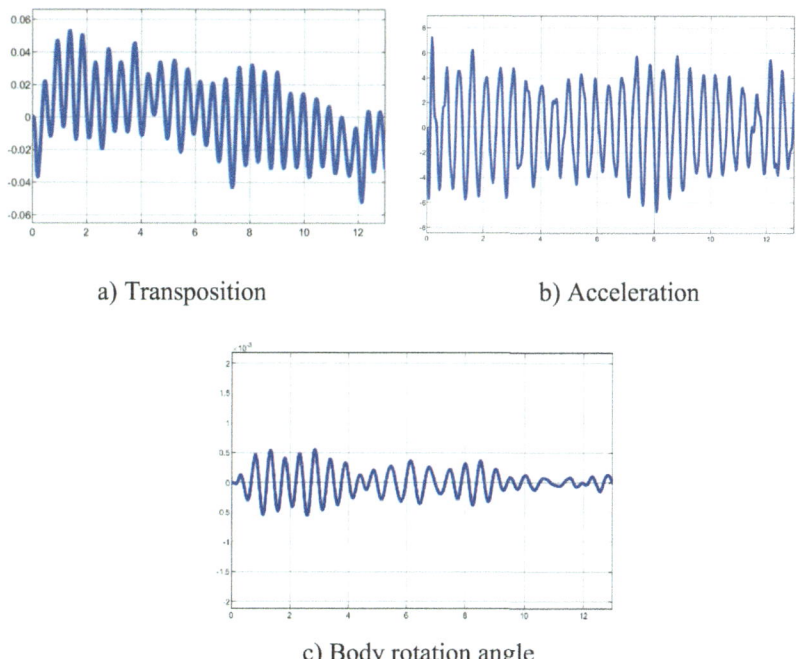

a) Transposition                              b) Acceleration

c) Body rotation angle

**Fig. 6.** Results when running at 60 km/h with damping coefficient of 1910 Ns/m

### 3.3 Discussion

In both damping coefficients, the vertical acceleration (with peak values of approximately 6 m/s$^2$) stabilizes after about 10 s, indicating that the suspension system can effectively absorb vibrations under non-ideal road conditions. The reduction in oscillations over time suggests that both damping coefficients can maintain comfort when the vehicle travels on uneven roads, while also minimizing the impacts on vehicle components caused by road-induced forces, potentially extending the vehicle's lifespan.

The application of ISO Class C Road surfaces remains relevant, as it accurately simulates the challenges suspension systems face in real-world conditions, such as those commonly encountered in Vietnam. This approach is particularly valuable for the design of suspension systems in electric vehicles, ensuring both comfort and durability on less-than-ideal road surfaces. The simulation results not only support the selection of optimal damping coefficients but also contribute to sustainability goals by reducing wear, tear, and long-term maintenance costs.

# 4   Conclusion

The shift from fossil-fuel-powered vehicles to electric vehicles not only reduces environmental pollution and improves air quality but also significantly contributes to Vietnam's sustainable development goals, particularly in achieving net-zero emissions by 2050. However, the increased mass and load of electric vehicles necessitate adjustments to the suspension system to ensure durability and stability in varying road conditions.

In this study, damping coefficients of 1910 Ns/m and 410 Ns/m were applied on an ISO Class C Road surface at a speed of 60 km/h to simulate typical road conditions in Vietnam. The ISO Class C Road surface, representing non-ideal conditions, induces significant vibrations and oscillations, which affect suspension performance and passenger comfort. The simulation results show that both damping coefficients effectively reduce oscillations after approximately 10 s, with substantial stabilization occurring earlier around 6–8 s. These results suggest that the selected damping coefficients can effectively minimize road-induced vibrations, meeting comfort requirements over time. The gradual reduction in oscillation amplitude indicates the suspension system's ability to handle rough surfaces, thereby reducing wear and tear on vehicle components, extending the vehicle's lifespan, and lowering maintenance costs.

Although the current damping coefficients ensure comfort under simulated conditions, further research is necessary to optimize the suspension system for real-world electric vehicle operations on uneven road surfaces, thus enhancing sustainability. Future studies should focus on developing more effective damping solutions for non-ideal road surfaces to improve vehicle stability and longevity while minimizing environmental impact and long-term maintenance costs.

# References

1. Vo, V.H.H., et al.: Automotive Dynamics. Vietnam Education Publishing House, Pham Van Hieu (2014)
2. Vo, V.H., Dung, N.T., Hung, T.T.: Modern Automobile Theory. Vietnam Education Publishing House, Pham Van Hieu (2021)
3. Quang, N.P.: Matlab & Simulink for Automatic Control Engineers. Science and Technology Publishing House (2006)
4. Kruczek, A., Stribrsky, A.: A full-car model for active suspension - some practical aspects. In: Proceedings of the IEEE International Conference on Mechatronics, ICM'04, Istanbul, Turkey, 3–5 June 2004, pp. 41–45 (2004)
5. ISO 8608:2016. Reporting Measured Data of Mechanical Vibration (2016)
6. VinFast., Vietnam Automotive Market Overview. Accessed Oct 2024. https://vinfastauto.com/vn_vi/
7. Hassaan, G.A., Mohammed, N.A.-A.: A study on suspension systems. Int. J. Comput. Techn. 2(1) (2015)
8. Akpakpavi, M.: Dynamics and vibration analysis. J. Multidisc. Eng. Sci. Stud. (JMESS) (2017)
9. Phu, N.M.T.T.V., Mai, P.X.: Soft sleeping environment of a male-sleeper bus. J. Sci. Technol. 11(108) (2016)

# Production of Energy Systems

# Leveraging Digital Twins and Dynamic Life Cycle Assessment for Sustainable Manufacturing: A Conceptual Framework

Yujia Luo[1]([envelope]) [iD], Rajesshkumar Madarkar[2] [iD], Xichun Luo[3] [iD], and Peter Ball[1] [iD]

[1] School for Business and Society, University of York, York YO10 5ZF, UK
yujia.luo@york.ac.uk
[2] School of Art, Design and Performance, Buckinghamshire New University, Buckinghamshire HP11 2JZ, UK
[3] Department of Design, Manufacturing and Engineering Management, University of Strathclyde, Glasgow G1 1XQ, UK

**Abstract.** The integration of digital twins (DTs) and dynamic life cycle assessment (LCA) offers a transformative approach to enhancing environmental sustainability in manufacturing systems. This paper proposes a conceptual framework to guide the development and implementation of DT-enabled dynamic LCA, emphasising system embeddedness – the integration of physical systems, virtual models, data, and performance objectives across multiple scales of manufacturing systems. This holistic approach enables the identification of environmental hotspots dynamically and the real-time monitoring of targeted optimisation strategies. The framework comprises three core pillars: (1) a multiscale DT architecture for real-time data collection and representation from diverse sources; (2) a dynamic LCA by leveraging real-time data feed from DTs and integrated with simulation modules in the DT system; and (3) a continuous feedback loop between DT's simulation and optimisation with LCA for predicative decision support and operational adjustments in the physical manufacturing system. This study contributes a roadmap for leveraging DT technology and dynamic LCA to drive real-time environmental improvements across the manufacturing system, empowering manufacturers to make real-time data-driven decisions that promote a more sustainable and resource-efficient system.

**Keywords:** Digital twins · Dynamic life cycle assessment · Real-time monitoring · Environmental sustainability

## 1 Introduction

The escalating consequences of climate change and resource depletion have underscored the urgent need for environmental sustainability in the manufacturing sector [1]. Traditional life cycle assessment (LCA) methodologies, while valuable for evaluating environmental impacts, often rely on static data and assumptions, failing to capture the dynamic nature of modern manufacturing processes [2]. These processes are characterised by

H. Kohl et al. (Eds.): GCSM 2024, LNME, pp. 285–293, 2025.
https://doi.org/10.1007/978-3-031-93891-7_32

fluctuations in energy consumption, material flows, and emissions, necessitating real-time monitoring and adaptive decision-making to achieve sustainability goals [3]. In this paper, we focus on LCA in manufacturing systems with broadening scope of the entire product lifecycle. This architecture would involve multiscale manufacturing processes and partnerships across the supply chain (SC).

Digital twins (DTs), as virtual replicas of physical systems, have emerged as a transformative technology in manufacturing [4]. By integrating real-time data from sensors, Internet of Things (IoT) devices, and other sources, DTs offer a comprehensive and dynamic representation of manufacturing processes [5]. This capability opens up new possibilities for data-driven optimisation, predictive maintenance, and, crucially, real-time environmental impact assessment [6]. Integrating DTs with dynamic LCA methodologies presents a promising pathway towards achieving environmental sustainability in manufacturing [7].

However, the integration of DTs and dynamic LCA faces several challenges, including the need for robust data collection and integration mechanisms, the development of accurate and computationally efficient LCA models, and the establishment of effective feedback loops between the DT and LCA systems [7]. Addressing these challenges requires a systematic approach that considers the complexities and interdependencies of modern manufacturing systems.

This paper proposes a conceptual framework to guide the integration of DTs and dynamic LCA in manufacturing. The framework aims to provide a roadmap for practitioners and researchers, outlining the key components, data flows, and integration mechanisms required for successful implementation. By bridging the gap between these two powerful technologies, this framework seeks to empower manufacturers with the tools and insights needed to achieve environmental sustainability in a dynamic and data-driven manner.

## 2   Literature Review

Dynamic LCA has emerged as a significant advancement over traditional static LCA methodologies, which rely on fixed datasets and assumptions. By incorporating real-time data, dynamic LCA aligns with the dynamic nature of manufacturing processes, providing a more accurate and up-to-date assessment of environmental impacts [8]. Several approaches, such as time series analysis [9], agent-based modelling, and system dynamics [10], have been proposed for conducting dynamic LCA, each with varying levels of complexity and granularity. However, challenges remain in terms of limited scope in existing studies, varying time horizons affecting accuracy, and scalability concerns, especially for complex manufacturing systems [11, 12].

The expansion of the Internet of Things (IoT) and sensor technologies, alongside the growth of cloud-based data platforms, has been instrumental in enabling the integration of DTs with LCA and facilitating real-time monitoring and control of manufacturing systems [13]. Notably, recent studies have demonstrated the potential of integrating real-time monitoring systems with LCA for more precise and responsive environmental impact assessments. For example, Schneider et al. [14] explored the use of real-time data from sensors to monitor the energy consumption and emissions of a manufacturing process, enabling a dynamic LCA that reflected the actual environmental performance of the system. Similarly, a study by Kim et al. [15] utilised real-time data from a smart factory to perform a dynamic LCA of a production line, demonstrating the feasibility of integrating LCA with Industry 4.0 technologies. These studies highlight the potential of real-time monitoring to enhance the accuracy and relevance of LCA in a manufacturing context.

The integration of DTs with LCA has been further propelled by the broader adoption of DTs in manufacturing. Recent trends demonstrate the increasing use for real-time production insights, enabling predictive maintenance and operational parameter optimisation [5]. As noted by Bitencourt et al. [16], DTs facilitate process optimisation via virtual experimentation and scenario analysis. Recent research also demonstrates the successful integration of DT and LCA in assessing carbon dioxide emissions, highlighting the potential for real-time environmental impact monitoring [17].

Despite advancements in DT and LCA integration within manufacturing [5, 16], a systematic framework for implementation is lacking [6, 11]. Existing research, while showcasing the potential of DT-enabled dynamic LCA, primarily focuses on specific applications or single-scale case studies [14, 15, 17], overlooking the potential of systematic data integration. This paper proposes a conceptual framework to address this gap, outlining key components, data flows, and integration mechanisms for successful DT-enabled dynamic LCA implementation, promoting environmental sustainability across the entire manufacturing system.

## 3   Conceptual Framework

Grounded in a System of Systems (SoS) perspective, a comprehensive conceptual framework is proposed to address the complexities of modern manufacturing and facilitate seamless DT and dynamic LCA integration. This framework recognises the interconnectedness of DT-enabled manufacturing subsystems, enabling a holistic understanding of their interactions and cumulative environmental impacts [12]. Crucially, the SoS perspective highlights feedback loops, cascading effects, and emergent behaviours within the manufacturing DT system, all of which influence the monitoring of system environmental outcomes [18]. This understanding is vital for DT-enabled dynamic LCA, as it reveals how localised changes can reverberate throughout the system, impacting environmental performance across multiple scales.

The proposed framework, centered around system embeddedness, creates a digital replica (the DT) of the physical manufacturing system, continuously updated with real-time data from various sources [5]. This DT serves as a digital proxy, enabling multi-scale analysis from individual processes to the extended production system [6]. The integration of diverse data sources pinpoints environmental hotspots and facilitates targeted interventions. Building upon this system embeddedness, the framework is structured around three core pillars (Fig. 1), each vital for integrating DTs and dynamic LCA for sustainable manufacturing.

**DT Architecture for Real-Time Data Integration and Synchronisation:** A robust real-time data architecture is crucial for effective DT-enabled LCA in dynamic manufacturing environments. Real-time, in this context, implies minimal data latency to facilitate

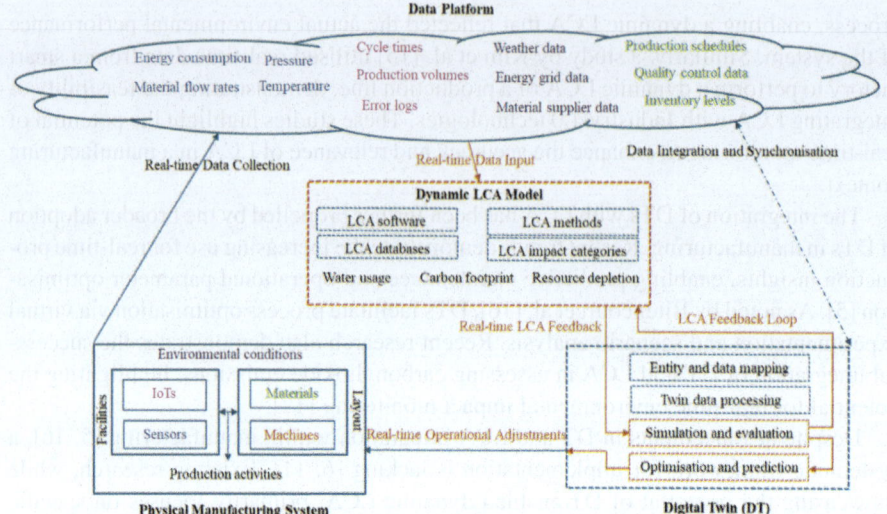

**Fig. 1.** Conceptual framework for integrating DTs and dynamic LCA.

prompt decision-making and control. The architecture integrates sensor networks (e.g., temperature, pressure, vibration sensors), communication protocols (e.g., open platform communications unified architecture), scalable data storage (e.g., time-series databases, distributed file systems), and powerful processing algorithms (e.g., stream processing, machine learning, and artificial intelligence). The complexity of such a DT architecture scales with system size and data diversity, potentially requiring distributed and modular designs.

**Dynamic LCA Methodology for Continuous Environmental Impact Assessment:** Integrating real-time manufacturing data into LCA via DT technology enhances traditional product-focused LCA, providing a comprehensive coverage of environmental impact assessment across the manufacturing system's various scales. By utilising the DT's computational capabilities for multiscale data handling and real-time acquisition, manufacturers can precisely identify and address environmental inefficiencies at the machine, work center, production line, and system levels in real time, enabling targeted process and resource optimisations. The identification of environmental bottlenecks also increases transparency by linking specific product batches to their actual production footprint over time, influencing design choices and informing consumer decisions. Furthermore, dynamic LCA monitoring at the manufacturing system level fosters prompt collaboration adjustments with suppliers on sustainability initiatives, ensuring compliance and simplifying reporting processes against regulatory standards.

**Real-time Feedback Loop Between DT and LCA for Decision Support and Operational Adjustments:** This pillar serves as the operational foundation of the framework, establishing a bidirectional communication channel between the DT and the dynamic LCA module with a feed-forward mechanism to the physical manufacturing system. The

DT provides real-time data to the LCA module for continuous assessment at various levels of granularity, from individual components to the entire system. In turn, the LCA module generates real-time environmental impact assessments for corresponding subsystems within the manufacturing system. These real-time LCA results then feed into the DT's simulation and optimisation modules, which generate proactive operational adjustments leveraging computational experimentation and AI-assisted evaluation. This real-time feedback loop mechanism is essential for translating insights from the DT system and LCA module into actionable adjustments at different scales of the manufacturing system.

In the subsequent section, we will discuss the specifics of data flow and integration within this framework, exploring how real-time data from various sources is harnessed to drive dynamic LCA analysis and ultimately enable sustainable decision-making in manufacturing.

## 4   Data Flow and Integration in DT-Enabled Dynamic LCA

The success of a DT-enabled dynamic LCA framework hinges on the seamless integration and flow of real-time data across multiple levels of the manufacturing system. This data-driven approach, facilitated by the DT, enables real-time monitoring, analysis, and optimisation of environmental impacts at various scales.

The foundation of this system is a network of interconnected data sources. Embedded sensors throughout the manufacturing environment monitor key parameters like energy consumption, temperature, pressure, and material flow, providing granular insights into individual processes. Simultaneously, connected IoT devices, such as machines and robots, transmit operational data, offering a view into the performance and health of both individual machines and the overall production line.

Beyond the factory floor, the system integrates data from external sources, including weather stations, energy grids, and material suppliers. This contextual data provides a broader perspective on environmental impact, incorporating factors like the carbon intensity of the energy grid or the footprint of raw materials. Integration with enterprise systems such as enterprise resource planning (ERP) and manufacturing execution systems (MES) further enriches the data landscape with production schedules, inventory levels, and quality control data.

This multiscale data integration, encompassing micro-level machine and process data, meso-level system data, and macro-level external contextual data, culminates in a comprehensive understanding of the manufacturing process's environmental impact, empowering effective decision-making and optimisation strategies.

Table 1 summarises the various data sources, their key parameters, collection methods, and scale level.

Following the identification of data sources and scales, the data flow is mapped to provide operational guidance within this conceptual framework. Figure 2 visually summarises the data flow and analysis stages in this DT-enabled dynamic LCA framework for multiscale manufacturing.

Raw data from various sources is transmitted to the central DT data platform, where it undergoes rigorous preprocessing, cleaning, and transformation. Machine learning

**Table 1.** Data sources and parameters for DT-enabled dynamic LCA.

| Data source | Key data parameters | Collection method | Manufacturing system level |
|---|---|---|---|
| Sensors | Energy consumption (electricity, gas, etc.), temperature, pressure, material flow rates | Direct measurement, continuous monitoring | Micro level (Process) |
| IoT Devices | Cycle times, production volumes, error logs, machine status | Data transmission from device APIs | Micro level (Machine) |
| Enterprise Systems | Production schedules, inventory levels, quality control data, material supplier data (environmental footprint of raw materials) | Integration with ERP/MES systems | Meso level (System) |
| External Sources | Weather data (temperature, humidity, solar irradiance), energy grid data (carbon intensity) | APIs, data subscriptions, manual input | Macro level (External) |

algorithms, trained on historical patterns, help eliminate errors and inconsistencies. The refined data is then aggregated, combining real-time sensor data with contextual information from internal and external sources. This enriched data pool forms the foundation for dynamic LCA analysis.

The integrated data fuels the DT, ensuring it accurately mirrors the current state of the manufacturing process. The DT, in turn, provides real-time data to the dynamic LCA module, enabling continuous measurement of environmental impacts across various categories, such as energy consumption, emissions, waste generation, and material flows. The LCA model leverages this real-time data, along with established impact assessment methodologies and data from sources like Ecoinvent (a leading life cycle inventory database), to comprehensively evaluate the environmental consequences of the manufacturing process. This evaluation can be conducted by employing LCA software tools such as SimaPro, GaBi, or openLCA.

The framework establishes dynamic feedback loops between the DT and LCA module, enabling real-time operational adjustments based on LCA results. This empowers the system to track environmental bottlenecks, provide feedback on operations, conduct scenario-based analyses, and suggest alternative action plans to achieve real-time and "all-time" (i.e., long-term) sustainability goals. For example, if a high-impact material is identified, the DT can simulate alternatives and process modifications using its built-in databases and capabilities like material substitution analysis and finite element analysis. Unknown material attributes can be estimated through historical data analysis, machine learning models, and expert knowledge. The integration of dynamic LCA results into the DT's visualisation interface creates a closed-loop human-machine interaction, offering both DT-generated actionable insights and allowing human input on performance monitoring and optimisation preferences across the manufacturing system.

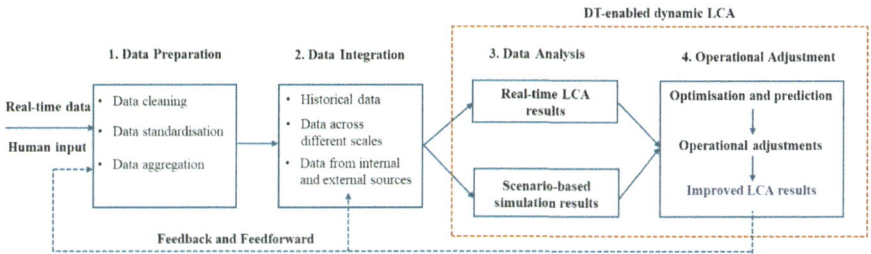

**Fig. 2.** Data across various stages in the DT-enabled dynamic LCA framework.

## 5 Discussion and Conclusion

This paper presents a novel framework for integrating DTs and dynamic LCA, offering a transformative approach to environmental sustainability in manufacturing. By enabling real-time, multiscale monitoring and assessment, the framework facilitates proactive decision-making and optimisation strategies across the entire manufacturing system, surpassing the limitations of traditional LCA methods [19]. Its flexibility and scalability suggest potentials beyond manufacturing, contributing to advancements in DT technology, LCA, systems engineering, and sustainable manufacturing [5].

Future research will focus on several key areas to fully realize the framework's potential. Standardization of data models and protocols is crucial for seamless interoperability between DT and LCA systems. Empirical validation across diverse industrial settings will provide valuable insights into the framework's efficacy and limitations. Exploring advanced simulation techniques within the DT can further enhance its predictive power. Additionally, a comprehensive analysis of the economic and social implications of DT-enabled dynamic LCA is essential [7].

While a detailed case study is beyond the scope of this paper, our subsequent research will investigate the application of the proposed DT-LCA framework within a solar panel production system, comparing it to traditional static LCA. This operationalisation will involve a systematic mapping of process, material, and data flows within both the physical production system and its DT. We will demonstrate the utilisation of time-series data (past, present, and predicted) within the DT for dynamic LCA, providing a ready-to-use data structure and processing procedures compatible with empirical manufacturing data. This real-world application will showcase the tangible advantages of the DT-LCA framework, including multiscale analysis, real-time monitoring, system-wide traceability, AI-powered computing, and adaptive control capabilities. While acknowledging the initial and ongoing economic investment required for implementation, the long-term benefits of enhanced sustainability, particularly in light of evolving EU regulations, are expected to outweigh the costs.

# References

1. Brundtland Commission: Our Common Future. Oxford University Press, Oxford (1987)
2. Finnveden, G., et al.: Recent developments in life cycle assessment. J. Environ. Manag. **91**(1), 1–21 (2009)
3. Matthews, H.S., Hendrickson, C.T., Weber, C.L.: The importance of carbon footprint estimation boundaries. Environ. Sci. Technol. **42**(16), 5839–5842 (2008)
4. Grieves, M.: Digital twin: Manufacturing excellence through virtual factory replication. White paper (2014)
5. Tao, F., Zhang, H., Liu, A., Nee, A.Y.C.: Digital twin in industry: state-of-the-art. IEEE Trans. Ind. Inf. **15**(4), 2405–2415 (2018)
6. Uhlemann, T.H., Lehmann, C., Steinhilper, R.: The digital twin: realizing the cyber-physical production system for Industry 4.0. Procedia CIRP **106**, 335–340 (2022)
7. Charnley, F., Tiwari, A., Brooker, P.: Barriers to the uptake of digital technologies for sustainability in manufacturing. J. Clean. Prod. **289**, 125649 (2021)
8. Zamagni, A., Pesonen, H.L., Swarr, T.E.: From digital twin to continuous LCA improvement: a conceptual framework. Procedia CIRP **80**, 695–700 (2019)
9. Heijungs, R., Huppes, G., Guinée, J.B.: Life cycle assessment: past, present, and future. Environ. Sci. Technol. **56**(19), 13632–13646 (2022)
10. Yu, F., et al.: Assessment of system sustainability: a critical review of the combined application of system dynamics and life cycle assessment. Energy Ecol. Environ., 1–10 (2024)
11. da Costa, T.P., da Costa, D.M.B., Murphy, F.: A systematic review of real-time data monitoring and its potential application to support dynamic life cycle inventories. Environ. Impact Assess. Rev. **105**, 107416 (2024)
12. Zhang, X., Jiang, D., Li, J., Zhao, Q., Zhang, M.: Carbon emission oriented life cycle assessment and optimization strategy for meat supply chain. J. Clean. Prod. **439**, 140727 (2024)
13. Babayigit, B., Abubaker, M.: Industrial internet of things: a review of improvements over traditional scada systems for industrial automation. IEEE Syst. J. (2023)
14. Schneider, D., Jordan, P., Dietz, J., Zaeh, M.F., Reinhart, G.: Concept for automated LCA of manufacturing processes. Procedia CIRP **116**, 59–64 (2023)
15. Kim, H., et al.: Smart factory transformation using Industry 4.0 toward ESG perspective: a critical review and future direction. Int. J. Precis. Eng. Manuf.-Smart Technol. **1**(2), 165–185 (2023)
16. Bitencourt, J., Wooley, A., Harris, G.: Verification and validation of digital twins: a systematic literature review for manufacturing applications. Int. J. Prod. Res. **63**, 1–29 (2024)
17. Yang, S., et al.: Coupling the digital twin technology and life cycle assessment: carbon dioxide emissions from polysilicon production. Sustain. Prod. Consump. **41**, 156–166 (2023)
18. Jamshidi, M. (ed.): Systems of Systems Engineering: Principles and Applications. CRC Press, Boca Raton (2017)
19. Chinnathai, M.K., Alkan, B.: A digital life-cycle management framework for sustainable smart manufacturing in energy intensive industries. J. Clean. Prod. **419**, 138259 (2023)

# HARDAT: Human Action Recognition Dataset for Manual Assembly Tasks

Lukas Büsch$^{(\boxtimes)}$, Mert Palazoğlu, and Thorsten Schüppstuhl

Institute of Aircraft Production Technology, Hamburg University of Technology,
Denickestraße 17, 21073 Hamburg, Germany
lukas.buesch@tuhh.de
http://tuhh.de/ifpt

**Abstract.** This paper introduces a novel Human Action Recognition (HAR) dataset designed to improve Human-Robot Collaboration (HRC) in green electrolyzer production. Recorded in a lab using RGB, depth, and skeletal data from Azure Kinect, the dataset focuses on assembly tasks, labeled with Methods-Time Measurement (MTM) primitives. The use of a green screen enables the study of background effects on HAR algorithms. The dataset addresses the challenges of data imbalance and limited training data in industrial HAR applications, offering standardized, mergeable, and extendable data for the research community. It aims to enhance the development of HAR algorithms in manufacturing contexts, with future plans for collaborative expansion and real-world application. The dataset and further information are available via a GitHub repository.

**Keywords:** Assembly · Human Action Recognition · Azure Kinect · Dataset · Artificial Intelligence · Methods-Time Measurement

## 1 Introduction

The Artificial Intelligence (AI) based Human Action Recognition (HAR) is used to detect and track activity by observing people's movements. While HAR is already used in various daily activities, its application in an industrial context is subject to challenges. HAR has various advantages in recognizing the progress of manual assembly tasks in a fine-grained manner [1] to enable Human Robot Collaboration (HRC) or other assembly assistance systems. This process optimization can, for example, make an important contribution to upscaling the production of large-scale electrolysers [2]. Here, the Methods-Time-Measurement (MTM) based action recognition is a promising approach [1,3]. In order to train AI applications to classify actions in an assembly context, a large amount of data is required. Due to privacy restrictions and company secrets, training data cannot be acquired in real industrial assembly lines and has to be recorded under lab conditions. This results in a lack of labeled training data. Since recognizing small nuances between assembly actions is of key importance, having high resolution training data for several action classes is required [4]. Hence, this work

H. Kohl et al. (Eds.): GCSM 2024, LNME, pp. 294–302, 2025.
https://doi.org/10.1007/978-3-031-93891-7_33

introduces a novel HAR dataset to contribute to the industrial application of HAR methods. The dataset was recorded in a lab setting in front of a green screen where numerous subjects perform generic assembly tasks. The recorded data is labeled by MTM-Primitives [1]. Furthermore, the data is recorded as skeleton, RGB, and depth data, to ensure its application for different state of the art HAR algorithms. Due to the utilization of a green screen setting, an investigation into background effects on the algorithms can be conducted.

First an investigation of state of the art training datasets is conducted (Sect. 2). Subsequently, the methodology is described (Sect. 3), followed by the analysis of the captured and labeled data (Sect. 4). Finally, the paper is concluded and an outlook for further research is given (Sect. 5). The dataset, as well as further information, is made available to the research community in the Git repository: https://github.com/LukasBuesch/HARDAT.

## 2    Related Work

Since HAR is not limited to industrial applications, there are numerous algorithms and methods available that are subject to continuous improvement. Popular networks for HAR are for example CNNs [4,5] or GCNs [1,6,7]. Common to all algorithms is that they require training data in the form of either RGB, depth or skeleton data. This data can be acquired by camera systems or, as for the skeletal data, by gyroscopic systems.

To train those HAR algorithms different datasets are available. An overview of HAR datasets with industrial context and their specifications are given in tab. 1. Besides the summarized aspects in the table, the datasets provide additional unique features. In ATTACH [6], the utilization of one or both hands per action is annotated. Also focusing on the collaboration of hands, in HA-VID [8] the impact of learning over assembly executions is included. Including a cobot, InHARD [9] depicts an HRC scenario. Assembly 101 [10], offers non-scripted tasks with detailed annotations, capturing natural variability and complexity in assembly. Furthermore, it offers 21 key-points per hand. While focusing on human -object interaction, IKEA-ASM [11] is recorded in different environments to increase the background diversity. Also focusing on human-object interactions, MECCANO [12] is recorded with egocentric views in an actual industrial setting. With large amount of multi-modal data, HA4M [13] challenges HAR algorithms with fine-grained actions. In OCA [14] and InHARD [9] IMU integrated vests were used to capture upper-body movements during assembly. Due to the high availability of data sets from other domains that can be used for HAR, these data sets are often used for the development of HAR algorithms and for transfer learning. Although these methods can be used as support, they do not replace the need for domain-specific datasets.

Datasets used for HAR in the manufacturing domain need to meet specific requirements [15], such as having multi-modal data with a variety of background and lighting conditions, having real-world data to capture the complexity of actual manufacturing processes, using remote sensors to minimize interference with workers, and also being reproducible to increase the amount of data

Table 1. Properties of the manufacturing datasets.

| Name | Data Type | Resolution | Size | Number of Action Classes |
|---|---|---|---|---|
| ATTACH [6] | RGB DEPTH IR | 2560 × 1440 320 × 288 – | 51.6 h | 51 |
| HA4M [13] | RGB DEPTH IR | 2048 × 1536 640 × 576 – | 4.1 TB | 12 |
| ASSEMBLY-101 [10] | RGB+ Monochrome cameras | 1920 × 1080 640 × 480 | 513 h | 1380 fine 202 coarse |
| MECCANO [12] | RGB | 1920 × 1080 | 7 h 64 349 frames | 12 |
| IKEA-ASM [11] | RGB DEPTH | 1920 × 1080 – | 35.27 h 3 million frames | 33 |
| HA-VID [8] | RGB DEPTH | 1280 × 720 512 × 512 | 86.9 h 1.5 million frames | 219 fine 81 coarse |
| InHARD [9] | RGB IMU | 1280 × 720 – | 2 million frames | 72 fine 14 coarse |
| OCA [14] | IMU | – | 6 h | 6 |

afterwards. Furthermore, balanced data positively enhances the performance of algorithms. Although some of the existing datasets have a variety of background and lighting conditions, these conditions cannot be modified for specific use-cases. While the nature of manufacturing leads to an unbalanced distribution of action classes, the state of the art datasets show an imbalance towards not assembly unique action classes. Furthermore, there is also a lack of standardization in the labeling of actions, particularly those of a fine-grained, causing issues combining and extending datasets.

## 3   Dataset Generation

For **Dataset Composition** two approaches were used for task selection. First, the most relevant actions for the manufacturing domain were selected by examining the tasks in the previous data sets. Secondly, actions performed with tools commonly used in manufacturing were selected based on the ALET [16] dataset. The commonly used tools are *hammer, screwdriver, drill, wrench, allen key, ratchet, pliers, file, tape measure, caliper, square, scissors, cutting knife, screws,* and *nails.* This resulted in the following list of tasks: *picking up* and *putting down* a wooden board, *measuring* distances and diameters, *hammering* the punch and nails, *drilling*, *screwing* and *unscrewing* with the hand, Allen key, wrench, ratchet, automatic screwdriver and screwdriver, *cutting* with knife and scissors, *changing* the head of the drill, *filing* the edge of the wood and *taking out* nails with the pliers.

With these tasks, five different assembly task lists were created to capture actions in a continuous flow, as online action recognition requires the use of continuous data [17]. When designing the task lists, it was ensured that tasks do not start and end with the same action and that the same two actions are not arranged in the same order more than twice across the lists. The aim is to reduce the weighting of the context of the individual actions in the continuous data set. Furthermore, to enhance the variability of the movements, the tools are randomly positioned prior to each participant's initial recording.

All tasks are divided into sequences by seven basic movements, the Methods-Time Measurement (MTM) primitives [1]: *reach, grasp, move, position, apply pressure* and *release*. Here, MTM standardizes the division of tasks by basic movements and enables the dataset to be extendable and mergeable.

The **Data Collection Process** was conducted on a workstation equipped with Microsoft's Azure Kinect as camera device. With the camera and the associated software, RGB and depth images can be capured, as well as skeletal data can be obtained by body tracking. Using body tracking information is beneficial for HAR due to independence of the data from background and changing performers [18]. The body tracking data is stored as spatial-temporal graphs [1]. These graphs consists of 32 body joints with position and orientation over time as well as a confidence level. RGB data can be used for tool tracking, which enhances the accuracy of HAR in assembly [3], as well as for improving the body tracking e.g. extracting the hand joints [7,18] to capture intricate motions. The RGB Video is recorded with 720p ($720 \times 1280$). Since some studies using depth image to capture 3D coordinates [7], the depth video is recorded with 480p ($480 \times 640$) as well. The whole data is recorded with 30 Frames Per Second (FPS). Although the body skeleton information is independent from the background [6], the background has an influence on the RGB and depth image. To enable the dataset for investigations of background effects or for setting up use-case specific backgrounds, a green screen is added as background of the workstation. The dimensions of the workbench as well as additional information on the used hardware are given in the GitHub repository. The RGB and the depth image of the workstation is given in Fig. 1.

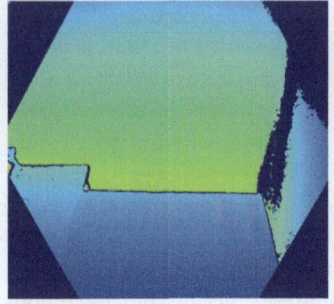 

**Fig. 1.** Depth (left) and RGB (right) image from camera at the workstation.

The above mentioned five assembly task lists were executed by every participant. Prior to the first recording all participants received an introduction. The tools as well as the recording procedure were explained. Subsequently, the participants had to self evaluate their handcraft skill level on scale from 1 to 5. Furthermore, each participant had to sign a data sharing consent agreement.

The **Data Annotation and File Generation** was carried out simultaneously to the recording of the data. Therefore, the supervisor was sitting in front of the workstation for live labeling and verbal assistance. At the start of every recording, gender and height of the participant and the chosen task list are written inside the body tracking data file. Within the annotation process, the start and end time and frame of each task and its subordinate MTM primitives were stored in a separate text file. Conducted actions that could not be matched with the defined MTM primitives were labeled as "none". Similarly, conducted actions between two tasks from the list were labeled as "none task". To increase the accuracy of the live labeling a post-correction was conducted for all data.

## 4    Analysis of the Generated Dataset

In total, 14 persons participated in the recordings, 11 men and 3 women. The height of the participants are ranging from 158 to 190 cm, the average is 180 cm. Furthermore, the average self skill rating is 3 out of 5. One participant asked his RGB data to be taken out from publication, which has an influence on the numbers for the RGB data. In the following only the numbers for body tracking and depth image are investigated. All numbers as well as additional information are given in the GitHub repository.

In total, all recorded data sum up to 1 271 797 frames, 11.78 hours and 99.7 GB of data. The total number of frames as well as a boxplot per task are depicted in Fig. 2.

As it can be seen in Fig. 2b, the task *screwing with hand* has the highest amount of frames, due to its occurrence before every other screwing motion. In contrast, *cutting* has the least amount of frames, as it was only included in three

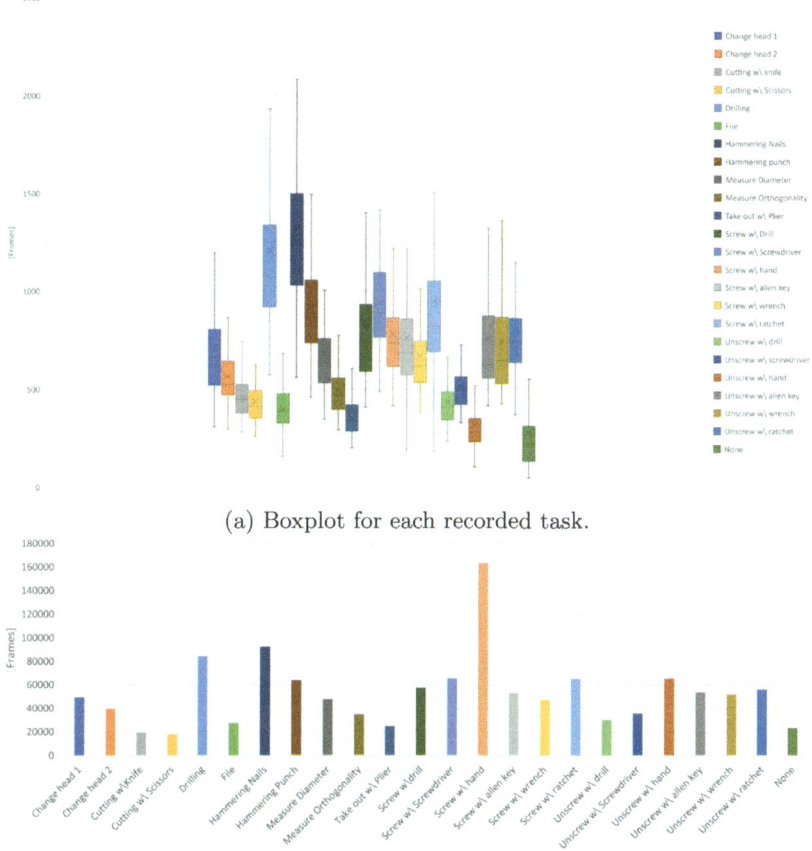

(a) Boxplot for each recorded task.

(b) Total number of frames for each recorded task.

**Fig. 2.** Data visualization for the recorded tasks.

task lists to create variety. While *hammering* in the nails took the most time on average because the nail had to be laboriously positioned correctly, *unscrewing* by hand took the least time because the screws had already been loosened with other tools. As can be seen from Fig. 2a, almost all assembly tasks have high standard deviations, with *hammering* the nails being the highest and *taking out* with pliers the lowest. This is mainly caused by the different skill levels of the participants as well as their physical strength when controlling the tools and applying force.

The total number of frames as well as a boxplot per MTM primitive are depicted in Fig. 3.

Figure 3b shows that the primitive *position* has the most amount of frames, while *release* has the least. This is because *position* took more time to execute on average than the other primitives, while *release* took the least

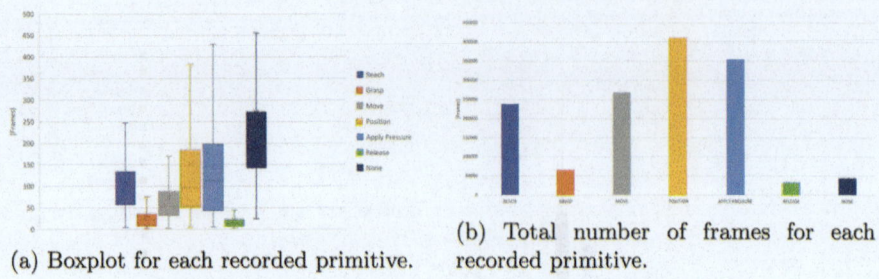

(a) Boxplot for each recorded primitive.    (b) Total number of frames for each recorded primitive.

**Fig. 3.** Data visualization for the recorded primitives.

time. For the same primitives, this difference can also be seen in the values of the standard deviation, as shown in Fig. 3a. One reason for the high standard deviation values is the limited range of MTM primitives. For example, in the hammering motion, each hit with the hammer is labeled as *apply pressure*, but also in the drilling task, the movement of the drill is labeled as *apply pressure* as this is the only primitive available for actions that require the application of force.

## 5    Conclusion and Outlook

The requirements for this dataset have been largely met, with notable achievements and some areas needing further improvement. Multi-modal data was successfully incorporated, utilizing remote sensors. A green screen was used to provide variable background conditions, though variety of lighting conditions was not achieved. The issue of data imbalance has been partly solved: while the overall imbalance persists, the specific imbalance towards non-assembly actions has been resolved. This dataset is standardized, making it both mergeable and extendable for future research. Compared to state-of-the-art datasets (see tab. 1), this dataset is of medium size. It also includes a medium number of action classes, with a strong emphasis on assembly-related activities.

For future work, we plan to extend the dataset collaboratively, inviting contributions from the research community to enhance its breadth and diversity. Further works should apply various HAR algorithms to evaluate their performance on the dataset. Additionally, extending the skeletal data using synthetic training data approaches could increase the number of training data. Addressing the current limitation, incorporating variable lighting conditions would better simulate real-world environments.

**Acknowledgements.** This work was supported by the German Federal Ministry of Education and Research (BMBF) under the grant number 03HY114F within the research project H$_2$Giga - HyPLANT100.

# References

1. Koch, J., Büsch, L., Gomse, M., Schüppstuhl, T.: A methods-time-measurement based approach to enable action recognition for multi-variant assembly in human-robot collaboration. Procedia CIRP **106**, 233–238 (2022)
2. Büsch, L., Jakschik, M., Syniawa, D., et al.: Hyplant100: industrialization from assembly to the construction site for gigawatt electrolysis. Hydrogen **5**(2), 185–208 (2024)
3. Büsch, L., Koch, J., Schoepflin, D., et al.: Towards recognition of human actions in collaborative tasks with robots: extending action recognition with tool recognition methods. Sensors **23**(12), 5718 (2023)
4. Myers, M.K., Wright, N., McGough, A.S., Martin, N.: Hand guided high resolution feature enhancement for fine-grained atomic action segmentation within complex human assemblies. In: 2023 IEEE/CVF Winter Conference on Applications of Computer Vision Workshops (WACVW), pp. 1–10. IEEE (2023)
5. Xiong, Q., Zhang, J., Wang, P., et al.: Transferable two-stream convolutional neural network for human action recognition. J. Manuf. Syst. **56**, 605–614 (2020)
6. Aganian, D., Stephan, B., Eisenbach, M., et al.: Attach dataset: annotated two-handed assembly actions for human action understanding. In: International Conference on Robotics and Automation (ICRA), pp. 1–7 (2023)
7. Terreran, M., Lazzaretto, M., Ghidoni, S.: Skeleton-based action and gesture recognition for human-robot collaboration. In: Petrovic, I., Menegatti, E., Marković, I. (eds.) Intelligent Autonomous Systems 17, vol. 577 of Lecture Notes in Networks and Systems, pp. 29–45. Springer, Cham (2023)
8. Zheng, H., Lee, R., Lu, Y.: Ha-vid: a human assembly video dataset for comprehensive assembly knowledge understanding (2023)
9. Dallel, M., Havard, V., Baudry, D., Savatier, X.: Inhard - industrial human action recognition dataset in the context of industrial collaborative robotics. In: 2020 IEEE International Conference on Human-Machine Systems (ICHMS), pp. 1–6. IEEE (2020)
10. Sener, F., Chatterjee, D., Shelepov, D., et al.: Assembly101: a large-scale multi-view video dataset for understanding procedural activities. In: 2022 IEEE/CVF Conference on Computer Vision and Pattern Recognition (CVPR), pp. 21064–21074 (2022)
11. Ben-Shabat, Y., Yu, X., Saleh, F., et al.: The ikea asm dataset: Understanding people assembling furniture through actions, objects and pose. In: Proceedings - 2021 IEEE Winter Conference on Applications of Computer Vision, WACV 2021 (2021)
12. Ragusa, F., Furnari, A., Livatino, S., Farinella, G.M.: The meccano dataset: understanding human-object interactions from egocentric videos in an industrial-like domain. In: 2021 IEEE Winter Conference on Applications of Computer Vision (WACV), pp. 1568–1577 (2021)
13. Cicirelli, G., Marani, R., Romeo, L., et al.: The ha4m dataset: multi-modal monitoring of an assembly task for human action recognition in manufacturing. Sci. Data **9**(1), 745 (2022)
14. Kuschan, J., Filaretov, H., Kruger, J.: Inertial measurement unit based human action recognition dataset for cyclic overhead car assembly and disassembly. In: 2022 IEEE 20th International Conference on Industrial Informatics (INDIN), pp. 469–476. IEEE (2022)

15. Rude, D.J., Adams, S., Beling, P.A.: A benchmark dataset for depth sensor based activity recognition in a manufacturing process. IFAC-PapersOnLine **48**(3), 668–674 (2015)
16. Kurnaz, F.C., Hocaoglu, B., Yilmaz, M.K., et al.: Alet (automated labeling of equipment and tools): a dataset, a baseline and a usecase for tool detection in the wild. In: ECCV2020 International Workshop on Assistive Computer Vision and Robotics (2020)
17. Rude, D.J., Adams, S., Beling, P.A.: Task recognition from joint tracking data in an operational manufacturing cell. J. Intell. Manuf. **29**(6), 1203–1217 (2018)
18. Aganian, D., Köhler, M., Stephan, B., et al.: Fusing hand and body skeletons for human action recognition in assembly. In: Artificial Neural Networks and Machine Learning – ICANN 2023: 32nd International Conference on Artificial Neural Networks, Heraklion, Crete, Greece, 26–29 September 2023, Proceedings, Part I, pp. 207–219. Springer-Verlag, Heidelberg (2023). https://doi.org/10.1007/978-3-031-44207-0_18

# Separation of Polymer Electrolyte Membrane Stack Components Using Sensor Integration for Non-destructive Disassembly

Dominik Goes$^{(\boxtimes)}$ ⓘ, David Kraus, Florian Kößler, and Jürgen Fleischer

KIT Karlsruhe Institute of Technology, 76131 Karlsruhe, Germany
`dominik.goes@kit.edu`

**Abstract.** Hydrogen technologies, such as polymer electrolyte membrane (PEM) electrolysis and fuel cells are considered the central and most promising technologies for the production and use of green hydrogen. End-of-life recycling is essential due to the presence of critical raw materials such as platinum group metals. Disassembly can improve the recycling outcome and enable other circular economy strategies such as reuse or remanufacturing of high value added components. The challenge in disassembling PEM stacks is to separate the stacked components non-destructively. This is due to the adhesion of the components to each other, as well as the low material thicknesses and component distances. The aim of this work is to identify suitable separation processes with a focus on mechanical processes. An industrial system concept will be developed and constructed. Sensors will be integrated into the system to enable accurate positioning of a cutting tool. Finally, the process is validated. The publication shows that non-destructive and automated separation of the individual components in PEM stacks using mechanical processes in combination with sensor-supported positioning enables the realisation of various circular economy strategies.

**Keywords:** Fuel Cell · Disassembly · Adhesion · Sensor integration

## 1 Introduction

### 1.1 Motivation

At the end of a product's life (EoL), different circular economy strategies can be pursued [1]: Recycling as a circular economy strategy involves the recovery of used materials. However, other strategies such as reuse and remanufacturing can retain the added value of an entire product or components. The prerequisite for reuse or remanufacturing of parts is a non-destructive extraction. The design of a PEM fuel cell stack is described in detail in [2]. Each stack is made up of up to hundreds of individual cells. Each cell features a bipolar plate (BPP), a membrane electrode assembly (MEA), followed by the next bipolar plate. There are

H. Kohl et al. (Eds.): GCSM 2024, LNME, pp. 303–311, 2025.
https://doi.org/10.1007/978-3-031-93891-7_34

different degradation mechanisms for the components. A typical cause of stack failure is perforation of the MEA [3]. The MEA or membrane is the component that limits the lifetime of the stack, whereas the BPP can have a longer lifetime. Regarding the cost structure, the main cost of the MEA is the cost of materials, whereas production costs predominate at BPP. [4]. Circular economy strategies can be derived from the duality between MEA and BPP in terms of degradation and cost structure: MEA needs to be recycled. Repair or remanufacturing should maintain the added value of BPP.

During operation, fuel cell stacks are exposed to conditions such as compression, acidic environments, overvoltage, temperature and humidity. After a stack has been used, BPP and MEA can adhere to each other in the sealing area and form a joint. In this area, the sealing material adheres to the polymer of the subgasket (e.g. polyethylene naphthalate) of the MEA. This adhesion, together with the low material thicknesses of the components (unshaped BPP around 100μm [5], laminated subgasket foils around 50μm [6]) of the components and the resulting small component distances (around 1mm or less) are the main challenges for disassembly. The causes and influencing factors (operating conditions see above) of adhesion have not yet been investigated. Tests such as peel and shear tests need to be carried out to evaluate the adhesive force. These analyses are not part of this work.

## 1.2  State of the Art

DIN/TS 54405 ([7]) specifies methods for separating and recovering adhesives and bonded parts from bonded component joints. A distinction is made between physical, chemical and mechanical methods. Chemical methods are not recommended due to the following reasons: (a) limited accessibility to the joint, (b) time consuming process, (c) damage to BPP or dissolving of catalyst from MEA not excluded, (d) possible contamination of BPP cooling channels. Aspects (a) and (d) can be justified by the structure of a stack and a BPP. Aspects (b) and (c) are process-related, as these methods usually result in the dissolution or decomposition of materials. Debonding by heating or freezing is a physical process. It can be used as an aid to debonding if the temperature limits are observed. Heating exceeds the sealant's glass transition temperature. Freezing embrittles the sealant. Both processes can lead to a reduction in adhesive force. Physical methods are not recommended for a repair scenario where only individual defined cells need to be separated. Targeted application to individual defined cells is difficult to achieve.

Non-destructive disassembly of PEM fuel cells has not been extensively studied in the literature. Component adhesion is also not addressed in the literature. The factors influencing adhesion have not yet been analysed. A solution for identifying the ideal position in relation to the Z-value (height) of the wedge is not proposed. In the literature, disassembly challenges have been identified and process chains elaborated ([8]), recycling strategies developed ([9]), manual disassembly procedures presented ([10]) and the need for automation highlighted ([11]). The patent landscape ([12–16]) provides more specific proposals for the disassembly of PEM fuel cells. The technical feasibility of the processes has not been demonstrated, there is no evidence of automation and some of the processes are destructive.

The state of the art shows that mechanical processes in particular are suitable for pursuing a variety of possible circular economy strategies. Furthermore, no automated solution has been proposed for the localisation of the joints and the positioning of a cutting tool. This work therefore focuses on the development of both, automated mechanical disassembly and localisation and positioning.

## 2    Materials and Methods

For reasons of availability and confidentiality, no commercial fuel cells can be used for the investigations and validations in this paper. Therefore, representative analogue components have been fabricated. The cell design is based on seven layer MEA and metallic BPP ([17]) including sealing with the following characteristics: Shaped and welded stainless steel foil (thickness 75μm, dimensions $219 \times 117$mm); bead sealing technology; screen printed elastomeric seal (width 1mm); flow field and manifolds. The investigations are not affected by the presence of CCM and GDL. Laminated subgasket foils (polyethylene naphthalate carrier film, 25μm thickness [6]) represent the MEA. A short stack consisting of alternately stacked laminated subgakset units and BPP was stacked and pressed (2,3MPa [18]). Adhesion in the sealing area occurred after a few weeks at room temperature and standard atmosphere in the compressed state.

## 3    Results and Discussion

Mechanical processes are suitable for disassembly, as described in Chap. 1.2. Different circular economy strategies can be achieved with mechanical processes. According to DIN/TS 54405, the options are peeling, cutting and stretching. Cutting is a possible solution in which the different circular economy strategies can be followed. One possible solution is cutting using a wedge as a separating tool. This process is analysed in detail below. A possible system design including sensor integration for position determination is described. The developed process is validated.

### 3.1    Construction of an Industrial System

An illustration of the system is shown in Fig. 1. The design includes two parallel linear axes with carriages. A load cell is mounted on the carriage of each axis. The axes are position controlled and are used to manipulate the cutting tool, in this case a knife-like steel wedge. Both axes can be driven synchronously or asynchronously. This means, that the angle of the cutting wedge around the Z axis can be varied. The lifting table is used to place and fix the fuel cell stack and is adjustable in height. Next to the lifting table, a laser triangulation line profile sensor is placed on its side so that it's measurement plane is perpendicular to the table surface. The fuel cell stack is placed on the lifting table with one of its short sides facing the profile sensor and the cutting wedge. Spring loaded anchor points, pneumatic cylinders or a vacuum clamping plate are all suitable for clamping the stack (not shown). An articulated arm robot with a suction gripper is integrated to handle the separated parts.

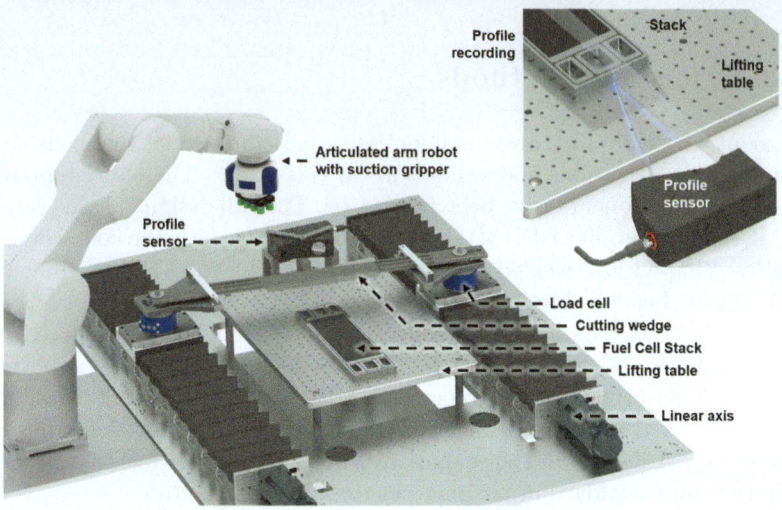

**Fig. 1.** Construction of the industrial system for disassembly of PEM fuel cells (left) with detailed view of the stack profiling setup (top right).

### 3.2    Separation Strategy and Parameters

To separate the top layer of the stack, the axes are driven so that the cutting wedge is inserted horizontally under this layer. To insert the wedge exactly between two layers and avoid collision with the stack, the height at which the wedge is moved must be precisely determined. This is achieved by using a profile sensor that detects the stack profile. This only demonstrates one of the options chosen for this system. The profile is analysed using a specially developed algorithm. Since the stack consists of periodically arranged layers the obtained profile also shows a pattern of recurring geometric features, i.e. the thin sides of the BPP, as the MEA doesn't tend to show up in the sensor profile due to its small width and transparency. An example profile is shown in Fig. 2. The setup for recording the profile is shown in Fig. 1. Notably each BPP is identifiable by it's two steep flanks. To identify the position of each BPP, the two flanks must be detected. The implemented algorithm achieves this by essentially calculating the second numerical derivative of the profile and finding inflection points. Since each BPP lies between two inflection points and always right of one rising and left of one falling inflection point, the position of each BPP can be calculated. The cutting position is then placed exactly between two plates. Minor features and noise in the profile are suppressed by only accepting inflection points correlating with sufficiently large slopes in the original profile.

The system operates in a closed information loop. First, the cutting wedge and lifting table are moved to their respective home positions. Then a measurement is taken by the profile sensor and the resulting profile is transferred to the custom evaluation software via an FTP server. There the cutting positions are determined using the above mentioned algorithm. The cutting coordinates are then sent back to the system control and the axes are driven accordingly in order to facilitate the cut. The separated layers can be removed manually or by an articulated arm robot. The process then starts again for the next layer.

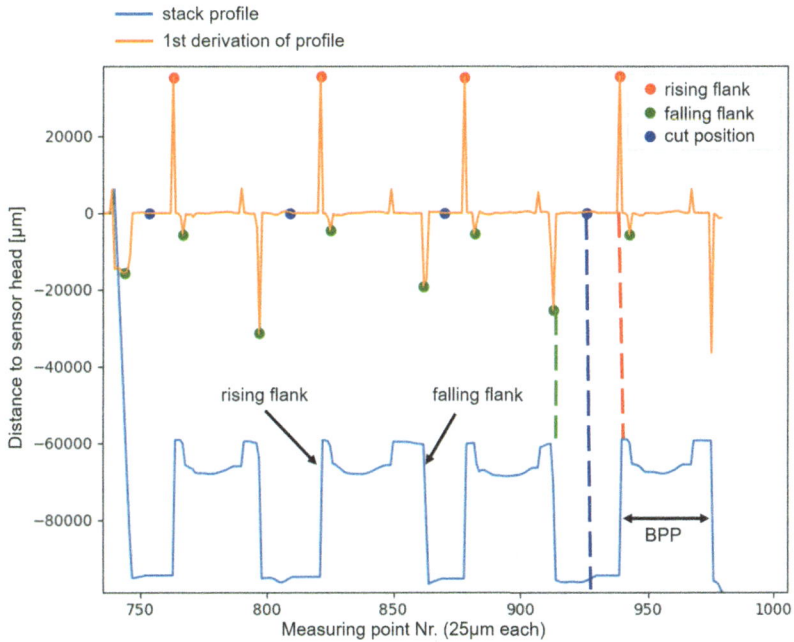

**Fig. 2.** The algorithm for analysing the cutting coordinates. The profile shows the cutting position between two BPPs. The MEA located between two BPPs cannot be seen on the profile.

## 3.3   Validation and Characterization

The analogue cells defined in Sect. 2 are used to validate the separation process. The validation test setup is shown in Fig. 3. The wedge can be set at different angles as it moves through the stack, allowing the influence of the cutting angle on the quality of the cut to be studied. By setting an angle, the cutting wedge is inserted over the corner of the stack rather than over the entire edge. This can simplify the insertion and eliminate height differences across the length of the stack in relation to the individual layers. The speed of the cutting movement can be adjusted to study the effect of cutting speed on the accuracy of the cut and the resulting damage patterns. Tests with the setup employing a cutting wedge with a continuous edge geometry at different cutting angles showed a good functionality of the algorithm described above. Out of twenty tests, only one profile showed an error in the detection of a BPP and did not lead to a correct cutting position. The requirement for the separation process is the non-destructive separation of the BPP. During disassembly, defects and damage can occur (see Table 1), which must be avoided.

**Table 1.** Failure patterns during separation process

| Failure patterns | Cause | Measures |
| --- | --- | --- |
| Edge deformation during insertion | Height differences over the length of the BPP | Insertion via corner instead of edge |
| Deformation of BPP corners | Algorithm errors | Algorithm optimisation |
| Deformation of BPP corners | Features in BPP corners (e.g. CVM connector or centring features) | Adaptation of cutting strategy or cutting wedge |
| Delamination of the seal or coating | Wedge thickness greater than cell distances | Acceptable |
| Damage to subgasket or delamination of GDL | Material thickness of the wedge greater than the distance between two cells | Acceptable |

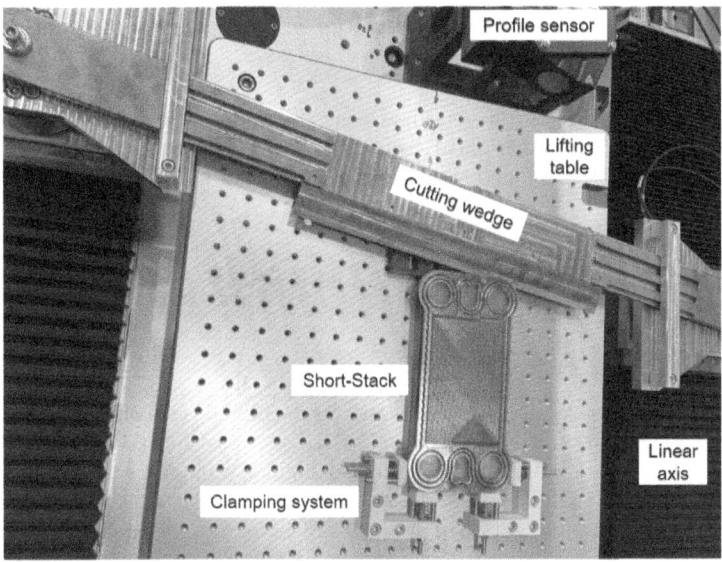

**Fig. 3.** Test setup for validation of the separation process with analogue short stack and angled wedge cutter.

### 3.4  Conclusion

Experiments using the setup as described in Sect. 3.1 showed the possibility of non-destructive disassembly of fuel cell stacks as BPP could be extracted undamaged. As the implemented process relies only on mechanical motion and force, and does not require any further thermal energy or material input, the process is energy and resource efficient. The implementation of a laser profile sensor enables the detection of cutting positions and promises to make the method transferable to other stack designs. Achieving undamaged disassembly is highly dependent on the geometry of the fuel cell stack and the cutting tool. Further research is required to improve the cutting geometry and tool response. There are several parameters that could be adjusted in future experiments to improve the results: (1) The geometry of the cutting wedge can be modified to avoid collision with small protruding geometric features; (2) The wedge could be agitated, e.g. by vibration or an ultrasonic pulse to further weaken adhesive bonds during cutting; (3) The stiffness of the wedge attachment could be reduced to allow small compensatory movement of the wedge; (4) A control loop could be implemented to monitor the cutting forces and adjust the cutting movement accordingly.

Other possible disassembly processes, such as peeling or alternative cutting processes, as well as the use of physical processes to reduce the adhesive force, must also be analysed. The adhesive force needs to be quantified and the factors influencing adhesion analysed. The process developed must be profitable in terms of cost and process time. These values cannot be quantified at present. The system is a proof of concept and needs to be optimised in terms of process time.

The automated system currently competes with manual disassembly because there is no proven process.

**Acknowledgment.** The authors acknowledge the financial support by the German Federal Ministry of Education and Research (BMBF) within the project "ReNaRe - Recycling - Nachhaltige Ressourcennutzung" under grant numbers 03HY111B.

# References

1. Ellen MacArthur Foundation: Towards the circular economy Vol.1: an economic and business rationale for an accelerated transition (2013). https://www.ellenmacarthurfoundation.org/towards-the-circular-economy-vol-1-an-economic-and-business-rationale-for-an
2. Schäfer, J., Allmendinger, S., Hofmann, J., Fleischer, J.: Genetic algorithm for the optimization of vision acquisition for on-the-fly position measurement of individual layers in fuel cell stack assembly. Procedia CIRP **104**, 1407–1411 (2021). https://doi.org/10.1016/j.procir.2021.11.237
3. Millet, P., Ranjbari, A., Guglielmo, F., Grigoriev, S., Auprêtre, F.: Cell failure mechanisms in PEM water electrolyzers. Int. J. Hydrogen Energy **37**, 17478–17487 (2012). https://doi.org/10.1016/j.ijhydene.2012.06.017
4. Kampker, A., Heimes, H., Kehrer, M., Hagedorn, S., Reims, P., Kaul, O.: Fuel cell system production cost modeling and analysis. Energy Rep. **9**, 248–255 (2023). https://doi.org/10.1016/j.egyr.2022.10.364
5. Song, Y., et al.: Review on current research of materials, fabrication and application for bipolar plate in proton exchange membrane fuel cell. Int. J. Hydrogen Energy **45**(54), 29832–29847 (2020). https://doi.org/10.1016/j.ijhydene.2019.07.231
6. CMC Klebetechnik GmbH - Technical data sheet subgasket foil CMC 61325. Accessed June 2024. https://www.cmc.de/wasserstofftechnik-brennstoffzellen
7. DIN/TS 54405:2020-12, Construction adhesives - Guideline for separation and recycling of adhesives and substrates from bonded joints (2020)
8. Al Assadi, A., et al.: Challenges and prospects of automated disassembly of fuel cells for a circular economy. Res. Conserv. Recycl. Adv. **19** (2023). https://doi.org/10.1016/j.rcradv.2023.200172
9. Wittstock, R., Pehlken, A., Wark, M.: Challenges in automotive fuel cells recycling. Recycling **1**(3), 343–364 (2016). https://doi.org/10.3390/recycling1030343
10. Férriz, A.M., Bernad, A., Mori, M., Fiorot, S.: End-of-life of fuel cell and hydrogen products: a state of the art. Int. J. Hydrogen Energy **44**(25), 12872–12879 (2019). https://doi.org/10.1016/j.ijhydene.2018.09.176
11. Uekert, T., Wikoff, H.M., Badgett, A.: Electrolyzer and fuel cell recycling for a circular hydrogen economy. Adv. Sustainable Syst. **8**, 2300449 (2024). https://doi.org/10.1002/adsu.202300449
12. Toyota Motor Group: Dismantling method of fuel cell, JP2005222818A, Patent granted (2011)
13. Toyota Motor Group: Fuel cell disassembly method, US7754371B2, Patent granted (2010)

14. Toyota Motor Group: Fuel cell disassembly method, JP4779345B2, Patent granted (2011)
15. Robert Bosch GmbH: Zelllagentrennvorrichtung zum gleichzeitigen Trennen von Zelllagen eines Zellstapels einer Brennstoffzelle, DE102022201566A1, Patent granted (2023)
16. Shenzhen Zhongwei Hydrogen Energy Technology Co ltd: Novel repair disassembly tool based on fuel cell stack, CN114346665A, Patent granted (2022)
17. MTZ extra: Entwicklung Brennstoffzellenantriebe - Die metallische Bipolarplatte, Dana Victor Reinz. Accessed June 2024. https://www.reinz.com/img_cpm/DANA-Reinz/Aktuelles/2108_MTZextra_Dana_mBPP_DE.pdf
18. Khetabi, E.M., Bouziane, K., Zamel, N., François, X., Meyer, Y., Candusso, D.: Effects of mechanical compression on the performance of polymer electrolyte fuel cells and analysis through in-situ characterisation techniques - a review. J. Power Sources **424**, 8–26 (2019). https://doi.org/10.1016/j.jpowsour.2019.03.071

# Making Sustainable Hydrogen Production a Reality: Scaling up the Production of Electrolysers and Fuel Cells

Mary Esther Ascheri(✉), Ulrike Beyer, Sören Scheffler, Stefan Lohberger, Sebastian Melzer, Jakob Arnold, and Samuel Rodrigo de Souza Cardoso

Fraunhofer IWU Institute for Machine Tools and Forming Technology, 09126 Chemnitz, Germany

mary.esther.ascheri@iwu.fraunhofer.de

**Abstract.** In the context of demanding sustainable energy production, hydrogen is known as the next green energy carrier. However, many challenges persist concerning the fulfillment of carbon-free hydrogen production. The central element for green hydrogen is the electrolyser, it separates water into hydrogen and oxygen using green energy. The task is to revolve around the urgent establishment of an efficient and economic amplification of electrolyser manufacturing on a large scale. The Referenzfabrik.H2 is actively addressing some challenges, linked to the upscaling of cost-efficient electrolyser and hydrogen systems production, employing technologies and development drawn from partners from industrial sectors and research institutes. Operating as a cooperative consortium, the Referenzfabrik.H2 functions as a value chain community of companies with different expertise, thereby enabling collaborative engagement and the dissemination of knowledge through a technological toolbox used to a sustainable ramp-up manufacturing of H2 system components. The Referenzfabrik.H2 wants also to serve as a connection between Germany and other countries. With partners in countries that have renewable energy potential, the Referenzfabrik.H2 can catalyze collaboration, through joint efforts. Among the initiatives are the development of manufacturing processes, automation and quality assurance solutions, and standardized techniques, driving the transformation of the hydrogen industry towards competitive hydrogen production and its acceptance.

**Keywords:** hydrogen · electrolyser · hydrogen systems · fuel cells

## 1 The Role of Hydrogen in Energy Transition

One of the key challenges in renewable energy utilization is intermittency, as sources like solar and wind are not constant. Hydrogen offers a solution by acting as a form of energy currency that can be stored for later use, effectively decoupling energy production from consumption. During periods of surplus renewable energy production, such as sunny days or windy nights, hydrogen production through electrolysis becomes an attractive option. This surplus energy is used to split water into hydrogen and oxygen, and the generated

H. Kohl et al. (Eds.): GCSM 2024, LNME, pp. 312–320, 2025.
https://doi.org/10.1007/978-3-031-93891-7_35

hydrogen is stored for later use. This capability transforms hydrogen into a valuable tool for balancing the intermittent nature of renewables and ensuring a continuous and reliable energy supply or feedstock [1, 2]. One of the significant advantages of hydrogen as an energy carrier is its exceptionally high energy density. Hydrogen reaches 2 up to 4 times higher energy content per unit mass compared to traditional fossil fuels [3–5]. This characteristic makes it an efficient and compact way to store and transport large amounts of energy. This high energy density makes hydrogen an attractive option for applications where space and weight considerations are critical, such as in FC vehicles or portable power systems. The ability to store energy opens a range of applications across various sectors [6, 7]. It is suitable for remote power generation and off-grid applications [8], where traditional energy infrastructure may be impractical. As a zero-emission fuel, hydrogen can play a transformative role in decarbonizing industries such as transportation, heating, and electricity generation [6, 9]. In the transport sector, FC vehicles powered by hydrogen can replace traditional internal combustion engines [10]. Additionally, hydrogen can be integrated into industrial processes [11], reducing the carbon footprint of heavy industries [12]. In the steel industry, hydrogen can replace fossil fuels in the production of iron [12]. Similarly, in fertilizer production offering a greener alternative replacing natural gas and in the chemical industry, known for its energy-intensive processes, can benefit with a cleaner feedstock, facilitating a shift towards sustainability [12]. Not to mention the current existing demand of hydrogen in those sectors which in 2021 less than 0,1% was made from electrolysis [13]. The development of environmentally friendly hydrogen production processes is challenging, hindering the establishment of a hydrogen economy worldwide. The key element for green hydrogen is the electrolyser. The high cost of this component and the hydrogen systems in general are the main challenges for the hydrogen market [14]. Therefore, the establishment of an efficient and economic amplification of these systems on a large scale is vital to meet the demand for green hydrogen. It is estimated that global demand will increase exponentially to 140 million tons of low-carbon hydrogen by 2030 [13]. The Referenzfabrik.H2 (translated as reference factory H2) from Fraunhofer IWU, brings the initiative, among others, of ramping up hydrogen systems manufacturing production by addressing the main challenges in this sector.

## 2 Electrolysers (EC)

Electrolysers are the fundamental component for green hydrogen generation. In this process, electrical energy generated from renewable sources is used to split water molecules into hydrogen and oxygen. The key principle involves passing an electric current through water, causing the water molecules to undergo a chemical reaction. This results in the release of hydrogen gas at the cathode and oxygen gas at the anode [15]. EC come in various designs, each with its own set of advantages and challenges concerning the component's material, component design, cell structure, and construction, directly influencing the operation temperature and conditions [16, 17]. There are four main types of ECs, the alkaline, the proton exchange membrane (PEM), the anion exchange membrane (AEM), and the solid oxide electrolysis cells (SOEC). Alkaline and AEM ECs operate with a basic electrolyte solution, the alkaline type is known for its reliability and mature technology [18]. The AEM is new in the market and the functionality is a combination of Alkaline and PEM types still in the low-maturity stage. SOEC operates at high temperatures, enables high-temperature heat integration, and can be integrated into industrial processes. The choice of EC type depends on factors such as application, scale, and availability of resources. The focus of research within this work is on the proton exchange membrane (PEM) EC type. PEMEC have advantages over other technologies, especially when coupled with renewable energies and the associated cyclical loads and short reaction times. They are also easy to operate, less corrosive, require less maintenance, have a low operating temperature, can be used reversibly, are more compact, and can work with lower cell voltages and higher current densities [19]. They can also be operated under high pressure, which leads to higher efficiencies (80–90%). The main disadvantages are high material costs [20], cross-permeation phenomena that increase with pressure, and the presence of water vapor together with the generated hydrogen, which requires dehumidification while the membrane must be constantly humidified. The PEM electrolysis cell (Fig. 1) consists of several components, including a proton-conducting membrane, which is usually made of a solid polymer electrolyte such as Nafion and is responsible for the diffusion of protons [19]. There is an anode and a cathode on both sides of the membrane, a catalytic layer to accelerate the electrochemical reactions, and a bipolar plate that serves as a structural support, conducts the electric current and facilitates the uniform distribution of reactants and products within the cell. The membrane can account for up to 40% of the total cost of the cell, and the bipolar plate up to 33% [20]. New materials and manufacturing processes need to be developed to enable industrial production and commercialization at lower overall costs. In terms of more effective integration of ECs with renewable energy sources, flexibility and efficiency of the system are decisive criteria. PEMECs are gaining favor due to their adaptability to dynamic energy inputs and reliability, making them suitable for coupling with intermittent renewables. The ability to quickly respond to the fluctuance of renewable energy positions PEMECs as favorable options for future hydrogen production systems.

## 3   Challenges Inside de Hydrogen Value Chain

A sustainable hydrogen production is still challenging. To replace the current demand for gray/black hydrogen, the green or low-carbon hydrogen production must be economically competitive. The infrastructure required for carbon-free hydrogen production (green hydrogen), including ECs, system design, and integration, demands substantial investment [21]. Without cost-effective solutions, the envisioned benefits of a hydrogen economy may remain out of reach for many regions and industries.

The cost challenge arises from various factors and in many scales, including the expense of advanced materials and alloys needed for EC components such as for porous transport layer and bipolar plates, and the risks associated with raw material availability, such as the catalysts materials, like platinum group metals. Besides, materials for the major components must withstand the harsh conditions of electrolysis, such as high temperatures and corrosive environments, and this poses an obstacle in achieving a balance between durability and cost-effectiveness, which opens scientific questions in terms of materials development [11]. Also, there is a need for a large-scale manufacturing capability. There are high costs involved along the whole production process, costs associated with low volume production and that could be minimized up to 2 times per $/kW with automatization, and digitalization in a scale effect. The increasing demand for low-carbon or carbon-free fuel and energy solutions pushes the hydrogen generated through electrolysis demand, which necessitates a substantial scaling up of EC production [22]. The EC capacity supply is very low due to the low manufacturing production volume available in the market. Developing advanced manufacturing technologies is a critical aspect of streamlining production processes, improving overall manufacturing efficiency, and production capacity, leading to final product cost reduction [22]. The technology maturity for commercial deployment and industrial implementation of such processes also plays a significant role in the market scale-up.

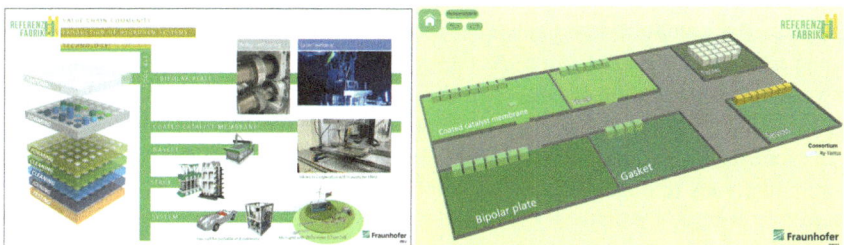

**Fig. 1.** (a) left: Referenzfabrik.H2 Toolbox schematic and technologies for EC and FC components production. (b) right: Referenzfabrik.H2 virtual platform

Furthermore, the efficiency of such systems directly impacts the energy balance of the hydrogen generation process. Integrating ECs into existing energy systems is sometimes difficult, particularly concerning grid compatibility and adaptation to variable renewable energy sources. The stability and reliability of the overall energy system are at a low maturity level, also the lack of regulatory frameworks for hydrogen technologies, which would facilitate the deployment of ECs and FCs, ensuring safety, interoperability, and greater market acceptance.

## 4   The Referenzfabrik.H2

The Referenzfabrik.H2 from the Fraunhofer IWU, represents a close connection between industrial companies and research resources, serves as a wide community for value creation that employs expertise and capabilities throughout the entire value chain for hydrogen systems. The aim is to provide orientation and an efficient network and thus accelerate the development of new solutions. To this end, unique EC and FC reference products and production systems (Fig. 1a) with innovative manufacturing technologies are being developed. One of the core strengths of Referenzfabrik.H2 lies in its collaborative efforts, offering a platform (Fig. 1b) where expertise from various sectors collaborates synergistically. The initiative brings together experts in design, testing, analysis, and quality monitoring, forming a well-rounded suite of services. This collaborative ecosystem extends beyond the confines of research institutions to include private sector entities from diverse industries. Through symbiotic partnerships, Referenzfabrik.H2 creates a multifaceted approach, enriching the initiative with insights and resources from different sectors. In essence, Referenzfabrik.H2 acts as a facilitator for the evolution of the hydrogen economy. It can engage in the breakdown and analysis of individual processes and components, provide targeted solutions, and offer several services inside the partner-platform to ensure the efficiency, reliability, and quality of these processes up to final product or entire systems. Various funding projects are the driving force behind Referenzfabrik.H2. This is funded by the Federal Ministry of Education and Research (BMBF) and is an integral part of the H2GIGA lead project (water electrolysis for green hydrogen on a gigawatt scale). FRHY (reference factory for high-rate electrolyser production) focuses on the development of novel roll-to-roll technologies. The National Action Plan for Fuel Cell Production (H2GO) funded by the Federal Ministry for Digital and Transport (BMDV) is a collaboration project representing impact as combines expertise of 19 Fraunhofer Institutes. The H2GO project focuses on the creation of innovative processes for FC production. As the transition from laboratory-scale innovations to practical implementation in large quantities is challenging, initiatives inside the Referenzfabrik.H2, demonstrate a strategic approach to bridge the gap between research and the industry up to international scale, emphasizing the importance of translating research outcomes into tangible solutions for the realization and adoption of hydrogen systems.

## 5   Referenzfabrik.H2 Technological Approach

Within Referenzfabrik.H2, the technological toolbox is an elaborate set of process technologies, and innovations designed to advance EC and FC manufacturing. Focused on the goal of ramping up production, Referenzfabrik.H2 employs a systematic approach to manufacture various components, such as the bipolar plate (BPP), catalytic coated membrane, gasket, etc. The process of fabricating each component involves a detailed procedure chain. Considering the BPP process production sequence as an example, it encompasses material selection, forming technology, cutting, cleaning, joining technology, coating processes, and finally submission for testing and quality control [11, 23, 24], see Fig. 2a below.

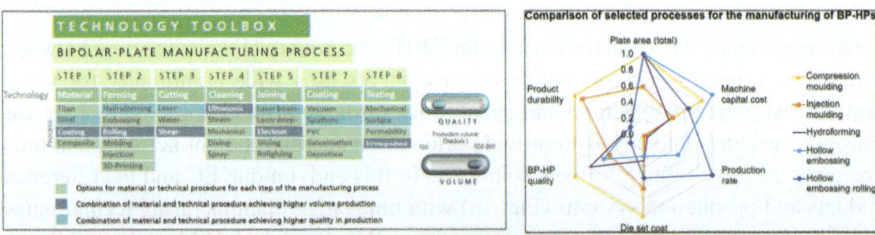

**Fig. 2.** (a) left: Schematic representation of the technological toolbox for bipolar plate manufacturing. (b) right: Example of how the data can be evaluated inside the toolbox for manufacturing of bipolar half plates (BP-HPs), with some of the forming processes included [11].

For example, in the forming process alone, Referenzfabrik.H2 can employ several techniques, such as hydroforming, stamping, rolling forming, 3D printing, or others. Each of these methods contributes uniquely with advantages and characteristics to the final product Fig. 3(b), which must be analyzed with other parameters such as cost, time, space need, and others. The goal of the Referenzfabrik.H2 toolbox is to categorize each process based on specific indicators or as desired and evaluate estimating not only the costs. This could range from choosing a process sequence to optimize the quality of the final product, prioritizing processes for high production volume, or applying a specific methodology due to equipment availability or preference. Inside Toolbox evaluates the trade-offs between different processes using a modeling tool. The choice between forming processes for the BPP can serve as an example, see Fig. 2b. While rolling forming can significantly increase production volume, the hydroforming process may offer superior BPP quality, in terms of surface homogeneity with affordable option comparing to molding, for instance. The evaluation can be carried out thanks a toolbox database which store several project results and parameters. This decision-making process ensures that diversity of the technological toolbox can be strategically applied, tailoring manufacturing approaches to meet specific objectives such as quality standards or production efficiency. Moreover, a detailed assessment of the entire process chain, evaluating sustainability indicators such as carbon footprint, waste generation, energy consumption, and overall resource utilization can provide a robust framework for the EC/FC value chain.

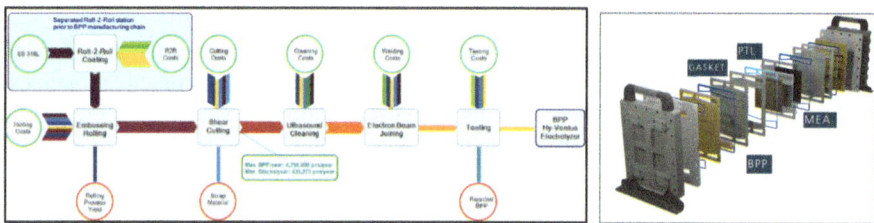

**Fig. 3.** (a) left: BPP manufacturing schematic simulation for material usage, costs, etc. (b) right: Example of a PEMEC; exploded view, designed by Referenzfabrik.H2; bipolar plate (BPP); porous transport layer (PTL); membrane electrode assembly (MEA)

By leveraging data-driven analysis, it is also essential to estimate environmental impact of the different manufacturing processes. The Fig. 3a is a schematic demonstration of the inputs and outputs for life cycle assessment modelling. This includes assessing the carbon emissions associated with each stage, identifying opportunities for waste reduction, optimizing energy use, and ensuring efficient resource management. Such process modeling and assessments contribute to building confidence among industries, investors, policymakers, and end-users with a common goal of reducing the environmental impact of their businesses more effectively. In this way, Referenzfabrik.H2 serves as an innovative value chain community, reducing the risks and uncertainties for industries interested in being part of the hydrogen value chain.

# 6  Collaborative Work: Bilateral and Multilateral

The need for strategic initiatives and collaboration is evident if we are aiming for a successful introduction of hydrogen in the market and society. Overcoming the economic barriers requires concentrated efforts in research, development, and large-scale manufacturing and is more effective if in cooperation work. For example, the Referenzfabrik.H2 has several projects with practical demonstrators, and hydrogen application prototypes in international collaboration projects in the southern African continent. In the HyTrA (Hydrogen Tryout Area) and HygO (Hydrogen & Oxygen Biotope Namibia) projects, hydrogen systems are being used in South Africa and Namibia, respectively. The central component of both projects is a microgrid, a power supply that is not yet able to compete with fossil alternatives in terms of cost. Other countries cooperating directly with Referenzfabrik.H2 are Norway and Czech Republic, where hydrogen systems prototypes are developed for the glass industry and transport application. The goal here is to enable an exchange of knowledge and demonstrate its functionality, leading therefore to more acceptance, and increasing demand that finally can result in cost reduction. As Referenzfabrik.H2 expands, existing collaborations between Germany and South America, e.g. with Uruguay, leverages the strengths of different regions, such as Germany's technological expertise and South America's renewable energy abundance, resulting in alternatives to scale up sustainable hydrogen production.

# 7  Conclusions

A keystone of Referenzfabrik.H2 is its engagement in a multitude of public-funded projects with a national scope, progressively advancing towards the practical and sustainable application of hydrogen systems on an international scale. This represents a step towards addressing the cost and technological challenges by focusing on efficient production processes with a technologic toolbox Referenzfabrik.H2 made to manufacture key components for green hydrogen production. Referenzfabrik.H2 can contribute to the strategic cooperation by sharing best practices, and technological insights, and promoting joint research and development projects that align with the hydrogen goals worldwide. In such a case, for more effective market growth bilateral and multilateral partnerships can be a crucial step forward for the efficient and economic production of electrolysis systems on a large scale.

# References

1. Borgschulte, A.: The hydrogen grand challenge. Front. Energy Res. (2016)
2. Møller, K.T., Jensen, T.R., Akiba, E., Li, H.W.: Hydrogen - A sustainable energy carrier. Progress in Natural Science: Materials International 27(1), 34–40 (2017)
3. Hydrogen Storage and Materials. In: Lu K, editor. Materials in Energy Conversion, Harvesting, and Storage, pp. 387–417. Wiley (2014)
4. Winter, C.: Electricity, hydrogen—competitors, partners? Int. J. Hydrogen Energy 30(13–14), 1371–1374 (2005)
5. Züttel, A., Remhof, A., Borgschulte, A., Friedrichs, O.: Hydrogen: the future energy carrier. Phil. Trans. A Math. Phys. Eng. Sci. 2010(368), 3329–3342 (1923)

6. Johnston, B., Mayo, M.C., Khare, A.: Hydrogen: the energy source for the 21st century. Technovation **25**(6), 569–585 (2005)
7. Sharma, S., Ghoshal, S.K.: Hydrogen the future transportation fuel: from production to applications. Renew. Sustain. Energy Rev. **43**, 1151–1158 (2015)
8. Rolo, I., Costa, V.A.F., Brito, F.P.: Hydrogen-based energy systems: current technology development status, opportunities and challenges. Energies **17**(1), 180 (2024)
9. Singla, M.K., Nijhawan, P., Oberoi, A.S.: Hydrogen fuel and fuel cell technology for cleaner future: a review. Environ. Sci. Pollut. Res. Int. **28**(13), 15607–15626 (2021)
10. Acar, C., Dincer, I.: The potential role of hydrogen as a sustainable transportation fuel to combat global warming. Int. J. Hydrogen Energy **45**(5), 3396–3406 (2020)
11. Porstmann, S., Wannemacher, T., Drossel, W.-G.: A comprehensive comparison of state-of-the-art manufacturing methods for fuel cell bipolar plates including anticipated future industry trends. J. Manuf. Process. **60**, 366–383 (2020)
12. Ramachandran, R.: An overview of industrial uses of hydrogen. Int. J. Hydrogen Energy **23**(7), 593–598 (1998)
13. IEA. Global Hydrogen Review 2022, IEA, Paris (2022). https://www.iea.org/reports/global-hydrogen-review-2022. Licence: CC BY 4.0
14. Proost, J.: State-of-the art CAPEX data for water electrolysers, and their impact on renewable hydrogen price settings. Int. J. Hydrogen Energy **44**(9), 4406–4413 (2019)
15. Kreuter, W.: Electrolysis: the important energy transformer in a world of sustainable energy. Int. J. Hydrogen Energy **23**(8), 661–666 (1998)
16. Guo, Y., Li, G., Zhou, J., Liu, Y.: Comparison between hydrogen production by alkaline water electrolysis and hydrogen production by PEM electrolysis. IOP Conf. Ser.: Earth Environ. Sci. **371**(4), 42022 (2019)
17. Godula-Jopek, A., Stolten, D. (eds.): Hydrogen Production: By Electrolysis, 1st edn. Wiley-VCH, Weinheim (2015)
18. Grigoriev, S.A., Fateev, V.N., Millet, P.: Alkaline electrolysers. In: Comprehensive Renewable Energy, pp. 459–472. Elsevier (2022)
19. Tellez-Cruz, M.M., Escorihuela, J., Solorza-Feria, O., Compañ, V.: Proton exchange membrane fuel cells (PEMFCs): advances and challenges. Polymers (Basel) **13**(18) (2021)
20. Mayyas, A.T., Ruth, M.F., Pivovar, B.S., Bender, G., Wipke, K.B.: Manufacturing Cost Analysis for Proton Exchange Membrane Water Electrolyzers (2019)
21. Lux, B., Deac, G., Kiefer, C.P., Kleinschmitt, C., Bernath, C., Franke, K., et al.: The role of hydrogen in a greenhouse gas-neutral energy supply system in Germany. Energy Convers. Manag. **270**, 116188 (2022)
22. Beyer, U., Porstmann, S., Baum, C., Müller, C.: Produktion der PEM-Systeme, Hochskalierung, Rollout-Konzept. In: Neugebauer, R. (ed.) Wasserstofftechnologien, pp. 297–330. Springer, Heidelberg (2022)
23. Saggiorato, N., et al.: A total cost of ownership model for low temperature PEM fuel cells in combined heat and power and backup power applications (2017)
24. Kampker, A., Kehrer, M., Heimes, H., Hagedorn, S.: Produktion von Elektrolyseursystemen (2023)

320     M. E. Ascheri et al.

# Augmented Reality Authoring for Efficient Inspections in Green Aviation: Evaluating Accuracy and Usability

Christian Masuhr[✉], Julian Koch, and Thorsten Schüppstuhl

Hamburg University of Technology, Institute for Aircraft Production Technology,
Denickestraße 17, 21073 Hamburg, Germany
christian.masuhr@tuhh.de
https://www.tuhh.de/ifpt

**Abstract.** This paper addresses the gap in AR-guided inspection by developing and evaluating an AR authoring tool tailored for green aviation. The tool aims to overcome the current limitations by assisting the inspector in the data acquisition process using hand-held sensors. The study introduces a sensor guidance prototype applicable to various inspection scenarios prone to human error. Usability assessments reveal positive feedback on real-time guidance and an intuitive interface, though challenges in tracking accuracy and managing complex trajectories are identified, particularly with the Microsoft HoloLens 2. Testing on the Magic Leap 2 (ML2) showed sufficient position accuracy of 1.39 mm, suitable for mobile authoring and tracking of trajectories. While results are promising, further optimization in tracking, user interface (UI), and real-world validation is needed.

**Keywords:** Augmented Reality Authoring · Augmented Reality for Inspection · Sensor Trajectory Guidance · AR Accuracy Analysis · Inspection Assistance

## 1 Introduction

In recent years, Augmented Reality (AR) has gained attention due to its potential to optimize manual industrial processes. Its application in maintenance, repair, and overhaul operations (MRO) remains relatively unexplored [1]. MRO relies on hand-guided sensors for inspections, where manual operation and sensor-specific guidance requirements (e.g., maximum velocity) directly impact data quality. Unlike assembly, where the final product offers a clear indicator of process quality, evaluating the manual data acquisition process is challenging. This is particularly crucial in aviation, where large-scale structures and the emergence of new, hydrogen-based propulsion technologies need reliable, mobile inspection solutions (e.g., leakage inspection) [2]. AR offers a promising solution

© The Author(s) 2025
H. Kohl et al. (Eds.): GCSM 2024, LNME, pp. 321–329, 2025.
https://doi.org/10.1007/978-3-031-93891-7_36

by providing real-time information on sensor guidance and sensor trajectory supervision, enhancing both the accuracy of inspections and, ultimately, passenger safety [3].

Though widely explored for assembly tasks, research on the application of AR in inspection remains limited [1]. Existing research demonstrates the effectiveness of AR in mobile inspection applications, using information overlay for improvements in inspection efficiency [4]. [5] introduces a successful AR application for operator training evaluating performance metrics from process execution but only focusing on assembly tasks. Similarly, [6] aims to provide operators with real-time information about tasks in a manufacturing environment and trajectories of mobile robots but suffers from limitations in tracking accuracy and comprehensive evaluation of real-time AR data. Despite these limitations, both [6] and [5] show promising research directions in providing feedback mechanisms, 3D model overlay, and intuitive interfaces to minimize cognitive load, indicating promising applications for enhancing the quality of manual sensor guidance. Robust and precise tracking remains a significant challenge in AR-guided inspections, directly impacting the accuracy of guidance [4]. A research gap exists in applying these tracking methods for sensor guidance applications compatible with various hand-guided sensors. [5] identifies the preparation time of AR applications as a significant drawback, while [6] emphasizes the importance of streamlining authoring processes to reduce deployment time and effort for wider adoption. [6] further demonstrates the positive impact of using AR also for authoring purposes on user experience, reduced time and errors. Advancements in AR robot programming [7] significantly improve speed and user confidence. Although AR shows excellent promise for tracking and authoring in sensor guidance applications, no known publications have specifically addressed this issue.

This research addresses the gap in AR-guided inspection by investigating a novel application for authoring inspection procedures, specifically within green aviation. This study will develop and evaluate an AR sensor guidance demonstrator applicable to multiple inspection use cases. Key design decisions will be outlined, and the usability of the proposed solution will be assessed through a user study. Crucially, a detailed accuracy analysis of tool tracking will be conducted, evaluating the technological foundation of the authoring solution and providing valuable insights for future research in AR-guided inspection.

This work is structured as follows: we examine two user roles and specific use cases to derive generalizations, compare their challenges, provide an overview of requirements for AR authoring, as well as methods (Sect. 2), for a prototype implementation of our modular authoring tool (Sect. 3). Main aspects such as usability and achievable accuracy are investigated (Sect. 4). Finally, we summarize our findings, discuss limitations, and identify areas for improvement (Sect. 5).

## 2    Approach and Methodologies

This section defines the main user roles, two relevant aircraft inspection use cases, and key requirements for the AR application and hardware. Methodologies for authoring and the conducted user study and accuracy analysis are presented.

The **User Role** *"Author"*, also referred to as the process expert, is primarily responsible for the adaption to specific environments and defining the parameters for inspection procedures, including the optimal trajectories, tolerance zones, speeds, and angles (Fig. 1[2]). The authoring tool is configured with the help of the author's knowledge and experience in the field, which includes path optimization and testing in accordance with sensor requirements and best practices, as well as preparing localization (e.g., position markers...). The user role *"Inspector"* utilizes the application generated by the author to carry out the inspection, missing the level of expertise or experience as the author. They follow the guidelines and procedures embedded in the application to ensure the inspection is performed accurately and efficiently (Fig. 1[1]). The inspector application visualizes sensor-specific trajectories for operators, providing live feedback on the current sensor pose, velocity, and acceleration. After starting the application, they have to maintain alignment with the defined trajectory inside a tolerance zone, performing manual adjustments to the sensor position.

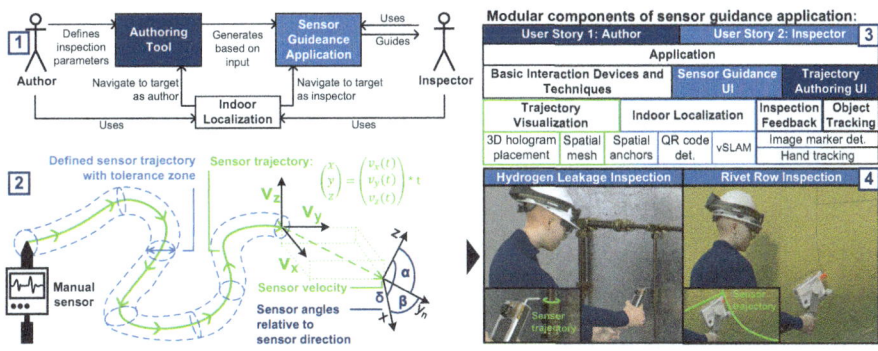

**Fig. 1.** Visualization of user roles flow chart [1], sensor trajectory [2], modular application components [3] and inspection use cases [4]

Two relevant **Inspection Use Cases** in aviation represent variability in manual sensor trajectories specific to sensor characteristics and emphasize the requirements of the proposed AR authoring solution (Fig. 1[4]):

*Hydrogen Leakage Inspection:* Aircraft with hydrogen-based propulsion systems require regular checks for leaks due to hydrogen's volatility and flammability. A hydrogen sniffer is used, requiring precise movement over potential leak surfaces. The sensor must maintain a specific distance and speed due to hydrogen's rapid dissipation. Challenges include homogeneity of measuring environment, gas invisibility, and sensor signal delay, all of which can be mitigated by AR support.

*Aircraft Rivet Row Inspection:* To ensure aircraft fuselage safety, rivet row inspections measure rivet placement and spacing for load distribution. Critical parameters include rivet pitch, edge distance, and height. A laser line scanner and encoder generate a point cloud, analyzed by software. Accurate data recording depends on maintaining specific angles, distances, and speeds, visualized through an AR application. Table 1 compares the specific sensor parameters for both inspection scenarios, highlighting key differences. While both scenarios require adherence to speed limitations, only rivet row inspection necessitates specific angle observations due to the nature of the laser scanning process.

**Table 1.** Definition of sensor trajectories for the seleted use cases

| Trajectory Requirements | | |
|---|---|---|
| use case | Leakage Inspektion | Riviet Row Inspection |
| Velocity | max. 10 mm/s | max. 20 mm/s |
| Trajectory Tolerances | 3 mm | 10 mm |
| Angle Tolerance | – | $\Delta\alpha,\ \Delta\beta,\ \Delta\delta = \pm 10°$ |
| Inspection Duration | 10 s for fitting; 1 m pipe 10 s | 1 rivet 2 s |
| Shape of Trajectory | Cylindrical (pipe), Torus (fitting sealing) | Cylindrical above riviet row |

We will now outline the **General Requirements** for an AR application design to support manual sensor guidance after fostering understanding using inspection examples. An effective authoring process should incorporate trajectory definitions that are adaptable to various inspection scenarios, utilizing standardized trajectories in three-dimensional Euclidean space. Authoring involves modular software adaptable to different scenarios and integrated with the application's functionalities. The user feedback and mechanisms for interaction must be non-intrusive, clear, and easily understandable to avoid disrupting the inspection process [8]. Precise and easy setup of hologram placement employing tracking is critical for authoring to facilitate trajectory use. Visual feedback indicating sensor positions or user interactions is vital due to workplace conditions and the need for language-free communication.

Following the presentation of the requirements, the methodological essentials are presented, which include choosing authoring methods, selecting hardware, and methods for user study and analyzing accuracy.

The **Method for Sensor Trajectory Definition** is a key aspect of designing the required authoring solution. A defined target trajectory consists of ordered points in space, describing motion, velocities, and accelerations while incorporating sensor- and task-specific tolerances (Fig. 1[2]). Based on robot programming methodologies, two primary categories for defining sensor trajectories are identified: Offline and Online Programming.

*Offline programming* leverages CAD software and model-based approaches to generate trajectories without requiring access to the physical object. Combined with formalized sensor-specific knowledge [5], this method presents a high potential for automation [9]. However, further research is necessary to evaluate its effectiveness, particularly regarding the usability of authoring tools and the impact of model accuracy on the resulting trajectory.

In contrast, *online programming* generates trajectories directly on the physical object, utilizing tracking methods to compensate for real-world tolerances, thus eliminating the reliance on precise 3D models. Integrating online programming within an AR application allows intuitive user interaction and definition of trajectories through hand-guided programming [7], addressing navigation and UI inefficiencies of 2D interfaces [5].

AR online authoring thereby enhances effectiveness and user experience by allowing direct trajectory modifications, flexibility and immediate testing within the AR environment. AR technology enables mobile tracking, crucial for inspecting large structures like aircraft. Using the same tracking hardware for authoring and inspection reduces setup effort. While offline authoring offers automation, the infrequent need for new trajectory authoring in aircraft inspection prioritizes quick sensor and inspection process utilization. Therefore, we adopt online AR authoring for its ease of use, adaptability, and missing precise models in various industries.

The chosen **Hardware Technology** are AR head-mounted displays (HMD) for hands-free functionality and flexibility. Despite alternatives like ML2 and Vuzix Blade offering features like higher tracking accuracy and lighter weight, the HL2 was selected for its strong community support and extensive software compatibility[5], making it the most suitable option for developing our application.

The initial **User Study** was focused on usability, expecting feedback for a follow-up study. Authoring efficiency and accuracy analyses were not conducted at this stage. The User Study employed the System Usability Scale to validate the developed prototype on the HL2, assessing its usability and limitations in a simulated inspection scenario (Fig. 2). A SUS questionnaire, combined with custom and open-ended questions, gathered detailed feedback on the tool's use case suitability and areas for improvement. The study, performed on real aircraft components, involved participants (10 male, two female) executing key functionalities, including localization setup and a combined author/inspector workflow.

A separate **AR Tracking Accuracy Analysis** was conducted using the ML2. The user study identified limitations in the tool tracking of the HL2 requiring millimeter-level precision. ML2 offers enhanced capabilities for tool tracking, including a handheld controller with independent SLAM-based tracking. To evaluate the ML2's accuracy, the AR HMD was placed in four static positions (P1-P4 HMD) (Fig. 3), while the controller, mounted to a UR10e robot for precise movement, was tracked at seven defined points (Fig. 3). The AR pose data was compared to a commercial tracking system for accuracy assessment. Metal pipes were used as a background to create a realistic environment.

## 3    Implementation

We selected the HL2 for our implementation leveraging its capabilities, using the Mixed Reality Toolkit (MRTK) for most functionalities, ensuring seamless integration and support. Our application combines marker-based and marker-less tracking to enhance accuracy and robustness (Fig. 1[3]). Vuforia was chosen for image marker detection because it offers superior performance in terms of accuracy and reliability, outperforming other options like ArUco markers, which are not supported anymore by HL2. The tool is designed to be intuitive for non-programmers, although specific environment adaptation requires training.

Authors set inspection parameters and environmental settings through a JSON file, using pre-measured inspection areas and strategically placed QR codes for localization, including marker IDs and previously defined trajectories. If no QR code is detected after starting the application, the system relies on vSLAM to update the user's position and orientation, ensuring consistent tracking. All user role functions, such as a live map, are selectable via a menu (Fig. 2[1]). Spatial Anchors, supported by MRTK, allow users to create trajectories by placing anchors on a spatial mesh. These anchors are linked to form a visual inspection path, with numerical labels guiding the user through the sequence (Fig. 2[2]). Trajectories are created via a pitch gesture or a tracked tool supported by raycasting. The author, as well as an inspector, can choose between tool or hand tracking, while hand tracking has a better performance. The designed tool features a cube on top, with each face holding Vuforia markers to ensure precise TCP determination.

This design supports real-time visualization of metrics such as speed, distance to trajectory, and angle, providing inspectors with immediate, actionable feedback (Fig. 2[3]). By integrating MRTK's Spatial Anchors, Vuforia for marker detection, and vSLAM for indoor localization, the application balances the precision of marker-based tracking with the adaptability of markerless methods, offering a robust solution tailored to industrial inspection tasks lag (Fig. 1[1][3]).

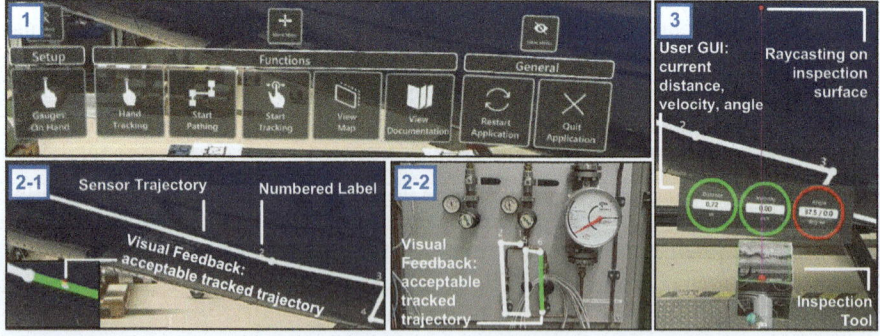

**Fig. 2.** Visualization of the developed AR user interface

## 4   Results

The **User Study** provides an average SUS score of 76 ($\sigma = 13{,}46$) for the standard questionnaire and 73.125 ($\sigma = 18{,}47$) for the custom one, both above the benchmark score of 70, indicating above-average usability but below the 80 threshold for excellent usability [10]. While the tool is deemed usable, feedback suggests areas for improvement. Issues include accuracy and efficiency performance due to lagging tool tracking features using Vuforia. Additionally, the menu UI requires a clearer feature presentation. Most participants feel the AR authoring tool performs better for rivet row inspection on aircraft fuselages, as the spatial mesh of the fuselage is more accurate than that of pipes. They note that the tool references the wall behind the pipes rather than the pipes themselves. Additionally, the tool's missing support for various trajectory types, such as curved or circular, makes inspecting pipe connectors for hydrogen leakage detection challenging. The results of the study suggest a further investigation of tracking accuracy.

We **Evaluate the AR Tracking Accuracy Using ML2**, based on the user-study findings. The ML2 is an alternative AR Hardware that promises improved performance with its additional controller for tool tracking. Our primary objective was to determine whether ML2's enhanced quantified tracking capabilities address the deficiencies observed with the HL2. Results show a maximum position repeatability of 0.0475 mm (Ø 0,2621 mm) and position accuracy of 1.39 mm (Ø 1,6172 mm), satisfying both use case requirements. Positioning the HMD at optimal angles and reduced distances from the pipes up to 300 mm affected position repeatability and accuracy (Fig. 3). Accuracies were higher after the initial measurement at the first HMD position due to SLAM tracking adjustments, as no calibration was carried out to reduce setup effort.

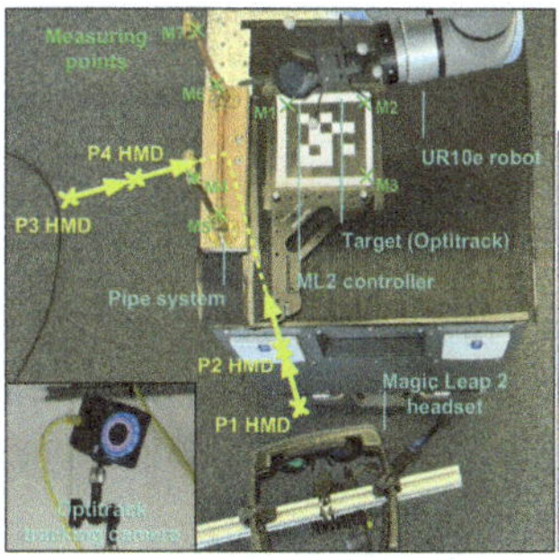

Analyses were carried out in accordance with EN ISO 9238

| Measurement series | HMD distance [mm] | AR Position Repeatability [mm] | | | AR Position Accuracy [mm] | | |
|---|---|---|---|---|---|---|---|
| | | mean | max | min | mean | max | min |
| 0 (initial) | 700 mm (P1) | 0,4136 | 6,3847 | 0,1063 | 1,6669 | 5,8435 | 0,4704 |
| 1 | 700 mm (P1) | 0,2327 | 1,9138 | 0,169 | 1,3979 | 3,6192 | 0,2174 |
| 2 | 500 mm (P2) | 0,1274 | 1,378 | 0,0475 | 1,4239 | 2,9949 | 0,4606 |
| 3 | 500 mm (P3) | 0,4393 | 2,5694 | 0,2468 | 1,8915 | 3,3774 | 0,3765 |
| 4 | 300 mm (P4) | 0,249 | 2,391 | 0,1012 | 1,7554 | 4,578 | 0,3901 |
| | Ø (1-4) | 0,2621 | 2,0631 | 0,1411 | 1,6172 | 3,6424 | 0,3612 |

**Fig. 3.** Measurement setup for determining the accuracy of the Magic Leap 2

## 5   Discussion

Current AR tools have limited inspection applications and face challenges with time-consuming setup processes. Applications are missing for effective manual sensor guidance to enhance efficiency and reliability. This study introduces a novel AR application designed to streamline inspections in green aviation. The core contribution is a dedicated authoring tool, validated through developing and testing a functional inspector application, empowering users to create mobile, sensor-guided inspection procedures. Our modular AR authoring tool, designed for hydrogen leakage detection and rivet row inspection, addresses these issues by integrating various tracking technologies. It supports customizable inputs and combines marker-based and markerless localization for increased robustness. The study of usability suggests positive user feedback and the potential to enhance both the speed and quality of inspections, aligning with findings from previous research in the field [7]. However, challenges remain, particularly in handling complex trajectories and tracking with the HL2, limiting its use in leakage inspection. Real-time tracking is constrained by computational demands, affecting hologram placement and AR scene generation. To overcome these challenges, we tested the tool on the ML2, which demonstrated sufficient position accuracy for authoring of 1,39 mm.

For a reliable sensor-guided application for inspectors, further updates are required—an aspect beyond the scope of this study. The application restricts trajectory variations to basic shapes, limiting its applicability to a wider range of inspection scenarios. Incorporating freeform splines or shape templates would enhance the system's versatility. The current evaluation lacks real-world testing by experts. It involves a small, male-dominated sample group with a technical background, causing potential statistical outliers. The accuracy assessment aimed to evaluate performance using pipes as a background, as unstable tracking near pipes was noted. We utilized DIN-standardized calculations for static positions by EN ISO 9238 but did not fully implement the measuring procedure for full trajectory accuracy. Our results confirm that the ML2's lowest position accuracy of 1,87 mm is still sufficient for sensor guidance - given that human handling typically achieves around a centimeter accuracy. We also demonstrated that accuracy depends on factors like distance and angle, which are influenced by user handling and should be considered in inspector feedback.

Future work should incorporate more diverse studies across industrial scenarios to evaluate and optimize tracking accuracy, robustness, and inspection performance under real-world conditions. This will involve testing with actual tools, advanced sensor trajectories, and expert evaluations. Combining depth cameras with trained neural networks improves markerless object tracking accuracy through superior pose estimation. Given expert scarcity and associated costs, evaluating the reduction of authoring effort is critical for adoption. Therefore, AR trajectory creation must be compared with appropriate offline approaches. Our findings can serve as a basis for further comprehensive research to improve inspection quality through detailed analysis of inspector actions.

**Acknowledgements.** This work was supported by the German Federal Ministry of Education and Research (BMBF) under the grant number 03HY114F within the research project H$_2$Giga – HyPLANT100.

# References

1. Palmarini, R., Erkoyuncu, J.A.: A systematic review of augmented reality applications in maintenance. R. CIM **49**, 215–228 (2018). https://doi.org/10.1016/j.rcim.2017.06.002
2. Moenck, K., Koch, J., Rath, J.-E., Büsch, L., et al.: Industry 5.0 in Aircraft Production and MRO (2024). https://doi.org/10.13140/RG.2.2.33842.00961
3. Koch, J., Lotzing, G., Eschen, et al.: A human-centered IIoT platform approach for manual inspections. Procedia CIRP (2023). https://doi.org/10.1016/j.procir.2023.04.089
4. Marino, E., Barbieri, L., Colacino, B., et.al.: An AR inspection tool to support workers in I4.0 environments. C. i. I.(2021). https://doi.org/10.1016/j.compind.2021.103412
5. Apostolopoulos, G., Andronas, D., Fourtakas, N., Makris, S.: Operator training framework for hybrid environments: an augmented reality module using machine learning object recognition. Procedia CIRP **106**, 102–107 (2022). https://doi.org/10.1016/j.procir.2022.02.162
6. Lotsaris, K., et al.: Augmented Reality (AR) based framework for supporting human workers in flexible manufacturing. Procedia CIRP **96**, 301–306 (2021). https://doi.org/10.1016/j.procir.2021.01.091
7. Ong, S.K., Yew, A., Thanigaivel, N.K., Nee, A.Y.: Augmented reality-assisted robot programming system for industrial applications. Rob. Comput.-Integrat. Manuf. **61**, 101820 (2020). https://doi.org/10.1016/j.rcim.2019.101820
8. Sekhavat, Y.A., Namani, M.S.: Projection-based AR: effective visual feedback in gait rehab. IEEE HMS **48**(6), 626–636 (2018). https://doi.org/10.1109/THMS.2018.2877850
9. Zwick, M., Gerdts, M., Stütz, P.: Sensor-model-based trajectory optimization for UAVs to enhance detection performance. Sensors **23**, 664 (2023). https://doi.org/10.3390/s23020664
10. Bangor, A., Kortum, P.T., Miller, J.T.: An empirical evaluation of the system usability scale. Int. J. Hum. Comput. Interact. **24**(6), 574–594 (2008). https://doi.org/10.1080/10447310802205776

# Energy Distribution

# Automated Control Cabinet Wiring Solution for Scalable Renewable Energy Systems

Milan Brisse$^{(\boxtimes)}$ ⓘ, Elias Milloch ⓘ, Johannes Prior ⓘ, and Bernd Kuhlenkötter ⓘ

Chair of Production Systems, Faculty of Mechanical Engineering, Ruhr-University Bochum, Industriestraße 38C, 44894 Bochum, Germany
brisse@lps.rub.de

**Abstract.** To successfully manage the transition to sustainable energy sources, the infrastructure for renewable energies such as photovoltaic systems, wind power, and hydrogen electrolyzers needs to be scaled up. Many of these systems rely on control cabinets, which are still manufactured manually. The shortage of skilled workers in electrical engineering requires rethinking toward automated control cabinet production to meet the increasing demand for energy infrastructure components. Almost half of the assembly time for manual control cabinets is spent on wiring. A reduction in cycle times using automation would lead to cost savings and an increased output in production. Furthermore, automated control cabinet wiring can reduce waste by minimizing errors during the wiring process. Current approaches to automate this process are inefficient, prone to errors, and time-consuming. In this paper, the processes for the automated wiring of control cabinets using different handling tools are presented and compared. Optimization potentials are identified, and solution approaches are presented. Furthermore, solutions are evaluated prototypically using a real research demonstrator for robot-assisted control cabinet wiring. The results of this paper demonstrate that the implementation of these optimizations leads to a flexible solution, contributing to a more efficient and cost-effective manufacturing process.

**Keywords:** Renewable energy infrastructure · automated wiring · manufacturing tools

## 1 Introduction

When it comes to successfully shifting towards the rising demand for renewable energies [1], it is essential to expand the energy infrastructure for technologies such as wind power, photovoltaic systems and hydrogen electrolyzers for storing energy. An essential component for the operation of these systems are control cabinets, which are still mainly manufactured manually [2]. However, the ongoing shortage of skilled workers is a hindering factor for the growth of its production [3]. A promising solution to meet the demand for control cabinets in energy infrastructure is the automation of production steps. Concerning the assembly of control cabinets, wiring represents almost half of the assembly time [2]. Automating this process has the potential to significantly reduce lead

© The Author(s) 2025
H. Kohl et al. (Eds.): GCSM 2024, LNME, pp. 333–341, 2025.
https://doi.org/10.1007/978-3-031-93891-7_37

times, resulting in cost savings and increased production output. Interest in automation is high, and several years of research have produced promising approaches for automated control cabinet wiring [3]. Nonetheless, there is still no turnkey and industry-proven solution commercially available that covers the complete wiring process, from wire supply to wire insertion and routing, for varying control cabinet sizes. This paper addresses the question of how processes in control cabinet wiring can be automated to increase the efficiency and cost-effectiveness of its production. The approach is based on the VDI 2221-1 standard [4] for the development of technical products and systems.

For this purpose, an overview of control cabinet construction basics and its production is given in the second chapter. Current approaches are then presented and com-pared in Sect. 3. The approaches are discussed and requirements for a new gripper for automated control cabinet wiring are derived. In Sect. 4, the proposed concept is presented. For this purpose, the gripper is split into individual modules each fulfilling a specific function. To evaluate the concept, a prototype of the gripper was tested in a realistic wiring scenario in Sect. 5. In Sect. 6 the results are evaluated, and future measures are presented.

## 2   Fundamentals

For the operation of renewable energy systems, electrical devices like relays and terminal blocks are needed. The devices are provided with either spring-loaded or screw connectors positioned at the front or at the top and the bottom. All devices are usually assembled inside control cabinets containing a mounting plate and assembly rails. [5].

The production of a control cabinet can be divided into the phases of planning, mechanical processing, assembly of mechanical and electric components, wiring, quality check, and logistics. The assembly is predominantly done manually, while the wiring takes up to 49% of the production time. It can be further subdivided into checking the circuit diagram, wire assembly, and the actual wiring, resulting in an average process time of 266 s for the wiring of two components. [2] Solutions for some production steps already exist in the market, e.g. for mechanical processing, or wire assembly. The solutions vary in their automation levels from manual tools to fully automated processing machines. However, there is no industry-proven solution purchasable that can flexibly cover the complete process, from automated wire supply to automated wire insertion and routing [6]. Current approaches from academia and industry will be further inspected in the following section.

## 3   Analysis

In this chapter, approaches for automated control cabinet wiring from industry and academia are first presented and analyzed regarding their limits and potential for improvement. Here, only solutions that represent a complete assembly of the wire with enough information available are presented. Based on this analysis, requirements for a new solution are derived.

## 3.1 State of Technology

The following concepts show promising approaches for automated cabinet wiring:

**Averex:** The engineering company Kiesling together with the manufacturer Rittal presented a fully automated wiring solution consisting of a portal robot called Averex [7]. It includes wire cutting, crimping, connecting the wires, and routing them inside the cabinet. The approach consists of a portal robot and a highly complex end-effector covering the complete process chain. The solution enables both spring-loaded and screw connections to be wired.

**Wirebot:** Polygon Technologies combined a portal robot with an articulated robot in their Wirebot solution [8]. The system includes vision as well as tactile sensors for high reliability. The articulated robot feeds the portal robot with previously assembled wires and plugs the wires into the terminals. The portal robot routes the wires inside the cable ducts and supplies the articulated robot with the wire ends. Both robots are equipped with complex end effectors enabling them to connect the wires to both spring-loaded and screw-type taking one minute process time per wire.

**Rittal:** The latest wiring solution from Rittal consists of an articulated collaborative robot equipped with a complex gripper that can store a pre-assembled wire, connect both ends to the terminals, and route the wire in the cabinet [9]. The gripper is covered by a housing so there is no further information about its mechanism. A key feature of this approach is the wire supply via a tube using compressed air. The tube connects the robot gripper to a fully automated wire assembly machine. To ensure reliable positioning of the wire ends, several cameras are used. During wiring, the approach does not guide the wire in the duct and allows the wire to fall free into the duct. So far, only the wiring of spring-loaded terminals from the top has been shown.

**RoboSchalt:** The approach presented by Spies et al. [10] also contains an articulated robot equipped with a single end-effector to store the pre-assembled wire and connect it to spring-loaded terminals. The end effector consists of two prismatic parallel grippers to hold the wire ends and a rotating wire reel to store the wire during the wiring process. Cameras are installed for wire and terminal detection. Regarding the wire supply, the authors state that a suitable wire magazine has been developed but needs further improvement.

All approaches theoretically achieve medium to high process reliability due to sensor monitoring and wire guidance. However, only long to medium cycle times are expected due to speed and additional steps like wire assembly. Industrial approaches tend to be costlier than academic ones due to fewer standard parts and fixed automation. There is a high variance in coverage of equipment, such as screw connectors or spring-loaded terminals. Approaches utilizing industrial robots often demonstrate greater flexibility in terms of control cabinet size compared to permanently automated approaches. However it should be noted, that collaborative robots move slower, therefore resulting in higher cycle times. The comparison shows three wire feed methods: wire assembly within the system, using a wire magazine, and by a pneumatic tube. Each approach supports only one type, making all systems inflexible in this aspect.

## 3.2  Deriving System Requirements

In the following section, the Functional Requirements (FR) for developing a new and improved system are defined based on the analysis of current system limitations. These requirements are then classified regarding their priority into Mandatory Requirement (MR), Essential Requirement (ER), or Desired Requirement (DR).

### FR1: Use of Standard Components (DR)

The system should utilize standard components instead of custom-made special parts to reduce costs and facilitate easier access to the technology for smaller companies.

### FR2: Flexibility in Cabinet Sizes (MR)

The system should be designed to cover various cabinet sizes, ensuring adaptability to different spatial constraints and requirements.

### FR3: Versatile Wire Handling (MR)

The system should be capable of covering the current methods of wire provision as well as promising new approaches from the industry, allowing for integration into diverse operational environments.

### FR4: Efficient Cycle Time (DR)

The system should achieve a cycle time of less than one minute to meet or exceed the performance of existing systems.

### FR5: Automation-Oriented Cabinet Design (DR)

The design of the control cabinet should ease automation in order to reduce the complexity of the process and increase process reliability.

### FR6: High Process Reliability (ER)

The system should ensure high process reliability by designing components and procedures that minimize the risk of errors and failures during wire routing and insertion.

# 4  Concept and Design

An overall system is to be designed based on the presented approaches and the defined requirements. To meet the FR5 a control cabinet suitable for automation and only spring-loaded terminals should be used, as described in [11]. Due to the FR1 for standardization and FR2 for flexibility regarding control cabinet sizes, an industrial robot is to be used. Therefore, a gripper is developed to realize the control cabinet wiring. The gripper is divided into four functional modules: the gripping unit for manipulating the wire, a swivel unit for positioning the gripper jaws, a storage unit for the wires, and a base module to connect the other modules. To meet the FR3 the system shall handle wires pre-assembled from a magazine, assembled within the system, or delivered via a pneumatic tube as shown in [9]. The solutions were developed using the morphological box method, which involves searching for partial solutions for each function and then combining them into a complete solution. To fulfill the FR1 only standard parts should be used in designing the gripper. The resulting solutions are displayed in Fig. 1. Function modules.

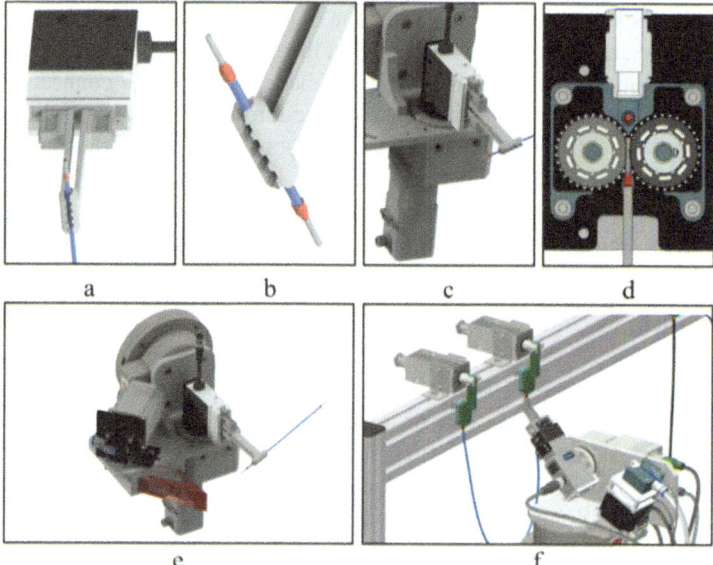

**Fig. 1.** Function modules.

The **Gripper Unit**, shown in Fig. 1a, comprises a parallel gripper and the associated gripper fingers. Besides grasping the wire, the gripper routes the wire in the cable duct for high process reliability (FR6). To guide the wire, the gripper fingers should be opened slightly. This requires a positionable parallel gripper and gripping surfaces that interlock to prevent the wire from slipping out during routing. The gripper fingers should be as long as possible for a high insertion depth. However, stiffness decreases with increasing finger length, reducing the gripping force. A T-piece can be used as gripper fingers for additional wire delivery by pneumatic tube, ensuring both ends can be joined and removed (Fig. 1b). However, this change reduces accessibility as the angled gripper jaws increase the distance between the gripper and the equipment.

The **Swivel Unit**, shown in Fig. 1c connects the gripper unit to the base module using a rotary drive. The rotary drive should be positionable to guide the second wire end out of the tube and to achieve greater flexibility when joining and guiding the wires. Although we use equipment optimized for automation as demanded by the FR5, the swivel unit enables the possibility of mounting the wire to varying connector positions.

The **Wire Feeder Unit (WFU)**, shown in Fig. 1d, enables the wires to be stored in a PTFE tube, which can be mounted into the robot's energy chain. This allows wires with a length of up to 3 m to be stored. To feed the wire into the tube, a system with two flexible rollers and a motor is utilized, capable of handling wires of different diameters, both inserting them into and retracting them from the tube. The rollers are driven by a stepper motor to enable precise positioning of the wire. The positioning of the wire is also supported by a light barrier, as can be seen at the top of Fig. 1d, which detects whether the wire has been inserted or if the end of the wire is just before the WFU rollers. The WFU also enables the possibility of delivering the wire by pneumatic tubing, thus supporting the FR3. To ensure that a cycle time of less than one minute can be achieved (FR4), it must be ensured that the feed speed of the rollers does not create a bottleneck.

The **Base Module**, shown in Fig. 1e, is the base frame, which connects the WFU and the swivel unit to the robot flange via a force-torque sensor. A laser triangulation sensor is also mounted on the underside. The sensors are not part of this paper and will be presented in detail in a future publication.

To meet the FR3 that wires can be removed from magazine, no changes need to be made. This was validated by a simulation in ABB RobotStudio ( Fig. 1f).

## 5 Evaluation

The developed modules are realized in the form of a prototype system. The gripper fingers are additively manufactured as part of a parameter study to investigate accessibility. These fingers are actuated by a positionable, electric parallel gripper to facilitate wire guidance in the cable duct. A stepper motor with an encoder is used for the swivel unit to ensure precise positioning of the gripper fingers. Standard plain bearings and gear wheels are used for the WFU. The rollers are additively manufactured from thermoplastic polyurethane (TPU), while the housing is made from polylactide (PLA) using the fused deposition modeling (FDM) process. In addition, the holder for the WFU is additively manufactured for test purposes. The carrier system is made of aluminum sheets. The finished structure can be seen in Fig. 2a.

### 5.1 Testing the Prototype in the Demonstrator

For testing the prototype, a test cell for robot-based wiring at the Chair of Production Systems of the Ruhr-University Bochum is used. Here, two industrial robots are in an open cell concept to test various concepts for automated control cabinet wiring. In this setup, one robot pre-assembles the wires while the second one joins the wires. The assembly robot is not in the scope of this work. A wiring frame from Lütze is used as the control cabinet to increase the automation of the process. A further description of the cell concept can be found in [11].

To install the wire in the control cabinet, the assembly robot first inserts the wire vertically into the WFU of the wiring gripper, partially feeding it into the storage tube (Fig. 2b). Subsequently, the robot provides the second wire end, which is held by the gripper, forming a loop between the gripper fingers and the WFU (Fig. 2c). Once the first wire end is positioned over the clamp, it is pushed into the clamp's opening until secured (Fig. 2d). At this point, the gripper jaws open slightly, guiding the wire without holding it tightly. The wire is then moved horizontally into the comb opening below the insertion position (Fig. 2e). As the gripper moves, the wire extrudes from the tube proportional to the distance traveled. For routing into another duct, the wire is guided around the wiring frame and into the outer wire guide while briefly closing the gripper jaws to tension the wire (Fig. 2f). To remove the second wire end from the tube, the gripper jaws first open and release the wire. Then the swivel unit guides the gripper towards the WFU, feeding the wire through until the second end can be grasped by the gripper fingers. Once secured, the gripper moves the wire end out of the removal position

(Fig. 2g). Following this, the second wire end is fixed in the clamp (Fig. 2h) and secured in the comb (Fig. 2i) as previously described. The process concludes with the release of the wire, and completing the installation.

The cycle time of the process presented was measured to be 71 s for a wire length of 90 cm and therefore not within the desired requirement of 60 s (FR4).

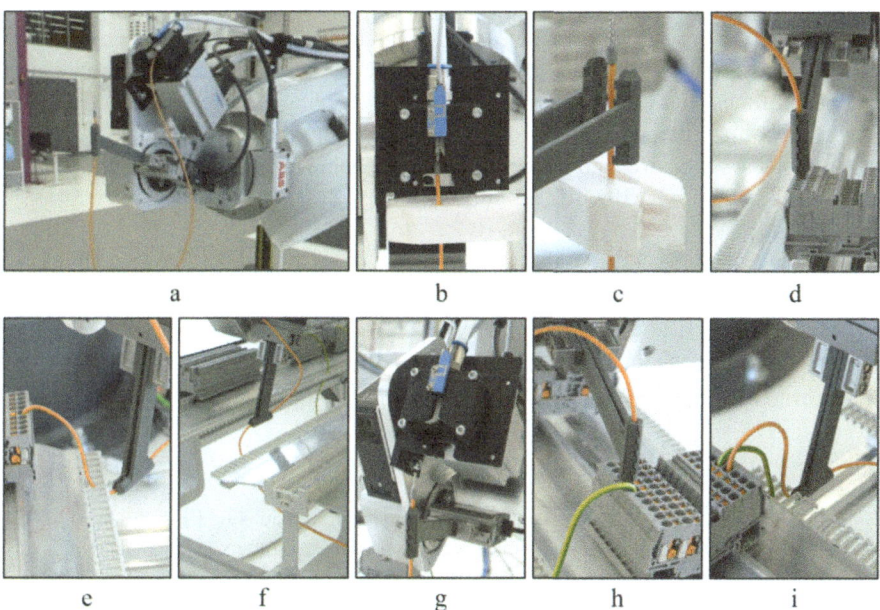

**Fig. 2.** Gripper and wiring process. Reproduced with Permission © 2025 ABB

## 6   Discussion

Although the approach was able to fulfill most of the requirements by design, it failed to achieve the desired cycle time of under one minute as demanded by the FR4. This is caused by poor trajectory planning and will be addressed in future work. The tests on the demonstrator show that the approach presented is promising but is subject to a certain degree of process uncertainty. The wire and the wire ends could be reliably positioned and guided. In addition, the wire storage proved to be a reliable solution for quickly inserting and removing the wire. The state of order of the wire can be determined and controlled at any time. The delicate design of the gripper fingers also made it possible to join several connectors next to each other without any problems. However, the process reliability during the first tests was low. The loop that forms after the wire is picked up (see Fig. 2a) can cause the wire to get caught in the wire combs or equipment. Additional guidance by the gripper or a different path planning is necessary here. In addition, the wire occasionally slipped inside the gripper fingers. This is mainly due to the flexibility of the additively manufactured plastic gripper fingers.

# 7  Conclusion and Outlook

The research question in this paper was how to automate processes in control cabinet wiring to increase production efficiency and cost-effectiveness. Therefore, the state of the art was analyzed, and requirements were derived. A modularized concept was designed and tested in a realistic wiring scenario. The proposed solution shows promise for fully automated and reliable control cabinet wiring but faces future challenges. In particular, the sensor technology for position detection should be addressed, and cycle times need to be optimized. Automated path planning needs to be implemented and optimized to reduce cycle times. Tests have also shown that the gripper fingers are currently too flexible and can only be manufactured additively. An approach must be found that combines high gripping forces, small size, and good wire guidance behavior. Additionally, future work needs to address automated wire assembly and supply.

**Acknowledgment.** Parts of this work were supported by the Federal Ministry of Education and Research (BMBF) under grant number 03HY113A within the project H2Giga – FertiRob.

# References

1. IEA Renewables 2023 IEA. https://www.iea.org/reports/renewables-2023. Accessed 21 Aug 2024
2. Tempel, P., Eger, F., Lechler, A., Verl, A.: Schaltschrankbau 4.0: Eine Studie über die Auto-matisierungs- und Digitalisierungspotentiale in der Fertigung von Schaltschränken und Schaltanlagen im klassischen Maschinen- und Anlagenbau, Stuttgart (2017)
3. Bründl, P., Stoidner, M., Bredthauer, J., Nguyen, H.G., Baechler, A., Franke, J.: Unlocking the potential of digitalization and automation: a qualitative and quantitative study of the control cabinet manufacturing industry. Prod. Manuf. Res. **12** (2024)
4. VDI 2221-1: Development of technical products and systems, Düsseldorf (2019)
5. Brecher, C., Weck, M.: Machine Tools Production Systems 3. Springer Fachmedien Wiesbaden (2022)
6. Großmann, C., Graeser, O., Schreiber, A.: ClipX: Auf dem Weg zur Industrialisierung des Schaltschrankbaus. In: Handbuch Industrie 4.0, 2nd edn. Springer Vieweg, Berlin (2017)
7. Kiesling, J., Koch, H.-R.: Automatisiert vom Plan zur Verdrahtung. https://www.industr.com/de/automatisiert-vom-plan-zur-verdrahtung-122902. Accessed 08 June 2024
8. Wirebot homepage. https://www.wire-bot.com. Accessed 13 June 2024
9. Rittal press section: SPS 2023. https://www.rittal.com/de-de/Unternehmen/Presse/Pressemeldungen/SPS2023_Preview. Accessed 12 June 2024
10. Spies, S., Bartelt, M., Hypki, A., Kuhlenkötter, B.: robot automation in control cabinet assembly. In: COMA'19 Proceedings: Knowledge Valorisation in the Age of Digitalization, p. 272 (2019)
11. Milloch, E., Bartelt, S., Brisse, M., Egel, R., Kuhlenkötter, B.: Electrolyzer control cabinet wiring as holistic approach. In: European Robotics Forum 2024: ERF - 15th European Robotics Forum (2024)

# Multi-agent Optimization of Industrial Microgrids Using Metaheuristics

Johannes Prior$^{(\boxtimes)}$ ⓘ, Simon Steinrötter, Milan Brisse ⓘ, Chris Taschelmayer,
and Bernd Kuhlenkötter ⓘ

Chair of Production Systems, Faculty of Mechanical Engineering, Ruhr-University Bochum,
Universitätsstraße 150, 44801 Bochum, Germany
prior@lps.rub.de

**Abstract.** Decentralized renewable energy (RE) generation and consumption through microgrids (MG), combined with short- and long-term storages and demand flexibility, present a promising approach for mitigating grid stress and reducing emissions in the industrial sector. This study aims to develop a simulation model framework that balances economic viability with the transition towards a sustainable industry, particularly regarding RE and its storage. The framework focuses on three key research areas: MG system sizing, optimal energy allocation, and demand flexibility. This framework is used for modeling multi-agent simulation models, which serve as the basis for metaheuristic-based optimization and enable a comprehensive approach to overall optimization. In an evaluation, a simulation model was established through a practical demonstration in a research factory environment, using historical data for MG system sizing and optimal energy allocation. The production was then scheduled based on RE availability. The results of the metaheuristic optimization demonstrate that the proposed methodology reduces costs by 18% and emissions by 3% compared to conventional, non-multi-agent optimization approaches.

**Keywords:** Microgrid · Demand Flexibility · Metaheuristics

## 1 Introduction

Progressive climate change is becoming an increasing concern. Previous measures remain inadequate, indicating significant deviation from the 1.5 °C maximum temperature rise target at the current pace [1]. Despite an increase in RE in recent years, emissions have not decreased proportionally. This is partly due to an increase in grid volatility and suggests saturation, indicating asynchrony between demand and availability during the peak periods. As a result, RE is occasionally restricted.[2].

Decentralized RE generation and consumption, coupled with storage and demand flexibility, represents promising avenues for mitigating the strain on public grids and reducing emissions. The industrial sector, as one of the largest consumers of electricity, possesses substantial capacity for photovoltaic (PV) systems on factory rooftops, demonstrating considerable potential in this regard.[3].

© The Author(s) 2025
H. Kohl et al. (Eds.): GCSM 2024, LNME, pp. 342–350, 2025.
https://doi.org/10.1007/978-3-031-93891-7_38

The transformation towards a $CO_2$-neutral industry requires methods for analyzing and evaluating sustainable investment measures. These methods should integrate sustainability aspects with economic assessment. Smaller production companies facing investment and financial constraints require support in their decision-making processes. This can be achieved through transparent cost analysis, ensuring effectiveness, and optimizing investment strategies.[4].

Common Battery Energy Storage Solutions (BESS) provide short-term storage for self-generated RE but face capacity and self-discharge issues. Hydrogen is a promising alternative for storing energy over extended periods in large quantities. This capability allows companies to store energy generated during sunny periods for use during periods with less sunlight. Optimizing the self-sufficiency share (SSR) through storage cuts costs, emissions, and infrastructure needs.[5].

Hydrogen storage components, particularly electrolyzers (EL), are currently experiencing a market ramp-up but remain costly and less efficient, especially in Power-to-Power (P2P) storage solutions with fuel cell (FC).[6].

Optimizing the adaptation to the specific needs of a company is essential to compensate for the low efficiency and excessive cost of the components, thereby ensuring economic operation. In addition to storage solutions, creating demand flexibility within a factory significantly boosts the SSR. This flexibility enables the adjustment of energy demand to supply fluctuations.[7].

In the current state of research, approaches often focus only on specific aspects like MG system sizing, optimal energy allocation, or demand flexible factories. Approaches such as [5] investigate MG system sizing using various metaheuristics, primarily based on evolutionary theory and swarm intelligence. These approaches typically optimize costs while often neglecting sustainability aspects. Furthermore, [8] dealt with the additional optimization of energy allocation strategies by incorporating additional parameters. These parameters included thresholds for the activation of further subalgorithms in the optimization process.

Conversely, studies by [7, 9] focus solely on optimizing energy allocation strategies, usually over short periods of one day. They also utilized metaheuristics with similar objectives, such as cost minimization. However, their variable parameters are the storage solutions' charge and discharge performances rather than the MG sizing.

The studies by [10, 11] are unique in presenting a partial integration of all three research areas, addressing flexibility measures, energy allocation strategies, and sizing of material and energy storage solutions, without sizing the entire MG.

In summary, it can be observed that there is currently no approach that fully integrates all three research areas. A comprehensive methodology for implementation in real companies is also lacking, and there are no metaheuristics known to simultaneously consider all three research areas and synchronize accordingly.

Owing to the absence of a comprehensive approach, a decentralized energy supply, particularly supported by P2P, remains economically unviable, delaying the widespread implementation of effective emission reduction. This study presents a multi-agent simulation and optimization model framework that considers all three areas. It was evaluated through a case study conducted in a research factory environment, enabling the derivation of MG system sizing, energy allocation strategies, and demand flexibility recommendations. To achieve these goals, the following research questions arise: 1) How can a comprehensive simulation model be designed to integrate the three research areas into a multi-agent framework? 2) What does a mechanism for synchronizing concurrently running optimizations of the respective agents look like for achieving comprehensive optimization of MGs?

The remainder of this paper is structured as follows: Sect. 2 offers a foundational overview of MGs, demand flexible factories, and metaheuristics. Section 3 introduces the realization of the simulation and optimization model framework, while Sect. 4 evaluates the application of these models to a case study. Finally, Sect. 5 concludes with a discussion and a conclusion.

## 2 Basics of Microgrids, Flexible Factories, and Metaheuristics

**MGs** are decentralized energy systems powered by RE such as PV or wind power, supplemented by energy storage solutions such as BESS or P2P. MGs can supply energy to a factory, thereby increasing its SSR.[2].

Despite the plethora of available RE sources, PV is particularly suitable for smaller factories because of its regulatory requirements. An MG includes an AC bus and a DC bus. The DC bus links PV systems with the BESS and P2P units. Electricity is inverted before entering the public grid or factory. A bidirectional inverter allows the MG to acquire and store electricity from the grid at advantageous prices, providing financial benefits to users. [6].

**Factories** can be divided into three entities. Machinery and workstations are categorized as *resources*. *Products* represent the raw materials, subassemblies, and finished goods manufactured within the factory, and must undergo a sequence of processes. Finally, *processes* that represent the processing of products require resources to be executed. This generic categorization was based on the *Product-Process-Resource* (PPR) notation proposed in [12]. The utilization of PPR notation can be employed to implement demand flexibility. Energy-consuming resources/processes can be scheduled within appropriate time windows.

**Metaheuristics** are abstract problem-solving techniques often characterized by their flexibility, adaptability, and efficiency, providing a generalized framework for optimizing various specific problem scenarios to which they can be tailored. Multi-objective metaheuristic optimizations for MGs, focusing on objectives such as monetary profit and emission reduction, present a complex optimization challenge owing to the multitude of input parameters. These parameters encompass the dimensions of the MG components and the configuration parameters for energy allocation strategies. Given the extensive range of solutions, deterministic methods are impractical because of computational demands. Instead, metaheuristics can efficiently identify promising solutions with comparatively few simulation runs.[5, 13].

An established metaheuristic is the *Non-dominated Sorting Genetic Algorithm II* (NSGA-II), which is like many other metaheuristics inspired by nature, particularly evolutionary theory. A simulation run can be regarded as an individual composed of *alleles* within its *DNA*, representing a unique combination of simulation run input parameters.

By iteratively selecting the best outcomes from a population of individuals and randomly combining these outcomes, improved results were obtained from generation to generation. The outcome is the *Pareto-Optimal Front*, which consists of multiple dominant solutions representing optimal outcomes for multiple objectives. A decision maker can select a compromise between the dominant solutions that offer better emission reductions and those that provide better monetary profit.[14].

## 3   Concept and Realization

The following section presents the implementation of the simulation and optimization model framework in the simulation environment AnyLogic [15]. Figure 1 presents a user interface from AnyLogic depicting the various interfaces and individual model agents for each component of the framework. Furthermore, there are two metaheuristic-based optimization models for MG system sizing and optimization of the energy allocation strategy. Based on a previous study [16], only new extensions are discussed here due to space limitations, preventing a detailed description of each functionality.

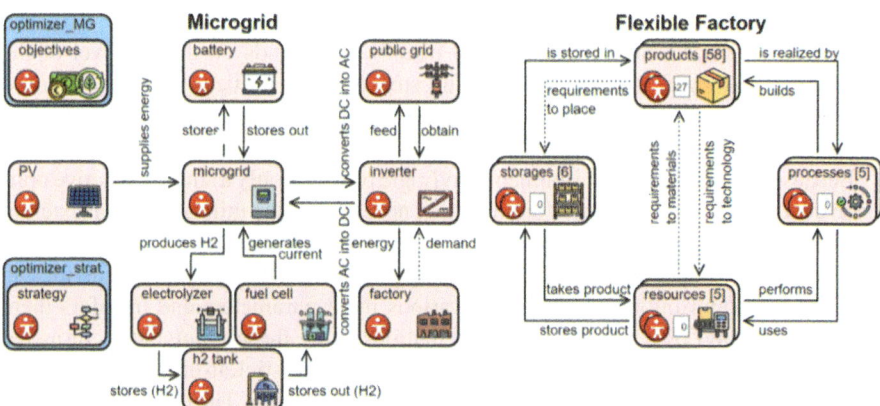

**Fig. 1.** Multi-agent simulation and optimization model framework in AnyLogic (Icons [17]).

A notable extension is the introduction of the *grid charging* **energy allocation strategy**, which enables the direct charging of storage solutions from the public grid at favorable electricity market prices. The strategy is based on the conventional approach to charging and discharging storage solutions, prioritizing, as described in [16], the BESS followed by the P2P storage. The grid charging strategy comprises three adaptive threshold values that can be specified. These include the value for identifying the base load of the factory outside operating hours, the value at which the electricity price is considered favorable, and the hydrogen storage level until it can directly source energy. The framework further considers dynamic time-of-use (ToU) electricity prices and fluctuating $CO_2$ emissions related to electricity consumption and emission reductions from feeding into the public grid.

A further extension is the **demand flexible factory** implemented in AnyLogic based on PPR notation. Figure 1 illustrates the structure on the right side, where each process, product, and resource represents an independent agent within an individual swarm of agents. In addition, each resource is allocated to a designated storage unit. At the beginning of a simulation run, the agents are initialized, with the dynamic creation and destruction of the product agents during the simulation upon the arrival of raw materials and the dispatching of finished products. Each product is either stored or currently being processed in a resource. The product is aware of its position in the queue within a storage and determines the next resource required for processing. After processing, the product autonomously transitions to the next storage or resource.

A combination of a resource and process is defined by an assigned cycle time, an assigned batch size, and an underlying load profile. Furthermore, the demand-flexible factory receives weekly order quantity and shift times for each resource.

The factory can adapt its energy consumption to the energy availability of the MG. Initially, it is verified whether the production rate, represented by the weekly order quantity, can be sustained. If this condition is met, the factory can adjust to the availability of RE. An internal algorithm monitors the energy levels every 15 min to detect any energy deficits or surpluses. In the event of a deficit, a resource is deactivated every 5 min, whereas in the case of a surplus, resources are reactivated every 5 min. The decision of which resources to activate or deactivate depends on the fill level of the corresponding storage. Resources with a full storage are prioritized and are either the first to be reactivated or the last to be deactivated. Finally, the adequacy of the production rate to meet the weekly order quantity was assessed hourly.

Because the simulation environment AnyLogic provides only a single-objective optimizer, conducting multi-objective optimization necessitates the utilization of an external tool. Following the concept proposed by [18], the optimization software *HeuristicLab* is bidirectionally linked with AnyLogic, utilizing the NSGA-II algorithm. HeuristicLab generates a random initial population of various input parameters and transfers them to AnyLogic. AnyLogic then conducts simulations using these inputs and returns the results, consisting of two objectives per simulation run. HeuristicLab utilizes these results to iteratively generate subsequent populations until termination. The overall result is the Pareto-Optimal Front.

Five input parameters were utilized for MG system sizing: PV power, BESS capacity, EL and FC power, and hydrogen tank ($H_2T$) capacity. The outcome comprises two objectives: net present value (NPV) and $CO_2$ emissions, where the former is maximized, and the latter is minimized. For the initial investment costs, the following values were adopted from a previous study [16]: PV 750 €/kWp, BESS 500 €/kW, EL 3500 €/kW, H2T 500 €/kW, and FC 2500 €/kW. According to [6], efficiencies of 90%, 75%, and 65% were achieved for BESS, EL, and FC, respectively. Ongoing operations and maintenance costs amount to 1% of the annual investment costs. In addition, there are other lump-sum costs, such as 10% of the investment costs for installation and electronics.

In addition to MG system sizing, another extension involves optimizing the energy allocation strategy by **employing a secondary metaheuristic**. This enables individual optimization of the three thresholds for the grid charging strategy.

To integrate both optimization approaches, the two metaheuristics were sequentially employed iteratively. Initially, system sizing was conducted, followed by optimization of the thresholds for the energy allocation strategy. After each optimization run, a decision maker selects the ideal compromise between emission reduction and profit maximization at their discretion. The parameter configuration from one optimization run was passed to the next until no significant changes were observed.

## 4  Evaluation

A case study was conducted based on the production of a wine bottle closure named *UniLokk* (Fig. 2), developed as part of a research factory. The machinery equipment, consisting of five resources, was retrofitted with electrical current clamps for live data transmissions. The factory was fully modeled based on the PPR notation. The production processes involved eight metalworking and assembly processes, each with a specific load profile. A demand of 800 UniLokks per week was established, with machinery operating five days a week in two shifts. The total energy consumption of production was composed of a generalized hourly standard load profile for manufacturing companies [19]. The energy consumption of the individual machines is added to this load profile. Over a period of 20 years, the energy consumption, with static electricity prices of approximately €0.42 for purchase and €0.06 for sale per kWh [20], discounted, resulted in an electricity cost of €2.174 million. During the same period, the $CO_2$ emissions amounted to 3116 tons (Fig. 2, yellow crosshair).

The following compares the results of three optimization runs. Each run involved 1000 iterations using the NSGA-II. Figure 2 illustrates the dispersion of the conventional optimization run with static electricity prices on the left (*Benchmark Results*). In contrast, the Pareto-Optimal Fronts from a run with *Demand Flexible* behavior and static electricity prices, as well as *Demand Flexible ToU* behavior with fluctuating electricity prices. Under *Adaptive ToU*, the results of the grid charging with demand flexibility strategy are depicted, wherein the fluctuating prices are also utilized.

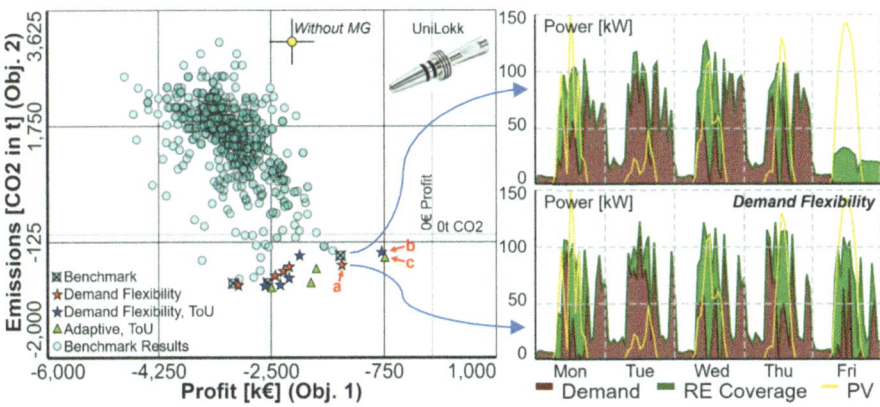

**Fig. 2.** Comparison of optimization runs and load profiles without/with demand flexibility.

Figure 2 illustrates on the right the energy consumption of production in a working week with and without demand flexibility with an MG configuration according to the marked Pareto-Optimal solutions. Without the demand flexibility approach, the weekly order volume is already reached by Thursday, whereas with demand flexibility, production continues until Friday. Concentrating production times around sunnier periods resulted in a higher SSR (in green). This necessitates a reduction in production output; hence, production on Fridays in this example. The improved scheduling of machine utilization during more productive hours for RE increases SSR, resulting in this comparison in a 1% reduction in costs and a 15% reduction in emissions.

The Pareto-Optimal solutions of the conventional energy allocation strategy with static electricity prices (*Benchmark*) demonstrate significant reductions in emissions compared to solutions without an MG, with solutions resulting in negative emissions, representing a difference of 3471 tons of $CO_2$ over 20 years. Furthermore, a few solutions indicate that the NPV has halved (configuration (b) and (c)).

The Demand Flexible configuration (a) featuring a 390 kWp PV, 240 kWh BESS, 30 kW EL, 40 kg $H_2$T, and 40 kW FC shows improved performance compared to benchmark results. Conversely, Demand Flexibility with ToU pricing (b) demonstrates even greater potential within the same MG configuration, potentially saving an extra one million euros over the next 20 years while achieving comparable emissions.

The Adaptive ToU (c) demonstrated the best performance after five iterations to optimize the three threshold values. Compared with the conventional strategy, the result shows a cost optimization of 18% and an additional emission reduction of 3%.

## 5   Conclusion

This study presents a multi-agent simulation and optimization model framework in the context of MGs, integrating system sizing, optimization of energy allocation, and flexible demand within a unified context. Using a research factory as a case study, the effectiveness of this approach was demonstrated through multiple simulated trial runs.

Regarding the first research question, the integration of many sub-models and functionalities into a single simulation model quickly results in exponentially increasing complexity. Therefore, the goal of the multi-agent approach was to allow each subcomponent to operate independently. This is evident in PPR notation, where each product autonomously knows when and how it is manufactured during production.

Regarding the second research question on synchronizing multiple optimization runs, it is important to highlight the strategy of decomposing optimization problems into simpler subproblems and utilizing simpler algorithms. This was accomplished using threshold values that incrementally fine-tuned the energy allocation strategy.

In the future, we plan to enhance our models with features like electricity prices and weather forecasting and implement them in industrial applications.

**Acknowledgment.** Parts of this work were supported by the Federal Ministry of Education and Research (BMBF) under grant number 03HY113A within the project H2Giga – FertiRob.

# References

1. IPCC. Climate Change 2023: Synthesis Report. Contribution of Working Groups I, II and III to the Sixth Assessment Report of the Intergovernmental Panel on Climate Change [Core Writing Team, H. Lee and J. Romero (eds.)], Geneva, Switzerland (2023)
2. Hafezi, R., Alipour, M.: Renewable energy sources: traditional and modern-age technologies. In: Affordable and Clean Energy 2021, p. 1085. Springer, Cham (2021)
3. Claußner, M., Huneke, F., Brinkhaus, M., et al.: Potentiale und Rahmenbedingungen für den Ausbau des Prosuming. Freiburg, Berlin (2022)
4. Peiseler, F., Runkel, M., Wettingfeld, M., et al. Nachhaltige Soziale Marktwirtschaft 2023
5. Zhang, Y., Campana, P.E., Lundblad, A., Yan, J.: Comparative study of hydrogen storage and battery storage in grid connected photovoltaic system: storage sizing and rule-based operation. Appl. Energy **201**, 397 (2017)
6. Quaschning, V.: Regenerative Energiesysteme: Technologie - Berechnung - Klimaschutz, 11th edn. Hanser, München (2021)
7. Schulz, J., Scharmer, V.M., Zaeh, M.F.: Energy self-sufficient manufacturing systems – integration of renewable and decentralized energy generation systems. Procedia Manuf. **43**, 40 (2020)
8. Crespi, E., Colbertaldo, P., Guandalini, G., Campanari, S.: Energy storage with Power-to-Power systems relying on photovoltaic and hydrogen: modelling the operation with secondary reserve provision. J. Energy Storage **55**, 105613 (2022)
9. Guo, Z., Wei, W., Bai, J., Mei, S.: Long-term operation of isolated microgrids with renewables and hybrid seasonal-battery storage. Appl. Energy **349**, 121628 (2023)
10. Bartolucci, L., Cordiner, S., Mulone, V., Pasquale, S.: Design of a multi-energy system under different hydrogen deployment scenarios. In: E3S Web of Conferences **238**, 2001 (2021)
11. Lombardi P., Liserre M.: Net-zero energy factory: exploitation of flexibility – a technical-economic analysis for a German carpentry. In: 2022 IEEE 21st Mediterranean Electrotechnical Conference (MELECON), p. 231. IEEE (2022)
12. Schleipen M., Drath R.: Three-view-concept for modeling process or manufacturing plants with AutomationML. In: IEEE Conference on Emerging, p. 1 (2009)
13. Dutta, P., Mahanand, B.S.: Chapter 9 - Affordable energy-intensive routing using metaheuristics. In: Cognitive Big Data Intelligence with a Metaheuristic Approach, p. 193. Academic Press (2022)
14. Deb, K., Pratap, A., Agarwal, S., Meyarivan, T.: A fast and elitist multiobjective genetic algorithm: NSGA-II. IEEE Trans. Evol. Comput. **6**, 182 (2002)
15. AnyLogic. https://www.anylogic.com/use-of-simulation/. Accessed 22 June 2023
16. Prior, J., Hypki, A., Kuhlenkötter, B.: Heuristische Optimierung von P2G2P Speichern: Ein Template basierter Ansatz zur Evaluierung von Speicherstrategien und Optimierungsalgorithmen. In: Erfolg durch nachhaltiges Energie- und Ressourcenmanagement, Nomos, Baden-Baden, p. 193 (2023)
17. Icons made by Freepik. www.flaticon.com
18. Bergmann, S.: Optimization of the design of modular production systems. In: Proceedings of the IEEE 2022 Winter Simulation Conference (WSC), p. 1783 (2022)
19. VDEW-Lastprofile. https://www.bdew.de/media/documents/1999_Repraesentative-VDEW-Lastprofile.pdf. Accessed 15 May 2024
20. Bundesnetzagentur. https://www.smard.de/home/downloadcenter/download-marktdaten/

# Application of Deep Reinforcement Learning for the Control of a Complex Industrial Energy Supply System

Heiko Ranzau[(✉)] [iD], Tobias Lademann[iD], Fabian Borst[iD], Oskay Ozen[iD], and Matthias Weigold[iD]

Institute of Production Management, Technology and Machine Tools (PTW), Technical University of Darmstadt, Otto-Berndt-Str. 2, 64287 Darmstadt, Germany
h.ranzau@ptw.tu-darmstadt.de

**Abstract.** Deep Reinforcement Learning (DRL) can optimize the operating strategies of industrial energy supply systems (IESS), enhancing energy efficiency and flexibility while reducing costs. This work conducts and evaluates a real-world application of Proximal Policy Optimization (PPO), a widely used DRL algorithm, on the complex ETA Research Factory's supply system. Our approach involves five steps: system boundary definition and data accumulation, modelling and validation, DRL implementation, simulation-based evaluation and finally, application of the DRL-based controller on the real system. In simulation experiments, the controller demonstrates a 43% reduction in operational costs with respect to the conventional controller while maintaining operational stability within safety limits, thus proving to be a cost-effective and reliable solution. During real-world application, the DRL-based controller upheld safety and robustness. Future research should aim at establishing comparability under various environmental conditions and closing the gap between simulation and reality.

**Keywords:** Deep Reinforcement Learning · Energy Efficiency · Industrial Energy Supply Systems

## 1 Introduction

To achieve a $CO_2$-neutral European economy by 2050, it is essential to reduce industrial energy consumption and accelerate the transition to renewable energy sources. Since the industrial sector causes about 20% of all European $CO_2$ emissions [1], it is essential to realize savings in this area. Industrial energy supply systems (IESS) provide factories with process heat, cooling, and electrical power and can significantly contribute to a more energy-efficient and flexible industry by intelligently optimizing their operating strategies [2]. Deep Reinforcement Learning (DRL) methods have been successfully applied for this purpose in simulation-based demonstrations, significantly reducing operating costs, energy consumption, and $CO_2$ emissions of IESS [3, 4]. However, they have not yet been applied to complex real-world IESS, which is the focus of this work. In Sect. 1.1

H. Kohl et al. (Eds.): GCSM 2024, LNME, pp. 351–359, 2025.
https://doi.org/10.1007/978-3-031-93891-7_39

and 1.2, we provide an overview of IESS and DRL. In Sect. 2 we introduce the methodology, the use case, and the implementation, before we evaluate the results in Sect. 3 and end with a discussion and outlook in Sect. 4.

## 1.1 Industrial Energy Supply Systems

A factory utilizes energy from diverse sources, including electricity, natural gas, and district heating [5]. These energy forms are either distributed directly or converted by IESS into other necessary forms, such as heating, cooling, or compressed air [3]. The IESS are part of the technical building equipment (TBE), which comprises the technical and usage-specific installations embedded in the building. The TBE is operated by building automation systems (BAS) across different system levels via standardized communication protocols. The relevant standard to this work, OPC UA, provides an independent, scalable solution for vertical and horizontal communication, including between programmable logic controllers (PLC) and external controllers [6]. Predominantly, however, the control and automation of IESS is realized through internal controllers such as two-point or PID controllers [3].

## 1.2 Applied Deep Reinforcement Learning

Fundamentally, Reinforcement Learning (RL) is one of the machine learning paradigms concerned with how intelligent agents take actions $(A)$ in an environment to achieve desired system states $(S)$ and maximize a cumulative reward $(R)$ [7]. As mentioned before, DRL is a promising approach for providing external controllers to increase energy efficiency and is a form of RL in which the value of a state, the policy or the system model is approximated by deep neural networks, enabling the control of much more complex problems than traditional RL methods. [3].

In terms of real-world utilization, few DRL applications are validated on real systems [8]. Our literature review aligns with these findings, as shown in Table 1. We compare algorithms, action space size, and indicate whether the addressed system is a complex IESS with different supply and storage systems, which we denote as hybrid energy systems (HES). We further investigate whether simulations are tuned using automated parameter identification (API), and if a real-world application (Real) takes place.

Notable real-world DRL applications primarily focus on room temperature control and small action spaces of 1–2 control signals. Schmidt et al. optimized a school's heating system using Q-Learning [9], while Zhang et al. and Du et al. applied Actor-Critic and DDPG algorithms to office buildings and residential Heating, Ventilation, and Air Conditioning (HVAC) systems, respectively [10, 11].

Most studies achieved significant energy savings but often relied on unvalidated simulations, except for Zhang et al. who used API to calibrate the simulation with real-world data. Although some studies explored HES, real-world DRL applications for optimizing HES control remain unaddressed, which is a research gap this work addresses.

**Table 1.** Overview of relevant literature.

| Work | Year | Algorithm | Actions | HES | API | Real |
|------|------|-----------|---------|-----|-----|------|
| [12] | 2016 | Q-Learning | n/a | ✓ | ✗ | ✗ |
| [9] | 2017 | fitted Q-iteration | 2 | ✗ | ✗ | ✓ |
| [13] | 2017 | TRPO, DDPG | 4 | ✓ | ✗ | ✗ |
| [14] | 2019 | AC | 10 | ✓ | ✗ | ✗ |
| [11] | 2019 | A3C | 1 | ✗ | ✓ | ✓ |
| [3] | 2019 | PPO, TRPO, A2C | 9 | ✓ | ✗ | ✗ |
| [2] | 2020 | PPO | 6 | ✓ | ✗ | ✗ |
| [15] | 2022 | DDPG | 1 | ✗ | ✗ | ✓ |
| [10] | 2022 | DDPG | 1 | ✗ | ✗ | ✓ |
| [16] | 2022 | SAC | 4 | ✓ | ✗ | ✗ |
| **Ours** | **2024** | **PPO** | **13** | ✓ | ✗ | ✓ |

## 2   Research Framework and Practical Application

### 2.1   Methodical Approach

The application of DRL for the control of IESS is a complex multi-step process, necessitating a methodical approach. We utilize the method first presented by Weigold et al. in [4], depicted in Fig. 1. The method assumes that the objective is to develop an optimized control strategy for an existing IESS.

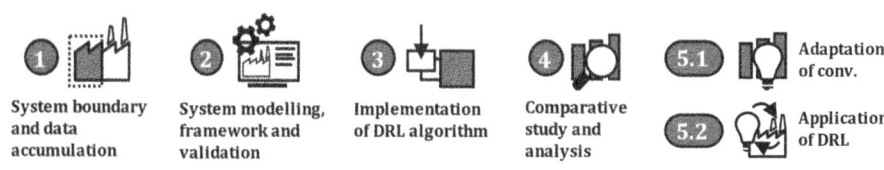

**Fig. 1.** Method for the application of DRL for IESS (compare [4]).

In **step 1**, the system boundary is defined, and necessary data, including technical documentation and historical data, is gathered to understand current conventional (conv.) control strategies. **Step 2** involves developing and validating a simulation model of the IESS. In **step 3**, a suitable DRL algorithm is selected, implemented, and trained using the simulation model, with defined performance indicators and hyperparameters. **Step 4** includes a comparative analysis between the DRL-derived strategy and the conventional approach, focused on operational safety and performance. Finally, the optimized DRL-based control strategy can either be analyzed and interpreted to adapt the conventional strategy (**step 5.1**) or directly applied to the IESS (**step 5.2**). Since the goal of this work is direct real-world application, option **5.2** is chosen.

## 2.2 Use Case

The use case of this work is the ETA Research Factory, which contains a representative process chain from the field of metal processing (see [17]). The HES of the ETA Research Factory is depicted in Fig. 2. The factory's thermal demand is met through three different thermal networks: the high-temperature heating network (HNHT), the low-temperature heating network (HNLT), and the cooling network (CN). The factory's HES interacts with external gas and electricity grids, allowing surplus electricity to be fed back into the grid. The HNHT is supported by two gas-based combined heat and power units (CHP) of different power rating and a condensing boiler (B), while the HNLT uses waste heat from production processes and includes a heat exchanger (HE), a heat pump (HP), and a thermally activated façade (AF) for cooling. The CN is used for cooling production machinery and spaces, utilizing a chiller (C) and the HP. Active thermal storage systems include a vacuum super-insulated storage (VSI) unit for the HNHT and high-volume fly ash (HVFA) concrete storage units for the HNLT and CN.

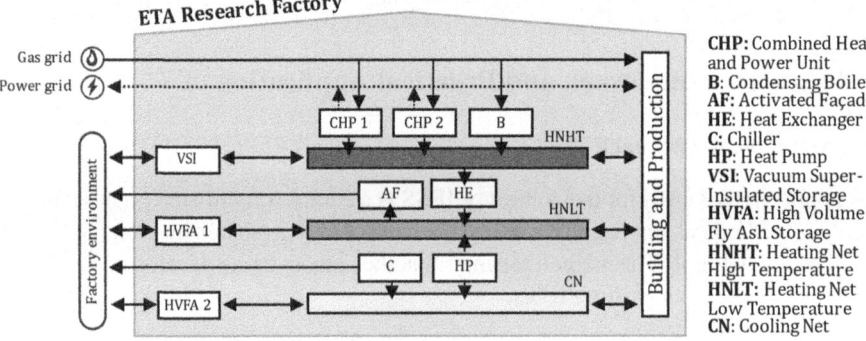

**Fig. 2.** Overview of the hydraulic energy supply network of the ETA Research Factory.

The conventional control strategy of the ETA Factory's thermal energy network is used as a base line in this work and utilizes a rule-based logic. The BAS employs *TwinCat* for control, with an overarching strategy managing 13 on/off signals and subordinate strategies controlling the respective subsystems such as pumps and valves. The system relies on hysteresis controllers based on temperature thresholds in various storage units, with different controllers for heating and cooling functions. The conventional control strategy can be overridden manually through a user interface, or by algorithmic control via OPC UA, which is the option we use to pass the DRL control signals.

## 2.3 Implementation

The implementation is carried out according to the method described in Sect. 2.1. Once the system to be optimized has been determined and the system boundary along with the conventional control strategy have been identified (**step 1**, see Sect. 2.2), a physical simulation model of the IESS is set up using *Modelica/Dymola*, manually parameterized at key points using measurements on the real system (**step 2**) and exported as an Functional Mock-up Unit (FMU). The *python*-based framework *eta-utility* (see [18], compare Fig. 3) is selected for training a DRL algorithm via a python-FMU interface and later for controlling the BAS via a python-OPC UA interface.

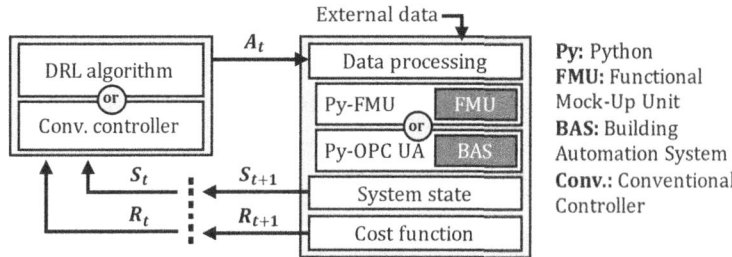

**Fig. 3.** The framework used for this study (compare [4]).

Next (**step 3**), Proximal Policy Optimization (PPO, see [19]) is selected as a suitable DRL algorithm and hyperparameters (Multi-Layer Perceptron with 500 shared neurons, 400 and 300 neurons for value and policy networks, minibatch size 256, 256 steps per episode, linear learning rate schedule from 0.0002 to 0.00002) are chosen based on experience from various test runs, previous work and the literature [3]. A cost function is selected and the DRL algorithm is trained on a server computer (Intel Xeon Silver 4110) with 8 environments running in parallel for 400 to 2000 training episodes of 8640 time steps each. The sampling time for all time steps is set to 30 s to fit the dynamics of the system. The cost function (see Eq. 1) is a sum of individual terms, e.g., costs for energy consumption (energy), temperature restriction violation costs (temp), switching costs (switch), and other costs (other) related to *policy shaping* (see [3]) and simulation abortions. Energy costs are the product of consumption and dynamic prices, temperature violation costs are fixed at 1€ per time step (a high, arbitrarily chosen value) and switching operation costs range from 0.5€ to 4€, depending on the system.

$$R_t = R_t^{energy} + R_t^{temp} + R_t^{switch} + R_t^{other} \tag{1}$$

The necessary external data includes factory demands, market, and weather-data, all based on historic data that in the case of the factory demands has been synthetically extended to generate a data set spanning 2 years. In **step 4**, a test series is designed to evaluate the DRL performance when applied to the IESS. For the final **step 5.2**, the conventional control strategy is implemented as a fallback mechanism to guarantee save operations in case predefined temperature restrictions are violated. Finally, the trained DRL is applied to the real system without any adaptation, referred to as *zero-shot* deployment. The test series and the corresponding results are listed in Sect. 3.

## 3 Evaluation

The settings described above were used to train an agent referred to as **Agent A** for a planned 2000 episodes (each episode is a randomly selected 3-day period from the training year and equals 8640 steps). Due to simulation instabilities and the resulting episode abortions, the term $R_t^{other}$ became dominant in the cost function Eq. 1, which destabilized the training due to the independence of simulation instabilities and the agent's actions. Thus, we aborted the training after only 1520 episodes and removed $R_t^{other}$ from Eq. 1 to ensure stable training. The observed simulation errors were used to enhance simulation robustness to further stabilize the learning progress. The training of Agent $\underline{A}$ is continued with the updated simulation and reward function for another 400 episodes. Figure 4 shows the application of Agent $\underline{A}$ on the simulation (left, 3-day period on historic data) and on the real system (right, 7-h real-world test run) with horizontal heat maps indicating an activated (black) or deactivated system (grey).

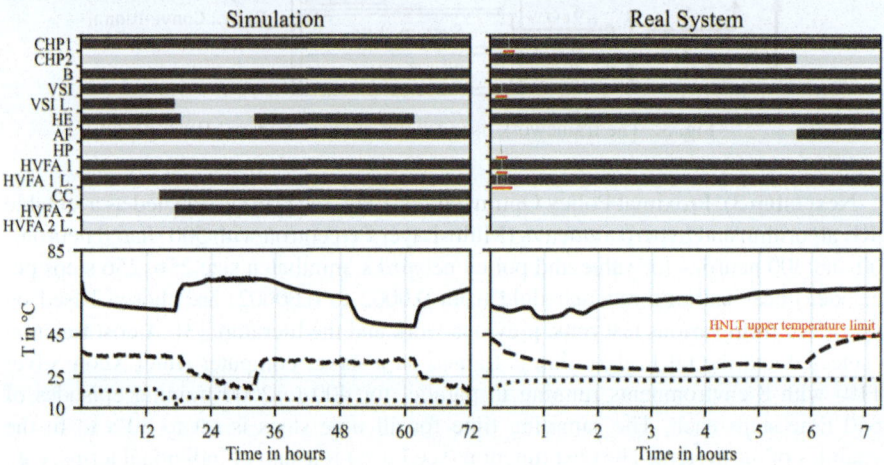

**Fig. 4.** Application of Agent $\underline{A}$ (HNHT: solid line, HNLT: dashed line, CN: dotted line). (Color figure online)

On the real system, Agent $\underline{A}$ exhibits sporadic undesirable high-frequency switching around 0.2h and exceeds the HNLT's upper temperature limit of 45 °C at 7.2 h (see red indicators), triggering the fallback mechanism and terminating the application. We attribute the high-frequency switching to the non-deterministic inference of the DRL-model and the temperature violations to a simplified simulation model of the AF, which neglects heat input from solar radiation, increasing the simulation-reality gap.

After further improvements to the simulation (the AF model was extended by an estimated characteristic curve incorporating solar radiation effects) and a study to determine the right training duration (we identified 800 episodes as a sufficient amount), **Agent B** is trained. As depicted on the right of Fig. 5, the temperature limits are successfully maintained and no switching operation takes place during the 8-h real application, indicating that the agent selects suitable and robust initial operating states to minimize switching frequency. Notably, in the 3-day simulation-based evaluation (left of Fig. 5) Agent $\underline{B}$ achieves an operating cost reduction of 43% compared to the conventional control strategy.

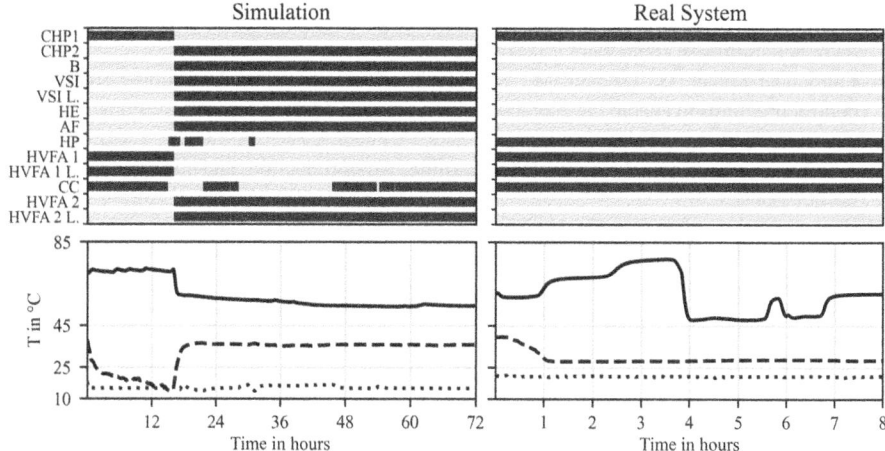

**Fig. 5.** Application of Agent $\underline{B}$ (HNHT: solid line, HNLT: dashed line, CN: dotted line) (Color figure online)

## 4 Discussion and Outlook

The evaluation successfully addresses the research gap and shows the application potential of DRL for complex IESS. Our DRL-based controller achieves operational stability and safety when applied on the real system and reduces operational costs by 43% in simulation-based evaluations. However, assessing cost-savings in real world applications is complex, as conducting experiments under identical external conditions is challenging due to external influencing factors such as weather conditions. Simulation based comparisons are a suitable approach, given the simulation-reality gap is sufficiently small. Future research should therefore find automated solutions to decrease the simulation-reality gap and establish comparability of control algorithms for diverse environmental scenarios.

**Funding.** The authors gratefully acknowledge the financial support of the Project KI4ETA under grant number 03EN2053A.

## References

1. European Environment Agency, 2024. EEA greenhouse gases—data viewer. https://www.eea.europa.eu/. Accessed 21 May 2024
2. Kohne, T., Ranzau, H., Panten, N., Weigold, M.: Comparative study of algorithms for optimized control of industrial energy supply systems. In: Proceedings of the 9th DACH+ Conference on Energy Informatics, vol. 3, no. 1 (2020)
3. Panten, N.: Deep Reinforcement Learning zur Betriebsoptimierung hybrider industrieller Energienetze. Dissertation. Shaker, Düren (2019)
4. Weigold, M., Ranzau, H., Schaumann, S., Kohne, T., Panten, N., Abele, E.: Method for the application of deep reinforcement learning for optimised control of industrial energy supply systems by the example of a central cooling system. CIRP Ann. **70**(1), 17–20 (2021)

5. Müller, E., Engelmann, J., Löffler, T., Jörg, S.: Energieeffiziente Fabriken planen und betreiben. Springer, Heidelberg (2009)
6. Fuhrländer-Völker, D., Lindner, M., Weigold, M.: Design method for building automation control programs to enable the energetic optimization of industrial supply systems. Procedia CIRP **104**, 229–234 (2021)
7. Sutton, R.S., Barto, A.: Reinforcement Learning: An introduction, 2nd edn. The MIT Press, Cambridge (2018)
8. Perera, A., Kamalaruban, P.: Applications of reinforcement learning in energy systems. Renew. Sustain. Energy Rev. **137**, 110618 (2021)
9. Schmidt, M., Moreno, M.V., Schülke, A., Macek, K., Mařik, K., Pastor, A.G.: Optimizing legacy building operation: the evolution into data-driven predictive cyber-physical systems. Energy Build. **148**, 257–279 (2017)
10. Du, Y., Li, F., Kurte, K., Munk, J., Zandi, H.: Demonstration of intelligent HVAC load management with deep reinforcement learning: real-world experience of machine learning in demand control. IEEE Power Energy Maga. **20**(3), 42–53 (2022)
11. Zhang, Z., Chong, A., Pan, Y., Zhang, C., Lam, K.P.: Whole building energy model for HVAC optimal control: a practical framework based on deep reinforcement learning. Energy Build. **199**, 472–490 (2019)
12. Sheikhi, A., Rayati, M., Ranjbar, A.M.: Dynamic load management for a residential customer; reinforcement learning approach. Sustain. Cities Soc. **24**, 42–51 (2016)
13. Bollenbacher, J., Rhein, B.: Optimal configuration and control strategy in a multi-carrier-energy system using reinforcement learning methods. In:2017 International Energy and Sustainability Conference (IESC), pp. 1–6 (2017)
14. Huang, X., Hong, S.H., Yu, M., Ding, Y., Jiang, J.: Demand response management for industrial facilities: a deep reinforcement learning approach. IEEE Access **7**, 82194–82205 (2019)
15. Svetozarevic, B., Baumann, C., Muntwiler, S., Di Natale, L., Zeilinger, M.N., Heer, P.: Data-driven control of room temperature and bidirectional EV charging using deep reinforcement learning: Simulations and experiments. Appl. Energy **307**, 118127 (2022)
16. Luo, Y., Liu, C., Lai, Q.: Optimal scheduling for a multi-energy microgrid by a soft actor-critic deep reinforcement learning. In: IEEE Power & Energy Society General Meeting (PESGM), pp. 1–5 (2022)
17. Abele, E., Schneider, J., Maier, A.: ETA - die Modell-Fabrik: Energieeffizienz weiter gedacht, Darmstadt, p. 64 (2018)
18. Grosch, B., et al.: A framework for researching energy optimization of factory operations. Energy Inf. **5**(1), 29 (2022)
19. Schulman, J., Wolski, F., Dhariwal, P., Radford, A., Klimov, O.: Proximal Policy Optimization Algorithms. arXiv-ID https://arxiv.org/abs/1707.06347 (2017)

# Scenario Study for Green Hydrogen in Sub-Saharan Africa

Yara Matschalow and Semih Severengiz(✉)

Bochum University of Applied Sciences, Sustainable Technologies Laboratory, Am Hochschulcampus 1, 44801 Bochum, Germany
semih.severengiz@hs-bochum.de

**Abstract.** Given the urgent need to transition to sustainable energy sources to address both environmental and socio-economic challenges, green hydrogen has emerged as a compelling option and a potential solution. This paper presents a scenario study for the development and integration of green hydrogen in decentralized energy systems in Sub-Saharan Africa, with a focus on Ghana. The research employs the scenario technique to explore the plausibility and feasibility of green hydrogen development in Ghana until 2035. In order to achieve this objective, 61 potential influencing factors were initially identified, and 20 key factors were subsequently selected. A total of 75 projections for the key factors were then developed, and three distinctive scenarios were created. The three scenarios, designated as "Corporate Dominance", "Decentralized Energy Supply", and "Lost Electrification", provide insights into potential future developments and their implications for energy policy, economic growth, and technological innovation. The study highlights the critical role of policy frameworks, technological advancements, educational programs, and economic incentives in shaping the future of energy systems. The findings offer valuable guidance for policymakers, businesses, and organizations to foster a sustainable and equitable energy future in Sub-Saharan Africa.

**Keywords:** Green hydrogen · decentralized energy systems · scenario study

## 1 Introduction

The SDG 7 addresses the necessity for clean, affordable, and reliable energy [1]. This objective is of particular significance in the context of African countries like Ghana. The insufficient and unreliable electrification of Ghana has resulted in an estimated five million individuals in the country lacking access to electricity, representing approximately 15% of the total population of 33.5 million in 2022 [2]. The inadequate power supply significantly impairs living conditions and hampers social and economic development [3]. As part of the investigation and evaluation of alternative power supply concepts, decentralized energy systems, such as solar systems and mini-grids, are being evaluated as a potential supplement to the national power grid. One advantage of such systems is that users are not dependent on the quality and availability of the electricity grid and are less affected by fluctuating electricity prices [4]. Furthermore, a significant proportion

© The Author(s) 2025
H. Kohl et al. (Eds.): GCSM 2024, LNME, pp. 360–367, 2025.
https://doi.org/10.1007/978-3-031-93891-7_40

of the population relies on environmentally detrimental practices, such as the use of diesel generators, to compensate for outages in the national power grid [5]. In 2022, 21 million tons of $CO_2$ were emitted through energy production, representing a 332% increase since the year 2000 [6].

The region is endowed with optimal resources to produce renewable energy, with solar energy representing a particularly promising avenue for the development of decentralized energy systems [7]. A crucial aspect of renewable energy is the storage of energy over both short and long periods of time to optimize the utilization of fluctuating sources. To date, batteries have been the predominant means of achieving this objective, although they are highly intensive in finite and rare resources and unsuitable for long-term storage. Green hydrogen exhibits optimal characteristics for long-term storage, rendering it an especially promising option for stabilizing the energy supply [8, 9]. Despite its promising properties, green hydrogen is not yet widely utilized in Ghana. This has led to uncertainty among investors, politicians, and other decision-makers regarding its potential use. To address the uncertainties of future developments, the scenario technique is particularly well-suited for the recording and presentation of a multitude of potential future scenarios. Furthermore, it provides a solid foundation for decision-making, taking into account a vast array of relevant indicators and underlying framework conditions. In this context, this study aims to examine the potential role of decentralized energy systems with green hydrogen in Ghana by 2035.

## 2   Scenario Technique

The scenario technique is an analytical method for developing complex visions of the future and serves as a basis for decision-making. The objective is to identify potential opportunities and risks [10]. As early as the 1960s, the scenario technique was employed for this purpose. As Kahn and Wiener aptly summarized, "scenarios are hypothetical sequences of events constructed for the purpose of drawing attention to causal processes and decision points."[11] The development of scenarios is based on assumptions about the present and trend studies, which employ both quantitative and qualitative analyses.[1] Finally, recommendations for action are formulated to support desirable visions of the future and to assist decision-makers.

The data utilized for the study was derived from surveys conducted as part of an master thesis, with an expert team comprising specialists from renowned universities, research institutes and enterprises in Ghana and Germany with extensive expertise in the fields of renewable energies, hydrogen technology, energy policy and sustainable financing[1]. The primary data obtained from the expert survey was initially collated and subsequently subjected to a structured processing procedure, encompassing the grouping of themes into higher- and lower-level categories. Secondary data was collected through a detailed systematic literature review, including reports, government evaluations and guidelines, market analyses and academic literature. The keywords employed were derived from the research question and the findings of the survey. The data collected

---

[1] For detailed analysis, insights and more information on the results of the study, please refer to [12]

was employed to identify potential influencing factors and prospective developments, thereby enabling the construction of realistic and comprehensive scenarios.

The use of scenario management software solutions with algorithms for clustering has become a well-established approach for the development of realistic scenarios. In our scenario development process, we have employed the ScMI scenario manager software. According ScMI, the approach is divided into four steps: scenario field analysis, scenario forecasting, scenario creation, and scenario evaluation and interpretation, which we also used in our work.

## 2.1 Scenario Field Analysis

In the scenario field analysis, the scope of the study was precisely defined and a catalog of 61 possible influencing factors was compiled from the primary and secondary data. The influencing factors were subsequently classified into the following areas of influence: energy supply, hydrogen, market, utilization, energy technology, locations, legal framework, substitution options, economy, society, the natural environment, and politics. A cross-impact analysis was conducted to identify the most influential key factors. The analysis is employed to evaluate the influence, plausibility, and consistency of two factors on one another by comparing them in isolation [13]. The result is an influence grid that enables an evaluation according to activity and passivity as well as the strength of the influence. The grid shown in Fig. 1 highlights the selected key factors in color on the left. The colors correspond to the respective areas of influence, which are listed on the right with the respective names of the key factors.

## 2.2 Scenario Forecasting

In the subsequent phase, a series of prospective projections are formulated for each of the selected key factors. These delineate a range of potential trajectories for the evolution of a given factor, encompassing both strategic relevance and distinctive characteristics. To this end, one or two dimensions of investigation are delineated for each factor, a spectrum of prospective developments within these dimensions is identified, and three to four future projections are formulated for each key factor, derived from a synthesis of the identified developments within the dimensions.

## 2.3 Scenario Creation

In scenario creation, the overall projections of all key factors are compared in a cross-impact analysis, which is also known as a consistency matrix. This serves to ascertain whether disparate projections of disparate key factors can occur in a consistent manner within a common scenario or are mutually incompatible. To generate the scenarios based on the calculations, firstly, the number of scenarios is determined by utilizing the second derivation of a scree diagram. The local maximum indicates that three scenarios should be considered. This approach achieves a balance between appropriate differentiation, obtaining measurable insights while minimizing information loss.

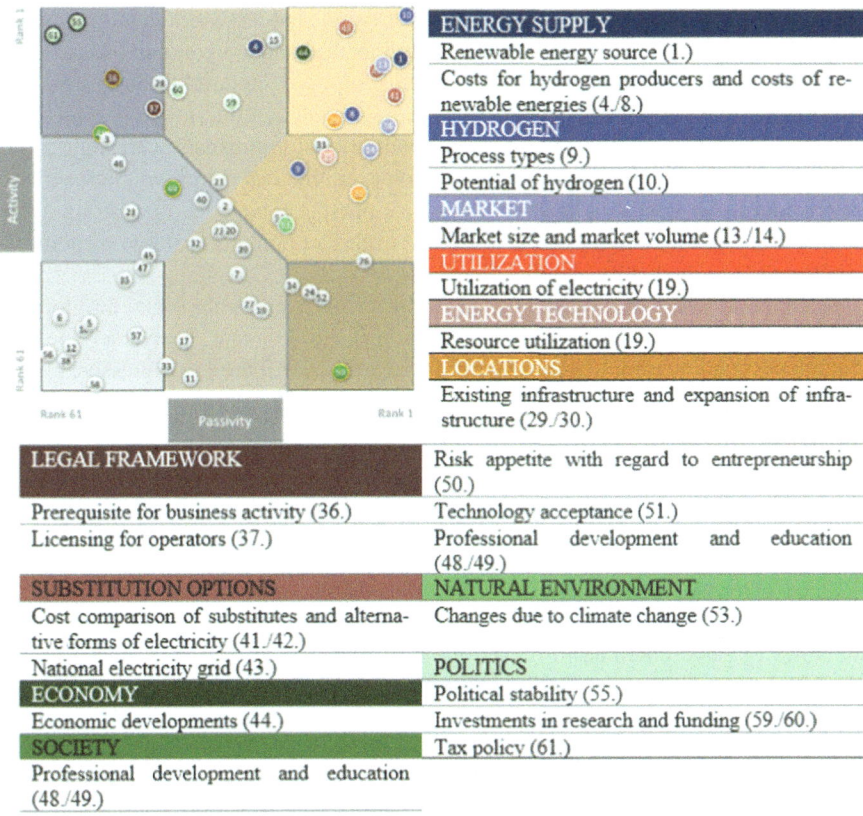

| | |
|---|---|
| **ENERGY SUPPLY** | |
| Renewable energy source (1.) | |
| Costs for hydrogen producers and costs of renewable energies (4./8.) | |
| **HYDROGEN** | |
| Process types (9.) | |
| Potential of hydrogen (10.) | |
| **MARKET** | |
| Market size and market volume (13./14.) | |
| **UTILIZATION** | |
| Utilization of electricity (19.) | |
| **ENERGY TECHNOLOGY** | |
| Resource utilization (19.) | |
| **LOCATIONS** | |
| Existing infrastructure and expansion of infrastructure (29./30.) | |

| **LEGAL FRAMEWORK** | Risk appetite with regard to entrepreneurship (50.) |
|---|---|
| Prerequisite for business activity (36.) | Technology acceptance (51.) |
| Licensing for operators (37.) | Professional development and education (48./49.) |
| **SUBSTITUTION OPTIONS** | **NATURAL ENVIRONMENT** |
| Cost comparison of substitutes and alternative forms of electricity (41./42.) | Changes due to climate change (53.) |
| National electricity grid (43.) | **POLITICS** |
| **ECONOMY** | Political stability (55.) |
| Economic developments (44.) | Investments in research and funding (59./60.) |
| **SOCIETY** | Tax policy (61.) |
| Professional development and education (48./49.) | |

**Fig. 1.** Influence grid with all influencing factors by passivity, activity and influence strength, key factors highlighted in color, grid exported from ScMI software

## 3    Results

The results of the study depend on the projections, including their dimensions and characteristics, of the relevant influencing factors. Due to the diverging interactions of the influencing factors, the evaluation in the ScMI software results in different characteristics of the projections for each scenario, which form the basis of the developed scenarios[2].

### 3.1    Scenario 1: Corporate Dominance

The "Corporate Dominance" scenario features large energy suppliers with rapid access to permits, stable green hydrogen prices, and robust law enforcement. Prices stagnate due to moderate renewable shares and isolated research barriers. Research is effective in some sectors due to medium to high investment, but particularly climate-friendly research faces low investment. Nevertheless, career paths are diverse and well supported

---

[2] The detailed projections for each key factor and scenario can be viewed in [12]

by extensive local training opportunities. Non-renewable processes dominate, exacerbating climate change and extreme weather conditions. Expanded hydrogen infrastructure ensures global accessibility and urban electricity supply is assured through simplified grid feed-in approvals, while remote areas remain excluded due to geographical and economic constraints. Policies favor large international companies with high capital requirements and complex tax exemptions. However, this has increased the scarcity of rare resources. Awareness, education and career prospects have led to a strong understanding and appreciation of hydrogen. But the exclusive benefits for large corporations exacerbate social inequality and division.

This scenario heralds a new era of green hydrogen, improving social development through better access to electricity, education and employment. Conversely, technological losses and environmentally harmful practices contribute to progressive climate change and the displacement of small and medium-sized enterprises (SMEs), which drives social division.

## 3.2  Scenario 2: Decentralized Energy Supply

In the "Decentralized Energy Supply" scenario, energy development is driven by many small, decentralized electricity producers. Renewable energy sources dominate the energy mix and contribute significantly to sustainability. Technological advances and economies of scale continue to reduce the cost of green hydrogen production, making it highly competitive, increasing efficiency and minimizing environmental impact. Hydrogen's potential as an energy carrier is well developed, due to favorable conditions like favorable policy, high investments and ongoing research. Decentralization leads to a wide distribution of electrification, making the energy system more independent and robust. Small energy systems consume fewer rare earths and metals and avoid fossil fuels. This ensures a stable and affordable energy supply, mitigates climate change and improves living conditions. There is a high level of public acceptance and understanding of the technology as well as a well-developed, cross-sectoral global infrastructure for green hydrogen. Policies with low capital requirements and fewer regulations encourage SMEs and obtaining feed-in permits is straightforward. The price competitiveness of hydrogen makes it an attractive energy option. The national grid is stable and reliable, but primarily covers urban centers, while remote areas are well served by decentralized systems. The development of the energy market has created a reliable constitutional framework that protects individual rights and upholds the rule of law. Social cohesion is promoted through cooperation in decentralized energy systems.

This scenario is characterized by innovation-driven growth. Technological advances and innovation-friendly policies lead to significant economic growth, creating numerous career and training opportunities in the energy and technology sectors, benefiting society.

## 3.3  Scenario 3: Lost Electrification

The "Lost Electrification" scenario presents a bleak outlook for renewable energy development, with a minimal contribution to total electricity generation and an unsustainable energy supply. Green hydrogen remains economically unattractive due to stagnating technological progress and lack of economies of scale. Innovation and investment efforts are constrained and the market for electrification is shrinking, characterized by high capital requirements, restrictive employment regulations, and exclusive licensing for operators. Hydrogen infrastructure is almost non-existent, and past reliance on fossil fuels has led to extreme resource scarcity. This exacerbates climate change and extreme weather events, further destabilizing the electricity supply. The national power grid lacks nationwide coverage and security of supply, resulting in persistent energy shortages. Weak rule of law and social fragmentation contribute to fragile political stability. Poor development of the energy market limits professional opportunities and academic education and training in green hydrogen and renewable energy. High taxation burdens business activity, and social acceptance of technology is low due to a lack of knowledge. Overall, the Lost Electrification scenario depicts a future of stagnation, limited resource use, and inefficient structures.

## 4 Discussion

### 4.1 Scenario Interpretation

Following the creation of the scenarios, a further analysis was conducted to evaluate all projections of the key factors in order to assess their proximity to the current reality, their desirability, and their feasibility. The results are presented in Fig. 2. It is evident that scenario 3 is the most proximate to the current situation in Ghana. Scenario 2 is the most proximate to the desired future state. Whereby scenario 1 is just above scenario 2 in terms of proximity to the current situation in Ghana.

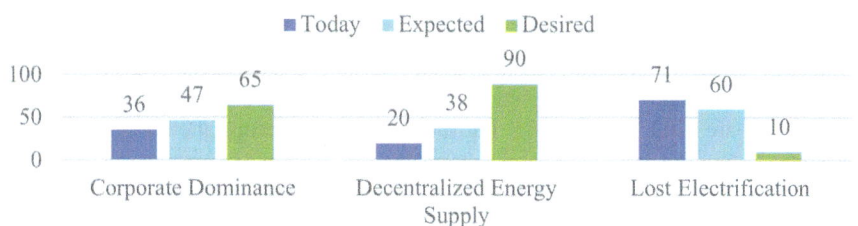

**Fig. 2.** Scenario interpretation according to their proximity to the current reality, their desirability, and their feasibility

This scenario study will be subjected to further investigation and quantitative analysis in subsequent studies, with the aim of corroborating the findings with additional data.

### 4.2 Recommendations for Action

Based on our scenario results we recommend the following:

- Support SMEs: Provide financial incentives, easier market access, and reduce bureaucratic hurdles.
- Improve market regulation: Ensure fair competition and prevent monopolies.

- Promote local energy producers: Strengthen community projects and create incentives for decentralized energy.
- Combat energy poverty: Use regulatory instruments to ensure equitable energy access.
- Adopt a regional approach: Consider local conditions and economic development in Sub-Saharan Africa.
- Invest in research and development as well as education: Focus on developing new technologies and climate change education.
- Enhance regulation and cooperation: Accelerate the transition to sustainable energy sources through robust regulation and international cooperation.
- Maintain flexibility: Be prepared to respond to unforeseen developments for a sustainable future.

## 5   Conclusion

This study examines the potential for green hydrogen development in Ghana and its implications for decentralized energy systems in Sub-Saharan Africa. Three scenarios illustrate different future paths for Ghana's energy sector by 2035.

The "Corporate Dominance" scenario highlights the risks of social inequality and environmental degradation. Conversely, the "Decentralized Energy Supply" scenario envisions a sustainable future with significant technological and economic benefits from renewable energy. The "Lost Electrification" scenario portrays a dismal future of stagnation and constrained advancement, underscoring the obstacles posed by inadequate investment and innovation.

The study highlights the pivotal role of policy frameworks, technological advancements, and economic incentives in shaping Ghana's energy future. To transition towards a desirable energy future, it is imperative to provide support to SMEs, promote local energy production, enhance regulatory mechanisms, and invest in research and development. Furthermore, international collaboration, educational initiatives, and robust governmental regulation are vital for fostering innovation and accelerating the adoption of green hydrogen.

**Acknowledgment.** The research work published in this paper was funded by the Federal Ministry for the Environment, Nature Conservation, Nuclear Safety and Consumer Protection under the on the funding program Export Initiative Environmental Protection. Funding reference: 67EXI6503A.

**Disclosure of Interests.** The authors have no competing interests to declare that are relevant to the content of this article.

## References

1. United Nations Department of Economic and Social Affairs: Goal 7 | Department of Economic and Social Affairs. https://sdgs.un.org/goals/goal7. Accessed 30 Aug 2024
2. Energy Progress Report: Ghana | Tracking SDG 7. https://trackingsdg7.esmap.org/country/ghana

3. Casati, P., Moner-Girona, M., Khaleel, S.I., Szabo, S., Nhamo, G.: Clean energy access as an enabler for social development. Energy Sustain. Dev. **72**, 114–126 (2023). https://doi.org/10.1016/j.esd.2022.12.003

4. Finke, S., Velenderić, M., Severengiz, S., Pankov, O., Baum, C.: Transition towards a full self-sufficiency through PV systems integration for sub-Saharan Africa. Renew. Energy Environ. Sustain. **7**, 8 (2022). https://doi.org/10.1051/rees/2021054

5. Seglah, P.A., et al.: Electricity generation in Ghana: evaluation of crop residues and the associated greenhouse gas mitigation potential. J. Clean. Prod. **395**, 136340 (2023). https://doi.org/10.1016/j.jclepro.2023.136340

6. International Energy Agency: Greenhouse Gas Emissions from Energy. https://www.iea.org/data-and-statistics/data-product/greenhouse-gas-emissions-from-energy

7. Economic and Social Commission for Asia and the Pacific: Decentralized energy system. https://www.unescap.org/sites/default/files/14.%20FS-Decentralized-energy-system.pdf

8. Modu, B., Abdullah, M.P., Bukar, A.L., Hamza, M.F.: A systematic review of hybrid renewable energy systems with hydrogen storage. Int. J. Hydrogen Energy **48**, 38354–38373 (2023). https://doi.org/10.1016/j.ijhydene.2023.06.126

9. Schneider, S., Velenderic, M., Staib, M., Küpper, E., Severengiz, S.: The Potential of Hydrogen-Based Storage Systems in Sub-saharan Africa. In:Presented at the International Renewable Energy Storage Conference (IRES 2022), 25 May 2023 (2023)

10. Schulte, A., Wittemund, T., Weber, P., Fink, A.: Preparing for an uncertain future: south west-phalia city scenarios 2030. In: Ahlemann, F., Schütte, R., and Stieglitz, S. (eds.) Innovation Through Information Systems, pp. 382–397. Springer, Heidelberg (2021)

11. Kahn, H., Wiener, A.J.: The Year 2000: A Framework for Speculation on the Next Thirty-Three Years. The Macmillan Company, New York (1967)

12. Matschalow, Y., Severengiz, S.: Additional information - Scenario study GH2GH (2024). https://www.hochschule-bochum.de/fileadmin/public/Die-BO_Fachbereiche/fb_e/Institute_und_Labore/LabNachhaltigkeitTechnik/GH2GH/AI1.pdf

13. Kosow, H., Gaßner, R.: Methoden der Zukunfts-und Szenarioanalyse Überblick, Bewertung und Auswahlkriterien (2008)

# The Role of the Digital Economy and Sustainable Factors: Economic, Social, Environmental in Renewable Energy Consumption in Southeast Asia

Nguyen Thanh Cong[1,2(✉)], Ho Thanh Truc[1], Nguyen Minh Nhut[1],
and Tran Tuan Ngoc[1]

[1] The Faculty of Economics and Public Management, Ho Chi Minh City Open
University, Ho Chi Minh City, Vietnam
cong.ngt@ou.edu.vn, thanhcong0420@gmail.com
[2] The Faculty of Economics and Public Management, Ho Chi Minh City Open University,
35 - 37 Ho Hao Hon Street, District 1, Ho Chi Minh City, Vietnam

**Abstract.** This study analyzes the role of the digital economy and sustainable factors: economic, social, and environmental in renewable energy consumption in Southeast Asia during the 2004 –2021 period. The aim of this research is to emphasize the important role of its in development and transformation of clean, environmental friendly energy. The contributions of this study support agencies of each country, including Vietnam, in determining correct and reasonable policies to achieve the goal of sustainable economic development. Based on panel data of 11 countries in Southeast Asia and regression model analysis techniques OLS, REM, FEM, FGLS to examine the relationship of the digital economy and sustainable factors: economic (GDP per capita, foreign direct investment, international trade), social (annual population growth), environmental ($CO_2$ emissions) for renewable energy consumption. Research results show that there is a relationship between the factors of digital economy, international trade, $CO_2$ emissions and renewable energy consumption; However, these factors have an impact on reducing renewable energy consumption, while the remaining factors are not statistically significant.

**Keywords:** digital economy · renewable energy consumption · environmental · sustainable · population growth

## 1 Introduction

The trend of energy transition to clean energy is a race between countries towards sustainable development. In particular, the renewable energy market of countries is forecast to growth strongly in the future [1]. Renewable energy is energy from natural resources such as sunlight, wind, rain, ocean waves,... and they are considered a clean energy source that can reduce harmful impacts to the environment [2].

© The Author(s) 2025
H. Kohl et al. (Eds.): GCSM 2024, LNME, pp. 368–377, 2025.
https://doi.org/10.1007/978-3-031-93891-7_41

Along with that, issues related to sustainable development such as transition and renewable energy consumption (REC) are topics that attract the attention of many scientists. The application of the digital economy plays an important role in promoting the development of renewable energy: The digital economy provides for the construction of energy internet platforms, thereby promoting the development and use of renewable energy [3]; Or research on the topic of renewable energy focuses on the impact of economic factors (economic growth, foreign direct investment, international trade,...) [4, 5]; Or some studies also consider social factors (population, urban population) [6], because when population increases rapidly, emissions will increase, thereby increasing demand for renewable energy will increase [7]; And there are some studies that also consider additional environmental impacts through CO2 [8, 9].

Although the above factors have been considered and evaluated in previous studies, research results on the impact of these factors on REC demand continue to be updated and concerned. Especially in Southeast Asia, there are not many studies related to this issue that evaluate the simultaneous impact of the digital economy and consider on a large scale all three aspects of economics, society, and environment. Therefore, this study hopes to contribute to the existing research literature by combining two theories: Environmental Kuznets curve (EKC) and Pollution Haven Hypothesis (PHH), and using the econometric models to clarify the impact of these factors on REC in Southeast Asia countries.

## 2 Hypothesis and Research Models

### 2.1 Research Hypothesis

**Correlation Between Digital Economy and REC**
Many empirical studies in recent times have demonstrated that the digital economy has a significant impact on the energy market, making the energy market smarter and more efficient [10]. Theoretically, the digital economy plays an important role in promoting renewable energy development: The digital economy provides for the construction of energy internet platforms, thereby promoting development and use of renewable energy [3]; The digital economy plays an important role in supporting the construction of smart grids, helping to optimize the consumption and integration of renewable energy [11]. Finally, the digital economy serves as a catalyst for the acceleration of green finance, which can promote green finance development and provide financial support for investment in renewable energy [12]. By taking advantage of digital advances, the expansion of renewable energy sources can be accelerated.

$H_1$: *Digital economy has a positive impact on REC*

**The Relationship Between the Group of Economic Factors and REC**
*GDP:* Many previous studies around the world have found a cause and effect rela-tionship between economic growth (Gross Domestic Product - GDP) and REC. As evidence, [13] studied over 80 countries, arguing that REC has a two-way causal relationship to economic growth in both the short and long term. In emerging economies research related to renewable energy has also increased significantly [14]. All these studies demonstrate that REC plays an important role for economic growth: higher economic growth requires

more energy consumption and for more efficient use of energy need higher economic growth.

*$H_{2a}$: GDP has a positive impact on REC.*

*FDI:* Researching the relationship between Foreign Direct Investment (FDI) and REC is also an issue of interest to many scientists. However, the research results are inconsistent: Specifically, a study analyzing data from 19 G20 countries suggests that FDI has no direct relationship with REC [15]; In Bangladesh, the research results of [16], show that there is a causal relationship between FDI and REC in the period 1980–2015, by testing Johansen cointegration and Granger causality test; Similar research by [17] in Kazakhstan and Uzbekistan during the period 1992–2018 also showed similar results that FDI has a positive impact on REC. On the contrary, some studies suggest that FDI has a negative impact on REC: Evidence is in South Asian countries, research by [18] suggests that FDI has a negative impact on REC demand.

*$H_{2b}$: FDI has a positive impact on REC.*

*IT:* Research results of [19] show that the impact of international trade (IT) on REC is very important. Research by [19] shows that trade openness has a positive impact on REC in the Balkans from 1998 to 2011. In addition, [20] with research in China has reported identified that import and export trade are important factors that will affect energy consumption. Exports can lead to more renewable energy production because increased export volumes will boost demand for and REC.

*$H_{2c}$: IT has a positive impact on REC.*

**The Factor of Social Have a Positive Impact on REC**
*POP:* According to [21], the urban population (POP) in ASEAN in 2018 was 49.25% and is expected to increase to 63.7% by 2050; It recommended incorporating renewable energy policies into the development process such as promoting smart cities, solar infrastructure, and water consumption. Because this policy is consistent with one of the United Nations Sustainable Development Goals (SDGs), the SDGs emphasize energy access towards sustainable development. The study by [6] examines the interrelationship between energy security, electricity consumption, population, and economic growth of Nepal, a resource-rich developing country. The authors could not confirm a long - run relationship between these variables. However, their findings supported the positive impact of population growth on electricity consumption.

*$H_3$: POP has a positive impact on REC.*

**The Factor of Environmental Have a Positive Impact on REC**
*CO2:* Research on the relationship between CO2 emissions (CO2) and REC has also received a lot of attention in recent times. Because understanding the relationship between CO2 and REC not only contributes to solving the problem of climate change but also promotes growth in the renewable energy industry, reducing dependence on imports fossil energy [22, 23]. Research by [24] confirmed the negative relationship between CO2 and renewable energy for G7 countries. Meanwhile, [19] conducted studies in 64 countries from 1990 to 2011 using the GMM estimator, the authors showed that CO2 have a positive impact on REC in high-income countries.

*$H_4$: $CO_2$ has a positive impact on REC.*

## 2.2  Research Models

Combining the two theories EKC and PHH with previous experimental research, the research model of the study is proposed as follows:

$$REC_{it} = \beta_0 + \beta_1 ICT_{it} \beta_2 gGDP_{it} + \beta_3 FDI_{it} + \beta_4 IT_{it} + \beta_5 POP_{it} + \beta_6 CO2_{it} + e_{it} \quad (1)$$

In which, REC denotes for Renewable energy consumption; Digital economy as ICT; Economic growth as gGDP; Foreign Direct Investment as FDI; International trade as IT; Population as POP; $CO_2$ emission is represented as $CO_2$; $\beta$ represents the coefficient ($\beta_1$, $\beta_2$, $\beta_3$, $\beta_4$, $\beta_5$, and $\beta_6$ are the coefficients of ICT, gGDP, FDI, IT, POP, and CO2); i denotes the countries involved in estimation (1, 2,…, N); t shows the period of analysis (2004 - 2021), and e denotes the residual.

# 3  Methodology

## 3.1  Data Sources

The research data of this topic is based on available data, 11 Southeast Asian countries were selected for research from 2004 to 2021. Table 1 presents a description of the variables and data sources.

**Table 1.**  Measurement of the variables and data sources

| Symbol | Variables | Units | Sources |
|---|---|---|---|
| REC | Renewable energy consumption | Share of renewable energy in total final energy consumption (%) | EIA |
| ICT | Information Communication Technology (Digital economy) | Individuals using the Internet (% of population) | WDI |
| gGDP | Economic Growth | GDP growth rate per capita, (constant 2015 US$) | WDI |
| FDI | Foreign Direct Investment | Net capital inflow (% GDP) | WDI |
| IT | International trade | Trade (% GDP) | WDI |
| POP | Population | Population growth (annual %) | WDI |
| CO2 | $CO_2$ emissions | Per capita $CO_2$ emissions | OWID |

## 3.2  Analytical Techniques

To evaluate the impact of digital economy and sustainable factors: economic, social, environmental in REC, it is important to first check multicollinearity in the model through the Variance Inflation Factor (VIF). Next, this study uses methods to estimate regression models such as Ordinary Least Squares, Fixed Effects Model and Random Effects

Model [25, 26]. In particular, after the testing process to select the appropriate estimation model and defect tests related to the model, if defects such as heteroscedasticity, autocorrelation,... appear, the method Generalized Least Square (GLS) [27] is used. The GLS method will basically overcome the shortcomings of the model, thereby making the regression model estimation results more reliable.

## 4   Empirical Results

### 4.1   Descriptive Statistics

The descriptive statistics results of the variables are presented in Table 2. The average REC is approximately 33%, there are significant differences between countries. Consumption is lowest in Brunei and highest in Myanmar; ICT averages at 36%, this data also shows that the level of digital economy application in countries is relatively positive in the coming time; gGDP is reflected through GDP growth rate per capita, Singapore is the leading country in terms of per capita growth, the lowest is Cambodia and Timor-Leste; The average values of FDI and IT are 5% and 124%, respectively. What is worth mentioning here is that the IT variable in Singapore reached the highest level of approximately 437%, which explains why Singapore is the an economic powerhouse in the region and the world; Average POP is 1.33%; In particular, $CO_2$ emissions per capita are approximately 4.0 tons, of which Cambodia and Timor-Leste have the lowest emissions and Brunei has the highest. In general, between countries in the region there are significant differences in values for several variables.

**Table 2.** Descriptive statistics

| Variables | Obs | Mean | Std. Dev. | Min | Max |
|---|---|---|---|---|---|
| REC | 191 | 32.76479 | 25.21331 | .01 | 85.77 |
| ICT | 196 | 36.2598 | 28.83855 | .0243374 | 98.08 |
| gGDP | 198 | 5.092021 | 4.649236 | −12.01637 | 31.96192 |
| FDI | 198 | 5.221555 | 6.924839 | −32.95523 | 32.69116 |
| IT | 198 | 124.2919 | 87.54675 | 4.821075 | 437.3267 |
| POP | 198 | 1.330962 | .7507094 | −4.170336 | 5.321517 |
| CO2 | 187 | 3.964261 | 5.117996 | .150382 | 21.70581 |

### 4.2   Pairwise Correlation

Table 3 presents the results of the correlation matrix of the variables in the model. The results show that the correlation between independent variables is low, the absolute value of the correlation coefficient between pairs of correlations is less than 0.5. However, there are two notable cases: the correlation between FDI and IT (correlation coefficient is 0.6915); ICT and $CO_2$ (correlation coefficient is 0.5852); This shows that there is a potential linear relationship between these independent variables and that multicollinearity may occur in the model. Finally, when combined with the results of the VIF, all VIF coefficients are less than 5 (VIF < 5.0). Therefore, it can be concluded that the model does not have multicollinearity phenomenon.

**Table 3.** Correlation Matrix

|       | REC     | ICT     | gGDP    | FDI     | IT      | POP     | CO2     |
|-------|---------|---------|---------|---------|---------|---------|---------|
| REC   | 1.0000  |         |         |         |         |         |         |
| ICT   | −0.7457 | 1.0000  |         |         |         |         |         |
| gGDP  | 0.3217  | −0.3665 | 1.0000  |         |         |         |         |
| FDI   | −0.0955 | 0.3007  | −0.0979 | 1.0000  |         |         |         |
| IT    | −0.4413 | 0.4716  | 0.0742  | 0.6915  | 1.0000  |         |         |
| POP   | −0.1037 | −0.1478 | 0.0672  | −0.0670 | 0.1875  | 1.0000  |         |
| CO2   | −0.6684 | 0.5852  | −0.3293 | 0.1314  | 0.3020  | 0.0839  | 1.0000  |

### 4.3 Regression Analysis

Looking at the results in Table 4, we see that the variables ICT, IT and $CO_2$ are almost all statistically significant in the models. This means that we will reject the hypotheses and accept the alternative hypotheses at the 1% level of significance. Meanwhile, the remaining variables gGDP, FDI and POP are not statistically significant. Another meaningful coincidence is that the statistically significant variables all have a negative relationship with REC (rejecting hypotheses $H_1$, $H_{2c}$ and $H_4$).

The digital economy has a negative impact on REC. Under the condition that other factors remain unchanged, if the digital economy increases by 1%, REC will decrease from 0.252%–0.424%. This shows that the application and transfer of clean technology and renewable energy-intensive technology to countries is still limited. Regarding IT, the results show a negative impact at the 1% significance level, meaning that a 1% increase in IT reduces the demand for REC by about 0.04%–0.1%. Finally, in the environmental aspect, the study discovered the negative impact of $CO_2$ on REC. REC will decrease in the range of 0.637%–1.394% when $CO_2$ increase by 1 ton per capita. This shows that although it is expected that increasing emissions will contribute to increased awareness of environmental protection issues, society's awareness of sustainability and the goal of climate change mitigation is still limited.

**Table 4.** Regression results

|  | (OLS) REC | (REM) REC | (FEM) REC | (GLS) REC |
|---|---|---|---|---|
| ICT | −0.424*** | −0.278*** | −0.252*** | −0.309*** |
|  | [−8.21] | [−10.19] | [−9.76] | [−8.69] |
| gGDP | 0.657** | −0.00849 | −0.0676 | −0.0262 |
|  | [2.59] | [−0.07] | [−0.60] | [−0.34] |
| FDI | 1.274*** | 0.0519 | 0.0381 | 0.0189 |
|  | [5.90] | [0.40] | [0.31] | [0.24] |
| IT | −0.107*** | 0.00852 | 0.0385** | −0.0444*** |
|  | [−5.25] | [0.51] | [2.24] | [−4.34] |
| POP | −2.306 | −0.754 | −1.124 | 0.302 |
|  | [−1.55] | [−0.87] | [−1.38] | [0.54] |
| CO2 | −1.394*** | −1.027*** | −0.637** | −0.855*** |
|  | [−5.60] | [−4.12] | [−2.51] | [−4.86] |
| _cons | 58.38*** | 45.15*** | 39.91*** | 47.25*** |
|  | [19.11] | [11.43] | [14.10] | [22.98] |
| R-sq | 0.718 |  | 0.437 |  |

Note: ***, ** and * represent significance levels of 0.01, 0.05 and 0.1 respectively.
The standard errors are reported in parentheses.

## 5   Discussion and Conclusion

The research results provide evidence that there exists a relationship between the digital economy, international trade, CO2 emissions and REC while the remaining factors GDP, FDI and POP are not statistically significant. However, the effects of statistically significant variables on REC are all negative. This shows that the application of the digital economy and the transfer of clean technology using a lot of renewable energy to other countries are still limited; Furthermore, the study results also show that promoting international trade helps reduce REC. This result, contrary to expectations, may be a sign that TI in countries has not yet pro-gressed to REC; Each country's CO2 need to be given more attention to promote REC: This is explained by society's awareness of sustainability goals and climate change mitigation is still limited, which may not be enough to promote the transition from alternative energy sources traditional energy to renewable.

To achieve the goal of promoting sustainable REC to limit the impact of climate change and move towards sustainable development, countries need to take action and have appropriate policies to achieve these goals. Firstly, encourage and promote the introduction of the digital economy into production and application of the latest advanced technology in industries related to renewable energy conversion; There is a need for economic policies that focus on promoting international trade and eliminating trade barriers between countries in the region. Policymakers need to ensure that trade between trading partners includes the transition or use of advanced renewable energy technologies; Finally, the government and industry experts need to pay attention to environmental policies that need to be more strictly enforced such as increased environmental fees, carbon taxes, etc. Encourage investment in the transition from high carbon technology environments to low carbon technology environments to reduce emissions and indirectly increase demand for REC.

Although the research has made certain contributions, limitations still exist. First of all, the study's analysis can only be collected up to 2021 and there is missing data on some variables such as $CO_2$ and REC in some countries. Therefore, future research requires data to be more complete and closer to reality which able to analyze more comprehensively and expand consideration of emerging issues such as the Covid-19 pandemic, recession economics, climate change,... In addition, research can consider the impact of factors on REC in a more multidimensional way by analyzing in the short and long term.

# References

1. Secretariat, R.: REN21 launches "Renewables Global Futures Report: Great debates towards 100% renewable energy". Energy **3** (2017)
2. Shahbaz, M., Nasir, M.A., Hille, E., Mahalik, M.K.: UK's net-zero carbon emissions target: investigating the potential role of economic growth, financial development, and R&D expenditures based on historical data (1870–2017). Technol. Forecast. Soc. Chang. **161**, 120255 (2020)
3. Lin, B., Huang, C.: Promoting variable renewable energy integration: the moderating effect of digitalization. Appl. Energy **337**, 120891 (2023). https://doi.org/10.1016/j.apenergy.2023.120891
4. Doytch, N., Narayan, S.: Does FDI influence renewable energy consumption? An analysis of sectoral FDI impact on renewable and non-renewable industrial energy consumption. Energy Econ. **54**, 291–301 (2016)
5. Fan, W., Hao, Y.: An empirical research on the relationship amongst renewable energy consumption, economic growth and foreign direct investment in China. Renew. Energy **146**, 598–609 (2020)
6. Nepal, R., Paija, N.: Energy security, electricity, population and economic growth: the case of a developing South Asian resource-rich economy. Energy Policy **132**, 771–781 (2019). https://doi.org/10.1016/j.enpol.2019.05.054
7. Cong, N.T., Lam, T.H., An, N.T., Hanh, N.D.D., Hong, T.C.: Determinants of household carbon emissions in Ho Chi Minh City in Vietnam. In: The Proceeding of the International Conference on Socio-economic and Environmental Issues in Development (ICSEED 2024), pp. 1548–1561. Finance Publishing House, Vietnam (2024)
8. Bhattacharya, M., et al.: The dynamic impact of renewable energy and institutions on economic output and CO2 emissions across regions. Renew. Energy (2017)
9. Mengal, A., Mirjat, N.H., Walasai, G.D., Khatri, S.A., Harijan, K., Uqaili, M.A.: Modeling of future electricity generation and emissions assessment for Pakistan. Processes **7**(4), 212 (2019)
10. Xu, Q., Zhong, M., Li, X.: How does digitalization affect energy? International evidence. Energy Econ. **107**, 105879 (2022). https://doi.org/10.1016/j.eneco.2022.105879

11. Shahbaz, M., Wang, J., Dong, K., Zhao, J.: The impact of digital economy on energy transition across the globe: the mediating role of government governance. Renew. Sustain. Energy Rev. **166**, 112620 (2022)
12. Mazzucato, M., Semieniuk, G.: Financing renewable energy: who is financing what and why it matters. Technol. Forecast. Social Change **127**, 8–22 (2022). https://doi.org/10.1016/j.tec hfore.2017.05.021
13. Apergis, N., Payne, J.E.: Renewable and non-renewable electricity consumption–growth nexus: evidence from emerging market economies. Appl. Energy **88**(12), 5226–5230 (2011)
14. Rahman, M.M., Sultana, N.: Impacts of institutional quality, economic growth, and exports on renewable energy: emerging countries perspective. Renew. Energy **189**, 938–951 (2022)
15. Lee, J.W.: The contribution of foreign direct investment to clean energy use, carbon emissions and economic growth. Energy Policy **55**, 483–489 (2013)
16. Khandker, L.L., Amin, S.B., Khan, F.: Renewable energy consumption and foreign direct investment: reports from Bangladesh. J. Account. **8**, 72–87 (2018)
17. Grabara, J., Tleppayev, A., Dabylova, M., Mihardjo, L.W., Dacko-Pikiewicz, Z.: Empirical research on the relationship amongst renewable energy consumption, economic growth and foreign direct investment in Kazakhstan and Uzbekistan. Energies **14**(2), 332 (2021)
18. Kang, X., Khan, F.U., Ullah, R., Arif, M., Rehman, S.U., Ullah, F.: Does foreign direct investment influence renewable energy consumption? Empirical evidence from south Asian countries. Energies **14**(12), 3470 (2021)
19. Omri, A., Nguyen, D.K.: On the determinants of renewable energy consumption: international evidence. Energy **72**(3), 554–560 (2014)
20. Chen, Y.: Factors influencing renewable energy consumption in China: an empirical analysis based on provincial panel data. J. Clean. Prod. **174**, 605–615 (2018)
21. Worldometers. Bangladesh population (2018). Accessed 30 May 2023. http://www.worldo meters.info/world-population/bangladesh-population/
22. Lavranos, N.: Challenges for foreign direct investment in the solar energy sector. In CESifo Forum, vol. 16, no. 2, pp. 23–27. ifo Institut-Leibniz-Institut für Wirtschaftsforschung an der Universität München, München (2015)
23. Raza, S.A., Shah, N.: Testing environmental Kuznets curve hypothesis in G7 countries: the role of renewable energy consumption and trade. Environ. Sci. Pollut. Res. **25**, 26965–26977 (2018)
24. Cong, N.T., Nuong, N.T.X., Thinh, V.H.H.: Financial inclusion, REC, and CO2 emissions: empirical research in Asia and insights for Vietnam. In: The Proceeding of the International Conference on Socio-economic and Environmental Issues in Development (ICSEED 2024), pp. 1425–1437. Finance Publishing House, Vietnam (2024)
25. Gujarati, D.N., Porter, D.C.: Basic econometrics. McGraw-hill (2009)
26. Hausman, J.A.: Specification tests in econometrics. Econometrica **46**(6), 1251 (1978). https://doi.org/10.2307/1913827
27. Hsiao, C.: Panel data analysis - advantages and challenges. Test **16**(1), 1–22 (2007)

# Technical Processes

# Optimization of the $CO_2$ Supply of a Cryogenic Minimum Quantity Lubrication System for Machining Processes

Andreas Röckelein[✉], Trixi Meier, and Nico Hanenkamp

Institute for Resource and Energy Efficient Production Systems, University Erlangen-Nuremberg, Dr.-Mack-Str. 81, 90762 Fürth, Germany
andreas.roeckelein@fau.de

**Abstract.** The reduction of lubricants plays an important role in improving sustainability of machining processes. However, the heat development in dry machining can lead to poor surface qualities. Therefore, cryogenic cooling with liquid carbon dioxide ($LCO_2$) has been introduced to adopt the cooling function of conventional lubricants. The density of the $CO_2$ stream varies depending on a number of factors, such as change in ambient temperature, the length of the supply line and filling level of the $CO_2$ riser bottle. This affects the cooling capacity as it determines the amount of $CO_2$ that can expand at the nozzle outlet to cool the cutting operation. In addition, density variations cause turbulence in the jet formation at the nozzle exit. This prevents a meaningful combination of this cooling strategy with minimum quantity lubrication, as the oil droplets cannot reach the cutting zone uniformly. Furthermore, an unstable cooling process directly affects product quality. Thus, the stabilization of the $CO_2$ density is a key factor in improving process controllability. Possible optimizations of the supply line regarding the $CO_2$ density as well as means to adjust the $CO_2$ flow rate dynamically are discussed in this paper.

**Keywords:** cryogenic cooling · carbon dioxide · machining

## 1 Introduction and State of the Art

One of the most important improvements in manufacturing with respect to sustainability is the reduction of lubricants. The energy usage for the compression and circulation of lubricants can be reduced drastically through the adaptation of cooling and lubrication strategies. However, substitutions such as dry machining and minimum quantity lubrication (MQL) have a lower performance especially with regards to cooling capacity. Thus, the cooling capacity of conventional lubricants has to be supplied by new means, e. g. cryogenic cooling. The application of $LCO_2$ represents a suitable alternative, since it has a high cooling capacity that is released only at expansion. Hence, the transport and storage of $LCO_2$ is more practical in comparison to liquid nitrogen. Recent discoveries by Pusavec et al. [1] showed that cryogenic machining as an alternative process offers improved surface roughness of the work piece. Courbon et al. [2] stated that cryogenic

© The Author(s) 2025
H. Kohl et al. (Eds.): GCSM 2024, LNME, pp. 381–389, 2025.
https://doi.org/10.1007/978-3-031-93891-7_42

cooling alone cannot prevent the sticking of titanium alloys. While the tribological behavior is improved by combination with MQL, a sticking of the material is also observed for this strategy. Overall, research regarding the development of cryogenic MQL (cMQL) is increasing. The current system design trend is moving towards 1-channel supply systems, as the $CO_2$-lubricant mix can also be provided through cooling channels in the tool. The lubricant is transferred into the $LCO_2$ stream by high-pressure pumps [3, 4]. Gross et al. [4] applied such a system to consider the influence of various oil combinations on the machinability of Ti6Al4V. Therein, the combination of bio-based oils with the cooling effect of $LCO_2$ represents an alternative to conventional cooling strategies, e.g. flood cooling. Grguras et al. [5] and Meier et al. [8] investigated the spraying and mixing behavior of various oils with $LCO_2$. Those system improvements include nozzle position [6] and nozzle geometry [7]. The former can lead to an improvement in tool life of around 80%. Pereira et al. [7] investigated variable pipe diameters and their influence on the $CO_2$ speed at the nozzle exit. It is conducted from these publications that the supply system is a central element for an efficient cooling strategy with $CO_2$. Meier et al. [8] mentioned that the pre-cooling of the $CO_2$ leads to an increased $CO_2$ density which is further investigated in the following. The optimization described in this research is performed on the one-channel supply system described by Hanenkamp et al. [3].

## 2 Objective, Materials and Methods

The properties of $CO_2$, such as density, flow rate and temperature, vary highly. This complicates a precise prediction as to how the machining process is impacted by cMQL. These variations originate from a fraction of the gaseous phase, which is dependent on the filling level of the riser bottles included in the $CO_2$ pack. Due to this, a density fluctuation from $\rho = 500$–$750$ kg/m$^3$ can be observed. Hence, the objective of this investigation is to optimize the density of the $CO_2$-stream and with that the cooling performance during machining processes. In [7] it is discussed that the density of $LCO_2$ can be enhanced by cooling. Thus, the goal is to achieve a medium density of $\rho = 800$ kg/m$^3$ combined with a pulsation-free beam at the nozzle exit. Additionally, it is desirable to adjust the flow rate dynamically during the process instead of exchanging parts of the supply system. Therefore, investigations regarding the realization of such a concept are performed. In the following section, the experimental setup is described.

### 2.1 Experimental Setup

The cooling of $CO_2$ is performed by the cooling device (type CWP100) from Lindr with three cooling pipelines that acts as a flow cooler by heat exchange. For these experimental purposes, a stainless-steel pipe with an inner diameter of $d = 1.7$ mm is inserted into the cooling basin in spirals and is attached to the housing. Its total length amounts to $l_{total} = 3.20$ m with 10 cm on each end reaching out of the water. The length over which the $CO_2$ is cooled thereby results in $l_{cooling} = 3$ m. The water basin is cooled by a ventilation on the back of the cooling unit. The filling capacity is 8 l of water. By means of a propeller unit just below the water surface, a circulation is generated which causes a uniform cooling of the water. The temperature of the cooling water is controlled by the usage of a PT 100 A measuring resistor. The unit allows a mode selection in seven levels which reach from a cooling temperature of T = 11.5 °C to −0.9 °C. This enables a $CO_2$ temperature of T = 3–4 °C dependent on the flow rate. The cooling unit switches off when the lower control limit for the temperature is reached and is reactivated at the upper control limit. These limits are specified in Table 1.

**Table 1.** Control levels of the cooling unit

| Cooling mode | Lower Control Limit (°C) | Upper Control Limit (°C) |
| --- | --- | --- |
| 1 | 9.8 | 11.5 |
| 2 | 7.4 | 9.2 |
| 3 | 5.9 | 7.6 |
| 4 | 4.3 | 5.7 |
| 5 | 2.6 | 4.6 |
| 6 | 0.9 | 2.8 |
| 7 | −0.9 | 0.4 |

This optimization is planned to be used in the cMQL mixing unit described by Hanenkamp et al. in [3]. This unit allows a separate supply of $LCO_2$ and lubricants and a distribution to the process zone in one channel after mixing. However, the objectives of this investigation are the characterization of the properties of the $LCO_2$. So, the system is used for the transport of the $CO_2$ without mixing it with MQL oil. Therefore, no statement about the interaction of the $CO_2$ in the optimized system with lubricants can be made in this research.

The properties of the $CO_2$, namely density, temperature, and flow rate, are measured with three Coriolis sensors during the experiments. The usage of three sensors enables the observation of the course of the density or the temperature of the $CO_2$ throughout the whole pipe system. The first sensor shows the density shortly after the cooling process to visualize the density at the lowest temperature possible. Over the course of the pipes the other two sensors are placed to observe the trend of the density. The evaluation of the system is performed using the data at sensor 3, since this measuring point is closest to the process zone where the $CO_2$ will be applied, see Fig. 1. The sensors are flow rate measurement sensors of the type M14 by Bronkhorst and are calibrated to $LCO_2$. The generated data is collected via a raspberry Pi that allows a sampling rate of 1 Hz. The statistical analysis of the experiment data is performed using the Minitab software.

## 2.2 Initial State

As this research discusses an improvement to an existing system, the details of this system must be considered first. At the beginning of this investigation, the $LCO_2$ flows from the $CO_2$ pack directly to the nozzle without elements in between to alter the density. By the use of three Coriolis sensors, the density, temperature and flow rate is visualized over the course of the pipeline system which is shown in Fig. 1. These distances are based on existing tubes produced by the company Swagelok that are isolated for reduced heating of the $CO_2$ by the environment throughout the system. The inner diameter of each pipe is $d_{pipe}$ = 1.7 mm. For this preliminary experiment a nozzle with a diameter of $d_{nozzle}$ = 0.2 mm is used. The experiment was performed over the course of t = 5 min, which leads to about 250 data points.

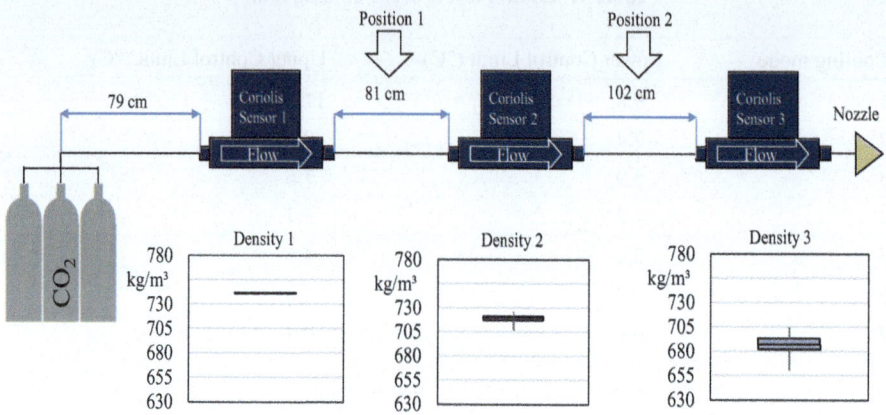

**Fig. 1.** Overview of the initial state and the experimental setup

## 2.3 Design of Experiment Based Optimization

The number of total experiments is decreased by applying the method of the design of experiment in which the investigated parameters are separated into a lower and a higher value. Applying statistical analysis via Minitab, more than one factor can be adjusted for each iteration. The order of experiments is randomized, so that the influence of the operator and the fill level of the $CO_2$ pack is reduced. The design of experiments is given in Table 2 below:

**Table 2.** Experimental plan

| Cooling mode | Nozzle diameter (mm) | Total pipe length (m) | Time (min) |
|---|---|---|---|
| 2 | 0.2 | 14.80 | 5 |
| 2 | 0.3 | 9.80 | 5 |

*(continued)*

**Table 2.** (*continued*)

| Cooling mode | Nozzle diameter (mm) | Total pipe length (m) | Time (min) |
| --- | --- | --- | --- |
| 5 | 0.3 | 9.80 | 2 |
| 5 | 0.2 | 14.80 | 2 |
| 2 | 0.3 | 14.80 | 2 |
| 2 | 0.2 | 9.80 | 2 |
| 5 | 0.2 | 9.80 | 5 |
| 5 | 0.3 | 14.80 | 5 |

### 2.4  Integration of a Dynamic Flow Rate Control

Conventionally, the flow rate in [3] is determined by the nozzle diameter, since it is the lowest inner diameter in the system. Thus, it acts as a bottleneck. In the following sections, possible means of integrating a dynamic flow rate control are discussed. The behavior of such a unit is simulated by a simple needle valve. This creates two alternatives for the simultaneous usage of a needle valve with a cooling unit: In option A, the needle valve is placed at Position 1 and the cooling unit at Position 2, while they are positioned vice versa in option B. The reduction of the inner diameter at the needle valve causes the CO$_2$ to expand thereafter inside the pipe system. This leads to premature cooling and a minimized cooling capacity in the machining process.

## 3   Results and Discussion

### 3.1  Influence of Pre-cooling of CO$_2$

As the investigation by [7] proves that an improvement of the CO$_2$ stream density can be achieved by cooling, it must still be considered to what extent the proposed cooling unit in this paper is capable. Similar to the analysis of the initial state, a short overview can be given by the use of boxplots in the following diagram, shown in Fig. 2.

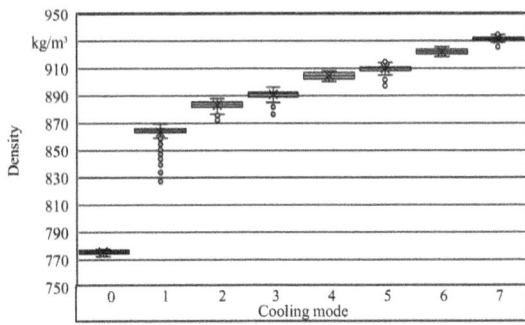

**Fig. 2.** Overview of the influence of the pre-cooling mode on the CO$_2$ density

This confirms the findings of other investigations and shows clearly that the density rises the lower the cooling temperature gets. Even cooling mode 1 which amounts to a water temperature of T $= 9.8$–$11.5$ °C depicts a distinct improvement of about $\rho = 100$ kg/m$^3$. As one can see by the position of the boxes, the whole data distribution rises for each cooling mode. However, no statement about the variation can be made, as the standard derivation does not show a clear trend. It is noticeable that the points below the boxes occur frequently. These represent extreme outliers. The explanation to this behavior lies within a ramp up phase (about t $= 60$ s) at the start of each experiment in which remaining $CO_2$ from the set up before leaves the pipe system. Another feature that cannot be recognized from the presentation of data is the quality of the $CO_2$ beam at the nozzle exit. It is observed that the beam starts to pulsate at very high densities. The flow rate and the nozzle diameter might also have an influence on that behavior. The pulsation is caused by the forming of dry ice at the outside of the nozzle, which restrains the beam from arriving at the process zone uniformly. This behavior complicates the process control. In this set of experiments with a nozzle diameter of d $= 0.2$ mm the forming of dry ice is noticeable at cooling modes above 4. Afterwards, the pulsation increases for lower pre-cooling temperatures.

## 3.2 Results of the Design of Experiment

The main objective of the design of experiment is to find the relevant factor influencing the behavior of the $CO_2$ stream density. Therefore, the factors flow rate (realized through different nozzle diameters), measuring time, total pipe length and cooling mode are analyzed. The relevance of these factors is evaluated by Pareto analysis of the median of the density using Minitab statistical software. The main contributions to the median density stem from the total pipe length and the nozzle diameter. The Pareto analysis assesses the influence of the cooling mode as low, since the cooling unit is turned on for both factor levels. Sect. 3.1 shows that a pre-cooling of the $CO_2$ is clearly beneficial compared to the uncooled state. The most promising combination of factors is derived from the Multi-Vari chart below in Fig. 3:

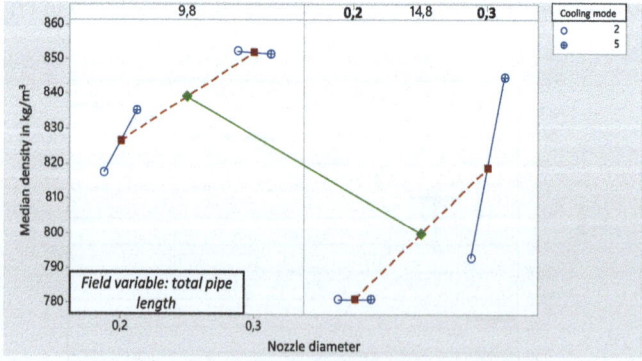

**Fig. 3.** Multi-Vari chart for median density

A local optimum is found at the combination of total pipe length l = 9.80 m and a nozzle diameter of d = 0.3 mm. If these two values are given, the difference between the two cooling modes does not seems to be significant in relation to the other combinations. However, this setting appears together with strong pulsation of the CO$_2$ beam due to the formation of dry ice, which does not guarantee a successful process control. The optimal configuration that meets both criteria is found at a nozzle diameter of d = 0.2 mm and a total pipe length of l$_{pipe}$ = 14.80 m. Further investigations regarding the effect of the cooling mode show optimal behavior at cooling mode 4. This leads to a median density of ρ = 804.0 kg/m$^3$ and a standard deviation of σ = 5.17. In the initial state, the median density is located at ρ = 683 kg/m$^3$. In comparison, the new supply system is considered a significant improvement with a confidence level of α = 0.05.

### 3.3  Results on the Realization of a Dynamic Flow Rate Control

The following measurements are performed with three Coriolis sensors. The first one is located directly after the CO$_2$ pack to monitor the initial state, while the second one is placed after the needle valve. The last sensor observes the CO$_2$ behavior after the cooling unit. As is shown in Fig. 4, the density drops at the valve and is increased afterwards. The experiment could not be fully executed because of the formation of dry ice, which lead to uncontrollable CO$_2$ emissions (cooling modes 3, 6 and 7). The importance of a stabilization of the CO$_2$ is highlighted by the variations of the golden bars. Nozzle diameter d = 0.2 mm achieves higher densities after cooling, since the flow rate is lower in comparison. Thus, the effect of the needle valve is reduced, as the nozzle remains the dominating bottleneck. The density reaches a comparable state to the initial values by pre-cooling and even surpasses them for high cooling modes. Even though the median density looks acceptable, this configuration is not recommended because it reinforces the formation of dry ice and turbulences. The standard deviation rises to above σ = 50. The alternative is to increase the density of the CO$_2$ stream before the expansion at the needle valve takes place.

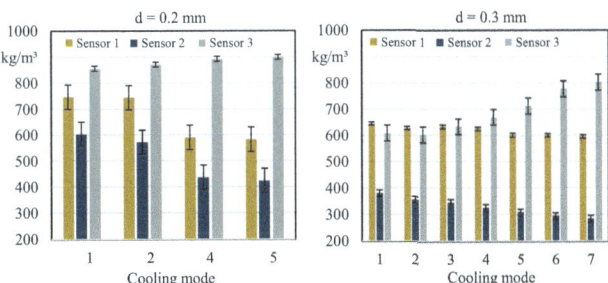

**Fig. 4.** Density medians for needle valve configuration A

In configuration B (Fig. 5), the densities at the initial state show a high variance with values from ρ = 724 kg/m$^3$ to ρ = 536 kg/m$^3$ and highlight the need for improvement. It is also clear that the initial density is increased through the application of the cooling unit and that higher cooling modes lead to higher densities. Furthermore, the theoretical concept is confirmed as the density decrease from sensor 2 to sensor 3 is considerably lower than at valve position one. It remains above the initial state. Additionally, a uniform beam is achieved for nozzle diameters of d = 0.2 mm and for a combination of cooling mode 1 and the larger nozzle diameter. In contrast to the set up before, the beam quality is not unusable for every factor configuration. When the pre-cooling is in place, the reduction of the density by the valve and therefore the loss in flow rate decreases decisively.

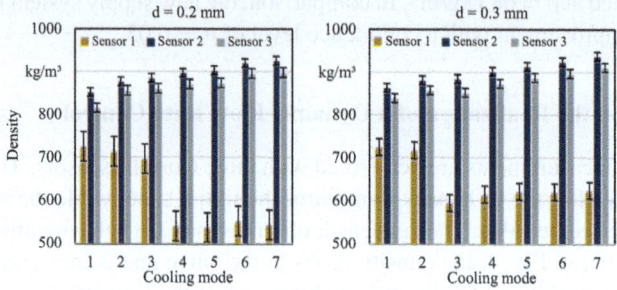

**Fig. 5.** Density medians for needle valve configuration B

# 4    Conclusion and Outlook

This research investigates the influences of supply line design on the density behavior of $LCO_2$. The following factors need to be considered, when installing a cryogenic cooling system in the cutting machine:

- In conventional cryogenic processes, the $CO_2$ density varies greatly depending on environmental factors, such as room temperature fluctuations, filling level or defects of the riser bottles.
- The density correlates to the pre-cooling temperature of the $CO_2$. A lower temperature relates to higher density levels. A higher density is considered beneficial for the cooling capacity of the process, since a higher amount of $CO_2$ expands at the nozzle.
- When a pre-cooling of the $CO_2$ is active, the main contributors to the median density are the nozzle diameter and the total pipe length. However, a higher median density can be unfavorable regarding the beam form and the building of dry ice.
- A dynamic flow rate control via a needle valve is problematic, as the $CO_2$ expands after the valve. This effect can be reduced by pre-cooling. However, the position of the pre-cooling unit and the needle valve play a vital role in the $LCO_2$ behavior.
- It is recommended to place the cooling unit in front of the valve, as fewer turbulences are observed in the beam formation. Additionally, the maximum density for high cooling modes is increased in this configuration.

# References

1. Pusavec, F., Hamdi, H., Kopac, J., Jawahir, I.S.: Surface integrity in cryogenic machining of nickel based alloy-Inconel 718. J. Mater. Process. Technol. **211**(4), 773–783 (2011). https://doi.org/10.1016/j.jmatprotec.2010.12.013
2. Courbon, C., Sterle, L., Cici, M., Pusavec, F.: Tribological effect of lubricated liquid carbon dioxide on TiAl6V4 and AiSI1045 under extreme conctact conditions. Procedia Manuf. **47**, 511–516 (2020). https://doi.org/10.1016/j.promfg.2020.04.139
3. Hanenkamp, N., Gross, D., Amon, S.: Hybrid supply system for conventional and CO2/MQL-based cryogenic cooling. Procedia CIRP **77**, 219–222 (2018). https://doi.org/10.1016/j.procir.2018.08.293
4. Gross, D., Blauhöfer, M., Hanenkamp, N.: Milling of Ti6Al4V with carbon dioxide as carrier medium for minimum quantity lubrication with different oils. Procedia Manuf. **43**, 439–446 (2020). https://doi.org/10.1016/j.promfg.2020.02.190
5. Grguras, D., Sterle, L., Krajnik, P., Pusavec, F.: A novel cryogenic machining concept based on a lubricated liquid carbon dioxide. Int. J. Mach. Tools Manuf **145**, 103456 (2019). https://doi.org/10.1016/j.ijmachtools.2019.103456
6. Bermingham, M.J., Palanisamy, S., Kent, D., Dargusch, M.S.: A comparison of cryogenic and high pressure emulsion cooling technologies on tool life and chip morphology in Ti–6Al–4V cutting. Journals of Materials Processing Technology **4**, 752–765 (2012). https://doi.org/10.1016/j.jmatprotec.2011.10.027
7. Pereira, O., Rodríguez, A., Barreiro, J., Fernández-Abia, A.I., López-de-Lacalle, L.N.: Nozzle design for combined use of MQL and cryogenic gas in machining. Int. J. Precis. Eng. Manuf.-Green Technol. **4**, 87–95 (2017). https://doi.org/10.1007/s40684-017-0012-3
8. Meier, T., Lermer, M., Gross, D., Hanenkamp, N.: Influence of carbon dioxide temperature on sprayability and solubility in cryogenic minimum quantity lubrication with bio-based lubricants. In: Kohl, H., Seliger, G., Dietrich, F. (eds.), Manufacturing Driving Circular Economy. GCSM 2022, pp. 202–210 (2023)

# A Study of a Metal Sintering Process Using Two-Way ANOVA and Bonferroni Correction

Anh-Duc Pham[1] (iD), Duy-Hoang Nguyen[2], and Trieu Khoa Nguyen[2](✉) (iD)

[1] Faculty of Mechanical Engineering, The University of Danang, University of Science and Technology, Danang, Vietnam
[2] Faculty of Mechanical Engineering, Industrial University of Ho Chi Minh City, Ho Chi Minh City, Vietnam
nguyenkhoatrieu@iuh.edu.vn

**Abstract.** This article presents a study of the metal sintering process using the two-way analysis of variance (ANOVA) method. Sintering technology, turning metal powder into products through shaping with molds and then sintering into shapes, is a technology that creates products differently than traditional machining methods. It offers advantages like the ability to create complex shapes with minimal waste, control metal density, and produce porous metal, making it increasingly popular. This study analyzed the two most important process parameters of the sintering process, including pressing density (A) and sintering mode (B). ANOVA results showed that both parameters were highly important and had no correlation between them. After that, the post-hoc tests (Bonferroni correction) also demonstrated that the hardnesses in pressing density A4 ($44.967 \pm 3.319$, $p < 0.0083$) were significantly higher compared with those in the three others. The post-hoc tests also revealed that the hardnesses in sintering mode B2 ($44.025 \pm 4.045$) were significantly higher compared with those in B1 ($39.150 \pm 2.998$, $p < 0.0167$) and B3 ($39.967 \pm 3.381$, $p < 0.0167$). The process is straightforward to implement and can be verified using Excel, making it highly suitable for sustainable industrial production environments.

**Keywords:** Sintering Technology · Metal Sintering · Pressing Density · Sintering Mode · ANOVA · Sustainable Production

## 1 Introduction

Metal sintering is a manufacturing technique employed to fabricate components from metal powders through heat application, causing particle fusion and the formation of the desired component [1, 2]. The resulting products exhibit increased density and strength, making them suitable for diverse applications across various industries to meet the difficult demands of sustainable production nowadays. Therefore, there have been many studies on metal sintering, especially the sintering mode. Yiwen Lei et al. [3] in their work researched sintering temperature and heat treatment in metal sintering. Abolfazl Malti et al. [4] studied the effect of sintering temperature. Milad Hojati et al. [5] researched the mechanical and physical properties of different alloyed sintered steels through temperature and internal atmosphere (Ar). Meanwhile, to optimize sintering conditions, the researchers studied more process factors. For example, S. Ahmad et al. [6] studied composition, sintering temperature, heating rate, and soaking time to optimize the sintering process of titanium foams using the Taguchi method. Of the three process factors, N.H. Mohamad Nor et al. [7] concluded that sintering temperature was the most influential variable contributing to the best final strength, followed by heating rate, dwelling time, and cooling temperature. Meanwhile, Mukesh Kumar et al. [8] studied abrasive concentration in ferromagnetic particles (AC) %, compacting pressure (CP) N/mm2, and sintering time (ST) min to optimize and predict the sintering process. Additionally, sintering temperature and dwell time were investigated for maximum nitrogen absorption, densification, and increased microhardness using response surface methodology (RSM) by Sadaqat Ali et al. [9]. More comprehensively, S.V. Zavadiuk et al. [10] studied four parameters including heating rate, sintering temperature, holding time, and subsequent heat treatment to optimize the process. In a follow-up study, Ahmad Gheysarian et al. [11] optimized the sintering temperature, cooling time, and grain size parameters to reduce residual stresses of copper-aluminum functionally graded material using response surface methodology. Similarly, Tahani A. Alrebdi et al. [12] optimized sintering time, sintering temperature, and compaction pressure using the Taguchi method. Based on the advantages of the Taguchi method [13], Navin Kumar et al. [14], in addition to CNT's wt.%, also studied sintering temperature and heating time.

From here, an important conclusion can be drawn that composition, pressing pressure, heating rate, sintering temperature, sintering time, and subsequent heat treatment are the most important process parameters of the sintering process that affect output product quality. However, several other important process parameters have not been studied such as air humidity, type, and ratio of protective gases during sintering, especially in the high-quality gear sintering process. Therefore, this study presents a work on the sintering process, focusing on the pressing mode and the sintering mode using ANOVA. This simple method, with verification from Excel software, can indicate the appropriate pressing mode and sintering mode and can be easily applied in our company's practice as well as in the manufacturing industry.

## 2 Materials and Methods

### 2.1 Sintered-Steel Gear

In the current procedure, once the powder alloy has been thoroughly mixed, it is compressed within a mold at a pressure of approximately 300 tons to form the initial gear blank. Subsequently, the gear blanks undergo an inspection to determine if they meet the criteria for sintering. This involves assessing their weight, volume, and pressing density. The pivotal stage involves placing the gear blanks into the sintering furnace, where specific conditions such as temperature, gas concentration, air humidity, sintering duration, and furnace capacity are meticulously defined. Following sintering, the resultant products are subjected to tests for surface finish, hardness, and dimensional accuracy. The gears are then sorted accordingly, with any substandard items being rejected and recycled promptly. Those meeting the necessary standards will continue to be subjected to surface hardening or quenching to increase hardness. After that, they are immersed in preservative oil before being dispatched to customers.

## 2.2  Metal Powder

The main raw material source of this technology is metal powder, which includes powders such as steel, nickel, copper, cobalt, aluminum, carbon,… Products such as inserts used for turning, and parts such as small gears in machines like in this case, are mostly mass-produced using metal powder. In this study, the material used is steel powder mixed with a small proportion of other metals and additives. The specific ingredients are not disclosed due to the technological secrecy of the host company.

## 2.3  Pressing Density

Pressing density represents the relationship between the mass m(g) and the volume V(ml) formed during and after the pressing process. A mass of m(g) of metal powder is put into the mold, then the mold is brought into the hydraulic press. The hydraulic press will press a force of about 200–300 tons into the mold cavity. At that time, the metal powder will be gradually heated up due to the compression pressure generating heat, from which the metal powder molecules gradually stick together to form a relatively durable body. Depending on the pressure, the level of bonding between metal powder molecules is large or small, and the amount of space between the molecular bond gaps of the product is more or less. From there, it will affect the product's volume V (ml).

## 2.4  Sintering Mode

Sintering mode is a set of factors such as temperature, time, humidity, protective gas, etc. For this study, the sintering mode was divided into three types, each with fixed parameters. For example, mode 1 will have a temperature of 1600°C, and 70% humidity, using 35% Argon + 65% Helium gas. Mode 2 and 3 parameters were different from mode 1.

## 2.5  Hardness Testing

When the product is released from the oven, it will be allowed to cool at room temperature and its hardness will be measured using a Matsuzawa Micro Vickers MMT-X Series hardness tester. The diamond indenter has a quadrangular pyramid in which the indenter approach's speed is 50μm/sec. The hardness in this study was unified using the HRC unit.

# 3  Results and Discussion

## 3.1  Hardness Results

Pressing density was denoted as A (g/ml) while sintering mode was denoted as B (−). Pressing density had 4 levels: 1.76, 1.8, 1.9, and 1.95 g/ml, denoted from A1 to A4 respectively. There were 3 types of sintering modes and were denoted from B1 to B3. The hardness results are presented in Table 1.

**Table 1.** The hardness results.

| Pressing density | Sintering mode | | | | | |
|---|---|---|---|---|---|---|
| | B1 | | B2 | | B3 | |
| | Value | Sum | Value | Sum | Value | Sum |
| A1 | 35.0 | 110.5 | 37.8 | 120.8 | 35.5 | 115.5 |
| | 36.9 | | 40.3 | | 37.1 | |
| | 38.6 | | 42.7 | | 38.9 | |
| A2 | 35.4 | 111.2 | 38.9 | 123.4 | 36.2 | 113.1 |
| | 37.1 | | 41.0 | | 37.6 | |
| | 38.7 | | 43.5 | | 39.3 | |
| A3 | 37.8 | 120.6 | 43.9 | 137.7 | 38.1 | 124.2 |
| | 40.1 | | 46.0 | | 41.3 | |
| | 42.7 | | 47.8 | | 44.8 | |
| A4 | 40.4 | 127.5 | 47.5 | 146.4 | 41.6 | 130.8 |
| | 42.5 | | 48.8 | | 43.6 | |
| | 44.6 | | 50.1 | | 45.6 | |

Figure 1 shows the sintered-gear before being subjected to hardness testing. The gear has an outside diameter of 30 mm, a thickness of 15 mm, and a total number of teeth of 9.

The visual comparisons in Fig. 2 show that level 4 of factor A, the pressing density, resulted in the highest average hardness. However, from the results of this analysis, it is still not possible to conclude whether this difference is statistically significant or not. The visual comparisons in Fig. 3 show that level 2 of sintering mode B resulted in the highest average hardness. However, as in the same for the pressing pressure A, from the results of this analysis, it is still not possible to conclude whether this difference is statistically significant or not. Therefore, further analyses were performed in the following sections.

## 3.2 ANOVA Results

The analysis of variance (ANOVA) [15, 16] was used to assess the pressing density and the sintering mode. Pressing density was denoted as A (g/ml) while sintering mode was denoted as B (−). Pressing density had 4 levels: 1.76, 1.8, 1.9, and 1.95 g/ml, denoted from A1 to A4 respectively. There were 3 types of sintering modes and were denoted from B1 to B3. The ANOVA results are presented in Table 2 with $\alpha = 0.05$. Furthermore, as can be seen in these results, the pressing density influenced the hardness result (FA = 21.47618 > F3; 24;0.95 = 3.008787), as the sintering mode influenced the hardness result (FB = 18.27144 > F2; 24; 0.95 = 3.402826). Whereas there was no interaction between A and B (FAB = 0.325537 < F6; 24; 0.95 = 2.508189). On the other hand, these results also can be verified by using Excel or other software solutions.

**Fig. 1.** The sintered-gear.

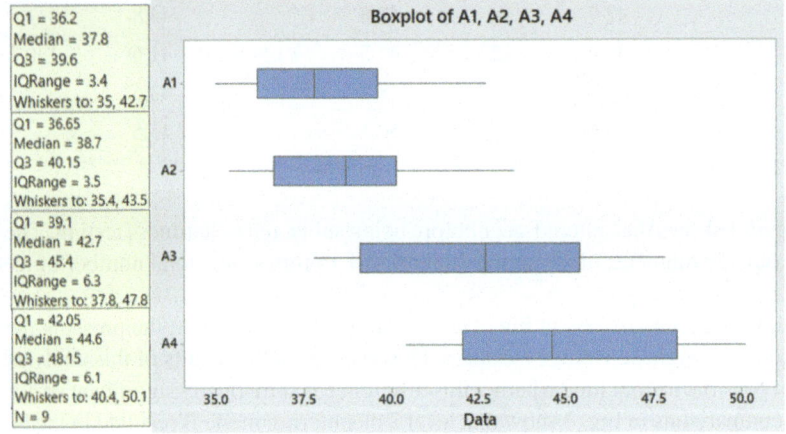

**Fig. 2.** The boxplot of the pressing density A.

### 3.3 Post-Hoc Tests Using Bonferroni Correction

When verifying with Excel, because ANOVA produced p-values that were less than the significance levels for both A and B ($5.572E-0.7$ and $1.506E-0.5$, respectively), post hoc tests were utilized to find out which group means differ from one another. Table 3 presents Bonferroni correction for both A and B.

Using Table 3, a post-hoc test was performed for compression density (A), presented in Table 4.

From Table 4, the post-hoc tests (Bonferroni correction) demonstrated that the hardnesses in pressing density A4 ($44.967 \pm 3.319$) were significantly higher compared with those in A1 ($38.086 \pm 2.392$, $p < 0.0083$) and A2 ($38.633 \pm 2.490$, $p < 0.0083$). Although the difference in mean hardness from A3 and A4 was not statistically significant, using transitivity nature as A3 was not different from A2 while A4 was different from A2 ($p < 0.0083$), the conclusion that the hardnesses in pressing density A4 ($44.967$

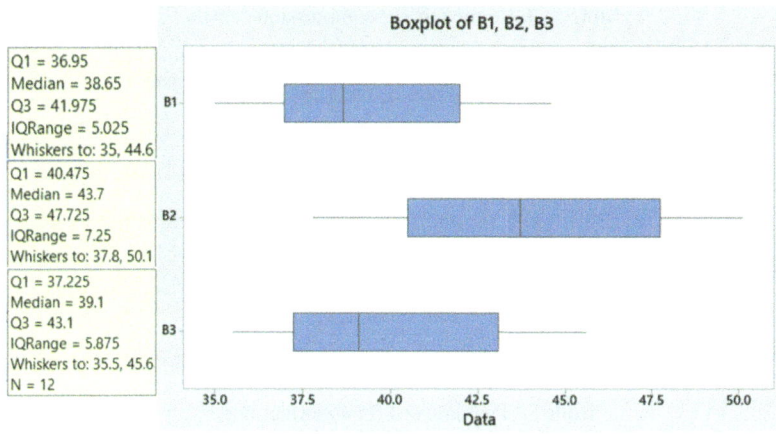

**Fig. 3.** The boxplot of the sintering mode B.

**Table 2.** ANOVA results.

| Source of Variation | SS | df | MS | F | F crit |
|---|---|---|---|---|---|
| Sample | 288.4608 | 3 | 96.1536 | 21.4762 | 3.0088 |
| Columns | 163.6106 | 2 | 81.8053 | 18.2714 | 3.4028 |
| Interaction | 8.745 | 6 | 1.4575 | 0.3255 | 2.5082 |
| Within | 107.4533 | 24 | 4.4772 | | |
| Total | 568.2697 | 35 | | | |

**Table 3.** Bonferroni correction.

| Test | Alpha |
|---|---|
| ANOVA | 0.05 |
| Post-hoc Tests (Bonferroni Corrected) A | 0.0083 |
| Post-hoc Tests (Bonferroni Corrected) B | 0.0167 |

$\pm$ 3.319) were significantly higher compared with those in three others could be given out.

Using Table 3, a post-hoc test was also performed for sintering mode (B), presented in Table 5.

From Table 5, the post-hoc test revealed that the hardnesses in sintering mode B2 (44.025 $\pm$ 4.045) were significantly higher compared with those in B1 (39.150 $\pm$ 2.998, p < 0.0167) and B3 (39.967 $\pm$ 3.381, p < 0.0167). Thus, pressing density A4 and sintering mode B2 are recommended for production to obtain the best possible hardness. From this research, the quality of sintered gear manufacturing has been raised to a new

**Table 4.** Post-hoc test for compression density.

| Groups | p-value (t-test) | Significant |
|--------|-----------------|-------------|
| A1 v A2 | 0.6426 | No |
| A1 v A3 | 0.0063 | Yes |
| A1 v A4 | 0.0001 | Yes |
| A2 v A3 | 0.0152 | No |
| A2 v A4 | 0.0003 | Yes |
| A3 v A4 | 0.1428 | No |

**Table 5.** Post-hoc test for sintering mode.

| Groups | p-value (t-test) | Significant |
|--------|-----------------|-------------|
| B1 v B2 | 0.0029 | Yes |
| B1 v B3 | 0.5377 | No |
| B2 v B3 | 0.0141 | Yes |

level in this particular company. In terms of production sustainability, sintered gears clearly outperform wrought machined gears [15]. Sintered gears align with global trends focused on enhancing production sustainability. This method can contribute to further popularizing this sustainable production process in Vietnam.

## 4 Conclusion

This study examined the two key process parameters in the sintering process: pressing density (A) and sintering mode (B). ANOVA results indicated the high significance of both parameters, with no observed correlation between them. Subsequent post-hoc tests (Bonferroni correction) confirmed that hardness values at pressing density A4 (44.967 $\pm$ 3.319) were significantly higher than those at the other three densities. Additionally, the post-hoc analysis revealed that hardness values under sintering mode B2 (44.025 $\pm$ 4.045) were significantly greater compared to B1 (39.150 $\pm$ 2.998, $p < 0.0167$) and B3 (39.967 $\pm$ 3.381, $p < 0.0167$). Therefore, it was recommended to use pressing density A4 and sintering mode B2 during production to achieve the highest hardness levels possible. By optimizing for pressing density A4 and sintering mode B2, manufacturers can achieve the highest possible hardness levels, which is crucial for the durability and longevity of metal gears. This not only enhances the product's performance but also contributes to sustainability by extending the gear's lifecycle and reducing the need for frequent replacements. Furthermore, the method is straightforward enough to be implemented using Excel, making it accessible for industrial production environments, thus promoting efficient and sustainable manufacturing practices.

# References

1. Fang, Z.Z.: Sintering of Advanced Materials. Woodhead Publishing (2010)
2. Kriba, A., Mechighel, F.: Three-dimensional numerical study of the behavior of thermoelectric and mechanical coupling during spark plasma sintering of a polycrystalline material. Arch. Mech. Eng. **70**(4), 497–529 (2023)
3. Lei, Y., Li, X., Sun, R., Tang, Y., Niu, W.: Effect of sintering temperature and heat treatment on microstructure and properties of nickel-based superalloy. J. Alloy. Compd. **818**, 152882 (2020)
4. Malti, A., Kardani, A., Montazeri, A.: An insight into the temperature-dependent sintering mechanisms of metal nanoparticles through MD-based microstructural analysis. Powder Technol. **386**, 30–39 (2021)
5. Hojati, M., Danninger, H., Gierl-Mayer, C.: Mechanical and physical properties of differently alloyed sintered steels as a function of the sintering temperature. Metals **12**(1), 13 (2022)
6. Ahmad, S., Muhamad, N., Sahari, J., Jamaludin, K.R.: Optimisation of sintering factors of titanium foams using Taguchi method. Int. J. Integrat. Eng. **2**(1), 1–6 (2010)
7. Nor, N.H.M., Muhamad, N., Ihsan, A.K.A.M., Jamaludin, K.R.: Sintering parameter optimization of Ti-6Al-4V metal injection molding for highest strength using palm stearin binder. Procedia Eng. **68**, 359–364 (2013)
8. Rai, M., Singh, S., Farwaha, H.: Optimization and prediction of sintering process parameters for magnetic abrasives preparation using response surface methodology. Int. J. Data Netw. Sci. **3**(2), 103–108 (2019)
9. Ali, S., et al.: Optimization of sintering parameters of 316L stainless steel for in-situ nitrogen absorption and surface nitriding using response surface methodology. Processes **8**(3), 297 (2020)
10. Zavadiuk, S.V., Loboda, P.I., Soloviova, T.O., Trosnikova, I.I., Karasevska, O.P.: Optimization of the sintering parameters for materials manufactured by powder injection molding. Powder Metall. Met. Ceram. **59**(1), 22–28 (2020)
11. Gheysarian, A., Honarpisheh, M.: Optimization of the sintering temperature, cooling time and grain size parameters to reduce residual stresses of copper-aluminum functionally graded material using response surface methodology. J. Strain Anal. Eng. Des. **58**(1), 26–37 (2021)
12. Alrebdi, T.A., et al.: Optimization on powder metallurgy process parameters on nano boron carbide and micron titanium carbide particles reinforced AA 4015 composites by Taguchi technique. J. Nanomater. **2022**, 3577793 (2022)
13. Mahmood, N.Y.: Prediction of the optimum tensile–shear strength through the experimental results of similar and dissimilar spot welding joint. Arch. Mech. Eng. **67**(2), 197–210 (2020)
14. Kumar, N., et al.: Optimization of sintering process parameters by taguchi method for developing Al-CNT-reinforced powder composites. Crystals **13**(9), 1352 (2023)
15. Nguyen, T.K., Hwang, C.J., Lee, B.K.: Numerical investigation of warpage in insert injection-molded lightweight hybrid products. Int. J. Precis. Eng. Manuf. **18**, 187–195 (2017)
16. Nguyen, T.K., Nguyen, V.-T.: Study of an electrospinning process using orthogonal array. Int. J. Precis. Eng. Manuf. **25**, 2153–2161 (2024)

# Control Loop Based Dimensional Error Compensation for Milling of Near-Net-Shaped, Thin-Walled Structures

Lasse Evers[1]([✉]) [iD], Matthias Müller[2] [iD], Carsten Möller[1] [iD], Alexander Brouschkin[1] [iD], and Jan H. Dege[1] [iD]

[1] Institute of Production Management and Technologie (IPMT), Hamburg University of Technologie (TUHH), Denickestr. 17, 21073 Hamburg, Germany
lasse.evers@tuhh.de
[2] FOOKE GmbH, Raiffeisenstraße 22, 46325 Borken, Germany

**Abstract.** Additive manufacturing has the potential to save resources in the production of lightweight aerospace structural components made of Ti-6Al-4V. Currently, these components are milled out of plate-material, resulting in up to 95% of the material being converted into chips that can only be downcycled. However, machining near-net-shape parts poses new challenges. For example, commonly used methods such as the "waterline"- path approach, which uses the residual stiffness of the plate-material to reduce the deflection of the thin-walls due to process forces, can no longer be applied. In this paper, a dimensional error compensation method is presented, that measures the deflection of the workpiece during helical end mill finishing using eddy current sensors. The sensor values are used within a control loop to adjust the toolpaths width of cut and inclination in real time to minimize dimensional error. Next to adjusting for different compliant states of the workpiece, this method adjusts for increasing tool wear states, that produce higher process forces and thereby larger dimensional errors. The presented compensation is compared to a conventional machining approach to demonstrate its capability to enable finishing of near-net-shape parts within tight tolerances while maintaining high material removal rates.

**Keywords:** milling · thin-wall machining · deflection · additive manufacturing · control loop · energy and resource efficiency · waste reduction

## 1 Introduction

Aerospace structural components made of Ti-6Al-4V are typically machined from plate-material, resulting in up to 95% of the material being converted into chips [1]. These chips, contaminated by cooling lubricants and coated with an oxide layer, are unsuitable for economic recycling and reuse in safety-critical structural components, thereby impeding a circular economy [2]. Additive manufacturing (AM), with its near-net-shape forming capability, presents significant potential for reducing the ecological footprint of this manufacturing process by minimizing the material that needs to be machined. However,

H. Kohl et al. (Eds.): GCSM 2024, LNME, pp. 399–407, 2025.
https://doi.org/10.1007/978-3-031-93891-7_44

components produced through AM fail to meet aerospace tolerances due to their inhomogeneous surfaces, necessitating processing through various methods, including helical milling [3]. Machining near-net-shaped, thin-walled parts introduces new challenges. Process forces during machining can deflect the components, leading to excess material and dimensional errors [4]. Traditional cutting strategies, such as the "waterline"-path approach, rely on the residual stiffness of the plate-material to mitigate deflection and are consequently no longer viable [5]. Therefore, innovative strategies are essential to achieve the necessary dimensional accuracy.

One common method to reduce process forces is to optimize engagement parameters [6, 7]. However, this method has limited effectiveness without compromising productivity, rendering the process less economical compared to conventional machining. In some cases, if components are too compliant, machining becomes impossible. Without new strategies, the significant $CO_2$ reduction potential of AM remains untapped. Current research is exploring the adaptation of tool paths to the flexibility of the components [6]. Based on expected dimensional errors, the tool is given an additional width of cut $a_e$ or inclination $\xi$ to compensate for the workpiece deformation. These additional engagements are determined in advance through simulations, analytical calculation or experiments [6, 8–11]. However, these approaches are time-consuming, require qualified personnel, and do not account for variable factors such as tool wear, making the true deformation prediction difficult [6].

This paper therefore presents an alternative method to achieve the required tool path. The deflection is measured using eddy current sensors, which are used within a control loop to adjust the tool path in real time in order to keep a constant sensor-workpiece distance and minimize the dimensional error.

## 2  Methods

To achieve an automatic process that continuously adapts to varying parameters such as workpiece stiffness and tool-condition-dependent process forces to avoid pre-calculation, the system must directly measure the current deflection. For this purpose, two Micro-Epsilon ES-S2 eddy current sensors connected to eddyNCDT 3060 measurement modules are moved along the thin-wall on the side opposite to the tool. Thereby, they record the workpiece deflection as closely as possible to the final contour-generating point. The sensor values $\Delta S1$ and $\Delta S2$ are fed into the 5-axis machining center and converted into a G-Code usable variable. This variable serves as a feedback path for the control loop, as illustrated in Fig. 1. The loop is implemented as a PI-controller via synchronous actions in Siemens Sinumerik 840D sl, operating at an IPO-cycle of $t_{IPO} = 2$ ms in real time. Desired sensor values $R_{\Delta S1}$ and $R_{\Delta S2}$, representing the target distance between the sensors and the thin-wall, must be input as a reference beforehand. A translational axis and a rotational axis of the machining center are the controlled system of the control loop - in this use case the translational Y-Axis and rotational A-axis. Using these actuators allows for compensation of both displacement and inclination to achieve the desired sensor distances $R_{\Delta S}$. Using this method, the tool engagement condition should be continuously maintained and thereby compensate for thin-wall deflection due to process forces.

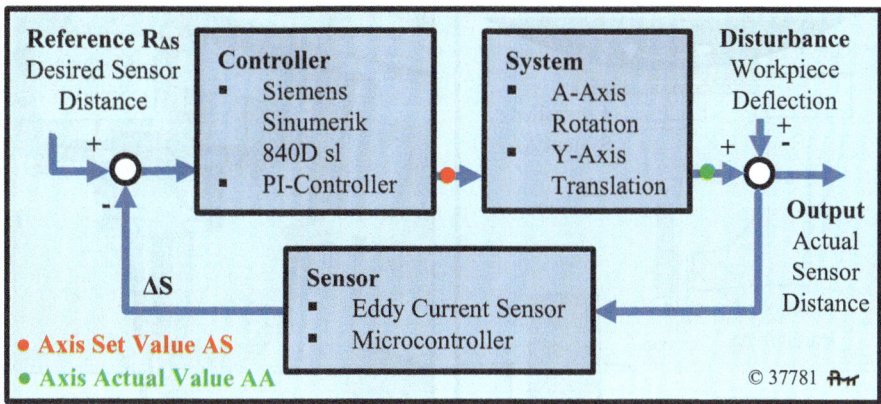

**Fig. 1.** Control Loop

**Sensor Data Preparation.** The process of peripheral milling involves a discontinuous cut, resulting in cyclic loading. Sensors therefor detect a deflection that fluctuates with the tooth engagement frequency. Since this fluctuation occurs too rapidly for the machine to compensate, it represents an unwanted disturbance in the signal and needs leveling. This is achieved by using an Arduino Due microcontroller positioned downstream of the eddy current sensor. The microcontroller calculates a moving average over the approximate tooth engagement wavelength $\lambda_z \approx 10$ ms before passing the signal to the PI-controller. By averaging the sensor data over this period, the rapid fluctuations are smoothed out, providing a stable signal for the feedback loop.

**PI-Controller.** The variable to be controlled is the distance $\Delta S$ measured by the eddy current sensors. Comprising the system to be controlled are the translational Y-axis and the rotational A-axis of the machining center, shown in Fig. 2. Before the machining trials, the sensors are positioned at an arbitrary distance within their measuring range $S_R = 2.0$ mm, and the necessary reference workpiece-sensor-distance $R_{\Delta S}$ to achieve the desired wall thickness is calculated. For example, with a desired wall thickness $t_d = 2.2$ mm, a tool radius $R_T = 8.0$ mm, and sensors-tool-rotation-axis-distance $S_d = 12.0$ mm, the desired sensor distance amounts to $R_{\Delta S} = 1.8$ mm.

**Y-Axis Control.** The control of additional width of cut $a_e$ via Y-axis feed is based on the deviation between the reference distance $R_{\Delta S2}$ and the actual distance $\Delta S_2$ of one sensor. An additional Y-axis set value $AS_Y$ required is calculated by the PI-controller, which adjusts the axis until the reference and actual value matches.

**A-Axis Control.** The control of additional inclination $\xi$ via A-axis feed is based on the difference between the actual distances $\Delta S_1$ and $\Delta S_2$ of the two sensors. Using the difference in distances and the vertical separation of the sensor positions $L_s = 14$ mm, as illustrated in Fig. 2, an angle $\alpha$ is calculated. The PI-controller calculates an additional A-axis set value $AS_A$ needed and adjusts the axis to bring $\alpha$ back to $0°$. Rotation is executed around the tool center point using the Siemens Sinumerik function TRAORI.

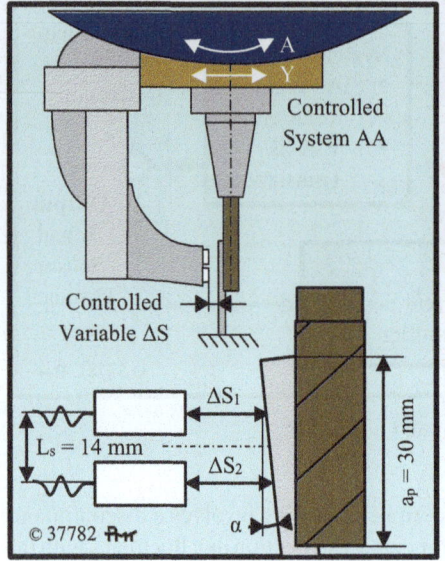 

**Fig. 2.** Test Setup

**PI-Gains.** The proportional and integral PI-controller gains $K_{PR}$ and $T_n$ were experimentally determined using the Ziegler-Nichols method. Following this, the integral time constant was manually adjusted to prevent excessive overshooting. This adjustment was necessary due to a high delay time $T_u = 70$ ms of the control loop, which resulted in poor controllability. Additionally, low maximum compensation speeds otherwise led to an integral windup effect.

**System Compensation Speeds.** The additional axis set values $AS_A$ and $AS_Y$ calculated by the PI-controller are transmitted to the machine. Due to the increased mechanical load on the tool by increasing feed and chip thickness, the speed at which the machine executes the set values gets limited. For achieving the desired additional width of cut $a_e$ via Y-axis, the maximum feed rate has been limited to $v_f = 20$ mm/min. Similarly, the maximum feed rate to increase the inclination $\xi$ via A-axis feed has been limited to $v_A = 38°/$min. Consequently, the axis actual additional values $AA_A$ and $AA_Y$ lag behind the axis set values $AS_A$ and $AS_Y$ of the PI-controller. Maintaining these controlled compensation speeds helps prevent undue stress and wear on the tool.

## 3   Test Setup

Experiments were conducted on a FOOKE ENDURA 711LINEAR 5-axis machining center. The workpiece was a cantilevered wall made of Ti-6Al-4V Grade 5, featuring a thickness of approximately $t \approx 2.6$ mm, a length of $l = 125$ mm, and a height of $h = 60$ mm. Using a solid carbide end mill with a diameter $D = 16$ mm, the workpiece was machined to a target width of $t_t = 2.2$ mm with two axial depths of cut $a_p = 30$ mm, following the process parameters depicted in Fig. 3. Six experimental series were conducted, as listed in Table 1. Next to comparing the presented control loop system with a conventional machining approach, the method gets compared using a worn tool with width of wear mark VB = 70 μm. To further demonstrate the current challenges with conventional strategies for machining compliant parts, two additional experimental series were conducted. After the first machining pass, an "air cut" was performed on the same tool path without further radial width of cut $a_e$ to remove any excess material left behind. This strategy was executed with both double and triple passes.

**Table 1.** Test Series

| Tool Condition | Milling Method | Repetitions |
|---|---|---|
| New | With Controll Loop | 3 |
| New | Conventional | 1 |
| Worn | With Controll Loop | 1 |
| Worn | Conventional | 1 |
| New | Conventional & 1x "air cut" | 1 |
| New | Conventional & 2x "air cut" | 1 |

During the machining with the control loop, the feedback system was activated after machining length $L_M = 10$ mm and deactivated after machining length $L_M = 115$ mm. The PI-controller axis set values $AS_A$ and $AS_Y$, as well as the actual axis values $AA_A$ and $AA_Y$ of the system, were recorded using the Siemens Sinumerik Trace function. Subsequently, the resulting surface profile was measured using a MarSurf XR 20 / GD 120 tactile measure instrument, and the maxima of the wall thicknesses were measured using an outside micro-meter.

## 4   Results

Figure 3 illustrates the control log, showing that the red PI-controller axis set values $AS_A$ and $AS_Y$ increase approximately 1.5 s after machining starts when the system is activated and decrease to zero after approximately 18 s when the system is deactivated. The blue and green actual axis values $AA_A$ and $AA_Y$ closely match the set values, showing good PI-controller setup.

At both the beginning and end of the machining process, the compensation values AA are higher due to the increased wall compliance at these points, with maximum inclination $\xi = 1.2°$ at $P1_A$ and additional width of cut $a_e = 0.4$ mm at $P1_Y$. In contrast, the stiffer middle section requires lower compensation values, with inclination $\xi = 0.9°$ at $P2_A$ and width of cut $a_e = 0.3$ mm at $P2_Y$. Additionally, during the machining of the lower of the two axial depths of cut $a_p$, even smaller compensation values are needed with inclination $\xi = 0.2°$ at $P3_A$ and width of cut $a_e = 0.05$ mm at $P3_Y$. The system's adaptation to tool wear is evident when comparing the compensation values used in the worn tool experiment. At $P4_A$, the inclination $\xi = 2.4°$ and at $P4_Y$ the width of cut $a_e$

= 0.8 mm show significantly higher compensation compared to the values with the new tool. This demonstrates the system's ability to adjust to changes in tool condition.

## 4.1 Dimensional Accuracy

Figure 3 A-A illustrates that with the conventional method, the dimensional error increases with the wall height due to workpiece deflection. The maximum dimensional error at the first pass amounts to $\Delta D = 330$ μm at $P5_{C.1}$. By applying a second and third "air cut" pass, this error can be reduced to $\Delta D = 270$ μm at $P5_{C.3}$. With the control loop activated, the maximum dimensional error is significantly reduced by 70% to $\Delta D = 80$ μm at $P5_{Loop}$. The maximum standard deviation between the 3 repetitions occurs at a wall height of $h = 30$ mm and amounts to $SD = 28$ μm.

Figure 3 B-B shows the dimensional error at machining length LM = 110 mm. The maximum dimensional error without the control loop is $\Delta D = 360$ μm at $P6_{C.1}$. A second and third pass are unable to significantly reduce this dimensional error further. With the control loop in place, the maximum dimensional error is reduced by 64% to $\Delta D = 130$ μm at $P6_{Loop}$. The maximum standard deviation between the 3 repetitions occurs again at a wall height of $h = 30$ mm and amounts to $SD = 35$ μm.

Figure 3 C-C displays the dimensional error in the middle of the wall when using a worn tool. Machining without the control loop shows that the wall was not machined beyond a height of $h = 50$ mm due to severe deflection, leaving the wall at its original stock dimension. With the assistance of the control loop, machining continued effectively, and the maximum dimensional error occurred at the lowest point of the wall, with $\Delta D = 180$ μm at $P7_{Loop}$, likely due to tool deflection.

## 4.2 Energy Savings

The findings indicate that machining compliant near-net-shaped thin-walled parts using conventional strategies is very time-consuming or nearly impossible. Even with two "air cuts" to remove excess material left behind after the first tool path, machining at very compliant positions (Fig. 3 B-B) left behind more than 85% of the wall thickness to be machined. This produces excessive machining time, thereby increasing energy, tool and resource consumption, which renders the AM structure unattractive for industrial usage and prevents the potential to reduce the to be machined and downcycled material by up to 95%. The proposed method demonstrates the ability to machine the compliant structures while maintaining high productivity, 66% less machining time and achieving up to 70% less dimensional error compared with the conventional approach. This makes AM structures attractive for industrial use and unlocks the full potential of material and energy savings in AM of thin-walled components.

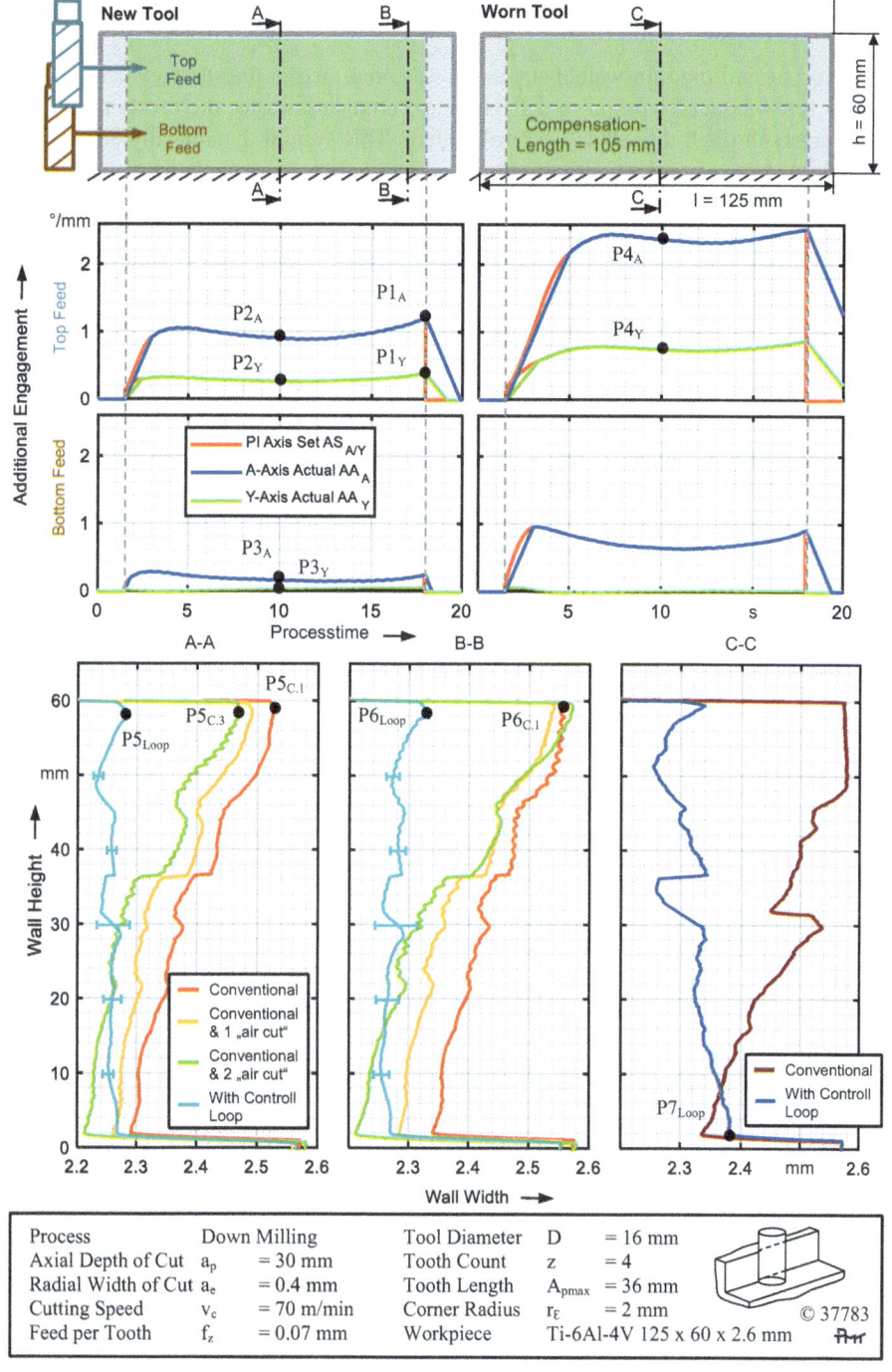

**Fig. 3.** Trace of the additional system feed & Profile of the dimensional error at various thin wall locations after machining

# 5 Conclusion and Outlook

A method for milling thin-walled structures was presented in the study, which records process force-induced workpiece deflection and compensates for the resulting dimensional errors through a feedback control system. This system dynamically adjusts the toolpath in terms of radial width of cut $a_e$ and inclination $\xi$ to counteract the typically arising dimensional deviations. Consequently, the system automatically adapts to the compliance of the workpiece and tool wear condition. Comparisons with conventional machining strategies demonstrate that the control loop reduces the maximum dimensional error by up to 70% while also reducing the machining time and associated energy and resource consumption by more than 66%. This advancement enables the effective machining of highly compliant parts, rendering AM for thin-walled parts more attractive for industrial usage. As a result, the full potential to reduce the to be machined and downcycled material of AM by up to 95% for thin-walled parts can be realized.

Future research should focus on the influence of different sensor positions, sensor data preparation methods, compensation speeds, PI-parameters and tools on compensation quality. Additionally, incorporating tool deflection into the compensation strategy offers significant potential for further dimensional error reduction [12].

**Acknowledgments.** The authors gratefully acknowledge the financial support of the Federal Ministry for Economic Affairs and Climate Action, Ministerium für Wirtschaft und Klimaschutz (BMWK) within the research project AMAvia, PNO 20W1902F.

# References

1. Herranz, S., et al.: The milling of airframe components with low rigidity: a general approach to avoid static and dynamic problems. Proc. Inst. Mech. Engineers Part B: J. Eng. Manuf. **219**(11), 789–801 (2005)
2. McDonald, D.T., Luo, P., Palanisamy, S., Dargusch, M.S., Xia, K.: Ti-6Al-4V recycled from machining chips by equal channel angular pressing. Key Eng. Mater. **520**, 295–300 (2012)
3. Alexander, I., Vladimir, G., Petr, P., Mihail, K., Yuriy, I.: Machining of thin-walled parts produced by additive manufacturing technologies. Procedia CIRP **41**, 1023–1026 (2016)
4. Budak, E., Altintas, Y.: Modeling and avoidance of static form errors in peripheral milling of plates. Int. J. Mach. Tools Manuf **35**(3), 459–476 (1995)
5. Scippa, A., Grossi, N., Campatelli, G.: FEM based cutting velocity selection for thin walled part machining. Procedia Cirp **14**, 287–292 (2014)
6. Del Sol, I., Rivero, A., López de Lacalle, L.N., Gamez, A.J.: Thin-wall machining of light alloys: A review of models and industrial approaches. Materials **12**(12), 2012 (2019)
7. Lassila, A.A., Svensson, D., Wang, W., Andersson, T.: Numerical evaluation of cutting strategies for thin-walled parts. Sci. Rep. **14**(1), 1459 (2024)
8. Wimmer, S.S.: Prognose und Kompensation von Formabweichungen bei der Fräsbearbeitung dünnwandiger Strukturen. utzverlag GmbH, München (2020)
9. Ratchev, S., Liu, S., Becker, A.A.: Error compensation strategy in milling flexible thin-wall parts. J. Mater. Process. Technol. **162**, 673–681 (2005)
10. Huang, N., Yin, C., Liang, L., Hu, J., Wu, S.: Error compensation for machining of large thin-walled part with sculptured surface based on on-machine measurement. Int. J. Adv. Manuf. Technol. **96**, 4345–4352 (2018)

11. Liu, H., et al.: State-space theory–based closed-loop control of machining error of thin-walled part modeling and application. Int. J. Adv. Manuf. Technol. **127**, 1721–1735 (2023)
12. Ma, W., et al.: Multi-stage error compensation with closed-loop quality control in five-axis flank milling of sculptured surface. Int. J. Adv. Manuf. Technol. **133**, 2891–2906 (2024)

# An Innovative Digital Liquid Metal Manufacturing Method for Aerospace Applications: Incorporating Life Cycle Assessment for Sustainability

Georgios Karadimas[✉], Emanuele Pagone, and Konstantinos Salonitis

Sustainable Manufacturing Systems Centre, School of Aerospace, Transport and Manufacturing, Cranfield University, Bedfordshire MK43 0AL, UK
george.karadimas@cranfield.ac.uk

**Abstract.** The Ultra Clean Cast (UCC) system presents an innovative approach to aerospace manufacturing by prioritizing component quality and manufacturing repeatability. It incorporates a cradle-to-gate life cycle assessment to highlight its additional environmental benefits, with a greater focus on enhancing sustainability. This novel approach improves upon traditional shape-casting by maintaining the high cleanliness of melt metal, critical for aluminum alloys, and difficult to achieve in general for aerospace parts, under varied conditions. By providing a sustainable, cost-efficient route for fabricating complex components, UCC is adaptable across aerospace platforms and evaluates the use of recycled aluminum, supporting the sector's shift towards a circular economy. This paper outlines the UCC system's integration of technological advancements with environmental responsibility, incorporating recycled aluminium raw material, material manufacturing, and product manufacturing stages. This system provides a new benchmark for environmentally friendly aircraft manufacturing by outlining process improvements and their implications for industry sustainability and efficiency. The findings highlight UCCs potential to affect aerospace manufacturing in the future, combining high-quality output with environmental considerations.

**Keywords:** Life Cycle Assessment · Sustainability · Ultra Clean Casting · Recycled Aluminum

## 1 Introduction

The manufacturing sector is making significant steps toward achieving net-zero emissions, with the aerospace industry leading these efforts. A major challenge in aerospace manufacturing is reducing carbon emissions. Aluminium, known for its high strength-to-weight ratio, is in high demand for various applications. However, despite advancements in manufacturing methods, aluminium casting is underused in aerospace due to strict standards and unpredictable quality [1, 2].

© The Author(s) 2025
H. Kohl et al. (Eds.): GCSM 2024, LNME, pp. 408–416, 2025.
https://doi.org/10.1007/978-3-031-93891-7_45

In aluminium casting, ingots are melted and poured into molds. Methods like die casting and investment casting are commonly used in aerospace. However, during melting, aluminium reacts with oxygen to form oxide films on the surface, which can mix with the molten metal and cause defects. These defects, such as porosity and leakage, weaken the cast components and can lead to cracks [3]. Porosity can reduce the fatigue life of aluminium components by 50% with just a 1% volumetric inclusion [4]. Despite regular cleaning practices, the quality of cast aluminium remains unpredictable, with few studies focusing on cleaning methods before casting [5]. This study introduces an Ultra Clean Casting (UCC) method for A356 aluminium, crucial for aerospace applications, supported by process maps and Life Cycle Assessment (LCA).

Furnaces are essential for the melting and holding stages of casting, but their design has changed little over the past decades [6]. Maintaining aluminium in a molten state is energy intensive. A study of 209 furnaces showed an average utilization rate below 39%, with 22% operating under 25% utilization, highlighting the need for more efficient use [7]. Different types of furnaces are used in the metallurgical industry, each suited to specific needs. Despite efficiency improvements, with some claiming up to 98%, sustainable practices in casting are still critical [8].

Aluminium is the most used non-ferrous material in the industry, with demand increasing by 3.9% from 104.1 million tons to 108.2 million tons [9]. There has been a slight decrease in greenhouse gas emissions from 1.13 GT $CO_2$e to 1.11 GT $CO_2$e and a 4.4% reduction in energy intensity of primary aluminium production [10]. Primary aluminium production consumes 168,000 MJ/t, while secondary aluminium uses only 11,200 MJ/t [11]. This significant difference highlights the importance of using recycled aluminium and promoting metal recycling. However, quality issues with recycled aluminium are still challenging for the casting industry [12]. This study aims to address this by exploring the use of 80% recycled aluminium in high-value casting applications.

Research to date has not identified a process capable of utilizing recycled aluminium to produce high-quality products without generating high emissions or being energy intensive. This study aims to address this gap by investigating a newly developed method. The research will consider scenarios using 100% virgin aluminium as well as combinations of recycled and virgin materials. A cradle-to-gate assessment will be conducted to evaluate the environmental and energy impacts of these scenarios.

## 2 Methods

This study's methodology involved a structured research approach to evaluate the environmental impacts of manufacturing processes, specifically focusing on the Ultra Clean Cast (UCC) system. The process included detailed process mapping, data collection, Life Cycle Assessment (LCA), and benchmarking.

The research focuses on detailed process maps to represent various manufacturing stages, essential for understanding inputs and outputs. Data was obtained from literature reviews, material databases, and company websites.

Life Cycle Assessment (LCA) was used to evaluate environmental impacts, following the international standard [13], and consisting of four phases. The phases included the goal and scope, the lifecycle inventory, the impact assessment, and the results analyses.

The study used a cradle-to-gate approach, covering environmental impacts from raw material extraction to the point where the product is ready for use or transportation out of the manufacturing plant. This assessment was conducted for both compared processes. The key environmental inputs measured include input and output materials, energy consumption, and waste [14]. The final stage involved benchmarking against conventional manufacturing methods, analyzing the effects on environmental outcomes. Changes in using different percentages of recycled and virgin materials for the novel UCC method provided insights into the influence of energy and waste variables, leading to recommendations for enhancing sustainability and promoting environmentally friendly practices [15].

## 3   Aluminium Processing Methods

### 3.1   Process Maps

The investment casting process, as seen in Fig. 1 starts with machining raw materials to create dies, producing swarf. Molten wax is cooled into wax patterns, which are coated with ceramic slurry to form molds. These molds are de-waxed, reclaiming wax and leaving empty ceramic molds. Aluminum alloy (virgin) is melted and poured into the molds. The castings are then cooled using electrical energy and a cooling medium. The final step is HIPing, where castings are treated with heat and pressure in an argon atmosphere, resulting in a high-quality final product.

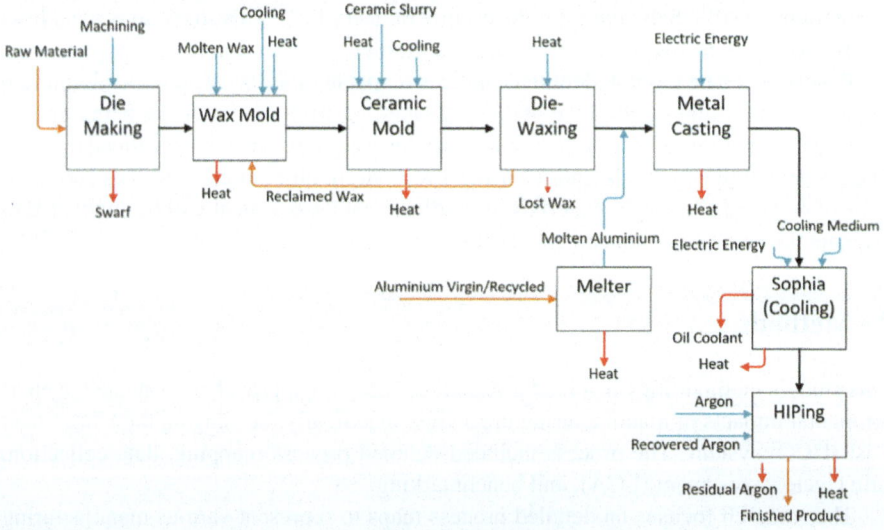

**Fig. 1.**  Investment casting process map.

In the UCC process, Fig. 2, aluminum alloy (virgin/recycled) is melted in a dry hearth melter using natural gas, producing dross and $CO_2$ waste. The molten aluminum is cleaned in an oxide sediment launder, removing impurities, and generating additional waste. The ultra-clean alloy is held in a dispensing furnace at 700 °C, creating more dross and $CO_2$ waste. Finally, the alloy undergoes counter gravity casting under vacuum conditions, producing the final product with minimal defects and some $CO_2$ waste.

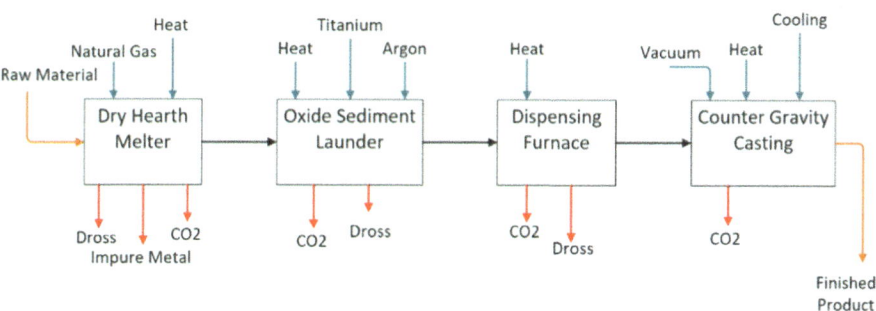

**Fig. 2.** Ultra Clean Cast process map.

Key differences between traditional methods and the UCC process include mold preparation, impurity removal, and final treatment. While investment casting relies on ceramic molds and HIPing for quality, UCC emphasizes ultra-clean alloy production and vacuum casting for defect minimization. In addition, this process efficiently utilizes recycled aluminum, achieving high-quality results through advanced impurity removal and controlled melting. Furthermore, this method reduces emissions and energy use, making it a sustainable option for producing high-quality components from recycled materials.

### 3.2 Inventory Analysis

In the lifecycle assessment of aluminium (A365) processes, it is crucial to identify and quantify the inputs and outputs at all manufacturing stages. This assessment involves not only direct measurements but also estimates derived from standard characterisation factors. These factors include the Material Utilization Factor, Energy Intensity, and $CO_2$ footprint, which are sourced from databases such as Granta EduPack 2024, Ecoinvent, and existing literature.

The functional unit the inventory calculations where made was for 1kg of A365 aluminium alloy.

To accurately calculate the environmental impacts associated with each stage of the aluminium casting processes depicted in the process maps, the following equations were employed:

$$Input\ Material\ Mass = Mass\ Produced \times Material\ Utilization\ Factor \qquad (1)$$

$$Consumed = Mass\ Produced \times Energy\ Intensity\ Factor \qquad (2)$$

$$Equivalent\ CO_2\ Emission = Mass\ Produced \times CO_2\ Intensity\ Factor \qquad (3)$$

The input and output data for both processes are showcased in Tables 1 and 2 below.

**Table 1.** UCC DLMM inventory.

| Process | Input Material (kg) | Input Energy (MJ) | CO2 Emissions (kg) | References |
|---|---|---|---|---|
| A356 Raw material | 1.2 | 119.5 | 8.6 kg | SimaPro9 |
| Dry Hearth | 1.2 | 3.5 | 0.4 | Dynamo Furnaces |
| Oxide Sediment Launder | 1.17 | 0.05 | 0.006 | Kanthal Furnaces |
| Dispensing Furnace | 1.16 | 0.05 | 0.006 | Kanthal Furnaces |
| Casting | 1.15 | Negligible for 1kg | Negligible for 1kg | EduPack |

**Table 2.** Investment casting inventory.

| Process | Input Material (kg) | Input Energy (MJ) | $CO_2$ Emissions (kg) | References |
|---|---|---|---|---|
| A356 Raw Material | 1.2 | 119.5 | 8.6 | SimaPro9 |
| Die Making | 1.2 | 1 | 0.05 | EduPack |
| Wax Mold Creation | 1.2 | 1 | 0.05 | EduPack |
| Ceramic Mold Creation | 1.2 | 1.5 | 0.1 | EduPack |
| De-Waxing | 1.19 | 0.5 | 0.03 | EduPack |
| Melting | 1.18 | 2 | 0.1 | EduPack |
| Pouring and Solidification | 1.17 | 0.5 | 0.03 | EduPack |
| SOPHIA Cooling | 1.17 | 0.5 | 0.05 | Signicast/ZOLLERN |
| HIPing | 1.16 | 1.1 | 0.22 | EduPack, Argon recovery method |

For the inventory analysis of both processes, material databases and recipes from SimaPro 9 software were utilized. For the investment casting method, the cooling step, known as SOPHIA, was estimated by assuming similar equipment to the cooling process, due to the lack of available literature on this specific process [16, 17].

## 4 Results

Inventory data were analyzed using SimaPro 9 and the Midpoint ReCiPe method for two scenarios for UCC process. Scenario 1 (S1) utilizes 100% virgin grade A356 alloy. The results (Fig. 3) show that raw materials dominate the environmental impact, contributing over 97% of the total energy (123.1 MJ/kg) and $CO_2$ emissions (9.01 kg/kg). Excluding raw materials, the dry hearth melter stage is the most impactful. Scenario 2 (S2) uses 20% virgin and 80% recycled alloy. This scenario significantly reduces energy use to 62 MJ/kg and $CO_2$ emissions to 4.8 kg/kg.

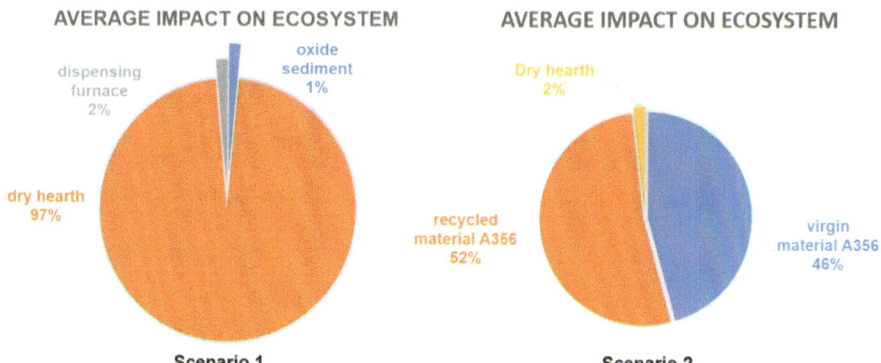

**Fig. 3.** Scenario 1 and 2, 100% Aluminium virgin material impact vs 80% recycled 20% virgin.

When UCC is compared with investment casting by implementing the inventory data to the LCA tool in SimaPro9 the outcome results show that UCC-S1 (100% virgin alloy) has the highest environmental impact, followed by Investment Casting, and UCC-S2 (80% recycled alloy) has the lowest impact, as seen in Fig. 4 below.

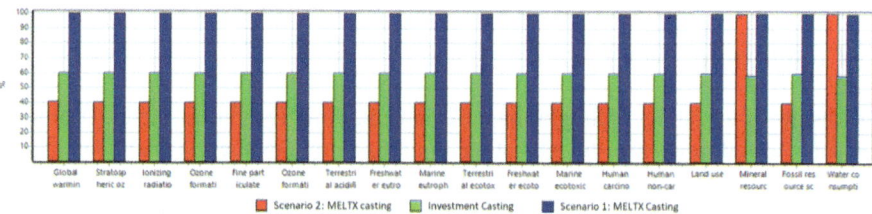

**Fig. 4.** Life cycle assessment comparison of midpoint results.

More specifically, UCC Scenario 1 (blue) consistently shows the highest impacts due to the use of virgin materials. Investment Casting (green) shows moderate impacts, while UCC Scenario 2 (red) demonstrates significantly lower impacts across almost all categories. This highlights the environmental benefits of using recycled materials in the UCC process, especially in reducing global warming potential, resource use, and other environmental burdens.

Finally, a comparative analysis was done to study the difference between energy consumption and $CO_2$ emissions for different % of virgin and recycled alloy as the initial charge in the process, as seen in Fig. 5 below.

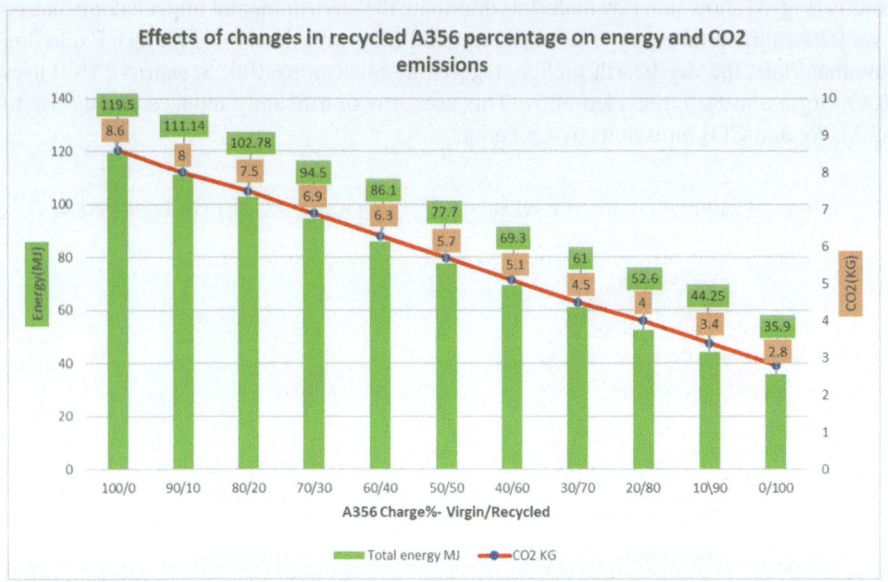

**Fig. 5.** Effects of changes in recycled A356 percentage on energy and CO2 emissions.

The figure shows that increasing the proportion of recycled A356 alloy in the charge significantly reduces both energy consumption and $CO_2$ emissions. At 100% virgin alloy, energy use is 119.5 MJ and $CO_2$ emissions are 8.6 kg/kg. Shifting to 100% recycled alloy lowers energy use to 35.9 MJ and $CO_2$ emissions to 2.8 kg/kg, illustrating the substantial environmental benefits of using recycled materials.

## 5  Conclusion

The results of this study highlight the environmental implications of the UCC process compared to traditional investment casting. Initial assumptions suggested a higher impact for UCC, but further analysis revealed that using a higher proportion of recycled aluminum significantly reduces its environmental footprint, making it a more sustainable option.

Proprietary data restrictions from foundries limited the accuracy of efficiency, energy consumption, and material loss estimates. Future research should incorporate cleaning methods in benchmarking, provide detailed information on casting processes, and use a functional unit for more precise comparisons.

**Acknowledgements.** The authors gratefully acknowledge the funding by the Ultra Clean Cast DLMM Program No 10065261.

# References

1. Cao, X., Campbell, J.: Defects in aluminium casting due to oxide film inclusion. J. Manuf. Process. **7**(1), 41–49 (2005)
2. Davis, J.R.: Aluminum and Aluminum Alloys. ASM International, Materials Park (1993)
3. Angeloni, L., Ruchert, C., Pommier, S.: Study on the effects of porosity on the fatigue life of aluminium components. J. Mater. Sci. **58**(2), 1245–1254 (2023)
4. Lee, P.D., Hunt, J.D.: Hydrogen porosity in directional solidified aluminium-copper alloys: in situ observation. Acta Mater. **48**(19), 4703–4713 (2000)
5. Bharambe, A., et al.: Analysis of defects in die casting on an industrial scale. Int. J. Cast Met. Res. **35**(3), 167–173 (2022)
6. Belt, C.: Utilization and efficiency of furnaces in the metallurgical industry. Metall. Eng. J. **10**(4), 299–305 (2015)
7. Belov, N.A., Eskin, D.G., Aksenov, A.A.: Multicomponent Phase Diagrams: Applications for Commercial Aluminum Alloys. Elsevier, Amsterdam (2002)
8. Srivatsan, T.S., Sivakumar, R.: Manufacturing Techniques for Materials: Engineering and Engineered. CRC Press, Boca Raton (2007)
9. International Aluminium Institute: Aluminium demand and greenhouse gas emissions. Global Industry Report, London (2024)
10. Hatch, J.E.: Aluminum: Properties and Physical Metallurgy. ASM International, Materials Park (1984)
11. Harborth, J.: Energy consumption in primary and secondary aluminium production. Sustain. Manufact. Process. 65–78 (2014)
12. Kaufman, J.G., Rooy, E.L.: Aluminum Alloy Castings: Properties, Processes, and Applications. ASM International, Materials Park (2004)
13. ISO 14040:2006. Environmental management — Life cycle assessment — Principles and framework. International Organization for Standardization (2006)
14. Norgate, T., Rankin, W.: The role of metals in sustainable development. Int. J. Life Cycle Assess. **7**(4), 230–234 (2002)
15. Schrijvers, D.L., Loubet, P., Sonnemann, G.: Critical review of guidelines against a systematic framework with regard to consistency on allocation procedures for recycling in LCA. Int. J. Life Cycle Assess. **21**, 994–1008 (2016)
16. Signicast: Introducing SOPHIA®: Aluminum Investment Castings for Aerospace, Defense and Special Applications. https://www.signicast.com/en/knowledge-center/webinars/introducing-sophia
17. ZOLLERN: Solid Metals. Fine Solutions. https://www.zollern.com/en/
18. PRé Sustainability: SimaPro 9. Life Cycle Assessment Software. https://simapro.com/
19. Dynamo Furnaces: Energy and Emissions Data. https://www.dynamo-furnaces.com/
20. Kanthal: Furnace Specifications and Energy Consumption. https://www.kanthal.com/
21. Granta Design: EduPack. Material and Process Information Database. https://www.grantadesign.com/products/education/edupack/
22. Argon Recovery Method for HIPing: High-Pressure Processes. https://www.example.com/argon-recovery

# Effect of Sustainable Hybrid Coolant Strategy on Machinability and Energy Consumption of NiTiHf High Temperature Shape Memory Alloy

Ozhan Kitay[1] , Emre Tascioglu[2] , and Yusuf Kaynak[3(✉)]

[1] Department of Machine and Metal Technologies, Bilecik Seyh Edebali University, 11100 Bilecik, Turkey
ozhan.kitay@bilecik.edu.tr
[2] Torun Bakır Alaşımları Metal San.ve Tic. A.Ş., Research and Development Center, 41480 Kocaeli, Turkey
[3] Department of Mechanical Engineering, Marmara University, 34854 Istanbul, Turkey
yusuf.kaynak@marmara.edu.tr

**Abstract.** Features of NiTiHf high temperature shape memory (SMA) alloy such as low thermal conductivity, high work-hardening tendency and superelasticity make machining operations difficult and cause high tool wear and high cutting forces. As industries move towards sustainability, it is vital to ensure the machining performance standards and energy efficiency of difficult-to-cut materials. Therefore, this study focuses on the energy consumption and in-depth analysis of machinability of NiTiHf alloy. The effects of minimum quantity lubrication (MQL) cooling and combination of both MQL + $CO_2$ cooling on machinability measures such as tool wear, cutting forces and surface quality were evaluated for various cutting speeds. Experimental findings show that MQL + $CO_2$ cooling reduces the cutting temperature by 30% more than MQL. MQL + $CO_2$ was found to be the finest effective cooling technique, reducing tool wear, and specific cutting energy (SCE) and creating the best possible surface. While the surface roughness (Ra) value in MQL is half of the dry cutting condition, the roughness in MQL + $CO_2$ is 69% less than the dry cutting condition. Even at the highest cutting speed, the effect of the MQL + $CO_2$ cooling regime is at maximum level and the tool flank wear is 20% less than MQL.

**Keywords:** High Temperature Shape Memory Alloy · Sustainability · Hybrid Machining

## 1 Introduction

Shape memory alloys (SMAs) are special smart materials that can change shape under the influence of parameters such as applied temperature, stress, etc. and return to their initial state after the effect is eliminated [1]. Shape memory alloys are used in many areas such as aerospace, automotive, fastener manufacturing, biomedical, oil and gas

© The Author(s) 2025
H. Kohl et al. (Eds.): GCSM 2024, LNME, pp. 417–425, 2025.
https://doi.org/10.1007/978-3-031-93891-7_46

industries [1, 2]. SMAs are considered as difficult to cut materials due to their properties such as high ductility, low thermal conductivity coefficient, and low elastic modulus [2, 3]. These disadvantages cause high wear on the cutting tool during the machining of the material, thus low tool life, high cutting forces, low chip breakability and burr formation as a result of the machining process [2–4].

The most widely used shape memory materials are NiTi alloys. Studies conducted to improve the machining performance of NiTi SMA material are available in the literature. Kaynak et al. investigated the effect of the phase state of NiTi alloy on machining performance under different cutting conditions. It was observed that cryogenic cooling applied during the machining process reduced tool wear, cutting forces and improved surface quality at high cutting speeds [5]. In another study using cryogenic cooling, tool wear and chip geometries formed during the machining of NiTi alloy were investigated and the obtained values were compared with the values obtained in dry cutting and minimum quantity lubrication (MQL) conditions [4]. In another study, the effects of cutting conditions (cryogenic, dry cutting and preheating) as well as cutting speeds on tool wear, cutting forces and chip thickness were investigated [6]. The results in the mentioned studies show that the cutting conditions improve the machinability of NiTi SMA material.

Different studies have been carried out to improve the machinability of NiTi alloys, and shape memory alloys such as NiTiZr, NiTiPd, NiTiPt and NiTiHf, which have better mechanical and shape memory properties, have been developed by adding various elements to the alloy [1, 7]. The NiTiHf alloy stands out with its high phase transformation temperature, superelastic behavior at high temperatures, and shape memory properties even under high loads [1]. Studies have been conducted on various deformation behaviors of NiTiHf alloy such as plastic deformation behavior [8], hot deformation [9], cold deformation [10] and manufacturing methods such as hot rolling [11] and extrusion [12]. There are many studies in different fields regarding NiTiHf shape memory material. However, it has been found that there are very few studies on NiTiHf machining.

It is an undeniable fact that machining plays a very important role in shaping the global economy [13]. Cutting parameters obtained by using personal experiences are usually far from the best value. This situation makes it necessary to determine the best cutting parameters. Because long tool life, low energy consumption and a clean environment must be provided for a sustainable machining industry. In this study, the machining performance of NiTiHf alloy at variable cutting speeds under MQL and MQL + $CO_2$ conditions instead of the traditional cooling method was investigated in terms of tool wear and energy consumption as well as cutting temperature and surface quality.

## 2    Experimental Setup

The work material used was a Ni50.3Ti29.7Hf20 high-temperature SMA. The Ni-rich alloy was produced utilizing a vacuum induction skull-melting technique with a heat size of ~27 kg (designated as FS#7). The molten metal was poured into steel molds, and the resulting cast ingots were vacuum homogenized for 72 h at 1050 °C followed by furnace cooling. The material was then canned and extruded at 900 °C through an area reduction ratio of 7:1. The extruded rods were centerless ground to a final diameter of 10 mm and cut into 60-mm-long pieces. The average hardness of the workpiece was $395 \pm 5$ HV. The NiTiHf alloy was in its martensitic phase at room temperature as determined by Differential Scanning Calorimetry (DSC) where the martensite start (Ms), martensite finish (Mf), austenite start (As) and austenite finish (Af) temperatures are 94 °C, 59 °C, 107 °C and 134 °C, respectively [14].

The machining experiments were carried out on a Puma GT2100 CNC turning center, which has a maximum spindle speed of 4500 rpm and 18 kW. The PCLNL 2525 M12 tool holder with a rake angle of $\alpha = -6°$ degree was used. A DNMG 11 04 04-MF cutting tool insert, 1105 grade coating, specified by the recommendation of the cutting tool companies, were used in the experiments. Cutting speeds ($V_c$) were selected as 60, 90 and 120 m/min. The feed rate, $f = 0.07$ mm/rev, and depth of cut, $a_p = 0.7$ mm, were kept constant. Machining experiments were carried out under the minimum quantity lubrication (MQL), MQL + $CO_2$(50 bar) hybrid and dry condition. Implementation of the machining tests is shown in Fig. 1. Cutting force ($F_c$) during the experiments were measured using a KISTLER 2129AA dynamometer. Cutting temperature was recorded using an Optris PI 400 infrared camera. Emissivity of the work material was determined as 0.63. The arithmetical average surface roughness ($R_a$) of the machined parts was measured with a Mitutoyo SJ210 surface profilometer. Tool wear and surface topography were captured using a Keyence VHX-6000 digital optical microscope.

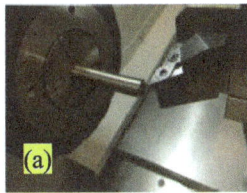  

**Fig. 1.** Experimental design for a) dry, b) MQL, c) MQL + $CO_2$

## 3   Results and Discussion

### 3.1   Cutting Temperature

Cutting temperature is a parameter that directly affects machining performance, especially tool wear, and therefore sustainable manufacturing. Therefore, apart from the dynamics of the machining itself, cooling/lubricating conditions are considered a sustainable manufacturing strategy. Figure 2 shows the maximum cutting temperatures measured at varying cutting speeds. As the cutting speed increased, temperature values increased in all conditions. The lowest temperatures occurred in the MQL + $CO_2$ hybrid cooling condition. The maximum temperature value under dry cutting condition at a cutting speed of 60 m/min was 925 °C. When the dry cutting condition is taken as a reference, the temperature was 7% lower in the MQL condition and 52% lower in the MQL + $CO_2$ condition. At the highest cutting speed of 120 m/min, the maximum temperature value was 1135 °C in dry cutting. In the MQL condition the temperature is 19%

less than in dry cutting and in the MQL + $CO_2$ condition it is 52% less. The austenite start temperature (As) of NiTiHf shape memory alloy is 107 °C, and this temperature, which is quite low for machining, was exceeded in all tests.

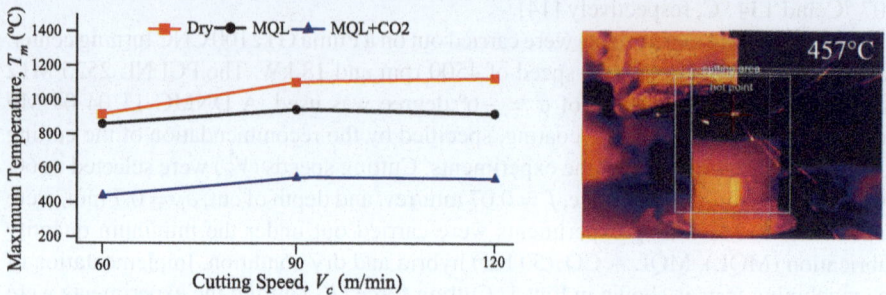

**Fig. 2.** Measured maximum cutting temperatures at different cutting conditions

## 3.2 Tool Wear Analysis

Cutting tool cost is an important factor in terms of sustainability in machining. Therefore, extending the life of tools against wear with cutting fluids contributes to sustainability. In this study, as stated in the ISO 3685 norm, the maximum flank wear value of 0.4 mm was taken as the criterion to determine the tool life. Tool flank wear images are shown in Fig. 3a. In general, wear on the cutting tool occurs as a result of the abrasive wear mechanism triggered by the tensions between the cutting tool and the workpiece. As the cutting speed increased, the amount of wear increased in all conditions. The percentage increase in wear amounts from 60 m/min cutting speed to 120 m/min cutting speed was 145% in dry cutting, 81% in MQL, and 140% in MQL + $CO_2$ condition. At a cutting speed of 60 m/min, the highest flank wear was in dry cutting condition with a wear amount of 281 μm. The amount of wear was 5% less in the MQL condition than dry and 41% less in the MQL + $CO_2$ cutting condition than dry. At the highest cutting speed of 120 m/min, the highest wear amount was again in the dry cutting condition. Tool flank wear was measured as 487 μm in MQL and 398 μm in MQL + $CO_2$. Cutting tool nose wear images are shown in Fig. 3b. As the cutting speed increased, the amount of nose wear increased in all cutting conditions. While the most wear at the highest cutting speed occurred in the dry cutting condition, the least amount of wear occurred in the MQL + $CO_2$ condition. At the lowest cutting speed of 60 m/min, the maximum amount of wear was 85 μm under dry cutting condition. The amount of nose wear in dry cutting condition is approximately 30% higher than other cutting conditions. At the highest cutting speed, the difference in the amount of wear between cutting conditions is clearly visible. At a cutting speed of 120 m/min, the highest nose wear occurred under dry cutting condition. The amount of nose wear under dry cutting condition was measured as 547 μm. In the MQL condition, the wear was measured as 457 μm, which is 16% less than in dry. In the MQL + $CO_2$ cutting condition, a wear amount of 418 μm was measured and the nose wear amount was 23% less than in the dry cutting condition.

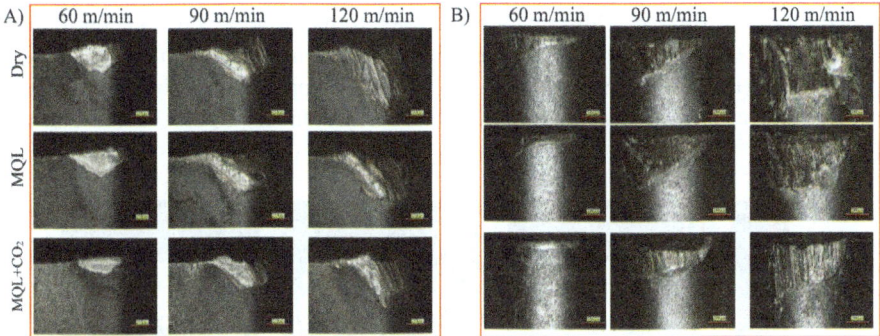

**Fig. 3.** A) Cutting tool flank wear images B) Cutting tool nose wear images

### 3.3  Surface Roughness and Topography

The surface quality of the workpiece material may vary depending on cutting tool wear, cutting speed, chip breaker form and type of coolant used. Roughness values measured at different cutting conditions are shown in Fig. 4 as a function of cutting speed. The highest roughness value is under dry cutting condition at a cutting speed of 60 m/min. The roughness value of the surface in dry cutting was measured as 0.67 μm. The MQL condition created a roughness value of 0.49, approximately 26% less than the dry cutting condition. In the MQL + $CO_2$ condition, 40% less roughness value was measured. The highest roughness value was measured under dry cutting condition with 5.6 μm at the highest cutting speed of 120 m/min. In dry cutting condition at a cutting speed of 120 m/min, the roughness is very high due to the high amount of wear of the cutting tool and the deterioration of the tool geometry. The roughness value in MQL is 53% less than the dry cutting condition. In MQL + $CO_2$, the roughness is 1.6 μm and is 69% less than the dry cutting condition.

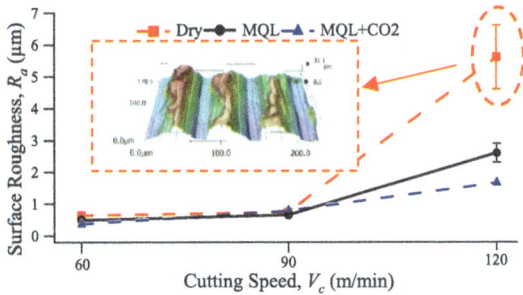

**Fig. 4.** Surface roughness values machined specimens

Surface topography images of machined samples are shown in Fig. 5. It is obvious that the surface quality decreases with increasing cutting speed. The decrease in surface quality with increasing cutting speed is explained by the wear of the cutting tool. The highest speed of the dry cutting condition, where the amount of wear was the highest, created the worst surface. This situation is related to the nose area of the cutting tool being exposed to serious wear. Tool wear and deterioration of tool geometry negatively affected the surface quality. Surface topography images support the roughness values and the best surface at the highest cutting speed was formed under MQL + $CO_2$ condition.

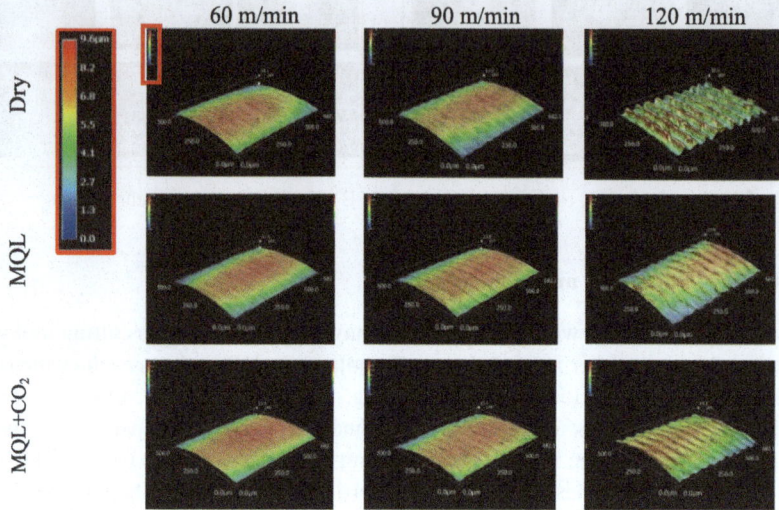

**Fig. 5.** Surface topography images of the machined specimens

## 3.4  Cutting Force and Energy Consumption

The cutting forces measured during machining are shown in Fig. 6a. In all cutting conditions, as cutting speed increased, cutting forces also increased. This increase can be explained by increasing temperature and therefore increasing cutting tool wear. The highest cutting force at 60 m/min was measured under dry cutting condition. The forces increased due to increasing temperature and cutting tool wear. In cutting conditions exceeding 600 °C, which is the softening point of the carbide cutting tool, the forces increased more due to wear. Since the cutting temperatures were lower in the tests carried out under cooling/lubricating conditions, the amount of wear occurred less, therefore the cutting forces were less. Considering the dry cutting condition, MQL also produced 9% less cutting force. This can be explained by the fact that the MQL condition reduces the friction forces and the cutting process is easier. MQL + $CO_2$ produced 36% less cutting force. This can be explained by the fact that the temperature in the cutting zone is much lower than in dry cutting and the cutting tool has not reached its softening temperature. The dry cutting condition created the highest cutting force at the highest cutting speed of 120 m/min. It can be seen that with the increase in cutting speed, cutting tool wear played a more dominant role, and the difference between cutting conditions decreased with the increase in cutting forces. The MQL condition created approximately 1.5% less cutting force than the dry cutting condition. In the MQL + $CO_2$ cutting condition, 6% less cutting force occurred.

Specific cutting energy (SCE) is a parameter that defines the machinability of the work material and measures the efficiency of the cutting process. The specific cutting energy is found by dividing the cutting power ($Pc$) by the material removal rate (MRR) as in Eq. 1 [15].

$$SCE = P_c/MRR \tag{1}$$

$$P_c = F_c \times V_c \tag{2}$$

$$MRR = V_c \times f \times a_p \tag{3}$$

where $F_c$ is the cutting force, $V_c$ is the cutting speed, f is the feed rate and $a_p$ is the depth of cut. The effect of cutting conditions on the calculated specific cutting energy is shown in Fig. 6b. The excessive increase in cutting tool wear with increasing cutting speed led to high cutting forces. As a result, the power and specific cutting energy required for machining have increased. The highest SCE is in the dry condition and the lowest is in the MQL + $CO_2$ condition. The impact of cutting conditions on energy consumption is clearly visible. The lubricating feature of the MQL condition gave effective results compared to the dry condition. As a cryogenic coolant, $CO_2$ gas can provide cooling down to -78 °C [16].When the lubricating/cooling feature of carbon dioxide combined with MQL, energy consumption decreased significantly under the MQL + CO2 condition.

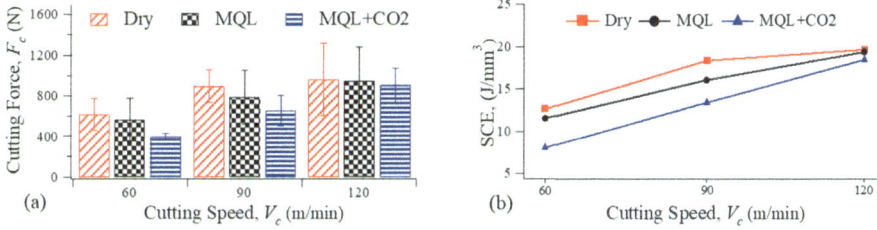

**Fig. 6.** a) Cutting force values at different cutting conditions, b) Calculated specific cutting energy values

## 4   Conclusions

This study examines the effect of different cutting conditions on the machinability of NiTiHf alloy in terms of sustainability. The results show that the lubricated condition already provides more sustainable machining compared to dry cutting, while the use of hybrid coolant further increases efficiency. When MQL and $CO_2$ are used together, longer tool life and better surface quality are achieved. In addition to reducing environmental pollution compared to conventional coolers, the power consumption for machining also decreases compared to dry cutting and MQL alone. These results also contribute to the machinability of the NiTiHf alloy, about which there are not many studies.

**Acknowledgments.** Financial support from TUBITAK (The scientific and Technological Research Council of Türkiye) under Project number 116M346 is greatly acknowledged.

# References

1. Karaca, H., Acar, E., Tobe, H., Saghaian, S.: NiTiHf-based shape memory alloys. Mater. Sci. Technol. **30**(13), 1530–1544 (2014)
2. Weinert, K., Petzoldt, V.: Machining of NiTi based shape memory alloys. Mater. Sci. Eng. A **378**(1–2), 180–184 (2004)
3. Huang, H.: A study of high-speed milling characteristics of nitinol. Mater. Manuf. Process. **19**(2), 159–175 (2004)
4. Kaynak, Y., Karaca, H., Noebe, R., Jawahir, I.: Tool-wear analysis in cryogenic machining of NiTi shape memory alloys: a comparison of tool-wear performance with dry and MQL machining. Wear **306**(1–2), 51–63 (2013)
5. Kaynak, Y., Karaca, H., Jawahir, I.: Cutting speed dependent microstructure and transformation behavior of NiTi alloy in dry and cryogenic machining. J. Mater. Eng. Perform. **24**(1), 452–460 (2015)
6. Kaynak, Y., Karaca, H., Noebe, R., Jawahir, I.: Analysis of tool-wear and cutting force components in dry, preheated, and cryogenic machining of NiTi shape memory alloys. Procedia CIRP **8**, 498–503 (2013)
7. Noebe, R., Biles, T., Padula, S.: NiTi-based high-temperature shape-memory alloys: properties, prospects, and potential applications. Mater. Eng.-New York **32**, 145 (2006)
8. Simon, A.A.: Shape memory response and microstructural evolution of a severe plastically deformed high temperature shape memory alloy (NiTiHf). Texas A&M University (2006)
9. Belbasi, M., Salehi, M.T., Mousavi, S.A.A.A.: Hot deformation behavior of NiTiHf shape memory alloy under hot compression test. J. Mater. Eng. Perform. **21**(12), 2594–2599 (2012)
10. Steiner, G., Peterlechner, M., Waitz, T., Karnthaler, H.: TEM investigation of severely deformed NiTi and NiTiHf shape memory alloys. In: EMC 2008 14th European Microscopy Congress, 1–5 September 2008, Aachen, Germany, pp. 489–490. Springer (2008)
11. Belbasi, M., Salehi, M.T., Mousavi, S.A.A.A., Ebrahimi, S.M.: A study on the mechanical behavior and microstructure of NiTiHf shape memory alloy under hot deformation. Mater. Sci. Eng. A **560**, 96–102 (2013)
12. Kockar, B., Karaman, I., Kim, J., Chumlyakov, Y.: A method to enhance cyclic reversibility of NiTiHf high temperature shape memory alloys. Scripta Mater. **54**(12), 2203–2208 (2006)
13. Sutherland, J., et al.: Challenges for the manufacturing enterprise to achieve sustainable development. In: Manufacturing Systems and Technologies for the New Frontier, pp. 15–18. Springer (2008)
14. Kaynak, Y., Tascioglu, E., Benafan, O.: Surface integrity characteristics of NiTiHf high temperature shape memory alloys. In: International Conference on Advanced Surface Enhancement, pp. 254–262. Springer (2019)
15. Younas, M., Jaffery, S.H.I., Khan, A., Khan, M.: Development and analysis of tool wear and energy consumption maps for turning of titanium alloy (Ti6Al4V). J. Manuf. Process. **62**, 613–622 (2021)
16. Kitay, O., Kaynak, Y.: The effect of flood, high-pressure cooling, and CO2-assisted cryogenic machining on microhardness, microstructure, and X-ray diffraction patterns of NiTi shape memory alloy. J. Mater. Eng. Perform. **30**(8), 5799–5810 (2021)

# Reducing Waste in Liquid Silicone Rubber Process Chains by New Injection Mould Process Chain

Eckart Uhlmann[1,2], Robert Bolz[1,2(✉)], Mitchel Polte[1,2], and Luiz Schweitzer[2]

[1] Institute for Machine Tools and Factory Management IWF, Technische Universität Berlin,
Pascalstrasse 8-9, 10587 Berlin, Germany
`bolz@tu-berlin.de`

[2] Fraunhofer Institute for Production Systems and Design Technology IPK, Pascalstrasse 8-9,
10587 Berlin, Germany

**Abstract.** The number of components made of liquid silicon rubber (LSR) is following a global increasing trend. Due to the low viscosity, the different LSR types are suitable for injection moulding, as they can reproduce even the most complex structures. However, the low viscosity of LSR results in challenges regarding the manufacturing of the injection moulds.

LSR injection mould manufacturers operate with highest effort to produce perfectly fitting mould inserts. In case of a perfect sealing, a venting of the cavities is limited, leading to defects in the moulded parts and consequently to an increase of production waste. This paper presents a far more effective process chain solving these contradictory requirements, describing new types of LSR moulding tools and an associated process chain for a resource efficient production. Experimental studies revealed the possibility of an electrical discharge dressing to generate a form fit between the mould inserts with a uniform material removal, leading to flash-free injection moulding at lowest closing forces. Thereby, the specific surface structure created by EDM presents a novel approach for venting the cavities. A resource-oriented customization of the injection moulding process can additionally reduce the energy consumption, leading to a more sustainable production.

**Keywords:** liquid silicone rubber; LSR · tooling; injection mould · EDM

## 1 Introduction

### 1.1 Increasing Demand for Liquid Silicone Rubber

The use of injection moulded parts by processing liquid silicone rubber (LSR) has been steadily increasing for years. The global LSR market size in 2023 was about USD 1,358.58 Million and is expected to perform a compound annual growth rate of about CAGR $\approx 8.5\%$ from 2023 to 2032 [1].

The increasing demand for silicone components is mainly due to its advantages, such as high temperature resistance, high resistance to shear forces $F_S$, low degradation over

H. Kohl et al. (Eds.): GCSM 2024, LNME, pp. 426–433, 2025.
https://doi.org/10.1007/978-3-031-93891-7_47

time t, high resistance to oxidation and hydrolysis, and resistance to oils and greases. In medical products such as cannulas, baby teats or breastfeeding aids, the biocompatible and hypoallergenic properties come into play, as does the bacterial resistance [2–4].

Due to the consistency and the type of crosslinking of LSR components, the possible applications are diverse and can be found in the automotive, electronics, consumer goods and food, as well as in the construction and healthcare sector [1, 2, 5].

## 1.2  Challenges in LSR Injection Moulding

The high and increasing application of LSR components calls for production on a large-scale using injection moulding processes in order to be economic. However, LSR places very high requirements on processing methods and consequently on LSR injection moulds. These requirements are handled with great effort and correspondingly high production costs $c_p$ by tool and mould making industry, applying manual polishing and similar finishing processes [2, 5].

The high-precision moulds required for LSR injection moulding are manufactured by specialized companies at high production costs $c_p$. Since mould and tool making is strongly characterized by single part and small series production, moulds are comparatively high-priced. In the plastics processing industry, moulds, especially those with high precision, account for over 60% of production costs $c_{prod}$ of the final product. Only by using advanced manufacturing technologies and achieving high quality, local companies can prevail over competitors from low-cost countries [6–8]. An example for a simple LSR injection mould geometry is given in Fig. 1.

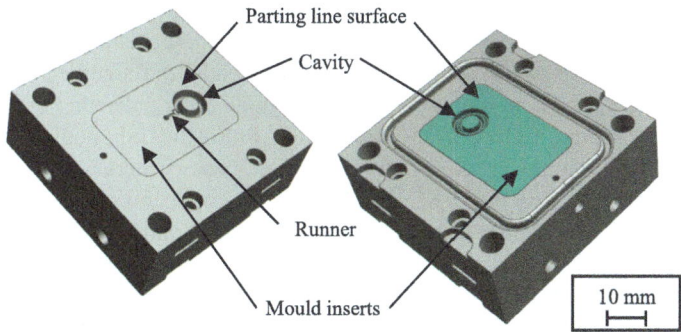

**Fig. 1.** CAD illustration of the LSR injection mould.

The LSR is guided through the runner into the cavity, where the LSR vulcanizes at temperatures $\vartheta \approx 180\ °C$ and consequently solidifies. The very low viscosity $\eta$ of the LSR during the injection process requires a sufficient sealing between the mould inserts in order to prevent LSR to enter the mould parting line. To achieve this, a form-fit of the applied mould inserts with a planarity $P_T$ of the mould parting line surfaces in the range of a few micrometres is targeted in recent production processes, even with large parting line surface areas $A_P$. Parting lines with a gap widths $s_P \geq 10\ \mu m$ between the mould inserts result in flash formation, see Fig. 2. In contrast to conventional thermoplastic processing, where parting lines with gap widths of 10 $\mu$m $\leq s_T \leq$ 20 $\mu$m are common, cost- and energy-intensive high-precision machining processes must be used in the production of LSR moulds [9–11].

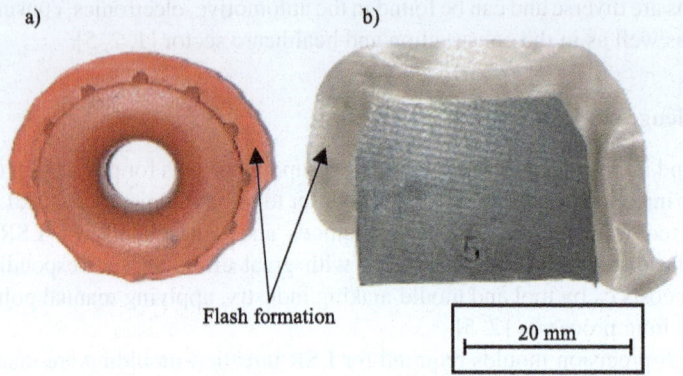

Fig. 2. Examples of flash formation on a) a sealing plug and b) a valve component.

The reworking of the components for removing the flashes accounts for up to 40% of the total production costs $c_p$ of an injection-moulded silicone component. Additionally, the removed flashes cannot be further processed or reused as the silicone is vulcanized, leading to an undesired waste of material. In addition to flash formation on the injection moulded part, there is a risk that LSR enters the mould parting lines, vulcanizes there, and successively destroys the mould insert with each subsequent injection process due to high closing forces $F_c$ of the injection moulding machine. As a result, the moulds have to be regularly overhauled or replaced, which leads to a significant increase in energy and resource consumption as well as production costs $c_p$. Another requirement is contradicting the aspired sealing in the mould parting line in the LSR mould. The ambient air contained in the mould cavities must be able to escape. Otherwise air pockets and geometric defects will occur on the injection-moulded components resulting in increased production waste and additional material and energy consumption [10, 12].

Therefore, improving the LSR mould making and LSR injection moulding by reducing manufacturing effort and avoiding flash formation is a significant issue. This paper aims at solving the above stated challenges by a new process chain for manufacturing of the tool inserts.

## 2   Innovative Injection Moulds with Adapted Parting Line Surface

Electrical discharge machining (EDM) and milling are the main manufacturing technologies applied in tool and mould making industry [11, 13, 14]. In order to meet the before-mentioned challenges with already existing machinery in most companies, a new process has been developed. In this, the parting line surfaces of the mould inserts are dressed to each other by applying an adapted EDM process in combination with new LSR mould insert parting line designs. This electrical discharge dressing leads to a form fit in the range of single micrometres and consequently enables a high sealing D of the parting lines of the LSR mould inserts.

In order to establish an EDM process with similar material removal volume $V_{W,E}$ on both electrodes, in this case mould inserts, several experimental studies have been executed. The most critical EDM processing parameters were investigated. The discharge current $i_e$, discharge duration $t_e$, polarity as well as discharge type have been varied according to a design of experiments setup with simplified electrode geometries. The effects of the different processing parameters have been described in a model, which was used to generate and further adapt a set of processing parameters meeting the requirements of low but similar surface roughness Rz on both electrodes as well as minimum machining time $t_{ero}$. The material removal volume $V_{W,E}$ and the surface roughness Rz were measured by the tactile measurement device Hommel-Etamic nanoscan 855, JENOPTIK AG, Jena, Germany, and multi-sensor measurement device Zeiss O-Inspect, CARL ZEISS AG, Oberkochen, Germany. After successfully meeting the requirements, the EDM process has been realised with a demonstrator geometry.

First demonstrators of these new mould inserts are shown in Fig. 3. The reduced parting line surface area $A_P$ simplifies the electrical discharge dressing of the parting line surfaces due to less material to be removed. Additionally, it reduces the required closing force $F_c$ to obtain a sealing in the parting line. These novel LSR moulds shall enable an almost flash-free production of injection moulded silicone components due to significantly reduced LSR in the parting lines. This results in less waste of materials. This increase of sustainability is added by significantly lower production cost $c_{prod}$.

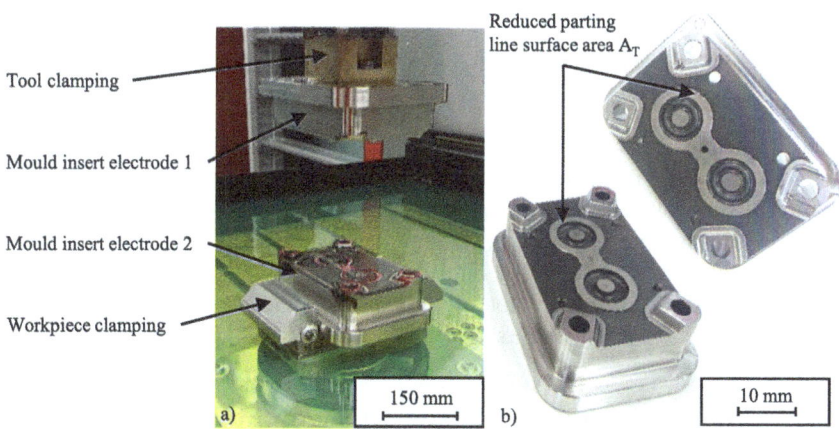

**Fig. 3.** Illustration of a) the setup for electrical discharge dressing and b) first demonstrators with reduced parting line surface area $A_T$.

In order to ensure sufficient venting of the LSR mould cavities despite the required sealing D of the mould inserts to each other, the aperiodic microstructures by the electrical discharge dressing were analysed. To enable sufficient venting, channels in LSR mould inserts with structure heights between $3\ \mu m \leq h_s \leq 6\ \mu m$ are recommended by DOW CHEMICAL COMPANY, Midland, USA [14]. Therefore, the generated aperiodic microstructures should be in a similar range to allow venting of the mould cavities and presenting a labyrinth seal for the LSR.

The surface roughness of the microstructures generated by the electrical discharge dressing is in the range of 10 μm < Rz < 22 μm. The ability of the venting of the mould cavities was proved in trials leading compressed air into the mould and fine powder applied in the parting line. However, these microstructures might be influenced by the closing forces $F_c$ and thereby generated mechanical stresses σ in the injection process. Consequently, the influence of closing forces $F_c$ on the microstructure was investigated as well. The measured values for the surface roughness Rz for four different parting line surfaces, samples 1–4, loaded with increasing induced pressures p are shown in Fig. 4.

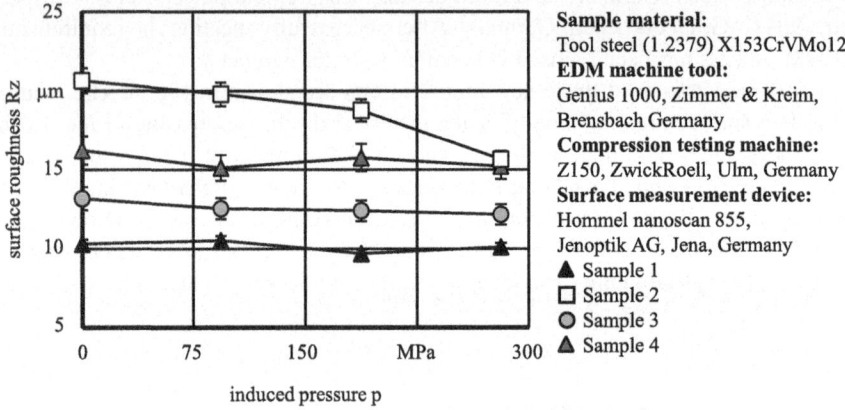

**Fig. 4.** Surface roughness Rz for differently prepared surface samples.

It can be seen that an increasing compressive load on the parting line surface causes a reduction for samples with initial surface roughness Rz > 20 μm, indicating a mechanical deformation of the peaks of the surface profile. The samples with surface roughness Rz < 20 μm show no significant effect by the induced pressure p due to higher mechanical strength f of the smaller structures. These results indicate that parting line surfaces with surface roughness Rz < 20 μm are not influenced by the mechanical stresses σ of the injection moulding process. The same should apply to the aspired venting of the cavities.

## 3   LSR Injection Moulding with New Injection Mould Inserts

The general functionality of the new LSR mould inserts with structured parting line surface has been successfully validated in LSR injection moulding process on a Boy 25E, DR. BOY GMBH & CO. KG, Neustadt (Wied), Germany, shown in Fig. 5.

The sampling of the LSR mould was initiated with the lowest closing force available on the applied injection moulding machine with $F_c$ = 50 kN. The material volume $V_m$ programmed for injection into the mould was increased step by step until the moulded

Structured parting line surface providing sealing and venting

Mould insert 1     Cavities     Moulded parts          Sprue          Mould insert 2

**Fig. 5.** Mould inserts after LSR injection moulding

parts were completely filled. An increase of the closing force $F_c$ was not necessary due to no existing flash formation on the moulded parts.

The proof of non-existing flashes on the moulded parts has been realised by optical measurements on the optical measurement device G4 Infinite Focus, ALICONA IMAGING GMBH, Gratz, Austria, shown in Fig. 6.

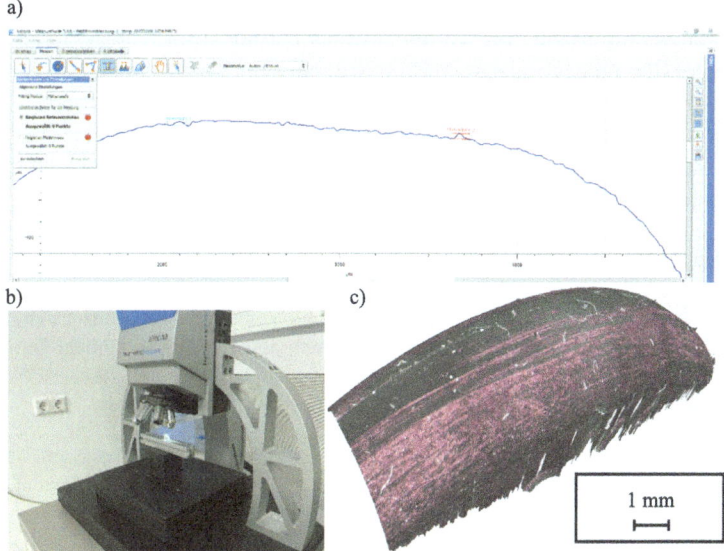

**Fig. 6.** Illustration of optical measurement of the moulded parts a) surface profile, b) measurement device, c) measured 3D surface data

The structures measured at the parting line of the moulded parts showed a maximum height $h_{max} = 5.6$ μm. Although this value is already sufficient for the majority of applications, further reduction might be possible by adapting the electrical discharge dressing as well as the injection moulding process.

At the end it could be proved that the new LSR mould inserts with its innovative parting line surface are able to significantly reduce effort in manufacturing this parting line area and still meet the high requirements on injection mould tools. In addition to the reduction of manufacturing effort in tooling, the injection moulding can be realised with low closing force $F_c$ and therefore reduced energy consumption of the injection mould-ing machine. The lack of post-processing for removing flashes additionally reduces production waste and additional costs.

## 4  Conclusion and Outlook

This paper presents the development of innovative LSR moulds and a new process chain for manufacturing them. The recently applied manufacturing strategies for LSR moulds include high effort in machining a perfect sealing between the mould inserts in order to avoid flashing and the implementation of vacuum and venting channels to allow air to escape from the cavity during filling to avoid trapped air [15].

The development enables a more sustainable processing of LSR by reducing the energy consumption $E_c$ of the injection moulding process. Additionally, the avoidance of flashes on the moulded parts reduces production waste and resource consumption as well as a post-processing of the parts. The reduced closing forces Fc, an effective venting by the microstructure on the surface and the prevention of flashes could be proved by a demonstrator LSR moulding tool. Due to the characteristics of the EDM processing, this approach is only usable for two-dimensional parting surface areas. In case of not plane surfaces, the process-related working gap between the electrodes leads to lateral shape deviations, which affect the sealing in the parting of the mould inserts.

Further research in this area will focus on even smaller parting surface areas $A_p$, the design of tool inserts for more complex parting line surfaces as well as energy-optimised injection moulding process.

**Acknowledgements.**

**EUROPEAN UNION**

**European Regional Development Fund**

The project "Development of innovative tools for LSR injection moulding" is co-financed by the Euro-pean Union and European Regional Development Fund in cooperation with ROEPER FORMENBAU GMBH, Berlin, Germany.

## References

1. Global Liquid Silicone Rubber Market 2024–2033, report published in June 2023, Liquid Silicone Rubber Market Size, Share & Trends Analysis Report 2018. https://www.custom marketinsights.com/report/liquid-silicone-rubber-market/. Accessed 07 June 2024
2. Eyerer, P., Hirth, T., Elsner, P.: Polymer Engineering Technologien und Praxis. Springer, Heidelberg (2008)
3. ProMed Molding Homepage. https://promedmolding.com/2019-beyond-what-the-lsr-mat erial-market-is-predicting/. Accessed 05 May 2024

4. Wintermantel, E.: Medizintechnik, 4th edn. Springer, Berlin (2008)

5. Johannaber, F., Michaeli, W.: Handbuch Spritzgießen, 2nd edn. Hanser, München (2004)

6. Boos, W., Salmen, M., Kelzenberg, C., Johannsen, L.; Helbig, J., Ebbecke, C.: Tooling in Germany 2018. Aachen (2018)

7. Schuh, G., Pitsch, M., Kelzenberg, C., Lieb, H., Ziskoven, H.: Getaktete Fertigung im Werkzeugbau, 1st edn. RWTH Aachen Werkzeugmaschinenlabor, Aachen (2015)

8. Klotzbach, C.: Gestaltungsmodell für den industriellen Werkzeugbau. Shaker, Aachen (2007)

9. Menges, G., Michaeli, W., Mohren, P.: Spritzgiesswerkzeuge, 6th edn. Hanser, München (2007)

10. Schlitt, C.: Thermoplasten und Flüssigsilikonen mit unterschiedlichen Mechanismen zur Initiierung der Vernetzung hergestellt im Mehrkomponenten-Spritzgießverfahren. Kassel University Press, Kassel (2018)

11. Uhlmann, E., Mullany, B., Biermann, D., Rajurkar, K.P., Hausotte, T., Brinksmeier, E.: Process chains for high-precision components with micro-scale features. CIRP Ann. **65**, 549–572 (2016)

12. Walde, H.: Beitrag zum vollautomatischen Spritzgiessen von Flüssigsilikonkautschuk, 1st edn. Augustinus-Buch, Aachen (1996)

13. Uhlmann, E., Domingos, D., Kühne, S., Zanatta, Z., Gomes, J.: Development and optimization of process chains for the micro mold industry: technological limits of the micro milling and micro die-sinking EDM technologies. In: euspen's 15th International Conference & Exhibition, Leuven, Belgium (2015)

14. Dow Chemical Company Homepage. https://www.dow.com/content/dam/dcc/documents/en-us/catalog-selguide/45/45-1581-01-liquid-si-rubber-prod-sel-guide.pdf. Accessed 05 May 2024

15. Plastics Technology: Getting Into LSR--Part IV: How LSR Tooling Is Different. https://www.ptonline.com/articles/getting-into-lsr--part-iv-how-lsr-tooling-is-different. Accessed 09 Aug 2024

# Modification of Edge Zone Properties of the Current Collector Foil During Calendering of Lithium-Ion Battery Electrodes Using Induction

Andreas Mayr[1]([✉]), Julian Link[1], and Rüdiger Daub[1,2]

[1] Technical University of Munich, TUM School of Engineering and Design, Department of Mechanical Engineering, Institute for Machine Tools and Industrial Management (iwb), Boltzmannstrasse 15, 85748 Garching, Germany
andreas.mayr@iwb.tum.de
[2] Fraunhofer Institute for Casting, Composite and Processing Technology (IGCV), Am Technologiezentrum 10, 86159 Augsburg, Germany

**Abstract.** Increasingly demanding requirements for lithium-ion batteries (LIBs) in terms of costs, energy density, and safety require the development of optimized manufacturing processes. In the electrode production of LIBs, the calendering process significantly influences the resulting volumetric energy density. Efforts to achieve higher energy densities lead to higher compaction rates during calendering, which increase electrode defects, such as wrinkles along the uncoated area of the current collector foil. These defects cause potential scrap in the subsequent electrode processing steps, increasing manufacturing costs. Therefore, a comprehensive understanding of the process-structure relationships and the associated wrinkle reduction during calendering is crucial for producing cost-efficient, high-quality electrodes. In this work, an induction heating unit is implemented to induce heat locally in the uncoated edge zone of the lithium-ion electrode. This influences the mechanical properties of the current collector foil and thus reduces the formation of wrinkles during the calendering process. The results show that the inductively applied heat effectively counteracts the formation of wrinkle defects. This study proves the capability of the induction principle to significantly reduce the scrap rate during calendering by improving the downstream processability of the electrodes.

**Keywords:** Lithium-ion battery · Calendering · Induction heating

## 1 Introduction

The ongoing shift from fossil-fueled transportation towards electromobility, driven by regulatory measures, is increasing the global demand for lithium-ion batteries (LIBs). The reduction of costs, the improvement of performance, and the increase in energy density represent key drivers of innovation in the field of LIB research [1]. Production

H. Kohl et al. (Eds.): GCSM 2024, LNME, pp. 434–442, 2025.
https://doi.org/10.1007/978-3-031-93891-7_48

research into new materials and innovative processes on a pilot scale is an essential part of establishing a cost-efficient and sustainable battery production. An in-depth understanding of the individual process steps in battery production is necessary for producing high-energy and cost-efficient LIBs [2].

Calendering is crucial in electrode production, significantly impacting the electrode quality by increasing the volumetric energy density and improving mechanical [3] and electrochemical properties [4]. However, achieving increased volumetric energy densities through calendering introduces process-related challenges. Higher compaction rates increase the volumetric energy density, however they can lead to defects and higher scrap rates, thereby creating a conflict between improving the energy density and avoiding scrap to reduce costs [5]. A thorough understanding of the process-structure relationships during calendering is an enabler to resolve the trade-off between increasing energy density and avoiding scrap. The calendering-induced defects can be divided into geometric, structural, and mechanical defect patterns [5]. The effect of the calendering process on different defect patterns [6, 7], the integration of inline measurement technology for monitoring these [8], and the interaction with subsequent process steps [9] have already been investigated on varying electrode materials. The geometric defect pattern of so-called wrinkles, which refers to the formation of folds due to different elongations in the transition area between the coating and the uncoated current collector foil [10], adversely impacts subsequent processing and can cause scrap [5]. The possibilities of counteracting defects, such as wrinkles, during calendering are limited. In previous research work, the temperature of the calendering rolls was identified as a relevant parameter influencing geometric defects [1, 5]. An alternative method to reduce wrinkles is the mechanical stretching of the uncoated edge area using additional pairs of rollers [10]. In addition, there are approaches for the application of inductive heat treatment in battery production. For example, as a pre-treatment for the recycling process of electrodes [11] or as an innovative drying process after the coating of electrodes [12]. To the best of the authors knowledge, there are no scientific studies that deal with the reduction of wrinkles through induction heating and the corresponding suitable process parameters.

In this work, the temperature treatment utilizing the induction principle was analyzed regarding the reduction of wrinkles by locally inducing heat into the current collector foil. This allowed the targeted modification of the current collector foil's mechanical properties. The objective of this study was to enhance the elongation of the current collector foil through annealing, thereby reducing the discrepancy in length between the uncoated and the coated areas of the electrode and the calendering-induced stresses. The experiments were performed on the calendering system at the *iwb* by integrating an inductive heating unit (IHU). As part of this technical feasibility study, an electrode was compacted to a reference thickness and the effectiveness of IHU to reduce wrinkles was compared at two power settings.

## 2  Experimental and Methodology

### 2.1  Electrode Materials and Calendering System

A commercially procured lithium nickel manganese cobalt oxide (NMC622) cathode was used for the proof of applicability and functionality of the IHU, given that compaction of cathodes is particularly constrained by the occurring wrinkle defects [5]. The double-sided coated electrode contained 95.5 wt.-% NMC622, 2.25 wt.-% carbon black, 0.75 wt.-% conductive graphite additive, and 1.5 wt.-% polyvinylidene fluoride binder material with a mass loading of around 18.3 mg/cm$^2$ per side. The foil width was 180 mm with a coating width of 150 mm. An aluminum foil (1100 alloy) with a thickness of 15 μm was used as the current collector.

The calendering system (GKL 600 MS, Saueressig Group, Germany) of the pilot production line at the *iwb* was used to carry out the experiments. It can operate at web speeds ranging from 0.3 to 30 m/min, with a maximum line load of 2000 N/mm for a coating width of 400 mm. Two calendering rolls, each with a width of 600 mm and a diameter of 550 mm, can be heated up to a maximum of 120 °C. The web tension can be set separately for the unwinder, two web tension units and rewinder in the range from 20 to 500 N. The experiments were performed in an ambient atmosphere.

The electrodes were compacted from an initial thickness of 167 μm to approximately 133 μm, which corresponds to a reduction in porosity from 44% to approximately 29%. A target roll gap of 85 μm was set, which led to a rolling force in the range of approximately 60 kN. The web speed $v_{web}$ was set at 1 m/min and the calendering rolls were not heated. The web tension in the experiments was 30 N at the unwinder and rewinder, with an increased web tension of 50 N set at the web tension unit after the roll gap.

### 2.2  Electrode Characterization

For monitoring the formation of the wrinkles, a laser triangulation sensor (LJ-X8060, Keyence, Japan) was installed after the roll gap, which was used to continuously capture monochrome intensity images and height data of the wrinkles. The measurements were carried out at a frequency of 3333.3 Hz. The laser line had a width of 16 mm. The so-called defect evaluation index (DEI) was used to quantify the wrinkles based on a previous study [8]. This represents the range between the lowest valley and the highest peak of the wrinkles along a profile line. The DEI was analyzed on a profile line at a distance of approximately 0.5 mm from the coating edge. In this study, the mean DEI value from five height images was compared for the reference without the use of the IHU and two IHU parameter levels, which are explained in the following subsection.

The porosity was determined using a gas pycnometer (Ultrapyc 5000 Micro, Anton Paar GmbH, Austria). A minimum of 15 electrode samples with a diameter of 10.95 mm were positioned vertically within the sample chamber. This ensured optimal exposure to the working gas argon. The electrode thickness was measured manually with a tactile dial gauge (ID-C112AX, Mitutoyo, Japan). To investigate the influence of the IHU on the mechanical properties of the current collector foil, electrode samples with a diameter of 15 mm were punched out at the edge zone of the electrode and decoated using a solvent (N-methyl-2-pyrrolidone) and an ultrasonic cleaning device. Subsequently, a laser scanning microscope (LSM; VK-X1000, Keyence, Japan) was employed to capture images of the particle indentations in the current collector foil at a 50-fold magnification. The determination of the number of particle indentations was based on the automated characterization from previous work [13].

## 2.3  Induction Heating Unit

The cathodes were inductively heated utilizing the IHU (HU2000+, Himmelwerk GmbH, Germany), which is a combination of pass-through inductors connected to high-frequency inverters. The IHU has a nominal power output of 2 kW and a frequency range of 0.55 to 1 MHz. The pass-through inductors, constructed from copper, are cooled internally by a water coolant. The IHU system is integrated upstream of the calendering rolls to apply heat precisely to the uncoated edge zone of the electrode. The mechanical integration of the IHU into the calendering system at the *iwb* is presented in Fig. 1. Since only one side of the electrode web was heated during the experiments, the heat-affected zone (HAZ) is shown schematically at the right edge of the electrode web. The HAZ represents the edge zone in which the product properties of the electrode or the current collector foil are influenced by the heat input. The pass-through inductor is displayed in detail in the bottom right-hand corner.

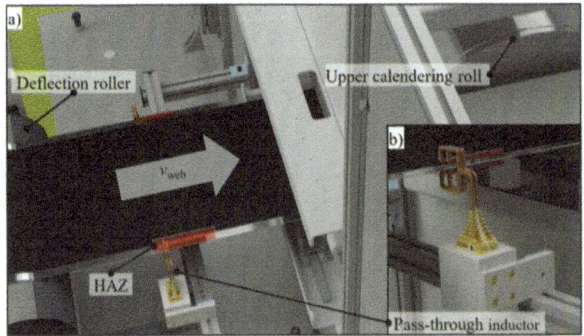

**Fig. 1.**  a) Mechanical integration of the IHU in front of the roll gap and schematic representation of the HAZ of the pass-through inductors; b) detailed visualization of the pass-through inductor.

A pass-through inductor (see Fig. 1b) with a width of approximately 1 cm, a length of approximately 10 cm, and an opening of approximately 1 cm was employed in this experimental setup. The current collector foil was fed through the center of the inductor opening. As the current collector foils used had a low thickness in the region of 15 μm, a low penetration depth is required to heat them. As the penetration depth decreases with increasing frequency, high frequencies are particularly suitable for the thin current collector foils to ensure efficient heating [12, 14]. The frequency of the alternating current passing through the inductor and the IHU power are therefore key variables in this context [11]. In this study three configurations were investigated: (1) a reference without IHU utilization, (2) 10% of the power respectively a nominal power of 0.2 kW, and (3) 11% of the power respectively a nominal power of 0.22 kW. The resulting frequency range was approximately 733 to 737 kHz. The settings were selected based on preliminary tests for the respective compaction from 44% to 29% porosity. To characterize the inductively induced temperature, irreversible temperature indicators (testoterm, Testo AG, Germany) were applied to the uncoated edge area of the current collector foil before passing through the inductor geometries, covering a temperature range from 116 to 154 °C.

The mechanical design and mounting of the inductors allow for flexible adjustment of their position relative to the uncoated edge area and overall electrode width by altering their angle on the profile mounting and shifting them longitudinally. Consequently, the system provides the flexibility to adapt to changing electrode formats in a short changeover time.

## 3   Results and Discussion

In-line image acquisition with laser triangulation sensors permits the evaluation of wrinkles as a function of the parameters of power of the IHU. Figure 2 depicts three exemplary images of the edge zone of the electrodes, representing the three configurations. The black area in the lower section of Fig. 2 represents the electrode coating. At the transition to the uncoated current collector foil, the wrinkles can be observed in varying intensity. The dashed line in Fig. 2a represents the schematic measurement location for the DEI evaluation for all three configurations.

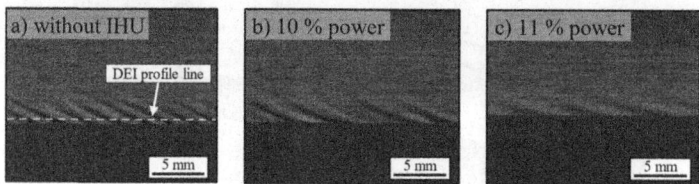

**Fig. 2.** Formation of wrinkles without use of the IHU and at varying IHU power levels.

Figure 2a illustrates the formation of numerous directional wrinkles on the current collector foil during compaction, without the application of localized heating to the edge zone. The height characteristics of these wrinkles vary within the millimeter range and were quantified with a mean DEI value of 0.111 mm, which was in accordance with other studies [10]. Utilization of the IHU with a power of 10% resulted in a reduction in the number of wrinkles (see Fig. 2b), although the mean DEI remained at a relatively elevated 0.088 mm (see Table 1). Upon further increasing the power of the IHU to 11%, only minor wrinkles were observed and the mean DEI decreased to approximately 0.065 mm (see Fig. 2c). The IHU facilitated a substantial reduction of the defect pattern as a result of the localized application of temperature, which increased the ductility of the current collector foil. Table 1 shows the calculated DEI and number of wrinkles in the edge zone area as well as the temperature ranges of the current collector foil as determined by temperature indicators for the corresponding configurations.

The temperature indicators demonstrated that, as anticipated, the temperature of the current collector foil increased correspondingly with the rise in IHU power. The

**Table 1.** Quantification of the IHU parameter regarding the indicated temperature range and the mean values for DEI and number of the detected wrinkles (images analyzed: 5 per configuration).

| Power setting in % | Temperature in °C | DEI in mm | Number of wrinkles |
| --- | --- | --- | --- |
| 0 | 23 (Ambient atmosphere) | 0.111 | ≈ 7 |
| 10 | 121–132 | 0.088 | ≈ 4 |
| 11 | 132–143 | 0.065 | ≈ 4 |

temperature indicators were solely utilized to assess the magnitude of heat generated by induction. Thus, they are not suitable for process monitoring purposes and can only provide an initial assessment of the temperature. It is noteworthy that the temperature values listed in Table 1 are therefore only an indicator for the respective temperature range, which may be different for other types of temperature measurements, e.g., using a pyrometer. The power setting is dependent on the throughput time of the electrode, which correlates with the set web speed. As the web speed increases, the power level of the IHU must also be increased. However, incorrectly set process parameters can lead to overheating of the current collector foil in the uncoated area, which may potentially damage the foil and the electrode materials [12].

To characterize the geometric properties of the HAZ, the particle indentations in the current collector foil were investigated. These are also an indicator of the mechanical stability of the current collector foil, which is essential for further processing in the roll-to-roll process. Samples were taken from the edge zone of the electrode at two distances from the electrode edge for this purpose. The characteristics of the particle indentations were then analyzed using the LSM. Figure 3a–f displays the images recorded for the three configurations at the two measuring points. Figure 3g schematically presents the position of the decoated sample image and the distance $d$ from the edge of the electrode. The top row of images (Fig. 3a–c) shows the samples assumed to be in the area of the HAZ ($d_1 \approx 20$ mm), while the bottom row (Fig. 3d–f) are the ones with larger distances from the IHU ($d_2 \approx 35$ mm).

The analysis of the particle indentations on the untreated current collector foil indicates that there were no apparent differences in the number or the intensity of the particle indentations between the two measurement positions (see Fig. 3a, d). Conversely, it was observed that following heat treatment with the IHU, regardless of the power level, the particle indentations at position $d_1$ become more intense, for instance regarding the number of indentations (see Fig. 3b, c). This indicates an increased ductility of the current collector foil, as the densification was identical for all configurations. The aforementioned finding does not extend to the more distant measuring position $d_2$, as there was no apparent intensification of the particle indentations (see Fig. 3e, f). The number of particle indentations in the LSM images shown at position $d_1$ increased from 162 to 243 for 10% power and to 262 for 11% IHU power, starting from the sample not heated by IHU. The analysis of the particle indentations using LSM supported the initial assumption that the heating of the current collector foil and thus the modification of the edge zone properties only occurred locally in the area of the pass-through inductor. This is advantageous as it prevents unintentional interference with the neighboring electrode

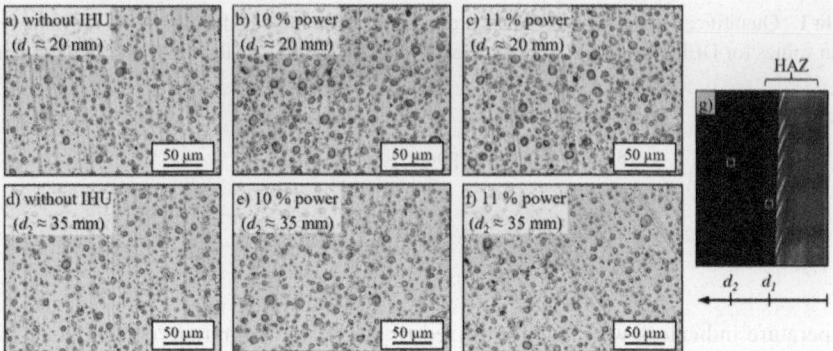

**Fig. 3.** a–f) Characterization of the HAZ regarding the mechanical deformation behavior of the current collector foil in the form of particle indentations; g) schematic representation of the measuring position of the LSM images as dashed lines.

properties. The results proved that the plastic deformation of the current collector foil can be modified by utilizing the IHU. In addition to the reduction of wrinkles, other defect patterns associated with the plastic deformation behavior of the electrode, such as the camber effect [15], may also be reduced by specifically modifying the edge zone properties. Further analysis is required to determine the extent of this interdependence.

## 4 Conclusion

Calendering has a significant influence on the physical properties of electrodes, increasing the volumetric energy density by reducing the coating thickness. High energy densities are associated with calendering-induced defects and therefore scrap, particularly when compacting cathodes. The objective of this study was to implement an inductive heating unit into an existing calendering system to reduce the occurrence of wrinkle defects commonly associated with calendering. The results demonstrated that the inductive heat treatment affected the material properties of the current collector foil, resulting in a reduction in height and number of wrinkles at elevated compaction rates. Consequently, high electrode densities can be achieved without negatively affecting the scrap rate due to defect patterns. The principle of the IHU appears to be a scalable solution compared to alternative methods, allowing high throughput rates and a standardized integration design, which is a prerequisite for industrialization. The integration of an IHU and the heating of the edge zone in a continuous process, combined with in-line and off-line analysis, provides a framework for investigating the formation and reduction of defect patterns.

**Acknowledgements.** We express our gratitude towards the Federal Ministry of Education and Research (BMBF) for funding the research. The presented content in this paper has been achieved within the scope of the projects "InteKal" (03XP0348) and "TUBE" (03XP0425). Furthermore, we would like to thank our research partners and project coordinator Juelich for their trustful cooperation and support within the project. Many thanks to Lucas Hille, Filip Dorau, and Yunhao Liang for their support during the experimental sessions.

# References

1. Kwade, A., Haselrieder, W., Leithoff, R., Modlinger, A., Dietrich, F., Droeder, K.: Current status and challenges for automotive battery production technologies. Nat. Energy **3**, 290–300 (2018)
2. Keppeler, M., Roessler, S., Braunwarth, W.: Production research as key factor for successful establishment of battery production on the example of large-scale automotive cells containing nickel-rich LiNi0.8Mn0.1Co0.1O2 electrodes. Energy Tech. **8** (2020)
3. Schreiner, D., Oguntke, M., Günther, T., Reinhart, G.: Modelling of the calendering process of NMC-622 cathodes in battery production analyzing machine/material–process–structure correlations. Energy Tech. **7** (2019)
4. Abdollahifar, M., Cavers, H., Scheffler, S., Diener, A., Lippke, M., Kwade, A.: Insights into influencing electrode calendering on the battery performance. Adv. Energy Mater. **13** (2023)
5. Günther, T., Schreiner, D., Metkar, A., Meyer, C., Kwade, A., Reinhart, G.: Classification of calendering-induced electrode defects and their influence on subsequent processes of lithium-ion battery production. Energy Tech. **8** (2020)
6. Wurba, A.-K., Altmann, L., Fleischer, J.: Analysis of longitudinal wrinkle formation during calendering of NMC811 cathodes under variation of different process parameters. Prod. Eng. Res. Devel. **18**, 497–506 (2024)
7. Wurba, A.-K., et al.: Methodology for the characterization and understanding of longitudinal wrinkling during calendering of lithium-ion and sodium-ion battery electrodes. Procedia CIRP **120**, 314–319 (2023)
8. Mayr, A., Schreiner, D., Stumper, B., Daub, R.: In-line sensor-based process control of the calendering process for lithium-ion batteries. Procedia CIRP **107**, 295–301 (2022)
9. Mayer, D., Schwab, B., Fleischer, J.: Influence of electrode corrugation after calendering on the geometry of single electrode sheets in battery cell production. Energy Tech. **11** (2023)
10. Bold, B.: Kompensation der Wrinkle-Bildung beim Kalandrieren von Lithium-Ionen-Kathoden. Dissertation, Karlsruhe Institut für Technologie (2023)
11. Wagner, M., Grießl, D., Hiller, M., Kwade, A.: Induction heating as a pre-treatment for the recycling of Li-ion battery cathodes – technical feasibility. J. Clean. Prod. **428**, 139338 (2023)
12. Horstig, M.-W. von, Schoo, A., Loellhoeffel, T., Mayer, J.K., Kwade, A.: A perspective on innovative drying methods for energy-efficient solvent-based production of lithium-ion battery electrodes. Energy Tech. **10** (2022)
13. Hille, L., Hoffmann, P., Kriegler, J., Mayr, A., Zaeh, M.F.: Automated geometry characterization of laser-structured battery electrodes. Prod. Eng. Res. Devel. **17**, 773–783 (2023)
14. Rudnev, V., Loveless, D., Cook, R.L.: Handbook of Induction Heating. CRC Press (2017)
15. Tran, H.-Y., Lindner, A., Menesklou, W., Braunwarth, W.: Toward calenderability of high-energy cathode based on NMC622 during the roll-to-roll process. Energy Tech. **11** (2023)

# Identification of Key Challenges and Optimization Options Along the Process Chain for the Manufacturing of a Thermally Optimized Injection Molding Tool Using the DED Process

Thore Gericke[✉], Moritz Jens, Nik Schwichtenberg, and Alexander Mattes

Kiel University of Applied Sciences (UAS), Institute for Production Technology and CIMTT, Schwentinestr. 13, 24149 Kiel, Germany

```
{Thore.Gericke,Moritz.Jens,Nik.Schwichtenberg,
Alexander.Mattes}@fh-kiel.de
```

**Abstract.** This paper describes the energy saving approaches along the complete path from the CAD model to the final post process step of a multi-material component manufactured with DED, using a thermally optimized injection tool as an example. The study focuses in particular on the possible optimization approaches for necessary iterations caused by instabilities during the production of complex components. These are different from approaches used in conventional manufacturing processes, as the special characteristics of additive manufacturing using DED must be considered. Improvements are achieved through geometry-specific process parameter optimization, NC-toolpath optimization strategies and a process monitoring through melt pool analysis. For this purpose, an existing CCD camera is used to detect deviations by means of AI-based image segmentation and classification. As a result of the optimizations carried out, the production time and the number of iterations for a comparable component were significantly reduced and considerable energy savings of 53.4% were achieved.

**Keywords:** Direct Energy Deposition · Process Chain · Multi-Material · Injection Molding · Conformal Cooling

## 1 Introduction and Motivation

Additive Manufacturing by powder-based Direct Energy Deposition (DED) has several advantages adverse to the widespread powder bed process: Higher build-up rates, ability of component repair by build-up on existing parts and enabling the connection of different materials cohesively with comparatively low effort. Furthermore, the high geometric degree of freedom with good accuracy thanks to the 5-axis guidance of the application nozzle gives the possibility to realize complex internal structures like cooling channels in combination with Multi-Material (MM) and Functionally Graded Material (FGM) [1, 2].

© The Author(s) 2025
H. Kohl et al. (Eds.): GCSM 2024, LNME, pp. 443–451, 2025.
https://doi.org/10.1007/978-3-031-93891-7_49

However, it is essential to consider the energy consumption per component when manufacturing additively using DED machines. Due to the long manufacturing cycles, unsuccessful production attempts in particular and the high energy consumption of DED machines have a significant impact on the overall energy consumption. Getting to the consumption of DED machines, the variable quantity, depending to large extent on the actual material processed, and the quantity, which requires constant power in proportion to the operating time - regardless of processing - have to be considered [3, 4].

In order to unlock the potential for energy savings, there exist two levers:

Firstly, often three to four experimental productions must be performed on average, until a new component meets the quality requirements. Consequently, a reduction in the number of unsuccessful tests conducted during the experimental determination of an optimal manufacturing strategy will directly result in a significant energy saving. Secondly, the average power consumption of the DED machine at Kiel UAS "Addup BeAM Modulo 400", with an electrical power rating of 50 kW, is still 86.9% of average power consumption during non-productive times, due to the large number of aggregates. Especially when an inert gas atmosphere is used, the machine cannot be switched off or reconfigured at any time during iterative process adjustments. This results in an average of 66.2% non-productive time, which was measured during the production of new components. This ratio demonstrates that reduction of non-productive time is a significant lever for reducing overall energy consumption. For the production of the first mold core, a production time of 4.57 h, using 208.5 kWh, per component and a total of 5 iterations, using 1,042.7 kWh, were required. The objective is a significant reduction in production time and iterations through the use of highly effective optimization methods for a comparable injection molding tool core (see Fig. 3a and 3b).

## 2  DED Optimization Methods

All steps of manufacturing by DED represent a complex system of process-related influencing variables that need to be considered. Particularly in the initial production of series components and prototypes, various sub-steps go through an iterative process that is associated with increasing effort along the process chain. Figure 1 illustrates the fundamental sub-steps from the CAD model to the error-free initial sample, as well as the iterative steps and optimization methods which are explained below.

Regarding the processing parameters, the laser power, feed rate, nozzle distance and powder mass flow are significant factors. Due to the diverse influences, these parameter sets must be determined separately for each material and for all mixing ratios when using multi-materials. Optimization of these parameters is the first method to reduce energy consumption per workpiece through reduced iterations (cf. Sect. 3). The determination of material-specific process parameters represents the first iterative step (loop 1) in the process chain and is the basis to ensure sufficient quality of the produced feature. Additional systematic instabilities are in most cases due to insufficient deposition strategy or the developed process parameters for combined geometries (loop 4). The process instabilities may result in either excessively high or low heat input, prevent a stable build-up and potentially leading to significant shape deviations [7].

All expertise gained during the development of an optimal deposition strategy for each geometric feature of the component is integrated during the toolpath optimization of the application nozzle in CAM Programming (cf. Sect. 4). [5] In order to guarantee an optimal NC program, it is crucial to consider the typical machine-related factors, such as kinematic limitations or the accessibility of the nozzle. The toolpath is optimized through the use of simulation of the additive material deposition (loop 2), which is employed to verify the programmed deposition.

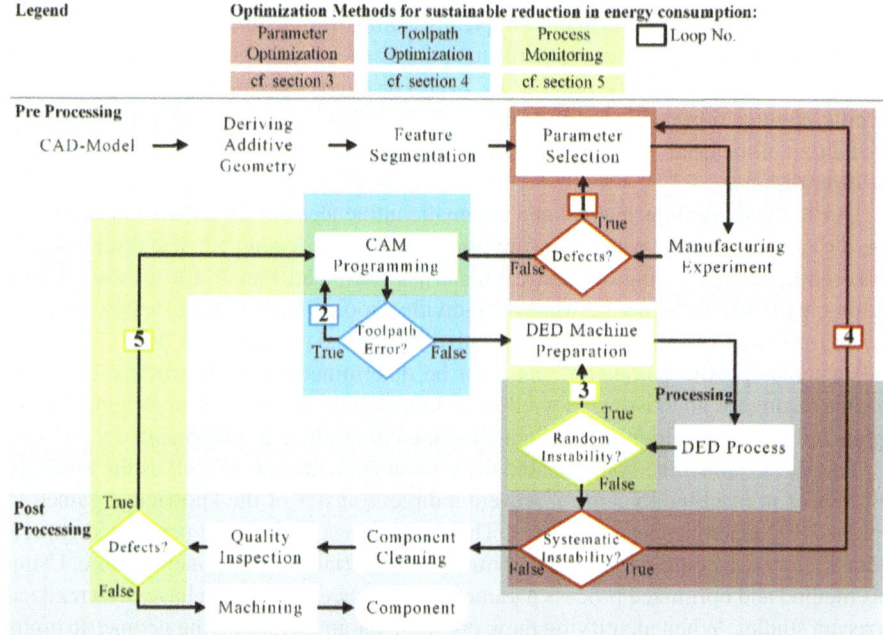

**Fig. 1.** Flow chart that illustrates the optimization methods during the DED process chain from the digital model (CAD-Model) to the physical part (component)

During the actual execution of the additive deposition complex interaction of all technical influencing variables can lead to instability, which need to be detected by a sufficient process monitoring. DED is always accompanied by thermally induced stresses due to the process and is therefore highly geometry-dependent. In order to conserve energy in the event of insufficient components due to random errors, a process monitoring system is currently being developed (cf. Sect. 5). The randomly occurring influences can be divided into two groups: process-interrupting (loop 3) or defect-causing (loop 5) disruptive influences in the process. In both cases, the components cannot be used afterwards. [6] Typical random errors include instabilities in the powder mass flow, misalignment of the laser focus lens or material build-up in the nozzle. This results in cracks at the component level and inherent defects such as flaws and porosity. The geometry of the machined component cannot have any of these defects which can, in most cases, only be detected using radiographic testing methods [8].

## 3   Parameter Optimization

In the research project presented here, a multi-material combination between the two materials 1.2311 (40CrMnMo7) and CW305G (CuAl10Fe1) is produced in order to optimize the thermal properties of a mold core for an injection molded component. Specifically, previously unfeasible, conformal cooling channels are integrated during additive manufacturing using DED. The manufacturing of these require a high level of geometric accuracy, as these channels typically cannot be machined subsequently. Comprehensive studies are therefore required to optimize the process parameters. The process time per component could be reduced by 22.3% through experimental and application-specific optimization of the production parameters. This was achieved in particular by a reduction of material-specific stabilization times and a feature-related optimization of the layer distance.

Due to the thermomechanical and kinematic influences during additive manufacturing, the optimal process parameters are highly geometry-dependent. The discrepancies observed in the processing of identical materials can be attributed, for instance, to the reciprocal thermal influence between the individual toolpaths. [9] These include the laser power, traversing speed, layer distance, path distance and powder mass flow. Geometry- and material-specific parameter sets must be determined using destructive or radio-graphic testing for all existing geometries resulting on the basis of feature segmentation, optimized iteratively (loop 1) and then combined to produce a component.

The various machine concepts and drive parameterizations, as well as the resulting differences in machine dynamics, prevent a direct transfer of the knowledge gained to similar DED machines and processes. Therefore, investigations are performed experimentally for each combination of DED machine, material and component feature. Using this method and optimized process parameters, tolerances of 0.1 mm have been realized in recent studies. When identifying these optimum parameters, existing geometric limits have to be considered. These include, for example, geometry-specific minimum toolpath lengths, layer thicknesses and bore diameters, as well as the achievable accelerations of individual traversing axes and for multi-axis movements (loop 4).

In addition to the geometric limitations, a number of material-, and process-specific constraints emerge since defects may occur during DED process with unsuitable parameters. Especially in multi-material applications complex mechanisms occur due to material mixing, particularly in material transition zones, which can only be detected using extensive analysis methods. The identified defects include cracks (Fig. 2a), pores (Fig. 2b) and localized material accumulations (Fig. 2c).

## 4   Toolpath Optimization

The simulation of the material deposition used here, which includes the physical machine characteristics and the existing machine kinematics, can be used to detect errors in the path generation at an early stage and derive appropriate optimizations (loop 2, see Fig. 3f). The toolpaths of the application nozzle are programmed and simulated using a commercially available computer-aided manufacturing (CAM) software initially used for machining processes. Therefore, adjustments have to be made manually, but they

**Fig. 2.** Cracking (a), pores (b) and material accumulation (c) during parameter optimization

are able to reduce the number of experimental production tests. For the mold core under consideration, the number of necessary iterations could be reduced from five to three by optimizing toolpaths at an early stage.

All adjustments require several different programming steps, which consist of specific programming strategies and have significant potential for improvement in the time required for programming. The programming strategies for the manufacturing of specific geometric features are being verified experimentally and documented individually in order to transfer them to other similar feature components. The conformal cooling channels are the most complex geometric elements manufactured for this component, as the required dimensional accuracy must be achieved throughout the DED process without any further post-processing. The surface of the outer geometry is ultimately processed by milling, thereby enabling the attainment of markedly reduced dimensional and surface tolerances.

Firstly, the additive geometry is derived from the 3D model according to the specific component features that are to be manufactured using the additive manufacturing process, considering the allowances for subsequent post-processing. The complete mold core is divided into layers, each of which is equally spaced vertically. This is a common approach in many other additive manufacturing processes. Intersection curves were calculated for each of these layers, which provide the geometry for subsequent programming steps (Fig. 3c). Secondly, the synchronization curves and reference paths are generated. The areas of the mold that represent a single closed volume are constructed using 3-axis toolpaths. To form the upper half of the cooling channels, it is necessary to create overhangs. For these areas, the movement position of the powder nozzle contains a variable angle of attack of up to 45°. 3 + 2-axis and 5-axis toolpaths are used. Derived synchronization curves are defined in terms of the requisite tool vectors at relevant points, which can then be interpolated (Fig. 3d).

Already available calculation cycles can be used to program the boundary curves of any equidistant plane, but do not consider the special requirements of the DED process (see Fig. 3e). Due to the limited machine dynamics and the strong effects of parameter adjustments, extensive iterative adjustments are required for sufficient path generation. The final stage of the process is therefore the calculation and verification of the additive paths. In order to realize the additive paths of each layer, a specific DED functionality is used within the CAM system. The toolpaths of the filling areas of each layer are derived from the previously created references.

Two additional adjustments must be made to create the overhangs. First, the coordinate system is tilted according to the calculated synchronization curve so that the Z-axis is once again in line with the nozzle installation direction. The increase in distance caused by changing the angle of attack must then be compensated. Another slicing with smaller layer distances must therefore be carried out in order to correctly cover the areas with sufficient toolpaths to create overhangs.

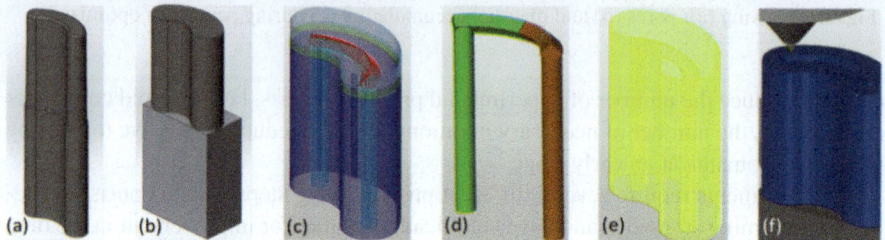

**Fig. 3.** 3D model of the mold core (a), additive geometry (b), calculated cutting curves (c) synchronization curves (d), reference paths (e) and NC-simulation (f) during CAM programming

## 5   Process Monitoring

The experienced machine operator is able to identify manufacturing defects and deviations as soon as they accumulate by observing the process. This enables the user to intervene in the process and avoid the wastage of several hours, thus reducing the overall energy consumption. Nevertheless, due to the extensive times required for larger components, it is not always feasible to manually monitor the entire process. The implementation of process monitoring has the potential to reduce energy consumption within the iterative DED process chain. This is because the time required for manufacturing could be significantly reduced if the production process were to be halted as soon as errors are identified in real time. It enables the necessary iteration to be performed at an earlier stage or to prevent it completely, thus preventing the wastage of material and energy. Consequently, a number of data-driven approaches are currently being developed with the objective of automating the process monitoring.

The foundation for a monitoring system capable of detecting both random process instabilities (loop 3) and defect-causing influences (loop 5) during the process, which frequently become visible only in the quality inspection after component production, is the acquisition of relevant process and quality data at sufficient frequency and quality. [10] The optimal stage for data collection in the DED process chain is during the manufacturing experiment, as this stage offers the highest probability of capturing instabilities. The DED machine is equipped with various sensors suitable as a data source. For example, integrated calibration tools, such as trace functions, are particularly suited for the recording of data related to the toolpath and laser power in order to detect systematic deviation from programmed parameters. Furthermore, an optical CCD camera system records the melt pool in real time, in a coaxial configuration with the laser beam to enable detection of random errors during melt pool formation. The data collected through these data sources during failed manufacturing experiments becomes a valuable resource for artificial intelligence (AI) models. This method leads to an expansion of the database, enabling the potential to improve the monitoring system over time in order to sustainably prevent the occurrence of energy and material waste.

The data collected can be utilized for AI approaches, that enable real-time defect detection. Specifically, process data (e.g., melt pool images) and quality data (e.g., deviation from optimal nozzle distance) are integrated to train supervised learning classification models. These models predict defects or process errors based on patterns in the process data. Especially for problems with diverse influencing factors, the real time images of the melt pool collected by the CCD camera enables detection of random process instabilities regarding the powder mass flow, misalignment of the laser focus lens or material build-up in the nozzle. [11] Two methodologies for utilizing the melt pool images in the context of process monitoring are presented below. Firstly, the YOLOv8 classification model [12] was fine-tuned using these melt pool images, achieving an accuracy of 0.2 mm in determining the distance between the nozzle and the component (Fig. 4b). Secondly, the YOLOv8 segmentation model [12] was optimized by additional training with manually segmented images to accurately retrieve the shape of the melt pool and calculate the characteristics (Fig. 4a). Current research focuses on detecting instabilities based on these characteristics.

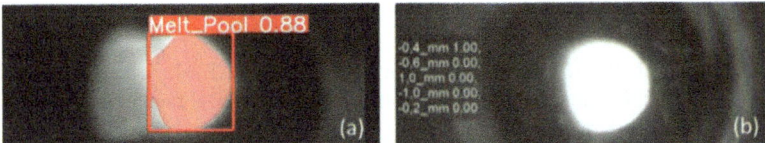

**Fig. 4.** Melt pool segmentation (a) and classification of melt pool distance with corresponding probability (b) on images coaxial to laser beam

## 6    Conclusion and Outlook

The case study presented here shows the optimization approaches during the entire process chain for the successful production of a thermally optimized injection mold core with conformal cooling channels made of multi-material using the DED process. However, the high average proportion of 66.2% non-productive time and the required number of iterations in the production of a new component demonstrates that there is considerable potential for reducing energy consumption. The return loops required during the process chain and the extensive reliance on manual adjustments and experimental investigations make application in an industrial environment considerably challenging.

Through experimental and feature-related optimization of the process parameters, the production time per component could be reduced, in particular by optimization of material-specific stabilization times and the layer distance. By optimizing the tool paths on the basis of a material deposition simulation, which is integrated into the CAM software, the number of production tests could be significantly reduced. Experimental process monitoring to detect process instabilities was achieved by means of melt pool analysis. Images of the melt pool captured using a CCD-camera were combined with AI-based data evaluation. This made it possible to detect random instabilities and defect-causing influences, thereby reducing the potential for energy wastage in the future.

By using the optimization approaches presented here, both the manufacturing time per part ($-22.3\%$) and the number of iterations ($-40\%$) for a comparable mold was significantly reduced. This resulted in energy savings of approximately 556.3 kWh or 53.4% for the production of the investigated components. The performance of the manufactured multi-material mold is currently being verified in an industrial application. If this proves to be successful, further optimization of molds by FGM will be pursued, which requires significantly more extensive experimental optimization. The study carried out here resulted in several other research topics and optimization approaches.

**Acknowledgments.** This Research is supported by the Federal Ministry for Economic Affairs and Climate Action (BMWK) on the basis of a decision by the German Bundestag.

 Federal Ministry
for Economic Affairs
and Climate Action

# References

1. Gericke, T., Rickerts, L.M., Mattes, A., Schimmelpfennig, T.-M.: Thermal optimization of injection molds using functionally graded materials. In: Fraunhofer Direct Digital Manufacturing Conference, Berlin (2023)
2. Ahn, D.-G., Park, S.-H., Kim, H.-S.: Manufacture of an injection mould with rapid and uniform cooling characteristics for the fan parts using a DMT process. Int. J. Precis. Eng. Manufact. (2010)
3. Schoenung, J.M., Lavernia, E.J., He, H., El-Azab, S.: Specific energy of metal am feedstock: a comparison. Add. Sustain. **5** (2024)
4. Holkup, T., Vyroubal, J., Smolik, J.: Improving energy efficiency of machine tools. In: 11th Conference on Sustainable Manufacturing, Berlin (2013)
5. Zhang, R., Jiang, F., Xue, L., Yu, J.: Review of additive manufacturing techniques for large-scale metal functionally graded materials. Crystals (2022)
6. Frazier, W.E.: Metal additive manufacturing: a review. J. Mater. Eng. Perform. **23** (2014)
7. Neubert, J., Weilnhammer, G.: Laserstrahl-schweißen: Leitfaden für die Praxis. DVS Media, Düsseldorf (2009)
8. Latif, K., Yusof, Y., Kadir, A.Z.A.: New method to utilize STEP-NC data interface model for 3D printing. Prog. Add. Manufact. **8**, 1677–1686 (2023)
9. Mazzucato, F., Aversa, A., Doglione, R., Biamino, S., Valente, A., Lombardi, M.: Influence of process parameters and deposition strategy on laser metal deposition of 316L powder. Metals **9**, 1160 (2019)
10. Schwichtenberg, N., Komkowski, T., Pongboonchai-Empl, T., Neustock, J., Mattes, A.: Data driven welding optimization. In: Advanced Course on Data Science & Machine Learning (2023)
11. Era, I.Z., Farahani, M.A., Wuest, T., Liu, Z.: Machine learning in directed energy deposition (DED) additive manufacturing: a state-of-the-art review. Manufact. Lett. **35**, 689–700 (2023)

12. Jocher, G., Chaurasia, A., Qiu, J.: Ultralytics YOLO. Ultralytics, version 8.0.0, released January 10, 2023. https://ultralytics.com. Accessed 20 Aug 2024

# Hybrid Manufacturing Process Chain for a Speedboat-Propeller: A Case Study on Process Sustainability

F. J. Matschnig, M. Hoffmann⬤, G. Mauthner(✉) ⬤, T. Trautner⬤, and F. Bleicher⬤

TU Wien, Institute of Production Engineering and Photonic Technologies, Karlsplatz 13, 1040 Vienna, Austria

mauthner@ift.at

**Abstract.** Additive-subtractive hybrid manufacturing has realized significant development in recent years. Utilizing such hybrid manufacturing process chains, complex-shaped part geometries can be manufactured near net shape in a resource- and energy-efficient manner. However, key aspects for a successful application of these process chains remain the systematic selection of additive and subtractive part features, integrated CAD/CAM toolpath planning functionality as well as the accurate identification of the in-process workpiece geometry. This paper investigates the impact of exploiting hybrid manufacturing over traditional approaches on metrics such as cycle time, raw material usage, and energy demand to further drive sustainability aspects in the manufacturing environment. An experimental case study has been performed utilizing a hybrid manufacturing cell at the Institute of Production Engineering and Photonic Technologies at TU Wien consisting of a 6-axis wire-arc additive manufacturing (WAAM) robot with 2-axis part manipulator, a 3D-laser-scanner, and a 5-axis machining center.

**Keywords:** Additive manufacturing · Energy efficiency · Manufacturing process

## 1 Introduction

In recent years, the awareness for more sustainable production technologies has significantly increased leading to additional research efforts to analyze and evaluate different manufacturing processes in regards to their environmental impact [1]. One promising technology towards more sustainable manufacturing is *Wire-Arc Additive Manufacturing (WAAM)*. WAAM, as one example of several existing metal additive manufacturing processes, has highlighted its potential for reduction in material waste in different studies [2] and has proven to be an alternative to conventional manufacturing techniques in selected applications [3]. Although numerous studies on this process exist in literature, there is still a recognizable lack of data regarding the application of WAAM for different geometries and materials, as well as its direct impact on material savings, energy demand and processing times compared to conventional manufacturing processes. In this study a beneficial use-case to compare a hybrid process chain of WAAM and subtractive manufacturing with the conventional purely subtractive process is carried out in order to provide additional data and insights on their environmental impact.

© The Author(s) 2025
H. Kohl et al. (Eds.): GCSM 2024, LNME, pp. 452–460, 2025.
https://doi.org/10.1007/978-3-031-93891-7_50

## 2   Literature Review

Additive manufacturing such as WAAM, or *Direct Energy Deposition (DED)* have proven to be advantageous to saving raw material and energy consumption due to the layered construction of near-net-shape metal components [4]. Researchers have analyzed these saving potentials and compared them with different additive manufacturing technologies such as *Selected Laser Melting (SLM)* as well as with conventional processes such as forging, hot-rolling or milling, commonly using *Life Cycle Assessment (LCA)* methodology. An overview of selected research results is given in this chapter.

Several researchers focused on comparing the environmental impact when using different additive processes such as WAAM and SLM. *Kokare et al.* compared the production of a simple steel wall using WAAM, SLM and conventional milling. Based on the provided data, SLM turned out to be the least sustainable process with its environmental impact being four times higher than with WAAM and even more compared to milling [5]. *Wurst et al.* showed that the production of a hollow sphere made of aluminum using WAAM results in a lower ecological impact compared to SLM [6]. *Jackson et al.* compared powder-based DED technologies with WAAM on the basis of energy consumption and showed that WAAM processes consume 85% less energy during deposition [7]. Another study has been conducted by *Campatelli et al.*, who compare WAAM with SLM highlighting the electric energy demand of WAAM being lower by more than one order of magnitude [8]. Beside SLM, WAAM processes have been challenged by researchers in comparison to various conventional manufacturing techniques. *Shah et al.* used a LCA to show that topology optimization and material savings through WAAM can reduce the environmental impact in the production of I-beams by up to 24% compared to conventional manufacturing using hot rolling [9]. Another study shows that WAAM enables a 40% reduction in energy consumption and 55% material savings compared to conventional forging of a Ti6Al4V component [10]. *Bekker and Verlinde* highlighted that the environmental impact of WAAM is comparable to that of sand casting. In their LCA CNC processes perform poorly compared to WAAM [11]. Further LCAs and comparative studies also show that WAAM is advantageous compared to various manufacturing processes and can lead to reductions in material usage, energy consumption and cost in some applications [12–16]. Based on the industrial relevance and wide usage of conventional subtractive technologies, a comparison of these manufacturing types with WAAM processes is of special importance. *Campatelli et al.* compare WAAM with subtractive manufacturing (milling). The study shows that the utilization of WAAM processes results in a reduction of energy demand, particularly in the raw material production phase of the LCA. Leaving raw material production out of the analysis, WAAM turns out to be more inefficient than conventional milling [17]. On the contrary, *Kokare et al.* highlights the opposite using a more complex propeller geometry and in his calculation comes to the conclusion that production using WAAM (including subtractive postprocessing) consumes less energy than purely subtractive production [18].

To summarize, there is already a large amount of research data on energy consumption and process sustainability when using metal additive manufacturing techniques such as WAAM. Various studies comparing WAAM to conventional manufacturing techniques indicate that there is a significant potential for reduced material and energy consumption. However, the results are dependent on specific use-cases and related input factors such as the baseline technology, the materials used or the final geometry. The results of the comparison vary in the presented state of the art. This indicates that there is a lack of sufficient process data from different use cases. Modern CAD/CAM software could be utilized to derive performance indicators like cycle time and material usage and get information regarding the economic performance of the process.

## 3   Hybrid Manufacturing Process Chain: A Use-Case

The aim of this paper is to evaluate a selected hybrid manufacturing process chain using WAAM in comparison to traditional milling processes and its impact on relevant performance indicators such as processing times, material usage and energy consumption. For this purpose, a demonstration part geometry of a speedboat propeller has been selected, and process definitions for conventional CNC milling as well as hybrid manufacturing strategies utilizing WAAM have been developed. The relevant performance indicators have been measured during the production of a demonstration part.

### 3.1   Demonstration Part: Speedboat Propeller

Today, WAAM is often used to manufacture complex part geometries in the energy or shipping sector. Especially blade geometries such as Pelton or Francis blades have frequently been selected to highlight WAAM capabilities. Hence, a *Formula Light Speedboat Propeller* was chosen as a demonstration part for this study. Figure 1 illustrates the selected demonstration geometry and the robot cell for the study in this paper.

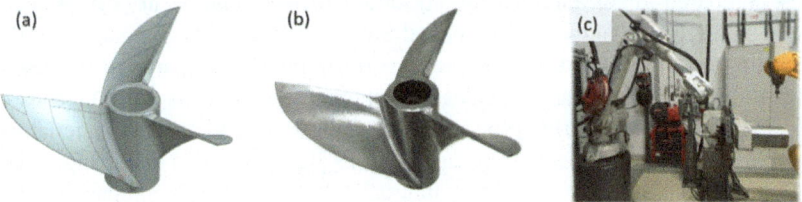

**Fig. 1.**   (a) CAD geometry; (b) Final part manufactured; (c) Robot Cell used at *IFT*

Such propeller geometries are conventionally manufactured in small quantities from solid material applying CNC milling, using a full bar/block of raw material. Complex curvatures and large amounts of raw material being machined provide an ideal baseline for comparing the WAAM process with conventional machining and to show its potential for more sustainable manufacturing overall. The chosen propeller has a diameter of 203 mm, a total height of 85 mm and consists of a total of three blades that converge to a blade thickness of 1 mm towards the maximum diameter. In the course of this study the propeller (final mass: 0.74 kg), was manufactured using a solid wire of grade G3Si1 which is usually used for welding structural steel, pipe steels and shipbuilding steels.

## 3.2   Experimental Setup

To evaluate potential advantages, two process chains will be considered in this study. As a baseline process, conventional milling strategies have been applied to produce the propeller out of a steel bar (D: 205 mm, H: 90 mm) solely using subtractive methods. In contrast, a hybrid process chain consisting of the WAAM part build-up, which is the near net-shape production of the three blades, and milling, which is the final machining to target geometry, has been designed and developed. Figure 2 summarizes the process chains which will be compared in this study.

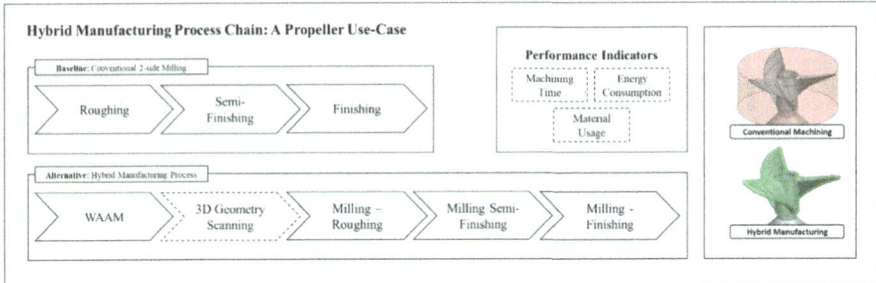

**Fig. 2.** Process chains investigated in the study.

For the *hybrid process*, a manufacturing strategy utilizing hardware and software at the laboratory of *Institute of Production Engineering and Photonic Technologies (IFT)* has been developed and utilized to execute selected cutting tests for performance measurement. First, a special fixture to hold a pre-turned cylindrical substrate body, which acts as raw material for the WAAM process, was machined. The fixture holding the substrate body is mounted on the part manipulator of an *IRB 2600* 6-axis welding robot, equipped with a *TPS 400i* welding power source. The fixture itself is designed to improve the accessibility of the welding torch as well as the cutting tool for final machining. Second, the following WAAM process consists of the layer-by-layer deposition of the blades onto the cylindrical blank. A total of 34 layers with an average layer height of 2.3 mm is applied for each of the three propeller blades. To manage part quality issues due to heat accumulation during the welding of the different blade layers, three counter measures have been integrated in the manufacturing strategy. (a) During the material build-up, the welding torch switches between the three blades after each layer; (b) After every third layer, a cooling time of 30 s is applied (no welding occurs); (c) Additionally, the welding parameters are continuously adjusted, Table 1 list those.

A CMT synergic line for steel developed by the manufacturer *Fronius* (synergic line 3148) was used throughout the entire build-up process. Third, in between the WAAM and following milling process of the demonstration part, the in-process workpiece is digitized using a manual 3D scanner. The resulting point cloud was converted into a STL object which was used for optimal CAD/CAM toolpath planning for the final machining step. The toolpaths for the additive as well as subtractive operations has been created utilizing modern CAD/CAM software (*Siemens NX 2212*), existing machine models of both, robot and CNC machining center, and in-house developed post-processors.

**Table 1.** Tools and cutting parameters used for subtractive machining.

| Layer Nr | Synergic line | Torch speed (mm/s) | Wire feed (m/min) |
|----------|---------------|--------------------|-------------------|
| 1–2      | 3148          | 6.66               | 3.4               |
| 3–4      | 3148          | 5.00               | 2.5               |
| 5–6      | 3148          | 4.16               | 2.1               |
| 7–34     | 3148          | 3.33               | 1.7               |

Lastly, the part is machined to the final geometry using a *DMGMori DMU75 monoBLOCK* 5-axis milling center with a *Fanuc 31i* numerical control. The baseline milling process has been designed with support of experienced technical personnel at *IFT* laboratory. Industrial relevant milling tools as well as cutting parameters have been selected. Table 2 lists the cutting parameters used in the experiments. In contrast to the hybrid manufacturing strategy, purely subtractive manufacturing required an additional roughing operation in the beginning (Ø25 end mill, 4-fluted, spindle speed = 2037 rpm, feed rate = 1222.2 m/min).

**Table 2.** Tools and cutting parameters used for subtractive machining.

| Milling Process | Tool | Spindle Speed (rpm) | Feed rate (m/min) |
|-----------------|------|---------------------|-------------------|
| Roughing        | Ø16 Endmill, 4 flutes  | 1592 | 636.8 |
| Semi-Finishing  | Ø16 Ballmill, 2 flutes | 1989 | 198.9 |
| Finishing       | Ø10 Ballmill, 2 flutes | 3820 | 382.0 |

Based on the described input parameters, a CAD/CAM cutting simulation provided relevant data about total cycle time and total material removal. When setting up the CAD/CAM process chain, particular attention was paid that the entire process is covered in one integrated software package, and functionalities for both, additive as well as subtractive machine program generation, exist. Utilizing the described CAD/CAM software and respective machine tools, the demonstration part can be manufactured and important performance indicators for total cycle time as well as material usage can be derived using the simulation capabilities of CAD/CAM system as well as various measurement equipment.

To retrieve data about the energy consumption of the two described process chains, *Siemens Sentron PAC 4200* multifunctional measuring devices were used. The energy consumption of both, the robotic system as well the milling machine, was measured directly at the feed-in to ensure that the data of the whole manufacturing cell is collected. The energy data of the robotic cell therefore includes the 6-axis robot, the 2-axis part manipulator, as well as the welding power source. The (aggregated) energy data is received via *ModbusTCP* using an in-house developed python script deployed on a network *EDGE device*. On the EDGE device, the data is converted to *MQTT* data format and transferred to a central timeseries. From there, another python script is used to analyze the data and calculate the overall power consumption by numerical integration methods.

## 4  Use-Case Validation

The baseline process as well as the hybrid process chain have been evaluated using the described manufacturing setup. Respective manufacturing data points have been gathered along the process chains using above mentioned data collection methods. In this chapter, the results comparing the two processes variants and the impact on productivity and environmental sustainability are highlighted. For this purpose, three important performance indicators have been selected: (1) machining time; (2) energy consumption of the machining systems; (3) material usage.

The manufacturing time measured during this study reflects the actual machining time on the machine. In the case of the hybrid process, this also includes the material build-up and interlayer cooling times. The time needed for re-clamping and scanning in between the two processes is excluded. The same applies to the purely subtractive process. Cycle time comparison showed that both processes have similar manufacturing times overall. Although larger amounts of material must be removed, the purely subtractive process turned out to be 2% faster compared to the near-net-shape approach. One reason for the conventional manufacturing approach being faster is the interlayer cooling time of the WAAM process which is necessary to prevent overheating. Although the part geometry allows to reduce those waiting times by switching from blade to blade in the build-up process, they make up almost half of the manufacturing time.

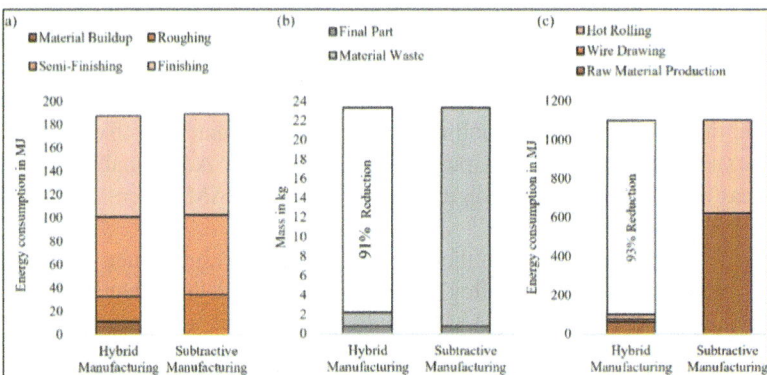

**Fig. 3.** Performance indicator process comparison of hybrid- and subtractive manufacturing: (a) Total energy consumption during manufacturing; (b) Material waste comparison; (c) Energy consumption of raw material production and pre-manufacturing

Figure 3 highlights three performance indicators relevant for sustainability comparison of the investigated demonstration part. Comparing total energy consumption of the hybrid and the solely subtractive machining process, both provide equal results (1.7% deviation). Even slight changes in either manufacturing strategy could lead to one being more energy efficient than the other. While manufacturing times as well as machining system energy consumption of the two processes are similar, the material consumption highlights significant benefits of WAAM processes over conventional subtractive milling. With the subtractive manufacturing approach there is a total amount of over 22 kg of steel waste produced to reach a final part mass of just 740 g. The hybrid approach produces just 1.5 kg of material waste in the production of the final part. This shows that a material waste reduction of over 91% can be achieved by switching to hybrid manufacturing with WAAM. Using the data collected by *Campatelli et al.* [17] and applying the characteristic values to the retrieved material waste, the energy consumption of raw material production and pre-manufacturing can be calculated. As shown in Fig. 3 WAAM has potential to reduce this energy consumption significantly by up to 93%. Combining the energy consumption of raw material production, pre-manufacturing and of actual machining, the solely subtractive process ends up consuming 4.6 times as much energy as the hybrid process. Additionally, it has been shown, that the production of such complex geometries is not economically feasible without the usage of integrated CAD/CAM software systems and their simulation capabilities.

## 5 Conclusion and Outlook

WAAM has been highlighted by researchers as a promising technology contributing to a more sustainable manufacturing environment. This study highlights that hybrid process chains show significant benefits regarding material usage and total energy demand when including raw material production steps. For the investigated demonstration part, the hybrid process saved around 78% of energy and 91% of material compared to the solely subtractive process. Regarding total machining time and energy consumption during machining, both process chains are at comparable level, indicating that advantages regarding near net-shape build up are weakened due to the additional WAAM process step. Further tests need to be carried out to determine how far the WAAM process can be pushed in terms of required cooling times, optimized material offsets and process parameters to further improve the performance indicators. Additionally, the baseline process has been evaluated using digital data from CAD/CAM system in combination with selected cutting test.

Future research could focus on the sensibility of these results when using different materials such as titanium or high alloy steel qualities. Overall, the gathered data supports that hybrid processes utilizing WAAM technology can contribute significantly to a more sustainable and resource efficient manufacturing environment. The reduction in material consumption, thus, reduction in overall energy usage justify intensive research work to support further industrialization of these processes.

## References

1. Gbededo, M.A., Liyanage, K., Garza-Reyes, J.A.: Towards a life cycle sustainability analysis: a systematic review of approaches to sustainable manufacturing. J. Clean. Prod. **184**, 1002–1015 (2018)

2. Agnusdei, L., Del Prete, A.: Additive manufacturing for sustainability: a systematic literature review. Sustain. Futures **4**, 100098 (2022)

3. Dias, M., Pragana, J.P., Ferreira, B., Ribeiro, I., Silva, C.M.: Economic and environmental potential of wire-arc additive manufacturing. Sustainability **14**(9), 5197 (2022)

4. Javaid, M., Haleem, A., Singh, R.P., Suman, R.: Rab, S: Role of additive manufacturing applications towards environmental sustainability. Adv. Ind. Eng. Polym. Res. **4**(4), 312–322 (2021)

5. Kokare, S., Oliveira, J.P., Santos, T.G., Godina, R.: Environmental and economic assessment of a steel wall fabricated by wire-based directed energy deposition. Add. Manufact. **61**, 103316 (2023)

6. Wurst, J., Steinhoff, T., Mozgova, I., Hassel, T., Lachmayer, R.: Aspects of a sustainability focused comparison of the wire arc additive manufacturing (WAAM) and the laser powder bed fusion (LPBF) process. In: Sustainable Design and Manufacturing (SDM 2022), pp. 88–97 (2023)

7. Jackson, M.A., Van Asten, A., Morrow, J.D., Min, S., Pfefferkorn, F.E.: A comparison of energy consumption in wire-based and powder-based additive-subtractive manufacturing. Procedia Manufact. **5**, 989–1005 (2016)

8. Campatelli, G., Campanella, D., Barcellona, A., Fratini, L., Grossi, N., Ingarao, G.: Microstructural, mechanical and energy demand characterization of alternative WAAM techniques for Al-alloy parts production. CIRP J. Manufact. Sci. Technol. **31**, 492–499 (2020)

9. Shah, I.H., Hadjipantelis, N., Walter, L., Myers, R.J., Gardner, L.: Environmental life cycle assessment of wire arc additively manufactured steel structural components. J. Clean. Prod. **389**, 136071 (2023)

10. Sword, J.I., Galloway, A., Toumpis, A.: An environmental impact comparison between wire+ arc additive manufacture and forging for the production of a titanium component. Sustain. Mater. Technol. **36**, e00600 (2023)

11. Bekker, A.C.M., Verlinden, J.C.: Life cycle assessment of wire + arc additive manufacturing compared to green sand casting and CNC milling in stainless steel. J. Clean. Prod. **177**, 438–447 (2018)

12. Mattos, B.M., de Souza Guimarães, C.: Life Cycle Assessment in Flange Part Production Comparing Wire and Arc Additive Manufacturing Method (WAAM) and Conventional Manufacturing in Terms of Energy Consumption, Greenhouse Gas Emissions and Solid Waste Generation (2023)

13. Pusateria, V., Olsen, S.I.: LCA and LCC of wire arc additively manufactured and repaired parts compared to conventional fabrication techniques. In: CIRP Conference on Life Cycle Engineering, pp. 491–496 (2024)

14. Sword, J.I., Galloway, A., Toumpis, A.: Analysis of environmental impact and mechanical properties of inconel 625 produced using wire arc additive manufacturing. Sustainability **16**(10), 4178 (2024)

15. Reis, R.C., Kokare, S., Oliveira, J.P., Matias, J.C., Godina, R.: Life cycle assessment of metal products: a comparison between wire arc additive manufacturing and CNC milling. Adv. Ind. Manufact. Eng. **6**, 100117 (2023)

16. Kokare, S., Matos, F., Oliveira, J.P., Godina, R.: Life cycle assessment of wire arc additive manufacturing process. In: Conference on Mechanical, Automotive and Materials Engineering (CMAME 2022), pp. 135–144 (2023)

17. Campatelli, G., Montevecchi, F., Venturini, G., Ingarao, G., Priarone, P.C.: Integrated WAAM-subtractive versus pure subtractive manufacturing approaches: an energy efficiency comparison. Int. J. Precis. Eng. Manufact.-Green Technol. **7**, 1–11 (2019)

18. Kokare, S., Oliveira, J.P., Godina, R.: Comparison of wire arc additive manufacturing and subtractive manufacturing approaches from an environmental and economic perspective. In: Flexible Automation and Intelligent Manufacturing: Establishing Bridges for More Sustainable Manufacturing Systems, pp. 868–878 (2023)

# Fiber Metal Laminates: From Production to Sustainability Assessment

Mariateresa Caggiano[1]([✉]), Maria Rosaria Saffioti[2], Giovanna Rotella[1,2], and Simonetto Enrico[3]

[1] Department of Engineering, University LUM Giuseppe Degennaro, 70010 Casamassima, BA, Italy
caggiano@lum.it

[2] Department of Mechanical, Energy and Management Engineering, University of Calabria, 87036 Rende, CS, Italy

[3] Department of Industrial Engineering, University of Padova, Via Venezia 1, 35131 Padova, Italy

**Abstract.** Fibre Metal Laminates (FML), a class of hybrid materials combining the advantages of metals and composites, have emerged as a promising alternative for producing lightweight structural materials. Consequently, research interest in FMLs is growing. This paper delves into the primary properties of FMLs, production techniques, potential applications, and functionalization strategies to optimize surface characteristics for superior FML quality. Additionally, the sustainability aspects of FML production are analyzed due to climate change and increasing environmental concerns. This comprehensive review aims to provide a detailed understanding of FMLs, highlighting their benefits and addressing the challenges associated with their use and production.

**Keywords:** Fibre Metal Laminates · Surface Functionalization · Sustainable Manufacturing

## 1 State of Art

Fibre Metal Laminate (FML) represents a composite structure comprising layers of metal sandwiching a Fibre-reinforced plastic layer. The metals commonly utilized include aluminum, magnesium, or titanium, while the fibre-reinforced layer typically consists of glass, carbon, or aramid fibres embedded in a composite matrix [1]. FMLs are classified according to the metal and constituent fibres or the structure of the laminate, thus the number of layers and fibre orientation [2].

At the Delft University of Technology, the first FML, aramid reinforced aluminum laminate, or ARALL, was produced in the 1980s. Afterward, carbon fibre-reinforced (CARALL) and glass fibre-reinforced (GLARE) aluminum laminates were created to enhance the mechanical properties of FMLs [3]. The orientation of the laminate, namely whether it is a cross-ply (woven) or a unidirectional hybrid laminate, should also be taken into consideration in addition to the metal and reinforcing components and the lay-up

H. Kohl et al. (Eds.): GCSM 2024, LNME, pp. 461–468, 2025.
https://doi.org/10.1007/978-3-031-93891-7_51

configuration as shown in Fig. 1, where the metal layers might be within or outside the multi-material [4]. However, combining metals with Fibre-reinforced polymers (FRP) to create FML can present various issues related to the formability of the fibres or the adhesion between materials.

**Fig. 1.** FML Configurations (reproduced with permission from [5], © Elsevier 2020)

FMLs combine the benefits of metals and fibre-reinforced composites, offering superior mechanical properties to traditional laminates made exclusively of fibre-reinforced composites or monolithic aluminum alloys [6]. The advantages of FMLs are high strength, high fatigue resistance, high impact resistance, high energy absorbing capacity, high fracture toughness, low density, excellent moisture resistance, excellent corrosion resistance, lower material degradation, and fire resistance [7–10]. Fewer parts may be required when building a component in FML than when building the identical element in metal alloy, which can result in significant cost savings. Labor savings from this may occasionally balance the increased cost of today's materials. Furthermore, FMLs require less maintenance and repair work due to their excellent fatigue resistance. These benefits lower the FMLs' maintenance expenses. In contrast, due to the long processing cycle due to the curing of the polymer matrix in the composite layers [11], the labor and overall cost of FMLs are high. This problem increases production time and decreases productivity.

## 2   Production of FMLs and Applications

The production of Fiber Metal Laminates (FMLs) primarily uses autoclave processing, involving several key stages [12]. The metal layer is initially pre-treated with chromic or phosphoric acid acids to enhance adhesion with composite laminates. The material is then prepared through cutting, resin application, and debunking. The curing process, which involves polymerization and bonding, generates residual stresses that require post-stretching to neutralize. Finally, quality is ensured through inspections such as ultrasound and X-ray testing. FMLs, including ARALL, CARALL, and GLARE, are known for their exceptional fatigue and impact resistance and are widely used in aerospace, automotive, construction, and shielding applications [13]. Their use in aerospace is growing due to their ability to replace traditional aluminum components, contributing to weight reduction and improved durability against high-speed impacts [14]. In the automotive industry, FMLs help reduce vehicle weight, lowering fuel consumption and $CO_2$ emissions [15]. Additionally, FMLs enhance crack resistance and durability in building applications, making them ideal for structures exposed to varying mechanical and environmental stresses. Their fatigue and corrosion resistance make them suitable for long-term use in harsh outdoor environments [16].

## 3 Role of Metal Surface Treatments for Improving FMLs Performance

The surface treatments of the metal to obtain sufficient adhesion at the metal-composite interface continues to be one of the most crucial problems in FMLs. FML laminates employ various surface preparation techniques for titanium and magnesium, including mechanical, chemical, and physical procedures. Each method's varying effects impact adhesion at the metal-composite interface on the morphology and structure of the metal. Each group has distinct methods that provide specific requirements for enhancing bonding strength and reducing delamination in FMLs. Mechanical treatments that physically alter the metal surface to increase roughness, thereby improving adhesion with polymer layers, include sandblasting, grinding, and polishing. Sandblasting involves blasting abrasive particles, like sand or glass beads, at high velocity onto the metal surface, creating a rough and clean texture that enhances adhesion by increasing the bonding area. Grinding and polishing prepare the surface by removing oxides and contaminants, and polishing also smooths out rough surfaces when necessary [17]. Other techniques, such as burnishing, shot blasting, and abrasion with sandpaper, similarly modify the surface topography, leading to a macro-roughened, more wettable surface that improves the interaction between the metal and composite layers [18]. Also, sandblasting with alumina or silica can create a peak-and-valley morphology, further enhancing adhesion. These mechanical treatments are often paired with degreasing, considered a minimum pretreatment for bonding, and have been shown to increase initial adhesion values effectively [19].

Chemical treatments that modify the metal surface's composition to enhance reactivity and bonding include chemical etching, anodization, phosphating, and coupling agents. Chemical etching involves applying acids or alkalis to remove a thin metal layer, revealing a more reactive surface and potentially creating microscopic roughness to improve adhesion [20]. Anodization, which applies an electrical current to the metal in an acidic electrolyte, forms an oxide layer that increases corrosion resistance and surface roughness, thus enhancing adhesion in FMLs. Phosphating creates a phosphate layer on the metal surface, further improving cohesion with composite materials. Traditional metal etching solutions, such as sulfuric-ferric acid (P2), Forest Product Laboratory (FPL) [21], and chromic-sulfuric acid (CAE), are commonly used to alter metal surfaces. Coupling agents, which form a chemical bridge between metal and polymer layers, also play a crucial role.

Lastly, the laser texturing process, which is categorized as a physical treatment, is still being studied by researchers. The morphology and microstructure of an aluminum substrate have been altered by laser texturing, which increases the FMLs bond strength and durability [22]. Table 1 shows the comparison of surface treatment methods. Each of these surface treatment groups offers distinct advantages, and the choice of method often depends on the specific requirements of the FML application, such as the type of metal used, the desired bond strength, and the environmental conditions the FML will be exposed to.

**Table 1.** Comparison of Surface Treatment Methods for Metal-Composite Interface in FMLs

| Method | Type | Surface Topography | Adhesion Improvement | Chemical State Alteration |
|---|---|---|---|---|
| Sandblasting | Mechanical | Increases roughness | High | Yes |
| Mechanical Abrasion | Mechanical | Macro-roughened surface | High | Yes |
| Phosphating | Chemical | Creates phosphate layer | High | Yes |
| Acid Etching | Chemical | Variable depending on acid type | High | Yes |
| Laser Texturing | Physical | Microstructural changes | High | No |

### Surface Treatment Methods on Magnesium Alloys for FLM Tests

As a reference case study, the study by Liu et al. [23] investigated the effects of surface treatments on the bonding strength and delamination behavior of fiber metal laminates (FMLs) based on AZ31B magnesium alloy. The FMLs consisted of two external magnesium alloy layers and a core of glass fiber-reinforced PA6. Six surface treatments were tested on the magnesium alloy sheets: three involved sandblasting at different pressures (1, 3, and 5 bar), and the other three combined sandblasting with annealing. The study employed a novel normal separation test to measure interlaminar properties in the normal direction (Mode I) and a single lap shear test for properties in the shear direction (Mode II). Results showed that bonding strength improved with increased surface roughness and that annealing enhanced the bonding. A new traction-separation model was developed to relate metal surface roughness to interlaminar performance in Modes I and II. Additionally, T-peel tests were conducted and numerically simulated to validate the proposed model, demonstrating its effectiveness in predicting the interlaminar behavior of the FMLs under combined normal and shear forces.

### Surface Treatment Methods on Titanium Alloys for Bonding Tests

As another reference case study, Rotella et al. [31] explored the surface modification of Ti6Al4V titanium alloy using pulsed Yb-laser irradiation to enhance adhesive bonding. The research focused on the effects of pulsed laser ablation on the alloy's surface morphology, chemistry, and mechanical properties, which were analyzed using SEM, XRD, wettability measurements, and instrumented indentation tests. The researchers identified optimal conditions for improving adhesive bonding by varying laser processing parameters. The experimental work involved treating 1.5 mm thick Ti6Al4V sheets with a pulsed ytterbium-doped fiber laser system. T-peel tests were then conducted on Ti6Al4V/epoxy joints to evaluate the effectiveness of the laser treatment. The results demonstrated that laser irradiation significantly enhanced surface roughness, wettability, and hardness, leading to a marked improvement in peel resistance. The total dissipated energy during T-peel tests was significantly higher for laser-treated samples than untreated ones, with scanning speed being a more critical factor than pulse fluence in generating surface modifications. The findings confirmed that laser irradiation is an effective surface preparation technique, promoting mechanical interlocking and improving the adhesive bonding strength of titanium alloys.

## 4   Sustainable Aspects of FMLs

In recent years, pursuing more sustainable and efficient materials has led to the development of innovative solutions such as FMLs. The hybrid composition of FML offers an excellent compromise between lightweight and strength. Table 2 reports the benefits and challenges of using FMLs.

Innovations in sustainable adhesives and biobased materials further improve the ecofriendliness and durability of FMLs. However, further studies are needed to promote sustainable production effectively.

**Table 2.** Summary of Fiber Metal Laminates (FMLs) Benefits and Challenges

| Aspect | Details |
|---|---|
| Sustainability | - Reduction in structural weight leads to lower fuel consumption and emissions [24] |
| | - Longer component life reduces the need for replacements, decreasing environmental impact [25]<br>- Polymer layers protect metals from corrosion, reducing maintenance [26] |
| | - New FMLs from wood and plant fibers offer lower carbon footprint and sustainable options [27] |
| | - Bio-based FMLs support the circular economy and sustainable industrial designs [28] |
| Performance | - Excellent compromise between lightweight and strength [24] |
| | - Reduced stress on vehicle/aircraft components, extending life [25] |
| | - Improved impact behavior and mechanical qualities with natural fibers [28, 29] |
| Environmental Benefits | - Lower fuel consumption and greenhouse gas emissions [30] |
| | - Decreased wear and tear on vehicle systems [25] |
| | - Reduction in VOC emissions with biodegradable adhesives [31] |
| Innovations | Use of biodegradable, biocompatible adhesives from renewable sources [31] |
| | - Development of adhesive-free processes like ultrasonic welding and hot pressing, enhancing recyclability and environmental safety [32] |
| Challenges | - Adhesives ' sustainability issues: petrochemical origin, VOC emissions, recycling difficulty, toxicity, and non-biodegradability [33] |
| | - Need for further studies to promote sustainable production [31] |

# 5 Conclusion

Fibre Metal Laminates (FMLs) are demonstrated to be a promising alternative for producing lightweight structural materials by combining the advantages of metals and composites. This paper analyzed the main properties of FMLs, manufacturing techniques, potential applications, and surface functionalization strategies to optimize the characteristics of FMLs. In addition, the sustainability aspects of manufacturing FMLs in response to growing environmental concerns were examined. FMLs offer significant environmental benefits, including reduced fuel consumption due to lower structural weight and longer component life, which reduces the need for maintenance and replacement. In conclusion, while FMLs show great potential as advanced structural materials, further research is needed to optimize manufacturing processes and improve overall sustainability. Future developments should focus on innovative production methods, including adhesive-free alternatives and biocompatible materials, promoting a more sustainable and environmentally friendly industry.

**Acknowledgments.** This research was supported by Ministry University Research within the PRIN 2022 program funded by the NextGenerationEU (project ADVANCE - ADhesiVe free Fibre Metal Laminates fabricatioN for aerospaCE applications - 2022W9SHCJ).

# References

1. Salve, A., Kulkarni, R., Mache, A.: A review: fiber metal laminates (FML's) - manufacturing, test methods and numerical modeling. Int. J. Eng. Technol. Sci. 3(2), 71–84 (2016). https://doi.org/10.15282/IJETS.6.2016.1.10.1060
2. El Etri, H., Korkmaz, M.E., Gupta, M.K., Gunay, M., Xu, J.: A state-of-the-art review on mechanical characteristics of different fiber metal laminates for aerospace and structural applications. Int. J. Adv. Manufact. Technol. 123(9), 2965–2991 (2022). https://doi.org/10.1007/S00170-022-10277-1
3. L. V.-R. LR-322 and undefined 1981: A new fatigue resistant material: Aramid reinforced aluminum laminate (ARALL). cir.nii.ac.jp. https://cir.nii.ac.jp/crid/1571980074004982528. Accessed 06 May 2024
4. Costa, R.D.F.S., Sales-Contini, R.C.M., Silva, F.J.G., Sebbe, N., Jesus, A.M.P.: A critical review on fiber metal laminates (FML): from manufacturing to sustainable processing. Metals 13(4), 638 (2023). https://doi.org/10.3390/MET13040638
5. Kazemi, M.E., Shanmugam, L., Yang, L., Yang, J.: A review on the hybrid titanium composite laminates (HTCLs) with focuses on surface treatments, fabrications, and mechanical properties. Compos. Part A Appl. Sci. Manuf. 128, 105679 (2020). https://doi.org/10.1016/J.COMPOSITESA.2019.105679
6. Chang, P.Y., Yeh, P.C., Yang, J.M.: Fatigue crack initiation in hybrid boron/glass/aluminum fiber metal laminates. Mater. Sci. Eng. A 496(1–2), 273–280 (2008). https://doi.org/10.1016/J.MSEA.2008.07.041
7. Vogelesang, L.B., Vlot, A.: Development of fibre metal laminates for advanced aerospace structures. J. Mater. Process. Technol. 103(1), 1–5 (2000). https://doi.org/10.1016/S0924-0136(00)00411-8
8. Gutowski, T.G.P.: Advanced composites manufacturing, p. 581 (1997). https://www.wiley.com/en-us/Advanced+Composites+Manufacturing-p-9780471153016. Accessed 20 May 2024

9. Asundi, A., Choi, A.Y.N.: Fiber metal laminates: an advanced material for future aircraft. J. Mater. Process. Technol. **63**(1–3), 384–394 (1997). https://doi.org/10.1016/S0924-013 6(96)02652-0

10. Cortés, P., Cantwell, W.J.: The prediction of tensile failure in titanium-based thermoplastic fibre-metal laminates (2006). https://doi.org/10.1016/j.compscitech.2005.11.031

11. Cortés, P., Cantwell, W.J.: The prediction of tensile failure in titanium-based thermoplastic fibre–metal laminates. Compos. Sci. Technol. **66**(13), 2306–2316 (2006). https://doi.org/10. 1016/J.COMPSCITECH.2005.11.031

12. Lee, D.W., Park, B.J., Park, S.Y., Choi, C.H., Song, J.: Fabrication of high-stiffness fiber-metal laminates and study of their behavior under low-velocity impact loadings. Compos. Struct. **189**, 61–69 (2018). https://doi.org/10.1016/J.COMPSTRUCT.2018.01.044

13. Sadighi, M., Alderliesten, R.C., Benedictus, R.: Impact resistance of fiber-metal laminates: a review. Int. J. Impact Eng **49**, 77–90 (2012). https://doi.org/10.1016/J.IJIMPENG.2012. 05.006

14. Alderliesten, R.C., Benedictus, R.: Fiber/metal composite technology for future primary aircraft structures. J. Aircraft **45**(4), 1182–1189 (2012). https://doi.org/10.2514/1.33946

15. Jiang, M., Shijie-Liu, Xiao, J.M.: Research on lightweight laminate for car body with excellent cushioning and energy absorption characteristics. Heliyon **8**(11), e11280 (2022). https://doi. org/10.1016/J.HELIYON.2022.E11280

16. Chang, Y.S., Hwang, D.G.: Composite laminate with reinforcement of metal mesh (2015)

17. Sinmazçelik, T., Avcu, E., Bora, M.Ö., Çoban, O.: A review: fibre metal laminates, background, bonding types and applied test methods. Mater. Des. **32**(7), 3671–3685 (2011). https:// doi.org/10.1016/J.MATDES.2011.03.011

18. Harris, A.F., Beevers, A.: The effects of grit-blasting on surface properties for adhesion. Int. J. Adhes. Adhes. **19**(6), 445–452 (1999). https://doi.org/10.1016/S0143-7496(98)00061-X

19. Critchlow, G.W., Yendall, K.A., Bahrani, D., Quinn, A., Andrews, F.: Strategies for the replacement of chromic acid anodising for the structural bonding of aluminium alloys. Int. J. Adhes. Adhes. **26**(6), 419–453 (2006). https://doi.org/10.1016/J.IJADHADH.2005.07.001

20. Park, S.Y., Choi, W.J., Choi, H.S., Kwon, H., Kim, S.H.: Recent trends in surface treatment technologies for airframe adhesive bonding processing: a review (1995–2008). J. Adhes. **86**(2), 192–221 (2010). https://doi.org/10.1080/00218460903418345

21. Kinloch, A.J., Little, M.S.G., Watts, J.F.: The role of the interphase in the environmental failure of adhesive joints. Acta Mater. **48**(18–19), 4543–4553 (2000). https://doi.org/10.1016/S1359-6454(00)00240-8

22. Critchlow, G.W., Cottam, C.A., Brewis, D.M., Emmony, D.C.: Further studies into the effectiveness of CO2-laser treatment of metals for adhesive bonding. Int. J. Adhes. Adhes. **17**(2), 143–150 (1997). https://doi.org/10.1016/S0143-7496(96)00037-1

23. Liu, Z., Simonetto, E., Ghiotti, A., Bruschi, S.: Experimental and numerical investigation of the effect of metal surface treatments on the delamination behaviour of magnesium alloy-based Fibre Metal Laminates. CIRP J. Manuf. Sci. Technol. **38**, 442–456 (2022). https://doi. org/10.1016/j.cirpj.2022.05.015

24. Harish Kumar, M., et al.: Investigation of natural fibre metal laminate as car front hood. Mater. Res. Exp. **8**(2), 025303 (2021). https://doi.org/10.1088/2053-1591/ABE49D

25. Zhu, L., Li, N., Childs, P.R.N.: Light-weighting in aerospace component and system design. Propul. Power Res. **7**(2), 103–119 (2018). https://doi.org/10.1016/J.JPPR.2018.04.001

26. Hamill, L., Hofmann, D.C., Nutt, S.: Galvanic corrosion and mechanical behavior of fiber metal laminates of metallic glass and carbon fiber composites. Adv. Eng. Mater. **20**(2), 1700711 (2018). https://doi.org/10.1002/ADEM.201700711

27. Rousseau, J., Donkeng, N.E.N., Farcas, F., Chevalier, S., Placet, V.: Thermal and hydrothermal ageing of flax/polypropylene composites and their stainless steel hybrid laminates. Compos.

Part A Appl. Sci. Manuf. **171**, 107582 (2023). https://doi.org/10.1016/J.COMPOSITESA.2023.107582

28. Germano Braga, G., Assunção Rosa, F., César dos Santos, J., del Pino, G.G., Panzera, T.H., Scarpa, F.: Fully biobased composite and fiber-metal laminates reinforced with Cynodon spp. Fibers. Polym Compos **44**(1), 453–464 (2023). https://doi.org/10.1002/PC.27109

29. Vieira, L.M.G., Dobah, Y., Dos Santos, J.C., Panzera, T.H., Campos Rubio, J.C., Scarpa, F.: Impact properties of novel natural fibre metal laminated composite materials. Appl. Sci. **12**(4), 1869 (2022). https://doi.org/10.3390/APP12041869

30. Lakshmi Kala, K., Prahlada Rao, K.: Synthesis and characterization of fabricated fiber metal laminates for aerospace applications. Mater. Today Proc. **64**, 37–43 (2022). https://doi.org/10.1016/J.MATPR.2022.03.488

31. Haina, Gul, S., Awais, M., Jabeen, S., Farooq, M.: Recent trends in preparation and applications of biodegradable polymer composites. J. Renew. Mater. **8**(10), 1305–1326 (2020). https://doi.org/10.32604/JRM.2020.010037

32. Bertolini, R., Stramare, A., Sorgato, M., Savio, E., Ghiotti, A., Bruschi, S.: On the role of metal surface modification and polymer matrix characteristics when drilling thermoplastic fibre metal laminates. CIRP Ann. (2024). https://doi.org/10.1016/J.CIRP.2024.04.090

33. Packham, D.E.: Adhesive technology and sustainability. Int. J. Adhes. Adhes. **29**(3), 248–252 (2009). https://doi.org/10.1016/J.IJADHADH.2008.06.002

# Investigation and Optimization of Surface Roughness in Grinding of AISI 52100 Using Response Surface Methodology

Pietro Andrea Miciaccia[1,2]([✉]), Domenico Monopoli[1], and Michele Dassisti[1]

[1] Department of Mechanical, Mathematics, and Management (DMMM), Polytechnic University of Bari, Bari, Italy
pietroandrea.miciaccia@poliba.it, pietro.miciaccia@skf.com
[2] SKF Industry Spa, Bari, Italy

**Abstract.** The surface quality of the raceway in ball bearing inner rings is crucial for preventing their premature failure. Surface roughness, quantified by parameters such as $R_a$ (arithmetic mean roughness) and $R_t$ (maximum height of the profile), is influenced by various cutting parameters. This experimental study has two primary objectives: first, to establish the correlation between cutting parameters and surface roughness during the grinding of AISI 52100 inner ring 6209 raceways; and second, to identify the optimal cutting parameters to enhance surface quality by minimizing $R_a$ and $R_t$ values. Machining tests were designed using a Central Composite Design (CCD). The data obtained were analyzed using Response Surface Methodology (RSM) and Analysis of Variance (ANOVA) to develop predictive regression models for $R_a$ and $R_t$. The optimization results indicated that the fine feed rate parameter has the most significant impact on surface roughness in grinding of AISI 52100. More specifically, the work conducted showed that the best value of Surface Roughness, in terms of $R_a$ and $R_t$, in the raceway's grinding of inner ring 6209 is achieved when the Cutting speed is 70.349 m/s and the Fine feed rate is 23.518 μm/s.

**Keywords:** Grinding · Roughness · Bearing

## 1 Introduction and Related Works

Bearings are critical mechanical components extensively utilized across various manufacturing industries. Their primary function is to support large loads while facilitating relative motion between mechanical parts in rotating machinery [1]. On the other hand, bearings are also a primary source of vibration in rotating machinery, often leading to premature fatigue failure. Consequently, enhancing bearing quality has garnered significant interest from both academia and industry. Bearing's raceway, its functional surface, performs two crucial roles: 1) absorbing radial and/or axial loads, and 2) facilitating the movement of rolling elements, which determines rotational accuracy. According to [2] raceway's surface quality is directly related to the bearing's performance and service life. Therefore, the grinding process of the bearing rings' inner and outer raceways is

H. Kohl et al. (Eds.): GCSM 2024, LNME, pp. 469–476, 2025.
https://doi.org/10.1007/978-3-031-93891-7_52

a fundamental step in the production process, as the surface roughness of the raceway depends on it. AISI 52100 stainless steel, also known as 100Cr6, is the most widely used material in bearing production due to its high hardness, excellent wear resistance, and dimensional stability [3]. Several works have explored the effects of various process parameters on the turning operations of AISI 52100. Allu [4] examined the influence of factors such as insert type, tool nose radius, cutting speed, feed rate, and depth of cut on surface roughness during the dry turning of AISI 52100, employing the Response Surface Methodology (RSM) approach.

Similarly, Bouacha [5] analyzed and modeled the relationship between cutting parameters (cutting speed, feed rate and depth of cut) and machining output variables (surface roughness and cutting forces) using RSM in the turning of AISI 52100 with CBN tools. Other authors [6] and [7] have examined the influence of cutting parameters on tool life, tool temperature, and chip formation in both dry turning and Minimum Quantity Lubrication (MQL) environments. Their findings indicate that MQL-based turning conditions significantly reduce the temperature at the tool tip, thereby improving tool performance and extending its operational life. However, none of these studies focus on the grinding process which, according to [8], produces different surface structures when compared to turning, leading to significantly different functional properties. The selection of input variables for constructing the regression model is informed by scientific literature. According to [9] and [10] cutting speed and feed rate are primary parameters in determining surface quality, particularly surface roughness, in machining processes. Generally, an increase in these parameters leads to an enhancement of the machined surface. On the other hand, other authors [11] and [12] have pointed out that cutting speed does not have a dominant effect on the quality of the machined surface. In this ambiguous context, the aim of this paper is twofold. The first objective is to determine the most influential grinding parameters on surface roughness using Analysis of Variance (ANOVA). The second objective involves identifying the optimal grinding condition to enhance surface quality, specifically surface roughness, across a range of cutting conditions, spanning from low to high grinding speeds.

## 2  Methodology

It is useful to clarify certain aspects of the nomenclature utilized in this research. Firstly, the term cutting speed ($V_c$) refers to the difference in speed $\Delta V$ (relative velocity) between the cutting tool and the surface of the workpiece it is operating on (see Fig. 1).

In the case of down-grinding, where the wheel motion and workpiece motion are in the same direction the module of the cutting speed is given by Eq. (1) [13].

$$\vec{V_c} = \vec{V_s} - \vec{V_w} \Rightarrow |V_c| = \frac{\pi d_s n_s}{60 \cdot 1000} - \frac{\pi d_w n_w}{60 \cdot 1000} = \frac{\pi}{60 \cdot 1000}(d_s n_s - d_w n_w) \quad (1)$$

where $n_s$ and $n_w$ are wheel and workpiece rotational speed (RPM), $d_s$ and $d_w$ are wheel and workpiece diameters (mm), while $\vec{V_s}$ and $\vec{V_w}$ are wheel and workpiece peripheral velocities at the contact point (m/s). It is also useful to consider that the inner ring (IR) raceway grinding process of a deep groove ball bearing involves three stages: rough grinding, fine grinding, and spark-out. Furthermore, the cutting parameters set during

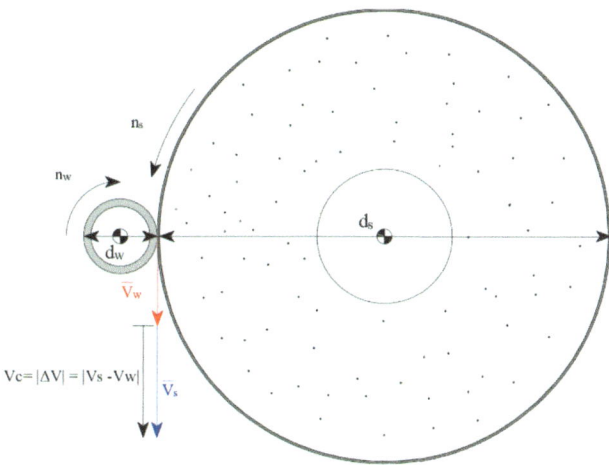

**Fig. 1.** Relative velocity between the wheel and the workpiece at the contact zone.

rough grinding have a negligible impact on the quality of the ground components [14]. Therefore, this study exclusively examines the process parameters of fine grinding (i.e. fine cutting speed and fine feed rate) to assess its influence on surface roughness. In this context, a Design of Experiments (DoE) with an extended range of process parameters was planned. The arithmetic average surface roughness $R_a$ and the maximum profile height $R_t$ were measured. Response Surface Methodology (RSM) was used to develop the predictive regression model, which was later validated with two tests using cutting parameter values outside the DoE. Analysis of Variance (ANOVA) was employed to identify the parameters with the most significant impact on surface roughness. Finally, based on the developed model, the optimal process parameter values that minimize surface roughness were identified.

## 2.1 Design of Experiment and Response Surface Methodology

The experimental plan was based on the face centered ($\alpha = 1$), Central Composite Design (CCD) with two factors and three levels. The three levels selected for Cutting Speed $V_c$ and Fine feed rate $f$ are presented in where $\beta_0$, $\beta_1$, $\beta_2$, $\beta_{11}$, $\beta_{22}$ and $\beta_{12}$ are the regression coefficients and $\varepsilon$ is a term that represents other sources of variability. RSM is based on regression analysis and often uses polynomial functions to potentially determine the correlation between input variables and output quantities, generally some performance measure or quality characteristic of product or process. In this work, a quadratic model with two design variables and six terms was utilized as shown in Eq. (2). Table 1 shows the three levels of cutting parameters selected for this work. Fourteen experiments were generated, including eight factorial and axial points and six center points (Table 2). ANOVA was performed using Design Expert® Software.

$$R_{a,t} = \beta_0 + \beta_1 V_c + \beta_2 f + \beta_{11} V_c^2 + \beta_{22} f^2 + \beta_{12} V_c f + \varepsilon \tag{2}$$

where $\beta_0$, $\beta_1$, $\beta_2$, $\beta_{11}$, $\beta_{22}$ and $\beta_{12}$ are the regression coefficients and $\varepsilon$ is a term that represents other sources of variability.

# 3   Experimental Procedure

In this study, AISI 52100 steel with a hardness of 60 HRC was utilized as workpiece material. Fourteen experimental runs were conducted by grinding the raceway and flanges of the IR of the single-row open radial ball bearing 6209, manufactured at SKF's plant in Bari, Italy. The grinding was performed using a Lidköping SGB 55 grinding machine, which can execute full-profile external plunge grinding on two rings per machining cycle. The dimensions of ring 6209 are a diameter of 57.6 mm ($d_w = 57.6$ mm) a width $B = 19$ mm and a bore diameter $d = 45$ m. According to Eq. (1), the experiments were conducted with $n_s$ values of 2000 (Low), 2350 (Medium), and 2700 (High) RPM, respectively; a constant $n_w$ value of 1300 RPM; a constant $d_w$ value of 57.6 mm and a constant $d_s$ value of 505.5 mm. For each test listed in Table 2, two inner rings were extracted from the machine, resulting in a total of twenty-eight inner rings examined. To ensure consistent grinding conditions, each ring was taken from the machine after the dressing operation, which was conducted using the same dressing parameters as the grinding wheel. This approach ensured that the cutting capability of the grinding wheel was considered constant for each ring. The Tyrolit's type 9GB120 9K4VCSSK100A sintered grinding wheel is used along with Tyrolit's REX diamond profile roller dresser. For surface roughness measurement it was used roughness tester named Mitutoyo CS-3000. The sampling length and the evaluation length was set to 0.25 mm and 10 mm using a Gaussian filter. The surface roughness parameters $R_a$ and $R_t$ were measured on three points at 120° for each ring in the axial direction. The triplet of measurements for $R_a$ and $R_t$ taken for each inner ring was next averaged to produce a single measurement of $R_a$ and $R_t$. This process yielded two average values for $R_a$ and two for $R_t$ for each run.

**Table 1.** The cutting parameters and levels affecting the grinding of AISI 52100 IR.

| Parameters | Symbol | Levels | | |
|---|---|---|---|---|
| | | Low | Medium | High |
| Cutting Speed (m/s) | $V_c$ (m/s) | 51,089 | 60,719 | 70,349 |
| Fine feed rate (μm/s) | $f$ (μm /s) | 10 | 20 | 30 |

# 4   Result and Discussion

## 4.1   ANOVA and Regression Analysis

As a result of the analysis, quadratic regression model equations for the optimal $V_c$ and $f$, which provided the best conditions for minimizing surface roughness $R_a$ and $R_t$ were developed, as shown in Eqs. (3) and (4).

$$R_a = -0.502967 + 0.023481V_c - 0.003670f + 0.000023V_cf$$

**Table 2.** Experimentation and measured responses.

| Test No. | Actual level of Variables | | Response | |
|---|---|---|---|---|
| | $V_c$ (m/s) | $f$ (μm/s) | $R_a$ (μm) | $R_t$ (μm) |
| 1 | 51,089 | 10 | 0,1471 | 1,9854 |
| 2 | 60,719 | 20 | 0,1444 | 1,9157 |
| 3 | 60,719 | 20 | 0,1480 | 1,9363 |
| 4 | 60,719 | 20 | 0,1429 | 2,0441 |
| 5 | 60,719 | 10 | 0,1544 | 2,2837 |
| 6 | 70,349 | 30 | 0,1184 | 1,6241 |
| 7 | 60,719 | 30 | 0,1407 | 2,0172 |
| 8 | 51,089 | 20 | 0,1319 | 1,5999 |
| 9 | 70,349 | 10 | 0,1301 | 1,7541 |
| 10 | 60,719 | 20 | 0,1427 | 1,9537 |
| 11 | 60,719 | 20 | 0,1433 | 1,9363 |
| 12 | 51,089 | 30 | 0,1263 | 1,6688 |
| 13 | 70,349 | 20 | 0,1180 | 1,4430 |
| 14 | 60,719 | 20 | 0,1496 | 1,9989 |

$$- 0.000203V_c^2 + 0.000038f^2 \tag{3}$$

$$R_t = - 12.57973 + 0.531534V_c - 0.125274f + 0.000484V_c f$$
$$- 0.004518V_c^2 + 0.002100f^2 \tag{4}$$

Analysis of Variance (ANOVA) was used to assess the contributions of $V_c$ and $f$ to the surface roughness variables during grinding. The results of the ANOVA for the quadratic model are presented in Table 3 and Table 4.

### 4.2 Validation Test

Additionally, Test No. 15 and 16 were conducted using the $V_c$ and $f$ values shown in Table 5 to validate the regression model. As indicated, there is good agreement between the predicted and measured values within a 95% confidence interval.

### 4.3 Optimization of $R_a$ and $R_t$

The optimal values of $V_c$ and $f$ for the raceway grinding process of the IR 6209, which maximize surface quality were determined using Design Expert Software and are listed in Table 6. Moreover, contour plots of $R_a$ and $R_t$ are shown in Fig. 2. In both cases, the lowest values of $R_a$ and $R_t$ occur when $V_c = 70.349$ m/s and $f = 23.518$ μm/s.

**Table 3.** The ANOVA for $R_a$.

| Source | Sum of Squares | df | Mean Square | F-Value | p-Value | |
|---|---|---|---|---|---|---|
| *Model* | 0.0017 | 5 | 0.0003 | 52.43 | < 0.0001 | significant |
| $V_c$ | 0.0003 | 1 | 0.0003 | 39.65 | 0.0002 | significant |
| $f$ | 0.0004 | 1 | 0.0004 | 56.03 | < 0.0001 | significant |
| $V_c f$ | 0.0001 | 1 | 0.0001 | 2.93 | 0.1253 | |
| $V_c^2$ | 0.0010 | 1 | 0.0010 | 154.94 | < 0.0001 | significant |
| $f^2$ | 0.0001 | 1 | 0.0001 | 6.35 | 0.0358 | significant |
| Lack of Fit | 9.44E-06 | 3 | 3.215E-06 | 0.3824 | 0.7706 | not significant |

$R^2 = 0.9704$, $R^2 (adjusted) = 0.8647$

**Table 4.** The ANOVA for $R_t$.

| Source | Sum of Squares | df | Mean Square | F-Value | p-Value | |
|---|---|---|---|---|---|---|
| *Model* | 0.6265 | 5 | 0.1253 | 70.91 | < 0.0001 | significant |
| $V_c$ | 0.0312 | 1 | 0.312 | 17.69 | 0.003 | significant |
| $f$ | 0.0848 | 1 | 0.0848 | 47.98 | 0.0001 | significant |
| $V_c f$ | 0.0087 | 1 | 0.0087 | 4.92 | 0.0573 | |
| $V_c^2$ | 0.4975 | 1 | 0.4975 | 281.55 | < 0.0001 | significant |
| $f^2$ | 0.1249 | 1 | 0.1249 | 70.70 | < 0.0001 | significant |
| Lack of Fit | 0.0025 | 3 | 0.0008 | 0.3630 | 0.7832 | not significant |

$R^2 = 0.9779$, $R^2 (adjusted) = 0.9641$

**Table 5.** Values and results used for the Validation Test.

| Test No | Level of Variables | | Response | | | | |
|---|---|---|---|---|---|---|---|
| | | | Predicted | | Measured | | |
| | $V_c$ (m/s) | $f$ ($\mu$m/s) | $R_a$ ($\mu$m) | $R_t$ ($\mu$m) | $R_a$ ($\mu$m) | $R_t$ ($\mu$m) | |
| 15 | 60.925 | 25 | 0.1411 | 1.9516 | 0.1451 | 1.9909 | |
| 16 | 55.632 | 15 | 0.1477 | 2.0050 | 0.1487 | 1.9426 | |

**Table 6.** Optimum cutting parameters and output response.

| Optimal Cutting Parameters | | Output Response | |
|---|---|---|---|
| Cutting Speed $V_c$ (m/s) | Fine feed rate $f$ ($\mu$m/s) | $R_a$ ($\mu$m) | $R_t$ ($\mu$m) |
| 70.349 | 23.518 | 0.1180 | 1.4680 |

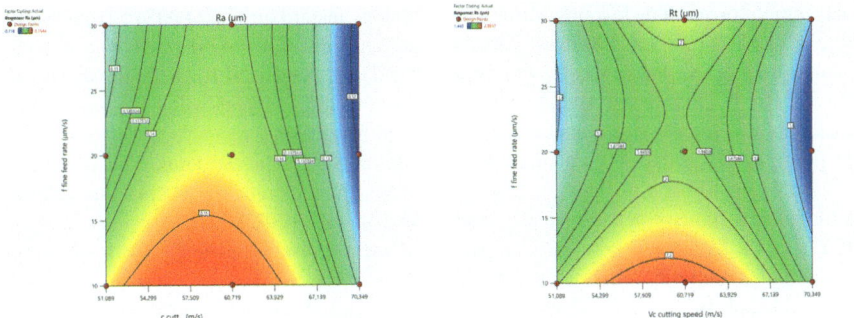

**Fig. 2.** The effects of the $V_c$ and $f$ on surface roughness $R_a$ (left) and $R_t$ (right).

## 5 Conclusion and Future Works

In this study, an experimental investigation was conducted to evaluate the influence of cutting parameters, specifically cutting speed ($V_c$) and fine feed rate ($f$), on surface roughness during the grinding of the raceway of an IR 6209 made of AISI 52100 steel. CCD combined with RSM was employed for the analysis. The results indicated that:

1. Surface roughness $R_a$ and $R_t$ are significantly affected by $V_c$ and $f$.
2. Fine feed rate $f$, as shown in Table 3 and Table 4, emerged as the most influential variable.
3. The cutting speed $V_c$ had a comparatively lesser impact on surface roughness.
4. Optimal surface roughness $R_a$ and $R_t$, were achieved at a cutting speed $V_c$ of 70.349 m/s and a fine feed rate $f$ of 23.518 $\mu$m/s.

These findings were derived from the grinding process of the IR 6209, conducted immediately after the dressing operation to restore the cutting properties of the grinding wheel. Future research should explore the degradation of surface roughness considering all rings ground between successive dressing cycles and examine the influence of cutting parameters on the surface quality of the inner ring's raceway, particularly in terms of waviness.

## References

1. Cheng, L., Ma, W., Xia, X., Wang, L.: Quality-achieving reliability assessment of rolling bearing based on bootstrap maximum entropy method and similarity method. Sci. Prog. **105**(2), 003685042211027 (2022)

2. Yan, K., Wang, Y., Zhu, Y., Hong, J., Zhai, Q.: Investigation on heat dissipation characteristic of ball bearing cage and inside cavity at ultra high rotation speed. Tribol. Int. **93**, 470–481 (2016)
3. Panda, A., Sahoo, A.K., Kumar, R., Das, R.K.: A review on machinability aspects for AISI 52100 bearing steel. Mater. Today Proc. **23**, 617–621 (2020)
4. Pradeep Allu, V., Linga Raju, D., Ramakrishna, S.: Performance investigation of surface roughness in hard turning of AISI 52100 steel - RSM approach. Mater. Today Proc. **18**, 261–269 (2019)
5. Bouacha, K., Yallese, M.A., Mabrouki, T., Rigal, J.-F.: Statistical analysis of surface roughness and cutting forces using response surface methodology in hard turning of AISI 52100 bearing steel with CBN tool. Int. J. Refract. Metals Hard Mater. **28**(3), 349–361 (2010)
6. Campos, P.H.D.S., Paes, V.D.C., Gonçalves, E.D.D.C., Ferreira, J.R., Balestrassi, P.P., da Silva, J.P.D.T.: Optimizing production in machining of hardened steels using response surface methodology. Acta Scientiarum. Technol. **41**(1), 38091 (2019)
7. Demirpolat, H., Binali, R., Patange, A.D., Pardeshi, S.S., Gnanasekaran, S.: Comparison of tool wear, surface roughness, cutting forces, tool tip temperature, and chip shape during sustainable turning of bearing steel. Materials **16**(12), 4408 (2023)
8. Grzesik, W., Żak, K., Kiszka, P.: Comparison of surface textures generated in hard turning and grinding operations. Procedia CIRP **13**, 84–89 (2014)
9. Cakir, M.C., Ensarioglu, C., Demirayak, I.: Mathematical modeling of surface roughness for evaluating the effects of cutting parameters and coating material. J. Mater. Process. Technol. **209**(1), 102–109 (2009)
10. Chou, Y.K., Evans, C.J., Barash, M.M.: Experimental investigation on CBN turning of hardened AISI 52100 steel. J. Mater. Process. Technol. **124**(3), 274–283 (2002)
11. Bartarya, G., Choudhury, S.K.: Influence of machining parameters on forces and surface roughness during finish hard turning of EN 31 steel. Proc. Inst. Mech. Eng. B J. Eng. Manuf. **228**(9), 1068–1080 (2014)
12. Bartarya, G., Choudhury, S.K.: Effect of cutting parameters on cutting force and surface roughness during finish hard turning AISI52100 grade steel. Procedia CIRP **1**, 651–656 (2012)
13. Marinescu, I.D., Hitchiner, M.P., Uhlmann, E., Rowe, W.B., Inasaki, I.: Handbook of Machining with Grinding Wheels. CRC Press (2006)
14. Tuan, N.A.: Multi-objective optimization of process parameters to enhance efficiency in the shoe-type centerless grinding operation for internal raceway of ball bearings. Metals **11**(6), 893 (2021)

# Exploring Magnetic Properties of Metal 3D Printed Products Manufactured by Laser Power Bed Fusion Method

Do Qui Duyen[1], Nguyen Phat Tai[1], Do Huu Minh Hieu[1], Nguyen Quoc Hung[2(✉)], and Ngo Chi Vinh[3]

[1] Faculty of Engineering, Vietnamese-German University (VGU), Bến Cát, Vietnam
[2] Institute of Engineering, HUTECH University, Ho Chi Minh City, Vietnam
nq.hung@hutech.edu.vn
[3] AlchLight LLC, Rochester, NY 14650, USA

**Abstract.** This study is centered on advancing sustainable manufacturing practices through the incorporation and assessment of magnetic properties in products produced via the laser power bed fusion technique, a prominent method in metal 3D printing. The investigation initiates with an in-depth exploration of metal 3D printing technology, particularly highlighting the power bed fusion process. Experimental procedures entail the creation of diverse specimens utilizing the power bed fusion method, followed by a comprehensive evaluation of their magnetic attributes. Subsequent analysis involves a comparison between these results and those of machine parts manufactured using conventional industrial fabrication techniques. The power bed fusion method is esteemed for its potential in metal 3D printing due to its ability to fabricate intricate components of exceptional quality, including magnetic properties. These discoveries hold significant importance for sectors like aerospace, automotive, and medical equipment manufacturing that necessitate high-quality, intricate elements. Furthermore, laser power bed fusion not only boosts manufacturing efficiency but also promotes sustainable practices by diminishing resource utilization, energy consumption, and industrial waste. This positions it as a vital technology for advancing both operational efficacy and environmental conservation objectives.

**Keywords:** Additive manufacturing · metal 3D printing · laser power bed fusion · magnetic properties

## 1 Introduction

Additive Manufacturing (AM), also known as 3D printing, has rapidly advanced as a technology offering significant opportunities for enhancing sustainability [1]. One of the primary advantages of AM lies in its capacity to fabricate intricate geometries that are often impractical or unattainable through traditional manufacturing methods [2]. Moreover, AM facilitates rapid prototyping and small-batch production, proving particularly beneficial in sectors like aerospace, automotive, and medical devices [3–6]. Fused

© The Author(s) 2025
H. Kohl et al. (Eds.): GCSM 2024, LNME, pp. 477–485, 2025.
https://doi.org/10.1007/978-3-031-93891-7_53

Deposition Modeling (FDM) remains a prevalent 3D printing technique, involving the extrusion of thermoplastic filaments through a heated nozzle to gradually build up layers of material and form objects. Recent developments in FDM have concentrated on bolstering the mechanical properties of printed components by integrating reinforced composite fibers and refining process parameters to enhance surface quality and dimensional precision [7]. Selective laser sintering (SLS) and selective laser melting (SLM), also known as laser powder bed fusion, represent notable 3D printing technologies. SLS employs a laser to selectively bond powdered materials, eliminating the need for support structures and enabling greater design flexibility [8]. Conversely, SLM utilizes a laser or electron beam to melt and fuse material powder layer by layer, resulting in fully functional parts [8]. Bound Metal Deposition (BMD) stands out as a metal fabrication technology that blends the flexibility of 3D printing with the capability to produce metal components. This process involves spraying a mixture of metal powder and binding agents onto a build plate, followed by heating to bind and bond the metal particles together [9]. The 3D-printed metal parts produced through BMD exhibit competitive mechanical properties compared to conventionally manufactured metal parts.

The magnetic properties of machine parts play a crucial role in ensuring their proper operation and reliability. Recent advancements in 3D printing technology have unveiled new avenues for designing and producing machine parts with customized magnetic properties. These properties are vital in applications such as motors, generators, sensors, and actuators, where they influence device efficiency and performance. In a recent study, 17-4 stainless steel parts were manufactured using the sustainable powder bed fusion method. The magnetic properties of the 3D-printed 17-4 stainless steel were evaluated and compared with those of conventionally fabricated 17-4 stainless steel and C-45 steel, a material commonly utilized in magnetic applications. The study delved into the underlying mechanisms behind the observed differences, indicating potential novel applications for 17-4 PH stainless steel in magnetic fields, potentially replacing traditional magnetic materials.

## 2   Material and Method

### 2.1   Material

Precipitation-hardened stainless steel powder 17-4, also known as 17-4PH, is a high-strength, corrosion-resistant material widely used in industrial applications [15]. In this study, the 17-4PH steel powder used was sourced from Oerlikon, USA. Table 1 describes the material composition.

**Table 1.** .

| Fe | Cr | Ni | Cu | Nb + Ta | C | Other |
|---|---|---|---|---|---|---|
| Balance | 17 | 4.5 | 4.0 | 0.3 | <0.07 | <1.0 |

The powder comprises spherical particles ranging from 500 nm to 35 μm in diameter, with an average size of about 20 μm. Sieving is essential to break up agglomerates of smaller spheres, ensuring a consistent and high-quality feedstock for Direct Metal Laser Sintering (DMLS) processes.

In this study, 17-4PH and hot-rolled C45 steel provided by ASSAB Vietnam Co., Ltd. were used.

Table 2 describes the material composition of hot rolled 17-4 ph steel.

**Table 2.** .

| Fe | Cr | Ni | P | S | Cu | Nb + Ta | Si | C |
|---|---|---|---|---|---|---|---|---|
| Balance | 17 | 4.5 | 0.04 | 0.03 | 4 | 0.45 | 1.00 | 0.07 |

Table 3 describes the material composition of hot rolled C45 steel.

**Table 3.** .

| C | Si | Mn | Ni | Cr | P | S |
|---|---|---|---|---|---|---|
| 0.42–0.5 | 0.15–0.35 | 0.15–0.35 | ≤0.4 | ≤0.4 | ≤0.003 | ≤0.035 |

## 2.2  Method

The 3D-printed samples were fabricated using the sustainable DMLS approach, which was carried out on a 3D metal printer (Creator, Coherent, Germany). This printer has a maximum laser power of 250 W. For this specific study, a laser power of 120 W and a focal beam diameter of 40 μm were selected. The scanning speed was set to 1000 mm per second, with a layer height of 0.025 mm and a 100-μm distance between scanning lines. The laser paths created visible stripes during the printing process. A sample with dimensions of 10 mm in diameter and 30 mm in height was printed.

The DMLS process began by creating a 3D CAD model of the desired part, which was then converted into an STL file and loaded into the DMLS machine software. High-quality, spherical 17-4 PH stainless steel powder was prepared by sieving to ensure a uniform particle size distribution. The 3D printer's chamber consists of four main components: a powder reservoir, a circulating coater, a build platform, and a container for powder overflow, as shown in Fig. 1. The powder was stored in the reservoir and fed into the printing process. During the process, the powder reservoir moved upwards (in the Z+ direction), while the build platform, which held the component carrier or 3D-printed part, moved downwards (in the Z-direction). The circulating coater transferred the powder from the reservoir to the build platform, spreading a thin layer of powder over the component in each layer during printing. The selected fabrication parameters, as mentioned earlier, were applied. The supporting structure was printed first, followed by the main printing process. A high-power laser selectively melted the powder according to the trajectory designed in the STL file, layer by layer, until the part was completed. The excess powder was collected in the replaceable container. After printing, the supporting structures were removed from the 3D-printed samples.

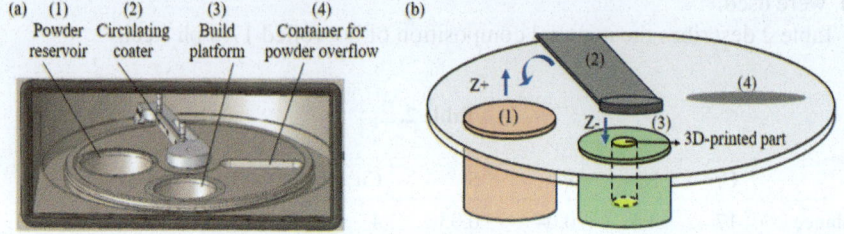

**Fig. 1.** (a) Main components in the printing chamber of Creator, and (b) Printing principle.

## 2.3 Characterization

The morphology of the 3D-printed samples was measured using scanning electron microscopy (JSMIT-500, JEOL, Japan). In addition, the magnetic properties of the 3D-printed 17-4 PH stainless steel sample, conventionally hot-rolled 17-4 PH stainless steel sample, and conventionally hot-rolled C-45 stainless steel sample were measured. To measure the magnetic properties, coils were wrapped around the samples to form magnetic field-generating coils. Specifically, 0.25-mm diameter copper wire was used to wrap 200 turns around each test sample, as illustrated in Fig. 2. The process of measuring the magnetic field generated by the coils involved using a current source to vary the supply current to the coil from 0 to 2.5 amperes in steps of 0.1 amperes. A magnetic field sensor was employed to measure and record the corresponding magnetic field values. This data is typically used to demonstrate the variation of magnetic field with current in a coil, showing the relationship between the input current and the resulting magnetic field.

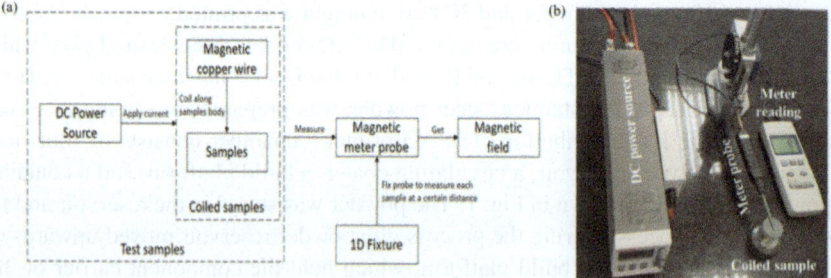

**Fig. 2.** (a) Arrangement diagram of experimental equipment, and (b) the measurement setup of magnetic property.

# 3   Result and Discussion

## 3.1   Morphology

Figure 3 shows the 3D-printed 17-4 PH stainless steel sample. The printed dimensions (Ø10x50 mm) matched the design model, demonstrating the precision of the 3D printing method. Additionally, conventionally hot-rolled 17-4 PH stainless steel and C-45 stainless steel samples were prepared for magnetic property measurements.

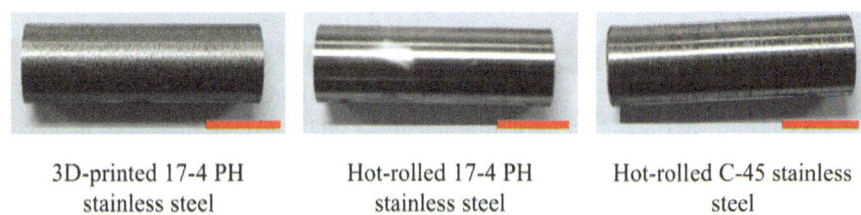

|  3D-printed 17-4 PH stainless steel  |  Hot-rolled 17-4 PH stainless steel  |  Hot-rolled C-45 stainless steel  |

**Fig. 3.** The 3D-printed 17-4 PH stainless steel sample compared with the conventionally hot-rolled 17-4 PH stainless steel and C-45 stainless steel samples. Scale bar: 1 cm.

In addition, the SEM (Scanning Electron Microscopy) images of the 3D-printed 17-4 PH stainless steel sample were analyzed, as shown in Fig. 4. The top surface was smoother than the body surface, which would be coiled with the copper wire for the magnetic property measurements. The top area interacted directly with the laser beam during the printing process. With sufficient laser power, the 17-4 PH stainless steel powders were completely melted and then bonded together to form a thin stainless steel layer. This process was repeated layer by layer until the desired dimensions were achieved. Consequently, the top surface, where the metal melted, appeared smoother. In contrast, the body surface was covered by unprocessed 17-4 PH stainless steel powder. During the printing process, when the laser beam interacted with the metal powders at the boundary area of the printed samples, the temperature was high enough to melt the powders. At that stage, the powders could be in a liquid or semi-liquid form and might adhere to the untreated 17-4 stainless steel powders around them. When the sample solidified and cooled down, those excess powders remained stuck to the body of the 3D-printed sample, causing a rougher surface compared to the top, as shown in Fig. 4b.

## 3.2   Magnetic Property

The 3D-printed 17-4 stainless steel sample, the conventionally hot-rolled 17-4 stainless steel sample, and the conventionally hot-rolled C-45 stainless steel sample were all coiled with copper wire, as shown in the inserted image in Fig. 5. This figure illustrates the measurement of the magnetic field (B) of the samples in response to varying current (I) from 0 to 2.5 amperes. At currents below 1.5 amperes, the magnetic field values for all three samples were nearly the same. However, as the current exceeded 1.5 amperes, distinct differences in the magnetic field values among the samples emerged, increasing with higher currents. At lower currents, the magnetic domains within the materials

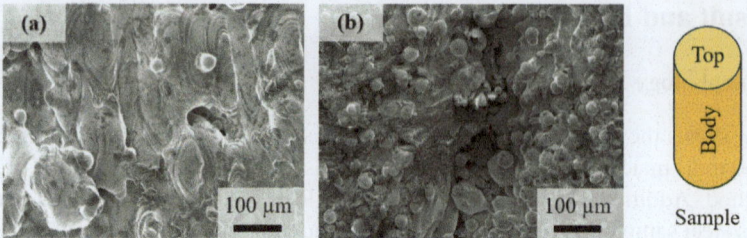

**Fig. 4.** SEM images of (a) the top, and (b) the body of the 3D-printed 17-4 PH stainless steel sample.

might not be fully aligned, and the differences in material properties such as grain size, porosity, and surface roughness have minimal effect on the overall measured magnetic field. However, when the current increases, the materials approach magnetic saturation, where the alignment of magnetic domains becomes more significant. At that stage, any imperfections or differences in material properties may create a clear impact on the overall magnetic field. Additionally, at higher currents, the thermal effects of the samples may also affect the magnetic properties of the investigated samples. At the maximum current of 2.5 amperes, the magnetic field values for the 3D-printed sample, the hot-rolled 17-4 PH stainless steel sample, and the hot-rolled C-45 stainless steel sample were 37.2 mT, 38.5 mT, and 39.8 mT, respectively.

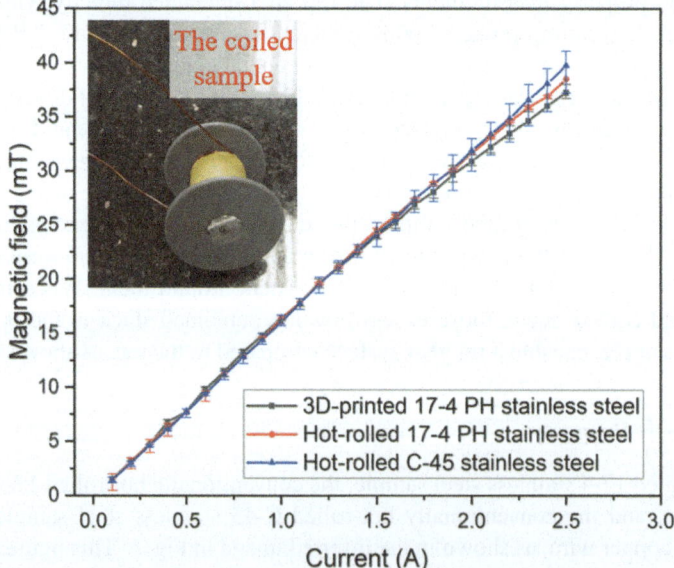

**Fig. 5.** The magnetic field B vs current I graph of the 3D-printed 17-4 stainless steel sample, the hot-rolled 17-4 stainless steel sample, and the hot-rolled C-45 stainless steel sample. The insert image is the sample after being coiled by the copper wire.

Compared to the hot-rolled 17-4 PH stainless steel sample, the magnetic field value of the 3D-printed sample showed a small reduction of approximately 4% at 2.5 amperes. This difference could be caused by the defects in the 3D-printed sample, such as the adhesion of unwanted powder to the body of the printed sample, which increases surface roughness, grain size, and porosity. The roughness from these adhered powders might act like tiny bumps or gaps, hindering the natural alignment of magnetic domains within the material, and leading to a weaker overall magnetic field [16, 17]. Porosity similarly disrupts magnetic alignment due to the presence of air and void spaces within the material [18]. Additionally, larger grain sizes may result in reduced responsiveness to magnetic fields [16, 17].

Compared to the magnetic field value of the hot-rolled C-45 stainless steel sample, the gap was approximately 7% for the 3D-printed sample, and approximately 3.4% for the hot-rolled 17-4 PH stainless steel sample. To apply 3D-printed 17-4 PH stainless steel in practical applications in Vietnam, where C-45 currently plays a significant role, the magnetic field value of the 17-4 PH stainless steel should be improved. While there may be limited room for improvement in conventionally hot-rolled 17-4 PH stainless steel, the 3D-printed version has potential for further development.

To reduce the gap between the 3D-printed 17-4 PH stainless steel sample and the hot-rolled 17-4 PH stainless steel sample, the researchers plan to investigate optimal laser processing parameters to produce better grain size and reduce porosity in the material. In addition, polishing the printed sample to remove unwanted adhered powders or altering the printing direction to achieve better morphology on the sample's surface could be effective strategies.

## 4   Conclusion

The magnetic properties of 3D-printed 17-4 stainless steel have been meticulously assessed and juxtaposed with those of conventionally hot-rolled 17-4 stainless steel and C-45 steel. This investigation has unveiled significant disparities in the magnetic attributes of the materials, accompanied by an in-depth analysis of the underlying mechanisms driving these differences. The study has elucidated the intricate interplay between the 3D printing process, microstructural characteristics, and magnetic performance.

Notably, the utilization of the 3D printing method for producing 17-4 stainless steel not only showcases the potential for enhanced magnetic capabilities but also underscores its alignment with sustainability objectives. This innovative technology optimizes material utilization, curtails waste generation compared to traditional manufacturing techniques, and reduces energy consumption in crafting intricate geometries. The outcomes of this research not only underscore the promise of 3D-printed 17-4 PH stainless steel as a viable alternative to traditional magnetic materials like C-45 but also advocate for sustainable manufacturing practices, thereby mitigating the environmental footprint associated with advanced magnetic material applications.

## References

1. Ford, S., Despeisse, M.: Additive manufacturing and sustainability: an exploratory study of the advantages and challenges. J. Clean. Prod. **137**, 1573–1587 (2016)

2. Wong, K.V., Hernandez, A.: A review of additive manufacturing. Int. Sch. Res. Not. **2012**(1), 208760 (2012)

3. Blakey-Milner, B., et al.: Metal additive manufacturing in aerospace: a review. Mater. Des. **209**, 110008 (2021)

4. Vasco, J.C.: Additive manufacturing for the automotive industry. In: Additive Manufacturing, pp. 505–530. Elsevier (2021)

5. Culmone, C., Smit, G., Breedveld, P.: Additive manufacturing of medical instruments: a state-of-the-art review. Addit. Manuf. **27**, 461–473 (2019)

6. Ngo, C.-V., Nguyen, Q.A., Le, N., Le, N.L.L., Nguyen, Q.H.: Fabrication of airy, lightweight polymer below-elbow cast by a combination of 3D scanning and 3D printing. In: Advances in Asian Mechanism and Machine Science: Proceedings of IFToMM Asian MMS 2021, pp. 628–637 (2022)

7. Crump, S.S.: Apparatus and method for creating three-dimensional objects. U.S. Patent No. 5,121,329, (1992)

8. Bhavar, V., Kattire, P., Patil, V., Khot, S., Gujar, K., Singh, R.: A review on powder bed fusion technology of metal additive manufacturing. In: Additive Manufacturing Handbook, pp. 251–253 (2017)

9. Hieu, D.H., Duyen, D.Q., Tai, N.P., Thang, N.V., Vinh, N.C., Hung, N.Q.: Crystal structure and mechanical properties of 3D printing parts using bound powder deposition method. In: Modern Mechanics and Applications: Select Proceedings of ICOMMA 2020, pp. 54–62 (2022)

10. Duda, T., Raghavan, L.V.: 3D metal printing technology. IFAC-PapersOnLine **49**(29), 103–110 (2016)

11. Thomas, D.S., Gilbert, S.W.: Costs and cost effectiveness of additive manufacturing. NIST Spec. Publ. **1176**, 12 (2014)

12. Shellabear, M., Nyrhilä, O.: DMLS-Development history and state of the art. Laser assisted netshape engineering 4, proceedings of the 4th LANE, pp. 21–24 (2004)

13. Aspden, H.: Power from Magnetism: Over-unity Motor Design. Sabberton Publications (1996)

14. Xu, J., et al.: Flexible, self-powered, magnetism/pressure dual-mode sensor based on magnetorheological plastomer. Compos. Sci. Technol. **183**, 107820 (2019)

15. Gülsoy, H., Özbek, S., Baykara, T.: Microstructural and mechanical properties of injection moulded gas and water atomised 17-4 PH stainless steel powder. Powder Metall. **50**(2), 120–126 (2007)

16. Aurongzeb, D., Ram, K.B., Menon, L.: Influence of surface/interface roughness and grain size on magnetic property of Fe/ Co bilayer. Appl. Phys. Lett. **87**(17) (2005)

17. Zhao, Y.-P., Gamache, R., Wang, G.-C., Lu, T.-M., Palasantzas, G., De Hosson, J.T.M.: Effect of surface roughness on magnetic domain wall thickness, domain size, and coercivity. J. Appl. Phys. **89**(2), 1325–1330 (2001)

18. Ternero, F., Rosa, L.G., Urban, P., Montes, J.M., Cuevas, F.G.: Influence of the total porosity on the properties of sintered materials—a review. Metals **11**(5), 730 (2021)

# Experimental Evaluation of Environmental Sustainability of Fused Deposition Modelling 3D Method

M. G. U. K. Priyamal[1], H. D. Ranasinghe[1], W. L. Raneesha Fernando[1(✉)], and Asela K. Kulatunga[2]

[1] University of Peradeniya, Peradeniya, Sri Lanka
raneesha@eng.pdn.ac.lk
[2] University of Exeter, Exeter, UK

**Abstract.** Additive manufacturing is an innovative technology across a wide range of industries and Fused Deposition Modeling (FDM) is the most common additive manufacturing method in the world. It is necessary to evaluate the sustainability of the additive manufacturing process due to global trends to move towards sustainable manufacturing. Environmental sustainability analysis of FDM is crucial because it uses a significant amount of electrical energy and resources. Therefore, this study aims to investigate the effect of process parameters on the mechanical properties of the FDM printed product and the environment. The experiments were designed using the half factorial design of experiment (DOE) method considering infill pattern, infill percentage, layer thickness, and nozzle temperature as changing variables. The tensile strength of the printed samples was measured using a tensile strength tester. Variation of tensile strength, material usage, electricity consumption, and $CO_2$ emissions with the changing process parameters was analysed. The results of this study reveal that the tensile strength of the products is influenced by all the key variables considered. Infill percentage is the most influencing parameter on material usage while nozzle temperature is that for the electricity consumption and $CO_2$ emissions.

**Keywords:** Additive manufacturing · tensile strength · process parameters · sustainable manufacturing

## 1 Introduction

In recent years, 3D printing, also known as additive manufacturing, has emerged as an innovative technology with far-reaching implications across a wide range of industries. Its ability to build complex structures layer by layer provides outstanding design flexibility and customization. However, as 3D printing technology advances, there is a growing awareness of the environmental implications of widespread adoption. Fused Deposition Modeling (FDM) is considered a widely practised additive manufacturing technology in the industry as well as academia [1]. A melted thermoplastic is fed through a nozzle to build the layers of the product on a printing bed. Even if FDM technology allows to

H. Kohl et al. (Eds.): GCSM 2024, LNME, pp. 486–493, 2025.
https://doi.org/10.1007/978-3-031-93891-7_54

manufacture of products with complex geometries, the mechanical properties of those products are not as good as products that are produced using traditional manufacturing methods [2].

Many studies have been conducted to identify the behaviour of mechanical properties of printed products following the FDM method with different operating parameters. One study has identified the optimal printing conditions; retraction speed, number of walls, and deposition angle, to obtain better visual quality and strength of the FDM-printed product using Polylactic acid (PLA) [1]. Another study has investigated the effect of annealing time, layer height and temperature on the dimensional change and tensile strength of the products printed using three different printing materials. The results of this study reveal that layer height has a significant impact on the tensile strength compared to the other two factors. Further, optimal parameters combination has been identified for each material to achieve the best tensile strength of the final product [2].

A comparative study has performed to investigate the effect of layer height, layer orientation, and infill density on mechanical properties of PLA and acrylonitrile butadiene styrene (ABS). The results of this study show that the infill percentage is the mostly influencing parameter on the mechanical properties of the products. Further, the products which are printed using PLA have higher tensile strength than the ABS products [3]. A recent study describes the sustainability approach for additive manufacturing technologies. It studied the effect of fan rate, nozzle temperature, printing speed, layer thickness and build plate temperature on the dimensional accuracy, cost of printing and carbon dioxide ($CO_2$) emissions. The results of this study show that layer thickness is highly affected by inner and outer width of the specimen and cost (material and labour). The build plate temperature is the significantly affecting factor on outer length, electricity cost and $CO_2$ emissions. Moreover, multi-objective optimization has been performed to identify the optimal combination of parameters to obtain the optimal results [4]. It is clear that there is a research gap in the sustainability evaluation of additive manufacturing technologies. Therefore, the purpose of this study is to bridge that research gap. This study aims to identify the effect of variation of process parameters in FDM technology on the mechanical properties of the final product and the environment.

## 2  Methodology

### 2.1  Design of Experiments

The design of experiment (DoE) was performed following the Half Factorial DOE method. By employing a half-factorial design, the experiment efficiently reduces the number of runs required for comprehensive testing, allowing for a more streamlined yet thorough investigation of the chosen parameters. Four key operating parameters; infill pattern, infill percentage, layer thickness and nozzle temperature, selected from the literature were considered as the changing parameters in DoE. The two levels of each parameter were selected and the resulting experimental plan is shown in Table 1.

### 2.2  3D Printing of Specimens

A 3D CAD model of the specimen (dog bone) was designed using SOLIDWORKS software following ISO guidelines. The dimensions of the specimen were selected according

**Table 1.** Design of experiments using half factorial method

| Test No. | Infill pattern | Infill Percentage (%) | Layer thickness (mm) | Nozzle temperature (°C) |
|---|---|---|---|---|
| 1 | Honeycomb | 80 | 0.2 | 230 |
| 2 | Cubic | 80 | 0.2 | 210 |
| 3 | Honeycomb | 100 | 0.4 | 230 |
| 4 | Cubic | 80 | 0.4 | 230 |
| 5 | Cubic | 100 | 0.4 | 210 |
| 6 | Cubic | 100 | 0.2 | 230 |
| 7 | Honeycomb | 80 | 0.4 | 210 |
| 8 | Honeycomb | 100 | 0.2 | 210 |

to the specimen size requirements of the Tensile testing machine as shown in Fig. 1(a). The designed samples were printed using FDM type 3D printer as illustrated by Fig. 1(b). The machine has a maximum build size of 305 × 305 × 305 mm and a maximum volume of 28.37 L. PLA filament with 1.75 mm diameter was used to prepare the samples. The main reason to select PLA material is its sustainability compared to other plastics because the carbon emission of 1000 single-use PLA products is 13%–63% lower compared to other plastic products [5]. A printed sample is illustrated in Fig. 2.

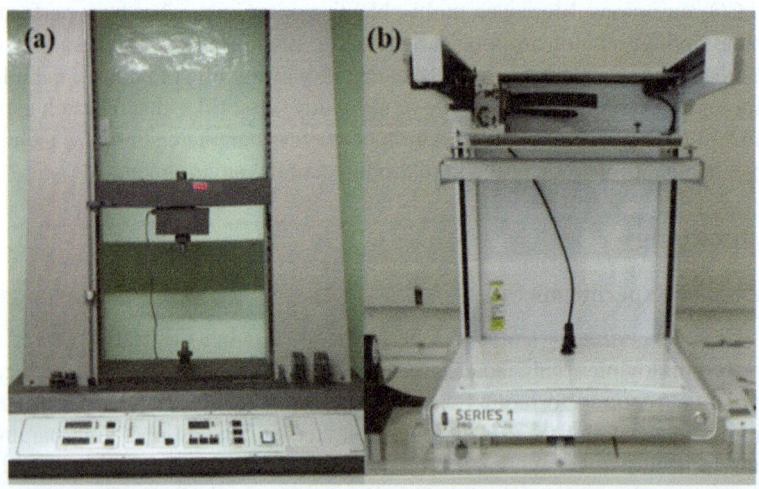

**Fig. 1.** (a) Tensile testing machine, (b) 3D printing machine

**Fig. 2.** 2D drawing of a specimen sample with dimensions

## 2.3  Data Collection

The dog bone samples were printed to test the variation of tensile strength with the changing level of process parameters. The samples printed using the 3D printer were tested using a tensile tester. To gather a wide range of elongation vs. tensile strength each dog bone was pulled till a fracture occurred. The rated power consumption of the printer is 400 W. Electricity usage was calculated referring rated power consumption of the machine and the printing time. Further, 0.71 kg of $CO_2$ is emitted to the atmosphere during the generation of 1kWh of electricity in Sri Lanka [6] and that was used to calculate the amount of $CO_2$ emitted during printing due to the usage of electricity.

## 3  Results and Discussion

A summary of printing time, amount of material used, tensile test results, electricity used, and amount of $CO_2$ generated is illustrated in Table 2. Variations of means of tensile strength, material usage, electricity usage and $CO_2$ emissions according to the levels of process parameters were analysed using the ANOVA method.

**Table 2.** Summary of Results

| Test No | Printing time | Material used (g) | Tensile strength (MPa) | Electricity (kWh) | Amount of $CO_2$ (kg) |
|---------|---------------|-------------------|------------------------|-------------------|-----------------------|
| 1 | 1 hr 16 min | 9.76 | 18.75 | 0.51 | 0.36 |
| 2 | 1 h 12 min | 9.82 | 18.29 | 0.41 | 0.29 |
| 3 | 48 min | 10.64 | 19.34 | 0.32 | 0.23 |
| 4 | 58 min | 10.14 | 15.07 | 0.39 | 0.28 |
| 5 | 1 h 45 min | 10.78 | 19.08 | 0.70 | 0.50 |
| 6 | 34 min | 11.08 | 20.4 | 0.23 | 0.16 |
| 7 | 57 min | 9.56 | 17.43 | 0.39 | 0.28 |
| 8 | 1 h 16 min | 10.97 | 21.4 | 0.51 | 0.36 |

### 3.1 Tensile Strength

Strength, a critical indicator of structural integrity, illuminates the specimens' ability to withstand external forces. According to the variation of tensile strength of the specimens with the change of process parameters, as illustrated by Fig. 3, the honeycomb infill pattern gives products with higher tensile strength. The ability to absorb the stress and vibration without deforming or cracking in a honeycomb infill pattern results in greater tensile strength compared to the cubic infill pattern [7]. The infill percentage of 100% also gives larger tensile stress due to the increase of resistance area [8].

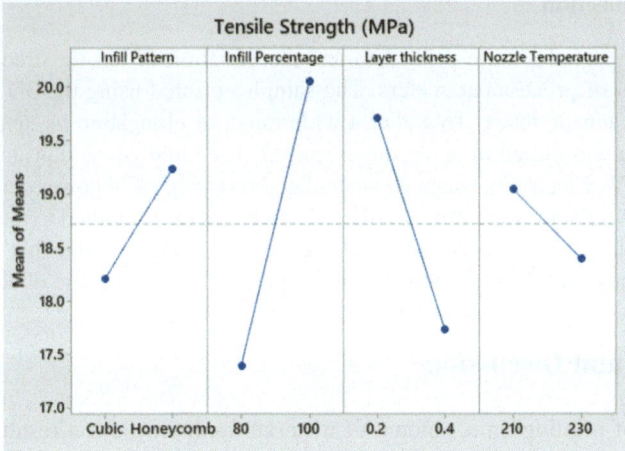

**Fig. 3.** Variation of tensile strength of the specimens with the change of process parameters

The lower level of the layer thickness (0.2 mm) again causes the high tensile strength. Decreasing layer thickness leads to increase the conjoined portion in between layers placed adjacently and less holes in the final product. Ultimately it results in greater cross-sectional area and tensile strength [9]. The nozzle temperature of 210 °C is responsible for a higher value of the tensile strength of the specimen than the 230 °C. The fluidity of the molten PLA material is increased temperatures after 210 °C [10]. This causes higher cooling time in order to ensure the perfect bonding between layers. However, insufficient cooling time results poor bonding between layers and ultimately lower tensile strength.

### 3.2 Material Used

Variation of material usage with the changing process parameters is illustrated in Fig. 4. According to that result the amount of material required to print one specimen is significantly affected by the infill percentage while the effect of other process parameters can be neglected. The infill percentage of 100% gives the higher tensile strength due to the creation of fewer holes and greater joining between layers.

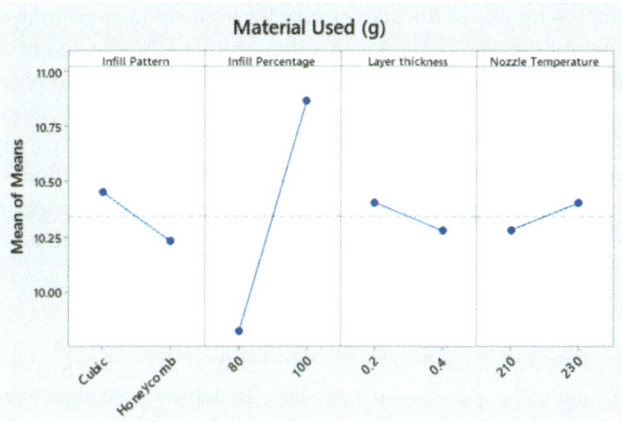

**Fig. 4.** Variation of material usage of the specimens with the change of process parameters

### 3.3 Electricity Used and Amount of CO$_2$ Emission

Variation in electricity usage with the changing process parameters is illustrated in Fig. 5. The amount of CO$_2$ generated during the printing of specimens, with the changing process parameters, also shows a similar behaviour as Fig. 5 because the amount of CO$_2$ emission in each test is directly dependent on electricity usage. There is a neutral effect of the infill pattern on the electricity usage and CO$_2$ generated while the effect of infill percentage is also very much less.

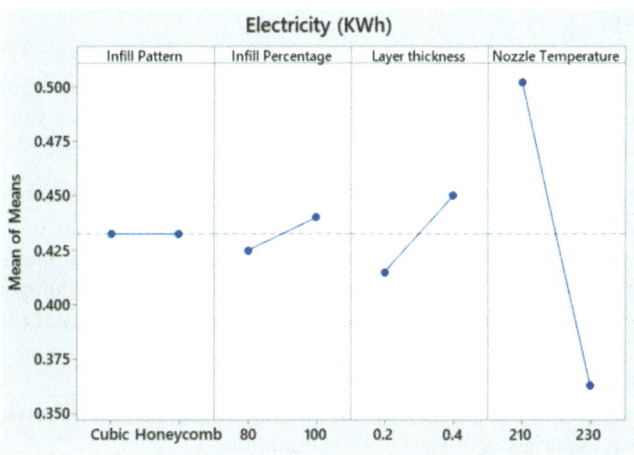

**Fig. 5.** Variation of electricity usage of the specimens with the change of process parameters

According to the literature, the printing time is increased when the layer thickness is decreased [9]. Higher printing time results in higher electricity consumption. However, according to the results of this study, it is evident that the mean values of electricity which resulted from the 0.4 mm layer thickness cause for high value of printing time and thus higher electricity consumption. The mean values are largely affected by the results of test no.5 which has comparatively high printing time. 100 of the infill percentage could be the reason for that. Individual comparison of other test results shows that lower layer thickness causes higher printing time and larger electricity consumption. Nozzle temperature is a significant influencing factor on electricity consumption. Lower nozzle temperature results in high printing time and high electricity consumption. It causes a high amount of $CO_2$ emissions and this result is confirmed by the literature [4].

## 4   Conclusion

This study was conducted to analyze the sustainability of 3D printing operations (FDM) without compromising the quality of the product. To achieve that objective variations of tensile strength, material usage, electricity consumption and amount of $CO_2$ emission were identified with the changing process parameters; infill pattern, infill percentage, layer thickness, and nozzle temperature. The following conclusions can be drawn based on the results of this study.

- Honeycomb infill pattern, 100 infill percentage, lower level of layer thickness, and 210 °C of nozzle temperature result in greater tensile strength of the printed samples.
- Material usage can be minimized by selecting a honeycomb pattern, lower infill percentage, higher layer thickness and 210 °C nozzle temperature.
- Smaller layer thickness and the 210 °C nozzle temperature cause higher printing time and thus result in greater electricity consumption and $CO_2$ emission during printing. However, the effect of infill pattern and percentage on electricity consumption and $CO_2$ emission can be neglected.

## 5   Future Work

This study has studied the effect of different process parameters on printing performances and the environment. The following areas can be considered for future studies based on the results of this study.

- A multi-response optimization can be performed to identify the best operating parameters to enhance the sustainability of 3D printing and those can be proposed for industries which use FDM for the businesses to enhance business productivity.
- Actual printing bed temperature can be deviated from the set-up temperature and it can be affected to the energy usage during printing and the quality of the final product. Deviation can be reduced by identifying better options while monitoring the real-time bed temperature.
- The sustainability of 3D printing operations can be enhanced by studying the effect of alternative materials on electricity consumption.

**Acknowledgement.** The authors would like to acknowledge the NRC15-151 Research Grant, Department of Manufacturing and Industrial Engineering and Civil Engineering, University of Peradeniya for providing the facilities to conduct the experiments and testing. The authors would further like to extend their thanks for the support given by Dr. Pramila Gamage and Eng. Avishka Maduranga.

# References

1. Jackson, B., Fouladi, K., Eslami, B.: Multi-parameter optimization of 3d printing condition for enhanced quality and strength. Polymers **14**(1586), 1–13 (2022)
2. Jelena, R.S., et al.: An experimental study on the impact of layer height and annealing parameters on the tensile strength and dimensional accuracy of FDM 3D printed parts. Materials **16**(4574), 1–19 (2023)
3. Adrián, R.P., Juan, C., Ana, M.C.: The influence of manufacturing parameters on the mechanical behaviour of PLA and ABS pieces manufactured by FDM: a comparative analysis. Materials **11**(1333), 1–21 (2018)
4. Yang, C.-J., Sin-Syuan, W.: Sustainable manufacturing decisions through the optimization of printing parameters in 3D printing. Appl. Sci. **12**(10060), 1–14 (2022)
5. Guanyi, C., et al.: Replacing traditional plastics with biodegradable plastics: impact on carbon emissions. Engineering **32**, 152–162 (2024)
6. Management, Presidential Task Force on Energy Demand Side. Carbon Footprint. https://www.energy.gov.lk/ODSM/Carbon-Footprint.html. Accessed 09 Jan 2024
7. Guan, R., Smith, D.: Influence of infill parameters on the tensile mechanical properties of 3D printed parts. J. Emerg. Invest. **2**, 1–5 (2020)
8. Alvarez C, K.L., Lagos C, R.F., Aizpun, M.: Investigating the influence of infill percentage on the mechanical properties of fused deposition modelled ABS parts. Ingenieria e Investigacion **36**(3), 110–116 (2016)
9. Yanping, L., et al.: Effects of printing layer thickness on mechanical properties of 3D-printed custom trays. J. Prosthet. Dent. **126**(5) (2020)
10. Fan, C., Shan, Z., Zou, G., Zhan, L., Yan, D.: Chin. J. Mech. Eng. **34**(1), 1–11 (2021)

# Life Cycle Assessment of Combined Additive and Subtractive Process Chains

Ralf Schlosser[✉], Knut Partes, Malte Schmidt, Helge Lorenzen, and Andre Wölk

Jade University of Applied Science, Friedrich-Paffrath-Straße 101, 26389 Wilhelmshaven, Germany
ralf.schlosser@jade-hs.de

**Abstract.** In this paper, the potential of a combined technology chain involving Wire Arc Additive Manufacturing (WAAM) and a subtractive milling process was analyzed in comparison to a conventional process chain, using a Cradle-to-Gate Life Cycle Analysis (LCA) conducted in the laboratories of Jade University of Applied Sciences.

The analysis revealed that the preferred technology chain, with respect to the evaluated Global Warming Potential (GWP), strongly depends on the specific application (component geometry). Consequently, the decision must be made on a case-by-case basis.

**Keywords:** Additive Manufacturing (AM) · GWP · LCA · Milling · WAAM

## 1 Introduction

Many companies have integrated sustainability as a core objective in their corporate goals. Nevertheless, a LCA according to DIN EN ISO 14040 and 14044 of products is not feasible with the existing company-wide reports since many relevant data are not available on manufacturing process level. Consequently, evaluations in impact categories such as Global Warming Potential (GWP) with methods like Product Carbon Footprints (DIN EN ISO 14067) are not possible with this data.

## 2 Goal and Scope Definition

The aim of this study is the LCA of a combined process chain of additive and subtractive milling processes in comparison to a conventional subtractive process chain starting with plate material. The study focuses on the GWP (CML2001-JAN2016, Global warming Potential 100 years) of the manufactured products with the intention to support the decision making of researchers and technical/ecological stakeholders involved in production planning. The framework of this study is designed to evaluate the environmental impact of the manufacturing processes on the basis of the constructed component volume and resulting weight. The material selected for this examination is SG2 (1.5125), known for its widespread applicability and relevance in welding and traditional manufacturing

H. Kohl et al. (Eds.): GCSM 2024, LNME, pp. 494–503, 2025.
https://doi.org/10.1007/978-3-031-93891-7_55

methods. The functional unit for this analysis is defined as the weight of material either added (in the case of additive processes) or removed (in the case of subtractive processes). This approach allows the direct comparison of the GWP associated with each technology chain.

In the LCA of this study, the system boundary is a cradle-to-gate analysis, focusing on the phases of raw material production and the manufacturing process within the production phase. This scope enables a coherent and comprehensive assessment of the environmental impacts associated with the early life stages of the product, from the extraction and processing of raw materials (cradle) up to, but not including, the point at which the product leaves the factory gate for distribution (gate). To achieve this, the study will systematically identify and quantify the relevant material and energy inputs and outputs for each of these phases. The environmental impacts of these inputs and outputs will be evaluated using the German life cycle inventory (LCI) data available in the GaBi software (version 9.2.1.68) as the investigation is carried out in Germany. The usage and recycling/disposal phases of the manufactured product are not considered further within the conducted LCA, as it is initially assumed that the additive/subtractive or conventionally manufactured components do not differ in terms of dimensions, functionality, and recyclability.

## 3   Inventory Analysis

For the comprehensive analysis of environmental impacts within the life cycle assessment of the study, differentiated approaches for the raw material extraction phase and the production phase are applied to ensure an accurate assessment.

For evaluating the environmental impacts of the raw material extraction phase, existing LCI databases are utilized. These databases provide extensive information on inputs such as energy and material consumption, as well as outputs, including emissions associated with the extraction and processing of raw materials. As the used material SG2 (1.5125) was missing in the available GaBi version, the existing dataset of 16MnCr5 (1.7131) was used estimating that the small difference of the alloying elements can be neglected (see Table 1).

**Table 1.** Comparison of the material composition (in wt%) showing the maximum values of the single alloying elements

| Material | Fe | C | Si | Mn | P | S | Cr | Ni | Cu | Mo |
|---|---|---|---|---|---|---|---|---|---|---|
| 1.5125 | 96.78 | 0.13 | 0.95 | 1.57 | 0.02 | 0.02 | 0.12 | 0.12 | 0.17 | 0.12 |
| 1.7131 | 96.94 | 0.19 | 0.40 | 1.30 | 0.03 | 0.04 | 1.10 | | | |

For a detailed evaluation of the production phase, the WAAM and milling processes are individually analyzed, as described within the next subchapters.

### 3.1 Wire Arc Additive Manufacturing Process

In AM, volumetric parts are generated by sequentially depositing layers of material on previously deposited layers. Fundamentally, the feedstock material in metal AM (e.g. powder or wire) is melted by an energy source (laser, electron beam or electric arc), deposited and solidifies again [8]. The WAAM process is based on the gas metal arc welding (GMAW) process. Here, an arc is generated between the tip of a consumable wire electrode and the underlying layer. Both components are melted and material from the wire electrode is deposited [9, 10]. The process chain of additive manufacturing is based on the following major steps consisting of multiple subtasks. First, trajectories and process parameters have to be generated (slicing) followed by the AM process. Manufactured parts are post processed, for example, with heat treatment or milling.

As WAAM has significant advantages in low equipment cost and high deposition rates, the shape stability (near-net shape) depends strongly on the heat input during the process [11]. Overheating as well as overcooling, caused by not controlling interpass temperatures, can led to shape distortion. Demanding matching process parameters to apply sufficient heat input is an additional quality factor [12].

GMAW process technologies like Cold Metal Transfer (CMT) are capable of reducing the heat input by modifying the droplet transfer and minimizing the arc burning time. Here, the arc is ignited and extinguished while the wire electrode is fed and retracted from the melt pool at high frequencies around 100 Hz [13, 14].

For the WAAM process a GMAW source with the CMT functionality (Fronius TPS 320i PULSE) is connected to an industrial robot arm (FANUC ARC Mate 100iC) with the weld gun attached to the robot flange. A volumetric part is manufactured by welding multiple, overlapping tracks with a defined distance of 4.25 mm (35% track-overlap) next to each other creating single layers. Those layers are stacked in a repetitive pattern, changing the weld direction by an angle of 90° to the previous layer, see Fig. 1.

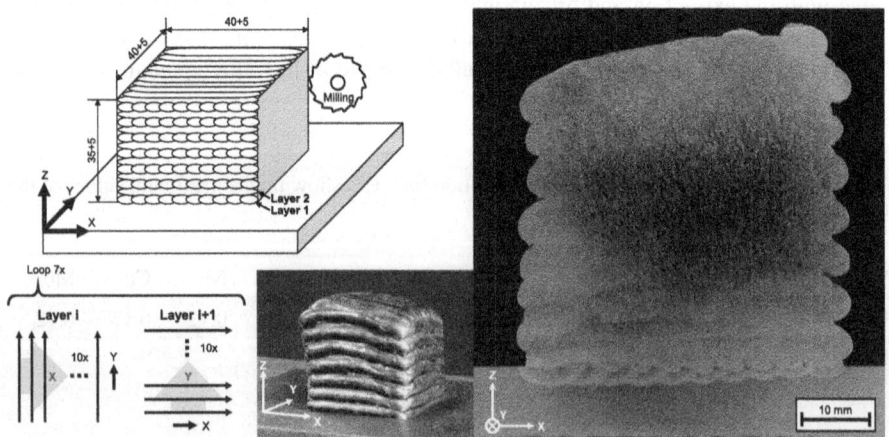

**Fig. 1.** WAAM deposition pattern and rough part dimensions

This pattern is repeated seven times resulting in 14 layers. Each single track has a length of 40 mm. Due to the initial deposition track geometry and subsequent milling process the part is oversized in width and height by approx. 5 mm. The weld gun is moved with a constant velocity of 25 cm/min. The weld source is set to constant current of 110 A (at 14.7 V) and a wire feed rate of 4 m/min using welding wire SG2 (1.5125) with 1.0 mm in diameter. A mixture of argon 82% and $CO_2$ 18% (M21-ArC-18) is used as active shielding gas with a flow of 15 l/min. Tracks are welded onto a substrate plate out of S235 mild steel.

The manufactured part fits the required oversize values with slight variations in Z-extension, strongly influenced by a sloped surface in the positive X- and Y-corner of the part. Although the required oversize was achieved, more material has to be removed in some areas. Figure 1 shows the manufactured part as well as a metallographic cross-section before the milling process. The part was cut in the Z-X-plane, grinded, polished and etched using a 3% alcoholic nitric acid. In the interaction zone between substrate plate and additive part a sufficient metallurgical bonding has been achieved. In the first layer's individual tracks can be differentiated, while towards higher layer a more homogenous microstructure can be seen.

The energy consumption of the complete WAAM System was measured with a frequency of 96 μs and the average welding power was identified with 2770 W. Only taking the main time into account the following specific values in Table 2 were created for the LCA within the experiments.

**Table 2.** WAAM input values for the LCA

|  | total Value | specific Value/per kg workpiece material |
|---|---|---|
| parts mass | 0.6529 kg | - |
| argon mass | 0.5435 kg | 0.8324 kg/kg |
| $CO_2$ mass | 0.1321 kg | 0.2023 kg/kg |
| Energy | 4.4 MJ | 6.7391 MJ/kg |

### 3.2 Milling

The Milling process as a cutting process is one of the standard processes in subtractive manufacturing of serial and single parts with high accuracy. Within industrial manufacturing cutting ratios of a view percent (automotive industry) up to 95–98% for integral structural components in aerospace technology are common. These milling operations are commonly applied on high automated CNC machine tools with complex supply functions and high energy demand.

Several investigations in cutting technology as well as in the LCA of cutting processes were conducted in the last years, showing that cutting processes have an ecological footprint depending on process parameter variations and the used machine tool and infrastructure. Several models were developed to forecast the required energy and supplies for energy consumption of machine tools and machining processes which Denkena et al. summarize in an extensive overview. [1].

Within this study a DMG-Mori Seiki DMU 50 machine tool and a HOLEX Pro Steel solid carbide steel High Perfomance Cutter with a diameter of 16 mm (202414 16) was used for the milling of the reference part.

For the evaluation of the milling process it was decided to neglect the influence of the machine tool production itself, as Bekker and Verlinden already proved the minor influence of the production phase compared to the use phase [7]. Although earlier investigations showed the influence of cutting parameters on the specific energy [15–17], it was also shown, that during machining operations peripheral aggregates as cooling for example have different non stationary levels of energy consumption independent of the chosen process [1, 18, 19].

To avoid a high complexity of the model and the experimental setup, it was decided to apply only a variation of the cutting speed with the constant process parameters depth of cut, width of cut and feed per tooth in accordance with the tool manufacturers recommendations. Table 3 shows the dataset which was applied for milling the cube. All parameter sets were applied under dry machining conditions.

**Table 3.**  Cutting parameters and specific energy $e_c$

| Parameter-set | $n$ [min⁻¹] | $v_c$ [m/min] | $f_z$ [mm] | $v_f$ [mm/min] | $a_e$ [mm] | $a_p$ [mm] | $D$ [mm] | $z$ | $P_{c,const}$ /W | $Q_z$ [mm³/s] | $e_{c,v}$ [Ws/mm³] | $e_{c,m}$ [MJ/kg] |
|---|---|---|---|---|---|---|---|---|---|---|---|---|
| 1 | 4000 | 201.1 | 0.08 | 960 | 6 | 2 | 16 | 3 | 2164 | 192 | 11.27 | **1.436** |
| 2 | 4250 | 213.6 | 0.08 | 1020 | 6 | 2 | 16 | 3 | 2169 | 204 | 10.63 | **1.354** |
| 3 | 4377 | 220.0 | 0.08 | 1050 | 6 | 2 | 16 | 3 | 1911 | 210 | 9.10 | **1.159** |
| 4 | 4500 | 226.2 | 0.08 | 1080 | 6 | 2 | 16 | 3 | 1998 | 216 | 9.25 | **1.178** |
| 5 | 4750 | 238.8 | 0.08 | 1140 | 6 | 2 | 16 | 3 | 2561 | 228 | 11.23 | **1.431** |
| 6 | 5000 | 251.3 | 0.08 | 1200 | 6 | 2 | 16 | 3 | 1967 | 240 | 8.20 | **1.044** |
| **Rounded Average for LCA** | | | | | | | | | | | | **1.250** |

Each parameter set was used three times and the power consumption of the machine tool was measured at the plug in. The constant cutting power value $P_{c,const}$ is the average power consumption of the complete machine tool incl. all peripheral aggregates during main time. The effects of auxiliary process and nonproductive times are not respected further on, as these times are very dependent on part geometry and manufacturing system. Within the experiments on the DMU 50 evolution machining center the total power consumption showed also the effects that, some power consumers of the machine tool which worked non stationary as already described by [1, 18, 19].

The volumetric and massic energy is calculated for each parameter set in accordance to Eq. (1) and (2) with the material removal rate $Q_z$ and the density $\rho$. The average specific massic energy for all six parameter sets was 1,267 MJ and is rounded to 1,250 MJ for the further implementation within the LCA. By applying the average about the six parameter sets the effect of not predicting the non-stationary power consumers can be lowered.

$$e_{c,v} = \frac{P_c}{Q_z} \tag{1}$$

$$e_{c,m} = \frac{e_{c,v}}{Q_z} \tag{2}$$

Within this study the tool wear and the wear of the machine tool itself are neglected. As [7] already showed for the manufacturing phase of a machine tool, that it can be neglected this assumption also make sense for the wear behavior of a machine tool. In context of the tool wear behavior process parameter sets were chosen for the steel milling which allow much more mass removal at the workpiece than the tool has itself. As shown in [10] it can be estimated that the factor between removed material and the influence of the cemented carbide on the LCA is low for moderate process parameter close to the economic optimum.

## 4 Impact Assessment

The inventory analysis for the WAAM and the milling process are used to set up a single parametric model within the GaBi-Software. Corresponding to the aim of the study the global warming potential will be assessed by the CML2001-JAN2016, Global warming Potential 100 years category. The model is shown in Fig. 2. The parametric model is based on the specific massic energy and the specific consumption of the shown resources and losses of workpiece material within the different processes.

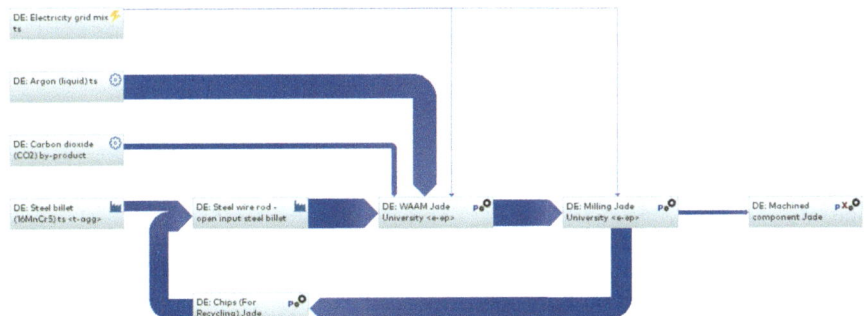

**Fig. 2.** LCA model in the GaBi software

The following existing GaBi processes and LCI-Data were used:

- DE: Electricity grid mix ts
- DE: Argon (liquid) ts
- DE: Carbon dioxide (CO2) by-product ammonia (NH3) (economic allocation) ts
- DE: Steel billet (16MnCr5) ts <t-agg>

- DE: Steel wire rod - open input steel billet ts <t-agg>

The existing process "DE: Steel wire rod - open input steel billet ts <t-agg>" is working on basis of a loss in material of about 5.2%. Comparing this data with [7] it is obviously that this deep investigation identified a loss of material from steel billet to rolled plate Material of 14,5% and to WAAM parts of 22,21%. As the first own parametrized model is used within the WAAM process, these ratios are lowed by the already implemented reduction within the "DE: Steel wire rod - open input steel billet ts <t-agg>" process to 9,8% and 17,9%.

For the evaluation of both process chains 3 scenarios (parts) were defined for the simulation. The three parts "wall", "thin walled cavity" and "cavity" were designed to respect a wide range of possible part geometries with a set of a few simple geometries.

The parts and the resulting mass are shown in Table 4. In addition to this the mass ratio resulting from the required part allowance (WAAM + 5 mm on each surface) or starting geometry. For Milling it was assumed for all three scenarios starting with 100 plate material with the contour geometry of 100 mm × 100 mm. This approach is respecting small manufactures working with a few starting geometries. In Serial production the plate material would be sawn to smaller blocks with a low allowance of <1 mm on each surface.

**Table 4.** Reference geometries for the scenario modelling

| | Finished Geometry | | | Starting geometry WAAM | | | Milling |
|---|---|---|---|---|---|---|---|
| | Wall | Thin walled cavity | Cavity | Wall | Thin walled cavity | Cavity | all |
| Contour/mm -length -height -width | 90 10 95 | 60 60 95 | 90 90 95 | 100 20 100 | 70 70 100 | 100 100 100 | 100 100 100 |
| Cavity/mm -length -height -width | | 50 50 95 | 50 50 95 | | 40 40 100 | 40 40 100 | |
| Ratio | 1 | 1 | 1 | | | | |
| Mass/kg | 0.671 | 0.820 | 4.176 | | | | |
| Ratio | | | | 0.428 | 0.317 | 0.633 | 0.086 0.105 0.639 |

Figure 3 shows the impact assessment of the three reference geometries. The three left columns show the result for the combined process chain of additive and subtractive manufacturing. The three columns on the right show the results of the conventional process chain of plate material and milling.

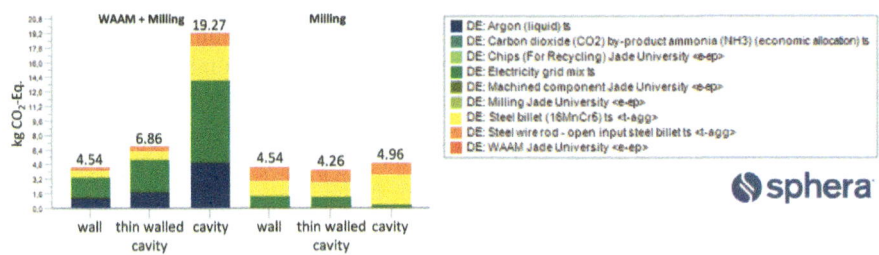

**Fig. 3.** LCA Results in CML2001-JAN2016, Global warming Potential 100 years

## 5   Interpretation

The simulation results show some expectable results due to the boundary conditions of the models. The driver for the GWP is the material in the conventional process chain and the electricity followed by argon for the additive subtractive process chain. The chosen reference parts showed a dramatic increase of the GWP with the increase of the mass of the reference part. Due to the same raw material geometry for milling and a high recycling rate of 90% of the chips the thin walled cavity shows the lowest impact. For the heaviest reference part, the conventional process chain has almost no influence by electrical energy but by the material.

Further on it can be seen, that only the geometry wall was possible to achieve the same low level of impact within the GWP.

### 5.1   Critical Discussion

The results for the different reference parts showed, that the conventional process chain showed a big advantage with increasing ratio. Only for small ratios (and bad starting conditions for the milling process comparable results were achieved. Considering that only main time was respected, it has to be estimated, that the milling processes with auxiliary and nonproductive times is less efficient. Further on, the tool wear behavior was neglected as well as the compressed air. Both last effects shouldn't increase the GWP to much. Figure 4 shows an estimation, that due to all 3 factors the milling process would require 2MJ/kg instead of 1,25 MJ/kg in the specific energy.

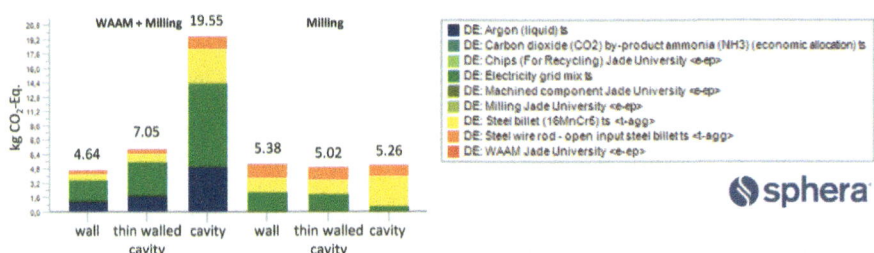

**Fig. 4.** LCA Results in CML2001-JAN2016, Global warming Potential 100 years estimating the specific massic cutting energy with 2MJ/kg

Even this bad estimation still delivers a high potential for two of the three reference parts. Nevertheless, it also shows, that the parts geometry has a major impact on the environmental business case. Also process capabilities like the productivity and quality have a big impact on the environmental business case. If it would be possible to reduce the required allowance on 1 mm instead of 5 mm more parts might be manufactured by combined additive and subtractive process chains in the future.

## 6  Summary and Outlook

In this paper, a LCA was conducted to compare the conventional process chain using steel plate materials and subtractive milling with an additive process chain that combines WAAM and milling. The analysis demonstrated that both process chains have their respective advantages, and there is no definitive answer as to which is superior. The geometry of the part plays a significant role and must be considered for a well-founded evaluation.

Additionally, this paper exclusively discusses the manufacturing of complete parts. However, in the context of the circular economy, the additive/subtractive process chain also shows significant potential for repair solutions instead of new manufacturing. Here too, the geometry of the parts needs to be carefully analyzed.

## References

1. Denkena, B., Abele, E., Brecher, C., Dittrich, M.-A., Kara, S., Mori, M.: Energy efficient machine tools. CIRP Ann. Manufact. Technol. **69**(2), 646–667 (2020). https://doi.org/10.1016/J.CIRP.2020.05.008
2. Zeulner, J.: Modellierung des Energiebedarfs von Zerspanungsprozessen zur Unterstützung des lebenszyklusorientierten Carbon Accounting für Unternehmen (2023). https://doi.org/10.26083/TUPRINTS-00024077
3. Kara, S., Bogdanski, G., Li, W.: Electricity metering and monitoring in manufacturing systems. In: Hesselbach, J., Hermann, C. (eds.) Glocalized Solutions for Sustainability in Manufacturing. 18th CIRP, pp. 1–10 (2011)
4. Hauschild, M.Z., Kara, S., Røpke, I.: Absolute sustainability: challenges to life cycle engineering. CIRP Ann. Manufact. Technol. **69**(2), 533–553 (2020). https://doi.org/10.1016/j.cirp.2020.05.004
5. https://www.grimm-edelstahlhandel.de/16mncr5-das-werkstoffdatenblatt/
6. https://pauly-stahlhandel.com/de/werkstoff-nr/1.5125
7. Bekker, A., Verlinden, J.: Life cycle assessment of wire + arc additive manufacturing compared to green sand casting and CNC milling in stainless steel. J. Clean. Prod. **177**, 438–447 (2018). https://doi.org/10.1016/j.jclepro.2017.12.148
8. DebRoy, T., et al.: Additive manufacturing of metallic components – process, structure and properties. Prog. Mater. Sci. **92**, 112–224 (2018). https://doi.org/10.1016/j.pmatsci.2017.10.001
9. Zhou, K. (ed.): Additive Manufacturing: Materials, Functionalities and Applications, 1st edn. Springer, Cham (2023)
10. Rodrigues, T.A., Duarte, V., Miranda, R.M., Santos, T.G., Oliveira, J.P.: Current status and perspectives on wire and arc additive manufacturing (WAAM). Materials **12**(7), 1121 (2019). https://doi.org/10.3390/ma12071121

11. van Le, T., Mai, D.S., Hoang, Q.H.: A study on wire and arc additive manufacturing of low-carbon steel components: process stability, microstructural and mechanical properties. J. Braz. Soc. Mech. Sci. Eng. **42**(9) (2020). https://doi.org/10.1007/s40430-020-02567-0

12. Kozamernik, N., Bračun, D., Klobčar, D.: WAAM system with interpass temperature control and forced cooling for near-net-shape printing of small metal components. Int. J. Adv. Manuf. Technol. **110**(7–8), 1955–1968 (2020). https://doi.org/10.1007/s00170-020-05958-8

13. Furukawa, K.: New CMT arc welding process – welding of steel to aluminium dissimilar metals and welding of super-thin aluminium sheets. Weld. Int. **20**(6), 440–445 (2006). https://doi.org/10.1533/wint.2006.3598

14. Posch, G., Chladil, K., Chladil, H.: Material properties of CMT—metal additive manufactured duplex stainless steel blade-like geometries. Weld World **61**(5), 873–882 (2017). https://doi.org/10.1007/s40194-017-0474-5

15. Schlosser, R., Klocke, F., Lung, D.: Sustainabilty in Manufacturing; Energy Consumption of Cutting Processes. Springer, Berlin (2011)

16. Yan, J., Li, L.: Multi-objective optimization of milling parameters – the trade-offs between energy, production rate and cutting quality. J. Clean. Prod. **52**, 462–471 (2013)

17. Draganescu, F., Gheorghe, M., Doicin, C.V.: Models of machine tool efficiency and specific consumed energy. J. Mater. Process. Technol. **141**(1), 9–15 (2003)

18. Kuhrke, B.: Methode zur Energie- und Medienbedarfsbewertung spanender Werkzeugmaschinen, Ph.D. thesis, Technical University of Darmstadt (2011)

19. Brecher, C. (ed.) Effizienzsteigerung von Werkzeugmaschinen durch Optimierung der Technologien zum Komponentenbetrieb - EWOTeK: Verbundprojekt im Rahmenkonzept „Forschung für die Produktion von morgen, Ressourceneffizienz in der Produktion des Bundesministeriums für Bildung und Forschung (BMBF). Report, RWTH Aachen University, Apprimus (2012)

20. Schlosser, R.: Methodik zur Prognose der Nachhaltigkeit des Energie- und Stoffeinsatzes spanender Fertigungsprozesse. Ph.D. thesis, RWTH Aachen, Aachen, Apprimus (2013)

21. Klocke, F., Schlosser, R., Sangermann, H.: Evaluation of the Energy Consumption of a Directed Lubricoolant Supply with Variable Pressures and Flow Rates in Cutting Processes. Sustainable Manufacturing, pp. 203–209. Springer, Berlin (2012)

# Maintenance, Repair and Lifecycle Extension

# Ontology-Based Framework for a Digital Twin in Maintenance

Sissy-Linh Nguyen[1(✉)], Svenja Nicole Schulte[2], and Kai Lindow[1]

[1] Fraunhofer IPK, Pascalstraße 8-9, 10587 Berlin, Germany
{sissy-linh.nguyen,kai.lindow}@ipk.fraunhofer.de
[2] Technische Universität Berlin, Pascalstraße 8-9, 10587 Berlin, Germany
svenja.schulte@tu-berlin.de

**Abstract.** With digital twins, manufacturers are able to use real-time data and simulations to optimize production and improve sustainability. The aim of this paper is to develop a twin architecture specially for the maintenance process of turbine blades. As a result, an ontology-based framework for a digital twin of turbine blades is presented that enables the connection of information and data during maintenance. The operational mechanisms of the digital twin are described, and the connectivity between the physical and virtual elements is demonstrated using the ontology model. The framework also supports analysis to optimize production, improve maintenance, and provide information to the services used throughout the lifecycle. Further research within ongoing projects will implement the digital twin and validate the use of the presented ontology in the digital twin.

**Keywords:** Digital twin · Ontology · Maintenance · Turbine blade · Lifecycle · Production · Sustainability

## 1 Introduction

The industrial sector is undergoing a profound transformation due to its global networking of markets and supply chains, impact on environment, increasing demand for customized products, need for qualified personnel due to the use of automation and robotics, fluctuating market demands, and political uncertainties. Approaches and technologies are therefore needed to create more efficient, sustainable, human-centered, and resilient manufacturing processes in order to maintain competitiveness and continuously improve production processes.

The digital twin is a cutting-edge technology that creates a digital representation of a physical object and consists of a linkage between data (digital shadow) and models (digital master) [1]. It enables a bidirectional flow of information and data between the virtual and physical components [1]. This enables simulation, monitoring, informed planning, and management of resources across the entire product life cycle and supports circularity in the economy through

© The Author(s) 2025
H. Kohl et al. (Eds.): GCSM 2024, LNME, pp. 507–515, 2025.
https://doi.org/10.1007/978-3-031-93891-7_56

enhanced maintenance and recycling [1–3]. Digital twins are applied in various industries, with the majority of applications being found in manufacturing, building environments, and information and communications technology [4].

At the same time, the use of knowledge graphs and ontologies is becoming increasingly popular for structured understanding of data, representation of flows, and automated reasoning in digital twins for manufacturing systems [5]. Ontologies, in particular, provide formal, machine-processable specifications to describe a set of concepts within a domain. The use of unique identifiers enables interoperability at the semantic level, as well as improved data management and reuse within large knowledge domains [6]. Semantic technologies are also associated with advancing cross-domain interoperability in digital twins [7]. While ontologies are used for structuring data, knowledge graphs are deployed for practical applications to provide insights from diverse data [6].

The objective of this paper is to identify possible architectures and models to address digital twins' key challenges in maintenance processes and to advance the semantic concept in the manufacturing industry by integrating environmental indicators. By integrating environmental indicators into the maintenance process, an environmental assessment can be included to evaluate the repair without having specialized expert knowledge in sustainability. This will facilitate the maturation of digital twins and promote the development of standards for enhanced interoperability, thereby improving decision-making and supporting sustainable manufacturing.

## 2   Related Works

This section reviews relevant literature on digital twins in maintenance and ontologies. A Web of Science search using "digital twin", "ontology" and "maintenance" in titles, abstracts, and keywords returned 29 results between 2019 and beginning of 2024. Of these, 10 articles were relevant to the manufacturing sector, while the rest focused on built environments, robotics, automation, infrastructure, and related topics.

In the field of maintenance in manufacturing, ontology-based digital twins have been developed for analytical purposes, with the aim of enabling the monitoring of health and decision-making support in products and manufacturing systems [7–10]. Four articles reused either existing ontologies or used manufacturing standards [7–9,11]. Two articles referred to the digital twin with semantic capabilities as Cognitive Twin, while two further articles developed domain ontologies integrating data from all lifecycle phases [7,8,11,12]. One paper linked the recyclability of materials, design decisions and waste reduction indicators in their model [12].

Furthermore, Karabulut et al. [6] found out that the use of knowledge graphs and semantics has become popular in digital twins, especially in the domains of manufacturing, smart cities, and infrastructure. Objectives of ontologies included system/data modeling, semantic interoperability, semantic relation extraction, and reasoning facilitation. With regard to digital twins, a semantic data model is used to contextualize digital shadow data with each other as well as other information provided by digital master models [13]. However, further research is required to determine the extent to which ontologies can be integrated into digital twins and to investigate the potential of knowledge graphs for reasoning [6].

Most digital twin models focus on monitoring and predictive maintenance, with few addressing corrective maintenance, which is carried out after fault identification. Data from fault detection can optimize engineering design. Challenges in digital twins include linking data and models from various sources and ensuring system interoperability, as well as addressing inconsistent and redundant data. The semantic approach effectively achieves interoperability and integrates digital twin capabilities. Despite manufacturing's role in reducing emissions and waste, sustainability is often overlooked, and maintenance lacks sustainability and waste reduction indicators.

The key contributions of this paper are: advancing digital twins with semantic capabilities, addressing interoperability through an ontology linking different data sources, and integrating environmental indicators for a broader understanding of sustainability into maintenances. This study identifies key aspects and relationships in maintenance to develop an ontology-based digital twin.

## 3    Methodology

The methodology proposed by Noy [14] was employed for the development of the ontology of this digital twin. It begins with the description of the domain and scope, as outlined in Sect. 3.1. Then, in Sect. 3.2, the potential for reusing existing ontologies is investigated. In Sect. 3.3, the concept is developed in order to provide an enumeration of the terms. In the final stage of the development process, the concept is formalized through the definition of classes, class hierarchies, and properties using a specific ontology language. This final stage will not be covered here, but it is worth noting that it forms part of further research in the future.

### 3.1    Domain Analysis

In order to ensure the effective development of the semantic data model, it is crucial to identify the necessary scope from the use case and to obtain necessary information sources relevant to the model. The findings are presented in Table 1.

### 3.2    Existing Ontologies

In order to identify existing ontologies, the domain of manufacturing was divided into three distinct categories: maintenance, digital twin and sustainability. A review of the literature revealed that, in many cases, ontologies were developed based on existing standards. In contrast, the concept of a digital twin is defined in a number of different ways, which also affects the choice of the developed

**Table 1.** Identification of scope, use cases and stakeholders

| Question | Answer |
|---|---|
| What is the domain that the ontology will cover? | Generation of work orders for corrective maintenance of a turbine blade |
| What is the ontology used for? | Knowledge abstraction; Support for work order generation |
| What type of question does the ontology answer? | Information structure of the maintenance of the turbine blade in a digital twin; Repair measure assessment depending on the defect type and criteria |
| Who will use and maintain the ontology? | Ontology expert and maintainer: service engineer; Data source: service engineer, worker, supplier, quality assurance specialist; User: service engineer, worker, supplier |

ontology. Similarities have been identified in the data management of digital twins, which can be applied as a generic data ontology for a variety of domains. The ontologies used for the conceptualization detailed in Sect. 3.3 are listed in Table 2.

**Table 2.** Reused ontologies

| Domain | Name | Description | Reference |
|---|---|---|---|
| Manufacturing | Reference Ontology for Industrial Maintenance (ROMAIN) | Basic Formal Ontology (BFO) compliant maintenance management ontology | [15] |
| | Ontology for maintenance procedure documentation (OMPD) | Web Ontology Language (OWL)-based ontology for maintenance procedures | [16] |
| Digital Twin | Digital Twin ontology model | Ontology for data management in a digital twin | [17] |
| Sustainability | Ontology-based system for supporting manufacturing sustainability | Ontology for supporting sustainable alternative manufacturing | [18] |
| | Ontology that relates sustainability terms to product and process | Ontology for product lifecycle management and sustainability in manufacturing | [19] |

## 3.3   Conceptualization of Knowledge

In the conceptualization stage, inputs are transformed into a semi-formal conceptual model of objects and concepts with taxonomic relationships. Two distinct

approaches have been employed to formalize this process. The bottom-up approach facilitated an understanding of the domain of the maintenance of turbine
blades. This was conducted in collaboration with subject matter experts. A number of terms have been compiled and discussed within this working group with
the objective of achieving consensus and facilitating their future utilization in
other working groups. By utilizing the top-level ontologies for maintenance and
the work plan specification, as presented in Table 2, the project content from
the working group can be compared with the research results and integrated
into a context for corrective maintenance. The detailed conceptualization will
be explained in Sect. 4.

## 4 Framework of the Ontology-Based Digital Twin for Maintenance

As illustrated in Fig. 1, the concept is divided into three layers, each of which is
responsible for answering specific questions from Sect. 3.1.

The domain for the work order generation is described in the service process. The initial step in the creation of the work plan in the product maintenance life plan is the identification and description of the defect, which is then
assigned to the relevant repair measure of the suppliers. The work specification
is derived from the totality of all defect types, their attributes (e.g. location,
inspection characteristics, measurements, inspection criteria) and the applicable
repair measures. Maintenance is carried out via an internal or external order.
Information that significantly influences the success rate, costs or customer satisfaction is sometimes fed back into the strategy in order to improve the work
plan generation of a product accordingly.

The next layer is the data process and describes the data management in
the digital twin. The function of the digital twin can be defined as follows: In
the data acquisition planned data are captured from strategy, 3D models or
documents. These data and models are referred to as digital master. In the next
step, this data must be tailored for the respective function (data manipulation).
State detection encompasses the identification and recording of the current state,
which is recorded in the digital shadow. The objective of the repair assessment is
to determine, with the assistance of the knowledge graph, the most appropriate
repair measure in accordance with the prevailing repair technologies that restores
the component to its target condition and impact parameters. It is essential that
the recommendation be made visually accessible to both the user, such as the
worker and supplier, and the ontology-based digital twin supervisor.

The concept of sustainability is presented in an entity relationship model
to illustrate interconnectedness of data relating to the product, materials, and
processes, and the work order generation of the repair assessment. This structure is also used in many other related works. It represents the basic classes
and taxonomy of the manufacturing sector. The concept is read as "product"
has "function". This implies that the product has a functional task. This information is incorporated into the "Repair Assessment" and the "Impact" which

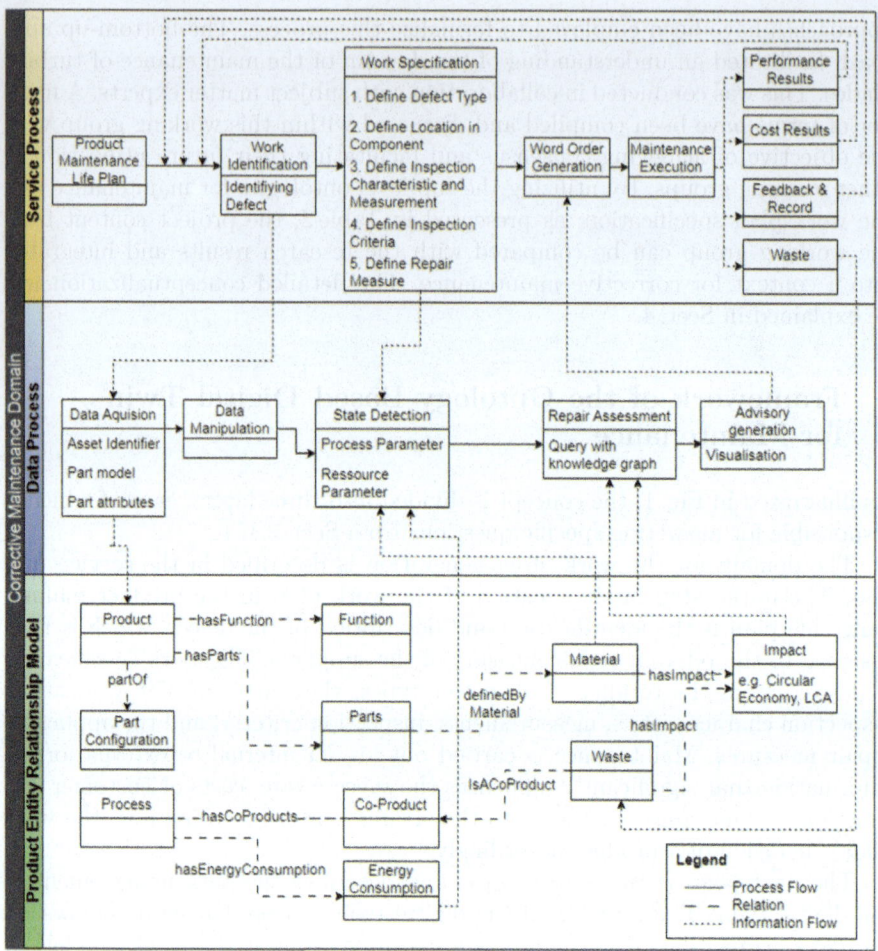

**Fig. 1.** Conceptualization of the corrective maintenance

addresses the non-functional requirement of an ecological assessment of a material, a resource used or a waste. A work plan is created that takes into account both functional and non-functional requirements.

## 4.1   Discussion

The utilization of existing standardized ontologies, as illustrated in Table 2, facilitates the transfer of activities and data models from this concept to other individual use cases within the manufacturing domain. The ontology has enabled the conceptualization of the digital twin of a turbine blade during maintenance, as well as the integration of sustainability measures.

When incorporating sustainability measures into repair assessments, a more comprehensive evaluation of potential repair measures becomes necessary. Company policies as well as legal requirements related to sustainability may necessitate the inclusion of such analyses. This entails addressing additional questions such as: Which technology results in the lowest $CO_2$ emissions? Which technology minimizes waste production? Nonetheless, it needs to be ensured that functionality of the respective turbine blade is restored and the remaining lifespan of the blade is maximized.

Technologies such as SPARQL and Large Language Models (LLMs) such as GPT-3 can be used to facilitate this: SPARQL queries can be used to retrieve information from linked data sources to assess the environmental impacts associated with different repair technologies. LLMs can be used to analyze and optimize repair strategies. They can also help generate detailed reports, analyze textual data related to environmental impact assessments, and recommend optimal repair measures based on predefined sustainability criteria.

Implementing the framework with SPARQL and LLMs enhances sustainability assessments in repair strategies, supporting informed decision-making and optimizing turbine blade environmental performance. This requires modeling an ontology from Fig. 1, involving engineers, software developers, and sustainability experts to ensure holistic trade-off evaluations, such as $CO_2$ reduction vs. product quality. A database with relevant data (operational sensors, maintenance records, environmental impact records) is needed for SPARQL queries. Additionally, integrating the ontology concept into the selected LLM is crucial for effective use.

## 5   Conclusion and Future Work

This paper presents an ontology-based digital twin of a turbine blade in maintenance, focusing on work order creation. While digital twins are common for condition monitoring and predictive maintenance, their use in manufacturing has been not fully used, despite the potential to optimize repair operations in terms of performance, cost-effectiveness, and sustainability.

Therefore, the paper proposed the ontology-based twin, which serves as a technical basis for the digital support of the repair process. As the knowledge base is primarily derived from experts in the field, the ontology facilitates a systematic and structured approach to knowledge documentation. Furthermore, additional domains, such as sustainability, can be incorporated into the repair assessment, thereby facilitating the incorporation of environmental sustainability considerations into maintenance decisions without the need for prior expertise. To integrate sustainability measures, the repair assessment must consider both the non-functional and functional requirements of the component and its repair measure. This was evaluated together with industry partners, fostering the development of strategies designed to minimize environmental impact and cultivate a sense of responsibility.

Future work will formalize, implement, and validate the framework in an ongoing research project. The service process has already been formalized in

Protégé and validated by the working group. The digital twin is currently in the implementation phase, developing instances and interfaces for information.

**Acknowledgments.** This research is part of the Pro FIT Maintenance, Repair and Overhaul (MRO) Phase II project inside the industry and science campus Werner-von-Siemens Centre, and is co-financed by the European Regional Development Fund (ERDF).

# References

1. Stark, R., Damerau, T.: Digital twin. In: Chatti, S., Tolio, T. (eds.) CIRP Encyclopedia of Production Engineering. Springer, Heidelberg (2019). https://doi.org/10.1007/978-3-642-35950-7_16870-1
2. Muench, S., Stoermer, E., Jensen, K., Asikainen, T., Salvi, M., Scapolo, F.: Towards a green and digital future. Publications Office of the European Union, Luxembourg (2022). JRC129319. 10.2760/54
3. Pal, S.K., Mishra, D., Pal, A., Dutta, S., Chakravarty, D., Pal S.: Digital twin - fundamental concepts to applications in advanced manufacturing, p. 12. (1. Aufl.) Springer, Cham (2020)
4. Lo, C.K., Chen, C.H., Zhong, R.Y.: A review of digital twin in product design and development. Adv. Eng. Inform. **48**, 101297 (2021). https://doi.org/10.1016/j.aei.2021.101297
5. Gogineni, S., Lindow, K., Nickel, J., Stark, R.: Applying Contextualization for Data-Driven Transformation in Manufacturing (2020). https://doi.org/10.1007/978-3-030-57997-5_19
6. Karabulut, E., Pileggi, S., Groth, P., Degeler, V.: Ontologies in digital twins: a systematic literature review. Future Gener. Comput. Syst. (2023). https://doi.org/10.1016/j.future.2023.12.013
7. D'Amico, R.D., Erkoyuncu, J.A., Addepalli, S., Penver, S.: Cognitive digital twin: an approach to improve the maintenance management. CIRP J. Manufact. Sci. Technol. (2022)
8. Meyers, B., et al.: Knowledge graphs in digital twins for manufacturing - lessons learned from an industrial case at atlas copco airpower. IFAC-PapersOnLine **55**, 13–18 (2022). https://doi.org/10.1016/j.ifacol.2022.09.361
9. Yan, W., et al.: Intelligent predictive maintenance of hydraulic systems based on virtual knowledge graph. Eng. Appl. Artif. Intell. **126** (2023). https://doi.org/10.1016/j.engappai.2023.106798
10. Pulikottil, T., Martínez-Arellano, G., Barata, J.: Immune system inspired smart maintenance framework: tool wear monitoring use case. Int. J. Adv. Manuf. Technol. **132**, 4699–4721 (2024). https://doi.org/10.1007/s00170-024-13472-4
11. Slee, D., Cain, S., Vichare, P., Olszewska, J.: Smart Lifts: An Ontological Perspective, pp. 210–219 (2021). https://doi.org/10.5220/0010690700003064
12. Guerra, V., Hamon, B., Bataillou, B., Inamdar, A., van Driel, W.D.: Towards a digital twin architecture for the lighting industry. Future Gener. Comput. Syst. **155**, 80–95 (2024). https://doi.org/10.1016/j.future.2024.01.028. ISSN 0167-739X
13. Stark, R., Schulte, S.N.: Determination of Digital Twin Maturity Levels Within Value Creation Networks. NAFEMS world congress (2021)

14. Noy, N.: Ontology Development 101: A Guide to Creating Your First Ontology (2001)
15. Karray, H., Ameri, F., Hodkiewicz, M., Louge, T.: ROMAIN: towards a BFO compliant reference ontology for industrial maintenance. Appl. Ontol. **14**, 1–24 (2019). https://doi.org/10.3233/AO-190208
16. Woods, C., French, T., Hodkiewicz, M., Bikaun, T.: An ontology for maintenance procedure documentation. Appl. Ontol. **18**, 1–38 (2023). https://doi.org/10.3233/AO-230279
17. Singh, S., et al.: Data management for developing digital twin ontology model. Proc. Inst. Mech. Eng. Part B J. Eng. Manufact. **235**(14), 2323–2337 (2021). https://doi.org/10.1177/0954405420978117
18. Giovannini, A., Aubry, A., Panetto, H., Dassisti, M., El Haouzi, H.: Ontology-based system for supporting manufacturing sustainability. Annu. Rev. Control. **36**(2), 309–317 (2012). https://doi.org/10.1016/j.arcontrol.2012.09.012
19. Borsato, M.: Bridging the gap between product lifecycle management and sustainability in manufacturing through ontology building. Comput. Ind. **65**(2), 258–269 (2014). https://doi.org/10.1016/j.compind.2013.11.003

# Enhancing Repairability in Digital Fabrication: Applying Lean Six Sigma for Optimized Design in a Local Fab Lab

Pham Minh Thanh[✉]

Vietnamese-German University, Bến Cát, Bình Dương, Vietnam
thanh.phaminh@gmail.com

**Abstract.** The growing adoption of digital fabrication technologies in localized production environments, such as Fab Labs (Fabrication Laboratories), presents opportunities for sustainable manufacturing practices. However, the integration of repairability principles into digital fabrication processes remains a challenge. This research develops a framework that combines Design for Repairability (DfR) principles with the Lean Six Sigma DMAIC (Define, Measure, Analyze, Improve, Control) methodology to optimize designs for enhanced repairability in digital fabrication. Focusing on additive and subtractive manufacturing technologies at a local Fab Lab, the study follows the DMAIC approach to define the problem of poor repairability, measure key metrics, analyze root causes, improve the design process, and control implementation. The framework leverages digital tools to identify potential failure points and optimize part geometry. Case studies are conducted to validate the framework and quantify the benefits in terms of reduced repair time, material savings, and extended product life. Results demonstrate significant improvements in repairability metrics, with repair time reduced by 37% and material waste decreased by 42% on average. This research contributes to the advancement of sustainable manufacturing by integrating DfR principles with Lean Six Sigma methodology in the context of digital fabrication.

**Keywords:** Digital fabrication repairability · Lean Six Sigma DMAIC · Design for Repairability

## 1 Introduction

Digital fabrication technologies have revolutionized localized production environments, enabling rapid prototyping and small-scale manufacturing of customized parts and products. Fab Labs, as open access workshops equipped with digital fabrication tools, play a crucial role in democratizing access to these technologies [1]. However, the focus on rapid production often overshadows considerations of product longevity and repairability, leading to increased waste and resource consumption [2].

Design for Repairability (DfR) is a design philosophy [1] that aims to create products that are easier to repair, maintain, and upgrade [3]. Integrating DfR principles into digital fabrication processes presents an opportunity to enhance the sustainability of Fab Lab

© The Author(s) 2025
H. Kohl et al. (Eds.): GCSM 2024, LNME, pp. 516–523, 2025.
https://doi.org/10.1007/978-3-031-93891-7_57

outputs. However, the implementation of DfR in digital fabrication contexts faces several challenges [2], including:

– Lack of standardized methodologies for assessing and improving repairability.
– Limited knowledge and tools for identifying potential failure points in digitally fabricated parts.
– Difficulty in balancing repairability with other design constraints such as cost and performance [3].

To address these challenges, this research proposes a novel framework that combines DfR principles with the Lean Six Sigma DMAIC methodology. The DMAIC approach provides a structured process for problem-solving and continuous improvement, making it well-suited for optimizing design processes in Fab Labs [4].

The primary objectives of this research are:

– Develop a comprehensive framework for integrating DfR principles into digital fabrication processes using Lean Six Sigma DMAIC methodology.
– Validate the framework through case studies conducted at a local Fab Lab.
– Quantify the benefits of the framework in terms of improved repairability metrics.

## 2  Methodology

This research was conducted at FabLab Thao Dien, a local Fab Lab equipped with various digital fabrication technologies including 3D printers, CNC machines, and laser cutters. The study followed the DMAIC phases of Lean Six Sigma, adapting each phase to address the specific challenges of integrating DfR into digital fabrication processes.

### 2.1  Define Phase

In the Define phase, we established the problem statement, project scope, and key stakeholders. A project charter was developed to outline the objectives, timeline, and expected outcomes. Voice of the Customer (VOC) [4, 8] analysis was conducted through surveys and interviews with Fab Lab users to identify critical-to-quality (CTQ) characteristics related to product repairability. This information was documented and referenced in the framework to ensure alignment with stakeholder needs.

### 2.2  Measure Phase

The Measure phase focused on identifying and collecting data on key repairability metrics. We developed a Repairability Index (RI) [5, 9] based on factors such as:

– Disassembly time.
– Number of specialized tools required.
– Availability of spare parts
– Modularity of design

Baseline data was collected for 50 randomly selected products which were consumer electronics, toys, and household items created at the Fab Lab over a 3-month period. The random selection aimed to provide a representative baseline across common product categories. The RI was calculated for each product using the following formula:

$$RI = (w1 * DT + w2 * ST + w3 * SP + w4 * MD)/(w1 + w2 + w3 + w4)$$

where:

DT = Disassembly Time score (1–10)
ST = Specialized Tools score (1–10)
SP = Spare Parts score (1–10)
MD = Modularity Design score (1–10)
w1, w2, w3, w4 = Weighting factors based on importance

The criteria for assigning scores were established through expert consultations and historical data analysis, ensuring that a score of 10 corresponds to optimal conditions while a score of 1 indicates severe limitations. The full scoring breakdown is (Table 1):

**Table 1.** Full scoring breakdown of RI's factors

| Scores | DT | ST | SP | MD |
|---|---|---|---|---|
| 10 | <5 min | No ST required | Readily available and easily sourced | Fully modular design; all components easily replaceable |
| 8 | 5–15 min | 1 ST required | Available but may require some effort to source | Mostly modular design; most components replaceable with some effort |
| 6 | 15–30 min | 2 STs required | Available but with limited options | Partially modular design; some components are replaceable, but others are not |
| 4 | 30–45 min | 3 STs required | Difficult to find; may require custom orders | Limited modularity; few components are replaceable |
| 2 | 45–60 min | 4 STs required | Rarely available; long lead times expected | Non-modular design; components are permanently joined |
| 1 | >60 min | >4 STs required | Not available; product must be replaced entirely | Completely monolithic design; no components are replaceable |

The multiplicative weights w1 to w4 were determined based on a Pairwise Comparison Analysis, where a panel of experts assessed the relative importance of each factor. This data-driven approach allowed us to quantify the importance of each variable objectively.

### 2.3  Analyze Phase

In the Analyze phase, we used various statistical and root cause analysis tools to identify the factors contributing to poor repairability. These included:

– Pareto analysis to identify the most significant contributors to low RI scores.
– Fishbone diagrams to explore potential root causes
– Failure Mode and Effects Analysis (FMEA) to assess potential failure points in digitally fabricated parts [6, 10]

Additionally, we conducted a Design of Experiments (DOE) to evaluate the impact of different design parameters on repairability metrics.

### 2.4  Improve Phase

Based on the insights gained from the Analyze phase, we developed a set of DfR guidelines tailored for digital fabrication processes. These guidelines were integrated into a digital design tool that provides real-time feedback on repairability during the CAD modeling process [7, 11].

Key features of the DfR-optimized design tool include:

– Automated identification of potential failure points.
– Suggestions for improving modularity and case of disassembly
– Material selection recommendations based on durability and availability

The tool was implemented at FabLab Innovation, and designers were trained on its use and the importance of DfR principles.

### 2.5  Control Phase

To ensure sustained improvements in repairability, we established a control plan [12–14] that includes:

– Regular monitoring and reporting of RI scores for new designs.
– Periodic audits of the design process to ensure adherence to DfR guidelines.
– Continuous training and education on DfR principles for Fab Lab users.

## 3  Results and Discussion

### 3.1  Baseline Repairability Assessment

The initial assessment of 50 products [5, 9] revealed a mean Repairability Index (RI) of 5.2 out of 10, indicating significant room for improvement. Table 2 shows the distribution of RI scores for the baseline products.

**Table 2.** Baseline Repairability Index scores

| RI Score Range | Frequency |
| --- | --- |
| 1.0–2.0 | 2 |
| 2.1–3.0 | 5 |
| 3.1–4.0 | 8 |
| 4.1–5.0 | 12 |
| 5.1–6.0 | 13 |
| 6.1–7.0 | 7 |
| 7.1–8.0 | 2 |
| 8.1–9.0 | 1 |
| 9.1–10.0 | 0 |

### 3.2 Root Cause Analysis

Pareto analysis identified the following key contributors to poor repairability:

– Lack of modularity in design (35%).
– Use of permanent joining methods (28%).
– Limited accessibility to internal components (20%).
– Absence of standardized parts (17%).

The percentages assigned to key factors in the Pareto analysis were calculated based on the frequency each factor appeared as a root cause across the FMEA and fishbone diagrams. The fishbone diagram and FMEA provided further insights into the root causes of these issues, highlighting factors such as time pressure, limited designer knowledge of DfR principles, and software limitations.

### 3.3 Design of Experiments Results

A two-level full factorial DOE was conducted to evaluate the impact of four design factors on the Repairability Index:

– Number of modules (2 levels: Low, High).
– Fastener type (2 levels: Permanent, Removable).
– Internal accessibility (2 levels: Limited, Open).
– Part standardization (2 levels: Custom, Standard).

The DOE results, analyzed using ANOVA, revealed that all four factors had statistically significant effects on the RI ($p < 0.05$). The most significant factor was modularity, followed by fastener type. Table 3 summarizes the DOE results.

### 3.4 Implementation of DfR-Optimized Design Tool

Following the implementation of the DfR-optimized design tool and associated guidelines, we observed significant improvements in repairability metrics. A follow-up assessment of 50 new products designed using the tool showed:

**Table 3.** DOE Results - Effect of Design Factors on Repairability Index

| Factor | F-value | p-value | Effect Size |
| --- | --- | --- | --- |
| Modularity | 87.3 | <0.001 | 2.4 |
| Fastener Type | 62.1 | <0.001 | 1.9 |
| Internal Accessibility | 41.5 | <0.001 | 1.5 |
| Part Standardization | 29.8 | <0.001 | 1.2 |

– Mean RI increased from 5.2 to 7.8 (50% improvement).
– Average disassembly time reduced by 37%.
– Material waste during repair processes decreased by 42%.

The results were calculated by comparing the average values before and after implementing the DfR-optimized design tool across the 50 products in the follow-up assessment (Table 4).

**Table 4.** Comparison of Repairability Index distributions before and after DfR implementation

| RI Score Range | Frequency (Before) | Frequency (After) |
| --- | --- | --- |
| 1.0–2.0 | 2 | 0 |
| 2.1–3.0 | 5 | 0 |
| 3.1–4.0 | 8 | 1 |
| 4.1–5.0 | 12 | 2 |
| 5.1–6.0 | 13 | 4 |
| 6.1–7.0 | 7 | 8 |
| 7.1–8.0 | 2 | 15 |
| 8.1–9.0 | 1 | 13 |
| 9.1–10.0 | 0 | 7 |

### 3.5 Case Study: Redesign of a 3D Printed Drone Frame

To further validate the framework, we conducted a case study involving the redesign of a popular 3D printed drone frame. The original design had an RI of 4.3, primarily due to its monolithic structure and use of permanent joining methods.

Using the DfR-optimized design tool, we redesigned the drone frame with the following improvements:

– Modular design with easily replaceable arms and central hub.
– Snap-fit connections for quick disassembly.
– Standardized mounting points for electronics.

– Improved internal accessibility for maintenance.

The redesigned drone frame achieved an RI of 8.7, representing a 102% improvement. Field testing showed that the average repair time for common issues decreased from 45 min to 12 min, and the need for complete replacement due to unrepairable damage was reduced by 78%. The redesigned drone frame maintained the same structural and functional characteristics as the original design, ensuring that performance was not compromised.

# 4  Conclusion

This research demonstrates the successful integration of Design for Repairability principles with Lean Six Sigma methodology in a digital fabrication context. The developed framework and DfR-optimized design tool led to significant improvements in product repairability at FabLab Innovation, as evidenced by increased Repairability Index scores and reduced repair times and material waste.

Key contributions of this work include:

– A structured approach for implementing DfR in Fab Lab environments.
– A quantitative method for assessing product repairability in digital fabrication.
– Identification of critical design factors influencing repairability.
– A validated design tool that provides real-time feedback on repairability during the CAD process.

By enhancing the repairability of digitally fabricated products, this research contributes to the broader goals of sustainable manufacturing and circular economy principles in localized production environments.

However, the current scope of the framework is limited to a specific Fab Lab environment with certain additive and subtractive manufacturing technologies, such as 3D printing and CNC machining. Hence, future work should focus on expanding the framework to address a wider range of digital fabrication technologies and materials, as well as exploring the potential for integrating artificial intelligence and machine learning to further optimize designs for repairability. Additionally, future studies should validate the framework across a wider variety of Fab Labs with different configurations and settings to ensure the generalizability of the findings.

# References

1. Gershenfeld, N.: How to make almost anything: The digital fabrication revolution. Foreign Aff. **91**(6), 43–57 (2012)
2. Prendeville, S., Hartung, G., Purvis, E., Brass, C., Hall, A.: Makespaces: from redistributed manufacturing to a circular economy. In: Sustainable Design and Manufacturing 2016, pp. 577–588. Springer, Cham (2016)
3. Diegel, O., Kristav, P., Motte, D., Kianian, B.: Additive manufacturing and its effect on sustainable design. In: Environmental Footprints and Eco-Design of Products and Processes, pp. 73–99 (2016)

4. Antony, J., Snee, R., Hoerl, R.: Lean Six Sigma: yesterday, today and tomorrow. Int. J. Qual. Reliab. Manag. **34**(7), 1073–1093 (2017)
5. Soh, S.L., Ong, S.K., Nee, A.Y.C.: Design for assembly and disassembly for remanufacturing. Assem. Autom. **36**(1), 12–24 (2016)
6. Despeisse, M., et al.: Unlocking value for a circular economy through 3D printing: a research agenda. Technol. Forecast. Soc. Chang. **115**, 75–84 (2017)
7. Zawadzki, P., Żywicki, K.: Smart product design and production control for effective mass customization in the Industry 4.0 concept. Manag. Prod. Eng. Rev. **7**(3), 105–112 (2016)
8. Esmaeilian, B., Behdad, S., Wang, B.: The evolution and future of manufacturing: a review. J. Manuf. Syst. **39**, 79–100 (2016)
9. Kohtala, C.: Addressing sustainability in research on distributed production: an integrated literature review. J. Clean. Prod. **106**, 654–668 (2015)
10. Bonvoisin, J., Halstenberg, F., Buchert, T., Stark, R.: A systematic literature review on modular product design. J. Eng. Des. **27**(7), 488–514 (2016)
11. Bocken, N.M., de Pauw, I., Bakker, C., van der Grinten, B.: Product design and business model strategies for a circular economy. J. Ind. Prod. Eng. **33**(5), 308–320 (2016)
12. Mourtzis, D., Doukas, M., Bernidaki, D.: Simulation in manufacturing: review and challenges. Procedia CIRP **25**, 213–229 (2014)
13. Lindemann, C., Reiher, T., Jahnke, U., Koch, R.: Towards a sustainable and economic selection of part candidates for additive manufacturing. Rapid Prototyping J. **21**(2), 216–227 (2015)
14. Kellens, K., Baumers, M., Gutowski, T.G., Flanagan, W., Lifset, R., Duflou, J.R.: Environmental dimensions of additive manufacturing: Mapping application domains and their environmental implications. J. Ind. Ecol. **21**(S1), S49–S68 (2017)

# Enhancing Sustainability in Manufacturing Through Life Cycle Optimization of Industrial Equipment

Nasser Amaitik[1]([⊠]) [iD], Ming Zhang[1] [iD], Kaifeng Wang[2] [iD], Yuchun Xu[1]([⊠]),
and Peihua Gu[2] [iD]

[1] Smart and Sustainable Manufacturing Research Centre, College of Engineering and Physical Sciences, Aston University, Birmingham B4 7ET, UK
{n.amaitik,y.xu16}@aston.ac.uk
[2] International Institute for Innovative Design and Intelligent Manufacturing of Tianjin University in Zhejiang, Shaoxing 312000, China

**Abstract.** In the drive towards sustainable manufacturing, extending the service life and improving efficiency of industrial equipment are crucial. Industrial equipment faces challenges of deterioration caused by wear, fatigue etc. effects. Effective maintenance is essential for minimizing the effect of these challenges. This research introduces an integrated model that facilitates the sustainable operation of industrial equipment through life cycle optimization. The model determines optimal maintenance schedules over life cycle by integrating multiple sub-models to analyze equipment. These sub-models include a reliability model to estimate deterioration, a cost model to evaluate the cost of maintenance scenarios, and a scheduling model to generate and select maintenance schedules that minimize life cycle costs while maximizing equipment reliability. By providing life cycle optimisation capabilities, this model supports the transition towards more sustainable manufacturing practices with minimized cost incurred.

**Keywords:** Sustainable Manufacturing · Life Cycle Cost · Predictive Maintenance · Reliability · Circular Economy

## 1 Introduction

In the context of rapidly evolving manufacturing technologies, life cycle optimization models with predictive maintenance capabilities have shown great potential for enhancing equipment lifetime performance and efficiency. The motivation for this study is the need for comprehensive models that effectively integrate cost-efficiency with enhanced equipment reliability, particularly in predictive maintenance strategies. Such comprehensive models integrate real-time condition data and life cycle cost (LCC) optimization to create proactive maintenance schedules, thereby preventing unexpected equipment failures and costly downtimes [1]. By focusing on both cost minimization and reliability maximization, such a model will contribute significantly to sustainable manufacturing

H. Kohl et al. (Eds.): GCSM 2024, LNME, pp. 524–533, 2025.
https://doi.org/10.1007/978-3-031-93891-7_58

practices, aligning with the industry's drive towards more economically sustainable operations. These are pivotal in developing sustainable manufacturing practices by predicting potential issues and optimizing the performance of industrial equipment and systems.

The use of life cycle optimization and predictive maintenance in industrial equipment has gained substantial attention in the academia and industry. Life cycle optimization involves managing equipment from its start to its end, with a focus on maximizing reliability and availability, as well as minimizing costs throughout its operational lifetime. Several studies have highlighted the importance of this approach in enhancing equipment reliability and performance. For instance, Bokrantz et al. [2] demonstrated that integrating life cycle assessment with maintenance strategies significantly reduces operational costs and environmental impact. Predictive maintenance, a critical component of life cycle optimization, utilizes real-time data and advanced analytics to predict equipment failures before they occur, minimizing downtime and extending equipment lifetime. Jardine et al. [3] introduced predictive maintenance models, which utilize techniques such as machine learning and statistical analysis, can significantly improve accuracy of failure predictions compared to traditional time-based maintenance. Additionally, Lee et al. [4] emphasized the role of predictive maintenance in fostering sustainable manufacturing by reducing waste and energy consumption.

While these studies confirm the transformative potential of life cycle optimization and predictive maintenance, there is still a need for a comprehensive model that accounts for economic aspects of life cycle optimization of equipment. This paper introduces an innovative framework of life cycle optimization for industrial equipment that integrates these aspects into a cohesive model. Applied to CNC (Computer Numerical Control) machine tools, the proposed framework enables manufacturers to enhance operational reliability and sustainability by optimizing LCC and maintenance schedules, thereby offering a more comprehensive approach to equipment management.

## 2 CNC Machine Tools in Manufacturing

CNC machine tools are pivotal in modern manufacturing due to their precision and efficiency in producing complex parts with minimal human intervention. They are widely used across various industries such as automotive, aerospace, and electronics, ensuring consistent and high-quality output. CNC machines integrate several key components essential for their operation and performance throughout their life cycle [5–7]. Figure 1 illustrates a schematic diagram of these components, including [8]:

- *Spindle (SD):* The spindle is the central rotating component that holds and drives the cutting tools or workpiece during machining operations.
- *Spindle Motor (SM):* It provides the necessary power to rotate the spindle and drive the cutting tools, ensuring precise and efficient material removal.
- *Cutting Tool (CT):* The cutting tool is attached to the spindle and is responsible for shaping the workpiece according to programmed specifications.
- *Holding Device (HD):* It securely grips and positions the workpiece during machining. This ensures stability and accuracy throughout the machining process.
- *Control Unit (CU):* The CNC controller serves as the brain of the machine tool, managing and coordinating all operations..

Maintaining these components in optimal condition is crucial for achieving high precision, minimizing downtime, and extending the machine tool's operational life. A comprehensive life cycle optimization framework, including predictive maintenance strategies, is essential to achieve these goals.

This paper focuses on proposing and testing a methodology of life cycle optimization with predictive maintenance for CNC machine tools. The subsequent section elaborates on the approach and findings, demonstrating its application in enhancing sustainability and efficiency in manufacturing operations.

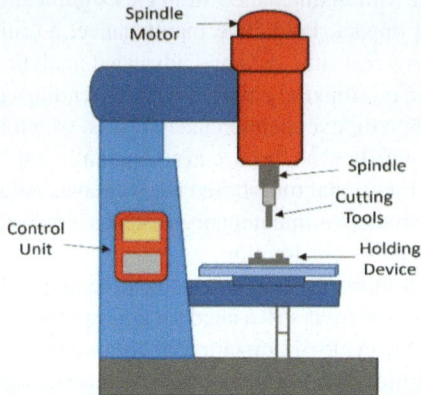

**Fig. 1.** CNC machine tool schematic diagram showing the main components.

## 3    Life Cycle Optimization Framework for CNC Machine Tools

In this section, we present the methodology and implementation of the proposed life cycle optimization framework for CNC machine tools.

### 3.1    Methodology

The proposed model determines optimal maintenance schedules over a desired planning horizon by integrating multiple sub-models to analyze equipment throughout its life cycle, as shown in Fig. 2. Specifically, the objective functions within these models are designed to minimize life cycle costs while maximizing equipment reliability. It introduces sub-models related to cost, reliability, and repair impact:

- *Degradation model*: This model uses the Weibull reliability function to model the deterioration of the machine tool components. It estimates the reliability until a component is likely to fail. Variability in machine operation intensity is accounted by incorporating the load factor through the parameter $\alpha$ in the Weibull function. This adjusts the failure rate based on the operational load, allowing the model to accurately reflect different intensities and complexities of parts.

- *Repair improvement model*: This model assesses the improvement in component condition after repair or replacement. It evaluates how different maintenance actions affect the performance and reliability of each component.
- *Maintenance cost model*: This model provides cost estimates for different repair strategies applied to each component.
- *Scheduling model*: This model uses a Genetic Algorithm (GA) to develop maintenance plans that minimize life cycle costs while maximizing component reliability. It considers the predicted degradation, the impact of repairs, and the associated costs to determine the optimal maintenance schedule.

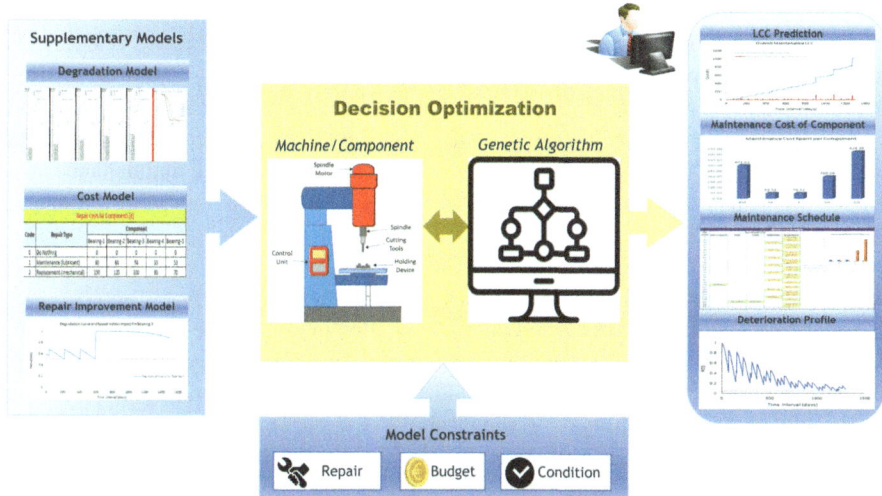

**Fig. 2.** Components of the life cycle optimization framework

**Problem Formulation and Algorithm**

Given a machine with known initial conditions for its components, and if a repair or no-repair action is taken, the degradation and improvement models can predict the components' conditions over any time interval. Each repair strategy has an associated repair cost. Therefore, it is possible to construct an optimization problem to determine the best set of repairs that achieve the maximum reliability/cost ratio.

The objective function and constraints for the machine $i$ at time interval $t$ to maximize reliability/cost ratio are formulated in Eqs. (1) through (6).

Objective function:

$$Maximize\ RT(i, t) \big/ PV(MC(i, t)) \tag{1}$$

Subject to:

$$RT(i, t) \geq Threshold_i(predefined\ minimum\ reliability\ for\ machine\ i) \tag{2}$$

$$RT(c, t) \geq Threshold_c \ (predefined \ minimum \ reliability \ for \ component \ c) \qquad (3)$$

$$\sum_{t=1}^{T} \sum_{c=1}^{C} PV(MC(c, t)) \leq Threshold_t \ (predefined \ max \ repair \ budget \ per \ t) \qquad (4)$$

$$RT(c, t) = \begin{cases} RT(c, t)^{new} & if \ c \ is \ selected \ for \ repair \\ RT(c, t) & otherwise \end{cases} \qquad (5)$$

$$RT(i, t) = \prod_{c=1}^{C} RT(c, t), \text{at any time t} \qquad (6)$$

where, $i$ denotes to machine, $t$ denotes to time interval (e.g., hours, days, years), $C$ is the number of components to be analyzed for the machine, $T$ is the number of time interval in planning horizon, $RT(i,t)$ is the overall reliability of the machine $i$ at the end of time $t$, $RT(c,t)$ is the reliability of the component $c$ at the end of time $t$, $MC(i,t)$ is total maintenance cost of the machine $i$ at the end of time $t$, $MC(c,t)$ is total maintenance cost of all components $c$ at the end of time $t$, $RT(c,t)^{new}$ is the new reliability of a component $c$ after maintenance at the end of time $t$, and $PV$ is the present value of costs.

The variables in this optimization problem for the CNC machine tools are the repair decisions for the five components SD, SM, CT, HD, and CU.

Separate optimizations are carried out for each time interval in the planning horizon. A solution for the problem is structured as a string of five elements, as shown in Fig. 3. Each variable can be assigned an integer value from 0 to 2, corresponding to one of the repair options (0 = do nothing; 1 = repair and 2 = replacement). More repair option can be added based on the case study analyzed.

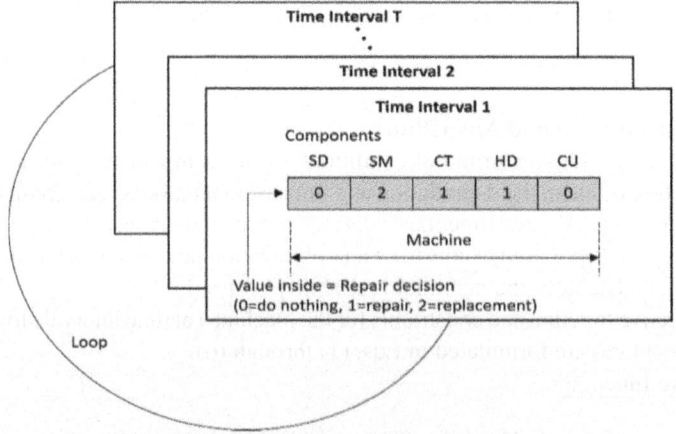

**Fig. 3.** Solution Structure for the schedule algorithm.

## 3.2 Implementation

### Data and Assumptions

Due to the limitation of accessing real-world data at this stage of development, we adopt a simulation-based approach for model experimentation. We use expert judgment supported by literature [7, 9, 10]. We estimate the model parameters, as summarized in Table 1, to experiment our approach. These parameters include components lifetime, Weibull reliability parameters ($\eta$-Eta and $\beta$-Beta) for degradation estimation, maintenance cost data, and percentages of resulting improvements for each component.

**Table 1.** Model parameters.

| Comp | Weibull Parameters | | Maintenance Cost and Resulting Improvement | | | |
|------|------|------|------|------|------|------|
| | $\eta$ | $\beta$ | Repair (£) | % Improvement | Replace (£) | % Improvement |
| SD | 800 Days | 1.5 | 500 | 40 | 1500 | 100 |
| SM | 1000 Days | 1.0 | 400 | 40 | 1800 | 100 |
| CT | 200 Days | 0.9 | 150 | 40 | 1000 | 100 |
| HD | 1200 Days | 1.2 | 300 | 40 | 2200 | 100 |
| CU | 1500 Days | 0.9 | 250 | 40 | 2500 | 100 |

The following assumptions have been considered for model implementation:

- *Spindle* has a high usage leading to faster wear, thus assigned a shorter $\eta$ and higher $\beta$, indicating aging-related deterioration.
- *Spindle Motor* is more robust with a constant failure rate, with $\eta$ and $\beta$ equal to 1.
- *Cutting Tool* is prone to early life deterioration due to frequent usage and high stress, thus assigned a low $\eta$ and $\beta$ less than 1.
- *Holding Device* is less prone to failure but still affected by usage over time, thus assigned a higher $\eta$ and moderate $\beta$.
- *Control Unit* is susceptible to early life failures due to electronic components, hence assigned a longer $\eta$ and $\beta$ less than 1.
- Two types of maintenance strategies are considered: Repair or Replacement.

The proposed model has been designed to minimize user input while delivering essential outputs necessary for informed and proactive decision-making regarding equipment maintenance. The required user inputs and outputs are outlined below:

| • User Inputs: | • User Outputs: |
|---|---|
| – Desired period of analysis | – Degradation curves |
| – Current age of components | – Maintenance schedule and cost analysis |

### Results and Discussions

To implement the proposed life cycle optimization model, a tool based on VBA programming language has been developed. It utilizes estimated data for each component (Table 1) and employs algorithms to simulate degradation, repair, and maintenance processes over the CNC machine tool lifetime. Considering inputs for 5-year (1825 days) as a desired period of analysis (also called planning horizon) with brand-new CNC machine tool components, the tool provides the following outputs:

***Degradation Curves.*** The tool provides degradation curves for components and the whole machine system throughout the analysis period (Fig. 4). These curves show the performance over time and analyze the impact of different repair and replacement strategies on component reliability. The Spindle component, for example, has experienced 5 interventions, with 4 repairs and 1 replacement throughout the planning horizon. The first repair intervention occurred on day 592, which was a part of maintenance activity along with other components. This resulted in an improvement in the Spindle's reliability to 0.743. This demonstrates the value of predictive maintenance interventions, as opposed to allowing components to run to failure and then replacing them.

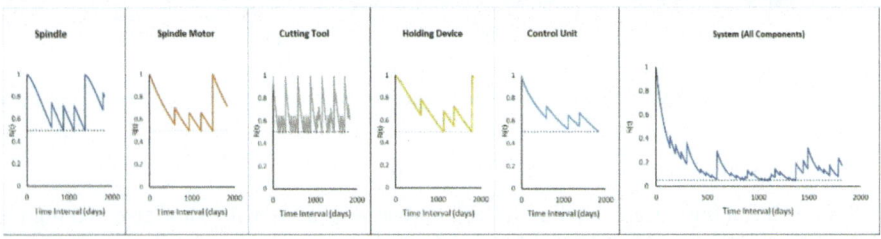

**Fig. 4.** Degradation curves for all components and system (CNC machine tool)

***Maintenance Schedule.*** The tool presents a predictive maintenance schedule for each component based on degradation and cost analysis so that manufacturers are informed in advance for taking the necessary arrangements for maintenance activities. Additionally, it shows statistics on repair and replacement implemented for each component (Fig. 5). For example, during the 5-years planning horizon, 41 repair/replacement interventions are expected to be occurred (32 repairs and 9 replacements).

***Maintenance Cost Analysis.*** The tool provides detailed analysis and insights into the maintenance cost of the machine. It shows cumulative costs, costs per intervention, and cost breakdown per component. For example, during the 5-year planning horizon, a cumulative cost analysis is conducted (Fig. 6-a) to provide manufacturers with insights into the intervals where higher costs are expected, enabling them to plan the required

**Fig. 5.** Maintenance schedule and statistics

budget for maintenance activities accordingly. Additionally, the cost incurred for each component is calculated (Fig. 6-b), highlighting the components that require the most expenditure. In our example, the Cutting Tool component demands the highest cost.

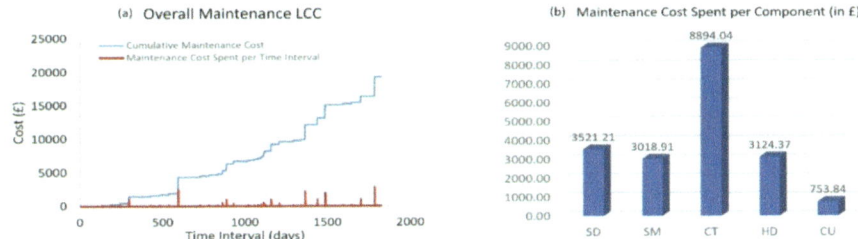

**Fig. 6.** Maintenance cost analysis

## 4 Conclusion

The application of life cycle optimization in managing industrial equipment, as seen in the CNC machine tools case study, offers substantial benefits in terms of operational reliability and sustainability. By continuously monitoring critical parameters and predicting the degradation of components, predictive maintenance within the life cycle optimization framework facilitates proactive planning. The life cycle optimization framework developed in this study employs advanced algorithms to optimize maintenance schedules based on degradation and cost analysis. Despite relying on simulated data, the model demonstrates potential to enhance CNC machine tools performance by minimizing downtime and extending component lifetime.

However, it is acknowledged that the use of simulated data in current study introduces some limitations, particularly in terms of directly translating these findings to real-world scenarios. When applied to real industrial environments, the model will need to account for the inherent complexities and variability in real-world operations. As empirical data becomes available, the accuracy and effectiveness of life cycle optimization models will improve, further supporting informed decision-making. Future research should focus on refining the model with real-world dat. Ultimately, this approach demonstrates its potential in fostering more sustainable manufacturing practices.

**Acknowledgement.** This work appreciates the support by Science and Technology Program Project of Shaoxing City under Grant 2023A14016, and previous support of Horizon Europe project Remanufacturing and Refurbishment Large of Industrial Equipment (RECLAIM) under grant No. 869884.

# References

1. Zhang, M., et al.: Predictive maintenance for remanufacturing based on hybrid-driven remaining useful life prediction. Appl. Sci. **12**(7) (2022). https://doi.org/10.3390/app120 73218
2. Bokrantz, J., Skoogh, A., Berlin, C., Wuest, T., Stahre, J.: Smart maintenance: a research agenda for industrial maintenance management. Int. J. Prod. Econ. (2020). https://doi.org/10. 1016/j.ijpe.2019.107547
3. Jardine, A.K.S., Lin, D., Banjevic, D.: A review on machinery diagnostics and prognostics implementing condition-based maintenance. Mech. Syst. Signal Process. **20**(7), 1483–1510 (2006). https://doi.org/10.1016/j.ymssp.2005.09.012
4. Lee, J., Ni, J., Djurdjanovic, D., Qiu, H., Liao, H.: Intelligent prognostics tools and e-maintenance. Comput. Ind. **57**(6), 476–489 (2014). https://doi.org/10.1016/j.compind.2006. 02.014
5. Altintas, Y., Brecher, C., Weck, M., Witt, S.: Virtual machine tool. CIRP Ann. **54**(2), 115–138 (2005). https://doi.org/10.1016/S0007-8506(07)60022-5
6. Groover, M.P.: Fundamentals of Modern Manufacturing: Materials, Processes, and Systems. Wiley (2013)
7. Kolokas, N., Ioannidis, D., Tzovaras, D.: DSF core: integrated decision support for optimal scheduling of lifetime extension strategies for industrial equipment. Sensors **23**(3) (2023). https://doi.org/10.3390/s23031332
8. Luo, W., Hu, T., Ye, Y., Zhang, C., Wei, Y.: A hybrid predictive maintenance approach for CNC machine tool driven by Digital Twin. Robot. Comput. Integr. Manufact. **65** (2020). https://doi.org/10.1016/j.rcim.2020.101974
9. Amaitik, N., et al.: Cost modelling to support optimum selection of life extension strategy for industrial equipment in smart manufacturing. Circ Econ Sus. (2022). https://doi.org/10. 1007/s43615-022-00154-0
10. Amaitik, N., et al.: Towards sustainable manufacturing by enabling optimum selection of life extension strategy for industrial equipment based on cost modelling. J. Remanufactur. **13**, 263–282 (2023). https://doi.org/10.1007/s13243-023-00129-w

# Disassembly and Remanufacturing

# Partial Automation of Traction Battery Cell Module Handling: Feasibility and Ergonomic Improvements

Andreas Lehner[1]([✉]), Cornelia Rutkowski[2], Alessandro Sala[1], Noel Scheder[1], Maximilian Haug[1], Astrid Arnberger[3], and Sebastian Schlund[1,4]

[1] Fraunhofer Austria Research GmbH, Theresianumgasse 7, 1040 Wien, Austria
`andreas.lehner@fraunhofer.at`
[2] Montanuniversität Leoben, Chair of Waste Processing Technology and Waste Management, Franz-Josef-Strasse 18/I, 8700 Leoben, Austria
[3] Saubermacher Dienstleistungs AG, Hans-Roth-Straße 1, 8073 Feldkirchen bei Graz, Austria
[4] TU Wien, Institute of Management Science, Theresianumgasse 27, 1040 Wien, Austria

**Abstract.** Associated with the high level of market placement of electric vehicles comes the challenge of the disposal of end-of-life (EoL) lithium-ion traction batteries (LITB). Recycling LITB is a multi-step process including the disassembly into cell modules (CM) and subsequent handling for further processing. Disassembly, currently conducted manually, poses an ergonomic strain and hazard potential for workers but is crucial to enhancing the ecology of downstream recycling processes by enabling selectivity. Based on requirements for the use-case handling during disassembly (process-related) and 13 design features of CM (product-related), a suitable lifting and handling system (LHS) was selected from six gripping options. An analysis of 50 CM models representing 90% of the Austrian market yielded an applicability of 75% for a pivot arm featuring lifting pins and electrically powered lifting capability. Both scenarios, manual handling and handling supported by the LHS were simulated in ema Work Designer and evaluated using the Ergonomic Assessment Worksheet (EAWS) method. The ergonomic assessment yielded an improvement from red to green in the EAWS scale for the simulated use-case featuring three different CM weights. The considerable ergonomic improvements call for further efforts to enhance partial or full automation to foster LITB disassembly for selective recycling.

**Keywords:** lithium-ion traction battery disassembly · partial automation · ergonomic assessment

## 1 Introduction

Lithium-ion traction batteries (LITB) have been touted as the most suitable energy storage solution for electric vehicles to accomplish the decarbonization of the transport sector [1]. The production of LITB is projected to grow exponentially over the coming decade due to demand for battery electric vehicles (BEV) [2]. Associated with the high level of

H. Kohl et al. (Eds.): GCSM 2024, LNME, pp. 537–545, 2025.
https://doi.org/10.1007/978-3-031-93891-7_59

market placement comes the challenge of the disposal of end-of-life (EoL) LITB. Recycling LITB is a multi-step process including, e.g., presorting, disassembly, and material recovery. The variety of possible steps is met by the variety of LITB models: cell chemistry, cell format, housing material, dimensions and weight are only some distinctive features which must be dealt with. Especially steps conducted manually, as disassembly, require human workers to manage this variety, posing a safety risk and ergonomic strain [3]. Yet this step is essential to increase the ecology of the subsequent recycling process: on the pyrometallurgical route, plastics or aluminium would be decomposed or slagged, whereas on the hydrometallurgical route, organics constitute an impurity and aluminium poses a selectivity problem when targeting lithium [1, 4–6]. By applying a hydrometallurgical approach, a reduction of 19% energy consumption, 40% water consumption, 18% GHG emissions, and 76% SOx emissions is possible, while only resulting in a 5% cost reduction compared to virgin material, pointing at low economic profitability of the ecologically meaningful LITB recycling [7]. In addition, disassembly (in contrast to dismantling) enables – besides recycling – reusing, refurbishing, remanufacturing, or repurposing parts [3]. Therefore, fostering the disassembly process is crucial for promoting sustainability and circular economy approaches [8]. This paper therefore quantifies the feasibility and ergonomic benefits of a lifting and handling system (LHS) to fulfil the two main functions of gripping/releasing a CM and lifting/handling a CM during disassembly.

## 2   State of the Art

### 2.1   Design of LITB

The design of LITB varies significantly due to automotive manufacturers optimizing the balance between features such as crash safety, performance, and space efficiency [9]. Yet certain generalities exist, as the division into three hierarchical levels: (1) battery system, containing several (2) CM, incorporating individual (3) cells (Fig. 1) [3].

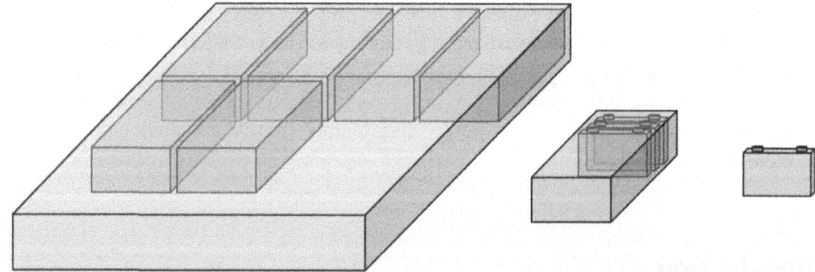

**Fig. 1.** Battery system (left), CM (mid), prismatic cell (right) (own illustration).

The battery system is usually designed as a flat pack, as it is mounted in the vehicle's underbody. Besides the CM, it comprises further components such as the battery management system and the battery system housing. Among the housing's main purposes is the protection of the CM from mechanical stress for safety reasons. Therefore, the CM are commonly bolted to the housing in which they are fully enclosed. For thermal control, a layer of heat-conducting paste is applied between the CM and the housing containing the cooling circuit. The CM incorporate groups of cells permanently mounted through e.g., adhesive bonds or welding. Typically, cells in a CM are not intended to be broken down as part of a servicing activity [10].

## 2.2 Disassembly and Handling of LITB CM

Manual disassembly is the state-of-the-art method to separate a battery system's components. The process of disassembling the battery involves removing the battery casing lid, battery management system, power electronics, and – if accessible – the thermal management system. Subsequently, all wires, cables, and connectors are detached. The final step involves disassembling the securing bolts and overcoming the highly adhesive effects caused by the heat-conducting paste (e.g., by using a chisel) to remove the CM. Currently, there is no commercial operation known to have reached full automation of handling CM for non-brand specific LITB models due to the wide variety in product design [1, 9, 11]. 83.3% of experts surveyed by Klohs et al. considered this step to be a challenge for automated process control [4].

As the subsequent processing is possibly conducted at a different location than the disassembly, transporting the CM may be necessary. Therefore, plastic containers filled with flame-retarding filling material are used, in which the CM is embedded after removal from the battery system.

## 3 Methodology

The methodology for assessing the feasibility and ergonomic improvement of a partial automation solution for CM handling is described in the following section. The use-case is the removal of a CM from a battery system and placing it in a nearby container for further processing. As EoL LITB are considered, the surface of CM may show signs of usage or contaminations as dust or liquids. Process-related targets comprise maximum safety for workers, usability, and minimum disruption of the workflow. This implicates a universal solution not tailored to certain CM models. The methodology is divided into the following four steps:

1. Selection of possible LHS: Initially, a screening of industrial methods for LITB CM gripping during assembly is conducted. Further possibilities shall be identified by capturing similarities of CM product design. These solutions of lifting and handling systems are collected under consideration of the process-related requirements and targets. Based on these solutions, a morphological box of relevant characteristics is derived.

2. Feasibility of selected LHS: To determine the feasibility of the selected LHS, product-related design features determined in step 1 are obtained through the evaluation of a set of LITB of BEV models. The models are selected according to the number of registrations in Austria in the years 2021, 2022, and 2023. A web search is conducted

to determine the quantitative and qualitative characteristics. This enables a quantitative conclusion on the applicability of the chosen LHS to the example of the Austrian waste management.

3. Simulation of use-case: Two versions (manual and partially automated) of the use-case removal of a CM and placing it in a transportation container are simulated in ema Work Designer.

4. Ergonomic Assessment: An ergonomic assessment for both versions of the use-case is conducted using the Ergonomic Assessment Worksheet (EAWS). EAWS is a tool used to evaluate physical strains at the workplace. It assesses various factors such as body postures, action forces, manual handling, and strain on the upper extremities to ensure a safe and healthy work environment. The difference in the resulting scores demonstrates the ergonomic benefit of partial automation.

# 4   Case Study

## 4.1   Selection of Possible LHS

The initial screening of LHS for CM assembly was conducted to identify renowned methods. These comprise vacuum gripping of the topside surface, parallel gripping of front or sideward surfaces (force-fitted or form-fitted using overhangs), and special versions of lifting pins and mechanical interlocks in holes. These solutions, proven on an industrial scale, however, may not work universally, as they are usually customized for the specific CM model. Furthermore, the process requirements include possibly polluted surfaces of the LITB (dry or wet), affecting force-fitting gripping methods. In addition, force-fitting might induce undesired, dangerous mechanical stress on the battery cells. Therefore, form-fitting methods are preferred.

An additional screening of state-of-the-art CM models yielded further typical design features suitable for gripping the CM – especially non-automated solutions not used in the prevalently highly automated production process. These comprise the usage of hooks in eyelets (e.g. for fixing the CM to the battery housing) and other suitable shapes as well as bolted eyelets or click screws in vertical threaded holes (e.g. for fixing surrounding components to the CM). The identified methods are listed in Table 1.

**Table 1.** Possible methods for CM gripping, type of connection and required design features.

| # | Method | Type of connection | Required design feature |
|---|--------|--------------------|--------------------------|
| 1 | Vacuum grip | Force-fitting | Clean, flat, even, air-tight topside surface |
| 2 | Parallel grip | Force-fitting | Flat, even surface (long or short side) |
| 3 | Parallel grip | Form-fitting | Durable overhang over topside surface |
| 4 | Lifting pins | Form-fitting | Min. of 3 vertical through holes |
| 5 | Bolted eyelets | Form-fitting | Min. of 3 vertical thread holes |
| 6 | Hooks | Form-fitting | Min. of 3 eyelets |

Along with the property mass, the requirements for the solutions were translated into the characteristics in Table 2 and superscripted by the number (#) in Table 1.

**Table 2.** Morphological box of relevant characteristics of LITB models.

| Section | Characteristics | Unification | | | | |
|---|---|---|---|---|---|---|
| | Mass [kg] | < 10 | < 20 | < 35 | < 55 | ≥ 55 |
| | Structural integrity | Yes[123456] | | No | | |
| CM | Accessible vertical through holes (min. 3) | Yes[4] | | No | | |
| | Accessible thread holes (min. 3) | Yes[5] | | No | | |
| | Accessible eyelets (min. 3) | Yes[6] | | No | | |
| Long side | Clearance CM to next CM or housing [mm] | < 15 | | ≥ 15[2] | | |
| | Planarity [mm] | < 5[2] | | ≥ 5 | | |
| Short side | Clearance CM to next CM or housing [mm] | < 15 | | ≥ 15[2] | | |
| | Planarity [mm] | < 5[2] | | ≥ 5 | | |
| | Durable, accessible overhang [mm] | < 10 | | ≥ 10[3] | | |
| Top side | Material | Plastic | | Metal[1] | | Mixture |
| | Planarity [mm] | < 1[1] | | ≥ 1 | | |
| | Airtightness | Yes[1] | | No | | |

## 4.2 Applicability of Selected LHS

The LITB models to provide a basis for determining applicability are the respective top 40 BEV models in Austria in the years 2021, 2022, and 2023 by registrations, comprising 57 different models in total [12]. The web search provided the necessary data (mass and visual footage of CM and LITB) to derive the required information (see Table 2) of 50 models, representing 90% of the combined registrations of the regarded period (105,205 of 116,433). Certain models were not reviewed as the necessary information could not be surveyed. Clearance and planarity were either estimated or gathered during disassembly analyses. The result of applying the characteristics in Table 2 to the 50 models is depicted in Fig. 2.

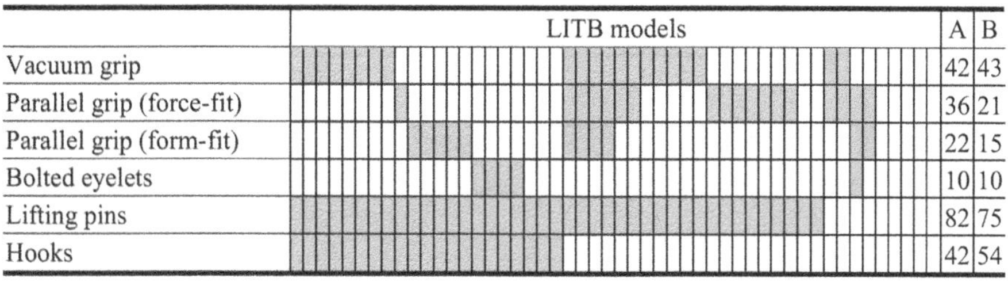

| | LITB models | A | B |
|---|---|---|---|
| Vacuum grip | | 42 | 43 |
| Parallel grip (force-fit) | | 36 | 21 |
| Parallel grip (form-fit) | | 22 | 15 |
| Bolted eyelets | | 10 | 10 |
| Lifting pins | | 82 | 75 |
| Hooks | | 42 | 54 |

**Fig. 2.** Suitability of gripping methods for the 50 reviewed models (grey mark indicates suitability for respective method); A: Share of reviewed models [%]; B: Share weighted by registrations in the chosen period in Austria [%].

82% of the reviewed models are considered suitable for lifting pins, covering a significant market share of 75%. Yet it must be noted that it may be necessary to slightly lift the CM for the lifting pin mechanism to expand after sliding it through the vertical hole. However, this is necessary in most cases, as the adhesion between the CM and the battery housing caused by the heat-conducting paste must first be loosened.

**Table 3.** Mass distribution of the 50 reviewed models.

| Mass [kg] | < 10 | < 20 | < 35 | < 55 | ≥ 55 |
|---|---|---|---|---|---|
| Share of reviewed models [%] | 4 | 50 | 18 | 14 | 14 |
| Share weighted by registrations [%] | 2 | 28 | 34 | 9 | 27 |

The evaluation of the CM masses (Table 3) yields a range of 9 kg to 89 kg. Considering process requirements for the functions of lifting, moving, and lowering the CM, a suitable, exemplary device is an electrically powered rope balancer on a pivot arm. Such devices cover the necessary load, can be equipped with quick-change systems for different grippers, and allow the usage of spacers for enhanced safety.

### 4.3 Simulation of Use-Case

The simulation of both use-cases – manual and partially automated – was conducted using ema Work Designer version 2.3.2.2. The modelled layout includes an exemplary LITB system containing a CM sitting on a table, a container placed on a table, and an exemplary vacuum gripper on an operating unit of a rope balancer (BINAR QLA 100i) for partially automated handling. The person was chosen according to DIN 33402–2: 2020–12 (German, age 40, male (95th percentile) and female (5th percentile)). The simulations were conducted with the parameters in Table 4. The CM masses represent the variety of CM models maximally portable by a single person. Respective loads to move these CM models with the LHS solution were chosen accordingly.

**Table 4.** Variations in the simulations.

| Handling | Gender | Mass of CM [kg] | Vertical load [N] | Horizontal load [N] |
|---|---|---|---|---|
| Manual | Male/Female | 10/20/35 | - | - |
| Partially automated | Male/Female | - | 5/5/5 | 30/45/67.5 |

**Manual Handling.** The simulated steps comprise: (1) walking to the table; (2) grabbing the CM and pulling it to the body; (3) carrying the CM to the container; (4) placing the CM on the bottom of the container; (5) returning to the initial position (Fig. 3).

**Partially Automated Handling.** The simulated steps comprise: (1) walking to the table while guiding the gripper to the CM; (2) attaching the gripper to the CM and lifting it; (3) walking/guiding the gripper plus CM to the container; (4) placing the CM on the bottom of the container and detaching the gripper; (5) returning to the initial position (Fig. 4).

**Fig. 3.** Shots of simulated steps (2), (3), (4) in ema Work Designer – Manual.

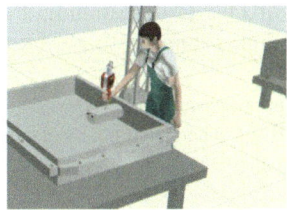

**Fig. 4.** Shot of simulated steps (2), (3), (4) in ema Work Designer – Partially automated.

### 4.4 Ergonomic Assessment Result

The EAWS scores of the simulated use-cases were automatically created by ema Work Designer to evaluate physical strains in the use-cases, listed in Table 5.

**Table 5.** Variations in the simulations.

| Mass of CM [kg] | Manual | | Vertical load [N] | Horizontal load [N] | Partially automated | |
|---|---|---|---|---|---|---|
| | Male | Female | | | Male | Female |
| 10 | 52 | 198.5 | 5 | 30 | 2 | 2 |
| 20 | 71.5 | 198.5 | 5 | 45 | 4.5 | 12 |
| 35 | 188.5 | 198.5 | 5 | 67,5 | 10.5 | 24.5 |

The reduction of whole body EAWS score from manual to partially automated handling resulted in an improvement from the colour scale red (score: >50, high risk – to be avoided) to green (score: <25, no risk or low risk – recommended) in every scenario in a traffic light scheme according to DIN EN 614-1:2009-06. The greatest benefit comes from avoiding high strain during lifting (score for 10–20–35 kg CM, male/female: 19.3/69.7–26.8/69.7–71.9/69.7) and placing (score for 10–20–35 kg CM, male/female: 23.2/90.3–32.2/90.3–86.3/90.3) the CM.

## 5 Conclusion and Outlook

The combined applicability of the considered gripping methods yields 90% of the surveyed LITB models – representing 81% of the market share. However, it must be noted that the used methodology does not allow a reliable statement regarding the applicability of a certain gripping method but rather indicates a plausible possibility. An implementation of the system to verify these findings is therefore needed. Moreover, different sizes or variants may be required for different CM models (e.g., different diameters of lifting pins). The variety in LITB product design will remain a challenge for both partial and full automation of CM handling and disassembly in general, as will upcoming trends like cell-to-pack integration or replacing bolting with adhesive bonding – pointing at the limitation of this work by only considering LITB represented on the Austrian market. Nevertheless, addressing these challenges has proven beneficial considering the substantial ergonomic improvements achieved through partial automation. Besides an implementation for verification, further research could be directed at determining the solution's economic viability and the transferability to full automation.

**Acknowledgement.** This work received partial funding from the Austrian Federal Ministry for Climate Action (BMK) under grant number 899505 (MoLIBity). The Austrian Research Promotion Agency (FFG) has been authorized for the program management.

# References

1. Wu, S., Kaden, N., Dröder, K.: A systematic review on lithium-ion battery disassembly processes for efficient recycling. Batteries **9**, 297 (2023)
2. Neef, C., Schmaltz, T., Thielmann, A.: Recycling von Lithium-Ionen-Batterien: Chancen und Herausforderungen für den Maschinen- und Anlagenbau (2021)
3. Gerlitz, E., Greifenstein, M., Hofmann, J., Fleischer, J.: Analysis of the variety of lithium-ion battery modules and the challenges for an agile automated disassembly system. Procedia CIRP **96**, 175–180 (2021)
4. Klohs, D., Offermanns, C., Heimes, H., Kampker, A.: Automated battery disassembly— examination of the product- and process-related challenges for automotive traction batteries. Recycling **8**, 89 (2023)
5. Peters, J.F., Baumann, M., Weil, M.: Recycling aktueller und zukünftiger Batteriespeicher: Technische, ökonomische und ökologische Implikationen (2018)
6. Doose, S., Mayer, J.K., Michalowski, P., Kwade, A.: Challenges in ecofriendly battery recycling and closed material cycles: a perspective on future lithium battery generations. Metals **11**, 291 (2021)
7. Gaines, L., Dai, Q., Vaughey, J.T., Gillard, S.: Direct recycling R&D at the ReCell center. Recycling **6**, 31 (2021)
8. Baazouzi, S., Rist, F.P., Weeber, M., Birke, K.P.: Optimization of disassembly strategies for electric vehicle batteries. Batteries **7**, 74 (2021)
9. Tan, W.J., Chin, C.M.M., Garg, A., Gao, L.: A hybrid disassembly framework for disas-sembly of electric vehicle batteries. Int. J. Energy Res. **45**, 8073–8082 (2021)
10. Thompson, D.L., et al.: The importance of design in lithium-ion battery recycling – a critical review. Green Chem. **22**, 7585–7603 (2020)
11. Hettesheimer, T.: Lithium-Ion Battery Roadmap. Industrialization Perspectives Toward 2030. Fraunhofer-Institut für System- und Innovationsforschung ISI (2023)
12. Statistics Austria. Registration of new motor vehicles. https://www.statistik.at/statistiken/tourismus-und-verkehr/fahrzeuge/kfz-neuzulassungen (2021–2023)

# Disassembly Production Design to Combat Information Security Risks in Remanufactured IoT Products

Yuki Kinoshita(✉) ⓘ

Kindai University, 1 Takaya Umenobe, Higashi-Hiroshima, Hiroshima 739-2116, Japan
y-kinoshita@uec.ac.jp

**Abstract.** For material circulation, End-of-life (EOL) products need to be collected and suitable lifecycle options, such as remanufacturing, and recycling determined, based on their conditions. Disassembly production planning involves decisions of lifecycle options for each component retrieved from the EOL products, and the number of EOL products or components for each recovery process, such as disassembly, inspection, remanufacturing, and recycling. Remanufacturing requires complete disassembly for inspection, and reassembling to keep product functionality. Meanwhile, although the spread of Internet-of-things (IoT) products has made our lives more convenient, information security risks have also increased because various types of IoT products can connect to the internet and store our personal data. Generally, for IoT products at an EOL stage, security updates should be conducted, and their ownership transferred. Additionally, material recycling can be a solution to address components with vulnerability. Therefore, remanufacturing IoT products requires dealing with information security risks by removing components with vulnerabilities for material recycling, and updating key/certification.

This study aims to design disassembly production to combat information security risks in remanufactured IoT products using a mathematical model. The proposed model determines the number of EOL products or components for each recovery process, such as disassembly, inspection, recycling, and key/certification updation.

**Keywords:** Remanufacturing · Sustainable Manufacturing · Circular Economy

## 1 Introduction

Circular economy, including remanufacturing and material recycling needs to be encouraged for sustainable manufacturing. The European Union aims sustainable products by establishing a new norm, Ecodesign for sustainable products regulation [1]. To retrieve values of end-of-life (EOL) products through remanufacturing and recycling, the collected EOL products are disassembled to obtain reusable and recyclable components or to dispose of hazardous components [2]. Based on the physical and value lifetime of each component, suitable lifecycle options, such as remanufacturing, material recycling,

H. Kohl et al. (Eds.): GCSM 2024, LNME, pp. 546–554, 2025.
https://doi.org/10.1007/978-3-031-93891-7_60

and disposal are preferred [3]. The removed components by non-destructive disassembly are inspected for remanufacturing or component reuse [4]. Meanwhile, components for material recycling and disposal are removed by destructive disassembly and inspected for recycling [4]. Thus, decisions and inspections are conducted based on the physical and value lifetimes.

On the other hand, spreading Internet-of-things (IoT) products increase information security risks, such as privacy leakages and DDoS attacks [5, 6]. Because, IoT products may always be monitoring personal data and connecting other products, information security risks of IoT products can cause critical damage [5, 6]. Thus, to remanufacture IoT products, in addition to their physical and value lifetimes, information security risk associated must be considered. Moreover, to transfer ownership through remanufacturing, security software and key/certification in IoT products should be updated to maintain the latest version, and prohibit a new owner from accessing old owner's data [5].

Kim et al. [2] addressed selective disassembly to retrieve the targeted component, with considering random disassembly operations. Tao et al. [7] proposed a 2-phase joint decision model to determine the optimal recovery option for each part using an automated disassembly system. Laili et al. [8] presented disassembly scheduling method to determine quantity, timing, and disassembly techniques. Tozanlı et al. [9] developed a disassembly-to-order (DTO) system to determine the optimal number for each process, such as collection, disassembly and inspections with blockchain technology to eliminate uncertainties related to the quality of the EOL products. Kinoshita et al. [4] also presented a DTO system to satisfy the demand for reused components and recycled materials with stochastic disassembly yields. These previous studies did not consider information security risks, even though information security risks might cause critical damage, such as privacy leakages.

This study proposes a disassembly production design that incorporates information security risks to determine the number of EOL products or components for each recovery process, such as disassembly, inspection, recycling, and key/certification updation. The proposed model evaluates the information security risks associated with each component to eliminate the vulnerabilities in remanufactured products by replacing components with information security risks with alternative ones. To validate the proposed model, numerical experiments are conducted. The novelty of the paper is consideration of information security risks such as expiration of security software support and personal information leak in remanufacturing. Security risks of each component are evaluated to reuse only secure components in remanufactured products, and security software and key/certification updates are introduced in remanufacturing processes.

## 2   Mathematical Model

### 2.1   Overview of Disassembly Production Design

The proposed disassembly production design determines the number of each process in the disassembly with an information security risk to satisfy the demand for remanufactured products and recycled materials. Figure 1 shows the structure of the proposed model.

First, EOL products with different statuses are purchased from different suppliers. The statuses determine the disassembly yields, to which a screening process is applied to decide whether the components can be remanufactured or not. Also, because components with information security risks cannot be used for remanufacturing, those components with information security risks are retrieved by destructive disassembly for material recycling. The components passing the screening are retrieved by non-destructive disassembly and sent to inspection for remanufacturing. Only those components passing the inspection, procured from outside and inventory are used for remanufacturing. Components with information security risks are replaced with alternatives. Security software update and key/certification updation are conducted for all the remanufactured products to combat information security risks. Expiration of security software support is considered as one of information security risks in the paper. The latest software will often require newer version of components. For example, Windows 11 needs a certain newer version of components such as a CPU, a computer memory, a trusted platform module.

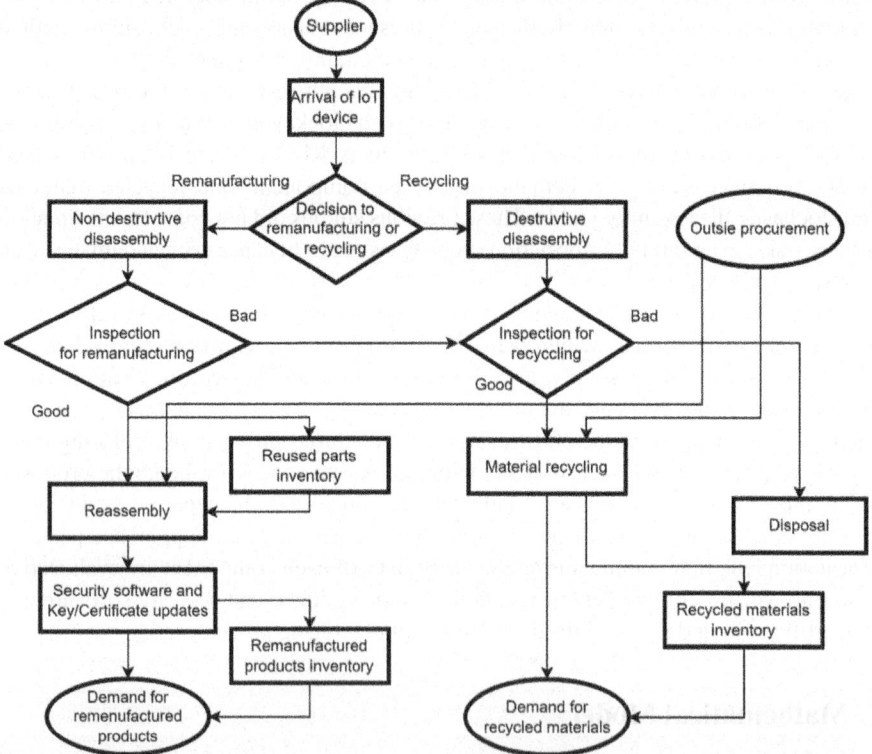

**Fig. 1.** Structure of proposed disassembly production design for information security risk

Components not passing the screening are retrieved by destructive disassembly operations and sent to inspection for recycling. The components disassembled destructively and those that failed to pass the inspection for remanufacturing are inspected for recycling. The components with good condition for recycling are recycled. Meanwhile, the components unsuitable for recycling are disposed of.

## 2.2 Formulation

The parameters and variables used in this study are listed as follows:

**Parameters**

$D_i^{RM}/D_m^{RC}$: Demand for remanufactured product and recycled material $m$

$sp_i^{RM}/sp_m^{RC}$: Selling price of remanufactured product $i$ and recycled material $m$

$prc_k/dpc_k$: Outside procurement and disposal cost of component $k$

$p_{ij}$: Purchase cost of EOL product $i$ from supplier $j$

$nddc_k/ddc_k$: Non-destructive and destructive disassembly cost of component $k$

$ac_i/supc_i/kupc_i$: Reassembly/Security software update
/Key Certification update cost of remanufactured product $i$

$hc_i^{RM}/hc_k^{RU}/hc_k^{RU}$: Holding cost of remanufactured product $i$, reused component $k$,
and recycled material $m$

$rcc_m$: Recycling cost of material $m$ per unit

$sr_k$: Information security risk of component $k$

$alt_{kk'}$: Alternative component $k'$ of component $k$

$w_{mk}$: Weight of material $m$ composed of component $k$

$Num_{ik}$: Number of components $k$ in EOL product $i$

$yld_{ijk}$: Disassembly yields of component $k$ in EOL product $i$
purchased from supplier $j$

$pct_k^{RU}/pct_k^{RC}$: Reuse and recycling percentage of component $k$

$CAP_{ij}^{EOL}$: Supply capacity of EOL product $i$ from supplier $j$

$Q_i^{SInvRM}/Q_k^{SInvRU}/Q_m^{SInvRC}$: Initial inventory quantity of remanufactured product $i$,
reused component $k$, and recycled material $m$

**Variables**

$Q_{ij}^{EOL}$: Quantity of purchased EOL product $i$ from supplier $j$

$Q_k^{RUPR}/Q_k^{RCPR}$: Quantity of outside procurement of component $k$
for remanufacturing and recycling

$Q_k^{DA}/Q_k^{NDA}/Q_k^{DDA}$: Quantity of disassembly/ non-destructive/destructive component $k$

$Q_k^{GNDA}/Q_k^{BNDA}$: Quantity of reusable/non-reusable component $k$

$Q_k^{GDDA}/Q_k^{BDDA}$: Quantity of recyclable/non-recyclable component $k$

$Q_i^{RM}/Q_k^{RU}$: Quantity of remanufactured product $i$/Reused component $k$

$Q_m^{RCW}/Q_m^{RC}$: Quantity of recyclable weight of material $m$/Recycled material $m$

$Q_i^{EInvRM}/Q_k^{EInvRU}/Q_m^{EInvRC}$: Final inventory quantity of remanufactured product $i$,
reused component $k$, and recycled material $m$

The proposed disassembly production design with information security risks maximizes the total profit computed from revenues from selling the remanufactured products and recycled materials, costs regarding procuring EOL products and components, disassembly, remanufacturing, recycling, inventory, and disposal; this optimization process is given by Eqs. (1)–(10).

Objective function:

$$TP = RMR + RCR - TPURC - TPRC$$
$$-TDAC - TRMC - TRCC - TINVC - TDPC \rightarrow Max \tag{1}$$

Constraint:

$$RMR = \sum_{i \in I} D_i^{RM} sp_i^{RM} \tag{2}$$

$$RCR = \sum_{m \in M} D_m^{RC} sp_m^{RC} \tag{3}$$

$$TPURC = \sum_{i \in I} \sum_{j \in J} p_{ij} Q_{ij}^{EOL} \tag{4}$$

$$TPRC = \sum_{k \in K} prc_k (Q_k^{RUPR} + Q_k^{RCPR}) \tag{5}$$

$$TDAC = \sum_{k \in K} (nddc_k Q_k^{NDA} + ddc_k Q_k^{DDA}) \tag{6}$$

$$TRMC = \sum_{i \in I} (ac_i + supc_i + kupc_i) Q_i^{RM} \tag{7}$$

$$TRCC = \sum_{m \in M} rcc_m Q_m^{RCW} \tag{8}$$

$$TINVC = \sum_{i \in I} hc_i^{RM} Q_i^{EInvRM} + \sum_{k \in K} hc_k^{RU} Q_k^{EInvRU} + \sum_{m \in M} hc_m^{RC} Q_m^{EInvRC} \tag{9}$$

$$TDPC = \sum_{k \in K} Q_k^{BDDA} dpc_k \tag{10}$$

The suppliers provide EOL products with different statuses under supply capacity as given by Eq. (11). The total obtained number for each component is calculated using Eq. (12). To consider the information security risks, parameter $sr_k$ is introduced, which is set to 1 if the component has information security risk, and to 0 otherwise. Because only components without information security risk can be used for remanufacturing, they can be retrieved by non-destructive disassembly as given by Eq. (13). The disassembly yield $yld_{ijk}$ indicates the component condition affecting the screening process, wherein it is decided whether the component is manufacturable or not. Equations (14) to (18) enforce the number of components in the divided flows due to screening and inspections to be matched.

$$Q_{ij}^{EOL} \leq Cap_{ij}^{EOL} \forall i \in I, \forall j \in J \tag{11}$$

$$Q_k^{DA} = \sum_{i \in} \sum_{j \in J} Num_{ik} Q_{ij}^{EOL} \forall k \in K \tag{12}$$

$$Q_k^{NDA} = \sum_{i \in I} \sum_{j \in J} Num_{ik} yld_{ijk} (1 - sr_k) Q_{ij}^{EOL} \forall k \in K \tag{13}$$

$$Q_k^{DDA} = Q_k^{DA} - Q_k^{NDA} \forall k \in K \tag{14}$$

$$Q_k^{GNDA} = pct_k^{RU} Q_k^{NDA} \forall k \in K \tag{15}$$

$$Q_k^{BNDA} = Q_k^{NDA} - Q_k^{GNDA} \forall k \in K \tag{16}$$

$$Q_k^{GDDA} = pct_m^{RC} \left( Q_k^{DDA} + Q_k^{BNDA} \right) \forall k \in K \tag{17}$$

$$Q_k^{BDDA} = Q_k^{DDA} + Q_k^{BNDA} - Q_k^{GDDA} \forall k \in K \tag{18}$$

Equation (19) means that the number of reusable components for remanufacturing. Because the components with information security risks cannot be used for remanufacturing, alternative components are used as given by Eqs. (20) and (21). The parameter $alt_{kk'}$ takes a value of 1 if component $k'$ is alternative for component $k$ with information security risks, and 0 otherwise, as given by Eqs. (22) and (23). Only one alternative can exist as an alternative for each component with information security risks.

$$Q_k^{RU} = Q_k^{SInvRU} + Q_k^{GNDA} + Q_k^{RUPR} \forall k \in K \tag{19}$$

$$Q_k^{RU} \geq \sum_{i \in I} Num_{ik}(1 - sr_k)Q_i^{RM} + \sum_{i \in I} \sum_{k' \in K/\{k\}} Num_{ik'} alt_{k'k} Q_i^{RM} \forall k \in K \tag{20}$$

$$Q_k^{RUPR} \leq M (1 - sr_k) \forall k \in K \tag{21}$$

$$\sum_{k' \in K/\{k\}} alt_{kk'} \geq sr_k \forall k \in K \tag{22}$$

$$alt_{kk} = 1 - sr_k \forall k \in K \tag{23}$$

Equations (24) and (25) represent the quantity of recyclable weight and recycled material $m$. Equations (26) and (27) ensure that the demands of the remanufactured products and recycled materials are satisfied. Inventory of the remanufactured products, reusable components and recycled materials are represented by Eqs. (28), (29), and (30) and respectively.

$$Q_m^{RCW} = \sum_{k \in K} (Q_k^{GDDA} + Q_k^{RCPR}) w_{mk} \forall m \in M \tag{24}$$

$$Q_m^{RC} = Q_m^{SInvRC} + Q_m^{RCW} \forall m \in M \tag{25}$$

$$Q_i^{RM} \geq D_i^{RM} \forall i \in I \tag{26}$$

$$Q_m^{RC} \geq D_m^{RC} \forall m \in M \tag{27}$$

$$Q_i^{EInvRM} = Q_i^{RM} - D_i^{RM} \forall i \in I \tag{28}$$

$$Q_m^{EInvRC} = Q_m^{RC} - D_m^{RC} \forall m \in M \tag{29}$$

$$Q_k^{EInvRU} = Q_k^{RU} - \sum_i Num_{ik}(1 - sr_k)Q_i^{RM}$$

$$- \sum_{k' \in K \setminus \{k\}} Num_{ik'} alt_{k'k} Q_i^{RM} \forall k \in K \tag{30}$$

$$\text{all variables are non} - \text{negative} \tag{31}$$

## 3  Numerical Experiments

To illustrate a design example of the proposed disassembly production with information security risk, numerical experiments were conducted. Three different types of IoT products were addressed. The IoT products were composed of eight different components and six different types of materials. Components B and H had information security risks, and components I and J were alternatives to them, respectively. The disposal cost of each component was 3USD. The initial inventory of remanufactured products, reusable components, and recycled materials were all set to 0. The disassembly yields of each component takes 0.00–1.00 values. The other parameters are listed in Tables 1, 2, 3 and 4.

**Table 1.** Capacity and purchased cost of EOL products.

| | Capacity | | | Purchase cost | | |
|---|---|---|---|---|---|---|
| | Product 1 | Product 2 | Product 3 | Product 1 | Product 2 | Product |
| Supplier 1 | 500 | 450 | 410 | 15 | 15 | 10 |
| Supplier 2 | 410 | 460 | 440 | 20 | 19 | 13 |

**Table 2.** Input data for remanufactured products.

| Product | Demand (unit) | Selling Price ($/kg) | Inventory Cost ($/kg) | Reassembly Cost ($) | Security Update Cost ($) | Key/Certification Update Cost ($) |
|---|---|---|---|---|---|---|
| Product 1 | 450 | 80 | 8 | 22 | 13 | 3 |
| Product 2 | 360 | 83 | 6 | 21 | 11 | 3 |
| Product 3 | 370 | 85 | 7 | 18 | 11 | 3 |

The numerical experiments were conducted on a laptop (Windows 11 with Intel(R) Core(TM) i5-10210U CPU @ 1.60 GHz), using an optimization solver named Gurobi

**Table 3.** A part of input data for components.

| Component | Contained number of comment (unit) | | | Reuse Percentage (%) | Recycle Percentage (%) | Disassembly Cost ($/unit) | | Outside Procurement Cost ($/unit) | Inventory Cost ($/unit) | Contained material weight (kg) | | | | | |
|---|---|---|---|---|---|---|---|---|---|---|---|---|---|---|---|
| | Product 1 | Product 2 | Product 3 | | | Non-Destructive | Destructive | | | α | β | γ | δ | ε | ζ |
| A | 2 | 0 | 3 | 88 | 86 | 1.7 | 0.7 | 14 | 3.5 | 1.0 | 1.5 | 0.0 | 0.0 | 0.7 | 0.0 |
| B | 2 | 2 | 0 | 81 | 85 | 1.3 | 0.7 | 11 | 3.5 | 0.0 | 0.0 | 0.0 | 0.0 | 1.2 | 0.5 |
| C | 0 | 2 | 2 | 97 | 96 | 1.6 | 0.9 | 14 | 3.9 | 0.2 | 0.0 | 0.0 | 1.5 | 0.0 | 1.0 |
| D | 2 | 1 | 0 | 84 | 96 | 1.3 | 0.9 | 13 | 3.8 | 0.0 | 0.0 | 0.1 | 0.8 | 1.4 | 0.6 |
| E | 2 | 3 | 2 | 92 | 92 | 1.7 | 0.9 | 11 | 3.9 | 0.8 | 1.5 | 0.9 | 1.5 | 0.0 | 0.0 |
| F | 3 | 2 | 1 | 74 | 88 | 1.2 | 1.0 | 11 | 4.0 | 0.0 | 0.0 | 1.3 | 0.4 | 1.1 | 0.0 |
| G | 1 | 3 | 2 | 75 | 100 | 1.2 | 1.0 | 14 | 3.8 | 0.3 | 0.0 | 0.0 | 0.0 | 0.2 | 0.0 |
| H | 3 | 0 | 0 | 74 | 95 | 1.3 | 0.9 | 13 | 4.0 | 0.7 | 0.0 | 0.0 | 0.0 | 0.4 | 0.8 |
| I | 0 | 0 | 3 | 98 | 85 | 1.6 | 0.9 | 14 | 3.4 | 0.0 | 0.0 | 0.0 | 0.0 | 1.6 | 0.8 |
| J | 0 | 3 | 2 | 88 | 96 | 1.1 | 0.9 | 11 | 3.4 | 0.6 | 0.0 | 0.0 | 0.0 | 0.5 | 0.7 |

**Table 4.** A part of input data for materials.

| Material Type | Demand (unit) | Selling Price ($/kg) | Inventory Cost ($/kg) | Recycling Cost ($) |
|---|---|---|---|---|
| α | 810 | 30 | 0.5 | 1.2 |
| β | 650 | 35 | 0.5 | 1.8 |
| γ | 700 | 30 | 0.5 | 1.4 |
| δ | 900 | 35 | 0.5 | 1.2 |
| ε | 780 | 32 | 0.5 | 1.0 |
| ζ | 930 | 31 | 0.5 | 1.4 |

[10]. The resulting profit was $68,553. To satisfy the demands for the remanufactured products and recycled materials, 1,383 units of EOL products were purchased and 5,171 units of components were procured for remanufacturing and recycling. All B and H components were recycled or disposed of owing to information security risks. As alternative components with information security risks, over 1,450 units of components I and J were procured from external suppliers in addition to retrieving the EOL products.

## 4    Conclusions and Future Studies

This study developed a disassembly production design incorporating information security risks to determine the number of EOL products and components for each recovery process, such as disassembly, inspection, recycling, and key/certification updation. Only remanufacturing factories, where security software and key/certification can update properly and absolutely, can utilize the proposed model, because these processes are critical for information security risks. To combat information security risks in IoT products, two novel parameters were introduced. One was security risk $sr_k$, which takes a value of 1 if security risks are present and 0 otherwise. The other was alternative component $alt_{kk'}$ which indicates that component $k$ with security risk can be replaced with component $k'$

without a security risk. However, the model did not consider the physical end of life of IoT devices, then, future studies should take account of it. Moreover, future studies shall adopt the proposed model to actual IoT products.

# References

1. European Commission: Provisional agreement for more sustainable consumer products. https://ec.europa.eu/commission/presscorner/detail/en/ip_23_6257, Accessed 16 Jul 2024
2. Kim, H.W., Park, C., Lee, D.H.: Selective disassembly sequencing with random operation times in parallel disassembly environment. Int. J. Prod. Res. **56**(24), 7243–7257 (2018)
3. Yoda, K., Irie, H., Kinoshita, Y., Yamada, T., Yamada, S., Inoue, M.: Remanufacturing option selection with disassembly for recovery rate and profit. International Journal of Automation Technology **14**(6), 930–942 (2020)
4. Kinoshita, Y., Yamada, T., Gupta, S.M.: Design of disassembly-to-order system for reused components and recycled materials using linear physical programming. Int. J. Sustain. Manuf. **4**(2/3/4), 121–149 (2020)
5. Yousefnezhad, N., Malhi, A., Främling, K.: Security in product lifecycle of IoT devices: a survey. J. Netw. Comput. Appl. **171**, 102779 (2020)
6. Schiller, E., Aidoo, A., Fuhrer, J., Stahl, J., Ziörjen, M., Stiller, B.: Landscape of IoT security. Comput. Sci. Rev.**44**, 100467 (2022)
7. Tao, Y., Meng, K., Lou, P., Peng, X., Qian, X.: Joint decision-making on automated disassembly system scheme selection and recovery route assignment using multi-objective meta-heuristic algorithm. Int. J. Prod. Res. **57**(1), 124–142 (2018)
8. Laili, Y., Li, Y., Fang, Y., Pham, D.T., Zhang, L.: Model review and algorithm comparison on multi-objective disassembly line balancing. J. Manuf. Syst. **56**, 484–500 (2020)
9. Tozanlı, Ö., Kongar, E., Gupta, S.M.: Trade-in-to-upgrade as a marketing strategy in disassembly-to-order systems at the edge of blockchain technology. Int. J. Prod. Res. **58**(23), 7183–7200 (2020)
10. Gurobi Optimization, The Leader in Decision Intelligence Technology, https://www.gurobi.com, Accessed 16 Jul 2024

# Reference Process for Production Program Planning in Re-Assembly

H. Neumann$^{(\boxtimes)}$, M. Riesener, S. Schmitz, and C. Hilger

Laboratory for Machine Tools and Production Engineering (WZL) of RWTH Aachen University,
Campus-Boulevard 30, 52074 Aachen, Germany
h.neumann@wzl.rwth-aachen.de

**Abstract.** The manufacturing industry is facing a rising demand of natural resources while being confronted with heightened global competition and increased pressure for sustainable products and production practices. In order to meet these diverse requirements, a transition from a linear to a circular economy is necessary. Within this new manufacturing paradigm, strategies such as Re-Assembly can be used to manufacture technologically upgraded products with significantly lower resource inputs and costs. However, also in Re-Assembly efficient production program planning is necessary to ensure economic viability. Methods used in linear production prove inadequate in Re-Assembly due to new influencing factors such as core conditions and availability. To address this, a reference process for production program planning in Re-Assembly is developed in this paper. Based on the linear production planning process, two literature researches were conducted to identify Re-Assembly-specific adaptions of the planning. The first literature review specifies necessary main tasks, while the second analyses planning modules and their input and output information. The identified tasks and modules are then used to create a new reference model for use in Re-Assembly.

**Keywords:** Re-Assembly · Production Planning · Circular Economy

## 1 Introduction

Due to the loss of biodiversity and the scarcity of natural resources, industry is at a crucial turning point [1]. The traditional linear economic model based on the production, consumption and disposal of products has left a clear ecological footprint [2]. With a share of one-third of the energy-related carbon emissions, the producing industry holds a high responsibility for the impact on the environment [1]. Given these pressing challenges, it is becoming increasingly clear that a paradigm shift is necessary [3]. This leads to an industry trend of more sustainable and circular production methods. The trend is supported by various political initiatives such as the EU´s new circular economy action plan [4].

The value-enhancing circular economy represents an essential new economic model for achieving long-term economic and ecological sustainability. In this paradigm, materials, components and products are reused and upgraded. This leads to minimized

H. Kohl et al. (Eds.): GCSM 2024, LNME, pp. 555–562, 2025.
https://doi.org/10.1007/978-3-031-93891-7_61

resource usage, while technological improvement in the product is still possible. A key enabler of this new economic model is the industrialized reprocessing in the form of Re-Assembly. However, successful Re-Assembly necessitates appropriate long- and medium-term planning in the form of production program planning (PPP). [5] Current planning approaches are no longer adequate due to changes in the type of used input resources and processes [6].

To address this, it is necessary to understand the fundamental tasks that need to be expanded upon from classical approaches. This paper conducts a literature review based on the reference tasks from the Aachener PPC model [7] to describe the new tasks and their relationship to classical production program planning.

## 2    Production Program Planning and Re-assembly

PPP is part of production planning and control (PPC). PPP is defined as the planning function that develops a long-term production plan, ensuring it is both marketable and feasible for the production. Additionally, this process yields a preliminary procurement plan, wherein the material requirements for production are broadly defined. [7] A well-recognized model for the PPC is the Aachener PPC model [7–9].

Following the Aachener PPC model, PPP for a linear producing company consists of the three main tasks: Sales planning, primary demand planning, and rough resource planning. Each main task consists of planning modules, which solve a subtask in the process. In sales planning the sales volume for the following time-periods are estimated. Primary demand planning uses this information in combination with further demands and the product inventory to calculate the production demand for products and components. Based on these demands the rough resource planning compares the resulting resource demand with the available production resources. If there are not sufficient resources to fulfill the demand, either the sales volumes or the resource availability has to be adapted [8].

In Re-Assembly the resource demands are not as predictable as they are in the linear economy as the conditions and quantities of the returned products are uncertain. The concept of Re-Assembly is similar to remanufacturing as defined in DIN Spec 91472 [10]. Re-Assembly includes the processes *product inspection, disassembly,* and *reassembly* but excludes the reconditioning of components. Re-Assembly is defined as a process in which a product is reconditioned to an as-good-as-new condition or better by replacing its components. The reconditioning of the components will either be done in a separate process or by the supplier. This leads to a short time in which the product is not in use or on the market. Products reconditioned through the Re-Assembly process can also be enhanced by the integration of upgrades. [5].

## 3    Literature Review and Derived Reference Process

To integrate the characteristics of Re-Assembly in the PPP process, a systematic literature review is conducted along the main tasks, planning modules, and the resulting information streams. The concept consists of two sub-sections aimed at addressing the research questions (RQ) (see Fig. 1.) Based on the basic structure of the reference model the sub-sections are detailed as follows: In Sect. 1 (*RQ 1*), the main tasks of PPP for Re-Assembly are defined by adapting or extending the known main tasks. In Sect. 2 (*RQ 2*), the level of detail of the adapted main tasks is increased by adding more specific planning modules to the main tasks and sequence these by comparing the input and output information (*RQ 3*).

### 3.1  Literature Review

The used systematic literature review (SLR) is based on the guidelines proposed by KITCHENHAM [11]. Initially, a protocol is created, detailing the framework and providing further specific information regarding the search. This protocol, shown in Fig. 1, includes the formulated research questions, the selection of databases, and the search terms. These terms are determined by breaking down the research questions into individual facets. Synonyms, abbreviations, and alternative spellings for these terms are utilized. Various search strings are created using boolean AND and OR connections. Furthermore, inclusion and exclusion criteria are defined. This helps to focus the literature search on the target area. Literature related to PPP in remanufacturing, ensuring relevance through the consideration of titles, abstracts, and keywords is included. Exclusion criteria filter out literature not available in electronic format, duplicates, works not written in German or English, and literature lacking full texts or abstracts.

| Research questions | | | |
|---|---|---|---|
| 1) | How must the main tasks of the PPP be expanded in order to be able to create the production program? | | |
| 2) | How must the planning modules be specified and expanded to consider all phases of the re-assembly process? | | |
| 3) | What data is required as input for the planning modules and what output do they provide for other modules | | |
| **Databases** | | | |
| IEEE Xplore (www.ieeexplore.com) | Springer (www.link.springer.de) | Web of Science (www.webofscience.com) | Scopus (www.scopus.com) |
| **Search terms** | | | |
| 1) production AND program planning AND remanufacturing material resource planning AND Remanufacturing multi period AND production planning AND remanufacturing | | 2) + 3) core acquisition management, sales planning, master production scheduling, material requirements planning, capacity requirements planning AND remanufacturing | |

**Fig. 1.** Protocol of the SLR

### 3.2  Main Tasks

Following the SLR review protocol, out of 2086 findings the most relevant findings for **RQ 1** were evaluated based on their titles, abstracts and full texts. The results were searched for planning methods in remanufacturing and Re-Assembly matching the planning horizon of production program planning. These were than compared with the linear PPP. The results demonstrate that the traditional tasks of PPP remain valid. The work of SCHELLER et al. [12], LAGE et al. [13] and BARBA- GUTIÉRREZ et al. [14] specifically emphasize the importance of sales planning, reinforcing its role as a central element in the reference process.

An additional task in PPP is primary demand planning, where the preliminary production program proposal is developed. This key task is also mentioned by SCHELLER et al. [12], LAGE et al. [13] and BARBA- GUTIÉRREZ et al. [14]. Another essential task within PPP, particularly in the context of Re-Assembly, is rough resource planning. This task is an indispensable part of PPP and has been discussed by DOH et al. [15], DENZIEL et al. [16], and HAN et al. [17] in the context of remanufacturing. In this process, the demand for products and components is broadly compared with the available production resources. Literature shows that in the planning for Re-Assembly an integration of the properties of the returned products is necessary [18–32]. From this literature the new main task *core supply management* is derived. The task integrates the impact of the uncertainties in quantity and quality of the cores. The task is extensively discussed in the literature, particularly by WEI et al. [18] and BARQUET et al. [19]. According to these authors, core supply management is responsible for managing the extensive data related to cores. It serves as an interface between reverse logistics and the traditional main tasks of PPP.

### 3.3 Planning Modules and Information Flow

In the next step the planning modules (RQ2) associated with the main tasks and the information flow between the modules (RQ3) are examined. Following the SLR review protocol for RQ 2, 14 results for the core acquisition management, 17 results for sales planning, seven results for the primary demand planning and 14 results for the resource planning were identified. Based on this literature the planning methods were clustered regarding their results and abstracted into planning modules. The sequence of the planning modules was defined using the information input and outputs of the modules.

The most relevant findings for core supply management, sorted by their focus, are shown in Table 1. The first planning module about core supply management deals with the forecast of core availability. The module aims to predict in which planning periods and to what extent cores will be available for remanufacturing.

The second planning module comprises the quality forecast of the cores and provides a quality forecast of the cores' components for further core supply management. The forecasting methods found in the literature by LIANG et al. [20] and TSILIYANNIS [21] work with the remaining service life of the cores and derive the quality of the cores when they are returned. Using usage data of the cores the quality forecasting can also be done before the cores are returned. The detailed method for the forecasting depends on the product and available product data.

In the third planning module, the condition of the cores is evaluated and categorized into quality levels. Several authors (see Table 1) consider the classification of cores into quality levels to be crucial for effective PPP in remanufacturing companies. According to MUTHA et al. [22] and YANIKOĞLU et al. [23], the systematic classification of cores into different quality levels can be used to determine their suitability for the remanufacturing process. The classification of cores range from simple distinctions such as "remanufacturable" and "non-remanufacturable" to finer gradations with several quality levels. To choose the right gradation, the detailed module needs to be adapted to the product and processes of the company.

**Table 1.** Literature focusing on Core Supply Management [14, 18, 20–32].

| Focus of the literature | Literature | | | |
|---|---|---|---|---|
| Forecast of Core availability | GUIDE et al. 2000 | HUSTER et al. 2023 | LIANG et al. 2014 | WEI et al. 2015 |
| | TSILIYANNIS 2020 | KRAPP et al. 2013 | AYDIN et al. 2018 | |
| Forecast of Core quality | GUIDE et al. 2000 | LIANG et al. 2014 | TSILIYANNIS 2020 | FERRER 2001 |
| Classification of Cores | GUIDE et al. 2000 | MUTHA et al. 2016 | LI et al. 2016 | WEI et al. 2015 |
| | AYDIN et al. 2018 | YANIKOĞLU et al. 2020 | CIPTOMULYONO et al. 2022 | |
| Determining use of components | GUPTA et al. 2000 | VOGELGESANG et al. 2015 | FERRER 2001 | BARBA-GUTIÉRRERZ et al. 2009 |

In the fourth planning module, the further use of the individual components of the cores is determined based on the core classification. The content of this planning module is derived from FERRER [33] and VOGELGESANG [31]. This planning module indicates probability for reuse for each component. Based on the quality determination of the components, the disassembly BOM is created.

The changes and the new main task inevitably result in new or modified planning modules. In the following, only the changes in the planning modules will be considered. In the area of sales planning, an additional planning module is derived based on the SLR. This planning module includes the coordination between the demand and return rate of cores. This is based on the work of GUIDE et al. [32], REDDY et al. [34] and INDEFURTH et al. [35].

In primary demand planning, the SLR did not result in any changes to the linear PPP process. In rough resource planning, new planning modules are necessary due to the new processes and uncertainties raised by Re-Assembly. In the first step, the reassembly and disassembly plans need to be created based on the information from primary demand planning and core supply management. The planning modules are derived from FERRER [33] and BARBA- GUTIERRERZ et al. [14]. The plans are created with the help of work plans and process times for the products as well as the demands and disassembly-BOM from the previous planning modules.

The next planning module follows from the literature sources of FERRER et al. [33] and GUPTA et al. [30]. In this module the product demand from the reassembly plan is transformed into a component demand. The component demand varies based on the condition of the returned products and is needed for the creation of the material demand. The last planning module is also derived from FERRER et al. and GUPTA et al. In this planning module, the net demands per component and period are determined from the gross demand per component and period. This is done by subtracting the stock of the respective component from the gross demand. The result is the rough material demand for Re-Assembly.

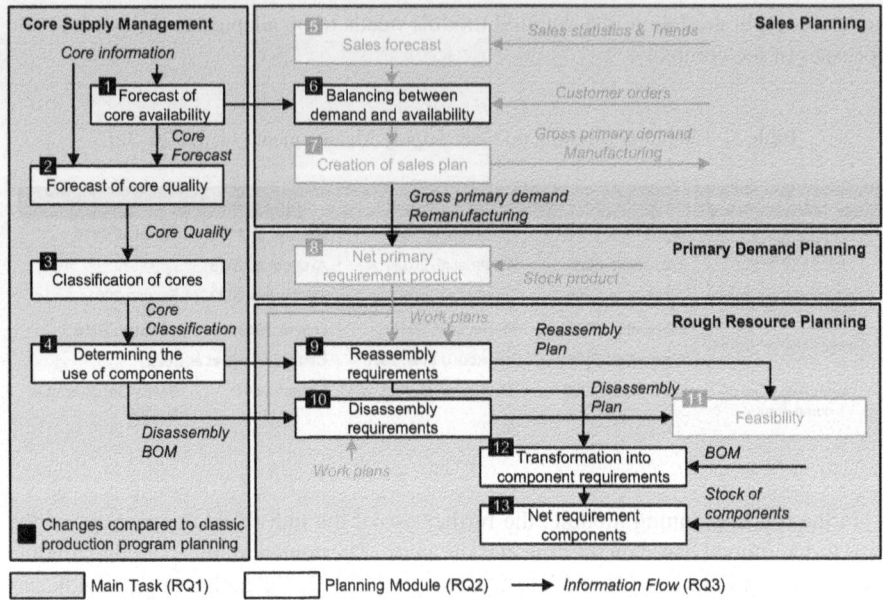

**Fig. 2.** Reference Process for Production Program Planning in Re-Assembly

Figure 2 illustrates the derived reference process for PPP in Re-Assembly. The changed process modules and information flows are highlighted in the figure. Additionally the results are matched to the research questions of the SLR.

## 4   Summary and Outlook

The conducted literature research in the work shows that there is a need for the adaption of the linear PPP. The changes are in the inclusion of a new main task, core supply management, and the redesign of the existing main tasks as well as their planning modules. Especially the uncertainties of quantity and condition of incoming cores lead to the necessity of the changes in the model. Next to the tasks and modules further information flows were derived.

In further research, the model needs to be detailed for specific company requirements and implemented in real-life use-cases. These use cases will show the effect of the model regarding the sustainability of the reconditioning process. For a company specific application, the planning modules and information demands need to be further specified and the sources for the demanded information need to be defined. As mentioned in Sect. 3.3 this includes the selection of specific quality estimation models for product and component conditions as well as detailed planning methods, which can handle the uncertainty in the process.

**Acknowledgements.** Funded by the Deutsche Forschungsgemeinschaft (DFG, German Research Foundation) under Germany's Excellence Strategy – EXC-2023 Internet of Production – 390621612.

# References

1. Sutherland, J.W., Skerlos, S.J., Haapala, K.R., Cooper, D., Zhao, F., Huang, A.: Indus-trial sustainability. reviewing the past and envisioning the future. J. Manuf. Sci. Eng. **142** (2020)
2. Ellen MacArthur Foundation Towards the circular economy. Economic and business rationale for an accelerated transition (2013)
3. European Commission, Directorate-General for Research and Innovation Circular Economy Research and Innovation. Connecting Economic & Enviromental Gains (2017)
4. European Commission: Directorate-General for Communication a new Circular Economy Action Plan, Brussels (2020)
5. Schuh, G., et al.: Grüne re-montage upgrade-Fabrik. In: Schuh, G., Brecher, C., Schmitt, R., Bergs, T. (eds.) Empower Green Production (2023)
6. Guide, V.R.: Production planning and control for remanufacturing: industry practice and re-search needs. J. Oper. Manage. **18**, 467–483 (2000)
7. Schuh, G., Schmidt, C. (eds.) Produktionsmanagement. Handbuch Produktion und Management 5. 2nd ed., Springer Vieweg, Berlin (2014)
8. Lödding, H.: Verfahren der Fertigungssteuerung. Springer, Berlin Heidelberg (2016)
9. Grabner, T.: Operations Management. Auftragserfüllung bei Sach- und Dienstleistungen. 4th ed., Springer Gabler, Wiesbaden, Germany (2019)
10. Remanufacturing (Reman). Qualitätsklassifizierung für zirkuläre Prozesse, Beuth Verlag, Berlin, June 2023
11. Kitchenham, B., Charters, S.: Guidelines for performing systematic literature reviews in software engineering **2** (2007)
12. Scheller, C., Schmidt, K., Spengler, T.S.: Decentralized master production and recycling scheduling of lithium-ion batteries: a techno-economic optimization model. J. Bus. Econ.. Bus. Econ. **91**, 253–282 (2021)
13. Lage Junior, M., Godinho Filho, M.: Master disassembly scheduling in a remanufacturing system with stochastic routings. Cent. Eur. J. Oper. Res. **25**, 123–138 (2017)
14. Barba-Gutiérrez, Y., Adenso-Díaz, B.: Reverse MRP under uncertain and imprecise demand. Int. J. Adv. Manuf. Technol.. J. Adv. Manuf. Technol. **40**, 413–424 (2009)
15. Doh, H.-H., Lee, D.-H.: Generic production planning model for remanufacturing systems. Proc. Inst. Mech. Eng. Part B: J. Eng. Manuf. **224**, 159–168 (2010)
16. Denizel, M., Ferguson, M., Souza, G.: Multiperiod remanufacturing planning with uncertain quality of inputs. IEEE Trans. Eng. Manage. **57**, 394–404 (2010)
17. Han, S.H., Dong, M.Y., Lu, S.X., Leung, S.C.H., Lim, M.K.: Production planning for hybrid remanufacturing and manufacturing system with component recovery. J. Oper. Res. Soc.. Oper. Res. Soc. **64**, 1447–1460 (2013)
18. Wei, S., Tang, O., Sundin, E.: Core (product) acquisition management for remanufacturing: a review. J. Remanuf. **5** (2015)
19. Barquet, A.P., Rozenfeld, H., Forcellini, F.A.: An integrated approach to remanufacturing: model of a remanufacturing system J. Remanuf. **3** (2013)
20. Liang, X., Jin, X., Ni, J.: Forecasting product returns for remanufacturing systems. J. Remanuf. **4** (2014)
21. Tsiliyannis, C.A.: Prognosis of product take-back for enhanced remanufacturing. J. Remanuf. **10**, 15–42 (2020)
22. Mutha, A., Bansal, S., Guide, V.D.R.: Managing demand uncertainty through core acquisition in remanufacturing. Prod. Oper. Manage. **25**, 1449–1464 (2016)
23. Yanıkoğlu, İ, Denizel, M.: The value of quality grading in remanufacturing under quality level uncertainty. Int. J. Prod. Res.. J. Prod. Res. **59**, 839–859 (2021)

24. Huster, S., Rosenberg, S., Glöser-Chahoud, S., Schultmann, F.: Remanufacturing capacity planning in new markets—effects of different forecasting assumptions on remanufacturing capacity planning for electric vehicle batteries. J. Remanuf. (2023)

25. Krapp, M., Nebel, J., Sahamie, R.: Using forecasts and managerial accounting information to enhance closed-loop supply chain management. OR Spectrum **35**, 975–1007 (2013)

26. Aydin, R., Kwong, C.K., Geda, M.W., Okudan Kremer, G.E.: Determining the optimal quantity and quality levels of used product returns for remanufacturing under multi-period and uncertain quality of returns. Int. J. Adv. Manuf. Technol. **94**, 4401–4414 (2018)

27. Inderfurth, K., Vogelgesang, S., Langella, I.M.: How yield process misspecification affects the solution of disassemble-to-order problems, pp 56–67

28. Li, X., Li, Y., Cai, X.: On core sorting in RMTS and RMTO Systems: a newsvendor framework. Decis. Sci.. Sci. **47**, 60–93 (2016)

29. Ciptomulyono, U., Mustajib, M.I., Karningsih, P. D., Anggrahini, D., Basuki, S.S.A.: A new multi-criteria method based on DEMATEL, ANP and grey clustering for quality sorting of incoming cores in remanufacturing systems under epistemic uncertainty: a case study of heavy-duty equipment. Cogent. Eng. **9** (2022)

30. Gupta, S.M., Veerakamolmal, P.: A bi-directional supply chain optimization model for reverse logistics. In: Proceedings of the 2000 IEEE International Symposium on Electronics and the Environment (Cat. No.00CH37082), pp 254–259. IEEE (2000)

31. Inderfurth, K., Vogelgesang, S., Langella, I.M.: How yield process misspecification affects the solution of disassemble-to-order problems. Int. J. Prod. Econ.. J. Prod. Econ. **169**, 56–67 (2015)

32. Guide, V.D.R., Jayaraman, V.: Product acquisition management: current industry practice and a proposed framework. Int. J. Prod. Res.. J. Prod. Res. **38**, 3779–3800 (2000)

33. Ferrer, G., Whybark, D.C.: Material planning for a remanufacturing facility. Prod. Oper. Manage. **10**, 112–124 (2001)

34. Reddy, K.N., Kumar, A., Velaga, N.R.: Scenario-based two-stage stochastic programming for a hybrid manufacturing-remanufacturing system with the uncertainty of returns. Qual. Demand Sādhanā **46** (2021)

35. Indefurth, K.: Production planning and control of closed-loop supply chains (2001)

# Cost Model for Evaluation of Industrial Re-Assembly Processes

A. Hermann[(⊠)], M. Riesener, S. Schmitz, and H.Neumann

Laboratory for Machine Tools and Production Engineering (WZL), Campus-Boulevard 30, 52074 Aachen, Germany
a.hermann@wzl.rwth-aachen.de

**Abstract.** Manufacturing companies must prioritize resource efficiency to remain competitive in the face of increasing demand for goods. Linear economic models, such as the 'Take-Make-Dispose' paradigm, are no longer adequate for meeting the demands of modern society, ecology, and regulation. The circular economy offers a resource-efficient alternative. In the circular economy, Re-Assembly is a promising strategy within this framework, involving industrial reprocessing of serial products to extend their lifecycle. However, the absence of optimized product designs for reprocessing hinders widespread adoption and economic viability of Re-Assembly processes. Therefore, integrated development across product design, production, and Re-Assembly is crucial to address this challenge. To achieve economic viability, it is necessary to evaluate costs throughout the product lifecycle. To enable companies to make informed decisions regarding the Re-Assembly process, a continuous cost assessment supported by sophisticated cost models is necessary. This paper presents a comprehensive cost model tailored to assess the economic feasibility of industrial Re-Assembly processes. The relevant cost drivers and necessary resources for the reprocessing are identified and used as input for the subsequently developed process-oriented cost model. This cost model is transferred into an optimization model. The model aims to enhance the sustainability and profitability of circular economy practices.

**Keywords:** Re-Assembly · Cost Modelling · Product and Production Development

## 1 Motivation

The manufacturing industry faces numerous challenges, including resource scarcity, stringent political regulations, and growing environmental awareness among consumers. This necessitates aligning economic activities with ecological, social, and regulatory requirements [1, 2]. Traditional linear economic models ("take-make-dispose") are inadequate for future challenges, while the circular economy offers a viable alternative [3]. The circular economy aims to decouple consumption and economic growth from natural resource use. The Re-Assembly approach, defined as the industrial reprocessing of complex series products, is key to advancing the circular economy in ecological and economic

© The Author(s) 2025
H. Kohl et al. (Eds.): GCSM 2024, LNME, pp. 563–571, 2025.
https://doi.org/10.1007/978-3-031-93891-7_62

ways. [4, 5] Out of an ecological perspective remanufacturing can reduce emissions by 90% and energy consumption by 56% compared to new production, according to the VDI [6].The reduced material and energy consumption can lead to margins two to three times higher, depending on the product [5]. Forecasts suggest that the economic potential of industrial reprocessing in Europe could reach €90 billion by 2030 [4]. However, the potential of Re-Assembly is hindered by insufficient product design for reprocessing [7]. Designing products for reprocessing maximizes benefits like waste reduction, material and energy savings [2]. Also the production process is not suitable to enable a disassembly of the products, and the process of the Re-Assembly is developed after the first lifecycle of the products, which leads to not applying a Re-assembly at all. Therefore, integrating Re-Assembly development into product development (PD) and production process development (PPD) is essential [8]. Furthermore, it is essential to implement a continuous monitoring and control of costs strategy. As development progresses, the potential for influencing costs diminishes. At the conclusion of the development phase, a considerable proportion of the cost price has already been established. Approximately 70% of the costs of a product are determined during the development and design phase. [9] In the next chapter the integrated development of process and production, the Re-assembly and the process based cost model (PBCM) and optimization modelling as the theoretical basics are described.

## 2   Theoretical Basics

This chapter will provide an overview of the theoretical basics of Re-Assembly, integrated PD and PPD, as well as PBCM.

### 2.1   Re-Assembly

SCHUH et al. propose that the remanufacturing process can be divided into two levels: the Re-Assembly and the refurbishment process. [5] The Re-Assembly comprises the following steps: disassembly, cleaning, inspection, Re-Assembly and final acceptance. In so-called "Re-Assembly factories," functional and value-enhancing measures are employed to upgrade and qualify used products for subsequent life cycles. [5].

### 2.2   Integrated Product and Process Development

The aim of the integrated development process is to harmonize PD and PPD. This approach is based on the iterative development process described by WLECKE et al. [10] and the industrialization of remanufacturing in HIP3E by HERMANN et al. [8]. The production and assembly processes are largely determined during the product development phase, while constraints on these processes influence PD. Interactions between PD, PPD, and Re-Assembly development must be identified and recorded to resolve conflicts early.

The integrated development process focuses on managing these interactions and is divided into four phases, based on product maturity and production/assembly flexibility. Transition to the next phase requires achieving a certain product maturity and reducing degrees of freedom in production and Re-Assembly. To accelerate development, numerous hypotheses are generated and re-evaluated across phases, leading to their expansion or rejection. [8, 10].

### 2.3 Process Based Cost Modelling (PBCM) and Optimization Modelling

PBCM is a tool for calculating production costs based on technical and operational parameters. It provides transparency regarding the impact of different parameters on overall costs and enables the monetary evaluation of potential optimizations. The model is primarily focused on cost estimation in early development phases and the assessment of alternatives in terms of materials, products and processes. For input product description are used which is transferred into technical parameters. These are modelled in terms of processing times, material and energy consumption, and machine failures. They are combined with company-specific parameters to create the input variables.[11, 12] Optimization models represent an abstraction of reality and are intended to find an optimal solution. The standard form of a linear optimization model comprises three key elements: decision variables, an objective function to be maximized or minimized, and constraints. The decision variables correspond to the decision space of the given decision situation and represent the degrees of freedom that exist and the decisions that are made by the model. [13].

## 3 State of the Art

The following section presents existing approaches from the literature that address the topic under consideration in this paper, followed by a brief evaluation of the approaches concerning the fulfillment of the objective of this paper.

DING et al. present a cost forecasting method for remanufacturing. Among other things, mathematical formulae for calculating remanufacturing costs are described. The remanufacturing costs are made up of the return costs, the costs in the process steps and the other costs. [14] DU et al. evaluate the recyclability of used machine tools, focusing on technological, economic, and ecological benefits. The initial step is to ascertain the technical feasibility of remanufacturing, followed by calculating costs, including those for returned cores, remanufacturing processes, and overheads. [15] XU et al. propose a procedure for forecasting remanufacturing costs, focusing on automotive components. They recommend a bottom-up approach, starting with delineating prerequisites and selecting a cost calculation method. The analysis focuses on cleaning, inspection, and reconditioning costs. [16] SCHAU et al. analyze life cycle costs for alternator remanufacturing, focusing on costs for both the remanufacturer and the user. This analysis aids in predicting costs, identifying savings, and supporting decision-making. [17] GHAZALLI and MURATA present an integrated evaluation system for selecting End-of-Life (EOL) strategies. The system first selects the EOL strategy at the product level and then determines strategies for individual components. The cost model includes labor costs for disassembly, sorting, cleaning, refurbishment, reassembly, and testing. [18] KRILL and THURSTON present a model for estimating the production, remanufacturing and disposal costs as well as the environmental impact of engine blocks. [19].

In summary, the reviewed approaches in Sect. 3 primarily focus on product remanufacturing, with varying degrees of detail and integration regarding the development of the Re-Assembly.

# 4  Concept

The cost model for evaluation of industrial Re-Assembly processes is based on the PBCM. It is used to validate the economics of different Re-Assembly hypotheses formulized in the development stage and must be embedded in the validation phase of the development. [8, 10] The results from the cost evaluation are used as a decision-making support. The model is divided in cost drivers and resources as input variables, the mathematical cost model, and the optimization model. For the cost model we assume that the number of returned products equals the number of sold upgrade products, assuming rejects occur only in reprocessing. Also costs which are not related to hypotheses or product manufacturing are excluded.

## 4.1  Cost Drivers and Resources

The input variables required to calculate Re-Assembly costs are divided into the categories "product-related input variables", "process-related input variables" and "dimensioning the Re-Assembly plant". Input variables derived from literature review [1–19] using the guidelines formulized by KITCHENHAM [20] and supplemented by expert interviews. The interview process includes a preliminary discussion in which the background to the question is specified. The interviews follow a guideline and includes intrinsic and extrinsic follow-up questions as well as interpretative questions [21].

### Product-related Input Variables
The product-related input variables refer to the respective product structure of the product hypothesis. The product structure is of significant importance in order to ascertain the proportion of components that can be utilized for reprocessing, the proportion that is generated as scrap following disassembly, and the proportion of purchased parts present within the product. This has a substantial impact on both the costs and the associated production process and the Re-assembly.

### Process-related Input Variables
In the development of the Re-Assembly, alternative process chains are defined. In order to evaluate these in economic terms, process input variables need to be defined. These include specification times, personnel and space requirements (workstation and machine based), and acquisition costs.

### Dimensioning the Re-Assembly Plant
The evaluation of alternative dimensioning concepts is an essential part of the economic analysis. For this purpose, dimensioning involves forecasting the number of units and different demand scenarios. The goal of dimensioning is to determine the production resource requirements in terms of production equipment, personnel and space required to remanufacture a planned number of items.

## 4.2  Cost Model

The cost model is applied after the input variables are defined. Periods for economic amortization are taken into account via parameter changes. Following the PBCM processes related costs for the processes in the Re-Assembly must be identified. The Re-Assembly includes the five process steps disassembly, cleaning, inspection, reprocessing, and assembly (see Fig. 1), as well as support processes like transport equipment costs, additional space costs, procurement, disposal, transport, and storage costs.

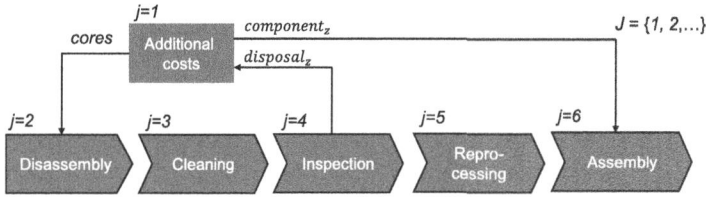

**Fig. 1.**  Processes in Re-Assembly

The control variable $j$ describes the process steps. Re-Assembly costs are calculated from the sum of the individual process costs (see Eq. (1)-(3)):

$$C_{Total} = \sum_{j=1}^{J} C_{Process}^{j} \tag{1}$$

$$C_{Total} = Total\ costs \tag{2}$$

$$C_{Process}^{j} = Tota,\ costs\ in\ j \tag{3}$$

The costs in $j$ are made up of variable and fixed costs. According to the PBCM framework  (see chapter 2.3), the variable costs in $j$ result from the sum of material costs, personnel costs and energy costs. Additionally, the fixed costs in $j$ result from the sum of equipment costs, area costs and maintenance costs for each process (see Eq. (4)-(6)):

$$C_{Process}^{j} = C_{variable}^{j} + C_{fix}^{j} \tag{4}$$

$$C_{variable}^{j} = C_{Material}^{j} + C_{Personnel}^{j} + C_{Energy}^{j} \tag{5}$$

$$C_{fix}^{j} = C_{Equipment}^{j} + C_{area}^{j} + C_{Maintenance}^{j} \tag{6}$$

For each of the considered process steps of the Re-Assembly (see Fig. 1), corresponding cost functions are developed to calculate the fix costs and variable cost in each respective area. As an example, the cost functions for the disassembly process as one of the process steps are derived in the following. A similar structure was also developed for the other four steps within the Re-Assembly.

**Variable Costs for Disassembly**

First, the variable costs of disassembly are considered (see Eq. (7)). Since there are no additional *material costs* incurred in the disassembly processes, these are not relevant here. The *personnel costs* consist of the number of stations, the personnel required per station, the annual costs per employee, and a safety factor, accounting for absences due to vacation and illness:

$$C_{Personnel}^{j=Dis} = NOS^{j=Dis} \times N_{Personnel/Station}^{j=Dis} \times S_{Personnel}^{j=Dis} \times P_{personnel} \tag{7}$$

$$N_{personnel/station}^{j=Dis} = Personnel\ requirements\ per\ station$$

$$NOS^{j=Dis} = Number\ of\ disassembly\ stations; \quad S_{Personnel}^{j=Dis} = Safety\ factor$$

$$P_{personnel} = Annual\ costs\ per\ employee\ (incl.\ non-wage\ labor\ costs)$$

The *energy costs* in the case of automated disassembly are calculated according to the following formula (8–9). Since the maximum power is not utilized over the entire usage time but typically at a rate of 0.6 to 0.8 [22], an efficiency factor is considered:

$$C_{Energy}^{j=Dis} = P_{Inst.Power}^{j=Dis} \times k_{Efficiency,Power}^{j=Dis} \times reqOT_{corrected}^{j=Dis} \times p_{Energy} \tag{8}$$

$$reqOT_{korrigiert}^{j=Dis} = ReVol \times CycleTime^{j=Dis} \times \frac{1}{z_G^{j=Dis} \times k_N^{j=Dis}} \tag{9}$$

$$ReVol = Quantity\ produced\ per\ Year; \quad reqOT^{j=Dis} = Annual\ required\ disassembly\ capacity$$

$$CycleTime^{j=Dis} = Disassembly\ target\ time\ per\ product; \quad z_G^{j=Dis} = Time\ degree\ in\ j$$

$$k_N^{j=Dis} = Degree\ of\ utilization\ in\ j; \quad C_{Energy}^{j=Dis} = Energy\ costs\ (here:\ Annual\ electricity\ costs)$$

$$P_{Inst.Power}^{j=Dis} = Maximum\ installed\ power\ [kW]; \quad p_{Energy} = Electricity\ costs\ [€/kWh]$$

**Fix Costs for Disassembly:**

Based on the number of required disassembly stations and the necessary space per disassembly station, the total space requirement is determined. The *space costs* are calculated as the product of the space requirement per station, the number of stations, and the annual space costs per square meter (see Eq. (10)):

$$C_{building}^{j=Dis} = R_{building} \times NOS^{j=Dis} \times Area_{NOS}^{j=Dis} \tag{10}$$

$$R_{building} = Annual\ floor\ space\ costs\ per\ m^2; \quad NOS^{j=Dis} = Number\ of\ disassembly\ stations$$

$$Area_{NOS}^{j=Dis} = Space\ requirement\ per\ station\ in\ m^2$$

The total *equipment costs* are calculated as the product of the annual costs per disassembly station and the number of disassembly stations. The equipment costs include both the costs for machinery and equipment as well as the costs for manual workstations:

$$R_{building} = Annual\ floor\ space\ costs\ per\ m^2$$

$$C_{Equipment}^{j=Dis} = Total\ equipment\ costs$$

$$R_{Equipment}^{j=Dis} = Annual\ equipment\ costs\ per\ disassembly\ station$$

The *maintenance costs* can be initially estimated by multiplying a maintenance factor with the annual equipment costs (see Eq. (11)):

$$C_{Maintenance}^{j=Dis} = C_{Equipment}^{j=Dis} \times Z^{j=Dis}$$

$$C_{Maintenance}^{j=Dis} = Maintenance\ costs;\ Z^{j=Dis} = Maintenance\ factor \qquad (11)$$

Subsequently, the total costs for each hypothesis must be calculated by applying the different input variables for all functions of the different process steps ($j = 1,...j = 6$) and for each hypothesis. The results allow for an economic comparison to be made, which in turn informs the decision as to whether a hypothesis should be continued or rejected for the next development phase.

**Optimization Model**

To support the cost modelling a linear optimization model in the direct comparison or integrated consideration of hypotheses formulated. First the cost model is adapted to form of a linear optimization model. Based on this, various decision situations are modelled. For example, the optimization model can be used to determine the profit-maximizing number of units. Binary decision variables are defined for each hypothesis and the cost functions are supplemented by the cost shares of all hypotheses so that a versatile solution space is spanned. The model, considering restrictions, delivers the optimal solution and determines which hypotheses should be developed further or rejected. The main difference between this model and the cost model is that several hypotheses are considered in an integrated manner.

# 5   Summary and Outlook

In this paper, a cost model is developed to evaluate the economic efficiency of industrial Re-Assembly processes as well as the economic validation of various alternatives. The fundamentals presented suggest that an integrated development of product, production process, and Re-Assembly, as well as early monitoring and control of costs, are important for the development of economical Re-Assembly processes. The cost model is based on input variables derived from the formulation of various development hypotheses. In addition, there is a mathematical cost model and an optimization model that supports the decision-making process. Further work is required to provide a more detailed account of the presented model and validated in industry.

**Acknowledgement.** Funded by the Deutsche Forschungsgemeinschaft (DFG, German Research Foundation) under Germany's Excellence Strategy – EXC-2023 Internet of Production – 390621612.

# References

1. Edinger-Schons, et al.: Sustainability transformation monitor (2023)
2. Ellen MacArthur foundation: towards the circular economy (2014)
3. Neugebauer, R.: Handbuch ressourcenorientierte Produktion. Hanser (2015)
4. Parker, et al.: Remanufacturing market study (2015)
5. Schuh, et al.: Green re-assembly upgrade factory (2023)
6. Verein Deutscher Ingenieure e. V.: Ressourceneffizienz durch Remanufacturing (2017)
7. Nasr, et al.: Remanufacturing: a key enabler to sustainable product systems (2006)
8. Hermann, et al.: Industrialization of remanufacturing in the highly iterative product and production process development (HIP3D) (2023)
9. Ehrlenspiel, K.: Kostengünstig Entwickeln und Konstruieren. Kostenmanagement bei der integrierten Produktentwicklung (2013)
10. Wlecke, et al.: Reducing time-to-market in the highly iterative and integrated product and production process development (HIP3D) (2022)
11. Fuchs, et al.: Process-based cost modeling of photonics manufacture (2006)
12. Johnson, M., Kirchain R.: Quantifying the effects of parts consolidation and development costs on material selection decisions: a process-based costing approach (2009)
13. Suhl, L., Mellouli T.: Optimierungssysteme. Modelle, Verfahren, Software, Anwendungen (2013)
14. Ding, et al.: A big data based cost prediction method for remanufacturing end-of-life products (2018)
15. Du, et al.: An integrated method for evaluating the remanufacturability of used machine tool (2012)
16. Xu, et al.: Remanufacturing and whole-life costing of lightweight components lightweight composite structures in transport (2016)
17. Schau, et al.: Life cycle costing in sustainability assessment—a case study of remanufactured alternators (2011)
18. Ghazalli, Z., Murata, A.: Development of an AHP–CBR evaluation system for remanufacturing: end-of-life selection strategy (2011)
19. Krill, M., Thurston, D.L.: Remanufacturing: impacts of sacrificial cylinder liners. J. Manuf. Sci. Eng. (2005)
20. Kitchenham, B.: Guidelines for performing systematic literature reviews in software engineering **2** (2007)
21. Meuser, M, Nagel, U.: Handbuch Frauen- und Geschlechterforschung. Theorie, Methoden, Empirie (2010)
22. Bauer, J.: Controlling für Industrieunternehmen. Kompakt und IT-unterstützt — Mit SAP®-Fallstudie (2006)

# Comprehensive Assessment of Remanufacturing Suitability and Enhancement

Tobias Lachnit[✉], Ida Vetter, Charlotte Braun, Thilo von Glasenapp, Jahn Feng, Martin Benfer, and Gisela Lanza

Wbk Institute of Production Science, Karlsruhe Institute of Technology (KIT), Kaiserstraße 12, 76131 Karlsruhe, Germany

tobias.lachnit@kit.edu

**Abstract.** For many companies, entering the field of circular value creation is a major challenge. They have difficulty in effectively assessing and quantifying the suitability and profitability of their products for remanufacturing and related strategies, which hinders their ability to make strategic decisions and implement necessary improvements. This work therefore proposes a comprehensive set of evaluation criteria to effectively quantify the suitability and profitability of products for remanufacturing and related circular economy strategies and to compare these within the product portfolio. These criteria focus on assessing the remanufacturability of various mechatronic products and evaluating their economic, ecological, and social sustainability. For easy and efficient assessment, a tool has been built that evaluates all criteria and product characteristics and provides recommendations for product improvement. With a remanufacturing score, the application makes it easy to analyse many products and compare their suitability. Additional options, such as the upgradeability of remanufactured products and the integration of remanufactured components into new production, can be used to investigate ways of increasing profitability. This research shows differences in sustainability and remanufacturability across different companies and product-specific potentials for improvement. The application provides companies with a simple assessment of their product's remanufacturing suitability and specific suggestions for improving efficiency.

**Keywords:** Sustainability Assessment · Circular Production · Remanufacturing · Product Upgrade

## 1 Introduction

The linear economic model's exclusive focus on efficiency is insufficient to address the finite nature of resources and conflicts with the goals of resource conservation and environmental protection. This highlights the urgent need for a transition to a circular economy [1]. Therefore, the importance of the R-strategies, particularly remanufacturing, to use resources more efficiently, extend the service life of products and minimize waste, is increasing. However, companies currently face challenges in objectively evaluating the suitability of their products for remanufacturing and in quantifying the economic,

© The Author(s) 2025
H. Kohl et al. (Eds.): GCSM 2024, LNME, pp. 572–579, 2025.
https://doi.org/10.1007/978-3-031-93891-7_63

environmental and social benefits. This paper presents a systematic methodology for assessing the remanufacturability of products, with an Excel tool serving as a central component.

## 2  State of the Art

The objective of a circular economy is to maximise value retention, if possible, first at the product level, then at the component level and finally at the material level [2]. The R-strategies represent the value-preserving processes that aim to maximise product, component and material value retention [3]. This paper generally focuses on the reprocessing of products as presented by Lickert et al. [4]. In detail, the R-strategies remanufacturing, remanufacturing upgrade, reintegration, and refurbishment are examined and enable value retention at the component level.

Remanufacturing refers to the reprocessing of used products or components. During the remanufacturing process, the products are completely dismantled, all components are inspected, worn components are replaced, and, if necessary, modernised. [5, 6] Remanufacturing promises the highest value retention at the component level and a quality level equivalent to a new product [3].

Remanufacturing upgrade combines the remanufacturing of a product with an upgrade process, thus improving product performance as part of the remanufacturing process [7]. In addition to the remanufacturing process, outdated components of the used product are brought up to date by incorporating technological advances and adapting the product to new customer needs [8].

As part of this paper, the R-strategy reintegration is introduced as a further value retention strategy of the circular economy. Reintegration contains the reuse of remanufactured components or assemblies within new production. The component or assembly undergoes a regular remanufacturing process and is afterwards integrated into new production of the subsequent product generation, a product within the same product family or a completely different product. The component or assembly can also be upgraded before reintegrating into new production. This approach can also be found in the work of other researchers. Hegedűs & Longauer [9] describe how, as part of product generation planning, both remanufactured and new components can be used within new production of a subsequent product generation and, according to this concept, the remanufactured components become the raw materials within new production. Following this, Sawyer-Beaulieu and Tam [10] recognise using "factory-approved" remanufactured components and new parts to produce new vehicles. Furthermore, IEC 62309 (DIN EN 62309 (VDE 0050) since 2005) [11], provides a concept to check the reliability, functionality and usage of "qualified-as-good-as-new" ("quagan") components within new products of the electronic, electro-mechanical and mechanical industry. According to DIN EN 62309 (VDE 0050) [11], a "new product" may contain one or more "qualified-as-good-as-new" components. The term "reintegration" for the R-strategy explained above has evolved from Kalverkamp et al. [12], who describe reusing used, but remanufactured components within new production. However, the term is not defined in more detail [12].

The R-strategy refurbishing differs from remanufacturing in terms of the achievable quality standard of the product after the process has been carried out. The quality standard for refurbished products is less stringent than for new products. [6].

Existing methods for determining the remanufacturing suitability of products are mainly used in product development. These are intended to identify necessary improvements that can still be implemented in the design process. The evaluation of methods within product development is usually qualitative, for example, using a three level scale (green, yellow, red), and no overall remanufacturing score is automatically displayed [13]. Existing remanufacturing scores based on quantitative key figures of the remanufacturing process steps [14, 15] are very mathematical and theoretical and neglect ecological, social and legal issues as well as market criteria, which are also important for calculating the remanufacturing suitability of a product. A related approach to analyse the suitability for a specific R-strategy based on decision trees is presented by Dvorak et al. [16]. A detailed analysis of methods, assessing the economic profitability of remanufacturing, it is referred to Vogt Duberg et al. [17]. Our approach differs from past assessment methodologies in terms of its comprehensive qualitative and quantitative assessment and its user-friendly application. An integrated financial calculation provides an initial evaluation of the economic viability of remanufacturing.

## 3   Research Scope

This paper aims to develop a holistic method for identifying and prioritizing products suitable for remanufacturing, including qualitative and quantitative criteria. The evaluation includes technical, economic, ecological, legal and social aspects of remanufacturing. Furthermore, the examination of remanufacturing is extended by evaluating the suitability for remanufacturing upgrade, reintegration and refurbishment. Comprehensive, precisely formulated product criteria enable a well-founded and reliable evaluation of the product suitability for the respective R-strategy. A separate financial calculation is provided within the tool to check the economic viability of the examined R-strategies. The method enables companies to assess existing products in their product portfolios with regard to their remanufacturing suitability. The automated output of the total scores for the various R-strategies contributes significantly to the tool's user-friendliness and understandability.

RQ1: Which technical, economic, ecological, legal and social requirements should a product fulfil to determine its suitability for remanufacturing, and how can companies evaluate these in an effective, quick and efficient way?

Existing methods do not provide specific instructions to increase the calculated remanufacturing score of a product. This deficit will be eliminated in the newly developed method by integrating a toolbox with recommendations for action. These suggestions are intended to enhance the generated remanufacturing score to improve economic and ecological sustainability in the long term.

RQ2: Based on the assessment, which specific guidance can be proposed to improve the remanufacturing suitability of a product?

## 4   Remanufacturing Assessment

A structured procedure is developed to evaluate the product concerning remanufacturing, as outlined in Fig. 1. Initially, scores for the analysed R-strategies are calculated based on the input data entered by the user. Technical, economic, ecological, legal and social criteria calculate a comprehensive remanufacturing or refurbishment score. If the user also evaluates the remanufacturing upgrade and reintegration criteria, additional scores for the remanufacturing upgrade and reintegration can be issued. The developed toolbox includes specific control variables based on product properties to improve the beforehand calculated remanufacturing score.

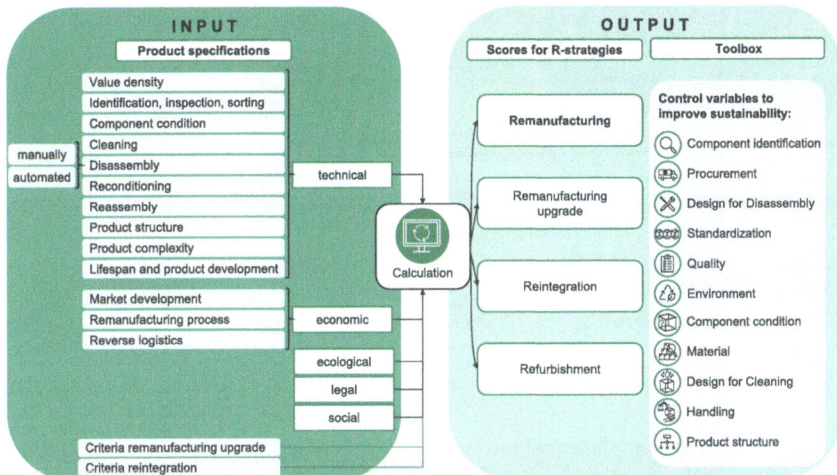

**Fig. 1.** Structure for the assessment of remanufacturing suitability

## 4.1 Methodology

Comprehensive evaluation criteria covering the entire process are essential to accurately assess a product's suitability for remanufacturing. A literature review was carried out to identify relevant references. Most of the criteria are based on the following contributions: Herrmann & Vetter [18], James et al. [19], Shabazi et al. [20] and Steinhilper & Hudelmaier [21]. In addition, further key criteria are added with the help of experts. The criteria are split into categories, as pictured in Fig. 1. These categories target the different steps of the remanufacturing process. Some of the criteria are exclusion criteria [22]. If those are not fulfilled, this product's respective strategy is impossible. The tool uses basic functions to assess product suitability for remanufacturing through utility analysis as described by Haag et al. [23]. The suitability of products for remanufacturing is calculated with the help of a utility analysis [23]. For every strategy, a score is displayed to predict the product's suitability. This score is calculated as sum of ratings of the different criteria multiplied by their weights. Finally, a score is calculated for remanufacturing, remanufacturing with product update and reintegration, that indicates the suitability, is calculated.

Furthermore, the tool provides a financial analysis, which allows the estimation of the overall profitability of the remanufacturing process. The key metrics are derived from the process parameters and the associated costs. Figure 2 provides an overview of all calculations, including specific input and output parameters.

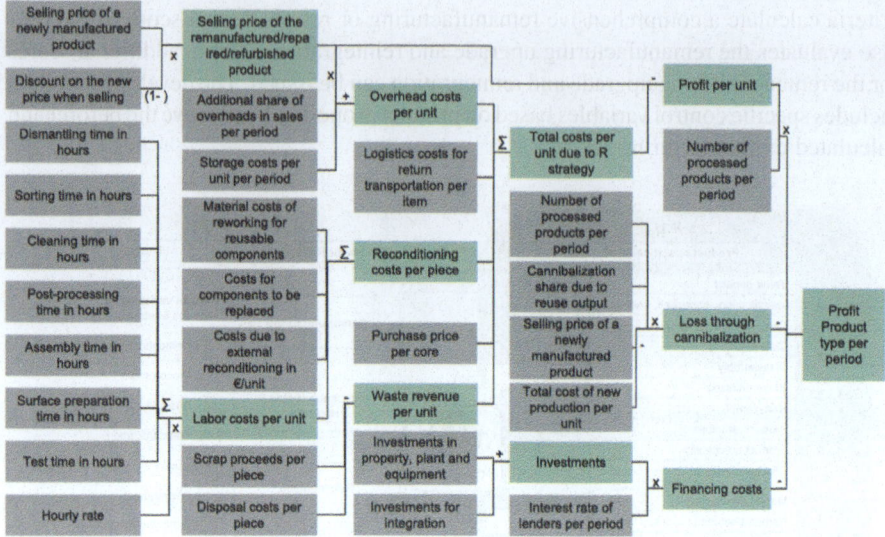

**Fig. 2.** Overview of financial calculation

After its development the excel tool was tested several times with different products. In this way errors that occurred were eliminated and iterative improvements were made. The exemplary implementation for a product is described in Sect. 5.

### 4.2 Tool Structure

Section 4.1 presented the methodology, which is then implemented in the spreadsheet program Microsoft Excel. The tool comprises eight worksheets: (1) an introduction and a brief manual about the usage of the tool, (2) a summary of the results, (3) & (4) sheets for the structured entry of the necessary data for the remanufacturing suitability evaluation as well as the financial calculations, (5) a detailed description of all control variables including goal and recommendations, (6) a sheet to adjust the weights of the criteria, (7) a sheet containing the detailed lists of the references to the criteria and (8) a sheet aiding in the functionality of the tool ("backengine") [24]. The input data is entered in the respective sheets. The user is guided in this process by remarks in the form of input messages. The utility analysis and the calculations are performed automatically with the help of worksheet functions and based on the values put in the weights sheet. The initial values for the weights are based on best practices but are adjustable at the user's discretion. The results are then consolidated and displayed on the results sheet. To aid the user, the most important values are also illustrated as graphs. A message warns that remanufacturing is impossible if one or more exclusion criteria are triggered. The criteria prohibiting product remanufacturing are consolidated in a recommendation section on the results sheets. In addition, improvable control variables are automatically determined and displayed on the results sheet. This is complemented by a function with which one of the control variables can be selected, and the relevant information (from the respective control variable fact sheet) is displayed. As a result, the user can be provided with targeted suggestions to improve the remanufacturing capability of the products assessed within the tool. The toolbox contains the following eleven control variables: Component identification, Component condition, Material, Procurement, Design for Cleaning, Design for Disassembly, Handling, Standardization, Product structure, Quality and Environment. All fact sheets describe the goal(s) of the respective control variable and key recommendations for action and measures to achieve the goal(s). Furthermore, literature recommendations are provided to deepen the selected improvement potentials, should the tool user require a more profound understanding. This toolbox thus supports the creation of a product with enhanced suitability for remanufacturing.

## 5  Case Study Validation and Results

For validation, the tool was used to assess electrical construction equipment. An expert of the company conducted a comprehensive evaluation of all product criteria and completed all parameters. This was followed by an evaluation in which any difficulties, problems and ambiguities were discussed with the responsible company employee. In addition, the results generated by the tool regarding the remanufacturability of the product were analysed and discussed with the manufacturer of the product. This process aimed to identify tool errors, user ambiguities and any missing remanufacturing aspects. The company's expert analysed the criteria using its products and validated that they effectively represent the remanufacturing process. The financial calculations proved reliable, closely matching detailed internal analysis results. All findings were subsequently incorporated into the tool. The tool suggests remanufacturing as the suitable R-strategy for the analysed product, excluding remanufacturing upgrade and reintegration due to unmet exclusion criteria. The remanufacturing score is 64.45%, indicating good performance but room for improvement. The product excelled in product criteria and reverse logistics (92%) and Product Structure (83%) but scored lower in identification, inspection, sorting and cleaning (20%). No control variables were flagged for improvement, suggesting that the product performed adequately in all categories with no significant deficiencies. To further improve the product, focus should be placed on areas with lower scores like e.g. the traceability of the product across the value chain. The financial analysis indicates that remanufacturing yields minimal profit, primarily due to the high cost of replacement components. No other anomalies were found in the results.

## 6  Summary and Outlook

The assessment application facilitates an efficient and quick determination of suitability for remanufacturing based on 188 individual criteria including in particular ecological, social and legal issues. The developed and publicly available tool [24] makes it easy to assess products' suitability for remanufacturing, product upgrade, refurbishment and integration of used components into the new production. The assessment offers specific guidance for improving a product's suitability for remanufacturing by integrating a toolbox proposing targeted recommendations designed to improve the overall remanufacturing score. The tool has been validated and improved in collaboration with several companies. In companies, the tool achieved equivalent results with significantly less effort than a comprehensive study. The extension to additional R-strategies could be incorporated to further improve the assessment. In addition, the specialization in specific product categories could improve the accuracy. Furthermore, there is a potential to recommend product improvements regarding remanufacturing suitability using artificial intelligence within the tool. Product improvement suggestions were generated in the first trial using a large language model and a substantial dataset. This approach has the potential to quickly provide companies with suggestions for product improvement.

**Acknowledgement.** This research work was undertaken in the context of the IGF-Project "IntroRemanNet" (Nr. 22626 N). The project is funded by the Federal Ministry for Economic Affairs and Climate Action on the basis of a decision by the German Bundestag.

# References

1. Ellen MacArthur Foundation, Towards the circular economy Vol. 1: an economic and business rationale for an accelerated transition (2013)
2. Bocken, N.M.P., Olivetti, E.A., Cullen, J.M., Potting, J., Lifset, R.: Taking the circularity to the next level: a special issue on the circular economy. J. Ind. Ecol. **21**(3), 476–482 (2017)
3. DIN SPEC 91472: Remanufacturing (Reman) - Qualitätsklassifizierung für zirkuläre Prozesse (2023)
4. Lickert, H., Görgens, S.J., Meyer, K., Dietrich, F.: Framework for mapping and developing closed loops in Urban Areas. Procedia CIRP **122**(2024), 360–365 (2024)
5. Sundin, E.: Product and process design for successful remanufacturing. Zugl.: Linköping, Univ., Diss. (2004)
6. Thierry, M., Salomon, M., van Nunen, J., van Wassenhove, L.: Strategic issues in product recovery management. Calif. Manage. Rev. **37**(2), 114–136 (1995)
7. Jukun, Y., Sheng, Z., Xiaoming, W., Peizhi, C.: Remanufacturing upgrade theory and technology system facing product multi-life cycle. In: 2014 Sixth International Conference on Measuring Technology and Mechatronics Automation (S. 472–475), IEEE (2014)
8. Peizhi, C., Jukun, Y., Sheng, Z.: Information-based remanufacturing upgrade study. In: 5th International Conference on Responsive Manufacturing - Green Manufacturing (ICRM 2010) (S. 32–37), IET (2010)
9. Hegedűs, D., Longauer, D.: Implementation of a circular supply chain model using reusable components in multiple product generations. Heliyon **9**(5), e15594 (2023)
10. Sawyer-Beaulieu, S., Tam, E.: Maximizing automotive parts reuse, remanufacturing, and recycling through effective end-of-life vehicle management: a different perspective on what needs to be done. SAE Int. J. Mater. Manuf. **8**(1) (2015)
11. DIN EN 62309 (VDE 0050): Zuverlässigkeit von Produkten mit wieder verwendeten Teilen Anforderungen an Funktionalität und Prüfungen (IEC 62309:2004) (2005)
12. Kalverkamp, M., Pehlken, A., Wuest, T.: Cascade use and the management of product lifecycles. Sustainability **9**(9), 1540 (2017)
13. Ahlstedt, E., Sundin, E.: Assessing product suitability for remanufacturing – a case study of a handheld battery-driven assembly tool. Procedia CIRP **116**, 582–587 (2023)
14. Bras, B., Hammond, R.: Towards design for remanufacturing: metrics for assessing remanufacturability (1996)
15. Chakraborty, K., Mondal, S., Mukherjee, K.: Analysis of product design characteristics for remanufacturing using Fuzzy AHP and axiomatic design. J. Eng. Des. **28**(5), 338–368 (2017)

16. Dvorak, J., Stanzl, L., Lachnit, T., Benfer, M., Balzereit, F., Lanza, G.: On the systematic selection of CE strategies for end-of-life products: a guide for practitioners (in press)
17. Circularity Days. Wolfsburg, Germany
18. Vogt Duberg, J., Sundin, E., Tang, O.: Assessing the profitability of remanufacturing initiation: a literature review. J. Remanuf. **14**, 69–92 (2024)
19. Herrmann, C., Vetter, O.: Ökologische und ökonomische Bewertung des Ressourcenaufwands - Remanufacturing von Produkten, VDI Zentrum Ressourceneffizienz GmbH (2021)
20. James, A.T., Kumar, G., Arora, A., Padhi, S.: Development of a design based remanufacturability index for automobile systems. Proc. Inst. Mech. Eng. Part D: J. Autom. Eng. **235**(12), 3138–3156 (2021)
21. Shahbazi, S., Johansen, K., Sundin, E.: Product design for automated remanufacturing—a case study of electric and electronic equipment in sweden. Sustainability **13**(16), 9039 (2021)
22. Steinhilper, R., Hudelmaier, U.: Erfolgreiches Produktrecycling zur erneuten Verwendung oder Verwertung: Ein Leitfaden für Unternehmer (IPA Projekt-Nr.: 104 850). Eschborn (1993)
23. Kühnapfel, J.B.: Nutzwertanalysen in Marketing und Vertrieb, Springer Fachmedien Wiesbaden; Imprint: Springer Gabler, Wiesbaden (2019)
24. Haag, C., Schuh, G., Kreysa, J., Schmelter, K.: Technologiebewertung in Technologiemanagement. In: Hrsg, G., Schuh, S., Klappert, G. Schuh, Springer, Berlin, pp. 309–366 (2011)
25. Lachnit, T., Braun, C., Vetter, I., Feng, J., von Glasenapp, T.: Remanufacturing Suitability Evaluation Tool. Zenodo (2024). https://doi.org/10.5281/zenodo.12483041

# Evaluation of Disruptive Business Models for Remanufacturing

Paul Molenda[1]([✉]) [iD], Jan Koller[2] [iD], Vihasith Ganga[1,2] [iD], and Anke Müller[1] [iD]

[1] Hof University, Alfons-Goppel-Platz 1, 95028 Hof, Germany
paul.molenda@hof-university.de
[2] Fraunhofer Institute for Manufacturing Engineering and Automation IPA, Universitaetsstrasse 9, 95447 Bayreuth, Germany

**Abstract.** The circular economy is an approach that aims to restore products and materials after their usage. Remanufacturing, as an element of the circular economy, enables sustainability through multiple product utilization cycles. The increasing digitalization, as well as the availability and exchange of information along the product life cycle, from its manufacture to the end of its life, offers new business opportunities for the remanufacturing industry. Instead of the traditional and established sale of remanufactured products, product-service systems are becoming more important. As a result, it is possible to open new markets and increase the acceptance of remanufactured products. This paper focuses on analyzing which disruptive business models will significantly impact the remanufacturing industry in the future. For this purpose, a systematic analysis of existing business models based on a literature review and the Business Model Navigator is first carried out. Disruptive business models for the remanufacturing industry are then evaluated with the help of the Business Model Canvas and PORTER'S Value Chain. Their applicability to the remanufacturing industry is then determined based on established criteria.

**Keywords:** Disruptive Business Models · Remanufacturing · Circular Economy

## 1 Introduction

Remanufacturers face growing challenges in a business environment characterized by dynamic framework conditions. Key factors driving this transformation include the ongoing progression of globalization and digitalization, demographic shifts, decarbonization of the industry, and the transition to a circular economy [1]. As a result, the business environment has become more volatile, uncertain, complex, and ambiguously, referred to as "VUCA" [2]. Succeeding in this VUCA environment requires remanufacturers to optimize and adapt their existing organizational structures, processes, and skills [1]. However, in most industries today, it is no longer enough to focus purely on product or process innovation [3]. In the long-term, remanufacturers must reassess their business strategies and actively pursue new opportunities, such as implementing disruptive business models, to foster innovation and resilience [1, 4].

H. Kohl et al. (Eds.): GCSM 2024, LNME, pp. 580–589, 2025.
https://doi.org/10.1007/978-3-031-93891-7_64

## 2   State of the Art

### 2.1   Remanufacturing

Remanufacturing is an industrial process in which used products are returned to at least the performance and quality level of the initial product with a full warranty, thus enabling an entirely new product life cycle [5–7]. Remanufacturing offers economic, ecological, and social advantages compared to other value retention processes, like repair or recycling [6]. Studies show that remanufacturing can lead to a resource requirement of only 10% compared to new parts production [7, 8]. Hence, by remanufacturing, up to 90% of energy and 55% of the material can be returned to the product life cycle [8, 9]. Furthermore, costs can be reduced by retaining the product shape and reducing the use of new materials. Therefore, the prices for remanufactured products are up to 40% lower than those of equivalent new products [9].

According to STEINHILPER, the standard remanufacturing process consists of five sequential steps, embedded in quality checks: (1) disassembly, (2) cleaning, (3) inspection and sorting, (4) reconditioning/replacement, and (5) reassembly [7]. The remanufacturing process can be performed by the original equipment manufacturer (OEM) or by a contracted or independent remanufacturer.

### 2.2   Disruptive Business Models

Disruptive business models are innovative concepts that fundamentally change the way industries operate. They often use new technologies, unique value propositions, and different cost structures to offer better or cheaper products and services. These business models challenge market leaders and can significantly change consumer behavior and market dynamics. New entrants can gain significant market share by offering superior or more affordable products or services while consumers enjoy easier access, lower prices, and greater variety. Table 1 shows an overview of selected disruptive business models. As these business models continue to evolve, they will shape the future of business and redefine the competitive landscape in various sectors. [10–13].

## 3   Review of Disruptive Business Models in Remanufacturing

First, a literature review is conducted to identify descriptions and implementations of disruptive business models in remanufacturing. Second, the 60 business models of the Business Model Navigator are examined for their applicability in remanufacturing.

### 3.1   Literature Review of Disruptive Business Models in Remanufacturing

The literature review aims to answer the research question: *Which disruptive business models in remanufacturing are described in the literature?* Therefore, the databases *IEEE Xplore, ScienceDirect, Scopus,* and *Google Scholar* were searched. The search is limited to the titles, as full-text searches often yield many unsuitable results [26]. Different spellings related to "Business Model" OR "Disruptive Business Model" OR "Innovative

**Table 1.** Overview of selected disruptive business models based on [3].

| Disruptive business model | Description | Examples |
|---|---|---|
| Platform-Based | Platforms connect consumers directly with service providers, cutting out traditional intermediaries | Amazon, Lyft, Uber, Airbnb, eBay, App Store |
| Subscription | Unlimited access to a product or service for a recurring fee | Adobe, Amazon Prime, Netflix, Spotify |
| Freemium | Basic services are provided for free, while advanced features or premium services are offered at a cost | Dropbox, LinkedIn, Duolingo, Zoom, Miro |
| Crowdsourcing and Crowdfunding | Leverage the collective power of a large group of people to fund or solve problems | Kickstarter, Indiegogo, Patreon |

Business Model" OR "Business Model Innovation" OR "Technology Innovation" OR "Technological Innovation" were combined by using the boolean operator AND with different spellings of "Remanufacturing" as the search query.

The literature review was conducted in May 2024 and initially resulted in 81 publications, including duplicates and those not meeting formal criteria. The formal criteria include considering only peer-reviewed publications in English with full-text access. After excluding duplicates (40), inaccessible publications (6), theses (6), and others (2), 27 publications remained.

The 60 disruptive business models of the Business Model Navigator were searched within full texts, excluding references. Additionally, use-oriented business models (Pool, Share, Lease, Rent, Pay-Per-Use, Performance-Based) were summarized into one category. This investigation resulted in 14 disruptive business models across 23 publications. The business models Customer Loyalty, E-Commerce, Licensing, Revenue Sharing, Subscription, Solution Provider, Push-to-Pull, Crowdfunding, and Reverse Engineering, as well as 12 publications were excluded due to less than five repetitions, indicating a lack of emphasis. Table 2 shows the result of the literature review.

As a result, the following five disruptive business models were identified for further evaluation regarding their applicability in remanufacturing industry and are briefly described below:

- **Auction**: Goods or services are sold to the highest bidder, with prices determined dynamically based on demand and competitive bidding among purchasers.
- **Digitization**: Physical goods, services, or business processes are transformed into digital formats, enhancing efficiency, accessibility, and scalability.
- **Mass Customization**: Combines mass production with individual customization, allowing personalized products/services while benefiting from economies of scale.

**Table 2.** Top 5 business models from the literature with the number of mentions.

| Disruptive business model | Bressanelli et al. [14] | Burggräf et al. [15] | Burggräf et al. [16] | Franceschi et al. [17] | Giacomo [18] | Koop et al. [19] | Lee & Kwak [20] | Mont et al. [21] | Shao et al. [22] | Souza [23] | Veleva & Bodkin [24] | Sum |
|---|---|---|---|---|---|---|---|---|---|---|---|---|
| Auction | | | | | | 1 | | | | | 18 | 19 |
| Digitization | 1 | 5 | | | | | | | | | | 6 |
| Mass Customization | | | | | | 6 | | | | | | 6 |
| Open Source | | | | | | 8 | | | | | | 8 |
| Use-oriented | 5 | 4 | 5 | 16 | 21 | 73 | 13 | 100 | 15 | 5 | 1 | 258 |

- **Open Source**: Products, usually software, are developed and made openly accessible to the public, often generating revenue from support and consulting services.
- **Use-oriented**: Provides access to a product rather than ownership, such as Leasing, Renting, Sharing, or Pooling, which can reduce costs and promote sustainability.

### 3.2 Expert-Based Review of Disruptive Business Models Based on the Business Model Navigator

An expert-based review was conducted to evaluate the potential relevance of additional disruptive business models of the Business Model Navigator to the remanufacturing industry, which were not yet addressed in the literature. Therefore, B2B-oriented business models were selected. Remanufacturing and business model experts rated them on a five-point scale from very low to very high based on criteria such as unique value propositions, innovative operational approaches, customer focus, and future-oriented characteristics. This expert-based review identified ten disruptive business models with "high" and "very high" potential for remanufacturing. According to the experts' assessment, five business models were excluded since they are already state-of-the-art practices or generic descriptions of the remanufacturing business model. This includes Integrator, Lock-in, Long Tail, Trash-to-cash, and Aikido. The remaining five disruptive business models were selected for further evaluation of their applicability in the remanufacturing industry and are briefly described below:

- **Barter**: Exchange of goods or services directly for other goods or services without the transaction of money, based on mutual agreement of value.
- **Franchising**: Franchisor grants a franchisee the rights to operate a business under the franchisor's brand and business system in exchange for fees and royalties.
- **Guaranteed Availability**: Service-level agreements often guarantee product or service availability to provide reliability and minimize downtime.
- **Leverage Customer Data**: Value is created from collecting and analyzing data, while revenue is generated by trading this data or utilizing it for internal purposes.
- **Sensor as a Service**: Deploys sensors to collect real-time data, which is then analyzed and provided to customers, enabling informed decision-making.

# 4  Evaluation of Disruptive Business Models in the Remanufacturing Industry

The final ten disruptive business models, identified through the literature review and the expert-based review, are then evaluated regarding their potential for remanufacturing. This evaluation is structured based on the Business Model Canvas for Remanufacturing by GUNASEKARA [25], and incorporated PORTER'S Value Chain [26] for a comprehensive company analysis. The PORTER'S Value Chain includes support activities, like "Firm Infrastructure" (FI), "Human Resource Management" (HR), "Technology Development" (T), and "Procurement" (P), and primary activities, like "Inbound Logistics" (IL), "Operations" (O), "Outbound Logistics" (OL), "Marketing and Sales" (MS), and "Service" (S). Each disruptive business model was assessed against these value chain activities to identify the focus areas. [26].

Subsequently, the experts' assessment provided examples of applying each disruptive business model within the remanufacturing context. The impact of these disruptive business models was assessed based on the proven criteria of time, costs, and quality, according to VDI-Guideline 2870 [27], extended by the dimensions of flexibility and sustainability [28]. Based on the expert assessments, ratings were assigned on a scale from low effect (1) to high effect (3). A zero effect (0) is assigned if the disruptive business model does not influence the criteria. This structured approach enabled a systematic expert-based evaluation and comparison of the potential of each disruptive business model in remanufacturing. Table 3 summarizes the results, detailing the assessments and examples for each identified disruptive business model.

The evaluation of the ten selected disruptive business models for remanufacturing shows a wide range of effects on time, cost, flexibility, sustainability, and quality. Disruptive business models emphasizing digitization and data usage yield broad benefits across all metrics, indicating significant potential for improving operational efficiency, sustainability, and product quality. Conversely, models like Auction and Barter, focusing on procurement and sales, offer limited benefits compared to digitally integrated models. Franchising and Guaranteed Availability show high efficiency and reliability, but Guaranteed Availability impacts sustainability differently.

Disruptive business models utilizing advanced technologies, such as sensor-based services, provide moderate to high benefits, particularly in real-time decision-making and operational optimization. Use-oriented models excel in flexibility, whereas Mass Customization, characterized by personalization, faces challenges in cost efficiency.

Impactful disruptive business models leverage digital technologies and data analytics, seamlessly integrating into various value chain activities to enhance operational efficiency, product quality, sustainability, and adaptability. To maximize benefits, businesses should prioritize adopting digital technologies and data-driven models, developing networks through franchising, leveraging customer data, ensuring product availability, and integrating sensor technologies. Additionally, use-oriented disruptive business models can improve cost-effectiveness and flexibility in targeted segments of the remanufacturing market. Focusing on these areas can enhance remanufacturing operations' effectiveness, sustainability, and adaptability, tailored to specific needs and value chain activities.

**Table 3.** Overview of identified disruptive business models in remanufacturing, their focus in the value chain, examples of their application, and evaluation.

| Disruptive business model | Value chain focus | Examples | Time efficiency | Cost effectiveness | Flexibility | Sustainability | Quality |
|---|---|---|---|---|---|---|---|
| Auction | P, MS | – Selling a bunch of cores<br>– Reverse auctions of remanufactured products<br>– Usage as an additional sales channel | 0 | 2 | 1 | 2 | 0 |
| Barter | P, MS | – Exchange of remanufactured products for used products or components<br>– Cooperation with a preferred supplier | 1 | 3 | 1 | 1 | 0 |
| Digitization | FI, T, P, IL, O, MS, OL, S | – Digital product passport of a turbo charger<br>– Digital product twin of an engine<br>– Digital production twin of the remanufacturing process | 3 | 3 | 3 | 3 | 3 |
| Franchising | FI, T, P, O, MS, S | – OEM provides a license for remanufacturing<br>– Network of independent remanufacturers | 2 | 3 | 2 | 2 | 2 |
| Guaranteed Availability | P, O, MS, OL | – Delivery of remanufactured products within 24 h<br>– Preferred customer service | 3 | 3 | 3 | 1 | 1 |

(*continued*)

**Table 3.** (*continued*)

| Disruptive business model | Value chain focus | Examples | Time efficiency | Cost effectiveness | Flexibility | Sustainability | Quality |
|---|---|---|---|---|---|---|---|
| Leverage Customer Data | T, P, IL, O, MS, OL, S | – Design for remanufacturing (use data for future product development)<br>– Use usage data for predictive maintenance and recommendations for exchange | 2 | 2 | 2 | 2 | 3 |
| Mass Customization | T, O, MS | – Personalization of remanufactured products<br>– Use of new materials or components in remanufacturing (e.g. upgrades) | 0 | 1 | 3 | 1 | 0 |
| Open Source | T, MS, S | – Collaborative design for remanufactured products<br>– Open-source platform for information sharing on product and process specifications | 2 | 2 | 2 | 1 | 2 |
| Sensor as a Service | T, P, IL, O | – Condition analysis of cores<br>– Optimize operating settings for machines | 2 | 2 | 2 | 2 | 3 |
| Use-oriented | T, P, MS, S | – Leasing a car and remanufacturing its components after return<br>– Shared logistic activities for used and remanufactured products | 2 | 2 | 3 | 2 | 0 |

## 5   Conclusion and Outlook

This paper identified the ten most suitable disruptive business models for remanufacturing through a literature review and an expert-based review. The focus of those disruptive business models in the remanufacturing value chain was analyzed, application examples were outlined, and their potential impact on time, cost, quality, flexibility, and sustainability was assessed. The results show the importance of adopting digital technologies and data-driven approaches across the remanufacturing value chain. Disruptive business models that integrate digitization and leverage customer data show the most promise in driving operational efficiency, sustainability, and product quality, suggesting a direction for future industry practices. In addition, disruptive business models established in new parts production, such as franchising or use-oriented business models, should be adapted to remanufacturing to maximize their advantages.

However, further research is needed to understand these disruptive business models' long-term impacts and scalability. In addition, the business models must be adapted to the specific needs and dynamics of the respective industry. Also, there is a need to explore the integration of new technologies such as artificial intelligence into remanufacturing and to examine the socio-economic implications, including user acceptance and regulatory challenges. Addressing these areas will help remanufacturing companies to remain resilient in a VUCA world.

## References

1. Gaubinger, K.: Hybrides Innovationsmanagement für den Mittelstand in einer VUCA-Welt. Springer, Berlin, Heidelberg (2021)
2. Mack, O., Khare, A.: Perspectives on a VUCA World. In: Mack, O., Khare, A., Krämer, A., Burgartz, T. (eds.) Managing in a VUCA World, pp. 3–20. Springer, Cham (2016)
3. Gassmann, O., Frankenberger, K., Choudury, M.: The business model navigator: 55 models that will revolutionise your business. Pearson, Harlow, England (2014)
4. Bartscht, J.: Why systems must explore the unknown to survive in VUCA environments. Kybernetes **44**(2), 253–270 (2015)
5. British Standards Institution: BS 8887-2:2009: Design for manufacture, assembly, disassembly and end-of-life processing (MADE). Part 2: Terms and definitions, London (2009)
6. Deutsches Institut für Normung e. V.: DIN SPEC 91472: Remanufacturing (Reman) - Qualitätsklassifizierung für zirkuläre Prozesse. Beuth Verlag, Berlin (2023)
7. Steinhilper, R.: Produktrecycling: Vielfachnutzen durch Mehrfachnutzung. Fraunhofer IRB Verlag, Stuttgart (1999)
8. Köhler, D.C.F.: Regenerative Supply Chains. Shaker Verlag, Aachen (2011)
9. Colledani, M., Battaïa, O.: A decision support system to manage the quality of end-of-life products in disassembly systems. CIRP Ann. **65**(1), 41–44 (2016)
10. Guarda, T.: Technology, Business, Innovation, and Entrepreneurship in Industry 4. 0, 1st ed. Springer, Cham (2023)
11. Prostean, G.I., Lavios, J.J., Brancu, L., Şahin, F. (eds.): Management, Innovation and Entrepreneurship in Challenging Global Times: Proceedings of the 16th International Symposium in Management (SIM 2021), 1st ed. Springer, Cham (2024)
12. Siedhoff, S.: Seizing Business Model Patterns for Disruptive Innovations, 1st edn. Springer, Wiesbaden (2019)

13. Slama, D.: The AIoT Playbook: A Practitioner's Guide to Smart, Connected Products and Solutions, 1st edn. Springer, Cham (2022)
14. Bressanelli, G., Saccani, N., Perona, M.: Remanufacturing for the circular economy: a business model analysis. In: Proceedings of the Summer School Francesco Turco (2022)
15. Burggräf, P., Dannapfel, M., Wagner, J., Heinbach, B., Föhlisch, N., Dackweiler, J.: "Re-LIFE": business models for data-based remanufacturing: adaptive remanufacturing for life cycle optimisation of networked capital goods. In: The Monetization of Tech-nical Data: Innovations from Industry and Research, pp. 507–520. Springer (2023)
16. Burggräf, P., Wagner, J., Heinbach, B., Wigger, M.: Design of a methodological framework for adaptive remanufacturing-based business models. Procedia CIRP 98, 547–552 (2021)
17. Franceschi, B., Bressanelli, G., Saccani, N.: Remanufacturing and Product-Service Systems for the Circular Economy: A Business Model Analysis, in: Smart Services Summit, Cham. Springer, Cham, pp. 133–141 (2023)
18. Giacomo Copani, S.B.: Remanufacturing with upgrade PSS for new sustainable business models. CIRP J. Manuf. Sci. Technol. 29, 245–256 (2020)
19. Koop, C., Grosse Erdmann, J., Koller, J., Döpper, F.: Circular business models for Remanufacturing in the electric bicycle industry. Front. Sustain. 2 (2021)
20. Lee, S., Kwak, M.: Consumer valuation of remanufactured products: a comparative study of product categories and business models. Sustainability 12(18), 7581 (2020)
21. Mont, O., Dalhammar, C., Jacobsson, N.: A new business model for baby prams based on leasing and product remanufacturing. J. Clean. Prod. 14(17), 1509–1518 (2006)
22. Shao, J., Huang, S., Lemus-Aguilar, I., Ünal, E.: Circular business models generation for automobile remanufacturing industry in China: barriers and opportunities. J. Manuf. Technol. Manage. 31(3), 542–571 (2020)
23. Souza, G.C.: Remanufacturing business models. In: Nasr, N. (ed.) Remanufacturing in the Circular Economy: Operations, Engineering and Logistics, pp. 61–84. Scrivener Publishing LLC (2020)
24. Veleva, V., Bodkin, G.: Emerging drivers and business models for equipment reuse and reman-ufacturing in the US: lessons from the biotech industry. J. Environ. Planning Manage. 61(9), 1631–1653 (2018)
25. Gunasekara, H., Gamage, J.R., Punchihewa, H.: Remanufacture for sustainability: a com-prehensive business model for automotive parts remanufacturing. Int. J. Sustain. Eng. 14(6), 1386–1395 (2021)
26. Porter, M.E.: Competitive advantage: Creating and Sustaining Superior Performance. Free Press, New York (1998)
27. VDI Verein Deutscher Ingenieure e. V.: VDI-Guideline 2870 Part 1: Lean production systems: Basic principles, introduction, and review. Beuth Verlag, Berlin (2012)
28. Herrmann, C.: Ganzheitliches Life Cycle Management: Nachhaltigkeit und Lebenszyklusori-entierung in Unternehmen. Springer, Heidelberg (2010)

# Digital Transformation in the Automotive Dismantling Industry: A Scenario Field Analysis, Recommendations for Action and Strategies

Janine Mügge[1]([✉]), Tobias Knauf[1], Joanna Steiner[1], Philip Staufenbiel[2], Tim Opalka[2], Sissy-Linh Nguyen[1], Theresa Riedelsheimer[1], and Kai Lindow[1]

[1] Fraunhofer Institute for Production Systems and Design Technology IPK, Pascalstraße 8-9, 10587 Berlin, Germany
janine.muegge@ipk.fraunhofer.de
[2] LRP-Autorecycling Leipzig GmbH, 04509 Krostitz, Germany

**Abstract.** The automotive recycling industry is part of a constantly and dynamically changing environment. In the course of the circular economy, the increasing demand for alternative sources of resources and materials and the advancing digital development in the industry, today's car dismantlers will face new challenges and opportunities in the coming years. In this paper, a scenario-based methodology is applied to assess possible developments in the industry and develop strategies to be prepared for upcoming changes. The methodology applied consists of developing possible scenarios, analyzing the opportunities and risks of a key scenario and formulating recommendations for action for the industry based on the predicted developments. The identified recommendations focus on six key topics to respond to the potential opportunities and risks in the future. Key topics for the development of the industry are: Cooperation along the value chain, adapting business models, integrating digital technologies, securing employees, expanding expertise and networks, as well as developing and establishing new standards.

**Keywords:** Scenario analyses · automotive · dismantler · end-of-life treatment · automotive trends · sustainability

## 1 Introduction

The linear economy threatens global ecosystems, leading the EU to adopt the circular economy [1]. This transition requires complex social, technical, economic, and environmental changes across many sectors, including the automotive industry. OEMs, suppliers and dismantlers must adapt to new challenges. Digitization and sustainability, closely linked to industry progress, present new opportunities and risks [2]. Digitization and sustainability are often closely intertwined with progress in this industry, presenting new business opportunities, possibilities, and risks that need to be leveraged or avoided [3]. While the goals of a carbon-neutral and a circular economy in the automotive industry

© The Author(s) 2025
H. Kohl et al. (Eds.): GCSM 2024, LNME, pp. 590–598, 2025.
https://doi.org/10.1007/978-3-031-93891-7_65

have been clearly defined, and corresponding secondary material quotas have been set, the practical implementation of these goals remains uncertain. In particular, the potential for the efficient treatment of end-of-life vehicles (ELV) is currently underutilized [2]. Defining the expectations of industry players regarding future developments is challenging due to the complex interdependence of influencing factors. This makes it difficult for vehicle dismantlers to make adequate preparations for future developments [4]. Future scenarios offer an opportunity to gain a better understanding of possible future developments and to develop strategies for probable challenges and opportunities [5]. In this paper, an analysis of future developments in the field of automobile dismantling is carried out. The following research questions (RQ) are intended to develop a comprehensive understanding of future vehicle dismantling:

- RQ1: Which key factors will influence future developments in the field of automotive dismantling?
- RQ 2: What could future vehicle dismantling scenarios look like?
- RQ3: What risks and opportunities will arise from future developments in the automotive dismantling industry?
- RQ4: What recommendations for action can be formulated based on the identified trends, opportunities and risks to prepare stakeholders in the automotive dismantling industry for future challenges?

To answer these research questions, a suitable methodology for the future development analysis was chosen and executed. This was conducted as part of the Digma-DT research project, which is funded by the German Federal Ministry for the Environment, Nature Conservation, Nuclear Safety and Consumer Protection. The analysis uses the German dismantling company LRP-Autorecycling Leipzig GmbH (LRP) as an example to examine assumptions about the current industrial situation of dismantling companies in Germany.

## 2   State of the Art

There are many suitable methods for forecasting future developments, such as the Delphi method, trend analysis, scenario technique, strategic foresight, bibliometrics or information retrieval [5]. The scenario technique, as developed by Gausemeier [5], offers a systematic and efficient approach to the analysis of future developments. It allows for the combination of several forecasting methods, such as bibliometrics for strategic early detection and for finding information on new future projections, or the use of the Delphi method to validate individual results of the scenario technique through expert interviews [5].

The scenario technique has been utilized since the 1960s as a methodology for analyzing future developments. A literature analysis by Bishop et al. [6] identified 23 methodological approaches that can be used in the identification of future trends and the development of scenarios. The scenario technique can be used to guide the strategic development of the company, as well as product creation processes [7]. Scenarios are plausible descriptions of possible futures based on assumptions about key drivers and their relationships. They represent different visions of the future through a coherent combination of conceivable developments (projections) of individual influencing factors and cover a spectrum ranging from expected to unlikely developments. They provide an overview of the potential effects of developments and measures [5, 6, 8].

# 3   Methodology

For the creation and evaluation of future scenarios, the method of Gausemeier and Plass [5] was chosen. The method is divided into five phases. The scenario preparation (1) includes determining the objective and the analysis of the design environment. In the scenario analysis (2), influencing factors for the design environment are identified. These factors are then compared and quantified in two matrices: the influence analysis and the relevance Analysis, where their mutual impact and relevance are assessed. The outcome of the second phase is the identification of the most relevant influencing factors, known as key factors. In the projection development (3), plausible future projections for each key factor are developed, discussed, and formulated. These future projections are compared in a pairwise matrix in scenario development (4) to evaluate their consistency. The most consistent future projections can be combined into potential unified future bundles (scenarios). Each key factor therefore occurs once in a projection bundle. The process of forming projection bundles can result in a multitude of possibilities, making it essential to reduce the number of projection bundles. For this purpose, a software is used, which performs a cluster analysis, a statistical data analysis method, to find homogeneous and maximally heterogeneous projection bundles and eliminate inconsistent futures. The highly consistent projection bundles are grouped into clusters based on their similarity, forming the possible future scenarios. In the last step, the scenario transfer (5), the possible future scenarios are examined. In this process, opportunities and risks are identified and evaluated based on their probability of occurrence and impact on the design environment. The aim is to select a robust reference scenario for the future and derive actionable recommendations from it [5].

# 4   Scenario Technique

## 4.1   Scenario Preparation

In the first step, the current status of vehicle dismantling in Germany is analyzed. The transfer, take-back and environmentally friendly dismantling of ELV in Germany is defined by the ELV Directive 2000/53/EC [9]. According to the yearly report from the end of 2023 on ELV dismantling rates in Germany, around 2.92 million vehicles reached the end of their life in 2021. However, around 2.3 million of these were exported to other EU countries as used vehicles and around 0.22 million to non-EU countries. Only 396,773 vehicles ultimately ended up with a certified ELV dismantler. This represents a statistical decrease of around 2.3% compared to the previous year [10] and is the lowest number ever recorded since 2004 [10]. The number of certified dismantling facilities has fallen in recent years, too. While there were still 1064 dismantling facilities in 2020, there were only 1030 in 2021 [10]. The majority of dismantling facilities are small and medium-sized enterprises (SMEs) [2]. The dismantling processes are characterized by a large number of manual processes and the degree of digitalization is low [11]. The future of vehicle dismantling will be characterized by new regulations, increased digitalization and sustainability demands. A proposal for a new ELV directive was published in 2023. In this, the current treatment of ELVs is assessed as suboptimal concerning the potential of a climate-neutral economy. This shows that the dismantling of ELV can be a decisive factor for a climate-neutral and circular automotive industry. Improved recovery can both increase reuse and recycling rates and ensure the availability of secondary materials [2]. The objective of the scenario technique was defined as the examination of potential developments of vehicle dismantling facilities concerning their digitalization and contribution to a circular economy in Germany over the next 10 years.

## 4.2 Scenario Analysis

In the first step of the scenario analysis, 39 influencing factors from the categories of regulation and politics, location and demographics, environment and surroundings as well as technological innovations and their applications are collected for the previously defined design environment. Subsequently, the relationships between the influencing factors are evaluated with the help of an influence matrix. This allows the strength of the influence of each factor on the others to be quantified (active sum) and the extent to which one factor can be influenced by the others to be determined (passive sum). In addition, a relevance analysis is carried out by comparing the influencing factors using a relevance matrix to determine their relative importance in comparison to each other (relevance sum). To select the key factors, the active sum, passive sum and relevance sum of each influencing factor are shown in a system grid diagram in Fig. 1. The active sum is shown on the y-axis, the passive sum on the x-axis and the relevance sum is shown in the form of a sphere diameter.

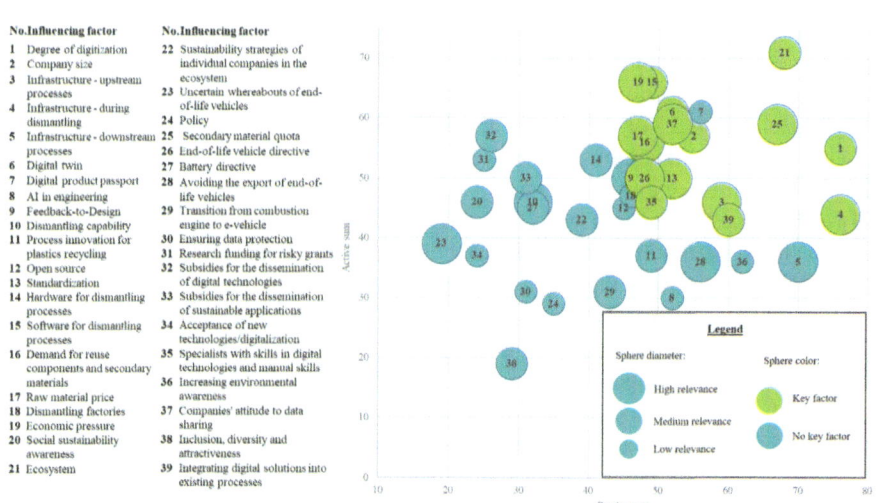

**Fig. 1.** System grid to identify the key factors.

Based on the quantified results of the system grid diagram, 15 key influencing factors were identified: Degree of digitalization, company size, digital twin, infrastructure - upstream processes and during dismantling, standardization, automation of dismantling, demand for reuse/remanufacturing components, ecosystem, secondary material quota, ELV directive, employees with expertise in digital technologies and craftsmanship, companies' attitude to data sharing, digital solutions, economy.

### 4.3 Projection Development

For each identified key influencing factor, various development options, known as future projections, are described in this phase. The time horizon is the next 10 years. This results in a total of 49 possible future projections. Using the key factor digital twin as an example, three possible future projections are presented: (1) "The digital twin plays no role in vehicle dismantling", (2) "The digital twin only performs simple tasks, such as displaying basic vehicle information" and (3) "The digital twin is a central component of vehicle dismantling. The digital twin enables the transparency of data flows throughout the value chain".

### 4.4 Scenario Development

The next step in the scenario technique is to create scenarios based on the defined projections. The defined future projections of the key factors are then compared with each other in a matrix and evaluated in pairs with numbers between 1 and 5 with regard to their consistency in a possible common future. Future projections can be evaluated from total inconsistency - meaning no possibility of a joint occurrence in the future - to neutrality in relation to each other and finally to strong mutual favourability. Each future projection in combination with one future projection of each of the other key factors is a possible description of the future and forms a projection bundle. Projection bundles that contain one or more inconsistent projection pairs can be excluded from the analysis. The large number of projection bundles is almost impossible to evaluate manually, which is why special cluster software is used, which compares each projection bundle with the other bundles and combines similar projection bundles in clusters. This process is carried out until three clusters remain. The limit of three clusters was chosen for this analysis because information was lost with each additional cluster, and the similarity between clusters was still too high with four clusters. Each of the three clusters indicates the probability with which the respective future projections of the key factors can occur, from which three scenarios were then formulated. The three scenarios are described in Table 1. In all scenarios, dismantling companies benefit from the increasing demand for remanufactured components and secondary materials. They are innovative, want to change the way of work, are starting to integrate new digital technologies into their processes and exchange data in the value chain to promote a circular economy. Standards for data exchange are already in place in many cases, but still need to be optimized. Cushioning the economic pressure caused by change in the industry, combating the shortage of skilled workers and people with digital know-how and the way the ELV directive is further developed are basic prerequisites for achieving these scenarios.

**Table 1.** Identified scenarios.

|  | Scenario 1 | Scenario 2 | Scenario 3 |
|---|---|---|---|
| Infrastructure | Expanded and standardised, joined forces with other players in the value chain, standardized digital network | Cooperation with other players in the value chain | Expanded and standardized collaboration in digital ecosystems with the other players in the value chain |
| Company size | Predominantly large companies | Evenly distributed between large and SME | Predominantly large companies |
| Level of digitization | High, digital twins as key technology, dismantlers will meet requirements | Advanced, dismantlers will keep up with increasing requirements, digital twin supports simple tasks | High, use of digital twins, large companies exceed requirements, risk that SMEs can't keep up |
| Level of automation | High | Low (pilot phase) | Moderate |
| Influence on the design field | Major | Moderate | Moderate |

## 4.5 Scenario Transfer

In the final phase, the scenario transfer, a reference scenario is derived, and opportunities and risks as well as recommendations for action are developed. First, the three developed scenarios are assessed according to their probability of occurrence and the strength of their impact and summarized in Table 2.

**Table 2.** Assessment of scenarios.

|  | Scenario 1 | Scenario 2 | Scenario 3 |
|---|---|---|---|
| Probability of occurrence | Low | High | Medium |
| Strength of their impact | Medium | High | Medium |

Scenarios with a high probability of occurrence and a high degree of impact on current vehicle dismantling are particularly significant for strategic development. As a result, scenario 2 is selected as the reference scenario. In the next step, the opportunities and risks for this scenario are analyzed and an excerpt is described in Table 3.

Based on the opportunities and risks, recommendations for action are derived for future strategy development. In the future, ELV dismantlers should focus on innovative strategies and adapt to industry changes to remain long-term economic players within the value chain. By increasing data exchange with other players along the value chain,

**Table 3.** Opportunities and risks analysis.

| Opportunities | Risks |
|---|---|
| Due to the increasing relevance of circular economy and demand for reuse, remanufacturing components and recycling materials, the dismantler is gaining importance in the value chain and becoming part of the ecosystem | Increased requirements for the treatment of ELV can pose a challenge for dismantlers, as they entail high costs for adapting treatment technologies in the short and medium term |
| Increasing demand for reuse, remanufacturing components and recycled materials | Smaller companies are unable to keep pace with digitalization and are being left behind |
| Increasing the degree of digitalization and exchange of information throughout the entire product life cycle can increase the efficiency of the dismantling processes | The introduction of increasing digitalization and automation in pilot studies requires employees to have digital skills that were not previously necessary |

ELV dismantlers can leverage synergies and enhance efficiency. To optimize cooperation with partner companies, new mechanisms for data exchange must be developed and the degree of digitalization should be increased. Standardization in data exchange and a willingness to share data and collaborate with other economic players are essential for achieving common goals and developing innovative solutions together. Employees should be involved in the changes caused by digitalization and partial automation at an early stage. Their digital skills should also be expanded. Dismantlers should integrate into the automotive supply chain, especially concerning new regulations and associated secondary material quotas. Adapting the main business model and customer base may become necessary. The focus will gradually shift from the small-scale spare parts business with reusable components to the supply of remanufacturing components to manufacturers and secondary materials for processing and recycling in the automotive value chain. Furthermore, in the future, the dismantlers must meet the sustainability requirements of an automotive industry supplier. At the same time, the dismantling processes must be adapted to comply with existing regulatory requirements, such as the separation of large materials like plastics.

## 5   Conclusion and Future Research Directions

In light of the new proposal for the ELV directive, the automotive value chain will adapt with a primary focus on increasing sustainability. The scenario technique offers insights into the future of vehicle dismantling in Germany in 10 years with a special focus on the degree of digitalization. As part of the scenario technique, 39 influencing factors were identified for the design field and the most relevant key factors (15) for the future of vehicle dismantling were derived (RQ 1). Future projections were developed for each key factor and combined into three consistent scenarios (RQ 2). These scenarios were evaluated, a reference scenario was derived and opportunities, risks (RQ 3) and recommendations for action (RQ 4) were formulated. ELV dismantlers must adapt to the changes in the industry and should try to shape them. The most important measures include working on secure data exchange with other players in the industry, expanding integration into the automotive supply chain, adapting business models towards secondary material suppliers and working closely with upstream and downstream partners. In addition, ELV dismantlers should leverage automation and digitalization technologies, involve their employees in the development and implementation of these processes and technologies and promote workforce diversification and training to increase efficiency and drive innovation. This approach will help them meet new regulatory requirements, optimize dismantling processes and ultimately become a key player in the evolving automotive industry.

# References

1. European commission: a new circular economy action plan. For a cleaner and more competitive Europe (2020)
2. Proposal for a Regulation of the European Parliament and of the Council on circularity requirements for vehicle design and on management of end-of-life vehicles, amending Regulations (EU) 2018/858 and 2019/1020 and repealing Directives 2000/53/EC and 2005/64/EC (2023)
3. Demartini, M., Evans, S., Tonelli, F.: Digitalization technologies for industrial sustainability. Procedia Manuf. **33**, 264–271 (2019)
4. Verordnung über die Überlassung, Rücknahme und umweltverträgliche Entsorgung von Altfahrzeugen. AltfahrzeugV (2020)
5. Gausemeier, J., Plass, C.: Zukunftsorientierte Unternehmensgestaltung. Strategien, Geschäftsprozesse und IT-Systeme für die Produktion von morgen. Hanser, München (2014)
6. Bishop, P., Hines, A., Collins, T.: The current state of scenario development: an overview of techniques. Foresight **9**, 5–25 (2007)
7. Haeberle, F., Parolin, G., Pigosso, D.C.A.: Scenario building guidelines for sustainable innovation. Proc. Des. Soc. **4**, 1289–1298 (2024)
8. IPCC: Workshop Report of the Intergovernmental Panel on Climate Change Workshop on the Use of Scenarios in the Sixth Assessment Report and Subsequent Assessments. Imperial College London, United Kingdom (2023)
9. The European Parliament and the Council of the European Union: Directive 2000/53/EC of the European Parliament and of the Council of 18 September 2000 on end of life vehicles (2002)
10. German Environment Agency, Federal Ministry for the Environment, Nature Conservation: Jahresbericht über die Altfahrzeug-Verwertungsquoten in Deutschland im Jahr 2021. nach Art. 7 Abs. 2 der Altfahrzeug-Richtlinie 2000/53/EG (2023)
11. Mügge, J., Seegrün, A., Faßbender, L., Riedelsheimer, T., Staufenbiel, P., Lindow, K.: Integrated Consideration of Data Flows and Life Cycle Assessment in Vehicle Dismantling pro-cesses. 2212–8271, vol. 122, pp. 1018–1023 (2024)

598     J. Mügge et al.

# The Automotive Industry - Potential and Strategy of Reusing Components for Improved Sustainability

Frank Riedel[1(✉)], Patrick Alexander Schmidt[1], Welf-Guntram Drossel[1],
Rafi Wertheim[1,3], and Eckard Lippmann[2]

[1] Fraunhofer-Institut for Machine Tools and Forming Technology (IWU), Chemnitz, Germany
frank.riedel@iwu.fraunhofer.de
[2] A.I.M. All in Metal GmbH, Markgröningen, Germany
[3] Braude College, Karmiel, Israel

**Abstract.** The automobile, produced in increasingly large quantities, significantly impacts global aspects of life. Consisting of thousands of high-tech parts, including high-quality aluminum and steel alloys, its production is resource intensive. Systematic recycling does not exist. The metallic components of a few vehicles are melted down and fed into the energy-intensive production process. This paper aims to demonstrate the potential of directly reusing these components and develop efficient manufacturing concepts for their implementation. By completing the material flow loop through remanufacturing and direct reuse, the project seeks to enhance resource efficiency in the automotive industry. Key sub-processes include identification, automated disassembly, and condition determination of components. A universal laser-based dismantling technology for almost all connections is presented, which makes it possible to separate components without damage.

**Keywords:** car industry · circular economy · disassembly · automated disassembly cell · universal dismantling laser technology

## 1 Introduction

Humanity is facing a critical turning point in the current era of increasing demand for manufactured goods. The per capita material footprint has risen from around 9 tons in 2000 to a worrying 12 tons in 2019. The Earth's resources are finite, which resulted in a global overuse crisis. The implications of this situation go beyond environmental concerns and raise an existential question for our species [1, 2].

This trend reveals the worrying reality, that despite governmental attempts to encourage resource conservation, society remains locked in a pattern of escalating consumption. The paper addresses the challenge of limited lifespan intentionally designed into industrial products, which perpetuates a culture of disposability. In addition, the generation of electronic waste is projected to increase by 24% between 2020 and 2030, which highlights the need for a fivefold increase in recycling rates [3].

© The Author(s) 2025
H. Kohl et al. (Eds.): GCSM 2024, LNME, pp. 599–606, 2025.
https://doi.org/10.1007/978-3-031-93891-7_66

A comparatively large industrial product with an extremely high resource consumption, is the automobile. It significantly impacts many areas of our lives like few other products. The automobile consists of up to 10,000 individual parts [4] on average and driven by innovation, high-tech components are installed with correspondingly high resource consumption. Components e.g., made of high-quality aluminum and steel alloys (boron-manganese, TRIP, dual-phase steels) are used in the body area.

Although a large part of the metallic components shows little property reduction such as aging or wear at the end of their lifecycle, these components are melted down. In extremely elaborate, energy and resource-intensive production processes, high-tech alloys are remelted, cast, and in numerous process steps such as heat treatment, forming, trimming, joining, and coating, almost identical components are produced. More energy efficient processes are needed in the future.

Our approach is to enable circular production cycles as depicted in Fig. 1. We aim to demonstrate the potential of the direct reuse of components for the mass product automobile and to develop manufacturing concepts that enable the efficient implementation of the direct reuse of components [5]. In terms of economic performance, remanufacturing can save 20% to 80% on production costs when compared to new product manufacturing [6].

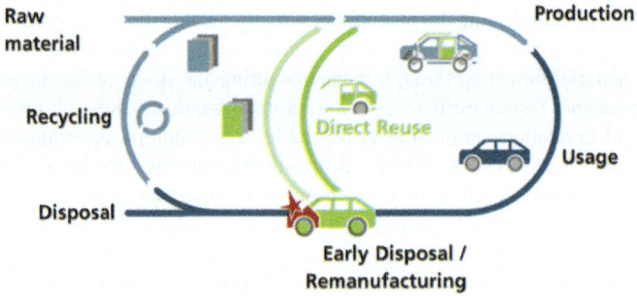

**Fig. 1.** Direct reuse to transform linear production towards circular production cycles.

In the current corporate sustainability landscape, there is an undeniable need for a paradigm shift towards circular economy (CE) principles. CE, as defined by [7], prioritizes reduction, reuse, recycling, and material recovery over the traditional 'end-of-life' concept. It aims to achieve sustainable development by balancing environmental quality, economic prosperity, and social equity for the benefit of present and future generations. Recognizing the urgency of this transition, industries, particularly the industrial sector, are facing challenges of resource scarcity, evolving market dynamics, and global competition, as articulated by [8]. Potential of component reuse and its current challenges.

Although, as described in the introduction, the relatively large technical industrial product automobile is a high-quality, resource-intensive product with a very large number of units, globally, hardly any notable recycling strategies are systematically implemented. In Europe, at the end of its lifecycle the automobile is disposed of or exported. During this process, spare parts sought after by small companies are recovered, which are supplied to a small circle of hobbyists on an insignificant scale. Mainly, the vehicles undergo de-fluidization and, to some extent, material separation. Non-metallic components, especially plastics, are predominantly disposed of and incinerated. The metallic automobile structure is scrapped and, if not stored, is subjected to a remelting process. Although remelting serves the fundamental recovery of raw materials, both predominantly used strategies today are to be assessed as critical in terms of energy and resources as shown in Fig. 2. The downward pyramid displays the recommended procedures with components at their end-of-life stage. The lower one goes the more energy, materials, and already performed work is lost. Thus, the upper procedures must be prioritized in a CE.

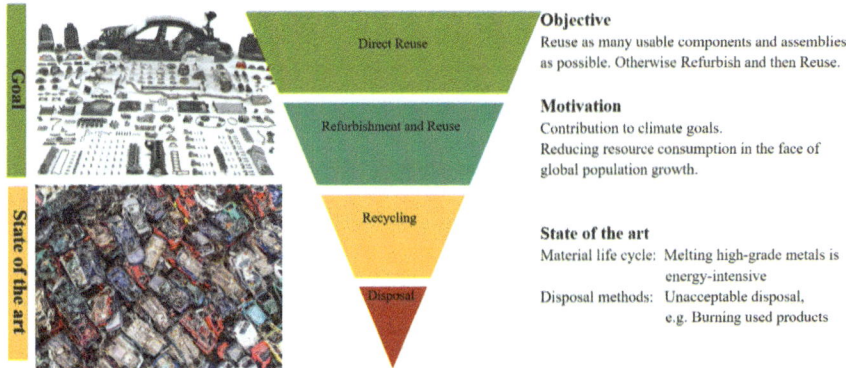

**Fig. 2.** Automotive product life cycle, state of the art and objective

The goal must be to start systematically developing and implementing concepts for the extremely large variety of automotive components in global circulation that enable the reuse or further use of usable components.

**Direct reuse** of components leads to a maximal effect for resources, energy, costs, and sustainability. If a significant number of undamaged components can be directly reused, then all materials, raw and auxiliary materials necessary for manufacturing, and especially the cost- and energy-intensive process chains are eliminated.

**Refurbishment** of reusable components leads to a large impact on resources and sustainability effects. The impact on energy and costs is variable. If the refurbishment of components requires fewer resources than new manufacturing, then the reuse of even refurbished components has a leverage effect corresponding to the delta, like the direct reuse of components.

Current obstacles to the implementation of strategies for the reuse or further use of usable components:

- The basic structure of automobiles, whether in the combustion or electric clusters, is relatively similar, but the global variety of brands and models means that the components differ in the materials used and geometric dimensions.

- Currently, the diversity of designs is growing due to the significant change of the automobile towards hybrid and electric vehicles and changes even within the scope of one brand and model due to short cycle times (~7 years)
- Due to the fierce competition in the automotive construction industry, there is a lack of opportunity for collaboration and availability of information and data.

Current positive aspects for the implementation of strategies for the reuse or further use of usable components:

- Large automotive global players sometimes unite various brands, leading to platform concepts, in which construction methods and components are similar.
- In the electrical and electronic sectors, similar components are already being used.

## 2   Methodological Procedure for Component Extraction

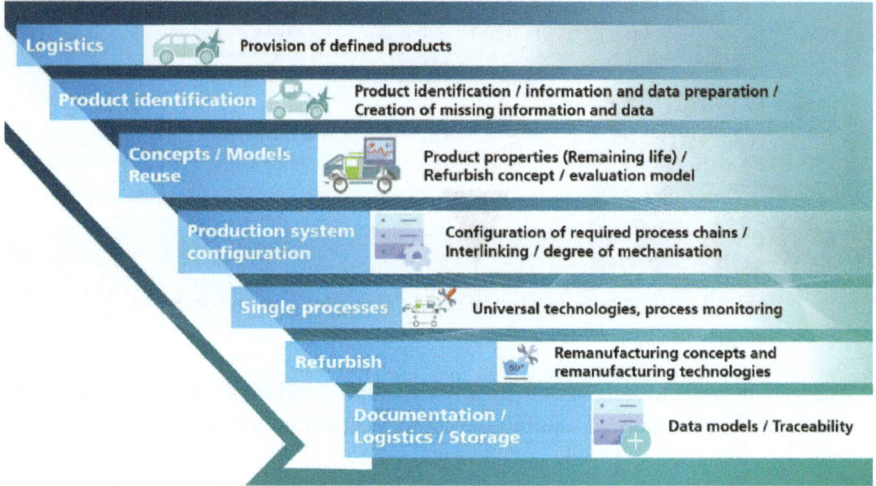

**Fig. 3.** Methodological procedures for component extraction

The actual extraction of components requires the consideration of a complex system of different processes, shown in Fig. 3. If all boundary conditions and the target values relating to the type of parts to be produced are known, the following process steps are required: product identification and digital twin, universal fully automatic disassembly cell, determining the properties of disassembly components, component refurbish for reuse, and documentation of component data for reuse. It should be noted that the development is aimed at fully automatic extraction of components from any vehicle. One area is outlined below.

The vision is a universal fully automatic disassembly cell with the following requirements:

- Technologies must be developed and integrated to enable highly automated disassembly for efficient remanufacturing. A universal automated disassembly strategy is required for the undamaged extraction and removal of defined components.

- Equipment required for highly automated disassembly and remanufacturing must be designed to be highly flexible to accommodate various products.

Around 100 different assembly and joining processes are used to assemble a car, which makes it unfeasible to build a reverse assembly process chain. Implementing this multitude of assembly and joining processes in such a universal fully automatic disassembly cell is not cost effective. The task is to reach the target component in a relatively unconventional way and to expose this target component without damaging it. We need universal dismantling technologies. Laser technology offers one solution. **Fehler! Verweisquelle konnte nicht gefunden werden.** Shows that laser technology can loosen the various connections of components without damaging the component.

## 3   Strategy for a Universal Fully Automatic Disassembly Cell with a Universal Disassembly Technology

To recover a significant number of parts for reuse, solutions that require a high level of manual disassembly will not be considered here. The vision is a fully automated disassembly cell in which any vehicle can be recognized and/or digitized. From this, the disassembly strategy is derived fully automatically. It is assumed that the entire vehicle will not be dismantled into its individual parts; instead, only those parts that can provide significant sustainability or value creation in the current situation will be recovered. This means that surrounding components will be removed destructively with minimal effort and then sent for material recovery.

The target components intended for reuse must be dismantled with as little damage as possible. The challenge here is that approximately 100 different joining methods are used in the manufacturing of new automobiles. In body construction alone, there are around 30 different joining methods (see Fig. 4).

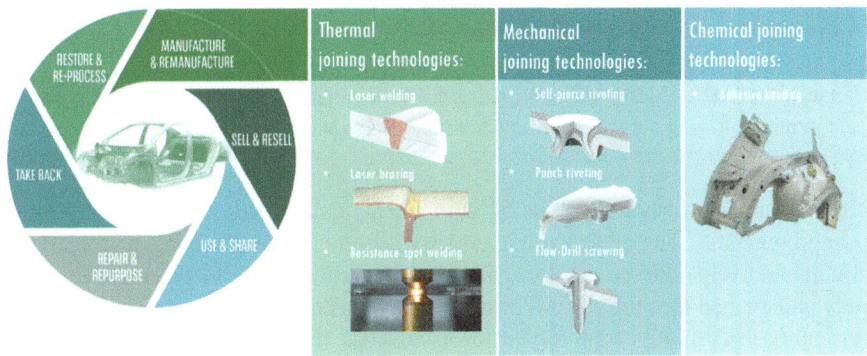

**Fig. 4.** Examples for joining technologies in car body construction.

It is logical that the variety of procedures cannot be fully implemented in a universal disassembly cell. A universal disassembly tool could be laser technology, which can separate nearly all connections with minimal damage. For example, when considering bonded joints, such as soldered connections, it is possible to melt the solder (CuSi3, $T_S$ = 1026 °C), which has a lower melting temperature than the base material being soldered (DC04, $T_S$ = ca. 1425 °C), and then vacuum it away. The separated parts of the base material remain completely undamaged (see Fig. 5). Mechanically joined connections can also be separated using a similar principle, where only the joining element is precisely melted, and the molten material is blown out (see Fig. 6). Naturally, holes remain in the component, but these can be used for rejoining.

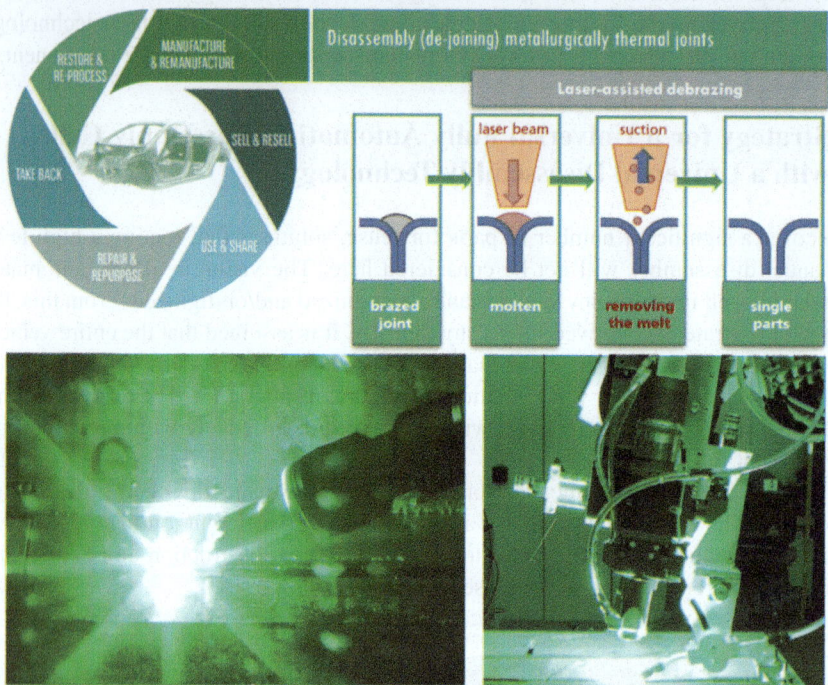

**Fig. 5.** Laser-assisted debrazing: Top: Principle; Bottom: Experimental proof of concept, steel brazed with CuSi3 ($T_S$ = 1026 °C), debrazed with a 3 kW green laser ($\lambda$ = 515 nm, $T_P$ < 1200 °C, $v_P$ = 4 m/min), blowing out the liquid solder with compressed air and vacuuming.

The separation of bonded metallic components using a laser tool is well-known and unproblematic. The adhesive is thermally decomposed by heating with the laser at a comparatively low temperature, allowing the components to be separated completely undamaged (see Fig. 7). All principles have been demonstrated in experimental investigations. Currently, a prototype for a universal disassembly cell with universal laser-assisted disassembly (de-joining) is being developed.

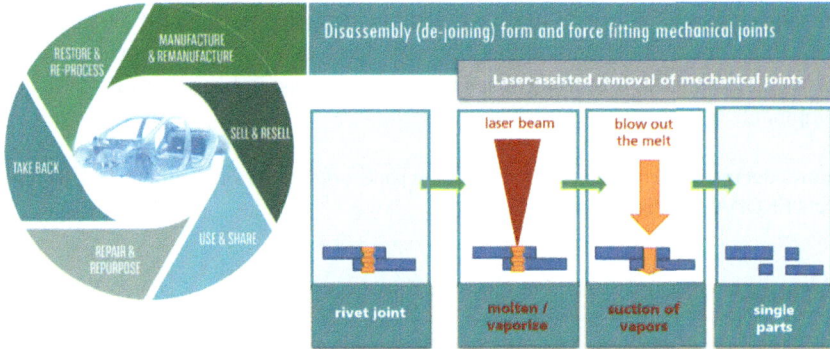

**Fig. 6.** Laser-assisted Disassembly (de-joining) form and force fitting mechanical joints.

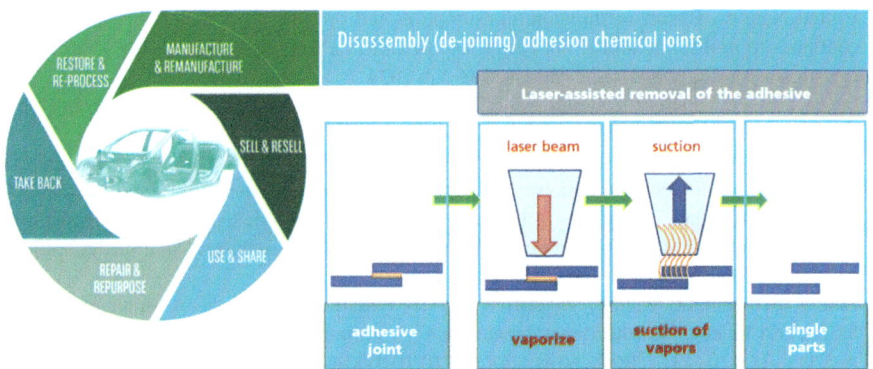

**Fig. 7.** Laser-assisted Disassembly (de-joining) adhesion chemical joints

## 4  Conclusion

There are approximately 2 billion vehicles worldwide. A true circular economy does not exist for these vehicles. The current practice for automobiles at the end of their life cycle involves, for a small portion of cars, the melting down of a mostly metallic fraction of materials. The manufacturing of a new component then undergoes the entire energy-intensive production process once again. However, in a used vehicle, there is a large number of components that are neither worn out nor aged and could be directly reused. This paper highlights the potential and current obstacles to the direct reuse of components.

The boundary conditions and the necessary process steps for extracting components from any vehicle up to their approval for direct reuse are briefly described. The focus of the paper is on the introduction of a fully automated disassembly cell in which vehicles are recognized or digitized, and from which the process steps for removing target components with minimal effort are fully automated. One proposed solution is a universal laser-based disassembly technology. The target components can be destructively cut free in the surrounding area. It is demonstrated that the direct joining connections of the target component can be separated without damaging the areas of the target component. Experiments have shown that this separation technology applies to almost all types of connections. A prototype of such a universal disassembly cell is currently under development.

**Acknowledgements.** The results presented in this paper originated from a current, BMBF-funded project "EKODA" (funding code 02J21E150).

# References

1. Department of economic and social affairs, World population prospects 2022. https://population.un.org/wpp/Graphs/DemographicProfiles/Line/900, Accessed 16 Jan 2024
2. Our world in data, Material footprint per capita, 2000 to 2019. https://ourworldindata.org/grapher/material-footprint-per-capita, Accessed 16 Jan 2024
3. United Nations, Ensure sustainable consumption and production patterns: electronic waste continues to proliferate and is not being disposed of in a responsible way, https://unstats.un.org/sdgs/report/2021/goal-12/, Accessed 16 Jan 2024
4. Focus, Aus wie vielen Einzelteilen besteht ein Auto? https://www.focus.de/auto/ratgeber/auto-abc/auto-aus-wie-vielen-einzelteilen-besteht-ein-auto_id_3514353.html, Accessed 25 Nov 2023
5. Patrick Alexander Schmidt, Ein zweites Leben für Batterie, Getriebe und Zahnräder: Kreislaufwirtschaft statt Recycling, https://www.iwu.fraunhofer.de/de/presse-und-medien/presseinformationen/PM-2023-ein-zweites-leben-fuer-batterie-getriebe-und-zahnraeder.html, Accessed 16 Jan 2024
6. Zhu, Y., Schwam, D., Wallace, J.F., Birceanu, S.: Evaluation of soldering, washout and thermal fatigue resistance of advanced metal materials for aluminum die-casting dies. Mater. Sci. Eng. A **379**(1–2), 420–431 (2004). https://doi.org/10.1016/j.msea.2004.03.020
7. Cantzler, J., Creutzig, F., Ayargarnchanakul, E., Javaid, A., Wong, L., Haas, W.: Saving resources and the climate? A systematic review of the circular economy and its mitigation potential. Environ. Res. Lett. **15**(12), 123001 (2020). https://doi.org/10.1088/1748-9326/abbeb7
8. Sharma, M., Kumar, A., Luthra, S., Joshi, S., Upadhyay, A.: The impact of environmental dynamism on low-carbon practices and digital supply chain networks to enhance sustainable performance: an empirical analysis. Bus Strat. Env. **31**(4), 1776–1788 (2022). https://doi.org/10.1002/bse.2983

# Reuse and Recycling

# Waste to Value Process Chain for Recycling of Fishing Gear Collected from Coastal Waters

Thomas Potempa$^{(\boxtimes)}$ ⓘ, James Henderson ⓘ, Julia Tetzner ⓘ, Sofie Hoang ⓘ, Max Ehleben ⓘ, and Max Juraschek ⓘ

Institute of Recycling, Ostfalia University of Applied Sciences, Wolfsburg, Germany
potempa@ostfalia.de

**Abstract.** Plastic waste in coastal waters has become a major environmental concern and a visible sign of ocean pollution. A significant share of the floating waste material originates from used and lost fishing gear. By establishing suitable recycling processing chains, used fishing gear can become a valuable resource for the manufacturing of products, fostering its extraction from the water and boosting local economies. Post-use fishing gear, however, is a very challenging input material for processing into recylcates. Understanding its degradation mechanisms is crucial for designing a recycling process that will yield high value materials. There are particular obstacles connected with each recycling step, starting with the collection of very heterogeneously aged and deteriorated fishing gear. Consecutively, shredding of the waste material requires the development of specific technological solutions. Marine dirt adds additional obstacles. Further processing steps that need to be adapted include extrusion, pelletizing injection molding and marketing of the products from recycled material in competition with ones using raw materials. This study presents a waste-to-value process chain design that addresses the specific challenges of recycling fishing gear collected from coastal waters. Data has been collected to identify and quantify sources of plastic waste from fishing activities, and solutions have been developed in local higher education through a joint German-Vietnamese effort.

**Keywords:** Recycling · Waste-to-value · Process chain · Fishing gear

## 1 Introduction

Due to its widespread and pervasive nature, marine litter has become a global problem [1]. Approximately 10 million tons of plastic enter the oceans annually, with fishing gear waste accounting for around 640,000 tons [2]. It doesn't respect national boundaries while affecting coastlines, seas, and oceans across the world. The interconnectedness of marine ecosystems allows litter to travel vast distances, carried by ocean currents and winds, resulting in the accumulation of debris in remote and diverse marine environments [3]. Marine litter originates from a variety of human activities such as shipping, fishing, tourism, and inadequate waste management practices [4].

To preserve the health of our oceans and the diverse marine life they contain, the need to combat marine litter is paramount. Marine litter poses a significant threat to

H. Kohl et al. (Eds.): GCSM 2024, LNME, pp. 609–616, 2025.
https://doi.org/10.1007/978-3-031-93891-7_67

ecosystems, marine animals, and coastal communities [5] [6]. Furthermore, it disrupts food chains, damages habitats, and harms marine species through ingestion and entanglement. As a consequence, the economic and cultural interests of coastal communities are affected [7].

This is especially true for developing countries like Vietnam, a country whose population lives at, on and from the sea. Vietnam is regarded as one of the countries with the highest amount of plastic waste discharged into the sea. According to a survey in 2015, the volume of plastic waste from Vietnam emitted to the sea is estimated to be between 0.28 and 0.73 million tons per year [8]. According to the Ministry of Natural Resources and Environment (2020), Vietnam does not provide national data to identify sources of plastic emissions from land and sea or create statistics on the amount of plastic waste around coastal areas [9].

Reducing plastic waste in the sea is a major challenge for Vietnam as the habit of throwing away plastic is increasing and the capacity for solid waste disposal is limited [10]. In addition, campaigns to change the habit of using plastic products and disposing of plastic waste have not achieved the desired results [11]. The proportion of plastic waste from nets and other fishing gear varies significantly from region to region, but it is clear that it makes up a large portion of the plastic waste discharged into the sea [12].

## 2   Fishing Gear and Ocean Waste

Several initiatives are underway in Europe to reduce the environmental impact of ALDFG (Abandoned, Lost, Discarded Fishing gear) through simply encouraging fishermen to take ashore the litter they encounter at sea while fishing. For example, BIM's Fishing for Litter initiative which was established in 2015 and is supported under the European Maritime and Fisheries Fund [13]. This project will build on the knowledge and best practice to explore the role of advanced technology in recycling and work to create a demand for ALDFG recycled materials.

The field of marine plastic waste management encompasses a broad spectrum of research areas, with a particular focus on recycling technologies and practices to mitigate ocean pollution. Within this domain, the recycling of fishing gear has gained significant attention as a crucial step towards sustainable fisheries and environmental conservation. Survey papers such as "GloLitter Fishing gear recycling technologies and practices" [14] provide comprehensive insights into the advancements in recycling processes for fishing gear, highlighting the challenges and opportunities in this area.

Several research works have endeavoured to address the issue of recycling post-use fishing gear, aiming to transform this waste material into valuable resources while reducing its environmental impact. Studies by Charter et al. (2020) [15] and Stolte et al. (2019) [16] have delved into the requirements for preparing unwanted fishing gear for recycling processes, emphasizing the importance of collection, sorting, disassembly, and cleaning techniques. Additionally, the exploration of different recycling methods, including primary and secondary (mechanical), tertiary (chemical), and quaternary (energy recovery) schemes, has been a focal point in the literature.

In evaluating existing approaches for recycling fishing gear, it becomes evident that challenges persist at each stage of the recycling chain. Issues such as heterogeneous aged gear, marine debris contamination, and the development of specialized techno-logical solutions for shredding and processing pose significant obstacles [17]. While some studies have reported advancements in extrusion, pelletizing, and injection mold-ing techniques for recycled fishing gear, the market competitiveness of products derived from recycled materials remains a key consideration.

A study on end-of-life fishing gear in Spain conducted by Basurko et al. [18] delves into the quantity and recyclability of discarded fishing gear, emphasizing the need for improved management practices within the circular economy framework. Despite chal-lenges such as complex material compositions and limited recycling technologies, the research underscores the importance of transitioning towards circular design and effec-tive end-of-life treatment for fishing gear. By identifying materials, establishing proper collection systems, and fostering recycling initiatives, the study advocates for tailored solutions to enhance the sustainability of fisheries and align with circular economy principles.

Findings presented by Liotta I. et al. [19] highlight the critical need for recycling fishing nets at the end of their life to mitigate plastic and microplastic pollution in marine environments. Their proposed recycling processes involve sorting, cleaning, grinding, dissolution, drying of polymeric matrix, and compression molding, leading to a decrease in fishing net waste in oceans and landfills. Additionally, the study emphasizes the environmental and economic benefits of recycling fishing nets, aligning with European directives on circular economy and sustainable waste management practices.

In conclusion, it is essential to acknowledge the limitations of current recycling approaches, keeping in mind socio-economic boundary conditions and highlighting the need for innovative solutions especially for developing countries.

## 3   Waste to Value Process Chain

To derive a waste to value process chain, it is necessary to understand the value chain/lifecycle of fishing gear in Vietnam. In general, based on interviews with different stakeholders within the value chain, the value chain is composed of 7 main steps:

After the production of the basic components such as nets, ropes and floats, they are assembled into the final fishing gear (1) and sold to the fishermen (fishing micro-enterprises) (2).It is common practice for fishermen to repair minor damage that occurs during use. (3). In the event of more extensive damage, the nets are returned to the commissioner for repair (4). The repaired net is reintroduced into the value chain and returned to the fishermen.

Once repairs have reached a certain threshold, the net will either be discarded or sold to a collector who specializes in the refurbishment of fishing gear. After manual separation, the actual processing occurs in neighboring countries (Laos, Cambodia) (5). The refurbished fishing gear might reenter the market at step 2 in a lower price segment. Nets that have been discarded are separated manually and are then transferred to waste handling facilities (6). The final stage in the life cycle of a fishing net is thermal disposal or incineration. (7) This final stage is regarded as the optimal scenario within the existing lifecycle of fishing gear. A proportion of the gear remains in the natural environment.

Only a very small amount is actually recycled, where the granulates made of recycled material are generally of low quality due to contamination.

Taking a closer look at the use-phase of fishing gear, a preliminary analysis of about 650 datasets from surveys performed by the REVFIN Project[1] in 2023/2024 indicate that fishing nets on average are used for 15 months, whereas ropes and buoys are in use for 25 and 28 month respectively. Moreover, the following average masses of fishing gear per boat/vessel have been identified in the preliminary analysis of the survey data (see Table 1). Data on fishing gear composition collected by RIMF[2] show that the plastic PA accounted for the highest proportion of fishing gear on average, at 22.5%, followed by PP at 15.3%, PE at 8.6% and PU, PVC and other plastics at 2.6%, 0.6% and 5.5% of the total mass, respectively, but the ratios are varying, depending on type, size and commissioning. Moreover, according to February 2023 statistics, the total number of fishing boats in Vietnam is 76,989 units.

**Table 1.** Masses, replacement times and annual replacements of different types of fishing gear per boat/vessel

| Type of fishing gear | Amount per boat | Uncertainty | Replacement after [month] | Annual replacement |
|---|---|---|---|---|
| Fishing nets | 480 kg | ±50 kg | 15 | 384 kg |
| Ropes | 240 kg | ±30 kg | 25 | 115 kg |
| Buoys | 6500 pieces | ±50 pieces | 28 | |
| Buoys [a] | 65 kg | ±0.50 kg | | 28 kg |
| Annual total mass | | | | **527 kg** |

[a]The mass of a single buoy was estimated to 10 gramms.

By extrapolating the data in Table 1, it is possible to estimate the amount and type of plastic material that could be recycled on an annual basis. The estimate yields approximately 9200 tons of PA, 6200 tons of PP and 3500 tons of PE and 3600 tons of other plastic types.

To define a waste to value process chain it is a prerequisite that the input material is almost free of impurities. Moreover, it should be clear, that such a process will neither be able to deal with EOL-plastics nor to recycle ALDFG, which has spent a longer period of time in the oceans or coastal waters. Ideally, plastic materials should not have reached their RIP (Recyclability-Inflection-Point). The RIP could be identical with the replacement periods listed in Table 1. But to achieve optimal quality of the recycled material (i.e. upcycling) the period for reaching RIP might be even shorter. Research on this topic is ongoing (as part of the REVFIN-project) and will be published in the near future.

As a result of the diverse experiments and recycling processes conducted as part of the REVFIN project, and in conjunction with the existing procedures for plastic recycling, a waste-to-value process chain has emerged. The following process chain focuses on the

---

[1] See: https.//www.revfin.asia.

[2] RIMF: Research Institute of Marine Fisheries, Hai Phong, Vietnam – project partner of the REVFIN project.

challenges and difficulties encountered in the recycling of fishing gear that is within the EOR and is shown in Fig. 1:

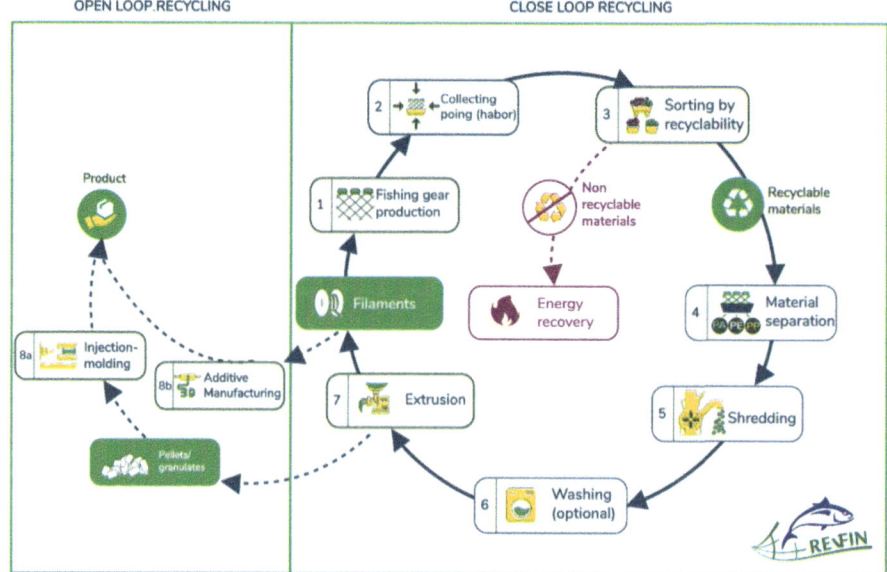

**Fig. 1.** Scheme of a Waste-to-value process chain for recycling of fishing gear

Step 1) Production of fishing gear and its usage

Step 2) Collect fishing gear at recycling directly at the harbor.

The introduction of standardized collection stations at ports allows nets to be imme-diately collected, sorted, and separated from other waste, improving visibility and increasing the amount of recyclable material collected.

Step 3) Sort by recyclability to extract recyclable materials

Step 4) Manual dismantling and separation of the different types of plastics;

To avoid impurities and contaminations a quality assurance process is highly requested.

Step 5) Shredding/Pre-cutting:

A shredder should be favored over a mill for fishing gear, since cutting mills tend to block much more quickly, especially when it comes to processing fishing nets. Shredders can automatically change the direction of rotation of the rotor to clear a blockage. Shredders run more continuously in terms of power peaks and volume. The shredder works at a significantly lower speed than a grinder/mill.

Step 6) Washing/cleaning if necessary

Step 7) Extrusion line with Twin-Screw Extruder and Water Strand Pelletizing

A twin-screw extruder is required in order to be able to modify the technical properties of the polymers in a targeted manner.

Step 8a) Injection molding

Step 8b) Filament extruding and 3D-Printing.

After pelletizing or filament production the process yields a first product for the market. To match possible quality control measures the introduction of a certain amount of new/unused material into step 7 could be considered.

# 4  Conclusion

As part of the REVFIN project the waste to value process chain has been engineered for the input material of fishing gear made from different plastic materials. In this context, different implementation options for the single process steps have been tested and validated for fishing nets and ropes.

Based on the preliminary implementation experience from the process chain development and surveys conducted as part of the REVFIN project, several crucial success factors could be identified [20]:

- The promotion of stakeholder cooperation facilitates the exchange of knowledge and resources, thereby enhancing the efficiency and acceptance of recycling processes.
- Awareness raising through public campaigns informs the public about the benefits of recycling and encourages greater support and participation.
- A standardized, optimized return system ensures the collection and reintroduction of used fishing equipment into the recycling process.
- Regular reports provide transparency and enable progress to be evaluated, thus ensuring continuous improvement of recycling processes.
- The creation of appropriate framework conditions supports the implementation and sustainability of recycling initiatives.
- Finally, design optimization improves the efficiency and quality of recycled materials through easy dismantling and sorting.

In addition to these success factors, the validated waste-to-value process chain for used fishing equipment will be installed at three higher education institutions in Vietnam to fulfil four main tasks:

- Support the educational process and increase practical skills of future employees
- Support ongoing education for employees of plastics enterprises
- Allow research and increase research capabilities of the partner universities
- Allow joint process development and research with Vietnamese enterprises

Against the background of the relevant waste volume connected to fishing gear, this will provide a building block towards creating awareness and establishing the means to increase the circularity of material flows. An important cornerstone is the emphasis on stakeholder cooperation as outlined above. The implementation activities will take this into account and further enhance the educational experience and practical skills of students while providing ongoing education for (future) employees in the plastics industry.

The objective of this study is to develop an effective process chain for recycling fishing equipment in Vietnam, contributing to the reduction of environmental pollution by plastic waste in coastal waters. By defining the current lifecycle of fishing gear in Vietnam and identifying challenges, an optimized waste-to-value process chain has been designed and various success factors impacting this process chain have been identified.

In addition to further research and analysis, it will be necessary to implement these findings in cooperation with stakeholders and universities and to implement them at the ports in order to establish a functioning, standardized recycling system for used fishing gear in Vietnam.

**Acknowledgements.** This research was conducted as part of the project "Prevention, reduction and recycling of fishnets polluting Vietnamese coastal waters (REVFIN)". The project is funded by the by the Federal Ministry for the Environment, Nature Conservation, Nuclear Safety and Consumer Protection (BMUV) on the basis of a resolution of the German Bundestag.

# References

1. 2017 Mediterranean Quality Status Report, United Nations Environment Programme, Barcelona (2017)
2. Thomas, K., Dorey, C., Obaidullah, F.: Ghost gear: the abandoned fishing nets haunting our oceans, Greenpeace Germany, Hamburg (2019)
3. Lebreton, L., Ferrari, L., et al.: Evidence that the great pacific garbage patch is rapidly accumulating plastic. Sci. Rep. (2018)
4. Veiga, J., Fleet, D., Kinsey, S., Nilsson, P.:Identifying Sources of Marine Litter - TGML Report, JRC Technical Reports, Luxembourg (2016)
5. Gilman, E., Musyl, M., Suuronen, P., Chaloupka, M., Gorgin, S., Wilson, J., Kuczenski, B.: Highest risk abandoned, lost and discarded fishing gear, g6 m iiiiii Scientific Reports (2021)
6. Gelabert, E., et al.: Multiple pressures and their combined effects in Europe's seas. In: ETC/ICM, Magdeburg (2019)
7. Korpinen, S., et al.: Multiple pressures and their combined effects in Europe's seas. In: ETC/ICM, Magdeburg, (2019)
8. Gilman, E.: Bycatch governance and best practice mitigation technology in global tuna fisheries, Marine Policy, Honolulu (2011)
9. Jambeck, J., et al.: Plastic Waste Inputs from Land into the Ocean. Science, Washington (2015)
10. Pham, T., Vu, B., Huynh, T., Vu, D.: Factors driving plastic-related behaviours: towards reducing marine plastic waste in Hoi An, Vietnam. J. Clean. Prod. (2023)
11. Dang, V.H., Gam, P.T., Son, N.T.X.: Vietnam's regulations to prevent pollution from plastic waste: a review based on the circular economy approach. J. Environ. Law **33**, 137–166 (2020)
12. Moss, E.: Reducing Plastic Pollution-Campaigns that Work, Moss & Mollusk Consulting, Tampa (2021)
13. Macfadyen, G., Huntington, T.: Abandoned Lost or Otherwise Discarded Fishing Gear, UNEP Regional Seas Reports and Studies (2009)
14. Barrett, C.: Fishing For Litter, KIMO International, https://fishingforlitter.org/. Accessed July 2024
15. Sala, A., Richardson, K.: Fishing Gear Recycling Technologies and Practices. FAO and IMO, Rome (2023)
16. Charter, M., Sherry, J., O'Connor, F.:Creating business opportunities from waste fishing nets. Opportunities for circular business models and circular design related to fishing gear, Blue Circular Economy (BCE), Norway (2020)
17. Stolte, A., Lamp, J., Schneider, F., Dederer, G.: A treatment scheme for derelict fishing gear (2019)

18. Van Meel, G.:Guidance document on conducting techno-economic feasibility studies for the establishment of port reception facilities for plastic waste. IMO, London (2023)

19. Basurko, O., et al.: End-of-life fishing gear in Spain: quantity and recyclability. Environ. Poll. **316**(2) (2022)

20. Liotta, I., et al.: Mitigation approach of plastic and microplastic pollution through recycling of fishing nets at the end of life. Process. Saf. Environ. Prot.Saf. Environ. Prot. **182**, 1143–1152 (2024)

21. Löw, C., Manhart, A., Prakash, S., Michalscheck, M.: Design-for-recycling (D4R) – State of play, Deutsche Gesellschaft für Internationale Zusammenarbeit (GIZ) GmbH, Freiburg (2021)

22. Hoang, S., Ehleben, M., Potempa, T., Henderson, J.: Sustainable approaches to fishing gear debris in Europe: effekctive management, reduction and recycling strategies, Ostfalia Hochschule für angewandte Wissenschaften (2023)

# Characterization of Virgin and Used Fishing Gear from Vietnam

Julia Tetzner[1](✉) ⓘ, Dinh Quynh Oanh[2], Nguyen Thi Tham[2], James Henderson[1] ⓘ, Thomas Potempa[1] ⓘ, and Max Ehleben[1] ⓘ

[1] Institute of Recycling, Ostfalia University of Applied Sciences, Wolfsburg, Germany
`ju.tetzner@ostfalia.de`
[2] Faculty of Environment, Ha Long University, Ha Long, Vietnam

**Abstract.** Marine plastic waste from fisheries has emerged as a significant environmental issue, posing a significant threat to marine biodiversity and ecosystem health. The material recycling of fishing gear is an effective way to reduce waste and promote a circular economy. This study examines the characteristics of virgin and used fishing nets and ropes to understand their recyclability before they are discarded. The used nets and ropes in this study were in service for up to two years. The material characteristics were identified based on Fourier-transform infrared (FTIR) spectroscopy while thermal properties were obtained using differential scanning calorimetry (DSC). Based on FTIR results, the main polymer of the fishing nets was found to be PE, while the fishing ropes were found to be a blend of PE/PP and a combination of PE/PP and PET. The aging and oxidation process produces new bands in the infrared (IR) spectra of PE and PP fishing nets and ropes. The significant differences between used and virgin materials were also obtained in terms of crystallization enthalpies. This suggests that the fishing nets and ropes were affected by environmental factors.

**Keywords:** Fishing gear · thermal analysis · aging · polymer degradation · material recycling

## 1 Introduction

In the course of the last few decades, marine plastic waste from fisheries and aquaculture has become a significant environmental problem. The fishing sector contributes significantly to marine plastic pollution through abandoned, lost or otherwise discarded fishing gear (ALDFG). It is estimated that the number of pieces of fishing gear that become marine plastic waste is increasing every year. A study by Richardson et al. [1] found that 2% of all fishing gear ends up in the ocean every year. A study by Kim et al. [2] estimated that 11 436 tons and 38 535 tons of trap and gillnets, respectively, are abandoned annually in the coastal waters of South Korea. Fishing gear lost or disposed into the ocean affect marine environments and coastal economies worldwide. It threatens the marine life cycle by entangling and killing marine mammals, fish and birds [3].

Vietnam is a country with a long coastline (3 260 km) with coastal fishing villages and major fishing ports in the coastal provinces from the north to the south of the country

© The Author(s) 2025
H. Kohl et al. (Eds.): GCSM 2024, LNME, pp. 617–625, 2025.
https://doi.org/10.1007/978-3-031-93891-7_68

(e.g., Hai Phong, Quang Ninh, Khanh Hoa, and Kien Giang). In recent years, the fisheries sector in Vietnam, including both marine fisheries and aquaculture, has made spectacular progress. However, Vietnam is one of the top countries contributing to marine plastic waste, generating between 0.28 and 0.73 million tons per year, according to a survey from 2015 [4].

Moreover, data on fishing gear composition collected by RIMF[1] as well as by several studies show that in average fishing nets and ropes are commonly made of different polymeric fibers such as polyethylene (PE), polypropylene (PP), and polyamide (PA) , 6]. Recent studies have shown that old fishing nets and ropes could be recycled using plastics recycling technologies to prevent the disposal of fishing gear. , 8]. However, the longevity of fishing gear and the environmental impacts on it can vary considerably depending on its exposure to marine environmental factors [3]. In addition, very little information is available on the composition and characteristics of plastic in fishing gear in Vietnam. Understanding the impact of these factors on new and old fishing nets is therefore crucial for sustainable fisheries management and marine conservation.

The main goal of this study is to compare the quality of virgin and used fishing gear polymers in order to identify the characteristics of fishing gear including nets and ropes in Vietnam. To achieve this goal, the thermal properties and chemical structure of virgin and used fishing gear polymers will be compared. By evaluating the characteristics of both virgin and used fishing gear, this research aims to assess and identify the properties of used fishing nets and ropes to know how to use their material for recycling, towards a circular economy, and to provide insights into the environmental sustainability of fishing practices in Vietnam's marine ecosystems.

## 2   Methodology

### 2.1   Materials

Four different types of fishing gear (Fig. 1) were collected in the coastal fishing village of Van Don district, Quang Ninh province. For each type of fishing gear, virgin and used samples were collected from a medium-sized enterprise that sells and also repairs fishing gear for reuse. These fishing gear samples were selected because of their widespread use by local fishermen in Quang Ninh. In order to estimate the age of the used fishing gear, the data from the survey conducted within the framework of the REVFIN project in 2023/2024 is used. This survey shows that the average lifespan of fishnets is 15 months, while that of ropes is 25 months. Two types of fishing nets (NET1 and NET2) and two types of fishing ropes (ROPE1 and ROPE2) were collected.

---

[1] RIMF: Research Institute of Marine Fisheries, Hai Phong, Vietnam – project partner of the REVFIN project.

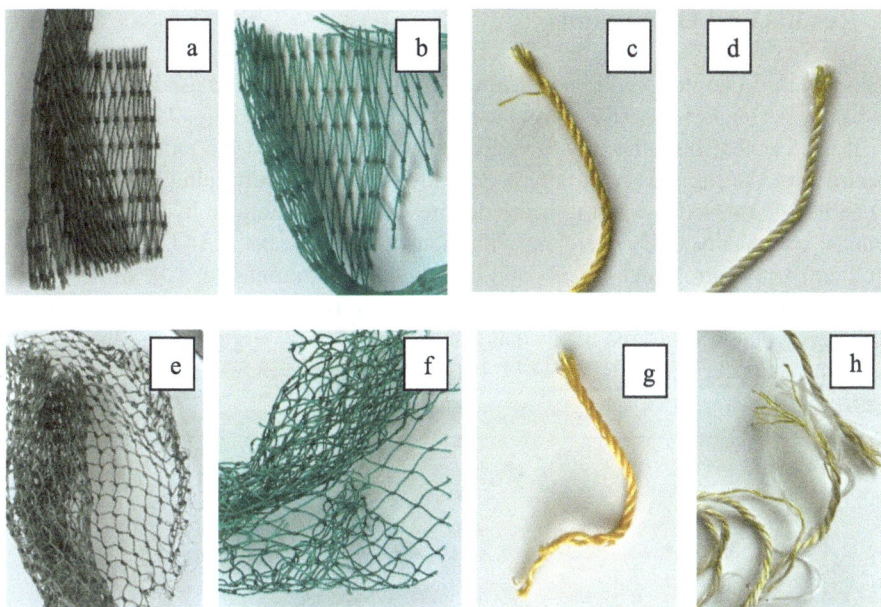

**Fig. 1.** Photos of fishing gear samples (virgin and used), collected in Quang Ninh Province, Vietnam; a: NET1 virgin; b: NET2 virgin; c: ROPE1 virgin; d: ROPE2 virgin; e: NET1 used, f: NET2 used; g: ROPE1 used; h: ROPE2 used

## 2.2  Methods

Individual polymer components contained in the fishing gear samples were identified using an Attenuated Total Reflection-Fourier transform infrared (ATR-FTIR) spectrometer. The IR spectra were recorded using the Bruker Tensor 27. The bonds and functional groups in the new and old fishing nets and ropes were identified. The data was processed and exported using Opus software. Measurements were made in a wave number range from $400 \text{ cm}^{-1}$ to $4000 \text{ cm}^{-1}$. Three measurements were carried out per sample. The types of polymers were obtained through the comparison with the Bruker ATR-FTIR spectral polymer database.

Thermal properties were investigated based on Differential Scanning Calorimetry (DSC). A detailed characterization of fishing gear materials such as the temperature of glass transition, crystallization temperature, melting temperature, and oxidation stability was obtained based on analyzing DSC curves. The measurements followed the DIN EN ISO 11357 standard. DSC measurements were performed using a NETZSCH DSC 214 Polyma. In the case of all the DSC measurements, the heating rate was 10 K/min under an oxygen atmosphere with isothermal steps at the start and stop to ensure instrument stability. The average weight of the sample was about 10 mg and were sealed in perforated aluminum vessels. The exothermic peak minimum and endothermic peak maximum were taken as crystallization and melting temperatures, ($T_c$ and $T_m$, respectively), whereas the area of each peak was considered as the crystallization and melting enthalpy, $\Delta H_c$ and $\Delta H_m$, respectively.

## 3  Results and Discussion

The spectra of samples NET1 (new and used) and NET2 (new and used) show similar IR-bands. A comparison of virgin and used NET2 can be seen in Fig. 2.

It is clear that both the green, virgin material spectra and the red, used material spectra have the same basic structure. The strong bands at wavenumbers 2914 cm$^{-1}$ (1) and 2847 cm$^{-1}$ (2) which indicate the presence of unsaturated hydrocarbon compounds, specifically C-H$_2$ stretching vibrations [9]. The peaks at 1472 cm$^{-1}$ (3) and 1462 cm$^{-1}$ (4) correspond to CH$_2$ and CH$_3$ bending vibrations [9; 10], while the IR bands at 730 cm$^{-1}$ (5) and 718 cm$^{-1}$ (6) are characteristic for C-H$_2$ rocking vibrations [10]. This pattern of infrared bands that can be seen in both, virgin and used materials, forms the characteristic spectrum of PE [9].

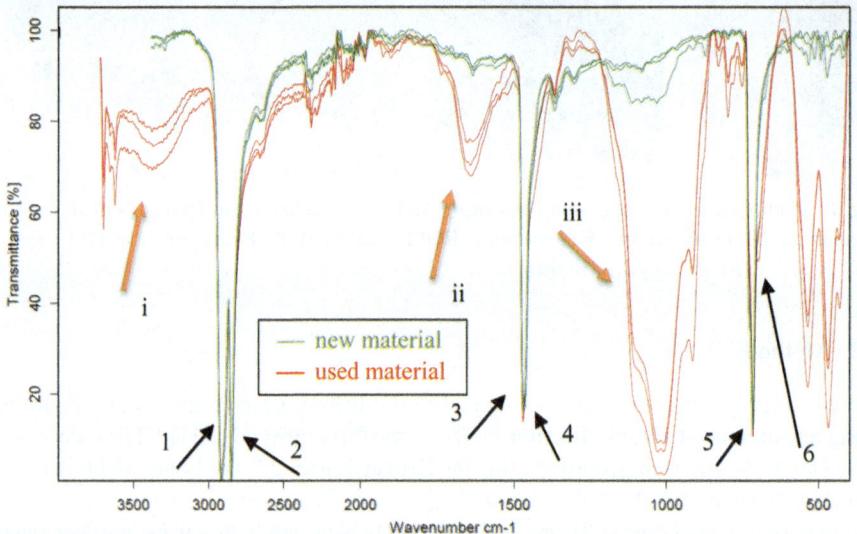

**Fig. 2.** Standardized IR-spectra of virgin (green) Vietnamese fishing net (NET2) and used (red) Vietnamese fishing net (NET2) in comparison, with markings of characteristic identification bands for HDPE (1-6) and markings of aging bands (i-iii)

In addition to these characteristic bands, the IR-spectra of both, new and used samples, further showed small peaks that can be seen at 3400 cm$^{-1}$ (i), 1636 cm$^{-1}$ (ii), around 1100 cm$^{-1}$ (iii) in all measurements (Fig. 2).

The absorption band around 3400 cm$^{-1}$ is indicative for an -OH group, while the slight band at 1636 cm$^{-1}$ is an indicator for a carbonyl group (-CO). The band at 1032 cm$^{-1}$ can also represent a -CO group [10]. These groups are caused by the aging of plastics, where UV radiation generates radicals which break the bonds in the polymer chain, thus changing the structure of the chain. In the presence of oxygen, the oxygen can be integrated into the polymer chain, resulting in the formation of peroxide and carbonyl groups. An increase in band intensity in these areas therefore indicates aging of the plastic [11].

Since each sample was measured three times (at different sample locations), Fig. 2 indicates that aging effects are dependent on the sample location tested. Height and structure of the aging bands i, ii, and iii are different at each measured location. Even for the virgin material, slight aging affects could be detected (see Fig. 2, aging band iii). The fact that slight signs of aging can also be identified in new samples suggests that the nets that were tested have already undergone slight aging - whether due to processing, incorrect storage and/or unsuitable material cannot be determined from these measurements. In combination with the study from Ranjan et al. [11], the signs of aging in new fishing nets might indicate improper storage, e.g., unprotected from UV light.

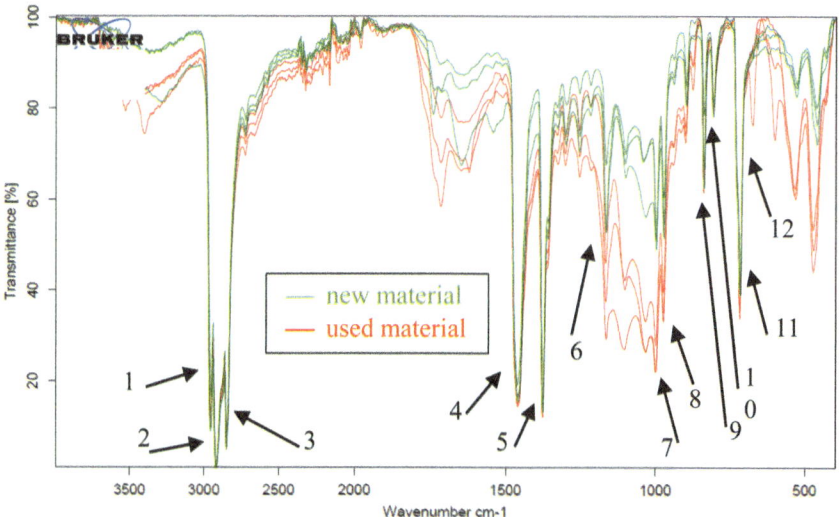

**Fig. 3.** Standardized IR-spectra of virgin (green) Vietnamese fishing rope (ROPE2) and used (red) Vietnamese fishing rope (ROPE2) in comparison, with markings of characteristic identification bands for PP (1–10) and PE (11;12)

The ROPE1 sample shows a single component material while ROPE2 consists of two different components: a white main component and a separately woven yellow fiber. To ensure that these were correctly identified, the different fractions were analyzed independently. The spectra of ROPE1 (new and used) and the yellow component of ROPE2 (new and used) show similarities to the net spectra (Fig. 2), but with an increased number of bands. In Fig. 3, a comparison of virgin and used Rope2 (yellow part) is shown. The analysis of the spectra showed that the materials were not a pure polymer but a PE-PP blend.

The following band combination is characteristic for PP: The bands at 2950 cm$^{-1}$ (**1**), 2915 cm$^{-1}$ (**2**) and 2848 cm$^{-1}$ (**3**) indicate a C-H stretching vibration [7]. The bands at 1461 cm$^{-1}$ (**4**) and 1377 cm$^{-1}$ (**5**) correspond to a CH$_2$ or CH$_3$ bending vibration, while the band at 1166 cm$^{-1}$ (**6**) can be assigned to a CH bending vibration, CH3 rocking vibration or a C-C stretching vibration. The bands at 997 cm$^{-1}$ (**7**), 972 cm$^{-1}$

(8), 840 cm$^{-1}$ (9) and 808 cm$^{-1}$ (10) all correspond to a CH$_3$ or CH$_2$ rocking vibration, but can also represent bending and stretching vibrations [11].

In addition to the assignment to PP, the bands at 2915 cm$^{-1}$, 2848 cm$^{-1}$, 1461 cm$^{-1}$ and 1376 cm$^{-1}$ can also be assigned to the spectrum of PE. While the bands at wave numbers 730 cm$^{-1}$ (11) and 719 cm$^{-1}$ (12) are characteristic for the CH$_2$ rocking vibrations of polyethylene [11]. On the basis of these band assignments and spectral comparisons in databases, it can be assumed that the rope materials ROPE1 and ROPE2 are a polymer blend of PE and PP.

As in the case of the NET spectra, there are also bands in the ROPE-spectra located around 3400 cm$^{-1}$, 1700 cm$^{-1}$ and 1100 cm$^{-1}$, which indicate degradation of the plastics [11]. However, a clear difference in intensity from virgin (green) to used (red) material can only be seen in the ROPE2 samples (Fig. 3).

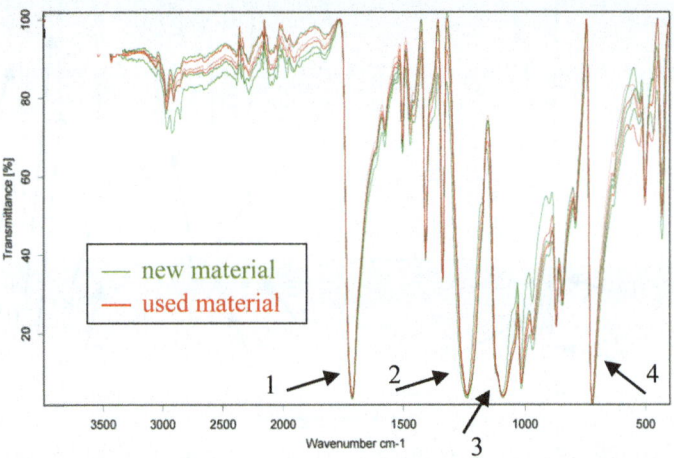

**Fig. 4.** Standardized IR-spectra of virgin (green) Vietnamese fishing rope (ROPE2) and used (red) Vietnamese fishing rope (ROPE2) in comparison, with markings of characteristic identification bands for PET (1–4)

The analysis of the spectra of the white main component of ROPE2 (Fig. 4) indicates PET material. This is supported by the characteristic band at the wavenumbers 1714 cm$^{-1}$ (1) which indicates a C = O stretch vibration, as well as bands at wavenumbers 1238 cm$^{-1}$ (2) and 1092 cm$^{-1}$ (3) which vibrate due to C-O stretch vibrations. The characteristic band at wavenumber 720 cm$^{-1}$ (4) is for an aromatic CH out-of-plane bend [9]. As the spectral curves are very similar a significant difference between virgin material (green spectra) and used material (red spectra) cannot be determined.

The DSC curves of virgin and used material of each NET-sample were analyzed in the temperature range of 50 - 180 °C. Table 1 summaries the thermal properties of the fishing nets under investigation (crystallization and melting temperatures, and melting and crystallization enthalpies for both virgin and used fishing nets). For the fishing nets, based on the DSC curves the peak temperature of melting was 134.4 °C for virgin PE and 137.1 °C for the used PE fishing nets.

The DSC showed a typical curve for HDPE, what is confirmed by the melting temperatures of approximately 134 °C, which is according to literature (135 °C) [12]. The crystallization temperature of the used PE material is slightly increased, possibly due to shortened polymer chains or nucleating agents that can cause this effect [12]. The small increase in crystallization temperature might indicate the degradation of the materials. On the other hand, there is little variation in the thermal properties of the fishing rope samples. Thus, polymer degradation and aging possibly occurred in the used fishing gear, which are interconnected processes driven by the exposure of polymers to environmental factors and the accumulation of damage over time. Understanding these degradation mechanisms is crucial for designing a recycling process yielding materials with high value.

**Table 1.** Averaged DSC-results with average error of each three virgin and used Vietnamese fishing nets (NET1 and NET2), with $T_M$, $T_C$, $\Delta H_M$ and $\Delta H_C$

|  |  | $T_M$ ($\pm$ SD) | $T_C$ ($\pm$ SD) | $\Delta H_M$ ($\pm$ SD) | $\Delta H_C$ ($\pm$ SD) |
|--|--|--|--|--|--|
|  |  | (°C) | (°C) | (J/g) | (J/g) |
| NET1 | Virgin | 134.4 ($\pm$0.7) | 112.6 ($\pm$0.5) | 188.6 ($\pm$5.8) | 196.4 ($\pm$9.2) |
|  | (n = 3) |  |  |  |  |
|  | Old | 136.7 ($\pm$0.3) | 114.6 ($\pm$0.4) | 196.2 ($\pm$1.4) | 195.0 ($\pm$11.6) |
|  | (n = 3) |  |  |  |  |
| NET2 | Virgin | 135.0 ($\pm$0.4) | 113.9 ($\pm$1.2) | 201.9 ($\pm$11.0) | 205.9 ($\pm$6.7) |
|  | (n = 3) |  |  |  |  |
|  | Old | 137.1 ($\pm$0.8) | 114.5 ($\pm$0.9) | 190.3 ($\pm$5.1) | 204.1 ($\pm$3.1) |
|  | (n = 3) |  |  |  |  |

# 4  Conclusions

The detailed sampling and characterization of fishing nets and ropes provided critical information for evaluating their recyclability. All net samples show the characteristic spectrum for PE and signs of aging, likely due to UV radiation. ROPE1 consists of a PE-PP blend, while ROPE2 is composed of PET and a PE-PP blend, both showing similar aging behavior to the PE nets.

If the recycling impact of fishing nets and ropes is to be increased, at least two factors have to be improved; firstly, the aging of the material during storage needs to be avoided and secondly, the mixing of materials to form composite products, such as ROPE2, needs to be prevented to increase the recyclability. Aging affects can be optimized by applying a suitable infrastructure for storage of the fishing gear during distribution and use. To increase recyclability, fishing gear manufacturers and commissioners need to be informed and trained not to mix different plastics into one product. To initiate a substitution process, commissioners of fishing gear should be aware not to purchase composite plastic products, such as ROPE2.

The distinct surface aging differences between old and new materials, which can be clearly identified, necessitate further investigations to understand their impact on the recyclate. These aging bands could be utilized to assess the quality of the recyclate and determine whether it is suitable for upcycling, where the material is converted into higher-value products, or downcycling, where it is transformed into lower-value products. Such insights are crucial for developing sustainable recycling processes and maximizing the efficiency of material reuse.

**Acknowledgements.** This research was conducted as part of the project "Prevention, reduction and recycling of fishnets polluting Vietnamese coastal waters (REVFIN)". The project is funded by the by the Federal Ministry for the Environment, Nature Conservation, Nuclear Safety and Consumer Protection (BMUV) on the basis of a resolution of the German Bundestag.

# References

1. Richardson, K., Hardesty, B.D., Vince, J., Wilcox, C.: Global estimates of fishing gear lost to the ocean each year. Sci. Adv. **8**(41), eabq0135 (2022). https://doi.org/10.1126/sciadv.abq 0135
2. Kim, S.-G., Lee, W.-I., Yuseok, M.: The estimation of derelict fishing gear in the coastal waters of South Korea: trap and gill-net fisheries. Mar. Policy **46**, 119–122 (2014)
3. Moore, C.J.: Synthetic polymers in the marine environment: a rapidly increasing, long-term threat. Environ. Res. **108**(2), 131–139 (2008). https://doi.org/10.1016/J.ENVRES.2008.07.025
4. Jambeck, J., et al.: Plastic Waste Inputs from Land into the Ocean. Science, Washington (2015)
5. Basurko, O.C., et al.: End-of-life fishing gear in Spain: quantity and recyclability. Environ. Poll. **316**(2), 120545 (2023)
6. Weißbach, G., Gerke, G., Stolte, A., Schneider, F.: Material studies for the recycling of abandoned, lost or otherwise discarded fishing gear (ALDFG). Waste Manage. Res. **40**(7), (2022)
7. Stolte, A., Lamp, J., Dederer, G., Schneider, F.: A treatment scheme for derelict fishing gear. Interreg Baltic Sea Region Programme 2014–2020, MARELITT Baltic Project report, 32 p. 2019
8. Wong, A.M.: Valued waste/wasted value: Waste, value and the labour process in electronic waste recycling in Singapore and Malaysia. Geogr. Compass **16**(4), e12616 (2022). https://doi.org/10.1111/gec3.12616
9. Jung, M.J., et al.: Validation of ATR FT-IR to identify polymers of plastic marine debris, including those ingested by marine organisms. J.M Lynch (2018)
10. Brunner, S.: Fouriertransformations-Infrarotspektroskopie (FTIR) in abgeschwächter Total-reflexion (ATR) und externer Reflexion (er) an Kunststoffen – Aufbau einer Spektren-datenbank, Identifikation der Zusammensetzung und Alterungserscheinungen mittels FTIR" –, TU München (2015)

11. Ranjan, V.P., Goel, S.: Recyclability of polypropylene after exposure to four different environmental conditions. Resour. Conserv. Recycl. **69**, 105494 (2021). https://doi.org/10.1016/j.resconrec.2021.105494
12. Braun, D.: Erkennen von Kunststoffen' 5. Auflg. München, Hanser (2012)

# AI-Based System for Enhanced Textile Recycling and Reuse

Jan Weimer[1], Karsten Pufahl[1], Marco Jagodzinski[1(✉)], Adam Cisowski[2], Birgit Kanngiesser[1], and Dirk Oberschmidt[1]

[1] Technische Universität Berlin, 10683 Berlin, Germany
m.jagodzinski@tu-berlin.de
[2] Freie Universität Berlin, 14195 Berlin, Germany

**Abstract.** The textile industry is highly competitive, with decreasing production costs and increasing disposal of clothes. Currently, such items are mostly combusted, disposed, or downcycled into lower value products like fleece, posing an ever-increasing demand on (fossil) resources and energy. Only a small share of garments is reused, as the sorting process is traditionally performed by trained personnel, resulting in significant time consumption. The method presented here employs computer vision and artificial intelligence for textile pre-sorting, significantly enhancing the proportion of reusable textiles. The method relies on a custom high-resolution camera system that captures images of textiles and garments on a conveyor belt. The system not only recognizes the type but also infers the quality as well as the fabric type of garments, ensuring high-quality sorting and substantially increasing the proportion of textiles that can be reused. The presented approach of fabric detection in visually inspected objects has potential applications beyond post-consumer sorting, such as in electronics and plastics recycling. It can also be applied to manufacturing tasks across various industries—including agriculture, food, and automotive—enhancing quality control, reducing waste, increasing efficiency, and thus supporting the principles of a circular economy.

**Keywords:** Textile recycling · Textile reuse · Fabric classification

## 1 Introduction

The textile industry is highly competitive, marked by declining production costs and a growing volume of discarded clothing. This sector, especially fast fashion, is a major contributor to greenhouse gas emissions and waste [1], with 92 million tons of textile wasted annually, and less than 1% being recycled into similar quality garments [2]. The increasing disposal of clothes poses significant environmental challenges, with most discarded textiles incinerated, sent to landfills, or downcycled into low-value products. This process wastes materials and increases fossil resource demand, worsening environmental degradation. State-of-the-art textile recycling uses mechanical and chemical methods. Both methods are limited by input quality and sorting efficiency. Traditional sorting is manual, time-consuming, labor-intensive, and inconsistent, especially with various fabric structures like plain weave and knitting patterns that require inspection.

© The Author(s) 2025
H. Kohl et al. (Eds.): GCSM 2024, LNME, pp. 626–634, 2025.
https://doi.org/10.1007/978-3-031-93891-7_69

This paper supports an automated textile sorting process, comprising garment collection, conveyor application, scanning, and sorting. The focus of this paper lies on automizing the scanning stage, as the manual inspection process poses the major bottleneck for larger scale applications of textile sorting. The application and sorting stages are beyond this paper's scope. The inspection process is conducted via a high-resolution camera and AI to analyze and classify textiles. The goal is to identify clothes for reuse, as those fractions of disposed textiles pose the major share of profit margins of textile sorting facilities. It is aimed to scan one garment every 1–5 s to be economically feasible. If higher throughput is demanded, it is aimed to parallelize the application and/or scanning stage. Yet, achieving low runtimes and high accuracy is crucial the computer vision AI system in order to meet throughput and quality demands.

Recent advancements in computer vision and AI offer new possibilities for automating textile sorting, analyzing visual data to accurately identify and classify objects. However, these technologies are still in early stages, needing improvements in accuracy, speed, and integration into recycling workflows. While textile texture analysis is a smaller field in computer vision, notable publications exist. Seçkin et al. [3] discuss the FabricNET dataset, focusing on statistical approaches and algorithms like XGBoost. This ensemble machine learning algorithm creates multiple decision trees to capture dataset structures and extracts properties such as textile mass and weft using regression models like Multilayer Perceptron (MLP). Schneider and Merhof [4], and Wang et al. [5], use algorithmic methods to determine material properties and weave structures through repetitive patterns and color contrasts. Wang et al. enhance color contrast in scanned images, convert them to greyscale, and use Crossed-Area segmentation to detect interlacing areas. By analyzing grey levels, they identify repeating patterns and divide images into sub-images. The orientation of these segments reveals the weave pattern, clustered using fuzzy c-means clustering (FCM). However, this method requires high-detail images, impractical for high-throughput industrial applications like conveyor belts. Therefore, a method to identify fabric types from lower resolution images, typically obtained with machine-vision cameras, is needed. The presented method uses computer vision and AI for textile pre-sorting with a high-resolution camera system capturing images on a conveyor belt. Advanced AI algorithms identify garments suitable for the second-hand market and detect defects, evaluating both type and condition. A neural network-based approach, trained on public and self-scanned image datasets, classifies fabric types, crucial for subsequent recycling processes.

## 2 Experimental Setup

### 2.1 Description of the Hardware

The self-scanned dataset was acquired by a self-built computer vision system (Fig. 1-a). The system includes a camera mounted above the conveyor belt, which is flat and wide to accommodate the large garments and ensure high throughput. The mechanical and optical parameters of the system are enlisted in Table 1. The conveyor belt is specifically designed for this application, featuring a flat and wide structure to support large garments. It operates at a high conveying speed to maximize throughput. Garments are placed flat on the front side of the conveyor belt to ensure optimal imaging conditions.

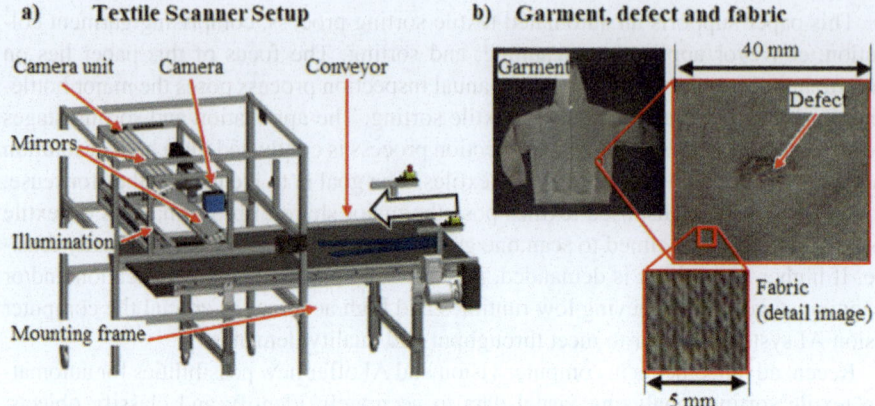

**Fig. 1.** a) CAD-Image of the textile scanning camera system. The covers of the camera-system are not depicted in the figure for better visibility. b) Exemplary image of a scanned garment and fabric defect

**Table 1.** Parameters and Dimensions of the textile scanning system and computing hardware

| Computer vision hardware | |
| --- | --- |
| Setup dimensions, length x width x height | 3 m x 1.5 m x 1.8 m |
| Conveyor belt dimensions, length x width x height | 2.5 m x 1.2 m x 0.9 m |
| Conveying speed | 1 m/s–3 m/s |
| Clearance between camera unit and conveyor belt | 250 mm |
| Distance camera-conveyor | 1.4 m |
| Camera FOV | 1.2 m |
| **Computer hardware used for model training and validation** | |
| CPU | 12$^{th}$ Gen Intel Core I9-12900K, 3.200 MHz, 16 Cores, 24 logical processors |
| GPU | NVIDIA GeForce GTX 1070, 1920 CUDA Cores, 1632 MHz (oc) |
| RAM | 4x 32 GB DDR4 RAM, 3.000 MHz |

The base construction of the mounting frame and camera unit is composed of aluminum extrusion bars, providing a lightweight yet robust framework. This construction facilitates easy adjustments and ensures the stability of the mounted components. A line scan camera is used to capture high-resolution images of the garments as they move along the conveyor belt. To make the camera unit more compact, two mirrors are used in a Z-shaped alignment, folding the optical path. The camera unit is fully enclosed to protect the components and ensure consistent imaging conditions. Dark field illumination is employed due to the lack of available coaxial systems suitable for bright field illumination within the required field of view (FOV). This lighting method enhances the contrast of surface features, aiding in the detection of defects and other pertinent characteristics of the garments.

## 2.2  Suppression of Vibrational Disturbances

Despite a rather stiff base-construction, it was observed the system incorporates measurable vibrations. These oscillations manifest as deviations in the camera's viewing angle, affecting both the rotational and translational vibrational modes, as depicted in Fig. 2-a.

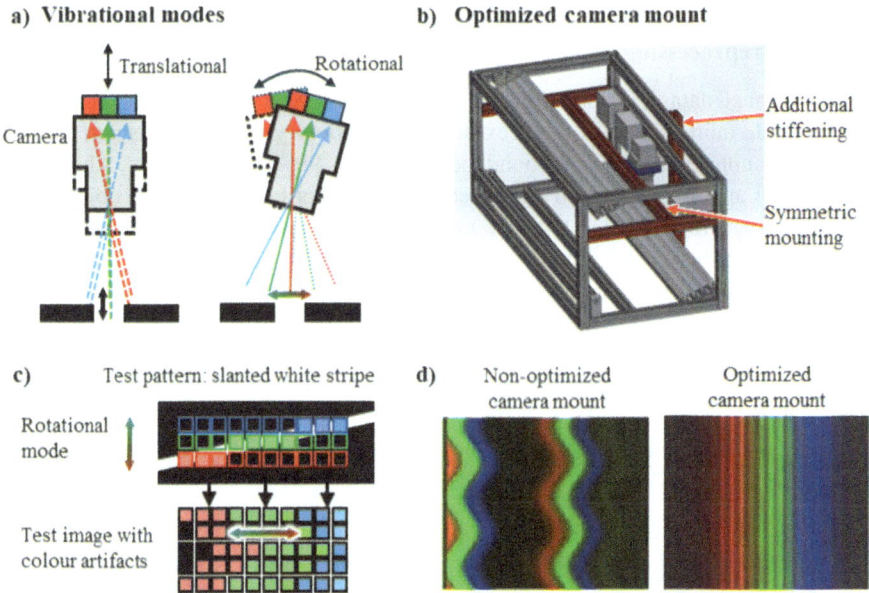

**Fig. 2.**  a) Potential vibrational modes of the camera system and related shift s of the optical axis and variations of the illumination in the RGB-channels. b) CAD image of the optimized mount. c) Depiction of image of a test pattern with a slightly slanted white stripe and wave-like artifacts. d) Acquired test images with and without the optimized mount.

The presence of these oscillations results in chatter-like artifacts and color deviations in the captured images due to the sequential line-by-line image acquisition of the line scan camera. As each line is recorded consecutively, any mechanical vibrations can cause misalignment between the lines. This misalignment results in inconsistencies in the captured image, appearing as chatter-like color artifacts. The color artifacts are particularly noticeable at black-white transitions. This phenomenon occurs because each color channel of the camera has its own optical axis, leading to each color pixel capturing different sections of the black-white transitions. Consequently, the misalignment among the color channels introduces visible color distortions, especially at high-contrast edges.

Furthermore, a test pattern, containing a slanted black-white transition mask, as depicted in Fig. 2-c can be employed to visualize the camera oscillations. This method allows for a clearer observation and analysis of the vibrational effects, enabling more precise adjustments to mitigate their impact on the camera system's performance. To address these issues, a local optimization of the camera's construction was implemented (see Fig. 2-b). By adjusting the camera mounting to a symmetric alignment (as well as additionally stiffening it), the oscillation modes can be aligned along the optical axis or line of sight of the camera. As depicted in Fig. 2-d, this alignment minimizes the deviation of the viewing angle caused by vibrational disturbances, thereby reducing the impact on image quality.

## 3   AI Based Fabric Detection

### 3.1   Data Preprocessing

The self-scanned dataset includes 584 images, each up to $3840 \times 5000$ pixels. The images contain textile fabric sheets of the "Textile materials library" from Re_fashion [6] on a mostly black conveyor belt background, occupying 80–90% of each image. About 10% of the images are either distorted due to scanning errors or lack discernible objects. The fabric sheets are fully labeled with the material composition, color, structure, source, and condition of the textiles. The second dataset, created by Seçkin et al. [3], consists of 130 images with a resolution of $640 \times 480$ pixels, categorized into "Plain" (50 images), "Satin" (41 images), and "Twill" (39 images). These images were taken with a digital handheld microscope, capturing single-layer fabrics made of 70–100% cotton yarn with thicknesses ranging from 0.1mm to 1mm. All images are defect-free and colored. For the training of the neural networks, to reduce training and prediction runtime and decrease disk space, the background of the self-scanned dataset needed to be removed, as depicted in Fig. 3.

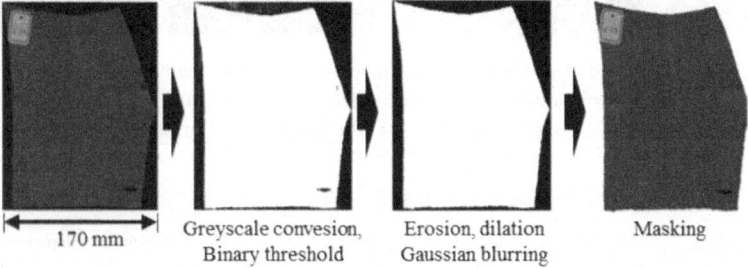

| | Greyscale convesion, | Erosion, dilation | Masking |
| 170 mm | Binary threshold | Gaussian blurring | |

**Fig. 3.** Example image of a textile sheet from the CRTX dataset and data preprocessing method

This involves converting the image to greyscale and applying binary thresholding with a value of 10, followed by morphological transformations like erosion and dilation to clean up the image. This approach, inspired by Zhang et al. [7], is used because the background is a mix of gray tones, and simple black pixel removal isn't effective. Data augmentation techniques are then applied to increase the quantity and diversity of the data.

To increase the variety and quantity of the dataset, various data augmentation techniques are applied. The first technique involves dividing the original image into smaller sub-images. An initial check ensures sub-images are acceptable by verifying that the average color level and the number of black pixels do not exceed predefined thresholds. Sub-images with irregular edges, which could negatively affect predictions, are also eliminated. Stratified k-fold cross-validation with 5 folds is employed to expand the training data, ensuring proportional class distribution and better representation. Classical augmentation techniques like rotating the image by up to 90 ° and flipping it horizontally and vertically, each with a 25% chance, are used. Both datasets are duplicated, and unsharp masking (USM) is applied to each duplicate to enhance image sharpness. USM involves applying Gaussian blur to the image, subtracting it to extract the unsharp mask, amplifying this mask, and recombining it with the original image to emphasize edge contrast, resulting in a visually sharper image.

### 3.2 Modeling

As baseline architecture the residual neural network (ResNet) proposed by He et al. [8] was used, which is known for its balance between computational complexity and performance. Variants of ResNets have achieved state-of-the-art accuracies [9, 10] in texture and defect classification tasks. A comparative study by Brigato and Iocchi [11] showed ResNets performed the best among examined algorithms. A pretrained ResNet50 with weights initialized to "ResNet50_Weights.IMAGENET1K_V2" [12] was used. The Adam optimization algorithm [13] is employed with a learning rate of 0.003, and the "CosineAnnealingWarmRestarts" scheduler adjusts the learning rate in a cosine decay pattern. As described by Hussain et al. [14], all layers except the last one are added with a fully connected layer with 256 neurons, a ReLU activation function, a dropout layer, and another fully connected layer with a softmax activation function.

The second architecture evaluated is the transformer-based ViT architecture by Dosovitskiy et al. [15]. The pretrained vit_l_16 architecture with weights set to "ViT_L_16_Weights.IMAGENET1K_V1" is used, with all layers except the last one being frozen. The head is replaced with a fully connected layer and a softmax activation function. All other parameters, such as the learning rate and optimizer, are kept consistent with the ResNet50 model.

### 3.3 Evaluation

The models were evaluated on the described datasets, trained for 50 epochs using 5-fold cross-validation with an 80% training and 20% validation split. Batch sizes were 6, and three different random seeds were used. Training was performed on a GPU with CUDA enabled. The results (Fig. 4) show that both models could generalize and classify the underlying structures of the images. The transformer-based architecture, ViT, outperformed ResNet50 in all metrics except runtime. Sharpening as a pre-processing step improved performance for the ViT architecture. However, ResNet50's performance declined on the sharpened FabricNet dataset, suggesting that sharpening removed significant features that the model would have otherwise learned.

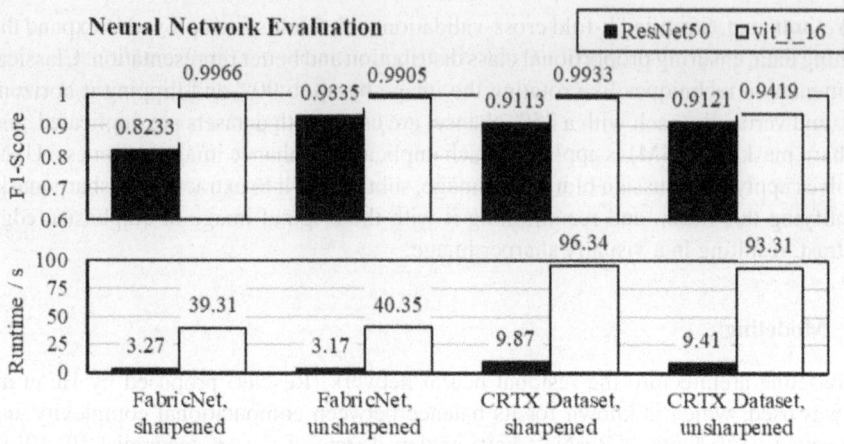

**Fig. 4.** F-Score and runtime of the evaluated neural networks

## 4 Conclusions

This paper implemented and evaluated several deep learning architectures on the Fabric-NET [3] dataset and a self-scanned dataset. Various preprocessing procedures and data augmentation techniques were applied to enhance the datasets. Both models could discern underlying structures in both datasets, but the transformer-based ViT outperformed ResNet50 in all metrics except runtime. Sharpening generally improved ViT's performance but worsened ResNet50's performance on the sharpened FabricNET dataset. To improve model accuracy and performance, increasing the amount of data is essential. The models were trained on relatively small datasets, and acquiring larger datasets is challenging. Manual dataset creation and labeling are time-consuming. Domain-adaptive pretraining on a textile dataset could further enhance performance, followed by fine-tuning for specific tasks. Higher base resolution images might improve accuracy but with higher runtime costs. Exploring other architectures, such as quantized deep learning architectures [12, 16] for reduced runtime and newer models like OmniVec [17] for better accuracy, could also be beneficial.

## 5 Outlook

The paper focuses on automating fabric structure identification using supervised learning algorithms and enhancing model performance with image sharpening and super-resolution techniques. Despite AI and computer vision advancements, issues with accuracy, speed, and integration in textile recycling persist. This approach addresses these gaps, supporting circular economy principles and sustainable practices, with the potential to revolutionize textile recycling. The method ensures high-quality items are selected for reuse, improving sorting efficiency and effectiveness, and increasing textile reuse. This defect detection approach also applies to electronics, plastics recycling, and various manufacturing industries, enhancing quality control, reducing waste, and increasing efficiency. In summary, this work aims to enhance textile recycling efficiency by automating fabric pattern recognition, reducing manual labor and errors.

# References

1. Khan, W., Ahmed, S., Dhoble, Y., Madhav, S.: A critical review of hazardous waste generation from textile industries and associated ecological impacts. J. Indian Chem. Soc. **100**(1), 100829 (2023)
2. Niinimäki, K., Peters, G., Dahlbo, H., Perry, P., Rissanen, T., Gwilt, A.: Author correction: the environmental price of fast fashion. Nat. Rev. Earth Environ. **1**(5), 278 (2020)
3. Seçkin, M., Seçkin, A.Ç., Demircioglu, P., Bogrekci, I.: Fabricnet: a microscopic image dataset of woven fabrics for predicting texture and weaving parameters through machine learning. Sustainability **15**, 15197, 10 (2023), https://doi.org/10.3390/su152115197 Author, F.: Contribution title. In: 9th International Proceedings on Proceedings, pp. 1–2. Publisher, Location (2010)
4. Schneider, D., Merhof, D.: Blind weave detection for woven fabrics. Pattern Anal. Appl. **18**, 08 (2014)
5. Wang, X., Georganas, N.D., Petriu, E.M.: Fabric texture analysis using computer vision techniques. IEEE Trans. Instrum. Meas. **60**, 44–56 (2011)
6. Publicly available pdf document, page 45, https://refashion.fr/rapport-activite/2021/public/pdf/regenerate_materials.pdf, Accessed: 2024/06/20
7. Zhang, L., Du, D., Li, L., Wu, Y., Luo, T:. Iterative knowledge distillation for automatic check-out. IEEE Trans. Multimedia 11 (2020)
8. He, K., Zhang, X., Ren, S., Sun, J.: Deep residual learning for image recognition. In: 2016 IEEE Conference on Computer Vision and Pattern Recognition (CVPR) (2015)
9. Zhang, L., Bian, Y., Jiang, P., Zhang, F.: A transfer residual neural network based on resnet-50 for detection of steel surface defects. Appl. Sci. **13**(9) (2023)
10. Shams, M.Y., et al.: Deep learning model based on resnet-50 for beef quality classification. Inf. Sci. Lett. **12**(1) (2023), https://digitalcommons.aaru.edu.jo/isl/vol12/iss1/24, Accessed 20 Jun 2024
11. Brigato, L., Iocchi, L.: A close look at deep learning with small data. In: 25th International Conference on Pattern Recognition (ICPR), pp. 2490–2497 (2020)
12. Pytorch. https://pytorch.org/, Accessed 17 May 2024
13. Kandel, I., Castelli, M., Popovič, A.: Comparative study of first order optimizers for image classification using convolutional neural networks on histopathology images. J Imaging **6**(9) (2020)
14. Hussain, M. A. I., Khan, B., Wang, Z.-J., Ding, S.: Woven fabric pattern recognition and classification based on deep convolutional neural networks. Electronics **9**, 1048, 06 (2020)
15. Dosovitskiy, A., e al.: An image is worth 16x16 words: transformers for image recognition at scale. In: International Conference on Learning Representations, Vienna, Austria (2021)
16. Jacob, B.: Quantization and training of neural networks for efficient integer-arithmetic-only inference (2017)
17. Srivastava, S., Sharma, G.: Omnivec: learning robust representations with cross modal sharing (2023)

634     J. Weimer et al.

# An Integrated Work Design Factors to Improve the Sustainable Manufacturing Performance of Garment Quality Inspection

Le Thi My Anh[1](✉) and Anh Phuong Ngo[2]

[1] Department of Mechanical and Manufacturing Engineering, Faculty of Engineering and Technology, Nguyen Tat Thanh University, 1165 National Route 1A, An Phu Dong Ward, District 12, Ho Chi Minh City, Vietnam
lemyanh79@gmail.com
[2] Department of Electrical and Computer Engineering, College of Engineering, North Carolina A&T State University, Greensboro, NC, USA

**Abstract.** This study examines the connection between sustainable manufacturing and ergonomic work design. It focuses on the impact of work-related musculoskeletal disorders (WMSDs) among garment workers in a Vietnamese SME factory who are responsible for quality inspection. The study uses the Rapid Entire Body Assessment technique and discovers medium to high-risk levels of WMSDs, indicating the need for comprehensive ergonomic intervention. The proposed ergonomic solutions are tailored to the anthropometric nuances of this demographic and include redesigning workstations and enhancing work methods and training. The study goes beyond ergonomics by advocating for significant reductions in waste through more effective work processes, promoting environmental sustainability. The research also highlights the economic benefits of optimized task sequencing and work design to increase value-added activities at the cellular manufacturing level. This holistic approach, which combines environmental sustainability with economic effectiveness and improved human health, is a paradigm shift toward sustainable manufacturing. The interventions aim to reduce stress, fatigue, and risks, leading to a healthier workforce and setting a benchmark for sustainable practices in industries that involve intense human-machine interactions.

**Keywords:** Sustainable Manufacturing · Ergonomic Work Design · Garment Quality Inspection

## 1 Introduction

The textile & garment industry, a cornerstone of the global economy, significantly contributes to employment and export income. In Vietnam, this industry, comprising around 6,000 companies, predominantly small and medium-sized enterprises (SMEs), employs approximately 1.6 million people, representing 12% of the industrial workforce and nearly 5% of the country's labor force [1]. This sector plays a crucial role in domestic and international markets, exporting products to over 180 countries and territories

© The Author(s) 2025
H. Kohl et al. (Eds.): GCSM 2024, LNME, pp. 635–645, 2025.
https://doi.org/10.1007/978-3-031-93891-7_70

[2]. A critical aspect of the garment industry is the inspection and grading of fabric, which is essential to maintaining the desired quality of the final products. Fabric inspection involves detecting defects that could compromise garment quality or lead to wasted resources. Methods used in fabric inspection include visual inspection, hand-feel inspection, and various testing procedures for colorfastness, shrinkage, weight, and width. This study focuses on the visual inspection method, where operators use their eyes and hands to assess fabric quality on an inspection machine, followed by labeling and packing at separate workstations.

Work-related musculoskeletal disorders (WMSDs) are common among garment workers due to repetitive tasks and poor postures. These disorders, which include injuries or dysfunctions of muscles, bones, nerves, tendons, ligaments, joints, cartilage, and spinal discs, result in significant economic costs due to lost wages, compensation expenses, and decreased productivity [3]. In the garment industry, WMSDs often manifest as pain in the shoulders, wrists, and fingers caused by repetitive arm movements and mandatory sitting positions [4]. The prevalence of WMSDs highlights the need for comprehensive ergonomic interventions to improve worker health, efficiency and effectiveness. Effective ergonomic practices enhance worker well-being, improve product quality, and reduce defects, thereby minimizing waste and fostering a more sustainable production process. Hence, addressing ergonomic issues can have a broader impact on the environmental challenges faced by the textile industry [5]. As a significant consumer of water, particularly in wet processing operations, the industry generates substantial effluent water that must be treated to avoid environmental harm. Additionally, carbon emissions from production processes are subject to stringent regulations, necessitating technological upgrades and process improvements to reduce emissions [6]. The industry can minimize its environmental impacts by implementing sustainable manufacturing practices, such as lean manufacturing principles and cleaner production approaches [7, 8]. This holistic approach promotes economic and environmental sustainability and reinforces ergonomics' importance.

This study focuses on ergonomic risks in fabric quality inspection in a Vietnamese SME factory. It identifies medium to high-risk levels of WMSDs and proposes interventions such as workstation redesign and targeted training. The research also emphasizes the importance of waste reduction and optimizing work processes for economic and environmental benefits. The remainder of this paper is organized as follows: Sect. 2 presents the research methodology, Sect. 3 analyzes the results and provides insights, and Sect. 4 concludes the paper.

## 2  Research Methodology

The REBA and similar tools enable practitioners to identify jobs at risk for WMSDs. They must first effectively and efficiently discriminate between at-risk jobs and jobs not at risk before attempting to discriminate levels of risk. An ideal assessment tool would have high sensitivity and specificity, minimizing false positives and negatives. However, before the predictive validity of REBA risk scores can be examined, it is necessary to establish that the measurements made by different observers are consistent and reliable. The reliability of a measure is a prerequisite for establishing validity [9]. Two types of validity are of interest: intra-rater and inter-rater reliability. Without some knowledge of the consistency of ratings within individual rates and between different rates, it is impossible to assess the predictive validity of REBA. The method for ergonomic risk assessment includes selection of workers and tasks, data collection, data analysis, results and discussion, and improvement [10]. Five levels of action are recommended according to REBA Scores (Table 1) [11].

**Table 1.** REBA Action Levels.

| Action Category | REBA Scores | Risk levels | Action needed (and further examine) |
|---|---|---|---|
| 0 | 1 | Negligible | No needed |
| 1 | 2–3 | Low | Maybe needed |
| 2 | 4–7 | Medium | Needed Soon |
| 3 | 8–10 | High | Needed Now |
| 4 | 11–15 | Very High | Needed Now |

## 2.1 Selection of Workers and Tasks

The company is a fabric dyeing and finishing factory. The shop floor comprises two operations: fabric inspection, labeling, and packaging. The factory has four fabric roll inspection machines and four labeling and packaging workstations staffed by eight workers. The study investigates the workstations used for fabric quality inspection. Table 2 shows detailed anthropometric data for four female workers involved in this task.

**Table 2.** Anthropometric data of four female workers in fabric quality inspection.

| Operation | Ages of workers | Elbow functional reach (cm) | Stature (cm) | Standing eye height (cm) | Standing elbow height (cm) | Length of arm (cm) |
|---|---|---|---|---|---|---|
| Fabric quality inspection | 40 | 31 | 153 | 143 | 91 | 54 |
| | 37 | 31 | 155 | 145 | 93 | 55 |
| | 35 | 31 | 155 | 145 | 93 | 55 |
| | 33 | 32 | 156 | 146 | 93 | 56 |
| **Average** | **36** | **31** | **155** | **145** | **92** | **55** |

The tasks used on this shop floor involve fabric roll inspection, labeling, and packaging. The fabric roll is held and hung at the back of the machine, and the fabric is pulled over the glass of the inspection table and wound onto another roll placed in front of the machine. The unwinding and winding of the fabric rolls are motorized. Any detected defects are marked immediately for further decisions. The fabric rolls are labeled with the brand name and packed into plastic bags.

**Table 3.** Summary of data inputs for REBA assessment.

| No | Task elements | Neck angle | Trunk angle | Leg | Force/ Load | Upper arm angle | Lower arm angle | Wrist angle | Coupling |
|---|---|---|---|---|---|---|---|---|---|
| 1 | Get tube | $10^0-20^0$ Twisted | $0^0-20^0$ Side bending | 1 leg standing | < 5 kg | $> 90^0$ Shoulder raised | $0^0 -$ $60^0$ | $> 15^0$ Twisted | – |
| 2 | Place tube | $10^0-20^0$ Twisted | $20^0 -$ $60^0$ | 1 leg standing | – | $20^0-45^0$ Abducted | $0^0 -$ $60^0$ | $> 15^0$ Twisted | – |
| 3 | Position fabrics to fit the tube | $10^0-20^0$ Twisted | $20^0 -$ $60^0$ | Straight legs | – | $20^0-45^0$ Abducted | $60^0 -$ $100^0$ | $> 15^0$ | – |
| 4 | Start the Electronic Code Meter | Extension | $20^0 -$ $60^0$ | 1 leg standing | – | $> 90^0$ Abducted | $0^0 -$ $60^0$ | $< 15^0$ | – |
| 5 | Inspect fabrics | $10^0-20^0$ | $0^0-20^0$ | Straight legs | – | $< 20^0$ | $60^0 -$ $100^0$ | $< 15^0$ | – |
| 6 | Pause the Electronic Code Meter | Extension | $20^0 -$ $60^0$ | 1 leg standing | – | $> 90^0$ Abducted | $0^0 -$ $60^0$ | $< 15^0$ | – |
| 7 | Place fabric roll on table of inspection machine | $10^0-20^0$ | $20^0 -$ $60^0$ | 1 leg standing | 5–11 kg | $45^0-90^0$ | $60^0 -$ $100^0$ | $< 15^0$ Bent | – |
| 8 | Get scissors | $> 20^0$ Twisted | $20^0 -$ $60^0$ Twisted | $30^0 -$ $60^0$ 1 leg standing | – | $20^0-45^0$ Abducted | $0^0 -$ $60^0$ | $< 15^0$ | Good |
| 9 | Cut fabric to create a finished fabric roll | $10^0-20^0$ | $0^0-20^0$ | Straight legs | – | $20^0-45^0$ Abducted | $0^0 -$ $60^0$ | $< 15^0$ Twisted | Good |
| 10 | Get a first label tape | $> 20^0$ Twisted | $0^0-20^0$ Twisted | $30^0 -$ $60^0$ 1 leg standing | – | $20^0-45^0$ Abducted | $0^0 -$ $60^0$ | $< 15^0$ Twisted | – |
| 11 | Place and stamp the first label tape on a finished fabric roll | $10^0-20^0$ | $0^0-20^0$ | Straight legs | – | $20^0-45^0$ | $60^0 -$ $100^0$ | $> 15^0$ Twisted | – |

(*continued*)

**Table 3.** (*continued*)

| No | Task elements | Neck angle | Trunk angle | Leg | Force/ Load | Upper arm angle | Lower arm angle | Wrist angle | Coupling |
|----|---------------|-----------|------------|-----|-------------|-----------------|-----------------|-------------|----------|
| 12 | Get a second label tape | $> 20^0$ Twisted | $0^0 - 20^0$ Twisted | $30^0 - 60^0$ 1 leg standing | – | $20^0 - 45^0$ Abducted | $0^0 - 60^0$ | $< 15^0$ Twisted | – |
| 13 | Place and stamp the second label tape on a finished fabric roll | $10^0 - 20^0$ | $0^0 - 20^0$ | Straight legs | – | $20^0 - 45^0$ | $60^0 - 100^0$ | $> 15^0$ Twisted | – |
| 14 | Get the finished fabric roll | $10^0 - 20^0$ Twisted | $0^0 - 60^0$ | $30^0 - 60^0$ 1 leg standing | – | $< 20^0$ Abducted | $60^0 - 100^0$ | $> 15^0$ Bent | – |
| 15 | Place the finished fabric roll on labeling and packaging workstation | $10^0 - 20^0$ Twisted | $0^0 - 20^0$ Twisted | $30^0 - 60^0$ | 5–11 kg | $20^0 - 45^0$ | $0^0 - 60^0$ | $> 15^0$ Bent | Fair |

## 2.2 Data Collection

Data collection was conducted using video recordings from a smartphone's camera. Four workers were recorded at four different workstations, each at least seven times while performing the same task elements. The camera was positioned at an appropriate distance to ensure that the working postures and process sequences were observed. The average values of criteria, including the angles of the neck, arm, wrist, trunk, and leg, were then calculated for REBA assessment (Table 3).

A time study established the standard time for fabric quality inspection. The sample size is 28. The observed times are shown on Table 4.

## 2.3 Data Analysis

The ergonomic assessment based on the REBA Employee Assessment Worksheet [12]Group A includes torso, neck, and leg postures recorded in Table A, adjusted for load weight. Group B consists of upper arm, lower arm, and wrist values entered in Table B. Then, Group A and B values are merged in Table C, adjusted for activity level, to calculate the final REBA score and risk category.

**Table 4.** Observed times of fabric quality inspection in seconds.

| Observed times | | | | | | |
|---|---|---|---|---|---|---|
| 56 | 59 | 59 | 60 | 60 | 61 | 62 |
| 62 | 62 | 63 | 63 | 63 | 64 | 64 |
| 64 | 64 | 64 | 65 | 65 | 67 | 67 |
| 67 | 68 | 70 | 70 | 71 | 71 | 74 |

## 3 Results and Discussion

Motion study is to evaluate fabric inspection, labeling, and packaging by selected workers. Task analysis and REBA scores for fabric inspection are shown in Table 5.

The worker sustains standing and bending postures for an average duration of 64 s, with a standard deviation of 4 s. The REBA results indicated that workers are exposed to medium to high levels of risk for WMSDs. Immediate ergonomic interventions are needed for 53% of the total task elements, categorized under Action Category 3. The tasks that required workers to reach far posed the highest risk for WMSDs (REBA scores ranging from 8 to 10). These tasks include getting tubes, starting rolls, getting scissors, and getting label tapes. The following postures increased the risk of WMSDs for workers and their ratio with total task elements:

- Trunk: $\geq 20^0$ flexion and/or twisted (100%)
- Neck: $\geq 20^0$ flexion/extension and/or twisted (80%)
- Leg position: knees flexion (73%)
- Upper arm: flexion with abduction, raised shoulder position (73%)
- Wrist: flexion and twisted (67%)

To reduce the risks of WMSDs, adjust the workstation height to avoid trunk, neck, and knee flexion. Then, train workers on the new methods and establish standard work and time. The pilot workstation is used for assessment. Modifying the workstation and work method involves minor adjustments to accommodate the worker better. The adjustments include raising the workstation height from 75 cm to 90 cm. The fabric rolls move at a speed of 58 m per minute, with the fabric measuring 32 m in length and 1.27 m in width. A female worker, who is 155 cm tall, received two days of training in the new method. This training is essential for the worker to master her skills before establishing a standard work sequence. Following this, a time study was conducted to determine the standard time for the new method.

The zone of convenience reach (ZCR) is the reach area allowing the worker to access objects while standing or sitting without bending forward.

$$ZCR = \sqrt{r^2 - d^2} \tag{1}$$

where: $r$ = reach fingertip or palm reach distance; $d$ = shoulder height above the horizontal work surface. Then, the ZCR is calculated to be 43 cm. The desk height is 90 cm, designed to accommodate the standing elbow height of female workers. This is based on anthropometric data shown in Figs. 1 and 2.

**Table 5.** Task analysis and REBA Score of fabric inspection.

| No. | Task elements | Left hand | Right hand | Time (s) | REBA score | Code of Action Category |
|-----|---------------|-----------|------------|----------|------------|-------------------------|
| 1 | Get tube | | x | 2 | 9 | 3 |
| 2 | Place tube | | x | 3 | 7 | 2 |
| 3 | Position fabrics to fit the tube | x | x | 4 | 5 | 2 |
| 4 | Start the Electronic Code Meter | x | | 2 | 9 | 3 |
| 5 | Inspect fabrics | x | x | 33 | 2 | 1 |
| 6 | Pause the Electronic Code Meter | x | | 2 | 9 | 3 |
| 7 | Place fabric roll on table of inspection machine | x | x | 2 | 8 | 3 |
| 8 | Get scissors | | x | 2 | 9 | 3 |
| 9 | Cut fabrics to create a finished fabric roll | x | x | 3 | 5 | 2 |
| 10 | Get a first label tape | | x | 1 | 9 | 3 |
| 11 | Place and stamp the first label tape on a finished fabric roll | x | x | 3 | 3 | 1 |
| 12 | Get a second label tape | | x | 1 | 9 | 3 |
| 13 | Place and stamp the second label tape on a finished fabric roll | x | x | 3 | 3 | 1 |
| 14 | Get the finished fabric roll | x | x | 1 | 7 | 2 |
| 15 | Place the finished fabric roll on labeling and packaging workstation | x | x | 2 | 10 | 3 |

**Table 6.** Comparison of REBA scores before and after improvement.

| No. | Task elements | REBA score | | Note |
|---|---|---|---|---|
| | | Before improvement | After improvement | |
| 1 | Get tube | 9 | 9 | Not changed |
| 2 | Place tube | 7 | 3 | Reduced |
| 3 | Position fabric to fit the tube | 5 | 3 | Reduced |
| 4 | Start the Electronic Code Meter | 9 | 5 | Reduced |
| 5 | Inspect fabric | 2 | 1 | Reduced |
| 6 | Pause the Electronic Code Meter | 9 | 5 | Reduced |
| 7 | Place fabric roll on table of inspection machine | 8 | 3 | Reduced |
| 8 | Get scissors | 9 | 9 | Not changed |
| 9 | Cut fabric to create a finished fabric roll | 5 | 3 | Reduced |
| 10 | Get a first label tape | 9 | 9 | Not changed |
| 11 | Place and stamp the first label tape on a finished fabric roll | 3 | 2 | Reduced |
| 12 | Get a second label tape | 9 | 9 | Not changed |
| 13 | Place and stamp the second label tape on a finished fabric roll | 3 | 2 | Reduced |
| 14 | Get the finished fabric roll | 7 | 4 | Reduced |
| 15 | Place the finished fabric roll on the labeling and packaging workstation | 10 | 10 | Not changed |

The REBA scores revealed that the most significant reduction was associated with operating the fabric quality inspection workstation, while those for other tasks, like retrieving the tube, scissors, tape, and placing the final products, remained unchanged (Table 6).

Before the improvement, the allowances were higher, including personal needs (5%), fundamental fatigue (4%), unavoidable delays (1%), standing posture (2%), bending posture (5%), and others (5%), totaling 22%. The average observed time is 64 s, with a performance rating of 100%, showing that the worker operates at an average pace. Consequently, the standard time is determined to be 78 s. After the improvement, the total allowance was reduced to 18% by reducing the allowance for bending posture

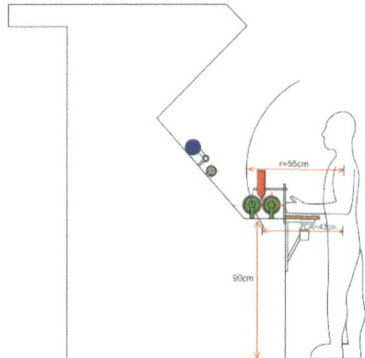

**Fig. 1.** Side view of fabric quality inspection workstation.

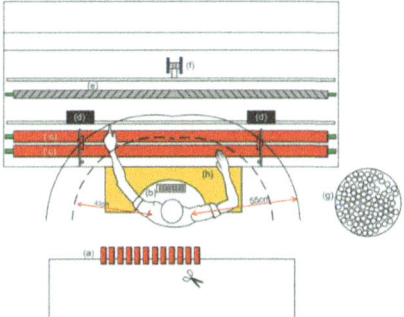

**Fig. 2.** Top view of fabric quality inspection workstation. (a) Tapes; (b) Control Panel; (c) Rollers; (d) Fabric Edge Detectors; (e) Rollers; (f) Electronic Code Meter; (g) Tubes; (h) Table.

to 1%. Then, the average observed time is 63 s, with a performance rating of 100%, indicating that the worker operates at an average speed. As a result, the standard time is 74 s. The planned throughputs were 340 fabric rolls per shift (8 h). However, before the improvement, the output of the fabric quality inspection workstation was 368 fabric rolls per shift. The throughput rate was 350 fabric rolls per shift, resulting in a defect rate of 5%. After the improvement, the output and throughput rate at the pilot workstation increased respectively to 387 and 368 fabric rolls per shift, leading to a 5.2% increase in effectiveness. The Energy Consumption per Unit of Output (ECUO) is calculated as:

$$ECUO = \frac{Total\ Energy\ Consumed\ (kWh)}{Number\ of\ Garments\ Inspected} \qquad (2)$$

A fabric quality inspection machine uses 6.4 kWh of energy per shift. The ECUOs were 0.018 kWh before improvement and 0.017 kWh after. Consequently, energy consumption decreased by 5.03% per fabric roll.

Further study could focus on placing tubes, scissors, and tape within easy reach of workers to reduce the need for excessive reaching, bending, and twisting. Additionally, arranging tubes and rolls at appropriate heights can help lower upper arm abduction and shoulder elevation. The next step would be to calculate the acceptable cost of defects to highlight the benefits of these improvements.

## 4   Conclusion

In conclusion, the scores from the REBA assessment indicate that workers' quality inspection is at medium to high-risk levels for developing WMSDs. Given these findings, it is imperative to implement ergonomic interventions to mitigate the potential risks of WMSDs. Practical solutions include the ergonomic redesign of workstations tailored to the anthropometric data specific to female Vietnamese workers in a garment factory. This redesign should ensure that workstations are appropriately sized and equipped to minimize strain. Furthermore, altering work methods to incorporate safer, more ergonomic practices can significantly reduce the physical demands placed on workers. In addition to these changes, providing comprehensive training for workers is crucial. Training should focus on proper body mechanics, the importance of regular breaks, and techniques for minimizing strain during repetitive tasks. To decrease WMSD risk, redesign workstations, improve work methods, and provide thorough training for a healthier work environment.

## References

1. Thuan, L.H.: Report on textile and garment industry, FPT Securities (2017)
2. Tot, B.V.: Report on textile and garment industry, FPT securities (2014)
3. Bridger, R.S.: The body as a mechanical system. In: Introduction to Human Factors and Ergonomics, CRC Press, pp. 31–63 (2018)
4. Das, S., Krishna Moorthy, M., Shanmugaraja, K.: Analysis of musculoskeletal disorder risk in cotton garment industry workers. J. Nat. Fibers **20**(1) (2023)
5. EU-OSHA, Work-related musculoskeletal disorders: prevalence, costs and demographics in the EU, European Agency for Safety and Health at Work (2019)
6. Bridger, R.S.: Standing and sitting at work. In: Introduction to Human Factors and Ergonomics, CRC Press, pp. 107–153 (2018)
7. Bridger, R.S.: Repetitive taks: risk assessment and task design. In: Introduction to Human Factors and Ergonomics, CRC Press, pp. 155–201 (2018)
8. Hussain, A., Case, K., Marshall, R., Summerskill, S.:Using ergonomic risk assessment methods for designing inclusive work practices: a case study. Hum. Fact. Ergon. Manuf. Serv. Ind. 337–355 (2016)
9. Kalkis, H., Roja, Z., Vaisla, G., Roja, I.: Causes of work related musculoskeletal disorders in the textile industry. In: Advances in Physical, Social & Occupational Ergonomics (2020)
10. Tayyab, M., Jemai, J., Lim, H., Sarkar, B.: A sustainable development frame-work for a cleaner multi-item multi-stage textile production system with a process improvement initiative. J. Clean. Prod. **246**, 119055 (2020)
11. Hignett, S., McAtamney, L.: Rapid entire body assessment (REBA). Appl. Ergon. **31**, 201–205 (2000)
12. Hedge, Ergonomics Plus, https://ergo-plus.com/wp-content/uploads/REBA.pdf. Accessed 22 Aug 2024

# End of Recyclability – Considering the Ageing of Plastic Materials and Its Implication on Potential Circularity

Thomas Potempa$^{(\boxtimes)}$ ⓘ, Max Ehleben ⓘ, and Max Juraschek ⓘ

Institute of Recycling, Ostfalia University of Applied Sciences, Wolfsburg, Germany
potempa@ostfalia.de

**Abstract.** The manufacturing of products is connected to environmental impacts due to material and energy demands. One central strategy towards reducing these impacts is establishing circular material flows by reintegrating end of life products as raw materials for production processes. Sustainable manufacturing aims for raising the share of recycled material, while primary resources are only covering losses during use phase and recycling. Hence, the transition towards circular economy requires considerations of end-of-life treatment and recyclability. However, the recyclability of materials depends on several influencing factors and typically degrades over time as ageing occurs. For this purpose, assessment methods for the actual quality of available materials are required providing decision support for identification of the best recycling route.

Against this background, a framework is presented for determining the recyclability of plastic materials over time, which incorporates decision criteria for up-, and down-recycling process routes and links this information with material quality and material degradation over time. Subsequently, it is shown that the recyclability of plastics is dependent on its age and that a point can be defined as EoR (End of Recyclability), which can support decision making on integrating recycled materials in manufacturing.

**Keywords:** Recycling · Circularity · Plastics · Sustainable Manufacturing

## 1 Introduction

In today's world, manufacturing and the making of products has become a central concern towards minimizing negative environmental impacts by human activities, especially given the urgent need to use available natural resources more efficiently. Implementing circular economy systems, in which recycling and recovery processes of materials are emphasized, can be one solution strategy. In this context, sustainable manufacturing aims at incorporating recycled materials in production processes to reduce the necessity of using primary resources.

Plastic materials, characterized by their durability and versatility, have been ubiquitous in various applications across sectors such as packaging, automotive, construction, and telecommunication for more than five decades. According to *Plastics Europe* [1],

H. Kohl et al. (Eds.): GCSM 2024, LNME, pp. 646–654, 2025.
https://doi.org/10.1007/978-3-031-93891-7_71

global plastic production rose to 390.7 Mt by the year 2021. However, the management of end-of-life plastics presents complex challenges. In Europe alone, approximately 29.5 Mt of plastic waste were collected in 2020, of which only 34% was recycled, while 42% was incinerated and 23% landfilled [1]. Significant gaps remain in our understanding of the properties of post-consumption plastic waste.

These gaps critically limit advancements in recycling technologies and overall waste management strategies [2]. The challenges are compounded by the inherent properties of plastics. As a large group of diverse polymers, each with unique chemical and physical properties, recyclability of plastics is influenced by factors such as polymer type - PET, HDPE, LDPE, PVC, PP, and PS being the most prevalent - and the presence of engineered polymers like ABS, PC, and PUR which exhibit 'enhanced' characteristics [3, 4]. The process of recycling is further complicated by the degradation of plastics over time due to exposure to environmental factors and usage patterns, which affects their physical and chemical integrity [5]. The degradation of plastic materials not only diminishes their mechanical and aesthetic properties but also raises concerns about the purity and quality of the recycled output material and consequently products. The immiscibility of different polymers and the presence of multi-polymer products in waste streams also pose significant barriers to achieving high-quality recycling outputs [6]. As such, a deeper understanding the ageing processes of plastics, their impact on recyclability, and the resultant implications for circular economy frameworks is crucial.

## 2  Plastics and Circular Material Flows

### 2.1  Product Development and Life Cycle Stages

Despite technological advances, recycled plastics are often considered inferior compared to virgin materials due to thermomechanical degradation during the recycling process and contamination. Furthermore, plastic production and disposal have led to significant environmental and health issues, including toxic exposures affecting both terrestrial and marine ecosystems. In this context, a deeper understanding of the material composition of plastic waste and its impurities is crucial to improve the efficiency of recycling processes contributing to a more sustainable society.

During the development of products, the benefits targeted are commonly defined with the customer demand in mind [7]. This can be achieved, for instance, through the development of a product profile. A product profile is a model of benefit that makes the intended provider, customer and user benefits accessible for validation and explicitly specifies the solution space for the design of a product creation process. As a consequence, user benefits are optimized, which leads to the assumption that the recyclability of plastics is determined by their useful lifespan. However, if an end-of-life (EoL) plastic can be recycled remains a subordinate question and mostly not in the focus of the product developing process.

Considering the life cycle stages of plastic products, as generically visualized in Fig. 1, sustainability principles at the end-of-life largely cover the well-known three R-strategies: reduce, reuse, and recycle. This framework has since evolved into more comprehensive concepts, as for instance the "10-R" set of principles, which integrate more holistically with the concepts of a circular economy [8].

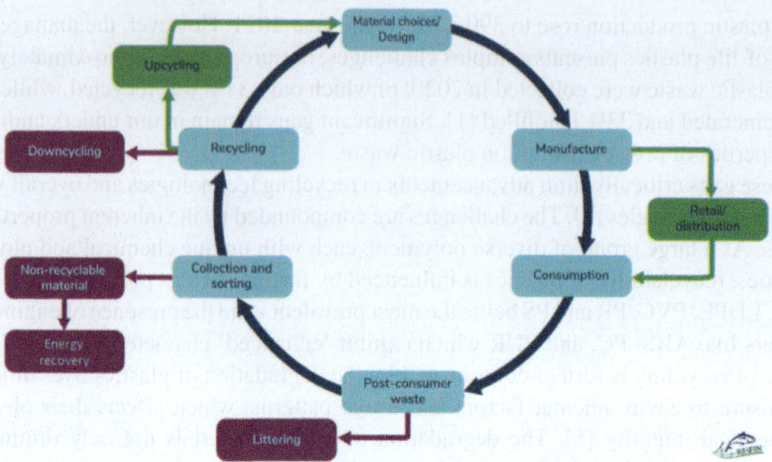

**Fig. 1.** Generic product life cycle for plastic products

## 2.2 Recyclability of Plastic Materials

The importance of recognizing EoL plastics that are of different composition and material quality is crucial for understanding the complexities involved in plastic recycling. Recycling is often promoted as a feasible solution towards eliminating plastic waste. However, significant challenges persist that hinder this goal. The inherent properties of polymer structures and product designs often reduce EoL recyclability. Although most thermoplastics used today are potentially recyclable, only a small portion actually enters these processes.

Taking a closer look at the polymeric materials, the effects of ageing and other degrading factors further limit their recyclability. Moreover, the degradation of polymer chains and bonds is altering its mechanical properties and reducing its future usability (or customer benefit). As a consequence, a lifespan extension is commonly introduced during material preparation via specific additives. But most additives are subject to degradation themselves as well and will decay over time making them to sacrificial material for the benefit of preserving the polymer characteristics for longer. The addition and degradation of additives vary depending on the polymer type and its intended use, as well as its chemical structure. An analysis of current literature in this field revealed insights on models to calculate recycling indicators, conduct material flow analyses or life cycle assessments [9–11]. A recent study, for instance, includes the adoption of a data-driven process that categorizes plastic waste based on its recyclability into different material recycling classes. This approach highlights the accumulation of contaminants, such as metal fillers and additives, throughout the polymer lifecycle, which may adversely affect both mechanical and chemical recycling processes [12]. The application of machine learning for automated classification further aids in defining key indicators of plastic waste recyclability [13].

Linking back to the product development process, the majority of available design methods for increasing circularity focus on post-consumer-plastics or end-of-life-plastics and thus follow a classical "end of pipe approach", neglecting the question of up to which point a "high quality" recycling is still possible. Environmental and ethical aspects are mostly considered in the context of the raw material, transportation and manufacturing processes. However, considering products during their use-phase individually and continuously over time would allow to select the environmentally or economically best performing pathway based on predefined targets and the actual material quality [14]. In this context, the need arises for an evaluation approach to identify the most beneficial (re-)utilization pathway for plastic materials to prevent or at least minimize negative environmental impacts while considering economic performance. The recyclability of materials is a decisive factor for actual material recovery success and depends on several influencing factors that change over time. A metric for recyclability is required to enable assessment methods, which would in turn enable the consideration of the quality of specific available materials to provide decision support for identification of the best utilization route for materials that already are or will become waste.

## 3  Determining Recyclability Based on Material Quality

Against this background, an approach is proposed for determining the recyclability of plastic materials over time by linking material quality and material degradation with the expected outcome of available utilization pathways for plastic products. This incorporates the corresponding process routes for recycling and recovery and allows to derive decision criteria for material utilization in recycling or energy recovery.

### 3.1  Material Quality and Value

The material quality of plastics can be described by its mechanical and aesthetical properties. In general, over time these properties in most cases deteriorate due to ageing. To prevent or minimize the effects of ageing and to ensure, e.g., mechanical functionality of a product, additives as for instance UV-protectors are used to stabilize polymers. The decay of additives depends on the environmental conditions the material is exposed to. Once the concentration of protecting additives falls below a specific threshold, the polymer chains themselves will deteriorate significantly. With the deterioration of the polymer, the material properties decline shown in Fig. 2 (top) and will be at some point in time not sufficient to ensure the product function. This functional boundary marks the end-of-life for a product. Further potential reasons for a product drop out of utilization, such as technological, psychological, economic or regulatory causes [15], might shift the end-of-life of a product towards higher remaining material quality.

A material product provides value, for instance by enabling economic value creation by its use. The value of the product can be based on the useful value creation it can provide until it reaches its end-of-life. Thus, the inherent value declines over time and usually a product is written off in this course. Alternatively, environmental or social performance can be considered as "value" (or combinations of all three dimensions).

Additional to the product value connected to its use, the resources a product is made of can also have a value as materials, for instance as input for manufacturing processes. This material value similarly declines over time with deteriorating material quality as the potential use in high-value recycling ("upcycling") as well in low-value recycling ("downcycling") is limited by the ageing of the polymers. Thermal energy recovery or the currently in development chemical recycling are final options. A schematic visualization of the value of the product and its materials is visualized in Fig. 2 in the lower graph. It is important to note that the relations of the value depicted are depending, among other factors, on material type, the specific process chains available for recycling, market prices as well as technological feasibility.

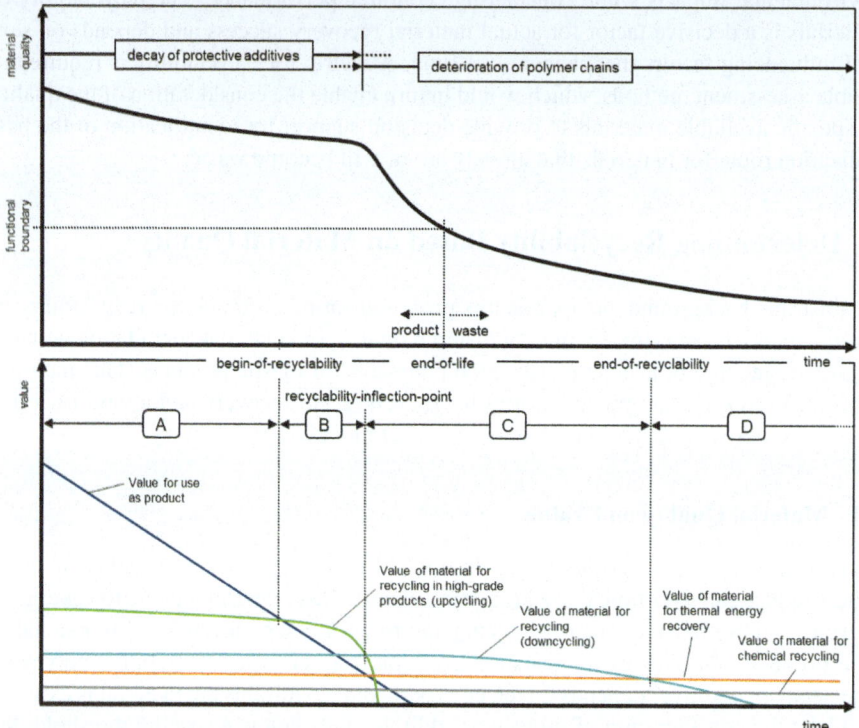

**Fig. 2.** Decay and deterioration of polymer materials over time (top) and an example for a thermoplastic material with corresponding course of value of product and material for different utilizations indicating begin and end-of-recyclability and inflection point (bottom)

## 3.2 Recyclability of Plastic Products

With the outlined modelling, a time period can be determined, in which the recycling of a plastic product is the action with the highest connected value. This time segment marks an area of *recyclability*. As shown in Fig. 2, the *begin-of-recyclability* is situated where the value of a product provided by continuously using it drops below the value of its materials to be used for creating new products. This point might lie before the functional end-of-life. Similarly, the *end-of-recyclability* is indicated by the intersection of the value of the materials for mechanical recycling with other options such as thermal energy recovery.

Over the deterioration of the polymer properties, a utilization in high-value recycling will increasingly incur a high effort in restoring the necessary material quality by adding higher shares of virgin raw materials or specific additives. If these efforts are exceeding the expected revenue, utilizing the materials for products with lower quality requirements will provide the more beneficial option. The transition between these options is defined as the *recyclability-infliction-point*.

With this concept of recyclability, a structured decision support can be developed for continuously selecting the best performing pathway for a product considering the ageing of plastic materials and its implication on potential circularity. In the schematic figure for an exemplary thermoplastic material, four areas can be indicated with each corresponding to the most favorable circularity options: (A) continuous use as product including repair or refurbishment, (B) high value material recycling, (C) low value material recycling and (D) energy recovery and chemical recycling.

## 4    Recyclability for Used Fishing Gear

The described concept of recyclability is exemplary applied to fishing gear used for catching fish in the ocean. Plastic waste in marine ecosystems has become a distressing global environmental challenge. This waste originates from a variety of human activities on land and sea. One significant waste stream in this context is used fishing gear, which can end up in the ocean due to, for instance, extreme weather conditions, mechanical failures, human error, vandalism or the lack of established waste treatment or recycling systems [16]. Fishing nets, as one of the most commonly used fishing gears, today in most cases are made of synthetic fibers like polyamide (PA), polyethylene (PE), and polypropylene (PP) due to their high strength and durability. The nets are often mixed with additives and coated with UV stabilizers to prevent ageing [16].

In a screening study, the material characteristics of 13 fishing nets made from PE or PA6 and 14 marine ropes made from a PE and PP mixture or PET have been analyzed with infrared spectroscopy and differential scanning calorimetry. In the results of the spectroscopy the ageing of the materials and their quality loss could be made visible. Aged material shows a lower intensity of the characteristic bands in comparison to new material due to degradation caused by UV radiation, which is corresponding with mechanical properties. The recyclability is calculated based on the material quality, which is determined in this case as the ratio of the current state (i.e. the intensity of the identified relevant spectroscopy bands) compared to reference values of unused equipment. Further mechanical testing is currently conducted to supplement the material quality calculation with further indicators. Based on these findings, the economic value and in turn the begin-of-recyclability, end-of-recyclability and recyclability inflection point can be defined by set boundaries relative values of the remaining material properties (for instance A: 100–80%; B: 79–70%; C: 69–25%; D: 24% and below). Examples for the boundaries of the fishing gear in or after its use phase sorted regarding their preferred further circular utilization as shown in Fig. 3.

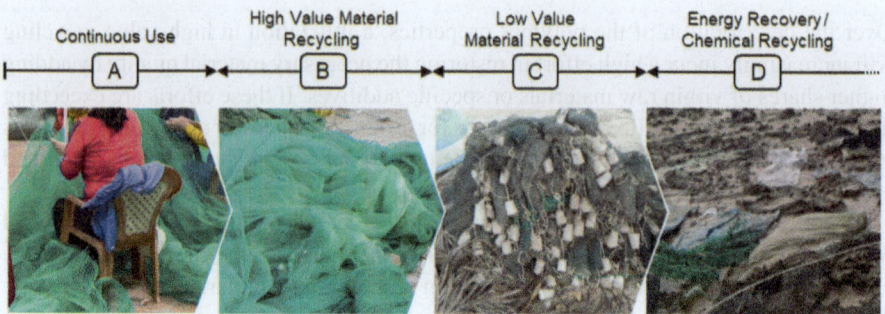

**Fig. 3.** Stages of material quality of fishing gear over its utilization time and mapping to corresponding circularity options with the highest value

## 5  Conclusion and Outlook

The recyclability of plastics is dependent on material quality and in turn on the ageing during the use phase. With the definition of a point as end-of-recyclability (EoR) a method is presented, which can support decision making on integrating recycled materials in manufacturing. The case study has shown the general applicability of the concept of EoR. A specific challenge was posed by the mixture of materials in used fishing gear. The data and measurements collected suggest that the varietal purity of a product is of crucial importance in its manufacture. If attention is paid to single-variety products or clear labeling of the materials during production, the material can be sorted more easily and better during recycling. Especially after the products have been used, the technical and economic costs related to mechanical recycling including sorting processes could be reduced. Furthermore, the analysis of material quality itself could be considerably simplified if polymers are not mixed or sufficiently labeled.

Currently, further research is conducted towards developing a robust and economical process chain for mechanical recycling of used fishing gear including methods for analyzing the material quality. Based on this, the calculation of the value functions will be refined with the collected data on the systems actual economic and environmental performance. This enables the development of the fundamental concept into a rapid decision support system for choosing the best circularity path for products in or at the end of use by taking the individual product properties as well as the background systems, as for instance the effort for recycling processes, into account.

**Acknowledgements.** This research was conducted within the project "Prevention, reduction and recycling of fishnets polluting Vietnamese coastal waters (REVFIN)". Funded by the Federal Ministry for the Environment, Nature Conservation, Nuclear Safety and Consumer Protection (BMUV) on the basis of a resolution of the German Bundestag.

## References

1. PlasticsEurope AISBL, Plastics - the facts (2022)

2. Ragaert, K., Delva, L., Van Geem, K.: Mechanical and chemical recycling of solid plastic waste. Waste Manag. **69**, 24–58 (2017). https://doi.org/10.1016/j.wasman.2017.07.044

3. Villanueva, A., Eder, P.: End-of-waste criteria for waste plastic for conversion. Technical proposals. EUR 26843, Luxembourg (2014). https://doi.org/10.2791/13033

4. Murphy, J.: Additives for Plastics Handbook. Elsevier (2001). https://doi.org/10.1016/B978-1-85617-370-4.X5000-3

5. Vilaplana, F., Karlsson, S.: Quality concepts for the improved use of recycled polymeric materials: a review. Macromol. Mater. Eng. **293**(4), 274–297 (2008). https://doi.org/10.1002/mame.200700393

6. Dahlbo, H., Poliakova, V., Mylläri, V., Sahimaa, O., Anderson, R.: Recycling potential of post-consumer plastic packaging waste in Finland. Waste Manag. **71**, 52–61 (2018). https://doi.org/10.1016/j.wasman.2017.10.033

7. Albers, A., Basedow, G.N., Heimicke, J., Marthaler, F., Spadinger, M., Rapp, S.: Developing a common understanding of business models from the product development perspective. Procedia CIRP **91**, 875–882 (2020). https://doi.org/10.1016/j.procir.2020.03.122

8. Kirchherr, J., Reike, D., Hekkert, M.: Conceptualizing the circular economy: an analysis of 114 definitions. Resour. Conserv. Recycl. **127**, 221–232 (2017). https://doi.org/10.1016/j.resconrec.2017.09.005

9. Huysveld, S., Hubo, S., Ragaert, K., Dewulf, J.: Advancing circular economy benefit indicators and application on open-loop recycling of mixed and contaminated plastic waste fractions. J. Clean. Prod. **211**, 1–13 (2019). https://doi.org/10.1016/j.jclepro.2018.11.110

10. Kuczenski, B., Geyer, R.: Material flow analysis of polyethylene terephthalate in the US, 1996–2007. Resour. Conserv. Recycl. **54**(12), 1161–1169 (2010). https://doi.org/10.1016/j.resconrec.2010.03.013

11. Simões, C.L., Pinto, L.M.C., Bernardo, C.: "Environmental and economic analysis of end of life management options for an HDPE product using a life cycle thinking approach. Waste Manag. Res. J. a Sustain. Circ. Econ. **32**(5), 414–422 (2014). https://doi.org/10.1177/0734242X14527334

12. Chin, H.H., Varbanov, P.S., Fózer, D., Mizsey, P., Klemeš, J.J., Jia, X.: Data-driven recyclability classification of plastic waste. Chem. Eng. Trans. **88** (2021). https://doi.org/10.3303/CET2188113

13. Nimmegeers, P., et al.: Extending multilevel statistical entropy analysis towards plastic recyclability prediction. Sustainability **13**(6), 3553 (2021). https://doi.org/10.3390/su13063553

14. Wilde, A.-S., Juraschek, M., Herrmann, C.: Data-driven decision support system enabling the circularity of products. Procedia CIRP - Accept. Publ., vol. 57th CIRP (2024)

15. Prakash, S., Dehoust, G., Gsell, M., Schleicher, T., Stamminger, R.: Influence of the service life of products in terms of their environmental impact: Establishing an information base and developing strategies against 'obsolescence': Final Report," Dessau-Roßlau (2020)

16. Hoang, S.V.A., Ehleben, M., Potempa, T., Henderson, J.: Sustainable approaches to fishing gear debris in Europe: effective management, reduction and recycling strategies (2023). https://doi.org/10.26271/opus-1735

# Circular Process for Reuse of Wind Turbine Poles as Secondary Raw Material

Barna Gal[1]([✉]), Lukas Kappis[2], Stefanie Eisl[3], Enrique Liesinger[1],
Pascal Froitzheim[2], Sebastian Schlund[1,3], and Wilko Flügge[2]

[1] Fraunhofer Austria Research GmbH, Theresianumgasse 7, 1040 Vienna, Austria
barna.gal@fraunhofer.at
[2] Fraunhofer Institut Für Großstrukturen in der Produktionstechnik, Albert-Einstein-Straße 30, 18059 Rostock, Germany
[3] Technische Universität Wien, Theresianumgasse 27, 1040 Vienna, Austria

**Abstract.** The demand for secondary steel resources in industrialised countries will increase significantly in the future as resources become scarcer. A 2 MW wind turbine contains over 170 tons of steel, whereby these quantities usually migrate to eastern regions at the end of the wind turbine's service life. In addition, there is a lack of knowledge of applications of mostly theoretical circular economy strategies and corresponding key technologies for wind turbines to ensure efficient material recovery of used components.

With regard to the identified research needs, our work contributes to the investigation of the reuse potential of steel derived from the towers of wind energy plants. Especially emphasizing its role as a secondary material within the circular economy paradigm. From a manufacturing perspective, our research focuses on evaluating the feasibility of bending pipe segments into a plane shape. To this end, the press-bending process is being investigated with the aid of FE simulations. The investigations are focused on examining the achievable flatness of the workpiece and analysing the effective process forces, which can be used to derive requirements for the machine tools. The study provides the foundation for the production-related realisation of the bending of tubular tower segments to a plane shape. Building on this, the remaining service life of the material can be investigated experimentally in future studies.

**Keywords:** Circular Economy · Wind Energy · Reuse · Steel

## 1 Introduction

The term circular economy (CE) is defined by the European Commission as 'a model of production and consumption, which involves sharing, leasing, reusing, repairing, refurbishing and recycling existing materials and products as long as possible.'[1]. The so-called R-principles are used to restrain and close resource flows [2, 3]. To differentiate a little bit more between the extension of product lifespans and the useful application of materials the following R-principles should be defined in detail [1]:

© The Author(s) 2025
H. Kohl et al. (Eds.): GCSM 2024, LNME, pp. 655–662, 2025.
https://doi.org/10.1007/978-3-031-93891-7_72

- Recycling: Processing used materials into new products to prevent waste of potentially useful materials. This is the last resort after other options have been exhausted and involves converting waste into reusable materials.
- Recovery (on a material level): Extracting energy or materials from waste that can't be recycled. This can include recovering energy through incineration or capturing valuable materials through advanced recovery processes.
- Reuse (on a functional level): Using products or components again, either for their original purpose or for a different function. This involves repairing, refurbishing, or repurposing items instead of discarding them.
- Remanufacturing: Rebuilding products to their original specifications using a combination of reused, repaired, and new parts. This can offer the performance and warranty of new products but with a reduced environmental impact.

Looking at a wind turbine as a resource potential, after reaching its End Of Life (EoL), it is important to understand that a wind turbine can basically be divided into four main components: the foundation, the tower, the nacelle, and the rotor blades [4, 5]. As wind energy continues to expand, with an estimated lifespan of wind turbines of around 20–25 years, the potential availability of raw materials from the dismantling of these turbines is imminent and current. All because the biggest and most mature onshore wind turbine markets in Europe are increasingly reaching their design lifetime [6].

Currently, there are only a limited number of concepts for the utilisation of components through R-strategies at the end of their service life [7]. Especially for this research, the tower of a wind turbine is mainly made of steel, followed by aluminum and copper [7, 8]. As these metals are highly recyclable, the EoL recycling option is very often used. This means that up to 90% of the towers can be recycled [5].

How these materials can be introduced into secondary cycles and that there are significant differences in CO2 emissions between the R-strategies should be the main discussion points of this research.

## 2 Literature Review

To determine the relevance and current state of research on this topic, a literature search and analysis was conducted on May 24th 2024, where a keyword string [('wind turbine' AND 'circular economy' AND 'tower') OR ('circular economy' AND 'steel' AND 'reuse') OR ('plate flattening' AND 'plate straightening')] was searched and resulted in 224 hits. To ensure the relevance and quality of the identified studies, a three-stage screening process was used, which included title, abstract and full text analysis. Following this process, 15 potentially relevant publications were identified.

Overviewing our state-of-the-art search, in terms of the circular economy, it was determined that up to 20% of the decommissioning costs are attributable to recycling, as evidenced by economic efficiency analyses [9]. Velenturf [3] discusses generally the use of steel in offshore wind turbines and its potential impact on resource availability and environmental sustainability. Further analyses have demonstrated that the materials steel, cast iron, and polymer materials contribute the most to CO2, SO2, and energy consumption. Additionally, it has been determined that modernisation and lifetime extension are more suitable than repowering concepts [7]. Nevertheless, the implementation of remanufacturing processes in companies is currently primarily driven by considerations of profitability and market protection [10]. Furthermore, it has been demonstrated that up to 90% of the components of a wind turbine can in principle be kept in circular processes [11, 12], and that a reuse concept for the steel tower and the concrete foundation could increase the concept of circular economy by 87% in terms of mass and 53% in terms of energy [13].

Despite the advantages of conventional recycling, the energy consumption of this process is still relevant: Resource consumption for remelting and subsequent grinding is estimated at an average of 9.54 MWh/t for steel parts [14]. For sheet reuse, the general consensus of the scientific community is that there is scope to replace the traditional remelting and subsequent refining of sheets with more innovative, efficient, and sustainable smelt-free recycling processes [15].

The first manufacturing approaches for straightening rounded plates were developed in an academic context in the field of materials research. Here, in some studies, pipe segments were cut into separate shell sections, which were then pressed flat in a press [16–18]. In these investigations, low-carbon micro-alloyed steels were considered, which are also used in the towers of wind turbines. The wall thicknesses of the analysed sheets ranged from 2 mm [16] to 17.4 mm [17], whereby the dimensions of the sheets were comparatively small with a maximum radius of 300 mm [17].

Adapting this approach to the large volume tubular towers of wind turbines with wall thicknesses of approximately 20 to 35 mm and radii of up to 4.5 m seems feasible. However, the press forces required must be analysed in relation to the plate material and plate dimensions. The machine tools must also be dimensioned. After pressing, the plate is expected to spring back elastically, leaving a residual curvature in the material. The extent of this springback effect needs to be analysed in relation to the plate properties and concepts to eliminate the residual curvature need to be developed to further ensure the integration of wind turbine tower components into a CE framework.

Based on these open analyses, the research question for this paper is:

What effects can be expected from the forming of a wind turbine tower for the recovery of secondary steel resources, and how could a forming process be designed for this purpose?

## 3   Methodology

The simulation of the flattening process of a pipe segment is carried out using the finite element method (FE) in the LS-DYNA software. It consists of the process steps shown in Fig. 1. In the first step, the roll bending process used to manufacture the tube is simulated. This process must be taken into account in the production chain as it has a significant influence on the plate properties due to work hardening and the formation of residual stresses. The plate is then prepared for bending: a uniformly rounded area is cut out and its elastic springback is simulated. The finished sheet is then placed in the model of a press and the press bending is simulated. Finally, the machine tool is removed to simulate elastic springback.

The workpiece is a plate of S355 with dimensions of 3000 × 1000 × 30 mm. The plate is meshed in the length, width and thickness directions with linear hexahedral

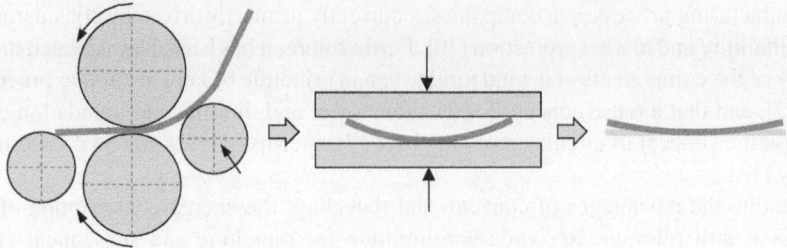

**Fig. 1.** Steps of the conducted FE analysis.

elements with an edge length of $5 \times 5 \times 3.75$ mm. To model the elastoplastic material behaviour, *MAT24 was used. It utilises an non-kinematic hardening law. An exemplary flow curve for S355 is provided to model the yielding behaviour. No failure model was implemented, and no fatigue was considered due to the lack of experimental data.

The FE model used for roll bending has been described in detail in [19]. Within this model, the rolls are idealised as cylindrical rigid bodies. The simulation of translational and rotational roll motion is performed with implicit and explicit time integration, respectively. As a result of the roller bending simulation, the plate is bent to an average outer radius of curvature of 1535 mm. In relation to wind turbine towers, this means a high wall thickness and a high curvature, resulting in high work hardening. This configuration can therefore be considered as a use case that places relatively high demands on the press-bending process and is therefore well suited for discussing the feasibility of the process.

The FE model for press-bending is shown in Fig. 2. Blocks of elastic steel with dimensions of $1000 \times 1000 \times 200$ mm are used as punches. The support on the machine tool is modelled using circular contacts with a diameter of 400 mm. The lowering of the upper punch is controlled by displacement and is calculated using explicit time integration. Several simulations were carried out, whereby the remaining gap between the closed punches was varied from 29.85 to 30 mm. Using symmetry in the Z direction, only one half of the test set-up is modelled.

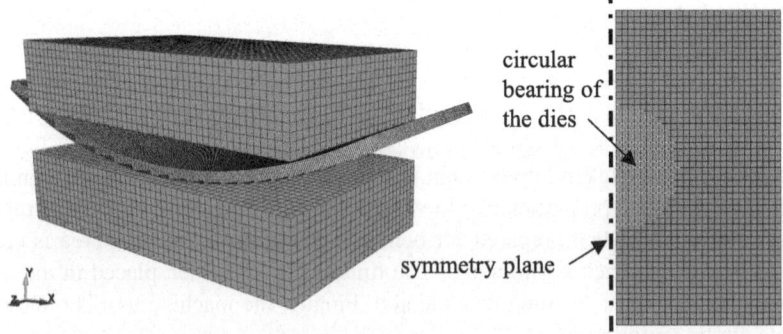

**Fig. 2.** FE-Modell of the press-bending process.

To simulate the subsequent elastic recovery, the tools are removed and a static equilibrium state is calculated using implicit time integration.

## 4  Results

Figure 3 shows the effective plastic strain in the plate material before (1) and after (2) press-bending with a gap of the closed punches of 30 mm. Initially it is homogeneous along the length of the plate, with a maximum of approximately 1.5% in the outer plate fibre. Due to the press bending, the plastic strain increases depending on the location and reaches up to 3.5% at the apex of the plate. It should be noted that a non-kinematic hardening model was used in the model. This model does not take into account the reduction in yield stress when the direction of loading is reversed between round and straight bending (Bauschinger effect) and therefore tends to underestimate the onset of plasticity. For the punches, the highest effective stress was found along the circular bearings. Here it reached up to 250 MPa.

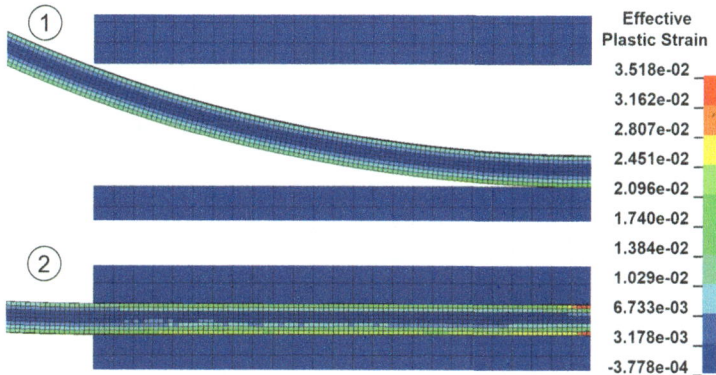

**Fig. 3.** Effective plastic strain of the plate before (1) and after (2) press-bending.

The investigation shows that the plastic strains that can theoretically occur during the flattening process are moderate. However, failure mechanisms could not be considered in the model. It therefore remains to be investigated experimentally, whether the material of real tower segments is suitable for press-bending after its service life and what residual capacity it has after such forming.

In Fig. 4, the spring-back plate contour after press-bending with a closed punch gap of 30 mm is visualised. The colour scale indicates the local height of the plate nodes relative to a 30 mm level, which corresponds to the surface of an ideally flattened plate. The contour achieves a good approximation of the desired flatness in the centre due to the strong plasticisation. However, there is significant springback at the edges. Above approximately 70 mm are the areas of the plate that were not in contact with the punches during the bending process and were therefore not formed. To compensate for the deviation from the desired flatness, various concepts can be considered. One possibility is to carry out the press-bending process continuously with a crane-controlled feed of the plate, so that it can be plasticised more uniformly over its entire length. In addition, the press tool can be designed to be concave so that individual areas are bent back more.

**Fig. 4.** Simulated spring-back plate contour after press-bending with a punch gap of 30 mm.

Figure 5 shows the influence of the punch feed respectively the gap between the closed punches on the required press force and the elastic springback of the plate. The displayed z-coordinate was determined after the elastic springback and refers to the highest workpiece point within the tool's engagement range. At a punch gap of 30 mm the required press force is 2.56 MN. It increases sharply with an increasing feed of the upper tool, whereby the elastic springback can only be reduced slightly due to this.

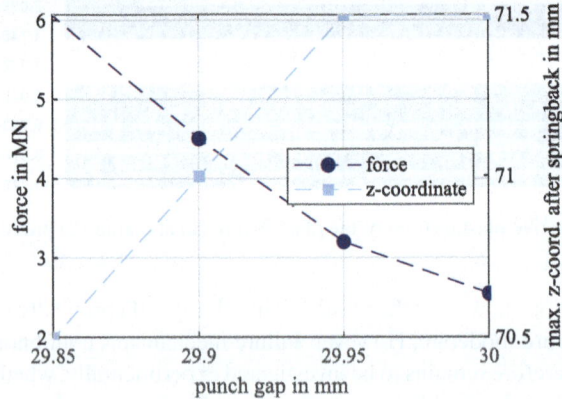

**Fig. 5.** Punch gap vs. resulting press force during press-bending and maximum z-coordinate after elastic springback.

The results show that powerful industrial presses with a maximum force of around 5 MN are suitable for press-bending tubular tower segments. It should be emphasised that the plate studied was a relatively thick plate with a high degree of pre-curvature and therefore a high degree of work hardening. In addition, the Bauschinger effect, which leads to a reduction in yield stress when the direction of loading is reversed, was not considered in the simulation. It can therefore be assumed that the process forces required for real tower segments would be lower.

# 5  Conclusion

Based on the problem of the lack of key CE technologies for wind turbine towers and the need to develop them due to the reaching of the EoL phase of the first large generation of wind turbines, this research project aims to recover as much steel as possible from the towers with as little energy as possible. As shown in the literature review, recycling approaches already exist, but are energy intensive. To enable a significantly less $CO_2$-intensive R-strategy for wind turbine towers, this paper discusses the possibility of flattening the steel towers to make them suitable for reuse without melting them down.

To investigate the feasibility of flattening from a manufacturing point of view, an FE analysis of flattening by press bending was carried out. The focus was on determining the required process forces and analysing the spring-back end contour to provide guidelines for the design of a manufacturing process. The results show that rebending of tubular tower segments can theoretically be achieved using a press-bending process with high-performance presses. To achieve flatness in a large area, the process can be carried out multiple times while the workpiece is fed forward. Additionally, a concave mould can be used to compensate for the elastic springback. In future work, it is planned to consider flattening by roll bending. This will presumably require lower process forces and enable a better springback compensation, but also involve a challenging progression, due to the inverse curvature of the workpiece.

It should be emphasised that pre-fatigue of the plate material and failure mechanisms were not considered in this study. For this reason, the feasibility in terms of material behaviour still must be proven experimentally on real pipe tower material. Furthermore, the mechanical properties of the straightened material must be determined to evaluate the possibilities of reuse.

Compared to the primary steel route, reuse approaches offer an 84% reduction in energy consumption and an 82% reduction in $CO_2$ emissions. Even compared to the recycled steel route, there is a 65% reduction in energy consumption and a 63% reduction in $CO_2$ emissions [20]. A final look into the future should investigate the real impact on $CO_2$ emissions, based on life cycle assessments, that could be achieved compared to state-of-the-art steel recycling processes.

## References

1. European Parliament. Circular economy: definition, importance and benefits (2023)
2. Kramer, K., Beauson, J.: Review existing strategies to improve circularity, sustainability and resilience of wind turbine blades-a comparison of research and industrial initiatives in Europe
3. Velenturf APM. A Framework and Baseline for the Integration of a Sustainable Circular Economy in Offshore Wind. Energies (2021)
4. Andersen, N., Eriksson, O., Hillman, K., Wallhagen, M.: Wind turbines' end-of-life: quantification and characterisation of future waste materials on a national level. Energies **9** (2016)
5. Zhang, W., Yu, H., Yin, B., Akbar, A., Liew, K.M.: Sustainable transformation of end-of-life wind turbine blades: advancing clean energy solutions in civil engineering through recycling and upcycling. J. Clean. Prod. (2023)
6. Zhao, F.: Global wind statistics (2023)

7. Kasner, R.: The environmental efficiency of materials used in the lifecycle of a wind farm. Sustain. Mater. Technol. (2022)
8. Cinar, S., Yildirim, M.B.: Reverse logistic network design for end-of-life wind turbines. optimization and dynamics with their applications: essays in Honor of Ferenc Szidarovszky (2017)
9. Topham, E., McMillan, D., Bradley, S., Hart, E.: Recycling offshore wind farms at decommissioning stage. Energy Policy (2019)
10. Jensen, J.P., Prendeville, S.M., Bocken, N.M.P., Peck, D.: Creating sustainable value through remanufacturing: three industry cases. J. Clean. Prod. (2019)
11. Demuytere, C., Vanderveken, I., Thomassen, G., Godoy León, M.F., Luca Peña, L.V., de, Blommaert C., et al.: Prospective material flow analysis of the end-of- life decommissioning: Case study of a North Sea offshore wind farm (2024)
12. Mello, G., Ferreira Dias, M., Robaina, M.: Evaluation of the environmental impacts related to the wind farms end-of-life. Energy Rep. (2022)
13. Gast, L., Meng, F., Morgan, D.: Assessing the circularity of onshore wind turbines: using material flow analysis for improving end-of-life resource management. Resources. Conser. Recycling (2024)
14. Haase, R.: Remanufacturing of metal components: reforming of sheet metal blanks (2020)
15. Cooper, D.R., Allwood, J.M.: Reusing steel and aluminum components at end of product life. Environ. Sci. Technol. (2012)
16. Moon, J., Jeong, H.J., Joo, S.-H., Sohn, S.S., Kim, K.-S., Lee, S., et al.: Simulation of pipe-manufacturing processes using sheet bending-flattening. Exp. Mech. (2018)
17. Zhang, W., Ding, D., Gu, M.: A model for predicting the yield strength difference between pipe and plate of low-carbon microalloyed steel. Metall. Mater. Trans. (2012)
18. Warwick, M., Vaka, H., Fang, C., Campbell, J., Dean, J., Clyne, T.W., et al.: Use of Profilometry-based indentation plastometry to study the effects of pipe wall flattening on tensile stress–strain curves of steels. Steel Res. Int. (2023)
19. Kappis, L., Cramon-Taubadel, E., von, Froitzheim P., Flügge, W.: Entwicklung eines empirischen KI-basierten Prognosemodells für das Umformergebnis beim Walzrunden von Grobblechen. In: Clausthaler Zentrum für Materialtechnik Clausthal, editor. Tagungsband, vol. 5, p. 574 (2023)
20. Werner, M., Haase, R., Hermeling, C.: reProd® – resource-autarkic production based on secondary semi-finished products. In: Kohl, H., Seliger, G., Dietrich, F., (eds.) Manufacturing Driving Circular Economy. Springer, Cham, pp. 51–59 (2023)

# Utilizing Waste Cast Stone: Compression Strength Analysis and Implication for Suitable Cast Stone Production

Devanshu Mudgal⬤, Emanuele Pagone$^{(\boxtimes)}$ ⬤, and Konstantinos Salonitis⬤

Sustainable Manufacturing Systems Centre, Faculty of Engineering and Applied Sciences, Cranfield University, Bedford MK43 0AL, UK
e.pagone@cranfield.ac.uk

**Abstract.** In recent years, the concept of circular economy has become increasingly popular as a sustainable approach to resource management. The present study explores the potential of reusing cast stone waste generated during the production process, in alignment with the principles of the circular economy. Cast stone waste typically goes to landfill sites discarded as a byproduct and poses environmental challenges due to its non-biodegradable nature. This work focuses on reusing this waste material as a viable resource, thus mitigating the environmental impact. Through a series of compression test experiments on cast stone, the presented study investigates the effectiveness of incorporating cast stone waste into production. The findings of this study contribute to cost reduction in raw material and waste minimization promoting economic and environmental benefits.

**Keywords:** Uniaxial compression test · Recycled aggregates · Cast stone

## 1 Introduction

Climate change and global warming are one of the major challenges being faced by this generation, which is largely driven by human activities such as industrial processes and construction. The construction industry is a significant contributor towards greenhouse gas emissions and environmental degradation due to the severe use of natural resources and production waste. As per the United Nations Environment Programme's report, the construction industry was responsible for 37% of global Greenhouse gas emissions in 2023 [1]. Raw material extraction such as aggregate mining causes damage to the riverbeds and mountains [2]. Aggregates are widely used in the cast stone production process. Cast stone is a popular construction material for architectural applications due to its ability to replicate the effect of naturally carved stone. It is also a highly durable material. The production and disposal of cast stone also result in a substantial amount of carbon dioxide emissions and contribute to landfill waste [3]. In recent years, there has been an increasing interest in sustainable construction practices that can mitigate these environmental impacts. One promising approach used in the industry is recycling of Construction & Demolition Waste (C&DW) as aggregates in production, hence reducing

H. Kohl et al. (Eds.): GCSM 2024, LNME, pp. 663–670, 2025.
https://doi.org/10.1007/978-3-031-93891-7_73

the amount of landfill waste [4]. Similarly, the waste cast stone can be used into in the production process. This approach not only reduces the need of virgin material but also minimizes the waste, hence aligning with the circular economy principles [5]. Within the construction industry, Portland cement is currently the first contributor to $CO_2$ emissions, hence the concrete industry is seeking to use more methods to reduce their emissions. The construction industry recently started using more environmentally friendly cementitious materials from other industries such as Ground Granulated Blast-furnace Slag (GGBS), Silica Fumes (SF) and Fly Ash (FA) in alignment with the BS EN 197–1 standard [6]. Currently, the $CO_2$ emissions from standardized concrete production in the UK are estimated at 76.3 kg $CO_2$ per tonne, which is a 26% reduction compared to 103.1 kg $CO_2$ per tonne agreed in the Kyoto Protocols and Climate Change Act [7]. However, these figures do not specify the strength class of concrete. For cast stone, the problem remains the same, Portland cement remains the main contributor to $CO_2$ emissions [3]. The cast stone industry is also moving towards greener alternatives to Portland cement; In a study by Mudgal et al. alternatives to Portland cement for cast stone have been identified for cast stone [8].

Another method to curtail relevant $CO_2$ emissions is by reducing the use of virgin raw materials, hence why the UK government as well as Denmark and Sweden has introduced an Aggregate Levy to prevent the use of natural resources and encourage the use of recycled material [9]. Therefore, the use of Recycled Aggregates (RA) in concrete is of significant interest because it promotes sustainable development and reduces the demand of mineral extraction reducing the amount of waste being sent to landfill. Currently, RA is used in lower-grade applications to conform with BS EN 12620 [10]. However, it can also be used for higher-grade construction applications if it conforms with BS 8500–1:2023 [11]. To promote a circular economy concrete industry has been utilizing its own waste as aggregate by using Recycled Concrete Aggregates (RCA) [12]. Most of the RCA are derived from C&DW. Compared to Natural Aggregates (NA), RCA has notable financial and environmental advantages [13]. In recent years, a vast majority of literature has focused on using RCA. In several works, it has been mentioned that Natural Aggregate Concrete (NAC) has superior compressive strength in comparison to concrete made from RCA [14–16]. In contrast to NA, RCA's surface is covered with aged mortar which leads to higher water absorption and porosity [14]. Higher water absorption leads to higher porosity in the concrete, hence reducing the strength of the concrete. The lower strength of concrete produced from RA has also been pointed to the parent concrete being weak [17]. However, some studies suggested that using a combination of recycled aggregates and admixtures (chemicals to enhance the properties of the concrete) will improve the mechanical properties of the concrete produced from the recycled aggregates [18, 19]. In a few studies, authors have argued that lower-strength parent concrete leads to higher-quality RCA as the low-strength mortar attached to the aggregates can be easily separated due to weaker bonds, creating stronger concrete [17]. Other studies suggest that concrete produced from a high-strength concrete parent RA would lead to better quality concrete as the bond between the mortar and the aggregate would be denser and stronger [20, 21].

In the concrete industry which closely resembles the cast stone industry, when the coarse Natural Aggregates (NA) have been replaced by 30% coarse RA or replaced by

20% fine RA, a marginal effect has been seen on the compressive strength of the concrete [22]. However, the strength starts to decrease gradually as more of the NA is replaced by the RA [23, 24]. It has been observed that the strength of concrete produced from either 100% coarse RA or 50% fine RA is typically 20% to 30% lower than from using NA [25, 26]. Where 100% of fine RA is used, a reduction in strength of up to 35% has been observed [27, 28]. Overall, more RA in the mix produces lower-strength concrete.

There are a lot of studies that have captured the effect on the strength of the concrete when using RA but no literature is currently available where cast stone waste has been re-introduced into the mix as RA and tested for compressive strength. To address the gap, this study aims to investigate the feasibility of using cast stone waste back as RA within the mix. It also investigates the effect of weight on compressive strength. Currently, there are three types of cast stones, Fibre Reinforced (contains Fibreglass) (FR), Semi-dry and Wet Cast (contains aggregates) (WC). For this study, only FR and WC can be considered as Semi-dry a powdery material. As per the UK Cast Stone Association, the acceptable cube strength for all cast stone should range between 35–50 MPa which also complies with BS EN 1992 and BS EN 1996 [29, 30].

## 2    Experiment

A few challenges faced by the CS industry is constant supply of NA. The RA used in this study comes from manufacturing waste such as rejected cast stone and spillovers. The waste from each cast stone was utilised to make the specimen for the same type. The test cubes made from the coarse RA were tested on the 14-day strength as they gained 90% of the strength during this curing period [31].

### 2.1    Aggregate Preparation

The aggregate preparation process involved the use of WC and FR production rejection waste. The cast stone waste was initially processed using a small crusher to break it down into different-sized aggregates. Following the crushing, aggregates were separated using sieved to achieve two different size ranges of 0–2 mm and 4–10 mm. This separation ensured that the aggregates met the specific size requirements for the mix design. During the sieving process, the crushed material was passed through a series of sieves resulting in the precise sorting of the particles [10]. This methodological preparation of aggregates is important for ensuring consistency and quality in the final specimen. The parent cast stone from which these aggregates were produced the batch had a strength of 40 MPa for WC and 42.5 MPa for FR cast stone.

### 2.2    Test Cube Preparation

The original mix design was used for both WC and FR. The mix design for these recipes is commercially sensitive, so it cannot be described this study. The mix design consists of ingredients such as sand, limestone and white Portland cement blended with admixtures. Once the mix was prepared it was poured into lubricated 150 mm × 100 mm × 150 mm moulds [32]. Within the mix composition for both cast stone materials, the

NA was replaced with RA by 0%, 10%, 20%, 40%, 50% and 100%. In total, 3 cubes were created for each composition equating to 21 sample in total per material. Wet cast test cubes are shown below in the mould for 0%, 10% and 10% recycled aggregates. The test cubes were left overnight within the moulds to cure, this initial setting process was completed under controlled conditions. The next, day the cubes were carefully de-moulded and weighed to record initial mass. The cubes were then submerged in a water tank to undergo further curing, which is initial for the hydration process that contributes towards the strength development of the cast stone. The test cubes remained within the water tank to gain 14-day strength, which is critical for cast stone to achieve 90% optimal performance and provide reliable data for subsequent testing [31]. Before the test, the wet weight of the cubes was noted to highlight any change in the density of the cubes.

# 3 Result

## 3.1 Wet Cast

The compression test was performed in compliance with BS EN 12390–3 [33]. The WC cubes were subjected to the compression test on the 14th day. As shown in Fig. 1, the cube samples S1, S2 and S3 with 0% recycled aggregates had the highest strength overall. Sample S3 had the highest strength, of 54.1 MPa. When 100% of coarse NA is replaced with the coarse RA, the strength is lowest between 33.6 MPa to 35 MPa. On average, the highest compression strength can be seen between 20% and 40% replacement of 12%.

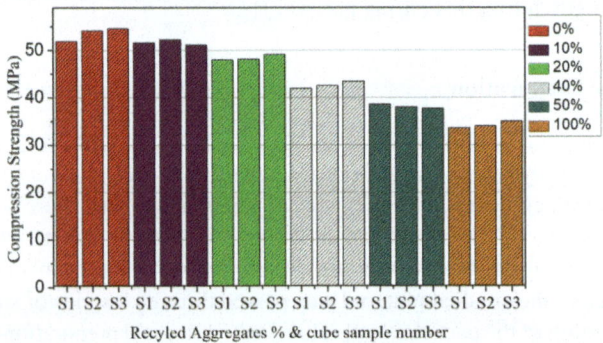

**Fig. 1.** Bar chart between Wet cast compression strength and % recycled aggregate replacement per sample

The correlation between the weight and compression strength can be seen in Fig. 2. Cube samples with the lowest weight have the lowest strength and the heavier cubes have more strength. For instance, cube samples S1 with 100% RA have the lowest compressive strength compared to 0% RA samples. At 40% RA, the samples weighed the same, however, all three samples have different compressive strengths of 32.2 MPa, 34.4 MPa and 33.7 MPa for S1, S2 and S2 respectively.

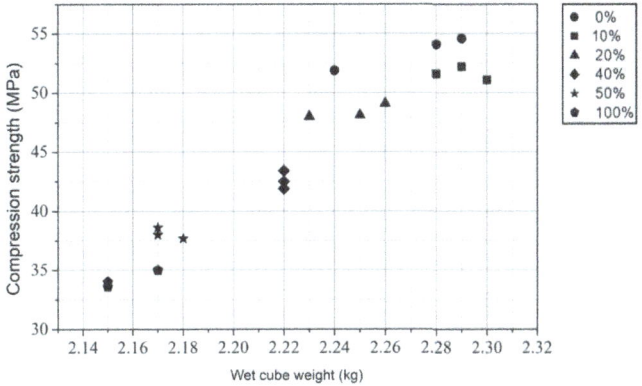

**Fig. 2.** Compression strength Vs wet cube weight (kg) for Wet cast

## 3.2 Fibre-Reinforced

Similar to Wet cast, a reduction in strength is observed as shown in Fig. 3. Sample S3 with 0% recycled aggregates has shown the highest strength whereas in sample S2 with 100% recycled aggregates, the lowest strength is observed. It can also be observed in Figure two that the cube strength on average drops drastically as soon as RA is introduced to the mix. On average, the reduction in the strength is 26% between 0% RA to 10% RA.

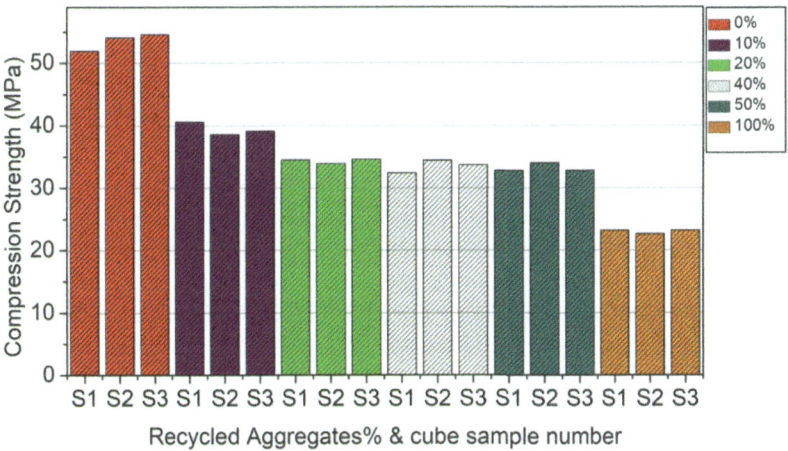

**Fig. 3.** Fibre compression strength Vs % recycled aggregate replacement per sample

A similar behavior to WC can be observed in the weight vs compressive strength chart. The lightweight sample with 100% RA had the lowest strength as shown in Fig. 4. Cubes with no recycled aggregates achieved the highest 14-day strength.

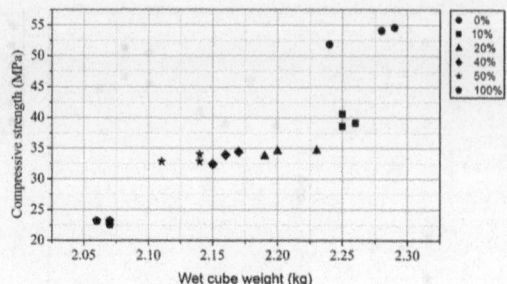

**Fig. 4.** Compression strength Vs wet cube weight (kg) for fibre-reinforced

## 4  Discussion

The results from Sects. 3.1 and 3.2 indicate that the cubes with higher 100% recycled aggregates exhibited the lowest compressive strength for both WC and FR cast stone. This reduction in strength can be attributed to the inherent weakness in recycled aggregates which often have pre-existing microcracks and a higher porosity compared to the NA. The weight and strength measurements further corroborate these findings with cubes with higher RA being lighter, suggesting higher void content. The presence of voids within the cast stone will reduce the density of the overall density and adversely affect the quality of the cast stone. Additionally, the fibre reinforcement was not able to bridge the gaps effectively. However, in both cases, the replacement of aggregates is still under the acceptable compressive strength limit of UK cast stone standards. In WC, from 0% to 50% RA is under the acceptable compressive strength. Similarly, in FR, 10% RA is under the acceptable limit.

## 5  Conclusion

To promote the circular economy and reduce overall greenhouse gas emissions, this study investigated the compressive strength of Wet cast and Fibre-reinforced cast stone with recycled cast stone aggregates. The initial findings reveal that with the increase in recycled cast stone aggregates within the mix, lower compressive strength is achieved. However, certain compositions of both materials are still acceptable by the UK Cast Stone Association standards as they were above 35 MPa strength. It also assures that a circular economy can be achieved in the cast stone industry by using construction and demolition waste. The primary objective of this study was to evaluate the feasibility of using recycled cast stone waste as aggregates. However, for future study a design of the experiment must be performed for a thorough testing of the samples for repeatability.

## References

1. Architecture, U.N.E.P.Y.C.f.E., Building Materials and the Climate: Constructing a New Future (2023)

2. Wu, B., Zhang, T.: Preparation of recycled sand from WRSG and compressive and flexural behavior of recycled aggregate concrete. Constr. Build. Mater.. Build. Mater. **414**, 134944 (2024)
3. Mudgal, D., et al.: Life-cycle-assessment of cast stone manufacturing: a case study. Procedia CIRP **104**, 624–629 (2021)
4. Aghayan, I., Khafajeh, R., Shamsaei, M.: Life cycle assessment, mechanical properties, and durability of roller compacted concrete pavement containing recycled waste materials. Int. J. Pavement Res. Technol. **14**, 595–606 (2021)
5. Colangelo, F., et al.: Comparative LCA of concrete with recycled aggregates: a circular economy mindset in Europe. Int. J. Life Cycle Assess.. J. Life Cycle Assess. **25**, 1790–1804 (2020)
6. Part, B.C., *1: Composition, specifications and conformity criteria for common cements.* 2011: European Committee for Standardization
7. Bostanci, S.C., Limbachiya, M., Kew, H.: Use of recycled aggregates for low carbon and cost effective concrete construction. J. Clean. Prod. **189**, 176–196 (2018)
8. Mudgal, D., et al.: Life-cycle-assessment of cast stone manufacturing: a case study. **104**, 624–629 (2021)
9. Söderholm, P.: Taxing virgin natural resources: lessons from aggregates taxation in Europe. Resour. Conserv. Recycl.. Conserv. Recycl. **55**(11), 911–922 (2011)
10. EN, B., 12620, Aggregates for concrete (2013). British Standards Publication
11. BSI, Concrete. Complementary British Standard to BS EN 206 - Method of specifying and guidance for the specifier, in BS 8500–1:2023 - TC. (2023). BSI
12. Alwaeli, M.: Waste utilization in concrete technology as a substitute for natural aggregates in the context of circular economy: an overview. Int. Multidisc. Sci. GeoConference: SGEM **19**(4.2), 151–158 (2019)
13. Zhang, L., et al.: Effective utilization and recycling of mixed recycled aggregates for a greener environment. **236**, 117600 (2019)
14. Qin, L., et al.: Experimental study on mechanical properties of coal gangue base geopolymer recycled aggregate concrete reinforced by steel fiber and nano-Al2O3. Rev. Adv. Mater. Sci. **62**(1), 20230343 (2023)
15. Zhang, H., et al.: Selection of nanomaterial-based active agents for packaging application: using life cycle assessment (LCA) as a tool. Packag. Technol. **30**(9), 575–586 (2017)
16. Pradhan, S., et al.: Comparative LCA of recycled and natural aggregate concrete using particle packing method and conventional method of design mix. J. Clean. Prod. **228**, 679–691 (2019)
17. Gebremariam, H.G., Taye, S., Tarekegn, A.G.: Parent concrete strength effects on the quality of recycled aggregate concrete: a review. Heliyon (2024)
18. Kurda, R., De Brito, J., Silvestre, J.D.: Combined economic and mechanical performance optimization of recycled aggregate concrete with high volume of fly ash. Appl. Sci. **8**(7), 1189 (2018)
19. Adessina, A., Fraj, A.B., Barthélémy, J.-F.: Improvement of the compressive strength of recycled aggregate concretes and relative effects on durability properties. Constr. Build. Mater.. Build. Mater. **384**, 131447 (2023)
20. Akbarnezhad, A., et al.: Effects of the parent concrete properties and crushing procedure on the properties of coarse recycled concrete aggregates. J. Mater. Civ. Eng. **25**(12), 1795–1802 (2013)
21. Xie, T., Gholampour, A., Ozbakkaloglu, T.: Toward the development of sustainable concretes with recycled concrete aggregates: comprehensive review of studies on mechanical properties. J. Mater. Civ. Eng. **30**(9), 04018211 (2018)
22. Dhir, R., Paine, K.A.: Suitability and practicality of using coarse RCA in normal and high-strength concrete. In: 1st International Conference on Sustainable Construction: Waste Management (2004)

23. Akbarnezhad, A., et al.: Microwave-assisted beneficiation of recycled concrete aggregates. Constr. Build. Mater.. Build. Mater. **25**(8), 3469–3479 (2011)
24. Mas, B., et al.: Influence of the amount of mixed recycled aggregates on the properties of concrete for non-structural use. Constr. Build. Mater.. Build. Mater. **27**(1), 612–622 (2012)
25. Dhir, R., et al.: Suitability of recycled concrete aggregate for use in BS 5328 designated mixes. In: Proceedings of the Institution of Civil Engineers-Structures Buildings, p. 257–274 (1999)
26. Etxeberria, M., et al.: Influence of amount of recycled coarse aggregates and production process on properties of recycled aggregate concrete. Cement Concrete Res. **37**(5), 735–742 (2007)
27. Evangelista, L., De Brito, J.: Mechanical behaviour of concrete made with fine recycled concrete aggregates. Cement Concrete Compos. **29**(5), 397–401 (2007)
28. Evangelista, L., De Brito, J.: Durability performance of concrete made with fine recycled concrete aggregates. Cement Concrete Compos. **32**(1), 9–14 (2010)
29. BSI, BS EN 1992-1-1:2004+A1:2014, in Eurocode 2: Design of concrete structures - general rules and rules for buildings (2014), BSI
30. BSI, BS EN 1996-1-1:2022 - TC, in Eurocode 6. Design of masonry structures - General rules for reinforced and unreinforced masonry structures (2023), BSI
31. Surana, S., Pillai, R.G., Santhanam, M.: Performance evaluation of curing compounds using durability parameters. Constr. Build. Mater.. Build. Mater. **148**, 538–547 (2017)
32. Standards, B.: BS 1881–108, in Testing concrete. Method for making test cubes from fresh concrete (1983)
33. Standard, B.: Testing hardened concrete, in Compressive strength of test specimens, BS EN, p. 12390–3 (2009)

# An Application of ARIMA Techniques to Enhance Circular Economy Manufacturing in the National Electricity Company in South Africa

Genevieve Bakam$^{(\boxtimes)}$, Khumbulani Mpofu, and Charles Mbohwa

Tshwane University of Technology, Pretoria, South Africa
bakamg@tut.ac.za

**Abstract.** Emerging economy strives for advanced manufacturing practices including waste management, resource optimisation and zero gas emissions. This study seeks to investigate and provide solutions enabling circular economy manufacturing in electricity production in South Africa. The Autoregressive integrated moving average (ARIMA) model was applied to predict the 2030 circular values using observation from 2017 to 2023. The Ljung-Box Q coefficients indicate the fitness of the time series models in comparison with the chi-square. Forecasting analysis highlighted a decrease in resource consumption and greenhouse gas emissions from 2024 to 2030 in alignment with the United Nations Sustainable Development Goals compared to an increase in waste generation. However, the target of reaching zero emissions in 2030 is far from realistic given ARIMA models. Eskom should fully adopt renewable energy sources like solar, wind and geothermal compared to fossil fuels for effective resource efficiency, zero greenhouse gas emissions and circular waste management to meet global requirements on pollution, climate change and finite resources preservation. The South African Government should update water infrastructure and regulations for technological innovation to achieve the circularity expectations without compromising the social corporate responsibility for inclusive growth.

**Keywords:** Circular economy · ARIMA · Forecasting · resource efficiency · Greenhouse Gas · waste management · Energy company and South Africa

## 1 Introduction

The traditional linear economy focusing on production and disposal has upgraded to circular economy (CE) involving all aspects of long-term sustainability such as reduce, repair, reuse and recycle (RE-X) and renewable energy sources. Circular manufacturing innovation involves value chains where business models, products, and services are designed, created and used to promote circularity and resource efficiency [1]. The concept of circularity addresses sustainability-related challenges associated with waste management, resource efficiency, responsible consumption, and renewable energy in

© The Author(s) 2025
H. Kohl et al. (Eds.): GCSM 2024, LNME, pp. 671–679, 2025.
https://doi.org/10.1007/978-3-031-93891-7_74

alignment with sustainable development goals (SDGs) [2, 3] posited circular manufacturing operations enable natural ecosystem resilience and the minimisation of greenhouse gas emissions. Dissociating environmental degradation and resource consumption from economic growth form part of the CE requirements enabling climate resilience [4], resource conservation, and pollution prevention [5]. Market demand and customer experience regarding sustainable products and services are driving innovation and competitiveness [5, 6] mentioned that businesses are incentivized by regulatory bodies such as waste management regulations and extended producer responsibility to implement eco-design principles and circular business models. Energy production represents the major contributor to global greenhouse gas emissions with 72% discharges and waste accounting for 3% [7]. The South African (SA) manufacturing sector relies on natural finite resources leading to overexploitation compromising future generations [8]. The SA energy mix includes 73.8% of total energy generated from fossil fuels and only 8.7% generated from renewable energy with 6% of energy demand not fulfilled given the constant load-shedding [9]. However, effective CE adoption is compromised by the primary goal of ensuring job creation and social corporate responsibility for inclusive growth [6]. Circular economy manufacturing (CEM) in electricity companies enables value creation through environmental protection, product durability and income maximisation for economic expansion [10]. Compared to previous studies based on the theoretical aspects of circularity [11], circular forecasting is applied for prediction analyses of the long-term impacts of non-circularity to provide preventive domestic environmental strategies. A further investigation of the synergies between resource efficiency, gas emissions and waste management to achieve successful transition to circularity in SA is provided. ARIMA techniques enable the prediction of waste generation and gas-emissions patterns in energy transformation for resources and production efficiency. This implies real-time adjustment of production processes, just-in-time materials acquisition, predictive waste reduction and efficient recycling processes. The integration of ARIMA techniques into CEM promotes the adoption of CE principles for environmental sustainability in energy generation companies.

## 2   Literature Review

Applying CE standards in sustainable manufacturing operations to achieve socioeconomic and environmental values resides on 4 pillars including applied materials, supply chain, design and production in addition to management and policymaking [12]. According to [6], the production creation and optimization process should be sustainable to reduce the environmental and economic impacts of waste residuals. They also argued that improved control systems and improved burners can be used to improve the combustion processes thereby reducing residues, ash volume, greenhouse gas emissions, and fuel consumption. Moreover, waste reduction from the start of the chain includes modification of processes, effective water management, and substitution of materials to attain green waste management as highlighted by [13]. CEM determines the product life cycle from innovation, design, creation, and consumption to remanufacturing [5]. For [8], market pressures, resource scarcity, and climate change are factors that force regulations, businesses, and policymakers to initiate the practice of resource efficiency.

An effective application of CE expectations depends on adopting industrial revolution technologies, upgrading organisational factors, and environmental protection targets [10, 14] reported that companies should be incentivized by regulations for relevant responsible resource consumption. The control of resource overuse, mitigation of greenhouse gases, and the conservation of biodiversity are some of the benefits that come with the use of resource efficiency practices [15]. Local electricity companies use fossil fuels like coal, oil, and natural gas as primary sources of power generation and power plants discharge a high volume of hazardous coal ash and oil residues increasing pollution, water contamination, and environmental degradation [13]. They further added that such companies must acquire licenses to perform environmental assessments before applying specific waste treatment methods for waste disposal. For [11], efficient and eco-friendly waste processing methods promote sustainable development by cutting down greenhouse gas emissions to protect biodiversity. [6] argued that power generation companies in SA have employed CE practices in their waste management systems to improve their resource efficiency and global footprint. Regulatory bodies like the Department of Environmental Affairs (DEA) controlling local electricity companies follow strict protocols to remain efficient in their waste management application methods [13]. For [15], electricity companies need improved technologies such as magnetic sorting, filtration systems and optical separation to improve recycling processes. Circularity in SA is limited by regulatory complexities, financial constraints, technological limitations, and skills shortages [16]. Manufacturing dynamic capabilities enable alignment with environmental demands while enforcing industrial resilience to circular manufacturing [17] based on CE data supporting real-time decision-making processes [18].

## 3  Methodology

The Autoregressive Integrated Moving Average (ARIMA) technique is used in time series forecasting to determine data patterns when data exhibit non-stationary with over-time variation [19]. In the ARIMA (p, d, q) model, (AR) is determined by the autoregressive order p, (I) is determined by d the degree of differencing d and (MA) is determined by q the moving average order q. The ARIMA (2,1,2) model is applied to determine the forecasting of 2030 circular economic indicators based on historical patterns. Secondary data was obtained from Eskom's annual reports published from 2017 to 2023 representing 28 observed quarters [9]. The model effectiveness is determined using the Ljung-Box Q statistics and significance value. CE indicators include resource efficiency measured by Coal and water consumption, gas emissions measured by Carbon dioxide ($CO_2$), Sulphur dioxide ($SO_2$) and Nitrogen oxides (NOx) and waste management measured by produced ash, asbestos and radioactive discharges. The Statistical Software for Social Sciences (SPSS) is used to determine the ARIMA model specifications from 2024 with 28 forecasted quarters following this prediction equation:

$$F_{t+1} = \alpha Y_t + (1 - \alpha)F_t \tag{1}$$

where: $F_{t+1}$ = forecast for period $t + 1$, $t$ = observation at period t, $F_t$ = forecast for period t, $\alpha$ = smoothing parameter (or constant) ($0 <= \alpha <= 1$). The model statistics use the Ljung-Box Q-test statistic to determine the best fit of the time series model based on the assumption of non-autocorrelation errors between lags. It also provides the Stationary R-squared, the R-squared and the Ljung-Box Q' coefficient and the significance. The Ljung-Box Q coefficient uses this formula:

$$Q = T(T + 2) \sum_{k=1}^{L} \left( \frac{\rho(k)^2}{(T - k)} \right) \tag{2}$$

where: T is the sample size, L is the number of autocorrelation lags, p(k) is the sample autocorrelation at lag k. The hypotheses are: The autocorrelations for the chosen lags are all zero in the chosen sample without serial correlation (H0) and the autocorrelations for the chosen lags are not zero in the chosen sample exhibiting serial correlation (H1).

The null hypotheses will be rejected if there is a significant p-value showing that the time series is not autocorrelated and if the $Q > X_{1-\alpha,h}^2$ where $X_{1-\alpha,h}^2$ represent the chi-square distribution value with h degrees of freedom. Although ARIMA has limitations on non-linearities, the model was validated at the optimal iteration with p, d and q generating model significance. Predictive techniques like Exponential Seasonal Smoothing (SES) for trend and seasonality and Long Short-Term Memory (LSTM) for sequential time series data can be applied for better results. The combination of the ARIMA and LSTM models on inventory forecasting increases demand and inventory reduction [19]. LSTM techniques enable predictive reliability and accuracy for waste management practices, suitable resource allocation and sustainable practices [20].

## 4   Discussion and Results

Table 1 displays the ARIMA model statistics for resource efficiency, gas emissions and waste as explained below.

**Table 1.** ARIMA model statistics

| Models | Model Fit statistics | | Ljung-Box Q(18) | | |
| --- | --- | --- | --- | --- | --- |
| | Stationary R-squared | R-squared | Statistics | DF | Sig. |
| Coal Burnt-Model_1 | ,892 | ,723 | 10,039 | 13 | ,691 |
| Water Consumption-Model_2 | ,821 | ,587 | 18,238 | 13 | ,149 |
| CO2 Emissions-Model_3 | ,879 | ,694 | 20,209 | 13 | ,090 |
| Sulphur Emissions-Model_4 | ,847 | ,709 | 10,147 | 13 | ,682 |
| Nitrogen Emissions-Model_5 | ,880 | ,735 | 15,836 | 13 | ,258 |
| Ash Produced-Model_6 | ,864 | ,638 | 16,683 | 13 | ,214 |
| Asbestos Disposed-Model_7 | ,214 | ,509 | 6,456 | 13 | ,928 |
| Radioactive Waste-Model_8 | ,056 | ,782 | 3,802 | 13 | ,993 |

### 4.1 Resource Efficiency Prediction Models

The model statistics in Table 1 indicate the best fit of the model given the coefficient of determination of 72.3% and 58.7% for coal and water consumption respectively. The Ljung-Box Q shows a significance of 0.691 and 0.149 for coal and water consumption highlighting that residuals of the model are independently distributed with no significant autocorrelation left by the models after specifications. Moreover, the statistic values of 10.039 and 18.237 are lower than the chi-square of 22.362 with 13 degrees of freedom at the significance level of 5% leading to the fitness and independence of the time series model. A trend of observed values from 2017Q1 to 2023Q4 being similar to the predicted values from 2024 quarter 1 to 2030 quarter 4 (see Fig. 1). However, there will be a decrease in coal usage from 28 Mt in 2024Q4 down to 24 in 2030Q4 and a decrease in water consumption from 69904ml in 2024Q4 down to 52719 in 2030Q4.

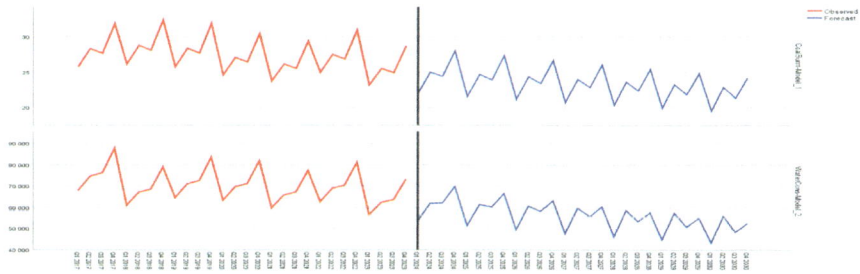

**Fig. 1.** Resource efficiency prediction models

### 4.2 Gas Emissions Prediction Models

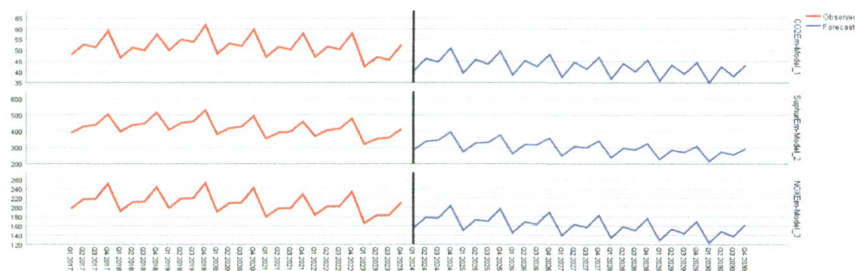

**Fig. 2.** Gas emissions prediction models

The model statistics in Table 1 indicate the best fit of the model given the coefficient of determination of 69.4%, 70.9% and 73.5% for coal and water consumption respectively. The Ljung-Box Q shows a significance of 0.09, 0.682 and 0.258 for carbon dioxide, sulphur and nitrogen gases highlighting that residuals of the model are independently distributed with no significant autocorrelation left by the models after specifications.

Moreover, the statistic values of 20.20, 10.147 and 15.836 Kt are lower than the chi-square of 22.362 with 13 degrees of freedom at the significance level of 5% leading to the fitness and independence of the time series model. A trend of observed values from 2017Q1 to 2023Q4 being similar to the predicted values from 2024Q1 to 2030Q4 (see Fig. 2). However, a slight decrease will be observed in $CO_2$, $SO_2$ and $NO_x$ gas emissions in the prediction period. The $CO_2$, $SO_2$ and $NO_x$ gas emissions will decrease from 51, 395 and 203 in 2024Q4 down to 43, 292 and 162 in 2030.

## 4.3 Waste Management Prediction Models

The model statistics in Table 3 indicate the best fit of the model given the coefficient of determination of 63.8%, 50.9% and 78.2% for coal and water consumption respectively. The Ljung-Box Q shows a significance of 0.214, 0.928 and 99.3 for ash, asbestos and radioactive wastes showing that residuals of the model are independently distributed with no significant autocorrelation left by the model after specifications. Moreover, the statistic values of 16.683, 6.456 and 3.802 are lower than the chi-square of 22.362 with 13 degrees of freedom at the significance level of 5% leading to the fitness and independence of the time series model. A trend of observed values from 2017Q1 to 2023Q4 and predicted values from 2024Q1 to 2030Q4 (see Fig. 3). Forecasting shows that the quantity of ash produced will slightly reduce from 8 in 2024Q4 to 7.5 Mt in 2030Q4 with weak variation during the year. Although a sharp increase in asbestos discharges reached more than 5000 tons each quarter in 2021 and dropped to 248 in 2022Q2, predicted values will increase from 600 in 2024Q4 to 1177 Mt in 2030Q4. The radioactive waste model displays stability from 2017 to 2022 followed by an increase in 2023. Previsions show that more radioactive waste will be deposited in nature from 95 in 2024 to 160 cubic metres in 2030. For all forecasting models, the null mull hypotheses will be accepted because the p-values are significant, and the Q coefficient is less than the chi-square at the significance level of 5%. This means that the autocorrelations for the chosen lags are all zero in the chosen model without serial correlation. Predictions revealed that the target of reaching zero emissions in 2030 is far from realistic given ARIMA models. Eskom should therefore enforce green solutions to meet company goals and global alignment to the sustainable development goals.

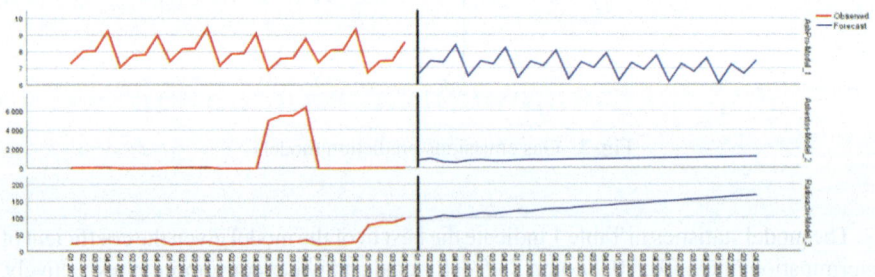

**Fig. 3.** Waste management prediction models

### 4.4 Forecasting Solutions for Circular Energy Production

Circular initiatives at Eskom include the just energy transition (JEST) strategies through partnerships and power purchase agreements (PPAs) to establish renewable energy development zones and the application of emissions abatement plans and high-frequency electrostatic precipitators and transformers [9]. To achieve forecasting environmental solutions, energy production companies should replace fossil-fuel-driven manufacturing with renewable sources, implement greenhouse gas-intensive and dry-cooled technologies, apply grid management strategies and system optimisation techniques. Solutions to minimise water waste include upgraded water infrastructure [16], long-term water storage through the Lesotho Highlands and Mokolo crocodile water projects, hydro-based features to repurpose decommissioned power stations and ash and oil-filtered tanks technologies. Future solutions for waste minimisation encompass new waste pollution prevention plans, updated waste governance, upgraded water infrastructure, circular water supplies with no contamination besides renewable energies sources like solar, geothermal, wind, hydro and nuclear. Reducing greenhouse gas emissions includes directing emitted gas underground, installing carbon footprint calculator, adopting less carbon-intensive fuels, adopting lower-emitting power plants, shifting to natural gas-firing boiler and combined-cycle turbines. It imperative to implement advanced technologies with specific circular manufacturing applications and advanced tracking systems. In the Middle East, Artificial Intelligence was coupled with predictive techniques to improve supply chain management, predictive maintenance and resource efficiency [21]. Following the European Union project on Resource Conservative Manufacturing, the modeling simulation techniques enhance decision-making processes in demand forecasting and remanufactured products [2].

## 5 Conclusion

Forecasting analysis indicates a decrease in resource consumption and gas emissions from 2024 to 2030 due to a few power stations closed down and the carbon tax incentives compared to a waste increase. Solving energy insecurity in the country suggests a continuous degradation of natural resources if circularity is not fully adopted in energy companies. Civil societies, industries, academia, and the SA government strive for integrated circularity innovation to promote environmental sustainability through the proliferation of resource-efficient practices, modernised power stations and upgraded manufacturing governance. To meet the SDGs agenda 2030, the SA government should ensure that the drivers of CE including global pressures, climate change, environmental concerns, industry 4.0 and regulatory requirements are well managed to safeguard the transition to circularity in the production sectors, and inclusive growth to ensure sustainable livelihoods. Future studies should consider the integration of the ARIMA with the SES and LSTM techniques to optimise forecasting results and production processes. Also, the ARIMA model can be applied to specific CE concepts after conducting linearity tests for comparative analysis.

# References

1. European environment agency. Circular economy in Europe – developing the knowledge base. https://www.eea.europa.eu/publications/circular-economy-in-europe Accessed 08 Apr 2024
2. Amir, S., Asif, F.M.A., Roci, M.: Towards circular economy: enhanced decision-making in circular manufacturing systems, Volume II. Palgrave Macmillan, Cham (2021). https://doi.org/10.1007/978-3-030-55285-5_12
3. Ponstein, H.J., Meyer-Aurich, A., Prochnow, A.: Greenhouse gases emissions and mitigation options for german wine production. J. Cleaner Prod. **212**, 800–809 (2019). https://doi.org/10.1016/j.jclepro.2018.11.206
4. World economic forum, towards the circular economy: accelerating the scale-up across global supply chains. https://www.weforum.org/reports/ Accessed 04 June 2024
5. DFFE. A circular economy guideline for the waste sector. www.dffe.gov.za/sites/default/files/docs/circulareconomy_guideline.pdf Accessed 27 Apr 2024
6. GreenCape. circular economy: market intelligence report: South Africa. https://greencape.co.za/sector/circular-economy/ Accessed 15 May 2024
7. Zero waste europe. zero waste - one of the solutions to ecocide, zero waste Europe, https://zerowasteeurope.eu/2013/07/zero-waste-one-ofthe-solutions-to-ecocide/ Accessed 13 May 2024
8. CSIR. Supporting the development of a globally competitive manufacturing sector through a more circular economy. https://www.csir.co.za/ Accessed 04 June 2024
9. Eskom. Integrated reports. https://www.eskom.co.za/wp-content/uploads/2023/10/Eskom_integrated_report_2023.pdf Accessed 10 May 2024
10. UNIDO. Circular economy, https://www.unido.org/sites/default/files/2017-07/Circular_Economy_UNIDO_0.pdf Accessed 03 June 2024
11. Frishammar, J., Parida, V.: Circular business model transformation: a roadmap for incumbent firms. Calif. Manage. Rev. **61**(2), 5–29 (2019)
12. Kazakova, E., Lee, J.: Sustainable manufacturing for a circular economy. Sustainability **14**, 17010 (2022). https://doi.org/10.3390/su142417010
13. DEA, South African state of waste report. https://soer.environment.gov.za/soer/UploadLibraryImages/UploadDocuments/141119143510_state%20of%20Waste%20Report_2018.pdf. Department of Environmental Affairs. Accessed 25 Apr 2024
14. Environmental protection agency network. monitoring progress in Europe's circular economy. https://epanet.eea.europa.eu/reports-letters/. Accessed 26 Apr 2024
15. Kristoffersen, E., Blomsma, F., Mikalef, P., Li, J.: The smart circular economy: a digital-enabled circular strategies framework for manufacturing companies. J. Bus. Res. **120**, 241–261 (2020)
16. Centre for renewable & sustainable energy studies (2023). Visualisation of South African energy data. https://www.crses.sun.ac.za/sa-energy-stats/ Accessed 08 June 2024
17. Díaz-Chao, Á., Ficapal-Cusí, P., Torrent-Sellens, J.: Environmental assets, Industry 4.0 technologies and firm performance in Spain: a dynamic capabilities path to reward sustainability. J. Cleaner Prod. **281**, 125264 (2021). https://doi.org/10.1016/j.jclepro.2020.125264
18. Acerbi, F., Sassanelli, C., Taisch, M.A.: Conceptual data model promoting data-driven circular manufacturing. Oper. Manag. Res. **15**(3), 838–857 (2022)
19. Wang, C.C., Chien, C.H., Trappey, A.J.: On the application of ARIMA and LSTM to predict order demand based on short lead time and on-time delivery requirements. Processes **9**(7), 1157 (2021)
20. Olawumi, M.A., Olawale, R.A., Oladapo, B.I.: Revolutionising waste management with the impact of long short-term memory networks on recycling rate predictions. Waste Manage. Bull. (2024)

21. Ronaghi, M.R.: The influence of artificial intelligence adoption on circular economy practices in manufacturing industries. Environ. Dev. Sustain. **25**, 14355–14380 (2023). https://doi.org/10.1007/s10668-022-02670-3

# Sustainable Management of Waste in the Olive Oil Sector

Guillermo Garcia-Garcia$^{(\boxtimes)}$ ⑩, Antonio Pérez⑩, and Mónica Calero⑩

Department of Chemical Engineering, Faculty of Sciences, Avda. Fuentenueva, s/n 18071 Granada, Spain
guillermo.garcia@ugr.es

**Abstract.** One of the challenges of olive oil production is the management of the large amounts of waste generated by the sector. Only 20% of the mass of the olive fruit ends up as olive oil. Most of the remaining biomass consists of the pulp and skin of the fruit, known as olive pomace. This olive pomace still contains oil, which is usually extracted with hexane and sold as lower-quality olive pomace oil. The biomass left over after extraction is called olive cake and is usually combusted to produce electricity or heat. Current waste management practices in the olive sector have a significant environmental impact and provide little economic benefit to olive oil producers. This work proposes a sustainable alternative to valorize olive cake, the main waste from the olive oil sector. The overall objective is to sustainably produce chemical compounds with market value, such as polyphenols, inositol, carbohydrates, bioethanol, tar and ash. The processes proposed also generate heat, that can be reused in the waste management alternative proposed. These processes can be integrated in a biorefinery that would be able to produce olive oil as the primary product, along with a wide range of other products with high market potential.

**Keywords:** Waste valorization · olive pomace · olive cake · hydrolysis

## 1 Introduction

Olive oil is valued by consumers for its unique health benefits and organoleptic properties. The Mediterranean region, particularly Spain, Italy, Greece, Turkey and Tunisia, is the world's largest producer of olive oil [1, 2]. Global production is largely dominated by Spain, which produces more than 40% of the world production and is the world's leading exporter of olive oil [3, 4]. In Spain, and to a lesser extent in the aforementioned Mediterranean countries, olive oil is an important product for the food sector and for the overall national economy. In fact, the price of olive oil has risen sharply in recent years, with prices in 2024 doubling those from the average of the previous 5 years, and currently reaching 7.94, 7.73 and 9.45 € per kg in Spain, Greece and Italy, respectively [5]. This is caused by the high consumer demand and recent poor harvests due to intense droughts.

It is therefore clear that the olive oil sector needs to improve its efficiency to remain competitive in the market. Inputs to the production process, particularly water, should be

H. Kohl et al. (Eds.): GCSM 2024, LNME, pp. 680–686, 2025.
https://doi.org/10.1007/978-3-031-93891-7_75

minimized as much as possible. Similarly, outputs that currently provide no value to olive oil businesses should be processed in order to enable these businesses to add value to their operations and generate additional income, which would then allow product prices to be reduced. Indeed, the olive oil sector currently generates large amounts of waste materials that not only provide no economic benefits, but also generates environmental impacts as well as economic costs. This is slowly changing with recent efforts to modernize the olive oil production processes and the entire olive oil supply chain. As a result, this industrial sector is currently demanding further advances to optimize its operations.

Research and development to propose more sustainable waste management treatments has been developing for years in the food sector [6, 7]. Several options have been proposed, such as anaerobic digestion to obtain bioenergy [8, 9], composting to fertilize soils [10, 11] and thermal treatments to obtain heat [12, 13]. However, a promising alternative is to use such food waste as a feedstock to obtain valuable compounds. Current research is investigating various techniques to recover useful materials and thus valorize food waste [14, 15]. Some researchers are investigating this approach in the olive oil sector (e.g. [16–18]), however this option is not yet implemented at industrial level.

The aim of this work is to find an alternative waste management solution to help the olive oil sector become more economically and environmentally sustainable. This paper reviews the most common way of managing waste in the olive oil sector and then proposes an alternative waste management solution that allows the recovery of several valuable compounds.

## 2  Current Waste Management Practice

One of the challenges of olive oil production is the management of the large amounts of biomass waste generated by the sector. Only 20% of the mass of the olive fruit ends up as olive oil. Most of the remaining biomass consists of the pulp and skin of the fruit, known as olive pomace (Fig. 1, left). This olive pomace still contains oil, so olive pomace is usually dried and the remaining oil is extracted with hexane. The resulting olive pomace oil is sold as lower-quality olive pomace oil. The biomass left over after extraction is called olive cake (Fig. 1, right). Olive cake is usually combusted to produce electricity or heat. Another residue from olive oil production is the olive stone, which is also combusted to produce energy. The heat generated in the combustion process is sometimes used in the drying process. The block flow diagram of the waste management practice described in this section is shown in Fig. 2.

The main advantage of the current waste management practice is the production of olive pomace oil, which has a good market position. However, the drying process requires a high energy input, which is not always available from the combustion process. In addition, both drying and combustion emit particles into the atmosphere, which generate an environmental impact. In addition to improving these two processes, there are opportunities to recover valuable compounds from the olive cake. This would also reduce the number of volatile solids emitted during combustion. Based on the authors' knowledge and discussions with waste managers from the olive oil sector, an alternative waste management solution is proposed in the next section Sect. 3.

**Fig. 1.** Left: olive pomace, right: olive cake.

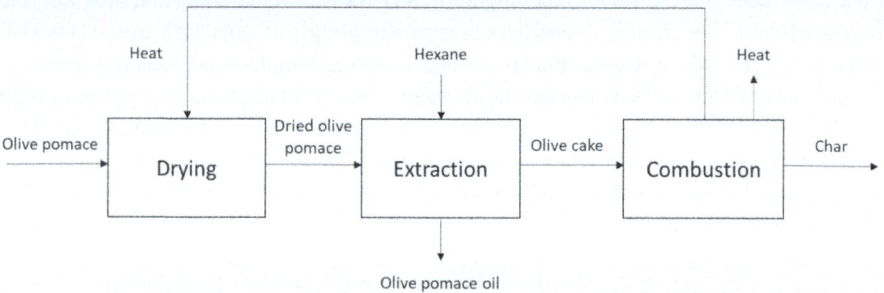

**Fig. 2.** Conventional waste treatment process for olive pomace.

## 3 Proposed Waste Management Alternative

This work proposes a waste management alternative to valorise olive cake, the main waste from the olive oil sector. The overall objective is to produce chemical compounds with market value, as well as heat that can be used in the treatment processes themselves. In addition, the processes proposed operate at low temperatures to save energy and use water as the only reagent. Therefore, the proposed waste management alternative is a sustainable solution that allows the recovery of valuable compounds, which saves the conventional production of such compounds, as well as uses minimal material and energy inputs.

The proposed waste management alternative is shown in Fig. 3. It starts with the drying of olive pomace and the extraction of olive pomace oil with hexane, as is currently done. This is unchanged due to the large share that olive pomace oil already has on the market. The resulting olive cake is then subjected to a hydrothermal treatment at 30°C, which produces a liquid phase that is filtered to extract polyphenols and inositol. Both compounds have been shown to be beneficial to human health and therefore have important applications in the food and pharmaceutical industries. Other carbohydrates, for instance sugars, can also be extracted and subsequently fermented to produce bioethanol.

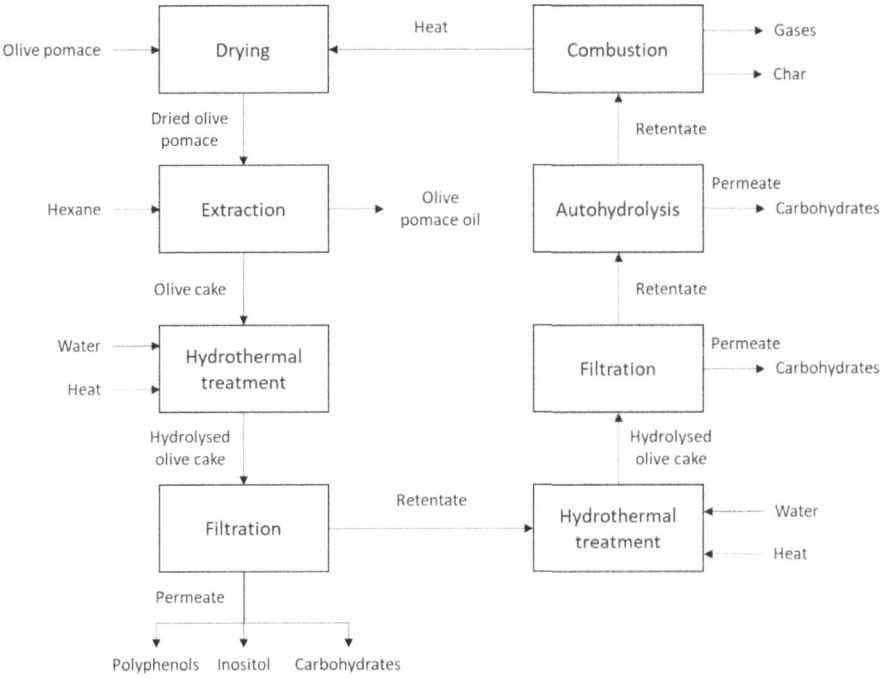

**Fig. 3.** Proposed waste treatment processes for olive pomace.

The solid phase remaining after the first hydrothermal treatment is then subjected to a further hydrothermal treatment at 70°C. The liquid phase obtained, as well as the previous permeate, is a good source of carbohydrates. The solid remaining after the second hydrothermal treatment is then subjected to autohydrolysis, which produces a sugar-rich solution and a solid that can be thermally treated for energy recovery. This solid is combusted and the heat generated is used in the drying process at the beginning of the process chain. Alternatively, future work will explore the gasification of the solid, with the objective of producing a gas that is combusted to recover the heat, and tar and ash as remaining liquid and solid phases, respectively. Tar has a wide range of applications, including in the production of varnishes, adhesives and preservatives. Ash can be employed as an adsorbent to replace activated carbon.

Preliminary tests have demonstrated the feasibility of the block flow diagram depicted in Fig. 3. A mass concentration of 25% w/w for the olive cake in water has been proven to be adequate for the hydrothermal treatment. Vacuum filtration was used for the separation of the permeate and retentate. Nine repetitions of the hydrothermal treatment at 30°C were conducted with samples of 500 g of dried olive cake, obtaining a homogeneous amount of permeate (mean: 1128.06 g, standard deviation: 79.54 g) and retentate (mean: 807.63 g, standard deviation: 71.41 g). Seven tests were carried out for the hydrothermal treatment at 70°C, also with 500 g of starting feedstock. The results obtained were also homogeneous for both the permeate (mean: 927.71 g, standard deviation: 93.52 g) and retentate (mean: 903.98 g, standard deviation: 61.17). Following the completion of both hydrothermal treatments, the solid retentates had a moisture content of 50–60%. The solid retentates were fully dried in an oven at 105°C prior to undergoing the subsequent process indicated in the block flow diagram shown in Fig. 3.

Autohydrolysis must be conducted at a considerably higher temperature. Several combinations of temperature, reaction time and concentrations are possible and are currently being tested. It is imperative to identify a compromise solution that strikes a balance between the maximum carbohydrates obtained in the permeate phase and the minimum temperatures and reaction times necessary to save energy. Ultimately, the retentate was subjected to drying and combustion. This process results in the emission of gases into the atmosphere; however, it does yield a solid char that can be activated and employed as an activated carbon substitute.

The proposed waste treatment processes described above allow the production of a comprehensive range of valuable chemicals that include polyphenols, inositol, carbohydrates, bioethanol, tar and ash. This contrasts with the current predominant approach in the sector, whereby the majority of outputs are released into the atmosphere or deposited in nearby soil. Moreover, we propose an innovative approach to optimizing the value of biomass management through the utilization of the biorefinery concept. A biorefinery is an industrial plant that utilizes biomass as a feedstock to produce a diverse range of valuable products, including chemicals, foodstuffs and animal feed, as well as energy in the form of electricity, heat or biofuels. While previous work has been conducted on the potential implementation of biorefineries in the olive oil sector, there is a lack of studies exploring the valorization of olive cake, particularly under mild conditions. Our proposed waste management alternative to valorize olive cake could be integrated into such a biorefinery, thereby enabling the production of olive oil as the main product, along with a diverse array of other valuable products that could also be commercialized.

## 4 Conclusions

The production of olive oil represents a significant industrial sector in numerous countries situated in the Mediterranean region. One of the most significant challenges currently facing this sector is the management of the waste generated throughout the production process, in particular the olive cake obtained from the processing of olive pomace. Presently, the majority of the olive cake is combusted. This work proposes a waste management alternative for olive cake to obtain polyphenols, inositol, carbohydrates, bioethanol, tar and ash. The heat released in the processes can be recovered and used in the treatment processes themselves. Further work could extend this approach to treat other materials generated in the sector, such as olive stones, leaves and branches, with a view to integrating them into a biorefinery. Consequently, a full-scale biorefinery would be capable of producing olive oil as its primary product, in addition to a multitude of other products that have a considerable market potential. Such an approach would have the dual benefit of increasing revenues for olive oil businesses and reducing the environmental impact of current waste management practices.

**Acknowledgements.** Guillermo Garcia-Garcia acknowledges the Grant 'Marie Skłodowska-Curie Actions (MSCA) Postdoctoral Fellowship' with Grant agreement ID: 101052284.

# References

1. European commission: factsheet olive oil. Agriculture and rural development (2020)
2. European commission: olive oil production. https://agridata.ec.europa.eu/extensions/Dashbo ardOliveOil/OliveOilProduction.html Accessed 06 Nov 2023
3. International olive council: the world of olive oil, https://www.internationaloliveoil.org/the-world-of-olive-oil/ Accessed 02 Feb 2024
4. FAOSTAT: Crops and livestock products https://www.fao.org/faostat/en/#data/QCL Accessed 09 May 2023
5. European commission. Agriculture and rural development: market situation in the olive oil and table olives sectors. Expert group for agricultural markets - arable crops and olive oil (2024)
6. Garcia-Garcia, G., Stone, J., Rahimifard, S.: Opportunities for waste valorisation in the food industry – a case study with four UK food manufacturers. J. Clean. Prod. **211**, 1339–1356 (2019). https://doi.org/10.1016/J.JCLEPRO.2018.11.269
7. Garcia-Garcia, G., Woolley, E., Rahimifard, S., Colwill, J., White, R., Needham, L.: A methodology for sustainable management of food waste. Waste Biomass Valorization **8**, 2209–2227 (2017). https://doi.org/10.1007/s12649-016-9720-0
8. Xu, F., Li, Y., Ge, X., Yang, L., Li, Y.: Anaerobic digestion of food waste – challenges and opportunities. Bioresour. Technol. **247**, 1047–1058 (2018). https://doi.org/10.1016/J.BIO RTECH.2017.09.020
9. Neri, A., Bernardi, B., Zimbalatti, G., Benalia, S.: An overview of anaerobic digestion of agricultural by-products and food waste for biomethane production. Energies 2023 **16**, 6851 (2023). https://doi.org/10.3390/EN16196851
10. Cerda, A., Artola, A., Font, X., Barrena, R., Gea, T., Sánchez, A.: Composting of food wastes: status and challenges. Bioresour. Technol. **248**, 57–67 (2018). https://doi.org/10.1016/j.bio rtech.2017.06.133
11. Awasthi, S.K., et al.: Changes in global trends in food waste composting: research challenges and opportunities (2019). https://doi.org/10.1016/j.biortech.2019.122555
12. Tang, Y., Dong, J., Chi, Y., Zhou, Z., Ni, M.: Energy and exergy optimization of food waste pretreatment and incineration. Environ. Sci. Pollut. Res. **24**, 18434–18443 (2017). https://doi. org/10.1007/s11356-017-9396-4
13. Kim, S., Lee, Y., Andrew Lin, K.Y., Hong, E., Kwon, E.E., Lee, J.: The valorization of food waste via pyrolysis. J. Clean. Prod. **259**, 120816 (2020). https://doi.org/10.1016/J.JCLEPRO. 2020.120816
14. Yadav, S., et al.: Valorisation of agri-food waste for bioactive compounds: recent trends and future sustainable challenges. Molecules 2024 **29**, 2055 (2024). https://doi.org/10.3390/MOL ECULES29092055
15. Liu, Z., de Souza, T.S.P., Holland, B., Dunshea, F., Barrow, C., Suleria, H.A.R.: Valorization of food waste to produce value-added products based on its bioactive compounds. Processes 2023 **11**, 840 (2023). https://doi.org/10.3390/PR11030840
16. Lozano, E.J., Blázquez, G., Calero, M., Martín-Lara, M.Á., Pérez-Huertas, S., Pérez, A.: Optimizing the extraction process of value-added products from olive cake using neuro-fuzzy models. Processes **12**, 317 (2024). https://doi.org/10.3390/PR12020317/S1
17. Moudache, M., Silva, F., Nerín, C., Zaidi, F.: Olive cake and leaf extracts as valuable sources of antioxidant and antimicrobial compounds: a comparative study. Waste Biomass Valorization **12**, 1431–1445 (2021). https://doi.org/10.1007/S12649-020-01080-8/TABLES/7
18. Fernández-Prior, M.Á., Fatuarte, J.C.P., Oria, A.B., Viera-Alcaide, I., Fernández-Bolaños, J., Rodríguez-Gutiérrez, G.: New liquid source of antioxidant phenolic compounds in the olive oil industry: alperujo water. Foods 2020 **9**, 962 (2020). https://doi.org/10.3390/FOO DS9070962

# Value Creation Networks and Supply Chain

# A Behaviourally Informed Heuristics-Framework for Net Zero Transformation in Manufacturing

Rashmeet Kaur, John Patsavellas[✉], and Konstantinos Salonitis

Sustainable Manufacturing Systems Centre, School of Aerospace, Transport and Manufacturing, Cranfield University, Cranfield, UK
John.Patsavellas@cranfield.ac.uk

**Abstract.** Transitioning to net zero emissions in manufacturing is fraught with challenges, from navigating uncertainties and making critical trade-offs, to overcoming biases and information asymmetry. Such behavioural challenges could potentially result in bounded rationality, where decision-makers operate under limited information and cognitive constraints. This paper introduces a framework that employs behaviourally informed heuristics to simplify the complexity of net-zero transformation. By incorporating behavioural model of rational choice and decision-making rules, the framework could help manufacturing decision-makers to manage uncertainties and cognitive limitations, thus broadening the toolkit for navigating the reduction of carbon emissions.

**Keywords:** Net zero manufacturing · Transformational framework · Bounded rationality

## 1 Introduction

The manufacturing sector plays a pivotal role in driving economic growth and development worldwide. However, its contribution to greenhouse gas emissions and environmental degradation has raised concerns about its long-term sustainability. As the global community intensifies efforts to combat climate change, the transition towards net-zero emissions in the manufacturing sector has become a critical imperative. This research paper explores the behavioural, cognitive, and organizational challenges associated with the net-zero transformation in the manufacturing sector. Resistance to change due to high costs, long investment cycles, and concerns about disrupting systems hinder the adoption of low-carbon technologies and processes, especially in energy-intensive industries fearing non-performance impact on production and low implementation of energy efficiency projects [1–3]. Overcoming this requires organizational culture shift, management commitment, employee engagement [4–6]. Lack of awareness, knowledge, skills among employees and managers [7–9] necessitates continuous education, training, knowledge transfer [10]. Aligning business-sustainability goals, fostering digital

© The Author(s) 2025
H. Kohl et al. (Eds.): GCSM 2024, LNME, pp. 689–698, 2025.
https://doi.org/10.1007/978-3-031-93891-7_76

transformation environment is crucial [6, 9, 10]. Short-term thinking, competing priorities constrain decarbonization strategies [11]. Trade-offs between economic gains-sustainability goals, resource constraints hinder net zero transition [8, 12]. Stakeholder engagement, collaboration engaging industrial communities to overcome resistance [13], industry-wide cooperation, sustainable value chain practices [14–16] is vital. Regulation, policy support like mandates, incentives can motivate shift from linear to circular approaches [1, 17, 18], but effectiveness depends on ambition level, company capacity, with greater challenges for smaller, energy-intensive firms [19]. To address these challenges, the paper presents a comprehensive review of existing models and frameworks for guiding the net-zero transformation and proposes a behaviourally informed strategic decision making. It critically evaluates current state of the art's strengths and limitations, highlighting the need for more holistic and adaptive approaches that integrate technological, economic, social, and behavioural dimensions to facilitate the transition towards a sustainable and resilient future.

## 2  Literature Review

### 2.1  Behavioural Challenges in Net Zero Manufacturing Transformation

The transition to net-zero manufacturing faces several behavioural challenges, as highlighted in multiple studies and summarised below in this section. Resistance to change emerges as a prominent obstacle, with employees and management often reluctant to adopt new technologies and practices [7, 12, 20]. This resistance is particularly evident in industries with long-established processes and infrastructures, such as the steel industry [7]. The lack of awareness, insufficient skills, and the discomfort caused by new net-zero practices further exacerbate this resistance [10, 12]. Organizational culture and leadership play important roles in overcoming these behavioural barriers. Studies emphasize the importance of top management commitment, continuous training programs, and the development of a sustainability-driven culture [10, 12]. The need for significant shifts in organizational behaviour and mindset is highlighted, particularly in aligning business objectives with sustainability goals [8, 22]. This cultural shift is essential for fostering an environment that supports and encourages sustainable practices and innovation. The challenge extends beyond individual organizations to industry-wide adoption of sustainable practices and innovative technologies [16, 23]. Many studies point out that achieving net-zero emissions will require significant changes in manufacturing processes and the adoption of new technologies by stakeholders across the value chain [16, 23]. This necessitates a collective effort and a shared commitment to sustainability goals, which can be challenging to cultivate across diverse industry participants. Overcoming these behavioural challenges requires coordinated efforts, supportive policies, and a commitment to long-term sustainability goals from all stakeholders involved in the manufacturing sector [10, 24]. The importance of employee engagement and training is frequently emphasized, as is the need for clear communication and collaboration between different sectors of the industry [12, 17]. Additionally, addressing concerns about job displacement and the potential economic impact of the transition is crucial for gaining widespread support and minimizing resistance to change [12, 22]. By addressing these behavioural challenges comprehensively, the manufacturing sector can navigate the complex path towards net-zero emissions. The behavioural challenges on the net zero journey in manufacturing can be summarized below in Table 1 in three categories of impact on decision making, namely complexity, information asymmetry and cognitive biases. These challenges compounded with the time bound pressure of achieving net zero goal contributes to bounded rationality.

**Table 1.** Summary of key behavioural challenges

| Category | Challenges | References |
|---|---|---|
| *Complexity* | High-emitting, fossil fuel-dependent infrastructures | [20, 25] |
| | Fundamental shifts in technologies, processes, and raw materials | [12, 26] |
| | Integration of renewables and energy efficiency | [11, 27] |
| | Multiple stakeholder involvement | [28] |
| | Diverse company characteristics | [29] |
| | Multi-sector, multi-region coordination | [13] |
| *Information Asymmetry* | Lack of accurate, comprehensive data | [20, 25, 26] |
| | Historical underreporting and underestimation of emissions | [27] |
| | Communication difficulties with broader audiences | [30] |
| | Challenges in integrating and analysing vast amounts of data | [29] |
| | Data availability, reliability, storage, and processing issues | [13] |
| *Cognitive Biases* | Status quo bias leading to resistance to change | [11] |
| | Short-term focus hindering climate change responses | [28] |
| | Underestimation of long-term decarbonization benefits | [20, 25, 28] |
| | Confirmation bias in seeking information | [20, 25] |
| | Herding influencing adoption of sustainable practices | [29] |
| | Anchoring effect on cost estimates and targets | [13, 24] |

## 2.2 Net Zero Manufacturing Transformation Models

The transition to net-zero emissions in manufacturing has led to various models and frameworks for this transformation. While providing insights, these models have limitations. The technology roadmap by [26] outlines a framework for net-zero emissions in emissions-intensive industries by 2050, emphasizing innovation-supporting policies and decarbonization. However, it assumes perfect rationality and does not address behavioural challenges or resistance to new technologies. Models integrating organizational and technological dimensions, like the ADO framework [21] and the theoretical model in [26], highlight enablers and barriers to decarbonization but often miss the complex stakeholder interrelationships and unintended consequences. Specific models like VENRA [31] for shipyard performance and I4.2-GiM [4] for integrating manufacturing with energy management offer structured approaches but lack generalizability. Roadmaps and step-by-step approaches, such as the bio steel cycle [1] for the steel industry and the Integrated Systems plus Principles Approach [3] for net-zero manufacturing, provide practical guidance but often rely on assumptions about technology availability and cost-effectiveness, overlooking economic and regulatory barriers. Sector-specific models for industries like cement [27], chemical and process [28], offer insights but ignore the interconnectedness between sectors and the need for a holistic approach. Many models fail to address social and behavioural dimensions, such as resistance to change, stakeholder engagement, and local contexts [13]. Models focusing solely on technological solutions and economic incentives may miss the complex decision-making processes and the role of values, culture, and emotions in adopting sustainable practices. The review of the current state of the art in various models and frameworks for net-zero emissions transformation in manufacturing reveals several limitations that can be summarised below in Table 2.

**Table 2.** Summary of limitations of current state of the art

| Challenge | Description |
|---|---|
| Rationality Assumption | Many models assume perfect rationality in decision-making, overlooking the cognitive biases and behavioural challenges that can hinder the adoption of new technologies |
| Complexity Oversimplification | Frameworks like ADO and other theoretical models often fail to capture the full complexity of stakeholder interactions and potential unintended consequences of decarbonization efforts |
| Limited Generalizability | Sector-specific models provide structured approaches but may lack applicability across diverse industrial contexts |
| Technological Optimism | Step-by-step approaches tend to rely on optimistic assumptions about technology availability and cost-effectiveness, potentially underestimating economic and regulatory barriers |
| Sectoral Silos | Industry-specific models for cement, chemical, and electronics often neglect the interconnectedness between sectors, missing opportunities for synergistic solutions |
| Social Dimension Gap | Many models inadequately address social and behavioural aspects, including resistance to change, stakeholder engagement, and local contextual factors |
| Holistic Perspective Lacking | Tendency to focus on technological solutions and economic incentives, overlooking the complex decision-making processes and the role of values, culture, and emotions in driving sustainable practices |

These findings highlight the need for more comprehensive, interdisciplinary approaches that integrate technological, economic, social, and behavioural dimensions to effectively guide the transition to net-zero emissions in manufacturing.

## 3   Discussion

Bounded rationality is the idea that when individuals or firms make decisions, they do so in a way that is purposeful but limited by constraints, rather than being fully informed and perfectly rational [31]. The concept was pioneered by Herbert A. Simon, who identified three key constraints on decision-making [32]. Firstly, only limited and often unreliable information regarding the alternatives and their consequences is available; the human mind has a limited ability to evaluate and process the available information; time constraints limit the ability to make decisions [33]. These factors contribute to bounded rationality for individuals. In organizations and firms, an inability to adequately deal with these constraints can lead to analysis paralysis and impaired decision-making. The behavioural model of rational choice that Simon put forth involved several key elements [34] – The set of all possible behaviour alternatives, subset of those alternatives considered by the individual, possible outcomes of each choice, pay-off function representing the value placed on each outcome, information (complete or incomplete) about which outcomes will occur for each choice and probability information for each outcome. Bounded rationality recognizes the real-world limitations and constraints on decision-making that make perfect rationality impossible [33]. Instead, individuals employ approximate methods compatible with the information, cognitive abilities, and time available to them. Understanding these methods can provide insight into how decisions are made by both individuals and organizations. In this research work, a behavioural model of rational choice has been made to support strategic decision making in net zero manufacturing transformations. This research work presents a conceptual framework that includes all the elements suggested by Simon in any behavioural model of rational choice. The framework begins with a list of all the decisions organisations need to take on their net zero journey as identified in literature and summarised below in Table 3 along with the heuristics to be mindful of while making these decisions.

**Table 3.** Decisions and heuristics on net zero journey.

| Decision | Heuristics | Challenge being tackled | References |
| --- | --- | --- | --- |
| Technology Adoption | Select low-carbon technologies, weighing perceived costs and disruption against long-term benefits | Cognitive Biases | [6, 7, 9, 11] |
| Data Management and Transparency | Establish robust systems for accurate data collection and reporting to ensure transparency, reliability, and informed decision-making | Information Asymmetry | [12, 21, 26, 27] |

*(continued)*

**Table 3.** (*continued*)

| Decision | Heuristics | Challenge being tackled | References |
|---|---|---|---|
| Stakeholder Engagement | Develop strong communication and collaboration strategies with all stakeholders, including co-producing decarbonization strategies with local communities for acceptance and success | Complexity | [13, 29] |
| Organizational Culture | Foster a sustainability-driven culture through leadership commitment, continuous training, and alignment of business objectives with sustainability goals | Cognitive Biases | [10, 22, 23] |
| Resource Allocation | Allocate resources, especially for SMEs, balancing investments in tangible assets, skills, culture, and knowledge management | Cognitive Biases | [10, 30] |
| Supply Chain Restructuring | Adopt circular economy principles, which requires significant restructuring of supply chains, product design, and end-of-life management | Complexity | [29, 31] |
| Energy Management | Integrate renewable energy sources and improve energy efficiency through sophisticated optimization and management. Consider on-site renewable energy systems where feasible | Complexity | [28] |

(*continued*)

**Table 3.** (*continued*)

| Decision | Heuristics | Challenge being tackled | References |
|---|---|---|---|
| Policy Engagement | Actively engage with policymakers to support the development of robust policy frameworks and financial incentives. Stay informed about evolving regulations and carbon pricing mechanisms | Cognitive Biases | [2, 10, 23, 25] |
| Long-term Planning | Develop strategies that account for the long-term impacts of climate change and decarbonization efforts, avoiding short-term focus | Cognitive Biases | [11, 29] |
| Innovation and R&D | Invest in research and development of new technologies and processes that can significantly reduce emissions, such as carbon capture and storage or alternative materials | Cognitive Biases | [1, 25] |
| Performance Measurement | Implement comprehensive performance measurement frameworks that integrate environmental, social, and economic factors | Information Asymmetry | [31] |
| Risk Assessment | Conduct thorough risk assessments, considering both the risks of inaction and the potential challenges of implementing new technologies and practices | Cognitive Biases | [31], [41] |

The heuristics identified from the literature review and summarised above can potentially support in limiting if not overcoming bounded rationality. The framework shown below in Fig. 1, does not have rationality assumption, in fact it is based on acknowledging its presence. The complexity is not oversimplified, rather it is accounted for by starting with a wide variety of decisions that could have a potential impact on overall progress. The framework is not sector specific and can be deployed in across manufacturing industry. It considers the entire manufacturing systems as the scope and also accounts for sector specific challenges in the second step with information and probability assessment.

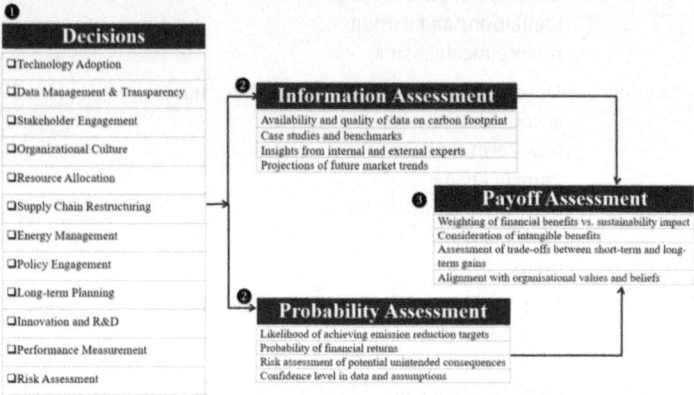

**Fig. 1.** Behavioural model of rational choice for net zero manufacturing.

# 4 Conclusion

This research paper delves into behavioural challenges, highlighting resistance to change, information asymmetries, cognitive biases, and the need for adaptive transformation models. The behavioural model of rational choice for net zero manufacturing transformation offers a framework for organizations to make better decisions for achieving net-zero emissions. By systematically evaluating choices, outcomes, and assessments, businesses can navigate sustainability complexities and ensure long-term success in a low-carbon future. However, this research has limitations. The proposed framework is based on existing literature and may not fully capture specific industrial contexts or regional variations. Rapid technological advancements and changing market dynamics may require periodic updates to the model. Future research should focus on empirical studies and case analyses to validate and refine the model. Practical insights from industry stakeholders and policymakers would enhance the model's applicability and relevance. Overall, this research supports efforts to decarbonize manufacturing and encourages further exploration and collaboration among researchers, policymakers, and industry professionals to accelerate the transition to a sustainable future.

# References

1. Kiessling, S., Darabkhani, H.G., Soliman, A.H.: The bio steel cycle: 7 steps to net-zero CO2 emissions steel production. Energies (Basel) **15**(23) (2022)
2. Vieira, L.C., Longo, M., Mura, M.: Are the European manufacturing and energy sectors on track for achieving net-zero emissions in 2050? an empirical analysis. Energy Policy **156** (2021)
3. Kissock, K., Pohlman, D., Smith, J., Chair, P.E.: Net-Zero Carbon Manufacturing At Net-Zero Cost. https://www.researchgate.net/publication/266220214
4. Tieng, H., et al.: I4.2-GiM: a novel green intelligent manufacturing framework for net zero. IEEE Trans. Autom. Sci. Eng. 1–21 (2024)
5. Sharma, M., Shah, J.K., Joshi, S.: Modeling enablers of supply chain decarbonisation to achieve zero carbon emissions: an environment, social and governance (ESG) perspective. Environ. Sci. Pollut. Res. **30**(31), 76718–76734 (2023)
6. Bonsu, N.O.: Towards a circular and low-carbon economy: insights from the transitioning to electric vehicles and net zero economy. J. Clean Prod. **256** (2020)
7. Mishra, R., Singh, R.K., Gunasekaran, A.: Adoption of industry 4.0 technologies for decarbonisation in the steel industry: self-assessment framework with case illustration. Ann. Oper. Res. (2023)
8. Blundel, R., Hampton, S.: How can smes contribute to net zero? an evidence review
9. Agrawal, R., et al.: Are emerging technologies unlocking the potential of sustainable practices in the context of a net-zero economy? an analysis of driving forces. Environ. Sci. Pollut. Res. (2023)
10. S. Bag, "From resources to sustainability: a practice-based view of net zero economy implementation in small and medium business-to-business firms," Benchmarking, 2023
11. Canal Vieira, L., Longo, M., Mura, M.: Responding to a wicked problem: how time, sense of place, and organisational boundaries shape companies. Decarbonisation Strat. Organ. Environ. **37**(1), 6–31 (2024)
12. Kumar, D., Singh, R.K., Mishra, R., Vlachos, I.: Big data analytics in supply chain decarbonisation: a systematic literature review and future research directions. Int. J. Prod. Res. **62**(4), 1489–1509 (2024)
13. Devine-Wright, P.: Decarbonisation of industrial clusters: a place-based research agenda. Energy Res. Soc. Sci. **91** (2022)
14. Malliaroudaki, M.I., et al.: Net zero roadmap modelling for sustainable dairy manufacturing and distribution. Chem. Eng. J. **475** (2023)
15. Glavič, P., et al.: Transitioning towards net-zero emissions in chemical and process indus-tries: a holistic perspective. Processes **11**(9) (2023)
16. Riahi, S., Mckenzie, J.A., Sandhu, S., Majewski, P.: Towards net zero emissions, recovered silicon from recycling PV waste panels for silicon carbide crystal production. Sustain. Mater. Technol. **36** (2023)
17. Helferty, P.H.: Getting to net zero through extended producer responsibility. J. Adv. Manufact. Process. **4**(3) (2022)
18. Lee, R.P., Keller, F., Meyer, B.: A concept to support the transformation from a linear to circular carbon economy: net zero emissions, resource efficiency and conservation through a coupling of the energy, chemical and waste management sectors. Clean Energy **1**(1), 102–113 (2017)
19. Buettner, S.M., et al.: How do german manufacturers react to the increasing societal pressure for decarbonization? applied sciences (Switzerland) **12**(2) (2022)
20. Talwar, S., et al.: Charting the path toward a greener world: a review of facilitating and inhibiting factors for carbon neutrality. J. Clean Prod. **423** (2023)

21. Kumar, R., Gupta, S., Rehman, U.: Circular economy a footstep toward net zero manufacturing: critical success factors analysis with case illustration. Sustainability (Switzerland) **15**(20) (2023)
22. Gajdzik, B., Piontek, B.: The strategic challenges of the decarbonisation of the manufacturing industry. Econ. Environ. **87**(4) (2023)
23. Underwood, R., et al.: Abundant material consumption based on a learning curve for photovoltaic toward net-zero emissions by 2050. Solar RRL **7**(8) (2023)
24. Kim, A., Miller, S.A.: Meeting industrial decarbonization goals: a case study of and roadmap to a net-zero emissions cement industry in California. Environ. Res. Lett. **18**(10) (2023)
25. Mani, A., Budd, T., Maine, E.: Emissions-intensive and trade-exposed industries: technological innovation and climate policy solutions to achieve net-zero emissions by 2050. RSC Sustainability **2**(4), 903–927 (2023)
26. Zhang, D.: CO2 utilization for concrete production: commercial deployment and pathways to net-zero emissions. Sci. Total Environ. **931** (2024)
27. Feng, R., Xu, X., Yu, Z.T., Lin, Q.: A machine-learning assisted multi-cluster assessment for decarbonization in the chemical fiber industry toward net-zero: a case study in a Chinese province. J. Clean Prod. **425** (2023)
28. Singh, R.K., Modgil, S.: Impact of information system flexibility and dynamic capabilities in building net zero supply chains. J. Enterp. Inf. Manag. **37**(3), 993–1015 (2024)
29. Rajbhandari, S., et al.: Thailand's net-zero emissions by 2050: analysis of economy-wide impacts. Sustain. Sci. **19**(1), 189–202 (2024)
30. Baihaqi, I., Lazakis, I., Supomo, H.: Integrated value engineering and risk assessment performance measurement framework in ship-manufacturing industry towards net zero emissions using fuzzy DEMATEL-AHP. Machines **11**(8) (2023)
31. Dequech, D.: Bounded rationality, institutions, and uncertainty. J. Econ. Issues **35**(4), 911–929 (2001)
32. Mallard, G.: Bounded rationality and behavioral economics. Bounded Rationality Behav. Econ. 1–132 (2015)
33. Kahneman, D.: Maps of bounded rationality: a perspective on intuitive judgement and choice. Nobel Prize Lect. (2002)
34. Simon, H.A.: A behavioral model of rational choice. Q. J. Econ. **69**(1), 99–118 (1955). Author, F.: Article title. Journal 2(5), 99–110 (2016)

# Industry and Consumer Views on Improving Sustainability and Circularity of Materials Intended for Single Use

Katri Salminen[1]([⊠]), Virpi Rämö[2], Piia Kanto[1], Jürgen Belle[3],
and Hanna Pihjalarinne[1]

[1] Tampere University of Applied Sciences, Kuntokatu 3, 33520 Tampere, Finland
katri.salminen@tuni.fi
[2] Aalto University, Otakaari 24, 02150 Espoo, Finland
[3] Hochschule München University of Applied Sciences, Lothstraße 34, 80335 München,
Germany

**Abstract.** Despite advances in more sustainable materials and plastics, the industry still faces challenges in commercializing them. The most critical factor for other businesses and consumers to make a purchase decision remains to be price. The current study aims at going beyond financial reasonings for the purchase decision and focuses on outlying industry ($N = 6$) and consumer ($N = 42$) views on the potential to improve the use of circularity and sustainability of materials. Both datasets were collected from Germany. The results detail the factors affecting the use of recyclable and biobased packaging materials in industry and further, the pricing of such products. The consumer study highlights the attitudes towards end-of-use of single-use plastics and aims pointing out the logic of a consumer when they decide to ignore the sorting instructions. Together, the two datasets create an overview on how the companies could be encouraged to use more sustainable materials in their products and how to ensure that the more sustainable products could be sorted and recycled accordingly, in order to boost circularity. The paper will give concrete recommendations for businesses in order to make their sustainable products more attractive to business-to-business markets and end user consumers.

**Keyword:** Circular Economy; Material Recycling; Sustainable business models

## 1 Introduction

In Europe, more than half of the plastics is thrown away as waste [1]. European Union (EU) is focusing on hindering the use of plastics [2] and overpackaging [3] as well as improving the communication of the importance of sustainability towards the consumers for behavior changes [4]. The ongoing policy efforts include a full transformation in the waste management in a manner that supports full recyclability [see, e.g., 5]. EU legislation is directly reflected to Germany's national legistlative landscape. Germany is already utilizing a mass of different environmental policies to regulate the amount of (plastic)

H. Kohl et al. (Eds.): GCSM 2024, LNME, pp. 699–707, 2025.
https://doi.org/10.1007/978-3-031-93891-7_77

waste [6, 7]. Despite increasing regulation, especially products intended for single use (e.g., packaging or hygienic products) continue to be a significant environmental burden globally [8] and create most of the plastic waste [9]. The barriers of better circularity of single-use products include short product lifetime, careless use by consumers, hygienic aspects, and gaps in waste management system [e.g., 10, 11]. The current study focuses on decision-making process of industry (i.e., companies) representatives and consumers regarding improving circularity and sustainability of products often inteded for single use. The study was explorative in nature and conducted in Germany in order to mit-igate the known effects of the regulation [see 6 for a comprehensive analysis of the effects of regulation within the context of Germany's dual system] and local recycling infrastructure [12] to the responses.

Companies have started to acknowledge the need for green transition and circularity on marketing and strategic level [13]. Factors like supply chain effectiveness, environ-mental capabilities and resources, instruments in a factory, cost savings, consumer pres-sure, competitive advantage, environmental impact and regulatory pressure affect the decision-making regarding sustainable packaging material development and selection by the end-user companies [e.g., 14, 15, 16, 17, 18]. Further, companies often focus only on the sustainability of their primary products. However, in EU, companies are expected to prepare for the reporting of indirect emissions inclusing purchases and services such as packaging by 2026 [19]. Unfortunately, creating feasible company-level strategies for decision-making processes affecting purchases such as industrial packaging materials is difficult and the whole research area is fragmented [20].

Company strategies alone are not sufficient to improve circularity of products intended for single use. Consumers play a large role in the success of the introduc-tion of sustainable single-use products [21]. However, they make the purchase decisions mostly based on price [22] which can become a barrier of the uptake of sustainable materials for single-use products. Further, previous studies have clearly shown that con-sumers lack in-depth understanding about how single-use products should be discarded [23] which is reflected in the incorrect disposal of single use products [24] and ability to understand labels [25]. Single use products are often disposed of as waste [26] or flushed [27] which can lead to significant environmental harm. Previous case studies have shown that numerous factors such as age [28], education [29] and regulations [30] affect to the attitudes and disposal behavior of single-use products.

In this explorative study, we focused on collecting qualitative data from German end-user companies (i.e., companies buying products intended for single use such as packaging) and consumer sector (i.e., citizens that are responsible for disposing or recy-cling such products) to study the potential to increase the circular economy of single use products. The aim of the current study was twofold: 1) To perform an explorative survey for the end-user companies regarding the landscape of factors affecting their selection of recyclable and biobased packaging materials (see 2.1. For methods and 4 for limitations) and 2) Study consumer ideas for improving the recyclability of products intended for single use (see 3.1. For methods and 4 for limitations).

## 2  Industry Study

### 2.1  Methods

Six respondents from German companies answered an online survey regarding their use of recycled and biobased packaging materials and factors that affects the material selection. Recycled and biobased materials were selected as both are identified as on-demand by industrial packaging end-user companies in Germany [31]. As the survey was exploratory in nature, it was distributed to end-user companies known to use at least recyclable materials in packaging meaning that the sample size remained small. The background information regarding the use of recyclable and biobased materials, company size, packaging materials used and existance of environmental strategies was collected. This background information is further used in the discussion to evaluate the suitability of the sample. After the background questions, the respondents were asked "What factors are important when considering whether recyclable (or biobased) materials can be used more?". The questionnaire was multi-choice and the answers given with a Likert scale varying from not important at all (1) to very important (5), 3 representing a neutral choice. Questionnaire was developed to test the importance of previously identified factors (i.e., stakeholders (Qs 1, 2, 10), price (Q4), strategies (Qs 5 and 8), regulations (Q7), and performance (Qs 3, 6, 9) [see, e.g., 14, 15, 16, 17, 18 for studies identifying the factors] in the German context in order to find out potential differences in the factors affecting the selection of recycled or biobased materials. The respondents were also asked about the product price formation to collect more information about pricing. Finally, the companies were asked to assess the importance of stakeholders in the packaging material selection (Liker scale, 1 to 5).

### 2.2  Results

Due to the explorative nature of the study, the survey N was low and therefore, no statistical tests were performed. Table 1 presents the results of background questions and Table 2 the questionnaire.

The companies were also asked pricing of the recyclable and biobased materials using the following questions. Q1: "How much more the company could invest to the packaging material costs (%) if the packaging material were recyclable or biobased" and Q2: "Do any additional costs for recyclable packaging materials increase the price of the product?" The answers to Q1 for recyclable materials were 0%, 30%, 5 to 20%, 5 to 10%, Do not know and 15%. The answers for the biobased materials were 0, 30 and 5 to 20% when three companies omitted the question. For Q2, four companies answered that the increased price of a recyclable packaging material would increase the price of the product (two noting that the price would be partially and two that it would be fully added to the cost of the product). For biobased materials the answers were the same. Finally, the respondents were asked about the importance of different stakeholder gourps. All but one reported consumer perspective as important or very important.

**Table 1.** Background information of the companies (Y = yes, N = No, percentages describe the rough amount of recyclable and biobased materials used, R = Recycling strategy, S = Social responsibility strategy, NA = not answered).

| Number | Company size | Material | Use of recyclable materials | Use of biobased materials | Strategy |
|--------|--------------|----------|-----------------------------|---------------------------|----------|
| 1 | LE | Plastic, Glass, Carton Paper, Metal | Y (10%) | N | R |
| 2 | SME | Plastic, Paper | Y (80%) | Y (50%) | R |
| 3 | LE | Plastic, Paper, Carton | Y (50%) | Y (NA) | R |
| 4 | LE | Glass, Carton, Plastic, Paper, Metal, Cork | Y (90%) | Y (50%) | S |
| 5 | LE | Paper, Cardboard | Y (85%) | N | R |
| 6 | SME | Paper, Carton | Y (95%) | Y (NA) | S |

**Table 2.** Factors affecting the choice of recyclable and biobased packaging materials.

| Factor | Average for recyclable materials | Average for biobased materials |
|--------|----------------------------------|--------------------------------|
| 1 Consumer shopping behavior | 3.5 | 3 |
| 2 External stakeholders | 3.83 | 4.17 |
| 3 Functionality of the packaging materials in their intended use | 4.83 | 4.17 |
| 4 Price of the packaging materials | 3.83 | 3.67 |
| 5 Environmental responsibility and related strategies and program | 4.5 | 4.5 |
| 6 Compatibility with production processes | 4.5 | 4.83 |
| 7 Legislation or similar regulations | 4.83 | 4.67 |

(*continued*)

**Table 2.** (*continued*)

| Factor | Average for recyclable materials | Average for biobased materials |
|---|---|---|
| 8 Available information related to, e.g., carbon footprint of materials | 4 | 4.17 |
| 9 Contaminants and other harmful substances in materials | 4.83 | 4.83 |
| 10 Material availability | 4.67 | 4.83 |

## 3   Consumer Study

### 3.1   Methods

42 university students (28 business students and 14 textile students) from Germany participated in a workshop where they were introduced a single-use product, packaging instructions related to its disposal and then asked questions regarding the products disposal and instructions itself. As previous studies show that age [28], regulations [29], and even education [30] affect to the disposal behavior of single-use products, we aimed to mitigate these effects by choosing a sample of university students exclusively from Germany. The single-use product given as an example for the participants was wet wipes as it is a known product that consumers find difficult to recycle [32]. During the workshop the participants were introduced to the theme (i.e., wet wipes and recyclability) and then asked following background questions: "When using wet wipes, do you look out for sorting instructions on packages of wet wipes?; and" Do you sort the wet wipe according to the instructions on the package?". The idea of the background questions was to ensure that the sample was in line with previous consumer studies regarding consumer difficulties in recycling SUPs [e.g., 23]. Then, the participants were asked ideas for improving the circularity of the product by a following question:"How could the consumers be engaged to sort better?".

### 3.2   Results

The results to the questions 1 and 2 are presented in Table 3 as percentages. For the third question, verbal explanation of the results is presented.

Finally, in total, 18 respondents provided ideas to engage the consumers more to the better disposal of the products. Out of these responses, eight mentioned to improved instructions, two illustration or hints on the package, four mentioned education or awareness-raising, one mentioned product prices, one better instruction on the waste bin, one country-wise sustainablity strategies also reflected to education and one encouragement to buy less.

**Table 3.** Amount of respondents reading the instructions and using the instructions to guide their disposal behavior.

|  | Yes | No |
| --- | --- | --- |
| Do you look for sorting instructions | 50% | 50% |
| Do you sort according to the instructions | 59% | 41% |

## 4  Discussion

The industry survey showed that there was a clear gap between the usage of recyclable (all used) and biobased (four respondents used) materials. This was partially reflected in the responses. The consumer opinions were an important factor in selecting recyclable materials but not as much when it came to the biobased ones. The selection of biobased materils was mostly motivated by other external stakeholders. This result is in contrast with previous studies [e.g., 17] identifying consumer pressure imperative in the transition towards sustainable materials. A reason for this could be the familiarity of the biobased materials in the sample [31]. Environmental strategies and legislation were evaluated to be very important factors affecting the use of both material categories, which is in line with previous studies [13]. However, the respondents put more emphasis on the functionality of the recyclable materials than biobased materials while highlighting compabitable with existing systems with biobased materials. This may indicate that a deeper analysis of the environmental strategies of the companies could reveal critical factors in the production systems or supply chain that should be refelcted also in strategies to facilitate the use of sustainable packaging materials. Finally, pricing of the materials is often swifted to the end-user companies having negative effect on the desirability of such materials [e.g., 14, 22]. On the contrary, the current data revealed that material cost increases especially for recycled materials were acceptable. Also, the respondents had a good understainding of the tolerable effect of the material cost to the cost of the final product but only when recycled materials were used. The reason behind the result might be in the sample itself (i.e., companies familiar with recyclable materials).

The consumer data showed that the respondents had difficulties in reading disposal instructions and complying them, as also noted previously [25]. This result was still a bit surprising. In the current study, age and education were partially controlled as they are known to affect the disposal practices positively [28]. Despite this, only half of the respondents expressed reading the instructions. The respondents stated that they would need detailed instructions and package markings to dispose the products better. Education can improve the correct disposal of single-use products [29]. The current results indicate university education itself may not be sufficient to make informed decisions about disposal. From industry perspective, this further means that better instructions are still needed to increase the circularity of the single-used products.

The study suffered from limitations. Both datasets were small and collected from Germany to control background factors potentially affecting the responses. Larger datasets from several countries are needed in order to address the problem from other perspectives (e.g., regulations). The current study provides findings that need to be researched further to achiece more holistic and less fragmented [18, 20] understanding of the factors affecting indusrial packaging and consumer behavior. First, the differences in the indusry results regarding recyclable and biobased materials should be mapped more systematically in order to clarify the role of the supply chain [e.g., 18]. Based on the results it also seems that the respondents may not have understood the concept of a biobased packaging material even though they were given examples. Respondents reporting their company using paper or carton were also answering that they do not use biobased materials. The results were also interesting from the perspective of pricing. Even though quite a few previous studies identify pricing as an independent factor affecting consumer and industry choices [e.g., 17], the reality can be more complex especially when different materials are compared and should be studied more. Finally, the consumers were not able to create innovative ideas to improve circularity of single-use products. Follow-up studies should consider carefully the structure of the workshops organised for data collection to gain novel ideas.

# References

1. Garcia, J.M., Robertson, M.L.: The future of plastics recycling. Science **358**(6365), 870–872 (2017)
2. Directive (EU) 2019/904 of the European Parliament and of the Council of 5 June 2019 on the reduction of the impact of certain plastic products on the environment
3. European Parliament and Council Directive 94/62/EC of 20 December 1994 on packaging and packaging waste (OJ L 365 31.12.1994, p. 10
4. EC (2023) Proposal for a directive of the European Parliament and of the council on substantiation and communication of explicit environmental claims (Green Claims Directive) EUR-Lex - 52023PC0166 - EN - EUR-Lex (europa.eu)
5. Communication from the commission to the European Parliament, the council, the European economic and social committee and the committee of the regions. a new circular economy action plan for a cleaner and more competitive Europe. COM/2020/98
6. Dehio, J., Janßen-Timmen, R., Rothgang, M.: Regulating markets for postconsumer recycling plastics: experiences from germany's dual system. Resour. Conserv. Recycl. **196**, 107048 (2023)
7. Simoens, M.C., Leipold, S.: Trading radical for incremental change: the politics of a circular economy transition in the German packaging sector. J. Environ. Planning Policy Manage. **23**(6), 822–836 (2021)
8. Chen, Y., Awasthi, A.K., Wei, F., Tan, Q., Li, J.: Single-use plastics: production, usage, disposal, and adverse impacts. Sci. Total. Environ. **752**, 141772 (2021)
9. Weghmann, V.: Waste management in Europe (2023)
10. Diggle, A., Walker, T.R.: Implementation of harmonized extended producer responsibility strategies to incentivize recovery of single-use plastic packaging waste in Canada. Waste Manage. **110**, 20–23 (2020)
11. Hahladakis, J.N., Iacovidou, E.: Closing the loop on plastic packaging materials: what is quality and how does it affect their circularity? Sci. Total. Environ. **630**, 1394–1400 (2018)
12. Zhang, Y., Wen, Z., Hu, Y., Zhang, T.: Waste flow of wet wipes and decision-making mechanism for consumers' discarding behaviors. J. Clean. Prod. **364**, 132684 (2022)
13. Maranesi, C., De Giovanni, P.: Modern circular economy: corporate strategy, supply chain, and industrial symbiosis. Sustainability **12**(22), 9383 (2020)

14. Ma, X., Park, C., Moultrie, J.: Factors for eliminating plastic in packaging: the European FMCG experts' view. J. Clean. Prod. **256**, 120492 (2020)

15. Olsmats, C., Kaivo-oja, J.: European packaging industry foresight study- identifying global drivers and driven packaging industry implications of the global megatrends. European J. Futures Res. **2**(1) (2014)

16. Monteiro, J., Silva, F.J.G., Ramos, S.F., Campilho, R.D.S.G., Fonseca, A.M.: Eco-design and sustainability in packaging: a survey. Procedia Manuf. **38**(2019), 1741–1749 (2019)

17. Afif, K., Rebolledo, C., Roy, J.: Drivers, barriers and performance outcomes of sustainable packaging: a systematic literature review. British Food J. **124**(3), 915–935 (2022)

18. Jagoda, S.U.M., Gamage, J.R., Karunathilake, H.P.: Environmentally sustainable plastic food packaging: a holistic life cycle thinking approach for design decisions. J. Clean. Prod. **400**, 136680 (2023)

19. Directive (EU) 2022/2464 of the European Parliament and of the Council of 14 December 2022 amending Regulation (EU) No 537/2014, Directive 2004/109/EC, Directive 2006/43/EC and Directive 2013/34/EU

20. Silva, N., Pålsson, H.: Industrial packaging and its impact on sustainability and circular economy: a systematic literature review. J. Clean. Prod. **333**, 130165 (2022)

21. Ketelsen, M., Janssen, M., Hamm, U.: Consumers' response to environmentally-friendly food packaging-a systematic review. J. Clean. Prod. **254**, 120123 (2020)

22. Laborda, E., Del-Busto, F., Bartolomé, C., Fernández, V.: Analysing the social acceptance of bio-based products made from recycled absorbent hygiene products in Europe. Sustainability **15**(4), 3008 (2023)

23. Jacobsen, L.F., Pedersen, S., Thøgersen, J.: Drivers of and barriers to consumers' plastic packaging waste avoidance and recycling-a systematic literature review. Waste Manage. **141**, 63–78 (2022)

24. Dagiliūtė, R., Žaltauskaitė, J., Sujetovienė, G.: Self-reported behaviours and measures related to plastic waste reduction: European citizens' perspective. Waste Manage. Res. **41**(9), 1460–1468 (2023)

25. Jóźwik-Pruska, J., Bobowicz, P., Hernández, C., Szalczyńska, M.: Consumer Awareness of the Eco-Labeling of Packaging. Fibres Text. Eastern Europe **30**(5), 39–46 (2022)

26. Köklü, R., Ateş, A., Deveci, E.Ü., Sivri, N.: Generic foresight model in changing hygiene habits with the pandemic: use of wet wipes in next generations. J. Mater. Cycles Waste Manage. **25**(1), 74–85 (2023)

27. Briain, O.Ó., Mendes, A.R.M., McCarron, S., Healy, M.G., Morrison, L.: The role of wet wipes and sanitary towels as a source of white microplastic fibres in the marine environment. Water Res. **182**, 116021 (2020)

28. Northen, S.L., Nieminen, L.K., Cunsolo, S., Iorfa, S.K., Roberts, K.P., Fletcher, S.: From shops to bins: a case study of consumer attitudes and behaviours towards plastics in a UK coastal city. Sustain. Sci. **18**(3), 1379–1395 (2023)

29. Walker, T.R., McGuinty, E., Charlebois, S., Music, J.: Single-use plastic packaging in the Canadian food industry: consumer behavior and perceptions. Humanit. Soc. Sci. Commun. **8**(1), 1–11 (2021)

30. Kasznik, D., Łapniewska, Z.: The end of plastic? the EU's directive on single-use plastics and its implementation in Poland. Environ Sci Policy **145**, 151–163 (2023)

31. Simoens, M.C.: Unpacking pathways to a circular economy: a study of packaging innovations in Germany. Sustain. Prod. Consumption **47**, 267–277 (2024)

32. Hadley, T., Hickey, K., Lix, K., Sharma, S., Berretta, T., Navessin, T.: Flushed but not forgotten: the rising costs and opportunities of disposable wet wipes. BioResources **18**(1), 2271 (2023)

# Digital Integration and Optimization Workflows in Timber Construction with Prefabricated Components: A Business Intelligence-Driven Application for Optimizing Carbon Footprints and Project Performance

Johannes Reinders[1]([✉]), Sebastian Orozco[1,2], Valentin Eingartner[1], Annika Feldhaus[3], Yaoning Yang[4], and Ignacio Borrego[1]

[1] Technical University of Berlin, 10623 Berlin, Germany
reinders@tu-berlin.de
[2] Costarican Institute of Technology, San José 10101, Costa Rica
[3] Fraunhofer Institute for Production Systems and Design Technology, 10587 Berlin, Germany
[4] Yunnan University, Kunming City 650504, Yunnan Province, China

**Abstract.** The construction industry is a major contributor to global carbon emissions and resource extraction, increasingly requiring sustainable alternatives. Timber Construction with Prefabricated Components (TCPC) presents a viable solution, yet it remains underutilized due to challenges in cross-domain integration and optimization. This study explores the potential of a Business Intelligence (BI)-based digital application designed to enhance TCPC project performance by synchronizing architectural design with prefabrication requirements and facilitating project optimization. The research introduces a Proof-of-Concept (PoC) application tested on a residential project in Berlin, Germany, showcasing the synchronization of building design data with prefabrication components and material performance metrics. The PoC allowed for real-time assessment of architectural variations and component optimization, significantly improving project carbon and cost efficiency. The results highlight the feasibility of this approach, with one test achieving a 57% reduction in embodied carbon of an optimized ceiling panel, through material optimization. This study underscores the relevance of digital tools in overcoming TCPC planning/execution challenges and emphasizes the need for further research integrating real-world manufacturing data and IoT capabilities. The findings suggest that the implementation of such digital tools could accelerate the mainstream adoption of TCPC, promoting more sustainable and efficient construction practices.

**Keywords:** Prefabricated Timber Construction · Cross-Domain Integration · Digital Optimization Workflow · Project Performance Enhancement · Sustainable Architecture

H. Kohl et al. (Eds.): GCSM 2024, LNME, pp. 708–715, 2025.
https://doi.org/10.1007/978-3-031-93891-7_78

# 1 Introduction

The construction industry is a key contributor to climate change. Accounting for approximately 40% of global carbon emissions [1], the sector is the largest emitter of carbon dioxide ($CO_2$) gases worldwide, and additionally responsible for 40 to 50% of all global resource extractions [1]. Timber construction with prefabricated components (TCPC) is widely regarded as a plausible alternative to curb these negative impacts [2]. TCPC has the potential to deliver substantial benefits towards the reduction of the ecological footprint and carbon emissions of the construction industry, while simultaneously enhancing its circularity [3]. Despite these advantages, TCPC has not yet been able to establish itself as a mainstream building technology on the global market.

One of the main challenges for TCPC project success is, that they still follow traditional planning and execution approaches that are not adapted to the specific requirements of prefabricated timber construction [4]. These traditional methods are insufficient when it comes to (1) synchronization of architectural design with timber prefabrication requirements and (2) effective project optimization workflows. Insufficient synchronisation of architectural design with prefabrication requirements causes a lack of standardised building components and processes. Without a minimum level of standardization, production is inefficient and more expensive, which reduces the benefits of serial production [5]. This subsequently leads to delays and additional costs. Moreover, conventional planning methods are based on a manual exploration of design proposals and alternatives, with a gradual modification of parameters through an iterative trial-and-error process [6]. This process is usually very time consuming and ineffective for finding optimized results [6].

# 2 Lean and Digital Methods in TCPC Planning Projects

## 2.1 Lean Principles for TCPC

Lean production revolves around the idea of doing more with less in terms of effort, material, personnel, equipment and space, while aiming to add value and eliminating waste throughout the value chain [7]. Numerous studies show that the application of lean principles in diverse industries led to improved production times, costs and quality [7]. Nonetheless, multiple authors indicate, that lean principles and practices in the construction industry have not reached a maturity level similar to that of the manufacturing industry [8]. So far, lean planning principles in the TCPC industry have been discussed thoroughly in the "leanWOOD" project at Technical University Munich [4]. The authors describe a TCPC-appropriate planning process, for which they identify the need for timber construction expertise from the early planning phases as an integral part of the project team and propose adapted cooperation models, which should lead to better project outcomes [4].

## 2.2 TCPC Digitalization Needs

The findings of "leanWOOD" help to set a conceptual framework and theoretical principles for an alternative planning approach. However, the authors do not provide a practical implementation method, that would integrate effective, digital planning. Yet, effective, digital planning methods render to be crucial for efficiency gains. Recently a review of prefabricated and modular timber construction, with a focus on evolution, trends and current challenges based on reviewed literature from 1990 to 2023 was published by the authors Gutiérrez, Negrão, Dias and Guindos (2024). The review reveals that only few studies addressed the practical implementations of digital technologies for prefabricated timber construction, although many studies coincide on the argument for digital technologies as crucial tools for efficiency gains [9].

### 2.3 Digital TCPC Planning Frameworks and Methods

In the construction industry Building Information Modelling (BIM) mainstream implementation is widely recognized as necessity to improve cost, time, value, and environmental performance for projects [10]. Indeed, BIM implementations have long been recognized as being successful for increasing project efficiency [11]. However, conventional BIM processes, as they are typically conducted in the construction industry, fall short on the integration of timber expertise and lack integration of prefabrication requirements as integral parts of the workflow. A recent review highlights a significant gap in the availability and adoption of BIM tools dedicated to timber construction and prefabrication [12]. Furthermore, the authors conclude that existing solutions, such as BIM and CAD, are underdeveloped and difficult to integrate into standard timber industry processes and suggest partnerships with wood producers [12]. This means that, although digital tools such as BIM exist, no digital methods or frameworks dedicated and tailored for TCPC in line with lean principles have been sufficiently established yet.

This research, therefore, aims to answer the question: How can seamless synchronization between planners and manufacturers be achieved in TCPC projects through a digital framework in order to enhance project performance? The goal is to develop a digital application that facilitates early integration of timber expertise and promotes continuous information flow through collaborative planning, while it incorporates lean principles. The expected outcome should significantly improve project times, reduce costs, and lower ecological footprints, adding value to TCPC projects. By optimizing these processes, the tool aims to strengthen TCPC's position in global markets, to ultimately lower the ecological footprint of the construction industry overall.

## 3  Setting a Framework and a PoC for the Digital Planning Application

### 3.1  Application Goals, Functionalities and Operational Principle

To address the need for integrating timber expertise and prefabrication requirements into digital planning in TCPC, the proposed application aims to enhance project coordination and performance. Therefore, effective aggregation, processing, and representation of data such as design specifications, component standards and environmental metrics are crucial. Through these functionalities, the proposed application can dismantle data silos and provide real-time access to project information across stakeholders, including developers, planners, and manufacturers. This approach has the potential to enable seamless synchronization and data-informed decisions from the early planning stages.

Furthermore, regarding project optimization, the application should facilitate real-time testing of variations. Diverse input operators, such as parametric slicers to adjust design specifications or material selectors, enable users to explore a multitude of options and receive instant feedback on critical metrics, such as costs, timelines, material consumption, and carbon footprint. By means of these data-driven forecasting capabilities, the application supports users in optimizing their projects efficiently, ensuring informed decisions.

Additionally, there is also a need for a user-friendly interface, tailored to the specific needs of different stakeholders. For instance, architects could experiment with design configurations while manufacturers explore optimal prefabrication components and materials usage. This collaboration would foster better communication and decision-making, reducing errors, delays, and waste.

As a premise operational principle, the digital application leverages BIM datasets as its key input and connects them with comprehensive component prefabrication data as well as other external datasets. Material LCA-information, market-driven cost data, and updated supplier registries can be included to allow for real-time operations and manipulations of this data within one collaborative workspace, which facilitates clear and quick, data-based project optimization (Fig. 1).

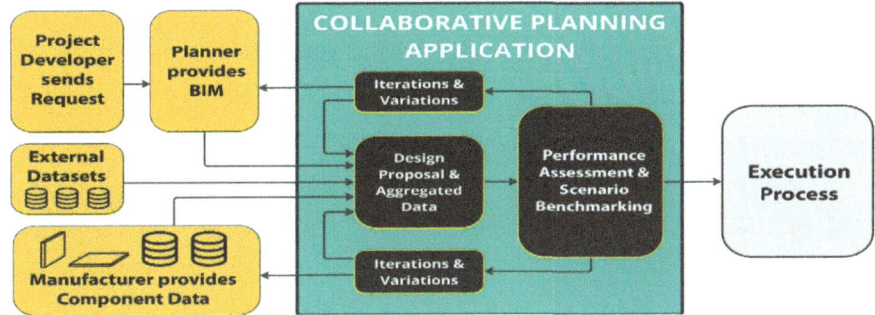

**Fig. 1.** Operational principle of the digital application.

## 3.2 Description of the PoC Application

To test the proposed digital planning framework, a PoC, Business-Intelligence (BI)-based application was developed and evaluated using an exemplary TCPC project in Berlin-Buch, am Sandhaus, in Germany as a case study. This case study was selected due to its complex urban context and varied building designs. The goal was to test the synchronization of architectural design with prefabrication requirements and the project optimization workflow. Two KPIs, investment cost and embodied carbon, were set as analytical metrics, to address both, financial and environmental considerations.

The selected case study consists of a masterplan featuring 155 residential buildings of various shapes and heights. Two apartment buildings, with seven floors each, were chosen as the test subjects. As for the component data, the team modeled a simple component catalog based on two different construction systems, a CLT-supported massive timber system, and a glulam-based timber skeleton system. The catalog was developed to serve as prefabrication components dataset for the PoC.

A prototypical data model was established for the PoC, consisting of BIM data, component data, and material Life Cycle Assessment (LCA) data. The BIM data sets are subdivided into their geometric and semantic parts, including parameters such as shape, footprint area, number of levels and height. Component data included component structure, materials, masses and costs. LCA data was drawn from the Ökobaudat platform, providing carbon footprint data for materials [13]. Through this data model, a link in between architectural design, components catalog, and material carbon footprints was effectively established. This enabled joint assessment of design and component parameters based on financial and environmental KPIs, such as investment cost and embodied carbon (Fig. 2).

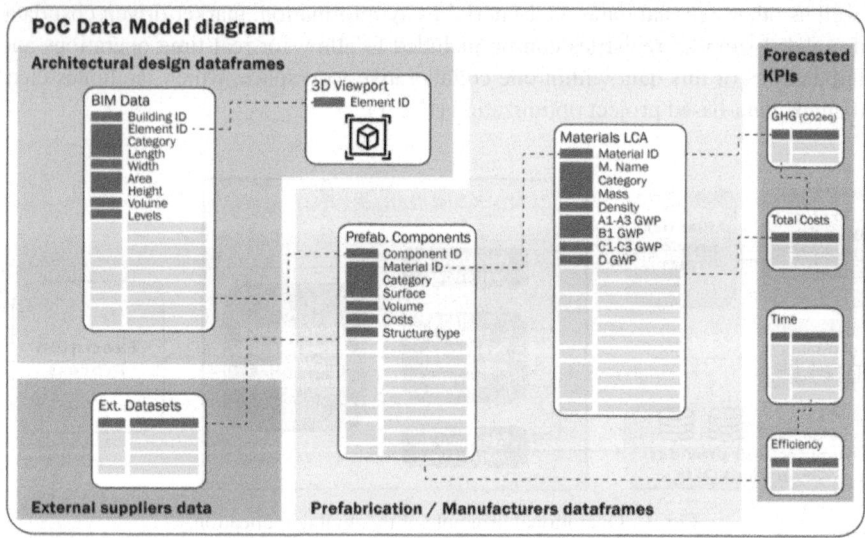

**Fig. 2.** Simplified PoC Data Model

As for the optimization workflow, two experimental scenarios were designed to test the PoC. In the first one, architectural design variations were tested based on the two different timber building systems. Floor plan, construction system and shape variations were generated through BIM and then uploaded to the PoC, allowing for real-time synchronization with prefabrication data. Users can visualize the selected design in a 3D-viewport and immediately see the impacts on cost and embodied carbon in numerical cards and interactive chart diagrams, with the corresponding prefabrication components and their embodied carbon footprints. This enables and facilitates rapid decision-making. In the second scenario, component optimization was tested on a single floor panel by adjusting its material composition. The goal was for the application to provide instant feedback on material costs and carbon footprints, to allow users to identify the optimal balance between cost and environmental impact.

## 4   Results

The PoC application demonstrated the feasibility of synchronizing architectural design data with prefabrication component data and improvement of project optimization workflows successfully. Although the PoC was limited in scope, it highlighted the potential for improved collaboration among stakeholders and provided valuable insights into the potential of the proposed digital planning framework for TCPC.

The PoC confirmed that real-time synchronization between BIM data and a prefabrication component catalog is achievable. Users could upload design variations, and the application responded instantly with updated performance metrics, including cost and carbon footprint.

- The PoC supported synchronization across multiple project levels, from building massing to individual prefabricated components. This multi-source integration enabled fast comparisons between different design options, with reduced manual adjustments. A process that usually consumes several working days for a project such as the case study was shortened to a few hours.
- Instant visualization of design changes and their impact on KPIs was enabled, helping to reduce revision time and provide early insights into project performance.

Concerning project optimization and decision-making the PoC facilitated rapid evaluation of design and component configurations using two experimental scenarios.

- Scenario 1: Variations in building shapes and construction systems were tested. The application provided immediate feedback on the cost and environmental impact of different options, streamlining the decision-making process.
- Scenario 2: Adjustments in material composition of a floor panel. The potential for optimizing the embodied carbon footprint without a significant cost increase was showcased. For example, replacing "rigid polyurethane" with "hemp wool" reduced embodied carbon by 57% with only a 3% cost increase (Fig. 3).

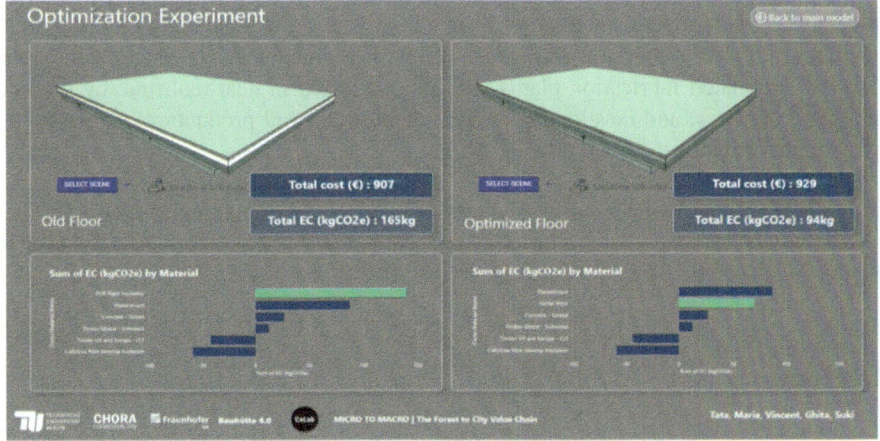

**Fig. 3.** Dashboard of the component optimization experiment.

## 5    Discussion

The PoC demonstrated the potential of the digital application framework to close the gap between project planning teams (developers, designers, planners) and execution teams (contractors, manufacturers) in implementing lean principles. The measurable improvements in time efficiency, decision-making, and environmental performance underscore the frameworks potential for driving more sustainable and cost-effective construction practices. While the PoC successfully demonstrated synchronization and optimization for a specific case study, further research is needed to assure scalability and reproducibility. However, the adaptability of the BI-based approach and data model already suggest potential for broader application.

Despite these benefits, challenges may exist in the context of the here introduced digital framework. First, the willingness of architects to limit design freedom may hinder widespread adoption. While prefabrication requires standardization, it may be seen as limiting creativity. Second, batch sizes could pose economic challenges for prefabrication in case of small-scale projects. Future research could address these challenges.

## 6    Future Outlook

The introduction of BI-based workflows for data modelling and visualization allows to bring diverse functionalities closer to the realm of architecture, engineering, construction, and operations (AECO)-industries, and within a short period of time, could transition from a niche data science experiment, into common praxis. After completion of the described demonstrator, the achieved functionality range triggers multiple pathways for applied research expansion.

Initially, the incorporation of real-world manufacturing data would enhance the utility and reliability of the component catalog. Industry-validated datasets would also improve tasks such as component optimization and LCA-phase-specific carbon balance breakdown. Secondly, integration of manufacturing IoT data from the prefabrication processes, could stimulate a transition from a relatively static planning tool into a real-time control platform over standardized, yet flexible, manufacturing. Shared databases from the automated fabrication plants could be analyzed in near-real-time by project designers, planners, and developers, as well as by industrial production engineers, to optimize and balance the main targeted KPIs.

**Acknowledgements.** We thank the Federal Ministry of Food, Agriculture and Consumer Protection, Berlin, Germany, as well as the Federal Agency of Renewable Resources (FNR) for funding of the research project DiKieHo (Funding Code 2221HV082B).

## References

1. United Nations Environment Programme, Yale Center for Ecosystems + Architecture (2023) Building Materials and the Climate: Constructing a New Future. Uneporg. 978-92-807-4064-6

2. Churkina, G., Organschi, A., Reyer, C.P.O., et al.: Buildings as a global carbon sink. Nat Sustain **3**, 269–276 (2020). https://doi.org/10.1038/s41893-019-0462-4

3. Van der Lught, P., Harsta, A.: Tomorrow´s Timber: Towards the next Building Revolution, 1st edn. Jeroen Van Oostveen, The Netherlands (2020)

4. Kaufmann, H., Huß, W., Schuster, S., et al.: leanWOOD Optimierte Planungsprozesse für Gebäude in vorgefertigter Holzbauweise (2017)

5. Kohl, H., Bunschoten, R., Oertwig, N., et al.: Studie zur Stärkung der Holzbauwirtschaft in der Metropolregion Berlin-Brandenburg (2023)

6. Touloupaki, E., Theodosiou, T.: Performance simulation integrated in parametric 3D modeling as a method for early stage design optimization - A review. In Energies **10**(5). MDPI AG (2017). https://doi.org/10.3390/en10050637

7. Gil-Vilda, F., Yagüe-Fabra, J.A., Sunyer, A.: From lean production to lean 4.0: a systematic literature review with a historical perspective. applied sciences **11**, 10318 (2021). https://doi.org/10.3390/app112110318

8. Babalola, O., Ibem, E.O., Ezema, I.C.: Implementation of lean practices in the construction industry: a systematic review. Build. Environ. **148**, 34–43 (2019). https://doi.org/10.1016/j.buildenv.2018.10.051

9. Gutiérrez, N., Negrão, J., Dias, A., Guindos, P.: Bibliometric review of prefabricated and modular timber construction from 1990 to 2023: evolution, trends, and current challenges. Sustainability **16**, 2134 (2024). https://doi.org/10.3390/su16052134

10. Alwan, Z., Jones, P., Holgate, P.: Strategic sustainable development in the UK construction industry, through the framework for strategic sustainable development, using building Information modelling. J. Clean. Prod. **140**, 349–358 (2017)

11. Sundquist, V., Domenico Leto, A., Gustafsson, M., Johansson, M., Roupé, M.: Bim in Construction Production: Gains and Hinders for Firms, Projects and Industry (2020)

12. Lobos Calquin, D., et al.: Implementation of building information modeling technologies in wood construction: a review of the state of the art from a multidisciplinary approach. In Buildings **14**(3). Multidisciplinary Digital Publishing Institute (MDPI) (2024). https://doi.org/10.3390/buildings14030584

13. BBSR ÖKOBAUDAT In: ÖKOBAUDAT informationsportal nachhaltiges bauen. https://www.oekobaudat.de/ Accessed 15 Apr 2024

# Integrating Digital Value Chains for Timber Construction

Valentin Eingartner[1]([✉]), Annika Feldhaus[2], Maxim Mintchev[1], Johannes Reinders[1],
Nicole Oertwig[2], Holger Kohl[2], and Raoul Bunschoten[1]

[1] Technische Universität, Berlin, Germany
`valentin.eingartner@tu-berlin.de`
[2] Fraunhofer Institute for Production Systems and Design Technology, Berlin, Germany

**Abstract.** The digital transformation of the construction industry offers opportunities to address its inherent complexity. Timber construction stands out as a sustainable alternative to traditional building methods, contributing to climate goals through carbon sequestration. However, the sector faces significant challenges, including fragmented processes, insufficient digital integration, and a lack of lifecycle-oriented approaches. This research presents a reference model that maps the entire timber value chain, from forest to city, to address these challenges. By fostering an understanding of existing processes, the model enables to navigate complexity, identify bottlenecks, and pursue lifecycle-oriented optimizations. Stakeholder workshops enrich process insights with practical experiential knowledge. Building on this foundation, a Business Intelligence tool, called Value Creation Digital Model (VCDM) was developed to support horizontal and vertical integration across stakeholders. While the VCDM is designed to show potential for optimizing regional value chains and to support circular construction, its adoption is limited by the slow pace of digitalization and data availability within the construction sector. Future scenarios envision the VCDM as a tool for planning, certification, and information dissemination, emphasizing the need for trust, participation, and standardized practices across the value chain.

**Keywords:** Integrative Planning · Digital Value Chain · Prefabricated Timber Construction · Circular Construction

## 1 Introduction

The construction sector is among the largest contributors to global carbon emissions [1]. Timber construction, utilizing a renewable resource, which is often locally available and acting as a carbon sink, offers a promising solution to decarbonize the industry. [2] However, achieving climate goals requires timber construction to overcome its current limited adoption in a traditionally mineral-based sector. Next to the conservative construction industry and therefore traditional market structures, this is hindered by the sector's high complexity and lack of digital integration. Simultaneously, the increasing focus on regional value chains and circularity introduces new challenges [3], which, if addressed correctly, could greatly benefit timber construction. This is addressed by

© The Author(s) 2025
H. Kohl et al. (Eds.): GCSM 2024, LNME, pp. 716–724, 2025.
https://doi.org/10.1007/978-3-031-93891-7_79

the Value Chain Digital Model (VCDM), which is intended to integrate information of different stakeholders and thus streamline processes along the value chain. [4]

This paper hypothesizes that by enhancing process understanding and leveraging digital solutions like the VCDM, complexity in timber construction can be handled. This will enable better integration of stakeholders, promote circularity, and improve overall efficiency. Specifically, it addresses the question on how digital integration of the timber construction value chain can be achieved to simultaneously promote sustainability and circularity. To answer this, a research project is presented that focuses on creating transparency and fostering collaboration along the timber construction value chain. This project adopts three interlinked approaches:

1) **A Value Chain Model from forest to city – Development of a reference model to support Process understanding:** To navigate the complexity of the timber construction value chain, a thorough understanding of existing processes is essential. This step focuses on clarifying obstacles and requirements for horizontal integration and digitalization by developing a comprehensive reference model. It was constructed through an extensive literature review and continuously refined with practical insights gained from workshops and project results.

2) **Information Flow and Digital Continuity – Stakeholder Engagement through workshops:** Through direct cooperation with stakeholders along the value chain, practical challenges and requirements for digitalization in timber construction were identified during workshops. Results were used to develop, validate and optimize a digital solution.

3) **A Solution-Oriented Approach - Enhancement of a digital application:** Insights derived from the process analysis and workshops were used to create a new iteration of the Value Chain Digital Model (VCDM), which facilitates horizontal networking for identified stakeholders along with the life cycle phases in construction. The VCDM was developed to enable practical applications as a response to the previously identified challenges.

## 2   Digital Integration in Timber Construction

In industries such as automotive, machinery and plant engineering, electronics, and food, efficient vertical and horizontal networking with suppliers and markets is well established [5]. In contrast, planning and production in forestry and the construction industries, typically do not occur within the same company, which makes planning activities as well as communication and data flows often poorly organized. Additionally, certification of construction projects is highly complex and is done mostly paper-based. Therefore, efficient planning and execution of construction projects poses a significant challenge, particularly in timber construction. The conventional project workflow, characterized by segmented stages of planning, tendering, production, and construction, presents a substantial obstacle for prefabricated timber construction. This is primarily due to the lack of integration of timber construction expertise into the planning process. [6] To address this, one approach discussed in the literature, *LeanWood*, focuses on streamlining planning processes by shifting critical decision-making to earlier stages, thereby increasing efficiency. The goal is to achieve integrative planning, which seeks to involve all relevant disciplines early in the process to maximize optimization potential. [7]

Digitalization plays a crucial role as an enabler for such integrative planning. Research into digital integration approaches such as Computer Integrated Construction (CIC) dates back to 1987, aiming to develop fundamental process models that define the scope of activities required for the provision of a facility [8]. The purpose of automated CIC processes is to automatically generate all information and associated documents for the design and construction of a built facility [9].

Today's approaches to implementing digitalized construction and planning processes are found in Building Information Modeling (BIM) but their implementation presents challenges. The unique, project-based nature of the construction industry often stands in opposition to the digitalization required for the consistent management of value chains. [10] This is especially true for timber construction and its associated complexity. The construction industry, due to many regulations, stakeholders and fragmented processes being a complex and disorganized sector, is still somewhat resistant of the adoption of BIM, while there is many analogue data created by many manual processes in timber construction in particular. The comprehensive literature review by Gutiérrez et al. reveals a lack of comprehensive digital approaches compatible with the unique characteristics of timber construction. Similarly, there is a lack of descriptive models that fully capture the specific aspects of timber construction across the entire value chain. [6] This gap is addressed through the development of a process model enriched with detailed information, serving as a reference framework for a digitally integrated timber construction process.

Connecting these ideas, the VCDM had been created previous to this work [4]. It provides different perspectives on the value creation network of regional timber construction, including forestry, architectural planning, urban planning and manufacturing. It provides a low-threshold entry point, enabling access even for users with limited

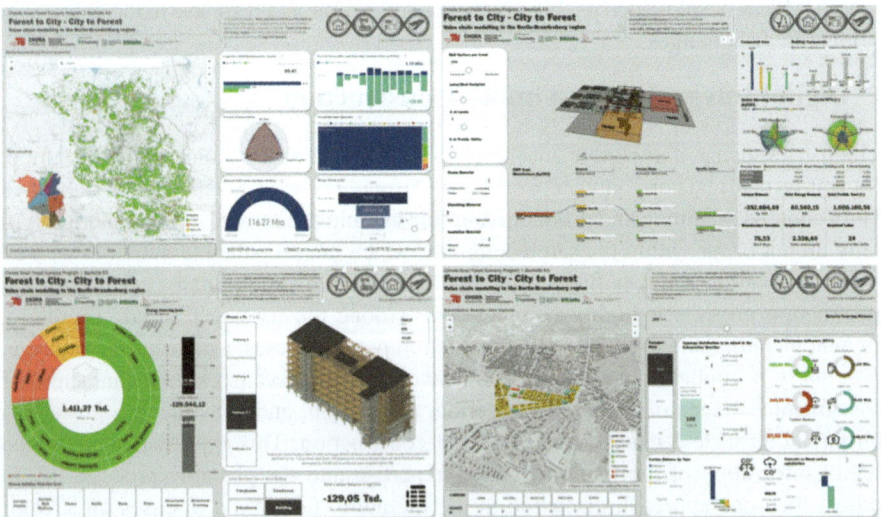

**Fig. 1.** Boards of the VCDM from forest (top left) through prefabrication (top right), building design (bottom left) and urban planning (bottom right).

technical or specialized knowledge. It gives an overview of different planning scenarios for building configurations, urban constellations or manufacturing planning. Users can adjust input parameters to compare the scenarios concerning a number of KPIs, such as cost, time, global warming potential, energy consumption, etc. to explore the effects of design and planning decisions in the very early planning stage of projects. The dashboards of the VCDM (Fig. 1) are realized through Microsoft Power BI and incorporate data from regional geographic information systems (GIS), LCA data [11] and forestry data [12] provided by the federal government of Germany, as well as urban planning data and BIM models of exemplary projects.

After showing the proof of concept, to be able to represent and transfer all life-cycle phases in the VCDM, the entire value chain has been analyzed and a process model as a reference was created, which is described in the following chapter.

## 3    A Value Chain Model from Forest to City

Using Integrated Enterprise Modelling (IEM) [13], a value chain model for urban timber construction was developed to support the horizontal networking of value creation. The goal of the model is to provide a method at the network level supporting digital information flows. Through input/output diagrams for each element, the system relationships, along with the generated and required information for planning, controlling, and evaluating production processes, were illustrated. Thus, the model promotes a holistic understanding, while uncovering obstacles to the implementation of digital solutions, and highlighting optimization potential. By detailing subprocesses within the model, the

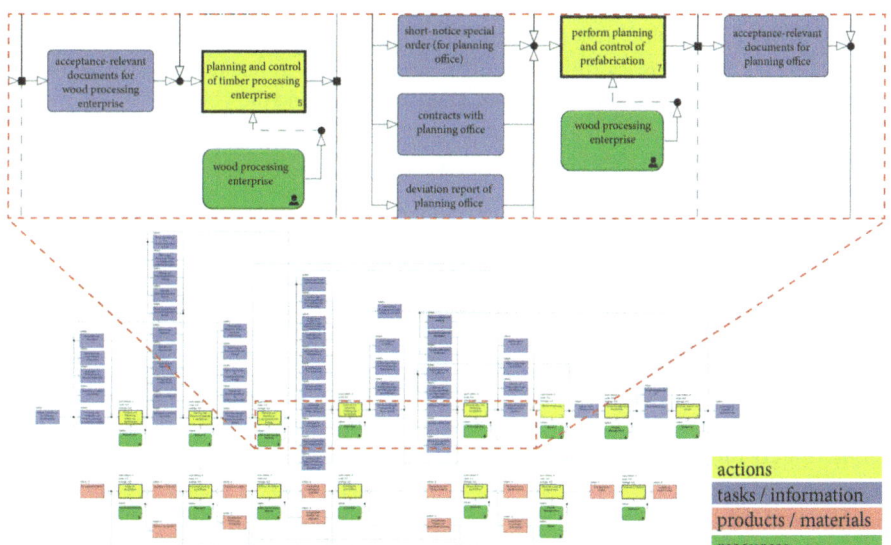

**Fig. 2.** Overview of the reference model for regional value chains in timber construction, showing the interconnections between actions (yellow), tasks/information (blue), products/materials (red), and resources (green).

complexity and dependencies among the individual activities and stakeholders become apparent. Fig. 2 provides an overview of the process model at the first level, with the underlying sub-processes embedded. It highlights the close interconnections between various actions, tasks/information, products/materials, and resources. In the figure, the overview over the entire value chain is given, while the exemplary actions (yellow) of "planning and control in wood-processing enterprise" and "planning and control of pre-fabrication" are shown a bit closer, emphasizing the information layer of the model. On the bottom of the figure, the material flows are indicated with the red elements.

Concerning the production processes, prefabrication in particular, is a focus subprocess due to its complexity and significant influence on upstream and downstream process phases. Thus, the model provides the mapping, analysis and comparison of prefabrication scenarios based on cost, time, and ecological criteria. This allows for the identification and analysis of emission-, cost-, or resource-intensive production processes and enables an economic-ecological assessment of processes.

# 4   A Solution-Oriented Approach –Value Chain Digital Model

These results formed the foundation for the enhancement of the Value Chain Digital Model (VCDM). To create an effective tool, however, it is crucial to understand the needs of the relevant stakeholders along the value chain as potential users, and to address their challenges and requirements. This was achieved in a subsequent step through a dedicated workshop, where stakeholder input was gathered to develop and optimize the VCDM's functionalities.

The workshop involving regional stakeholders of the timber construction value chain, such as representatives of the Brandenburg State Forestry Office, an urban planning office responsible for the development of a new timber-built neighborhood, and of a company offering coating solutions for timber construction, aimed to gain a real understanding of the challenges and requirements to the VCDM, to understand the potential of the model and to adapt it to the needs of the stakeholders, and to elicit solution-oriented suggestions for improving the VCDM. The workshop was divided into three parts.

## 4.1   Challenges and Requirements to the VCDM

The first part of the workshop consisted of an open discussion among the participants regarding the challenges and requirements of such a model. Specifically, the interfaces between the project developer and the planner, and between the planner and the manufacturer were discussed.

With regard to the interface between the project developer and the planner, the former has to rely on the latter's assurance that minimum building requirements have been met during the planning process. Automated parameter checks and instant requirement changes with the help of digital twins and Artificial Intelligence respectively are technically feasible, but underutilized. However, the biggest factor of uncertainty and time sink mentioned in the early planning stages is the building approval process, which can be digitalized, but public administrative staff is often not trained to use it. Participants pointed out the importance of viewing administrative personnel as key stakeholders in a digitally integrated value chain.

The interface between planners and manufacturers is characterized by the redundancy of planning processes. Architects and engineers often work with 2D models and old standards, which impedes the implementation of digital twins and BIM in planning, as well as collision checks, which are much simplified when modelling projects in 3D. This results in uncertainty for the manufacturers, which often carry out the entire planning process from the start so that planning mistakes are avoided. The aspects of data management and sovereignty are also relevant for the interface between the construction planner and the prefabricator, since it is an often unclearly allocated responsibility. It was thus suggested that the developer should delegate this task in large building projects.

## 4.2 Digital Integration of the Value Chain

The VCDM is currently in its development phase and is not yet fully adapted for industrial applications. To develop the Model towards more realistic framework conditions, the second section of the workshop focused on collecting feedback on the tool's functionalities as well as to propose potential scenarios. In the discussion round, there were doubts for which stakeholders the VCDM would be most relevant and who could mostly benefit from using the tool. For example, until now it is unclear if there is demand by sawmills and how they could profit from using the existing tool. Furthermore, the tool is designed for different scales that might not fit the needs of every stakeholder along the value chain. Therefore, more functionalities must be added to the VCDM.

First, even if the VCDM is initially intended for use in the metropolitan area of Berlin-Brandenburg, the model is only in English. Thus, adding German to the available languages will presumably extend the circle of users. Additionally, facilitating the native import of BIM models, and interaction with BIM, as well as the implementation of a saving and printing function of dashboard configurations can enable independent stakeholders to use the VCDM as an actual planning tool for their projects.

To enhance the correctness and accuracy of the data, it is planned to include as much data from each stakeholder with up-to-date information. Doing so, data interfaces for each level must be used and integrated into the VCDM. Similar to BIM, the benefit of the VCDM is higher, the more stakeholders provide data and use the system. This requires the stakeholders to collect the relevant data, according to the workshop results, which is challenging, as the digitalization in the building sector in Germany is progressing quite slowly [14]. Moreover, further improving the visualization of the data, as well as the knowledge gain of using the VCDM, stakeholders of value creation steps as well as the data they provide, are planned to be displayed more specifically in the dashboards. Implementing this will add to the accuracy of the VCDM, which mostly includes estimates based on experience and previous projects. The resulting data volume can potentially overflow the abilities of Microsoft PowerBI, therefore alternatives should be examined.

Improving the usability of the VCDM, more suggested functionalities in the workshop were to implement a suggestion system for timber supply on the basis of forestry data, and to separate the visualization of the material flow of massive timber, particle composites, and parts.

Addressing all limitations and exploiting the potentials by adding the functionalities described above, the VCDM will be a valuable contribution to the planning and control of the value creation of timber construction, and it will include many different features which require a documentation and user manual, which is planned as well.

### 4.3 Future Application Scenarios

Within the third part of the workshop, application scenarios were developed. The most important ones are the VCDM as a tool for information, VCDM as interface, and VCDM for certification and planning aid.

The VCDM as a tool for information is a scenario, which does not introduce the tool into operative processes along the value creation chain during the detailed planning or the execution phase. The aim in this scenario is to create interest and provide information in timber construction among the construction industry. As timber construction is yet a construction principle, it can inform and excite critics for the benefits of timber construction. Doubts for timber construction often include cost, security, and durability of the projects. [15] The VCDM provides much information about timber construction, showing that it is a feasible alternative to traditional construction with concrete and steel.

Using the VCDM as an interface aims to provide relevant KPIs to other software, using standardized exchange formats of the data. In this scenario, stakeholders use the tool as a basis to find correct information for their processes. The VCDM does not provide actual planning functionality, but it is part of actual value creation. To realize such a scenario, the software must include specific project data for each project and must be used by all relevant stakeholders.

The VCDM for certification and planning aid is intended to integrate planning and certification of projects and use it as the basis for certification. In this scenario, administrations who are responsible for certifying projects are using the software together with the usual stakeholders along the value chain. The VCDM must provide all relevant information for the certification, which includes a long and highly complicated process. Using planning and manufacturing data and transforming into certification documents would speed up certification immensely, which is still one of the longest processes in entire construction processes. However, many attempts to digitize the certification process have already failed, which makes this scenario the most challenging one.

## 5  Discussion and Conclusions

The development of the reference model and the VCDM represent a solution to support the digital integration of the timber construction value chain while incorporating sustainability parameters and circular economy potential early in the planning process. To fully unlock the potential of such solutions, the construction industry must embrace openness and a willingness to pursue such innovative approaches. A fundamental requirement for improving information flow among value chain partners is the availability of necessary data – ideally in real time provided by the process participants. However, even today, it is evident that information crucial for seamless workflows is often insufficiently available. Particularly in the context of sustainable construction, information about the methods, materials, and processes used across the value chain is crucial to effectively assess environmental impacts throughout the entire building lifecycle, including upstream and downstream processes. Additionally, accurate data on economic parameters, such as cost and time estimates in construction processes, is essential to mitigate the high risks often associated with timber construction. Consequently, digital integration depends on the availability of data from the industry, necessitating a foundation of trust and participation. Additionally, the consistent implementation of standardized tools, IT solutions, and practices is crucial to reducing redundancies and re-design phases between planning and execution stages, thereby fostering long-term collaboration and information exchange among stakeholders.

**Acknowledgements.** We thank the Federal Ministry of Food, Agriculture and Consumer Protection, Berlin, Germany, as well as the Federal Agency of Renewable Resources (FNR) for funding of the research project DiKieHo (Funding Code 2221HV082A-B).

# References

1. Wiest, M.: Ressourceneffizienzpotenziale des Bauhauptgewerbes im Hochbau. VDI ZRE Publikationen: Kurzanalyse Nr. 34 (2024). https://www.ressource-deutschland.de/fileadmin/user_upload/1_Themen/h_Publikationen/Kurzanalysen/VDI_ZRE-Kurzanalyse_34_RE-Potenziale_Bauhauptgewerbe_Hochbau_C1.pdf
2. Churkina, G., et al.: Buildings as a global carbon sink. Nat Sustain **3**, 269–276 (2020). https://doi.org/10.1038/s41893-019-0462-4
3. Bundesministerium für Wohnen, Stadtentwicklung und Bauwesen (BMWSB), Bundesministerium für Ernährung und Landwirtschaft (BMEL): Holzbauinitiative. Strategie der Bundesregierung zur Stärkung des Holzbaus als ein wichtiger Beitrag für ein klimagerechtes und ressourceneffizientes Bauen, Berlin (2023)
4. Orozco, S., Oertwig, N., Feldhaus, A., Eingartner, V., Mintchev, M., Kohl, H.: Planning and control of value creation networks in timber construction. In: Kohl, H., Seliger, G., Dietrich, F., Mur, S. (eds) Sustainable Manufacturing as a Driver for Growth. GCSM 2023. Lecture Notes in Mechanical Engineering. Springer, Cham (2025). https://doi.org/10.1007/978-3-031-77429-4_48
5. Wildemann, H. (ed.): Supply chain management, 1st edn. TCW Transfer-Centrum-Verl, München (2000)
6. Gutiérrez, N., Negrão, J., Dias, A., Guindos, P.: Bibliometric review of prefabricated and modular timber construction from 1990 to 2023: evolution, trends, and current challenges. Sustainability **15**(5), 2134 (2014). https://doi.org/10.3390/su16052134
7. Kaufmann, H., Huß, W., Schuster, S., Stieglmeier, M.: leanWOOD Optimierte Planungsprozesse für Gebäude in vorgefertigter Holzbauweise, München (2017)
8. Sanvido, V.: An Integrated Building Process Model, Technical Report No. 1 (1990). https://www.engr.psu.edu/ae/cic/publications/TechReports/TR_001_Sanvido_1990_IBPM.pdf. Accessed 12 June 2024
9. Warszawski, A., Sacks, R.: Computer Integrated Construction, Automation and Robotics in Constructio XXII (1995)
10. Schober, K.S., Hoff, P., Lecat, A., de Thieulloy, G., Siepen, S.: Turning point for the construction industry. The disruptive impact of Building Information Modelling (BIM), Roland Berger GmbH, Munich (2017)

11. Federal Ministry for Housing, Urban Development and Building. (2024). ÖKOBAUDAT 2024-I. https://www.oekobaudat.de/en.html. Accessed 30 Jan 2025
12. Federal Ministry of Food and Agriculture. (2022). Bundeswaldinventur. https://bwi.info/impressum.aspx. Accessed 30 Jan 2025
13. Mertins, K., Jochem, R., Jäkel, F.W.: A tool for object-oriented modelling and analysis of business processes, Computers in Industry, 33 (2–3) (1997). https://doi.org/10.1016/S0166-3615(97)00040-7
14. PricewaterhouseCoopers GmbH: The construction industry in times of crisis: progress on ESG, standstill on digital transformation (2024). https://www.pwc.de/de/risk-regulatory/risk/capital-projects-and-infrastructure/bauindustrie-unter-druck.html
15. Kohl, H., et al.: Studie zur Stärkung der Holzbauwirtschaft in der Metropolregion Berlin-Brandenburg (2023)

# Modelling and Analysis of Circular Production Network Structures

Tobias Lachnit[(✉)], Ariane Deckert, Moritz Hörger, Kevin Gleich, Finn Bail, Martin Benfer, and Gisela Lanza

Wbk Institute of Production Science, Karlsruhe Institute of Technology (KIT), Kaiserstraße 12, 76131 Karlsruhe, Germany
tobias.lachnit@kit.edu

**Abstract.** Circular business models have the potential to generate value for consumers, benefit producers and positively impact the environment. Despite its potential, designing a circular production network remains a significant challenge. Recognising multiple factors that influence the design of production networks, the proposed approach emphasises the importance of product, process and location attributes in the design and selection of a suitable production network. Five archetypes of reverse networks, specifically tailored to reverse logistics and circular production, are introduced. These network types are analysed and compared in a simulation. The research evaluates the performance of these network types in a case study from the construction machinery industry, taking into account factors such as production volumes, regional-specific transport costs and times, labor costs de-pendent on the country, varying return volumes across the European Union. The research shows how simulation can be used to develop and evaluate suitable network types for circular production.

**Keywords:** Reverse Logistics · Circular Production · Remanufacturing · Network Configuration · Return Volume

## 1 Introduction

The circular economy represents an important opportunity to contribute to future climate goals by decoupling the production of goods from the use of resources [1]. This transition is reflected in the growing number of companies integrating circular economy principles into their operations. A critical aspect of achieving these goals is the strategic design of production networks, which remains a significant challenge for globally operating companies [2]. However, the return of a significant number of used products from customers continues to pose considerable difficulties [3]. Therefore, this paper examines these challenges and explores strategies for effective product remanufacturing, aiming to provide insights into circular global production network (GPN) structures and suitable reverse network configurations for the return of cores for remanufacturing and refurbishment.

© The Author(s) 2025
H. Kohl et al. (Eds.): GCSM 2024, LNME, pp. 725–733, 2025.
https://doi.org/10.1007/978-3-031-93891-7_80

## 2   State of the Art

As GPNs are often the result of historical growth, they are usually not designed to cope with today's requirements in terms of volatility, complexity or new constraints such as increased closed-loop management [4]. Continuous adaptation of the network configuration, i.e. the structuring of GPNs concerning the geographical distribution of their sites and resources (network structure), is therefore of central importance for maintaining the competitiveness of GPNs. In this context, Abele et al. 2008 [5] define network phenotypes - local-for-local, hub-and-spoke, world factory, sequential and web structure - which are differentiated based on allocating production scopes and associated resources and the sequence of production steps. Each phenotype has different characteristics, such as local customisation, transactional costs, economies of scale and scope. This research uses these network configurations in reverse logistics and analyses their impact on costs through discrete event and agent-based simulations, building on Klenk's work in reverse supply chain simulation [3]. Consequently, this article focuses on the following primary research question:

RQ1: How do different network configurations and varying return volumes influence total costs per remanufactured product in reverse logistics operations?

Linear optimisation problems are often used to analyse and optimise reverse logistics. In contrast to linear programming (LP), simulations are particularly valuable for analysing complex systems, especially when dealing with uncertainty, non-linear relationships, and dynamic changes over time. They enable the evaluation of multiple design variants with relatively low effort, making it possible to explore a wide range of configurations [6]. It is crucial to comprehensively analyse the various cost types in reverse logistics in GPNs. These cost types include remanufacturing, transportation, handling and storage costs. Thus, the second research question arises as follows:

RQ2: Which cost components significantly impact total costs per remanufactured product in reverse logistics, and how do these costs vary across network configurations and return volumes?

## 3   Circular Production Network Structures

The following chapter provides the methodology and detailed description of circular production networks' implementation and structure process.

### 3.1   Methodology

The methodology of this study is based on the guidelines of VDI 3633, part 1 [6]. The methodological process begins with defining the objectives, followed by a system analysis, whose results are presented as a conceptual model. In this work, the conceptual model includes a system overview diagram, target values, input variables, and control variables. Developing an activity diagram and a mathematical formula describing the calculation of transport costs leads to a formal model. Both models are explained in

Sect. 3.2. Subsequently, the formal model is implemented into a software model in Python using the SimPy library. After implementing the model, experiments are conducted, and the results are analysed. A full factorial design is used in the experiments. Data collection and processing, crucial for accurate simulation input, are conducted case-specifically and in parallel with the other steps. This is detailed in the case study in Sect. 4 [6].

The applied simulation method encompasses several approaches, including discrete event simulation (DES) and agent-based simulation (ABS). DES is a process-oriented method that models events at specific points in time and represents systems as a series of discrete events, including queue modelling. ABS models the behaviour of individual agents and their interactions [7].

The simulation model used in this study is dynamic, stochastic, and discrete. Dynamic simulation models represent systems that change over time and analyse temporal processes. Stochastic simulation models involve random input components that lead to random outputs, allowing the estimation of system properties through exponentially distributed arrival times [8].

### 3.2 Implementation

This simulation aims to analyse the impact of return volumes in reverse logistics on selecting the economically optimal network type. The conceptual model of this work, shown in Fig. 1, is based on the work of Klenk [3] and forms the basis for the simulation. The modules are general functionality, participants, logistics and procurement, market, and product. The information agent stores all relevant information at discrete points in time in the simulation, which is important for subsequent cost calculations. The market module comprises the generation of cores and demand and matching supply and demand. The product module contains the specific product characteristics. The core arrives at the workshop, is transported to the collection centre, and finally to the remanufacturing plant. New components are also transported to the remanufacturing plant. At the end of the process, remanufactured finished goods are produced.

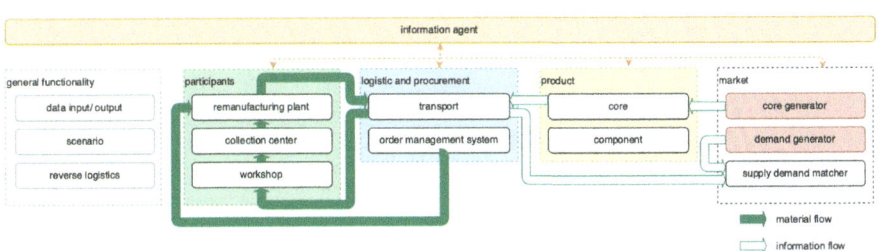

**Fig. 1.** Conceptual framework of the simulation modules and the material and information flow

The input variables for modelling reverse logistics processes include rejection rates, inventory and procurement parameters, and specific details for workshops, collection centres, and remanufacturing plants. Core-related variables are rejection rates, return parameters, and bills of materials. Component variables include inventory, procurement parameters, and rejection rates. At workshops, collection centres, and remanufacturing plants, variables encompass inspection, handling (loading, unloading, packaging), disassembly, and remanufacturing times, as well as depreciation and headcount. Transport parameters cover core and component capacities per pallet, transportation cost structures, and transport trigger levels. The control variables define the configurable parameters within the model and are detailed in Sect. 4. The target values serve to evaluate the cost-effectiveness of the reverse logistics processes. These include logistics, remanufacturing, and depreciation costs of the remanufacturing plant. Logistics costs cover direct transport from the workshop to the remanufacturing plant, indirect transport via collection centres, and handling and storage costs. Remanufacturing costs encompass inspection, remanufacturing, reassembly, disassembly, and depreciation.

The formal model of the reverse logistics simulation includes demand and core generation, stock checking, and supply-demand matching. Cores are inspected, palletised, and sent directly to the remanufacturing facility or through a collection centre. At the facility, cores are inspected, disassembled, and remanufactured. When all parts are available, reassembly occurs. The order management system monitors stock levels and places orders as needed. Finished products are assigned to customers on a FIFO basis.

## 4   Case Study and Network Structures

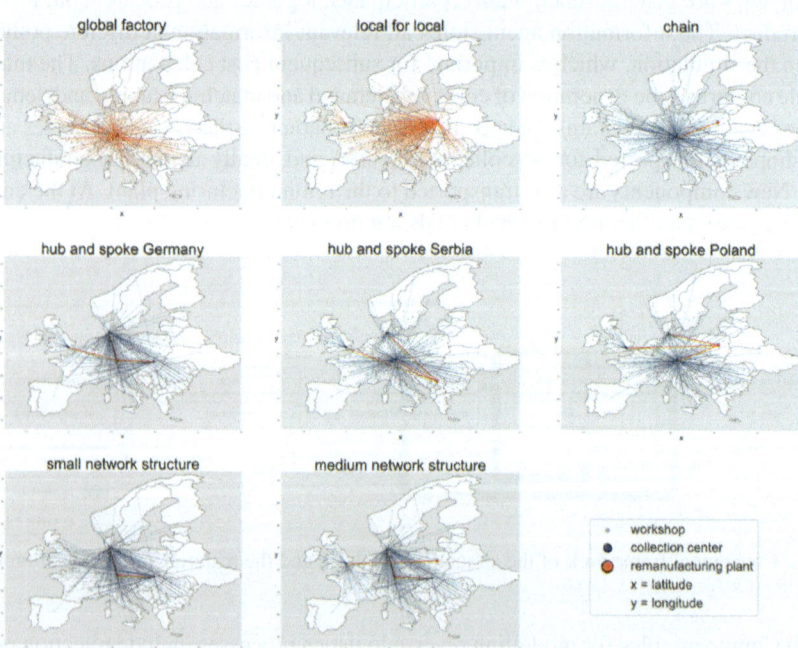

**Fig. 2.** Network configurations for reverse logistics in the construction equipment industry

The case study examines the potential reverse logistics of a company in the construction equipment industry. It explores variations in network configurations and return volumes for over 400 workshops, as visualised in Fig. 2. These scenarios were selected heuristically on the basis of discussions with industry experts. The network configurations include a "global factory" with a central remanufacturing facility in Germany, "local for local" with three remanufacturing facilities in the UK, Poland, and Croatia, various "hub and spoke (h&s)" scenarios with three collection centres and a remanufacturing facility in Germany, Serbia, or Poland, and a "chain" with a collection centre in Germany and a remanufacturing facility in Poland. The "small network" comprises two collection centres in Denmark and Slovakia and a remanufacturing facility in Germany. The "medium network" includes an additional collection centre in the UK and another remanufacturing facility in Poland. Additionally, return volumes vary from very low (800 cores per year) to low (1100 cores per year), medium (1400 cores per year), high (2000 cores per year), and very high (15000 cores per year).

In this case study, the following costs are considered: transportation costs, remanufacturing costs, storage costs, and depreciation of the remanufacturing plants. Ordering and penalty costs for unserved customers due to a specific scrap rate of cores and components are neglected. The following section explains the data collection and calculation of transportation, storage and processing costs for reverse logistics.

The transportation costs were systematically recorded by submitting price inquiries through a publicly accessible online portal of a logistics service provider [9]. An analysis of the correlation between transportation costs and distance revealed only a weak relationship. However, the cost analysis for varying pallet quantities identified two distinct linear trends: a steeper increase in costs was observed for the first pallets, while a more moderate linear increase was noted for the remaining pallets, up to a quantity of 20. Consequently, the transportation costs for each route are automatically determined via the portal for the first pallet. The costs for additional pallets on the same route are then calculated using a formula that accounts for these two linear trends. When querying the country-specific individual routes, deviations from standard dimensions for load carriers and means of transport are taken into account. The required storage area is increased by a factor of 1.11 to account for additional operational spaces, with rental costs set at 6.75 euros per square meter based on E&G Real Estate GmbH [10]. A five-level high-bay warehouse is assumed. Processing costs are determined based on publicly available labour rates sourced from various sources, including the Federal Statistical Office, Trading Economics, and the Austrian Federal Economic Chamber [11–13]. The following section describes the statistical distributions and model assumptions used, as well as additional specifications of the simulation. This simulation used exponential distributions to model customer demand and arrival processes, with process times following a truncated normal distribution. The maturity phase is simulated, where supply meets demand without direct coupling. Additionally, the t,S-order policy for order management is implemented.

The simulation setup includes a 90-day warm-up period and runs 365 days, with eight-hour workdays. To ensure statistical robustness, each simulation is replicated 20 times. An information agent monitors reverse logistics data. Pseudo-random numbers are consistently generated using a fixed seed value for reliability, and the Numpy random number generator transforms uniformly distributed random numbers to match desired distributions. Experiments employ a full factorial design to test all scenarios systematically. Simulation results are evaluated following VDI 3633, part 3 guidelines, averaging data and calculating confidence intervals. Statistical analysis includes the creation of a heatmap to visualize correlations, bar charts to show cost allocation, and interaction plots to illustrate variable relationships. Additionally, Tukey HSD tests were performed to identify significant differences between scenarios.

## 5 Results

The network configurations are visualized in terms of total costs per sold product using a heatmap in Fig. 3 and are analyzed with a Tukey HSD test. In response to RQ1: The results indicate that the "global factory," "hub and spoke Germany," "medium network structure," and "small network structure" configurations consistently incur higher costs compared to the "chain" configuration, with mean differences ranging from 235% to 299%. The "hub and spoke Poland" configuration showed higher and lower costs. The "local for local" configuration showed no significant difference compared to the "chain" configuration. "hub and spoke Serbia" proved to be the most cost-effective configuration in all categories, with mean differences ranging from 74% to 374%. The "local for local" scenario without a central hub as well as the "hub and spoke Serbia" scenario with a central hub are among the more cost-effective configurations. Interestingly, "hub and spoke Germany", despite having a similar configuration to "hub and spoke Serbia," performs significantly worse, necessitating a closer look at the cost distribution, which is crucial for answering RQ2.

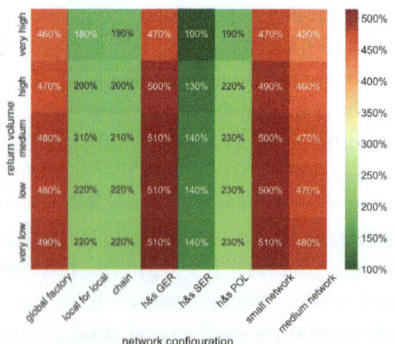

**Fig. 3.** Heatmap of transport costs per re-manufactured product for reverse logistics

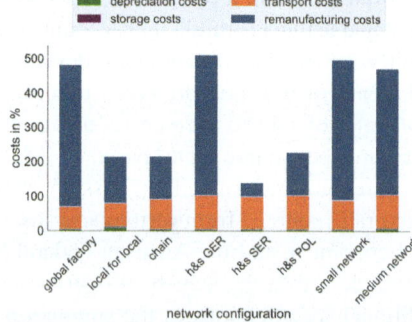

**Fig. 4.** Cost allocation for low return volume for reverse logistics

The cost distribution is visualised in Fig. 4. The average share of each cost type in the total costs per sold product is as follows: remanufacturing costs share 70%, transport costs share 28%, storage costs share 1%, and depreciation costs share 1%. This distribution indicates that remanufacturing costs constitute the majority of the total costs, followed by transport costs, while storage and depreciation costs have minimal impact. A look at the transportation costs in the interaction plot in Fig. 5 shows that these costs decrease with higher return volumes. This is because a fully loaded pallet can be transported from the workshop, distributing the transportation costs of a pallet over more cores. Additionally, it becomes clear that scenarios without a central hub incur lower transportation costs. This could be due to logistics companies already optimising highly competitive logistic networks. Internal logistic networks might disrupt rather than support them in terms of cost efficiency.

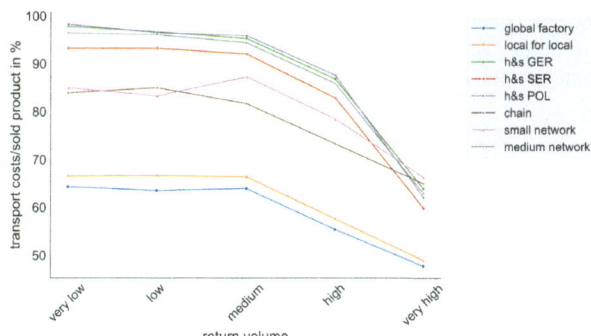

**Fig. 5.** Interaction plot of total transport costs per remanufactured product

The Tukey HSD test confirms that the "global factory" configuration is more cost-effective than the "local for local" configuration in most cases concerning transportation costs. Additionally, it is more cost-effective in all return volumes compared to all other network configurations. A closer look at remanufacturing costs in Fig. 3 shows that they significantly influence total costs. Since remanufacturing costs are labor-intensive in this case, they are significantly influenced by the labor costs of the respective country. This explains the differences in total costs between the various "hub and spoke" configurations.

## 6   Summary and Outlook

This study investigates the impact of different network configurations and return volumes on total costs per remanufactured product in reverse logistics (RQ1 and RQ2). It highlights remanufacturing costs as pivotal, particularly influenced by labour wages. The use case demonstrates cost advantages by strategically locating remanufacturing plants in low-cost locations such as Poland, Croatia and Serbia. Direct transportation to remanufacturing sites reduces costs in logistic networks like the "global factory" and "local for local." Higher return volumes mitigate differences in transportation expenses across configurations, underscoring the significance of remanufacturing costs. Future research may focus on optimising facility numbers across locations to enhance scalability and efficiency, particularly in competitive labour markets. Further opportunities lie in refining logistical strategies and inventory management for new components. Investigating the correlation between CO2 emissions and transportation costs, alongside the dynamics of donor and recipient markets within the EU, represents promising directions for future inquiry.

**Acknowledgement.** This research work was undertaken in the context of the IGF-Project "In-troRemanNet" (Nr. 22626 N). The project is funded by the Federal Ministry for Economic Affairs and Climate Action based on a decision by the German Bundestag.

# References

1. Ellen MacArthur Foundation: Towards the circular economy Vol. 1: An economic and business rationale for an accelerated transition. https://www.ellenmacarthurfoundation.org/towards-the-circular-economy-vol-1-an-economic-and-business-rationale-for-an. Accessed 23 June 2024
2. Steier, G.L., Jaspers, M.C., Peukert, S., Benfer, M., Lanza, G.: Strategic fit in global production networks – a decision support model for strategic configuration of global production networks. Procedia CIRP **120**, 1059–1064 (2023)
3. Klenk, F.: Transparenzsteigerung in der Rückführungslogistik zur Verbesserung der Mate-rialbedarfsplanung für das Remanufacturing. Forschungsberichte aus dem wbk, Institut für Produktionstechnik, Karlsruher Institut für Technologie (KIT), 272 (2023)
4. Peukert, S., Hörger, M., Lanza, G.: Fostering robustness in production networks in an increasingly disruption-prone world. CIRP J. Manuf. Sci. Technol. **41**, 413–429 (2023)
5. Abele, E., Meyer, T., Näher, U., Strube, G., Sykes, R. (eds.): Global Production: A Handbook for Strategy and Implementation. Springer, Berlin Heidelberg (2008)
6. VDI Verein Deutscher Ingenieure e.V. Gesellschaft Produktion und Logistik: VDI 3633 Blatt 1: Simulation von Logistik-, Materialfluss- und Produktionssystemen – Grundlagen. VDI-Verlag (2000)
7. Sumari, S., Ibrahim, R., Zakaria, N.H., Ab Hamid, A.H.: Comparing three simulation mod-els using taxonomy: system dynamic simulation, discrete event simulation and agent based simulation. Int. J. Manag. Excellence **1**(3), 55–57 (2013)
8. Law, A.M., Kelton, W.D.: Simulation modeling and analysis. 2nd edn. McGraw-Hill, Inc., 28–29 (1991)
9. Cargoboard GmbH & Co. KG Verwaltung & Zentrale. https://www.cargoboard.com. Accessed 21 June 2024
10. E & G Real Estate GmbH: Region Stuttgart Industrie- & Logistikimmobilien (2024). https://www.eug-immobilien.de/assets/main/Berichte/industrie-und-logistikmarkt-stuttgart/eug-ind ustrie-logistik-stuttgart-2024.pdf. Accessed 17 Aug 2024
11. Statistisches Bundesamt (Destatis). https://www.destatis.de. Accessed 21 June 2024
12. Trading Economics. https://tradingeconomics.com. Accessed 21 June 2024
13. Wirtschaftskammer Österreich. https://www.wko.at/statistik/eu/europa-arbeitskosten.pdf. Accessed 21 June 2024

# Leveraging Digital Logistics Twins to Harness Sustainability and Resilience in Manufacturing Networks

Frank Straube, Benjamin Gorgas[✉], and Angelica Coll

Institute for Technology and Management, Berlin University of Technology, 10623 Berlin, Germany
{frank.straube,gorgas,angelica.coll}@tu-berlin.de

**Abstract.** The current global business environment poses significant challenges for companies to maintain efficient and sustainable operations. This study focuses on the role of self-imposed restrictions in enhancing the sustainability and resilience of logistics networks. Using the automotive industry as an example, this research analyzes the impact of such restrictions on the overall sustainability and resilience of international logistics networks. Building on a case study research approach, qualitative and quantitative analyses are performed to present a comprehensive understanding of the current situation while developing practical solutions. By examining the threshold between different categories of restrictions, the study provides actionable recommendations for companies to enhance their process planning capabilities and leverage Digital Logistics Twins as a supportive environment. Specifically, by shifting the focus towards more proactive management of self-imposed restrictions, companies can improve their efficiency and sustainability. Furthermore, the insights gained from this study provide a valuable framework for organizations seeking to optimize their operations, leverage the potential of DLT, and contribute to the research on supply chain resilience and the role of digital technologies.

**Keywords:** Digital Logistics Twins (DLT) · Sustainability · Resilience · Automotive Industry · Supply Chain Management · Production Planning · Constraints · Logistics Networks

## 1 Introduction

In the contemporary global business environment, companies face increasing pressure to maintain efficient and sustainable operations. This pressure is driven by a combination of regulatory requirements, market demands, and environmental considerations. Sustainability, defined as the ability to meet present needs without compromising the ability of future generations to meet their own needs, has become a central focus for many organizations. Concurrently, resilience, which refers to the ability of a system to absorb disturbances while retaining its basic function and structure, is critical for navigating the complexities and uncertainties of global supply chains, especially within the context of changing environmental and social conditions [1–3].

© The Author(s) 2025
H. Kohl et al. (Eds.): GCSM 2024, LNME, pp. 734–741, 2025.
https://doi.org/10.1007/978-3-031-93891-7_81

Adaptability in this context refers to a company's ability to adjust to market fluctuations, supply interruptions due to supplier issues, and variability resulting from different planning environments. The literature review aimed to identify classes of restrictions that enhance adaptability, thereby increasing the sustainability and resilience of international value networks [4].

In this context, the automotive industry serves as an illustrative example due to its complex and globally dispersed supply chains. The industry is characterized by high demand variability, stringent regulatory requirements, and the need for precise coordination among multiple stakeholders [5]. To address these challenges, companies have increasingly adopted self-imposed restrictions, i.e., rules or constraints that go beyond regulatory requirements, to enhance operational efficiency and cope with the real conditions of the production and logistics network [6].

This study aims to analyze the role of self-imposed restrictions in the resilience and sustainability of logistics networks within the automotive industry. By examining various categories of restrictions and their thresholds, this research provides actionable recommendations for improving process planning and leveraging digital technologies such as Digital Logistics Twins (DLTs). Specially by shifting the focus towards more proactive management of self-imposed restrictions, companies can improve their efficiency and sustainability. Furthermore, the insights gained from this study provide a valuable framework for organizations seeking to optimize their operations and leverage the potential of DLT and contribute to the research on supply chain resilience and the role of digital technologies.

## 2 Methodology

This study employed two distinct methods to focus on enhancing sustainability and resilience through the reduction of internal restrictions. The primary aim was to identify restrictions, which are more feasible to adjust and thus enhance adaptability and resilience in manufacturing environments.

First, a systematic literature review (SLR) was conducted following the methodology outlined by Durach et al. (2017) to identify key restrictions in manufacturing environments [7]. A comprehensive search string was developed using terms related to constraints, production or manufacturing planning programs, and the automotive industry. This search yielded over 100 articles, which were subsequently reviewed and filtered based on their relevance. The content analysis of the selected papers resulted in the identification and categorization of several restrictions that impact manufacturing environments.

Following the SLR, the findings were validated through a semi-structured survey involving two industry representatives. This survey aimed to identify the types of restrictions from the SLR present in the production environments of the interviewed experts and determine the thresholds at which these restrictions are considered hard (non-negotiable) or soft (flexible). The focus was particularly on soft restrictions due to their easier implementation and potential for enhancing sustainability and resilience. The industry experts surveyed each possess over 10 years of experience in the field and are employed by an OEM in the automotive sector, ensuring that the insights gathered reflect a deep understanding of the industry's specific operational challenges and constraints.

The survey was designed with a combination of open-ended and close-ended questions, gathering detailed information on various aspects of the previously identified restrictions. The industry representatives were selected based on their extensive knowledge in production program planning, providing a comprehensive perspective necessary for a robust analysis. The empirical data collected included quantitative measures and qualitative insights.

The data were then analyzed to identify patterns and establish a coherent framework for defining thresholds. A comparative analysis was conducted to validate the restriction categories from the SLR.

By identifying and managing these soft restrictions, the study provides actionable recommendations for improving process planning and leveraging DLTs to simulate scenarios and optimize processes. DLTs are particularly suitable for this purpose because they create a virtual replica of the logistics network, enabling real-time monitoring, predictive analysis, and scenario testing [8]. This capability allows companies to anticipate potential disruptions, evaluate the impact of different constraints, and explore various strategies for enhancing efficiency and resilience. The ability of DLTs to provide detailed insights and foresight into logistics operations ensures that companies can make informed decisions and implement changes swiftly and effectively [9]. Through continuous simulation and optimization, DLTs help maintain a balance between operational demands and sustainability goals, ensuring a resilient supply chain capable of adapting to changing market conditions and environmental challenges [9, 10].

## 3   Results

The identified self-imposed restrictions and their implications derived from the SLR were clustered in 20 different categories (shown as an example in Fig. 1). From these categories, two have been the main focus of the research from the past decade, namely inventory management (14 articles) and production planning (12 articles). Following in relevance are demand and production capacity management (7 and 5 articles respectively) as well as logistics and transport and risk and uncertainty management (4 articles each). These categories have distinct roles and implications for the automotive industry which will be described in the subsequent discussion.

**Inventory Management:** These restrictions primarily involve limitations on storage capacity, safety stock levels, and inventory turnover. Effective inventory management is crucial for maintaining smooth operations and preventing stockouts. For instance, companies must balance the need for sufficient safety stock to buffer against demand variability while minimizing holding costs [11]. DLTs can enhance inventory management by providing real-time visibility and predictive analytics to optimize stock levels, such as the avoidance of stockouts and overstocking [9, 12].

**Production Planning:** These restrictions include setup and production costs, production capacity limits and sequence-dependent setup times. Effective production planning ensures that production schedules align with demand forecasts, optimizing resource use and minimizing downtime. Studies emphasize the importance of managing setup carryover to maintain the integrity of the production process [13]. Recent advancements in

**Categorization and frequency of identified restrictions**

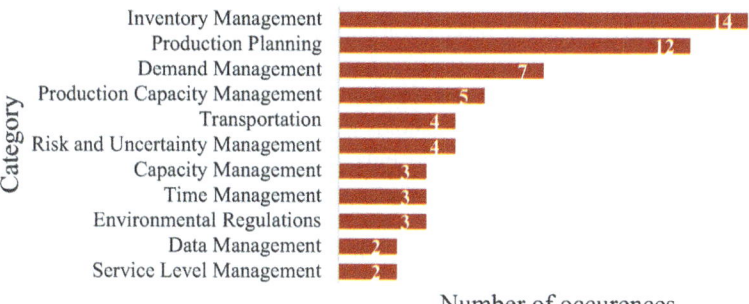

**Fig. 1.** Categorization and frequency of identified restrictions

smart manufacturing and the integration of lean principles further enhance production planning by reducing setup times and improving workflow efficiency [14].

**Demand Management:** This type of restrictions involves handling demand variability and stockouts. Robust forecasting and flexible production planning are essential for meeting customer demands without incurring excessive costs. Research highlights the importance of managing demand uncertainty to ensure timely product delivery and maintain high service levels [15]. Modern techniques, such as machine learning algorithms, have been employed to improve demand forecasting accuracy and responsiveness to market changes [16].

**Production Capacity Management:** These restrictions focus on balancing production quantities with available resources, including machine utilization and maintenance schedules. Ensuring that production does not exceed the capabilities of production lines is critical for maintaining efficiency [17].

**Risk and Uncertainty Management:** This type of restrictions involves preparing for potential disruptions in the supply chain. This includes incorporating safety stock levels, demand forecasting accuracy, and contingency planning. Robust optimization techniques can mitigate the impact of uncertainties on production schedules [18, 19]. Advanced methodologies, such as chance-constrained programming, offer robust solutions to address complex uncertainties in production and logistics [20].

**Transportation:** These restrictions address the efficient movement of goods through the supply chain, which includes vehicle capacity, route planning, and delivery schedules. Efficient logistics are essential for minimizing transportation costs and ensuring timely deliveries [21]. Innovative approaches, such as integrated vehicle transportation models, have been developed to optimize operations and improve overall supply chain efficiency [22].

The results obtained from the expert survey provide an empirical delimitation of the thresholds that delineate the transition from soft to hard restrictions. Additionally, the industry survey assessed the relevance of the restriction groups identified in the SLR, as evaluated by industry experts.

It is important to note that restriction groups can be interdependent, meaning that the management of one group can impact others. For instance, adjustments in inventory management may influence production planning and vice versa. These groups were selected based on their frequency of mention in the literature, highlighting their critical importance in the context of logistics. Parting from the quality analysis of the survey responses, we determined that the thresholds for transitioning from soft to hard restrictions are primarily defined by two components: *time* and *cost*. A restriction is considered hard if it cannot be adjusted within a short period (typically three months) and if modifying it would negatively affect the business case. Conversely, a restriction is deemed soft if it can be altered within this timeframe without compromising cost-efficiency. These definitions align with the findings of recent studies which emphasize the significance of temporal and financial parameters in determining operational flexibility [23, 24].

Moreover, the survey responses strongly indicate that certain restriction categories were deemed irrelevant by the experts, as these categories do not fall within the scope of tasks typically managed by the respondents. Specifically, categories related to transportation, vehicle management, and route planning were identified as irrelevant. The irrelevance of these categories highlights the fact that the professionals surveyed are not primarily engaged in these areas, and therefore, these restrictions do not pertain to their operational decision-making processes. This exclusion is essential to ensure that the analysis remains focused on the most pertinent aspects of logistics management as experienced by the survey participants.

## 4   Discussion

The aim is to present a cohesive framework for understanding and managing self-imposed restrictions in logistics networks within the automotive industry.

The integration of the findings from the different data collection methods suggests several strategic actions for enhancing supply chain resilience. Identifying hard restrictions enables companies to allocate resources to areas requiring stringent control, ensuring compliance and stability [17]. Recognizing soft restrictions allows for greater flexibility and adaptability in operations. Companies can implement dynamic strategies to adjust these restrictions based on market conditions and operational needs, aligning with the importance of demand management [13]. Proactive management of soft restrictions can lead to significant improvements in efficiency and cost-effectiveness. This involves continuously monitoring and adjusting safety stock levels, production schedules, and capacity utilization to respond to market changes and demand variability [18]. Implementing flexible production planning and robust demand forecasting can enhance a company's ability to meet customer needs while minimizing costs [13, 17].

The integration of DLTs can further enhance a company's ability to manage self-imposed restrictions. DLTs provide a virtual environment for real-time monitoring and simulation of logistics operations, enabling companies to identify potential bottlenecks and optimize processes [10]. This capability allows companies to anticipate potential disruptions, evaluate the impact of different constraints, and explore various strategies for enhancing efficiency and resilience.

For instance, in previous chapters, we discussed how DLT can simulate various disruption scenarios and optimize response strategies in real-time. A specific example is the reduction of procurement management restrictions. Traditionally, companies might restrict their procurement to a select small group of suppliers to minimize costs; however, this approach can limit flexibility and increase vulnerability to supply chain disruptions. By enabling additional suppliers, companies gain a significant lever on delivery reliability and planning capability, even during crises, which in turn has a higher financial benefit for companies than the costs incurred by enabling suppliers beyond the minimum cost threshold [25]. DLT can facilitate this flexibility by providing real-time visibility into supplier capabilities, inventory levels, and logistics constraints, enabling more informed and agile decision-making. This increased flexibility fosters the resilience of production programs, ensuring they remain well-planned and efficient, thereby avoiding waste and supporting sustainability efforts. Thus, DLT not only strengthens the resilience of logistics networks but also ensures that sustainability goals are not compromised during disruptions. By integrating specific sustainability metrics into DLTs and simulation methods, companies can measure and uphold sustainability alongside resilience.

In today's volatile world, it is imperative to question whether the determination of thresholds for restrictions should solely rely on the components of *cost* and *time*. Following this thought, the results also underscore the importance of incorporating environmental constraints into production planning. While specific $CO_2$ emission limits were not a primary focus for the surveyed companies, energy consumption and waste management are critical considerations. Adopting sustainable practices and reducing environmental impact can enhance a company's reputation and comply with regulatory requirements [21]. By integrating considerations such as environmental impact, social responsibility, and long-term strategic goals into the threshold-setting process, organizations can better navigate uncertainties and uphold their commitment to sustainable practices.

## 5    Conclusion

This study provides a comprehensive analysis of the impact of self-imposed restrictions on the resilience and sustainability of manufacturing and logistics networks in the automotive industry. Understanding and defining these thresholds is vital for developing strategies that enhance supply chain resilience and sustainability. By identifying which restrictions are hard and which are soft, companies can prioritize their management efforts, allocate resources more effectively, and adapt more swiftly to market changes.

The findings underscore the importance of proactive management of these restrictions to enhance operational efficiency, resilience, and sustainability. The integration of DLTs can further augment a company's capacity to adapt to environmental challenges, optimize operations, and enhance overall resilience. By effectively managing various categories of restrictions, companies can develop more resilient and sustainable networks. This increased resilience directly contributes to sustainability, as flexible and adaptable operations are better equipped to respond to and mitigate environmental impacts. Moreover, these findings are appliable not only for companies in the automotive industry, but also for any company working with highly customized products and/or handling several product portfolios.

Future research should concentrate on refining the thresholds that define hard and soft restrictions across different industry contexts. This involves further exploration to ensure these thresholds are universally applicable and provide a precise understanding of their impacts. Additionally, expanding empirical studies across various sectors and regions is crucial for validating the findings and enhancing their generalizability. Investigating the role of emerging digital technologies in logistics and resilience is also essential. This entails exploring the integration and impact of innovations such as DLTs to further enhance supply chain resilience, adaptability and sustainability.

# References

1. Straube, F., Nitsche, B.: Heading into "The New Normal": potential development paths of international logistics networks in the wake of the Coronavirus pandemic. Int. Verkehrswesen **72**(3), 31–35 (2020)
2. Tukamuhabwa, B.R., Stevenson, M., Busby, J., Zorzini, M.: Supply chain resilience: definition, review and theoretical foundations for further study. Int. J. Prod. Res. **53**(18), 5592–5623 (2015)
3. Hohenstein, N.O., Feisel, E., Hartmann, E., Giunipero, L.: Research on the phenomenon of supply chain resilience: a systematic review and paths for further investigation. Int. J. Phys. Distrib. Logist. Manag. **45**(1/2), 90–117 (2015)
4. Mishrif, A. (ed.): Business resilience and market adaptability: pandemic effects and strategies for recovery. In: The Political Economy of the Middle East (2024)
5. Marković, D., Mijušković, V.: Disruptions and competitive strategies in the automotive industry. In: Rezaei, N. (ed.) Transdisciplinarity, vol. 5, pp. 207–227 (2022)
6. Helmold, M., Küçük Yılmaz, A., Dathe, T., Flouris, T.G.: Supply chain risk management: cases and industry insights. In: Management for Professionals (2022)
7. Durach, C., Kembro, J., Wieland, A.: A new paradigm for systematic literature reviews in supply chain management. J. Supply Chain Manag. **53**(4), 67–85 (2017)
8. Gerlach, B., Zarnitz, S., Nitsche, B., Straube, F.: Digital supply chain twins—conceptual clarification use cases and benefits. Logistics **5**(4), 86 (2021)
9. Zarnitz, S.: Digital Supply Chain Twin: Framework for Planning and Control of Logistics Systems, Tech. Univ. Berl. (2023). https://depositonce.tu-berlin.de/items/05fafb7e-d875-4566-8ccf-2e0b041f9cdc
10. Ivanov, D., Dolgui, A., Das, A., Sokolov, B.: Digital supply chain twins: managing the ripple effect resilience, and disruption risks by data-driven optimization, simulation, and visibility. Handb. Ripple Eff. Supply Chain **276**, 309–332 (2019)
11. Digiesi, S., Mascolo, G., Mossa, G., Mummolo, G.: New Models for Sustainable Logistics: Internalization of External Costs in Inventory Management. Springer International Publishing, Cham (2016)
12. Liu, J., Zhang, K.: Design and simulation debugging of automobile connecting rod production line based on the digital twin. Appl. Sci. **13**(8), 4919 (2023)
13. Lohmer, J., Lasch, R.: Production planning and scheduling in multi-factory production networks: a systematic literature review. Int. J. Prod. Res. **59**(7), 2028–2054 (2021)
14. Tripathi, V., Chattopadhyaya, S., Mukhopadhyay, A.K., Sharma, S., Li, C., Di Bona, G.: A sustainable methodology using lean and smart manufacturing for the cleaner production of shop floor management in Industry 4.0. Mathematics **10**(3), 347 (2022)
15. Rexhausen, D., Pibernik, R., Kaiser, G.: Customer-facing supply chain practices—the impact of demand and distribution management on supply chain success. J. Oper. Manage. **30**(4), 269–281 (2012)
16. Spoor, J.M., Weber, J.: Evaluation of process planning in manufacturing by a neural network based on an energy definition of hopfield nets. J. Intell. Manuf. **35**(6), 2625–2643 (2023)

17. Slack, N., Chambers, S.: R, 6th edn. Johnston, Operations management (2010)
18. Nitsche, B.: Development of an assessment tool to control supply chain volatility (2019)
19. Paul, S.K., Agarwal, R., Sarker, R.A., Rahman, T., (eds.): Supply Chain Risk and Disruption Management: Latest Tools, Techniques and Management Approaches Flexible Systems Management (2023)
20. Hui, J., Wang, S., Bin, Z., Xiong, G., Lv, J.: Chance-constrained programming with robustness for lot-sizing and scheduling problems under complex uncertainty. Assem. Autom. **42**(4), 490–505 (2022)
21. Christopher, M.: Logistics & supply chain management, 4. ed. Harlow: Financial Times, Prentice Hall (2011)
22. Akbarpour, N., Kia, R., Hajiaghaei-Keshteli, M.: A new bi-objective integrated vehicle transportation model considering simultaneous pick-up and split delivery. Sci. Iran. **28**(6), 3569–3588 (2020)
23. How to Improve Supply Chain Resilience, SafetyCulture. www.safetyculture.com/topics/supply-chain-resilience/. Accessed 25 June 2024
24. Resilience in transport and logistics. www.mckinsey.com/capabilities/operations/our-insights/resilience-in-transport-and-logistics. Accessed 25 June 2024
25. Schlegel, G.L., Trent, R.J.: Supply Chain Risk Management: An Emerging Discipline. CRC Press (2014)

# Sustainable Sugarcane Supply Chain Network Design with Multiple Capacity Candidate in Facility Units

Keisuke Nagasawa[✉], Katsumi Morikawa, and Katsuhiko Takahashi

Hiroshima University, Kagamiyama 1-4-1, Higashi-Hiroshima 739-8527, Japan
nagasa-kei@hiroshima-u.ac.jp

**Abstract.** The sugarcane supply chain is rapidly gaining attention because unstable and rising fossil fuel prices, environmental limitations, and the energy needs of growing economies in developing countries are driving demand for clean, renewable energy. Bioenergy continues to attract attention from researchers because of its various advantages over other renewable energy sources. This paper developed a mixed-integer linear programming model (MILP) to design a sustainable sugarcane-based bioenergy supply chain network to optimize a facility location and transportation flow amount for decision variables. Candidate areas will be integrated into the mathematical optimization model as candidate sugarcane fields and sugarcane factories, which consist of a sugar unit, bioethanol unit, and bioelectricity unit, will comprise the proposed supply chain design. Demand locations and amounts of sugar, bioethanol, and bioelectricity are set for each, and the supply chain is designed to satisfy or shortage those demands. Each unit has candidates and lower limits rate of capacity from which to choose appropriately and design the supply chain. The results confirm the usefulness of considering the capacity candidate in facility units. It was observed that the introduction of a lower limit capacity constraint changes the composition and costs of the supply chain.

**Keywords:** Supply chain management · Sugarcane · Sustainability

## 1 Introduction

Sugarcane production is actively conducted in India, where the population and economy are growing rapidly, and also in Vietnam [1]. In addition to processed sugar, which is the main product of the sugarcane industry, by-products such as bagasse, molasses, and filter cake are produced in the processing process at factories. Currently, these by-products are often disposed of as is due to the lack of sufficient facilities. However, bagasse and molasses can be reused as raw materials for other industries, and filter cake can be reused as fertilizer for agricultural products, leading to a reduction in the total cost of the entire supply chain, effective use of resources, and improved environmental measures [2].

In fact, sugarcane bagasse and bagasse-derived bioethanol are being used as renewable biofuels in Brazil and other countries under the initiative of the national government.

H. Kohl et al. (Eds.): GCSM 2024, LNME, pp. 742–751, 2025.
https://doi.org/10.1007/978-3-031-93891-7_82

As the study by Husam et al. [3], some researches exist that consider Sugar-cane supply chain with scenarios and design for robustness.

Sugarcane juice and molasses (a coproduct derived during processing sugarcane) have been widely used for producing transportation fuel. In Brazil, fuel from sugar-cane juice covers about 40% of light vehicles' fuel needs (Leal et al., [4]). Also, the sugarcane residues are used for bioelectricity generation by utilizing co-generation technologies to satisfy the demand in the sugarcane factory and the surplus is sold to increase the economic returns (Fuess et al., [5]; Abdali et al., [6]).

In this study, the sugarcane supply chain network (SSCN) is treated which considering multiple capacity candidate based on mathematical formulation model of Abdali et al. [3]. In this study, additional conditions are taken into consideration. Cost aspects are important when operating a factory, but costs and minimum capacity rates need to be introduced to maintain operation. Abdali et al. [3] proposed SSCN model only considering capacity level of every unit without lower limit capacity. Their study treated the capacities of every facilities independently, it would have been better to treat each capacity simultaneously and list them as candidates, which would have reduced the number of candidate combinations to be treated. In addition, they did not set a lower limit capacity constraint that is set when a facility is in operation. Therefore, we proposed mathematical model, which considered capacity candidate and lower limits of capacity.

## 2 Problem Definition

In this study, we propose a supply chain design model that considers the costs and unsatisfied demand from the opening and operation of the fields, facilities and unit for demand zone shown in Fig. 1. Sugarcane fields and factories will be considered for construction from among the candidate location. After harvesting the sugarcane in the selected locations, it will be sent to the sugarcane factories. In the factories, the sugarcane is cleaned and prepared for crushing to extract the sugar juice from a dry pulpy residue called bagasse (Tables 1, 2, 3, 4, 5, and 6).

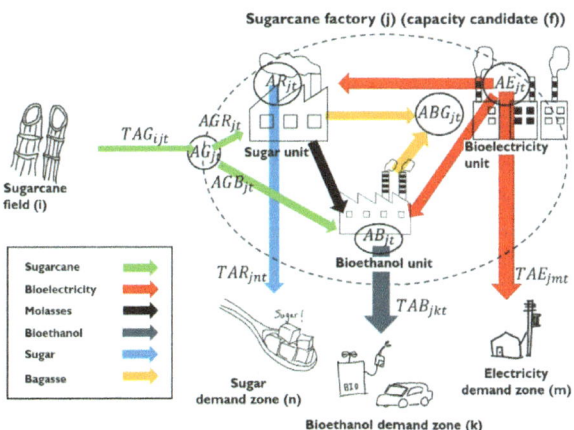

**Fig. 1.** The structure of the sugarcane supply chain network

**Table 1.** Index and Sets

| Index and Set | Definition |
|---|---|
| $i \in I, j \in J$ | Candidate locations of sugarcane fields and factories |
| $f \in F(, F^j)$ | Capacity candidate of bioethanol production, electricity generation and sugar production unit at factory j |
| $n \in N, k \in K, m \in M$ | Demand zones of sugar, bioethanol and electricity |
| $t \in T$ | Time periods |

**Table 2.** Parameters

| | Definition |
|---|---|
| $\partial$ | Yield of sugarcane |
| $\theta, \varphi, \omega, \Phi$ | Conversion factor of sugarcane to sugar, sugarcane to molasses, sugarcane to bioethanol and sugarcane to bagasse |
| $\varpi$ | Conversion factor of molasses to bioethanol |
| $\beta$ | Conversion factor of bagasse to bioelectricity |
| $CAP^1_{jf}, CAP^2_{jf}, CAP^3_{jf}$ | Maximum capacity of bioethanol production, bioelectricity generation and sugar production of the unit at sugarcane factory j of candidate f |
| $r_{ethanol} r_{electricity} r_{sugar}$ | Lower limit capacity rate of the unit at sugarcane factory for bioethanol production, Bioelectricity generation and Sugar production |
| $CAP5_j$ | Storage capacity of sugarcane at sugarcane factory j |
| $CAP6_j$ | Storage capacity of bioethanol at sugarcane factory j |
| $CAP7_j$ | Storage capacity of sugar at sugarcane factory j |
| $CAP8_j$ | Storage capacity of bagasse at sugarcane factory j |
| $DSC^1_{ij}$ | Distance between sugarcane field i and sugarcane factory j |
| $DSC^2_{jk}$ | Distance between sugarcane factory j and bioethanol demand zone k |
| $DSC^3_{jn}$ | Distance between sugarcane factory j and bioethanol demand zone n |
| $hlss$ | Proportion of harvesting loss of sugarcane |
| $glss$ | Proportion of weight loss of sugarcane during storage |
| $D1_{kt}$ | Bioethanol demand of demand zone k at period t |
| $D2_{mt}$ | Electricity demand of demand zone m at period t |
| $D3_{nt}$ | Sugar demand of demand zone n at period t |
| $IE_j$ | Amount of electricity consumed for processing sugarcane into products at sugarcane factory j |
| $LT_i$ | The available arable land at location i |
| $U$ | Big number |

**Table 3.** Economic Parameters

|  | Definition |
|---|---|
| $C_{1jf}$ | Cost of establishing sugarcane factory j at candidate f |
| $C_2$ | Land rental cost for sugarcane cultivation |
| $C_3, C_4, C_5, C_6$ | Cost of sugarcane cultivation, bioethanol production, bioelectricity production and sugar production |
| $C_7, C_8, C_9$ | inventory holding cost of sugarcane (with bagasse), bioethanol and sugar |
| $C_{10}, C_{11}, C_{12}$ | Transportation cost of sugarcane, bioethanol and sugar |
| $C_{13}$ | Cost of diesel consumption |
| $C_{14}$ | Cost electricity consumption |
| $C_{15}$ | Cost of water consumed for sugarcane cultivation |
| $C_{16}$ | Cost of water consumed for processing sugarcane |
| $C_{17}$ | Cost of CO2-eq emissions |
| $P_1, P_2, P_3$ | Sale price of bioethanol, bioelectricity and sugar |

**Table 4.** Environmental parameters

|  | Definition |
|---|---|
| $e_1$ | Diesel fuel utilization to cultivate sugarcane per unit area |
| $e_2, e_3$ | Diesel fuel utilization to transport sugarcane/sugar, bioethanol |
| $e_4$ | Electricity utilization to process sugarcane into final products |
| $E_1$ | Energy intensity of diesel fuel (MJ/ton) |
| $E_2$ | Energy intensity of diesel fuel (MJ/MW) |
| $g_1, g_2 g_3$ | CO2-eq emissions coefficient of cultivation, diesel fuel and processing sugarcane |
| $w_{1i}$ | Sugarcane water requirement at field i |
| $w_{2j}$ | Amount of water required for processing sugarcane at factory j |

**Table 5.** Binary variables

|  | Definition |
|---|---|
| $Y_{jf}$ | 1 if location j is selected for establishing sugarcane factory with capacity candidate f; 0 otherwise |
| $X_i$ | 1 if location i is selected for opening sugarcane field; 0 otherwise |

# 3   Proposed Model

The objective function and constraints in this research are as follows

**Table 6.** Continuous variables

| | Definition |
|---|---|
| $LA_i$ | sugarcane cultivated area at location i |
| $HG_{it}$ | Amount of harvested sugarcane at location i at period t |
| $TAG_{ijt}$ | Amount of sugarcane transported from sugarcane field at location i to sugarcane factory at location j at period t |
| $TAR_{jnt}$ | Amount of sugar transported from sugarcane factory j to sugar demand zone n at period t |
| $TAE_{jmt}$ | Amount of energy supplied from sugarcane factory j to electricity demand zone m at period t |
| $TAB_{jkt}$ | Amount of bioethanol transported from sugarcane factory j to bioethanol demand zone k at period t |
| $AG_{jt}$ | Amount of sugarcane used for production at sugarcane factory j at period t |
| $AGR_{jt}$ | Amount of sugarcane allocated for sugar production at sugarcane factory j at period t |
| $AGB_{jt}$ | Amount of sugarcane allocated for bioethanol production at sugarcane factory j at period t |
| $ABG_{jt}$ | Amount of resulted bagasse at sugarcane factory j at period t |
| $AR_{jt}$ | Amount of produced sugar at sugarcane factory j at period t |
| $AB_{jt}$ | Amount of produced bioethanol at sugarcane factory j at period t |
| $AE_{jt}$ | Amount of produced bioelectricity at sugarcane factory j at period t |
| $AM_{jt}$ | Amount of extracted molasses at sugarcane factory j at period t |
| $IAG_{jt}$ | Amount of inventory sugarcane at sugarcane factory j at period t |
| $IG_{jt}$ | Amount of inventory bagasse at sugarcane factory j at period t |
| $BG_{jt}$ | Amount of bagasse allocated for bioelectricity generation at sugarcane factory j at period t |
| $IB_{jt}$ | Amount of inventory bioethanol at sugarcane factory j at period t |
| $IR_{jt}$ | Amount of inventory sugar at sugarcane factory j at period t |
| $E$ | Total energy consumption |
| $W$ | Total water consumption |
| $G$ | Total CO2-eq emissions |
| $SDB_{kt}$ | Amount of unsatisfied bioethanol demand in demand zone k at period t |
| $SDE_{mt}$ | Amount of unsatisfied electricity demand in demand zone m at period t |
| $SDR_{nt}$ | Amount of unsatisfied sugar demand in demand zone n at period t |

min.

$$\alpha(CN + CE + CW + CG - RE) + (1 - \alpha)SD$$

The objective function is a weighted sum of the sum of network costs (CN), energy consumption costs (CE), water consumption costs (CW) and CO2 emission costs (CG), minus revenue (RE), with total shortages (SD).

$$CN = \sum_{j,f} C_{1f} Y_{jf} + \sum_{i} (C_2 + C_3) LA_i + \sum_{j,t} C_4 AB_{jt} + \sum_{j,t} C_5 AE_{jt} + \sum_{j,t} C_6 AR_{jt}$$

$$+ \sum_{j,t} C_7 (IAG_{jt} + IG_{jt}) + \sum_{j,t} C_8 IB_{jt} + \sum_{j,t} C_9 IR_{jt} + \sum_{i,j,t} C_{10} TAG_{ijt} DSC_{ij}^1$$

$$+ \sum_{j,k,t} C_{11} TAB_{jkt} DSC_{jk}^2 + \sum_{j,n,t} C_{12} TAR_{jnt} DSC_{jn}^3$$

$$CE = \left( \begin{array}{c} \sum_i e_1 LA_i + \sum_{i,j,t} e_2 TAG_{ijt} DSC_{ij}^1 \\ + \sum_{j,n,t} e_2 TAR_{jnt} DSC_{jn}^3 + \sum_{j,k,t} e_3 TAB_{jkt} DSC_{jk}^2 \end{array} \right) C_{13} + \sum_{j,t} e_4 AG_{jt} C_{14}$$

$$CW = \sum_i C_{15} w_{1i} LA_i + \sum_{j,t} C_{16} w_{2j} AG_{jt}$$

$$CG = \left( \sum_i g_1 LA_i + \left( \sum_{i,j,t} e_2 TAG_{ijt} DSC_{ij}^1 + \sum_{j,n,t} e_2 TAR_{jnt} DSC_{jn}^3 + \sum_{j,k,t} e_3 TAB_{jkt} DSC_{jk}^2 \right) g_2 + \sum_{j,t} g_2 AG_{jt} \right) C_{17}$$

$$RE = \sum_{j,n,t} P_1 TAR_{jnt} + \sum_{j,k,t} P_2 TAB_{jkt} \sum_{j,m,t} P_3 TAE_{jmt}$$

$$SD = \sum_{k,t} SDB_{kt} + \sum_{m,t} SDE_{mt} + \sum_{n,t} SDR_{nt}$$

subject to

$$LA_i \leq LT_i X_i \ \forall i \tag{1}$$

$$\sum_t HG_{it} = \partial LA_i \quad \forall i \tag{2}$$

$$HG_{it}(1 - hlss) = \sum_j TAG_{ijt} \quad \forall i, t \tag{3}$$

$$AG_{jt} = AGR_{jt} + AGB_{jt} \quad \forall j, t \tag{4}$$

$$AR_{jt} = \theta AGR_{jt} \quad \forall j, t \tag{5}$$

$$AM_{jt} = \varphi AGR_{jt} \quad \forall j, t \tag{6}$$

$$AB_{jt} = \varpi AM_{jt} + \omega AGB_{jt} \quad \forall j, t \tag{7}$$

$$ABG_{jt} = \Phi (AGR_{jt} + AGB_{jt}) \quad \forall j, t \tag{8}$$

$$AE_{jt} = \beta BG_{jt} \quad \forall j, t \tag{9}$$

$$\sum_i TAG_{ijt} + IAG_{jt-1}(1 - glss) = IAG_{jt} + AG_{jt} \quad \forall j, t \tag{10}$$

$$AR_{jt} + IR_{jt-1} = IR_{jt} + \sum_n TAR_{jnt} \quad \forall j, t \tag{11}$$

$$AB_{jt} + IB_{jt-1} = IB_{jt} + \sum_k TAB_{jkt} \quad \forall j, t \tag{12}$$

$$ABG_{jt} + IG_{jt-1} = IG_{jt} + BG_{jt} \quad \forall j, t \tag{13}$$

$$AE_{jt} - \left(IE_j AG_{jt}\right) = \sum_m TAE_{jmt} \quad \forall j, t \tag{14}$$

$$AB_{jt} \leq \sum_f CAP1_{jf} Y_{jf} \quad \forall j, t \tag{15}$$

$$AE_{jt} \leq \sum_f CAP2_{jf} Y_{jf} \quad \forall j, t \tag{16}$$

$$AR_{jt} \leq \sum_f CAP3_{jf} Y_{jf} \quad \forall j, t \tag{17}$$

$$IAG_{jt} \leq CAP5_j \quad \forall j, t \tag{18}$$

$$IB_{jt} \leq CAP6_j \quad \forall j, t \tag{19}$$

$$IR_{jt} \leq CAP7_j \quad \forall j, t \tag{20}$$

$$IG_{jt} \leq CAP8_j \quad \forall j, t \tag{21}$$

$$\sum_f Y_{jf} \leq 1 \quad \forall j \tag{22}$$

$$\sum_j TAB_{jkt} + SDB_{kt} \geq D1_{kt} \quad \forall k, t \tag{23}$$

$$\sum_j TAR_{jnt} + SDR_{nt} \geq D3_{nt} \quad \forall n, t \tag{24}$$

$$\sum_j TAE_{jmt} + SDE_{mt} = D2_{mt} \quad \forall m, t \tag{25}$$

$$E = E_1\left(\sum_i e_1 LA_i + \sum_{i,j,t} e_2 DSC_{ij}^1 TAG_{ijt} + \sum_{j,k,t} e_3 DSC_{jk}^2 TAB_{jkt} + \sum_{j,n,t} e_2 DSC_{jn}^3 TAR_{jnt}\right) \\ + \sum_{j,t} e_4 E_2 AG_{jt} \tag{26}$$

$$W = \sum_i w_{1i} LA_i + \sum_{j,t} w_{2j} AG_{jt} \tag{27}$$

$$G = \sum_i g_1 LA_i + g_2\left(\sum_{i,j,t} e_2 DSC_{ij}^1 TAG_{ijt} + \sum_{j,n,t} e_2 DSC_{jn}^3 TAR_{jnt} + \sum_{j,k,t} e_3 DSC_{jk}^2 TAB_{jkt}\right) \\ + \sum_{j,t} g_3 AG_{jt} \tag{28}$$

$$AB_{jt} \geq r_{ethanol} \sum_f CAP1_{jf} Y_{jf} \forall j, t \tag{29}$$

$$AE_{jt} \geq r_{electricity} \sum_f CAP2_{jf} Y_{jf} \forall j, t \tag{30}$$

$$AR_{jt} \geq r_{sugar} \sum_f CAP3_{jf} Y_{jf} \forall j, t \tag{31}$$

Constraints (1–4) indicate the relation of sugarcane cultivation area, harvested with loss of sugarcane and transported amount. Constraints (5–9) calculate the amounts of produced sugar, molasses, bioethanol, bagasse, and electricity at sugar cane factories. Constraints (10–14) balances inventory, production amount and transported demand at sugar factory units. Constraints (15–22) represents capacity constraints each sugar factory units. Constraints (23–25) represents the relation of total amount of production and shortage larger than the demand. Constraints (26–28) defines the consumption of energy, water and $CO_2$ emission. Constraints (29–31) represents lower limit capacity rate constraints of each units.

## 4  Results and Conclusion

For solving mathematical model of the last section, the mathematical optimization solver Gurobi Optimizer is used. Several numerical example can be solved. As in previous studies [3], geographical data for Iraq were used to show the composition of the supply chains obtained.

As a first experiment, when only the lower limit rate of capacity of factories unit ($r_{ethanol}$, $r_{electricity}$, $r_{sugar}$) was changed, the revenue in the supply chain were identified while parameters such as candidate capacity and cost were not varied. As shown in Fig. 2, when the lower limit of capacity constraint is introduced and made tighter, the revenue will decrease. If the lower limit rate of capacity is not related to the composition of the supply chain or to revenue, then the bar chart of revenue should remain unchanged. Therefore, the fact that revenue vary depending on the lower capacity limit means that the lower capacity limit must also be taken into account at the stage of selecting candidate capacities.

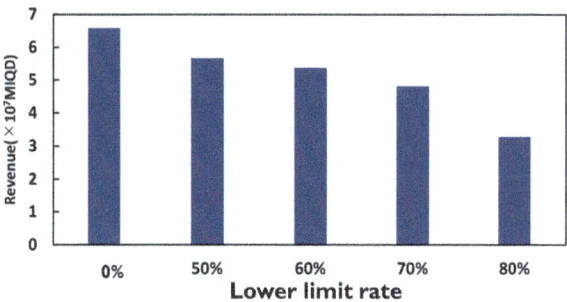

**Fig. 2.** Lower limit rate of capacity and Revenue

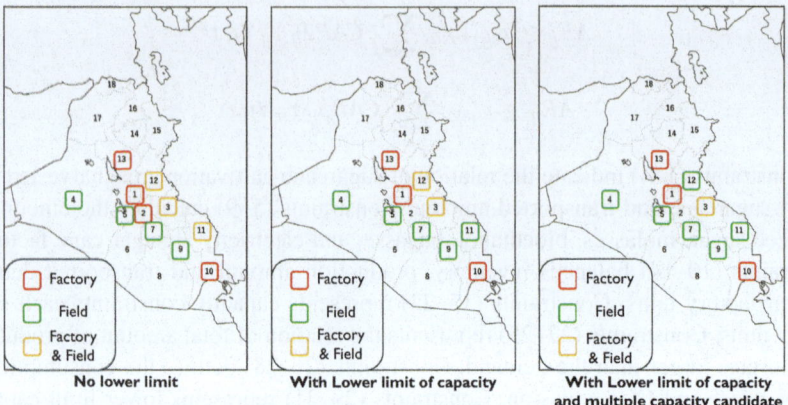

**Fig. 3.** Facility location with Lower limit of capacity and multiple capacity candidate

Figure 3 shows the results of the optimization for three cases on a map. The number from #1 to #18 are candidate of factory, sugarcane field or both factory and field. The numbers with red boxes represent location of factory, the green boxes represent location of sugarcane fields and the yellow boxes represent location of both factories and sugarcane fields constituting of each optimization results at supply chain in each case. As shown in Fig. 3, considering lower limit of capacity constraints changes the arrangement of sugarcane fields and facilities. For example, when there was no lower capacity limit, location #2 resulted a locating factory, but when there was a lower capacity limit, nothing being located. Also, when multiple candidate facility capacities are used, factories will be concentrated in some areas and no factories will be built in some areas. The result was that location #12 had both a factory and a sugarcane field built, but when there was a consideration of lower limit capacity and capacity candidate, it should only be a sugarcane field. The results suggest that the relationship between the lower limit capacity of the plant and the potential location should be carefully considered when selecting a candidate.

A future work is to use some indicators more realistic or considerable values. The exact parameters values of economic, environmental indicators, carbon footprint and water consumption in sugarcane fields, were not available in optimization. It is further important to consider how to select candidate and construct supply chains so that the sugarcane supply chain has a sustainable and resilient configuration.

**Acknowledgement.** This research was partially supported by the Japan Society for the Promotion of Science (JSPS), KAKENHI for Grant-in-Aid for Scientific Research (B) JP23K26329 from 2023 to 2026.

## References

1. Nguyen, T.T., et al Research and development prospects for sugarcane industry in Vietnam. Sugar Tech, **24**(5), 1330–1341 (2022)

2. Chouhan, V.K., Khan, S.H., Hajiaghaei-Keshteli, M.: Metaheuristic approaches to design and address multi-echelon sugarcane closed-loop supply chain network. Soft. Comput. **25**(16), 11377–11404 (2021)
3. Abdali, H., Sahebi, H., Pishvaee, M.: A sustainable robust optimization model to design a sugarcane-based bioenergy supply network: a case study. Chem. Eng. Res. Des. **180**, 265–284 (2022)
4. Leal, M.R.L., Walter, A.S., Seabra, J.E.: Sugarcane as an energy source. Biomass Convers. Biorefinery **3**(1), 17–26 (2013)
5. Fuess, L.T., de Araújo Júnior, M.M., Garcia, M.L., Zaiat, M.: Designing full-scale biodigestion plants for the treatment of vinasse in sugarcane biorefineries: how phase separation and alkalinization impact biogas and electricity production costs? Chem. Eng. Res. Des. **119**, 209–220 (2017)
6. Abdali, H., Sahebi, H., Pishvaee, M.: The water-energy-food-land nexus at the sugarcane-to-bioenergy supply chain: a sustainable network design model. Comput. Chem. Eng. **145**, 107199 (2021)

# Factors Affecting Halal Supply Chain Implementation in Food SMIs to Achieve Sustainable Performance

Meuthia Meuthia[1] ⓘ, Nilda Tri Putri[1(✉)] ⓘ, Rini Rahmahdian S[1] ⓘ,
Iwan Vanany G[2] ⓘ, and Muhammad Zaky Almazidan[1]

[1] Universitas Andalas, Padang, Indonesia
nildatp@eng.unand.ac.id
[2] Institut Teknologi Sepuluh Nopember, Surabaya, Indonesia

**Abstract.** The purpose of this research is to analyze the factors that support the application of halal supply chain management to sustainable performance (SP). The background of this research is due to the low knowledge and commitment of food SMEs to implementing the halal supply chain. The research method used is Structural Equation Modeling (SEM) with the Partial Least Squares (PLS) approach. The sample used in the study was food SMEs. The study used a significance level of 0.1% with a t-statistic value of 1.289 (df = 102). The results showed a positive and significant influence of the halal product demand factor and the integration of supply chain partners in the implementation of the halal supply chain on sustainable performance. In addition, the results also show that the most influential aspect of the sustainable performance dimension is the economic aspect. The contribution of the economic aspect is because halal food products will increase profits for food businesses. This research is expected to be a solution in implementing the halal supply chain to ensure halal products are marketed and develop businesses to support sustainable performance.

**Keywords:** Halal · IKM · Structural Equation Modeling-Partial Least Square (SEM-PLS) · Sustainable Performance (SP) · Supply Chain Management (SCM)

## 1 Introduction

Indonesia, with the world's largest Muslim population, has a significant demand for halal food. In 2020, 1.51 million small and medium industries (SMIs) focused on food products [1]. Research on halal food emphasizes the relationship between purchasing, marketing, and production processes [2]. In West Sumatra, 2,961 halal-certified products are produced by SMIs in Padang, with food SMIs being dominant. The global demand for halal products has led to the development of Halal Supply Chain Management (HSCM) by food SMIs. Factors related to implementation include the demand for halal products, marketing, quality improvement, management support, supply chain integration, competitive pressure, and government support [2]. However, the current focus is on halal products in general rather than specific products like food and beverages.

H. Kohl et al. (Eds.): GCSM 2024, LNME, pp. 752–759, 2025.
https://doi.org/10.1007/978-3-031-93891-7_83

The research by [2] focused on India-based companies trading Halal products. It gathered data from professionals in the Halal industry, especially those in the food and beverage sector. The study emphasized the importance of halal certification bodies, resources, production systems, ICT integration, and human resources management for halal supply chain management. Government support and integrated networks are key for halal supply chains in small and medium enterprises (SMIs), aiming for sustainability. Ensuring halal compliance throughout production and labeling is critical. While Halal Supply Chain Management (HSCM) offers growth opportunities and affects Halal product demand, achieving consistent sustainable performance remains a challenge for SMIs [2]. Our research examines the impact of HSCM-supporting factors on the sustainable performance of food SMIs and provides statistical information for relevant policy-making.

## 2 Literature Review and Hypothesis Development

### 2.1 Demand for Halal Products (DHP)

Halal and healthy foods avoid substances like blood and carrion, catering to the 24.9% of Muslims worldwide, as per Pew Research Center [3]. This demand grows with the Muslim population, setting higher standards for food safety and hygiene globally [4]. Interestingly, halal products are gaining popularity among non-Muslims too, boosting demand and enhancing business sustainability [2, 5].

H1. There is a significant and positive correlation between DHP and SP.

### 2.2 Halal Marketing (HM)

Halal Marketing (HM) involves promoting products and services following Islamic Sharia laws using strategies like word-of-mouth, email, SMS, and social media [6]. The aim is to foster responsible consumption and positive relationships between customers and companies [7]. It's crucial for both Muslims and non-Muslims to grasp Halal principles to devise effective marketing and production plans [8]. In the global market, the role of food labeling and packaging is significant [9]. Employing Halal sustainable marketing and encouraging Halal food adoption are key strategies for drawing in new customers and enhancing consumer trust [10].

H2. There is a significant and positive correlation between HM and SP.

### 2.3 Process Quality Improvement (PQI)

The Quality Management System (QMS) benefits halal food producers by building trust, reducing costs, and boosting profitability [7]. Quality improvement leads to better products and services, satisfying consumers [11]. Sustainable practices enhance product quality and contribute to waste management. Research by [7] shows the impact of Process Quality Improvement (PQI) on company sustainability.

H3. There is a significant and positive correlation between PQI and SP.

## 2.4  Management Commitment and Support (MCS)

Halal supply chain management adoption and management commitment and support are imperative to ensure the success and integrity of a halal supply chain. It also encompasses the determination and involvement of the top management in ensuring that the products and services produced by the company are aligned with halal principles [4]. The management must actively monitor and evaluate halal compliance in the company's supply chain and take immediate actions once finding an infringement [12].

H4. There is a significant and positive correlation between MCS and SP.

## 2.5  Integration of Supply Chain Partners (SCI)

Supply Chain Integration (SCI) reduces the risk to halal food integrity by minimizing waste, promoting efficiency, and advocating eco-friendly practices [13–16]. Successful integration of supply chain partners enhances operational efficiency, productivity, and business sustainability (research from [17]).

H5. There is a significant and positive relationship between SCI and SP.

## 2.6  Competitive Pressure (CP)

Competitive pressure is essential for companies to survive and thrive. Managers should implement ethical practices to maintain competitiveness. The government needs to consult with key stakeholders when developing regulatory policies. This will help ensure company sustainability and long-term survivability [2].

H6. There is a significant and positive correlation between CP and SP.

## 2.7  Government Support (GS)

Halal Supply Chain Management (HSCM) is an emerging area that benefits from government support, such as regulation, planning, and promotion of halal practices [2]. Indonesia, for example, has created certification bodies to facilitate halal compliance. Government incentives like financial aid and tax incentives can drive the adoption of HSCM, fostering economic, social, and environmental development.

H7. There is a significant and positive correlation between GS and SP.

## 2.8  Sustainable Performance (SUSTP)

Sustainable Performance (SUSTP) aims for a balance between economic, social, and environmental aspects to ensure sustainability [18]. Integrating sustainability into business operations not only boosts financial outcomes, lowers production costs, and expands market share but is also crucial in the food industry for enhancing supply chain efficiency and reducing costs [19–24]. Specifically, for Halal Supply Chain (HSC) practices, aligning economic activities with halal standards, especially in logistics, is essential [25]. Sustainability in the food sector focuses on social responsibilities such as fair labor and human rights [26–28], while in Halal Supply Chain Management (HSCM), environmental sustainability is crucial for maintaining product integrity and reducing environmental impact [25, 29].

## 3   Theoretical Framework and Methodology

Literature shows a positive effect of DHP, HM, PQI, MCS, SCI, and GS factors on SUSTP. Figure 2 suggests a research framework.

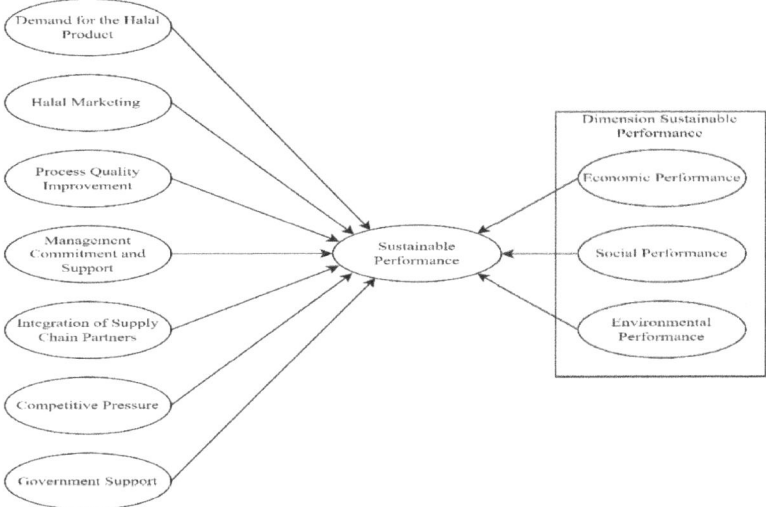

**Fig. 2.**  Theoretical Framework

A quantitative method was used to gather data from 103 small and medium-sized food industries (SMIs) in Padang. The data was collected through questionnaires distributed using purposive sampling. The variables were measured on a 1–5 Likert scale, and the data was analyzed using SmartPLS to test hypotheses on adopting a halal supply chain and sustainability. The study focused on SMEs in Indonesia, which lack the infrastructure for testing and logistics to separate halal and non-halal components, making it difficult to prevent contamination.

## 4   Findings

### 4.1   Measurement Model

A measurement model was used to test the correlation between indicators and the latent variable [30]. Table 1 demonstrates the composite reliability and convergent validity used to evaluate the measurement model. The discriminant validity implied "the degree to which the measure could be adequately differentiated from related constructs in the nomological net" [30]. Fornell and Larcker proposed that the discriminant validity was determined using the square root of the latent construct (Table 2).

The R-squared, or coefficient of determination [30], shows how much variation in a variable is explained by predictor factors. An R-squared of 0.951 in PLS-SEM indicates that the research model explains 95% of the variance in SUSTP.

**Table 1.** Item Loadings, Composite Reliability (CR), and Average Variance Extracted (AVE)

| Variables | Code | Outer loading | AVE | CR | Variables | Code | Outer loading | AVE | CR |
|---|---|---|---|---|---|---|---|---|---|
| Demand for Halal Products (DHP) | A1 | 0.743 | 0.63 | 0.836 | **Competitive Pressure (CP)** | F1 | 0.78 | 0.55 | 0.785 |
| | A2 | 0.864 | | | | F2 | 0.711 | | |
| | A3 | 0.769 | | | | F3 | 0.731 | | |
| | A4 | deleted | | | **Integration of Supply Chain Partners (SCI)** | E1 | 0.727 | 0.56 | 0.789 |
| Halal Marketing (HM) | B1 | 0.591 | 0.51 | 0.754 | | E2 | 0.69 | | |
| | B2 | deleted | | | | E3 | deleted | | |
| | B3 | deleted | | | | E4 | 0.815 | | |
| | B4 | 0.852 | | | | E5 | deleted | | |
| | B5 | 0.677 | | | **Economic Performance (EP)** | H1 | 0.807 | 0.59 | 0.811 |
| Process Quality Improvement (PQI) | C1 | deleted | 0.6 | 0.812 | | H2 | 0.867 | | |
| | C2 | 0.955 | | | | H3 | 0.613 | | |
| | C3 | 0.507 | | | | H4 | deleted | | |
| | C4 | deleted | | | | H5 | deleted | | |
| | C5 | 0.801 | | | **Social Performance (SP)** | I1 | deleted | 0.7 | 0.816 |
| Management Commitment Support (MCS) | D1 | 1.000 | 1.000 | 1.000 | | I2 | deleted | | |
| | D2 | deleted | | | | I3 | deleted | | |
| | D3 | deleted | | | | I4 | 0.676 | | |
| | D4 | deleted | | | | I5 | 0.967 | | |
| Government Support (GS) | G1 | 0.762 | 0.7 | 0.822 | **Environmental Performance (ENVP)** | J1 | deleted | 0.61 | 0.756 |
| | G2 | deleted | | | | J2 | 0.704 | | |
| | G3 | 0.904 | | | | J3 | deleted | | |

**Table 2.** Fornell-Larcker's Criterion

| | 1 | 2 | 3 | 4 | 5 | 6 | 7 | 8 | 9 | 10 |
|---|---|---|---|---|---|---|---|---|---|---|
| DHP | **0.794** | | | | | | | | | |
| HM | 0.404 | **0.715** | | | | | | | | |
| PQI | 0.272 | 0.446 | **0.777** | | | | | | | |
| MCS | 0.261 | 0.239 | 0.287 | **1.000** | | | | | | |
| SCI | 0.398 | 0.340 | 0.416 | 0.141 | **0.746** | | | | | |
| CP | 0.379 | 0.338 | 0.333 | 0.296 | 0.315 | **0.741** | | | | |

*(continued)*

**Table 2.** (*continued*)

|       | 1     | 2     | 3     | 4     | 5     | 6     | 7         | 8         | 9         | 10        |
|-------|-------|-------|-------|-------|-------|-------|-----------|-----------|-----------|-----------|
| GS    | 0.147 | 0.229 | 0.122 | 0.140 | 0.164 | 0.351 | **0.836** |           |           |           |
| EP    | 0.322 | 0.329 | 0.299 | 0.144 | 0.352 | 0.255 | 0.237     | **0.770** |           |           |
| SP    | 0.297 | 0.405 | 0.296 | 0.053 | 0.310 | 0.180 | 0.142     | 0.139     | **0.834** |           |
| ENVP  | 0.630 | 0.365 | 0.156 | 0.182 | 0.292 | 0.202 | 0.035     | 0.212     | 0.279     | **0.781** |

## 4.2  Structural Model

The Q-square (Q2) method evaluates how accurately a structural model predicts observed data, using a blindfolding technique [30]. For the sustainable performance factor, the Q2 value is 0.656. This study also used bootstrapping with 5,000 subsamples and 103 cases to assess the path coefficient's significance The findings are demonstrated in Table 3.

**Table 3.** Hypothesis Test Results

| H  | Relations       | β      | t-value | p-value   | Findings |
|----|-----------------|--------|---------|-----------|----------|
| 1  | DHP - > SUSTP   | 0.060  | 1.692   | **0.046** | Accepted |
| 2  | HM - > SUSTP    | −0.026 | 1.011   | 0.156     | Rejected |
| 3  | PQI - > SUSTP   | 0.026  | 0.845   | 0.199     | Rejected |
| 4  | MCS - > SUSTP   | 0.028  | 1.083   | 0.140     | Rejected |
| 5  | SCI - > SUSTP   | 0.062  | 2.077   | **0.019** | Accepted |
| 6  | CP - > SUSTP    | −0.012 | 0.538   | 0.295     | Rejected |
| 7  | GS - > SUSTP    | 0.001  | 0.045   | 0.482     | Rejected |
| 8  | EP - > SUSTP    | 0.977  | 60.314  | **0.000** | Accepted |
| 9  | SP - > SUSTP    | −0.002 | 0.076   | 0.470     | Rejected |
| 10 | ENVP - > SUSTP  | 0.002  | 0.058   | 0.477     | Rejected |

# 5  Discussion and Conclusion

Various research studies have shown that the Demand for Halal Products (DHP) and Integration of Supply Chain Partners (SCI) significantly affect the sustainable performance of food SMIs in Padang [2]. Additionally, variables like halal marketing did not affect sustainable performance [17; 7]. The study found that 41.7% of business units sought halal certification in 2023, emphasizing halal practices. Government support is crucial for halal supply chains, with economic interests often outweighing social and environmental ones. The lack of environmental awareness leads to few sustainable practices being adopted.

Two crucial factors for achieving sustainable performance in small and medium-sized food enterprises through holistic supply chain management (HSCM) are the Demand for Halal Products (DHP) and the Integration of Supply Chain Partners (SCI). The study found that only economic performance significantly impacted HSCM implementation by food SMIs. These results provide valuable insights for improving sustainable performance in food SMIs. Additionally, this research, focused specifically on food SMIs, yielded results different from those of a study on general products.

**Acknowledgement.** We gratefully acknowledge financial support from Faculty of Engineering and the Institute of Research and Community Service, Universitas Andalas as well, which funded this research under the grant of Penelitian Unggulan Jalur Kepakaran 2024 no. 360/UN16.19/PT.01.03/PUJK/2024.

# References

1. BPS I. Statistik Penyedia Makanan dan Minuman (2020)
2. Khan, M.I., Haleem, A., Khan, S.: Examining the link between halal supply chain management and sustainability. Int. J. Prod. Perform. Manage. **71**(7), 2793–2819 (2022)
3. Nurlatifah, H., Saefuddin, A., Nanere, M., Ratten, V.: Muslimpreneur: entrepreneur potential characteristics in Indonesia as the country with the largest Muslim population in the world. In: Entrepreneurial Innovation Springer, Cham (2022)
4. Bakar, A., Hamid, A., Syazwan, M., Talib, A.: Halal logistics: a marketing mix perspective. Intellect Discourse **22**(2), 191–214 (2014)
5. Bashir, A.: Awareness of purchasing halal food among non-Muslim consumers: an explorative study with reference to Cape Town of South Africa. J Islam Mark. **11**(6), 1295–1311 (2019)
6. Temporal, P.: Islamic Branding and Marketing: Creating a Global Islamic Business. Wiley, New York (2011)
7. Mabkhot, H.: Factors affecting the sustainability of halal product performance: Malaysian evidence. Sustainability **15**(3) (2023)
8. Shah Alam, S., Mohamed, S.N.: Applying the theory of planned behavior (TPB) in halal food purchasing. Int. J. Commer. Manag. **21**(1), 8–20 (2011)
9. Talib, M.S.A., Johan, M.R.M.: Issues in halal packaging: a conceptual paper. Int. Bus. Manag. **5**(2), 94–98 (2012)
10. Haleem, A., Khan, M.I., Khan, S.: Understanding the adoption of halal logistics through critical success factors and stakeholder objectives. Logistics **5**(2), 38 (2021)
11. Birch-Jensen, A., Gremyr, I., Halldorsson, A.: Digitally connected services: improvements through customer-initiated feedback. Eur. Manage. J. **38**(5), 814–825 (2020)
12. Silalahi, S.A.F., Fachrurazi, F., Fahham, A.M.: Factors affecting intention to adopt halal practices: case study of Indonesian small and medium enterprises. J Islam Mark. **13**(6), 1244–1263 (2022)
13. Ali, M., Tan, K., Pawar, K., Makhbul, Z.: Extenuating food integrity risk through supply chain integration: the case of halal food. Ind. Eng. Manage. Syst. **13**, 154–162 (2014)
14. Raman, R., Sreenivasan, A., Ma, S., Patwardhan, A., Nedungadi, P.: Green supply chain management research trends and linkages to UN sustainable development goals. Sustainability **15**(22), 15848 (2023)
15. Lee, R.: The effect of supply chain management strategy on operational and financial performance. Sustainability **13**(9), 5138 (2021)

16. Trivellas, P., Malindretos, G., Reklitis, P.: Implications of green logistics management on sustainable business and supply chain performance: evidence from a survey in the greek agri-food sector. Sustainability **12**(24), 10515 (2020)
17. Khanuja, A., Jain, R.K.: The mediating effect of supply chain flexibility on the relationship between supply chain integration and supply chain performance. J. Enterp. Inf. Manage. **35**(6), 1548–1569 (2021)
18. Seuring, S., Müller, M.: From a literature review to a conceptual framework for sustainable supply chain management. J. Clean. Prod. **16**(15), 1699–1710 (2008)
19. Adeyeye, A.R., Owen, N.: Sustainable performance in organizations: exploring the key drivers. J. Sustain. Dev. World Ecol. **20**(6), 491–505 (2013)
20. Schaltegger, S., Wagner, M.: Managing the Business Case for Sustainability: The Integration of Social, Environmental and Economic Performance. Routledge (2017)
21. Moreno-Camacho, C.A., et al.: Sustainability metrics for real case applications of the supply chain network design problem: a systematic literature review. J. Clean. Prod. **231**, 600–618 (2019)
22. An, K., Ouyang, Y.: Robust grain supply chain design considering post-harvest loss and harvest timing equilibrium. Transp. Res. Part E: Logistics Transp. Rev. **88**, 110–128 (2016)
23. Ojo, O.O., et al.: Potential impact of industry 4.0 in sustainable food supply chain environment. In: 2018 IEEE International Conference on Technology Management, Operations and Decisions (ICTMOD), pp. 172–177. IEEE (2018)
24. Rejeb, A., Rejeb, K., Zailani, S.: Are halal food supply chains sustainable: a review and bibliometric analysis. J. Foodserv. Bus. Res. **24**(5), 554–595 (2021)
25. Kurniawati, D.A., Cakravastia, A.: A review of halal supply chain research: sustainability and operations research perspective. Clean. Logistics Supply Chain **6**, 100096 (2023)
26. Prashar, A., et al.: A bibliometric and content analysis of sustainable development in small and medium-sized enterprises. J. Clean. Prod. **245**, 118665 (2020)
27. Zhu, L., Hu, D.: Sustainable logistics network modeling for enterprise supply chain. Math. Probl. Eng. **2017**(1), 9897850 (2017)
28. Catlin, J.R., Luchs, M.G., Phipps, M.: Consumer perceptions of the social vs. environmental dimensions of sustainability. J. Consum. Policy **40**(3), 245–277 (2017)
29. Ahmad, S., Wong, K.Y.: Development of weighted triple-bottom line sustainability indicators for the Malaysian food manufacturing industry using the delphi method. J. Clean. Prod. **229**, 1167–1182 (2019)
30. Hair, J., Hult, T., Ringle, C., Sarstedt, M.: A Primer on Partial Least Squares Structural Equation Modeling (PLS-SEM) (3th ed.). Sage, Los Angeles (2022)

# Factory Planning and Production Management

# Systematic AI Potential Analysis for Sustainable Rough Factory Planning

Dominik Kürpick[1]([⊠]), Jan-Philipp Disselkamp[1], Jonas Lick[1], Aschot Hovemann[1], and Roman Dumitrescu[2]

[1] Fraunhofer IEM, Zukunftsmeile 1, 33102 Paderborn, Germany
dominik.kuerpick@iem.fraunhofer.de
[2] University Paderborn - HNI, Fürstenallee 1, 33102 Paderborn, Germany

**Abstract.** Current megatrends are influencing industrial production and leading to ever shorter innovation cycles. The resulting fast pace of production requirements requires an accelerated development of production systems and an associated increase in efficiency in factory planning. Due to its knowledge-intensive activities, rough factory planning promises great potential to be supported in its activities by innovative technologies such as artificial intelligence. However, industrial companies face the challenge to recognize the potential of artificial intelligence (AI) in rough planning and to evaluate possible applications in their business context. As a result, a systematic approach for analyzing AI potential in rough factory planning was developed as part of this work. The system includes a procedural model and several artefacts used in it, which support the identification and evaluation of AI potential in organizations. This approach not only streamlines the planning process but also aligns with sustainable manufacturing principles by enhancing resource efficiency, promoting intelligent system design, and fostering innovation in product development and manufacturing processes.

**Keywords:** Rough Factory Planning · Artificial Intelligence · Potential Analysis

## 1 Introduction

The megatrends of globalization, digitalization and sustainability are forcing industrial companies to change their attitudes and practices. The manufacturing industry is subject to increasing market dynamics and is being forced into ever shorter product innovation cycles to meet changing conditions and fast-moving requirements [1]. As a result, the associated production systems have to be adapted to new product generations and conditions at a similar speed while maintaining the same level of resource utilization. Rough factory Planning is regarded as an elementary lever for significantly accelerating the development of production systems [2]. The high complexity and knowledge-intensive nature of the activities involved in rough planning means that rough planning is very time-consuming, highlighting the need to support factory planners in their tasks with innovative approaches and digital technologies [3]. Approaches from the field of artificial intelligence (AI) offer a promising solution to meet the requirements and make rough

H. Kohl et al. (Eds.): GCSM 2024, LNME, pp. 763–771, 2025.
https://doi.org/10.1007/978-3-031-93891-7_84

planning more efficient and sustainable. The targeted use of AI applications in rough planning promotes sustainability in the planning process by shortening development times and conserving resources. Additionally, it addresses the factory and the improvement of its manufacturing processes through higher-quality planning outcomes. The exponential speed at which AI technologies has developed in recent years, particularly through digitalization and Industry 4.0, requires all areas of the company to recognize and exploit potential to remain competitive [4]. For this reason, a systematic approach for analyzing AI potential in rough planning was developed as part of this paper. This is intended to support companies in identifying suitable AI application options in their rough planning and evaluating them regarding realization. Section 2 presents the current framework and challenges in the context of rough planning and AI in relation to a potential analysis. This is followed by the research method (Sect. 3) and existing approaches in the literature (Sect. 4). Section 5 presents the systematic approach and its application to the industrial example within the research project DATENFABRIK.NRW. Finally, the results are summarized and the need for further research is identified (Sect. 6).

## 2 Problem Analysis

The demands on industrial production have changed dramatically in recent decades due to increasing market dynamics and the fast pace of technology and require a considerable increase in efficiency in the planning of production facilities. Factory planning, also known as production system development, is part of the product development process [5] and comprises a systematic, target-orientated process carried out in successive phases for the construction or replanning of a factory [6]. **Rough planning** plays a decisive role within factory planning, as the essential boundary conditions of the factory are defined in this section, considering all restrictions and project objectives. According to GRUNDIG, rough planning can be divided into the four phases of function determination, dimensioning, structuring and design as shown in Fig. 1, which in turn are subject to specific planning tasks [7].

During the *functional determination* phase, the product structure and the underlying work flow diagram must be analyzed in order to derive an initial functional diagram of the material flow. In the *dimensioning phase*, the necessary quantities in the areas of operating resources, personnel and space are calculated for the design of the factory or production system. The aim of *structuring* is to analyze the material flow in more detail and to derive an initial ideal layout from this. Within the *design process*, the ideal layout is adapted to real area and room structures, which leads to the derivation of a so-called real layout, which describes the final layout [8].

For companies to remain globally competitive, the challenge is to use technologies such as AI to produce more efficiently. **Artificial Intelligence** describes a branch of computer science that can realize cognitive abilities like a human being, such as learning, planning or problem-solving in computer systems [9]. The technology is based on the input of large amounts of data (big data) to derive patterns, correlations or rules and use them to make decisions [10]. AI technologies are already being used in product planning by supporting activities such as technology foresight, market and competition analyses and business forecasts. Product development is also supported by AI-controlled

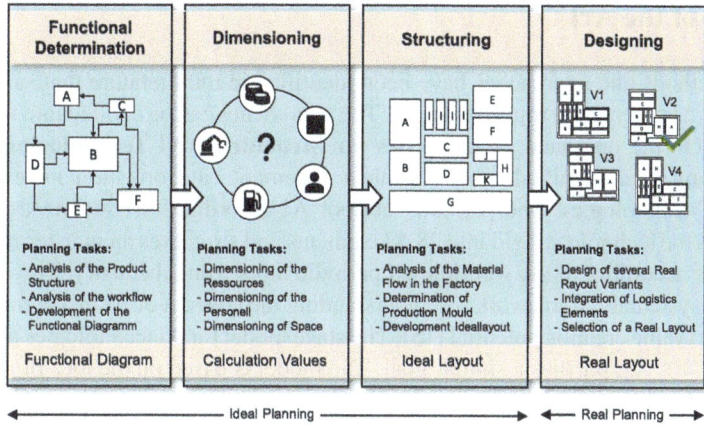

**Fig. 1.** The phases and planning tasks of rough planning [7]

requirements analyses or generative development of product designs [11], whereas the transparency of AI potential within factory or rough planning is still largely unexplored. One of the reasons for this is the lack of structured communication of existing best practices in this area [12]. The AI technology field also includes a broad spectrum of technologies and algorithms for various applications such as machine learning, natural language processing or computer vision [9]. As a result, industrial companies are confronted with an opaque and, above all, unstructured technology field when it comes to identifying potential, which makes it much more difficult to select suitable AI solutions for existing problems. In addition, the implementation of an AI application requires a structured and interdisciplinary project approach to bring in different skills and the necessary knowledge at the right time [13].

## 3   Research Method

The Systematic AI Potential Analysis was developed as part of a research project using **Österle and Otto's Consortium Research Methodology** [14]. The Consortium Research Methodology is divided into four phases: Analysis, Design, Evaluation and Dissemination. In the *analysis* phase, the research objective is defined, and a consortium is formed. As part of this research, the consortium consists of a research institute and two user companies. A research plan was drawn up, including a *literature review*. Different AI potentials were listed and filtered in the design phase. An initial setup of the AI potential catalogue was then carried out. In the third phase, the *evaluation* was carried with both companies in the consortium. The *results* were then summarized in expert workshops. Dissemination began in the final phase. To this end, regular rounds of best practice exchange were set up in the companies. In addition, the results were presented in expert workshops. This paper concludes the scientific exploitation.

## 4   State of the Art

Several methods and approaches have been identified in the literature that support the analysis of AI potential in rough planning. The approaches can be divided into two fields of action. On the one hand, **Approaches for Structuring AI Technologies and AI Applications** were identified, which enable a systematic categorization and evaluation of existing technologies. The periodic table of AI according to HARTMANN ET AL. categorizes the technology field into 28 AI elements and structures them in terms of their capabilities according to the well-known periodic table from chemistry. The approach provides a systematic framework for understanding the concept of AI in a company and tapping into value creation potential [15]. The stage model for AI technologies according to SEIFERT ET AL. pursues a similar goal. The model is based on the idea of assigning existing technologies to their typical application possibilities along the value creation stages, which allows so-called technology clusters to be developed. Besides to that AI technologies are divided into behavior-oriented and rationally inspired approaches [16]. To structure concrete AI applications, the approach according to KAISER ET AL. is suitable. The structuring framework of AI assistants in engineering contains a matrix that is spanned by two dimensions and thus brings use cases into a suitable structure [17].

On the other hand, **Approaches and Methodologies to support the Methodical Analysis of AI Potentials** in Rough Planning were identified. STEIREIF ET AL. have published a participatory procedure for identifying AI potential within five phases and evaluating it regarding realization and its potential. The method is characterized by the early involvement of relevant stakeholders [18]. The Aachener factory planning procedure also offers a proven approach to support the identification of AI potential in factory planning by describing the tasks as planning modules with input and output information as shown in Fig. 2 [3]. This specification technique offers a promising approach in the identification of AI potentials in rough planning, as AI applications can also be described by converting input variables into output variables.

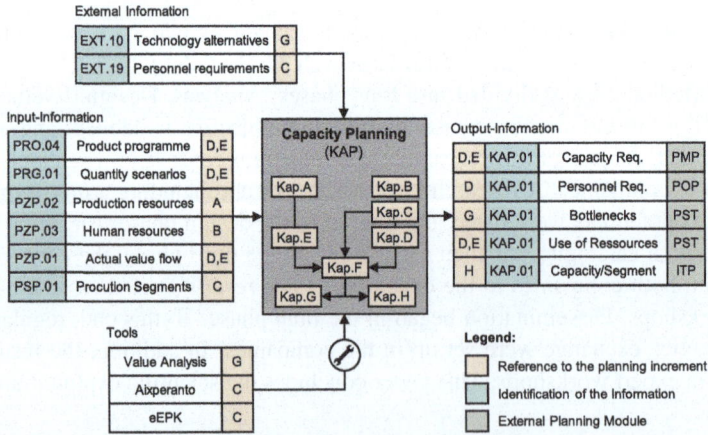

**Fig. 2.** Capacity Planning Module with Input and Output Variables [3]

Another instrument is the method for evaluating and selecting digital tools in factory planning according to HIRSCH ET AL. The decision on the use of a new tool like an AI solution is made regarding the necessary adaptation effort in the company [19]. The cost-benefit assessment portfolio according to POKORNI ET AL. is suitable for evaluating identified AI potential [20]. Furthermore the first Generative AI use cases in factory planning have been analysed [21].

## 5  Development and Application of the Systematic Approach

The systematic for analyzing AI potential in rough planning was developed as part of the research project DATENFABRIK.NRW and successfully applied in an industrial example. The systematic approach comprises six phases from analyzing the initial situation to identifying AI potentials and evaluating AI applications (Fig. 3). Several artefacts are used within the phases to provide the user with methodological and technical support. The phases with its output and artefacts are presented chronologically below.

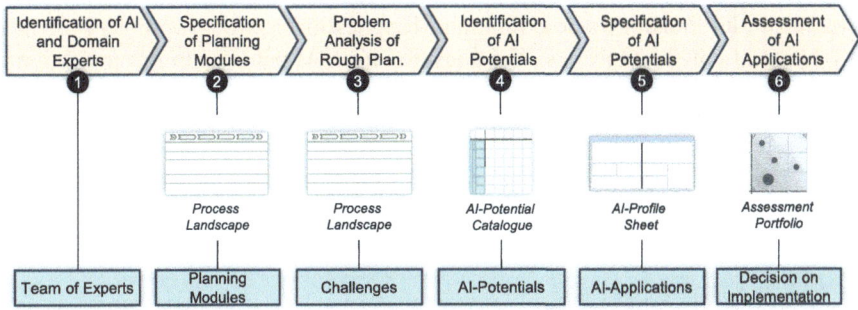

**Fig. 3.** Systematic for Analyzing AI Potentials in the Rough Planning of the Factory

The **First Phase** aims to select AI and domain experts relevant to the project. On the one hand, the necessary knowledge from the rough planning must be covered by employees directly involved in the process. On the other hand, know-how about AI is required to analyze suitable AI-based solutions and estimate their cost. Depending on the company's existing expertise, it is advisable to involve external AI experts. In the industrial example, factory planning participants and AI researchers were identified.

In the **Second Phase**, the initial situation is recorded in the rough planning and the activities are specified in the form of planning modules based on the Aachener factory planning procedure. A *process map* was developed for this purpose, in which the planning modules with their input and output information (data), IT systems and methods are recorded in a structured manner. This step helps those involved to understand the process and the activities of rough planning and to establish a common understanding. Using the industrial example, four planning modules (1. Analyzing the range of components, 2. Determining space requirements, 3. Analyzing the material flow, 4. Layout design) were considered and specified in the process map. Here it became clear that the planning modules have high data interfaces to product development, such as parts lists, which are required for analyzing the range of components.

In the **Third Phase**, the problem analysis takes place, in which existing challenges are recorded for the previously specified planning modules with which the factory planners are confronted in their rough planning activities. The challenges are identified in the *process map* and form the starting point for possible AI potential. In the industrial context, a total of eight challenges were identified for the four planning modules analyzed. One example of this is the considerable amount of time required to analyze make-or-buy decisions as part of the analysis of the component spectrum.

In the **Fourth Phase**, the findings from the process mapping in the rough planning are transferred to AI potential identification. The aim here is to find suitable and initially solution-neutral AI potential for the previously recorded challenges. For this purpose, an *AI potential catalogue* can be used to view possible potentials for the planning modules of rough planning in a two-dimensional matrix for the planning modules. The catalogue was developed as part of the project by means of a literature analysis and several expert discussions and produced several AI potentials for rough planning.

In the **Fifth Phase**, the collected AI potentials are specified and formulated into detailed AI applications. The documentation is carried out using an *AI profile*, which summarizes all the information for describing an AI application from both a problem-oriented and solution-oriented perspective. The transition from AI potential to AI application is illustrated in Fig. 4.

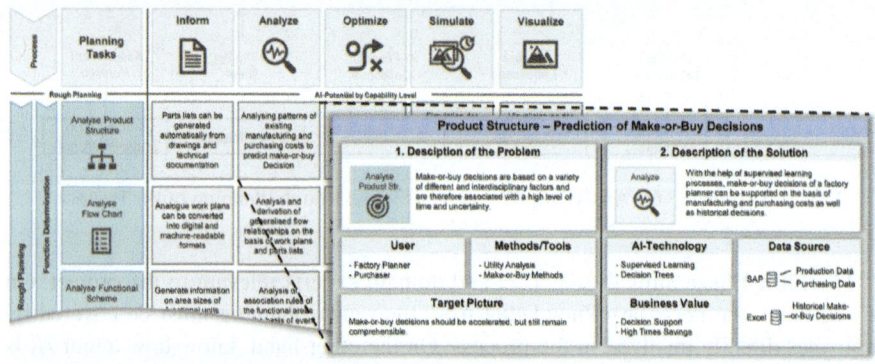

**Fig. 4.** Specification of an AI-Potential from the AI-Catalogue in the AI-Profile

Several AI profiles were formulated in the industry example. One example is the use of supervised learning processes to automate and continuously monitor the prediction of make-or-buy decisions based on historical decisions as well as purchasing and manufacturing costs.

The **Sixth Phase** of the process model involves evaluating and prioritizing the AI applications using an evaluation portfolio shown in Fig. 5. The classification and prioritization of the AI applications in the portfolio is based on an evaluation catalogue in which the applications are evaluated in the evaluation fields of business benefits, challenges, and strategic conformity, which are subject to subordinate categories. An example of a category in the business benefits assessment field is the time savings and improvement in planning quality associated with the AI application. The evaluation and prioritization offer companies the opportunity to take strategic measures and derive meaningful recommendations for action. The AI application "prediction of make-or-buy decisions" was rated as having the best cost-benefit ratio in the industrial context, particularly due to its high time savings, and should therefore be prioritized as a top priority. The required steps for implementation should be initiated as follows.

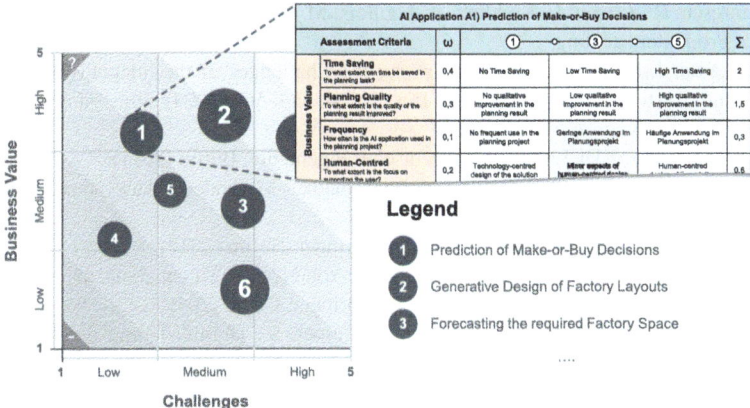

**Fig. 5.** Assessment of AI-Applications along the Categories in the Evaluation Portfolio

## 6 Summary and Outlook

In this article, a systematic approach for analyzing AI potentials in rough planning was developed and successfully applied as part of DATENFABRIK.NRW research project. The systematic comprises six phases, which are supported in their implementation using various artefacts. By applying the approach in the business context, several AI potentials in rough planning could be identified and prioritized. This paper is aimed at industrial companies that want to make their rough-cut planning processes and the resulting manufacturing more efficient, sustainable and future-oriented through the use of artificial intelligence. While there is existing research on AI potential analysis as shown in Sect. 4, it has not yet addressed the specific area of factory planning. This study seeks to bridge that gap and contribute to the advancement of knowledge in this field. Nevertheless, the results of this work reveal a need for further research. Future approaches must take greater account of the area of product development and its interfaces with rough planning. Both areas are generally pursued in parallel and in coordination with each other, resulting in a considerable exchange of information and data between the activities of both disciplines. Furthermore, research is needed to evaluate the economic benefits of AI applications, a major challenge for industrial companies, as highlighted by discussions with AI experts in this study. These challenges need to be addressed through methodological support.

# References

1. Ködding, P., Dumitrescu, R.: Forschungsfelder für Künstliche Intelligenz in der strategischen Produktplanung. In: Hartmann, E.A., (ed.) Digitalisierung Souverän Gestalten, pp. 59–73. Springer, Heidelberg (2021)
2. Disselkamp, J.-P., Cieply, J., Dyck, F., Grothe, R., Anacker, H., Dumitrescu, R.: Integrated product and production development - a systematic literature review. Procedia CIRP **119**, 716–721 (2023). https://doi.org/10.1016/j.procir.2023.06.198
3. Burggräf, P., Schuh, G.: Fabrikplanung. Springer, Heidelberg (2021)
4. Dumitrescu, R., Özcan, L. Ködding, P., Foullois, M., Bernijazov, R.: Künstliche Intelligenz in der Produktentstehung
5. Disselkamp, J.-P., Schütte, B., Dumitrescu, R.: Challenges of the integrative product and production system development. Proc. Des. Soc. **4**, 553–562 (2024). https://doi.org/10.1017/pds.2024.58
6. Factory planning, VDI 5200, Verein Deutscher Ingenieure, Berlin (2011)
7. Grundig, C.-G.: Fabrikplanung: Planungssystematik - Methoden - Anwendungen, 7th edn. Hanser, München (2021)
8. Pawellek, G.: Ganzheitliche Fabrikplanung. Springer, Berlin (2014)
9. Corea, F.: An Introduction to Data: Everything You Need to Know about AI, Big Data and Data Science (Studies in Big Data Ser v.50). Springer, Cham (2019)
10. Stich, V., Stroh, F., Abbas, M., Frings, K., Kremer, S.: Digitalisierung der Wirtschaft in Deutschland: Technologie- und Trendradar (2022)
11. Skilton, M., Hovsepian, F.: The 4th Industrial Revolution. Springer International Publishing, Cham (2018)
12. Kerkenberg, T.: Innovative und effiziente Fabrikplanung. Zeitschrift für wirtschaftlichen Fabrikbetrieb **109**(6), 435–438 (2014). https://doi.org/10.3139/104.111164
13. Röhler, M., Haghi, S.: Leitfaden Künstliche Intelligenz – Potenziale und Umsetzungen im Mittelstand
14. Österle, H., Otto, B.: Konsortialforschung. Wirtschaftsinformatik **52**(5), 273–285 (2010). https://doi.org/10.1007/s11576-010-0238-y
15. Hartmann, E.A.: Digitalisierung Souverän Gestalten. Springer, Berlin (2021)
16. Seifert, I., et al.: Potenziale der Künstlichen Intelligenz im produzierenden Gewerbe in Deutschland
17. Kaiser, L., Schräder, E., Bernijazov, R., Foullois, M.: Ein Ansatz zur Strukturierung von KI-Assistenzen im model-based systems engineering. Proc. Tag des Syst. Eng. **2022**, 113–117 (2022)
18. Steireif, N., Kranz, M., Langhanki, J., Imorde, J., Maetschke, J., Mütze-Niewöhner, S.: Potenzialanalyse von KI-Anwendungen in der Produktion. Zeitschrift für wirtschaftlichen Fabrikbetrieb **118**(4), 258–264 (2023). https://doi.org/10.1515/zwf-2023-1052
19. Hirsch, B., Klemke, T., Wulf, S., Nyhuis, P.: Digitale Werkzeuge in der Fabrikplanung. Zeitschrift für wirtschaftlichen Fabrikbetrieb **105**(3), 151–156 (2010). https://doi.org/10.3139/104.110273
20. Pokorni, B., Braun, M., Knecht, C.: Menschzentrierte KI-Anwendungen in der Produktion (2021). https://doi.org/10.24406/publica-fhg-300817
21. Disselkamp, J.-P., Kurpick, D., Schutte, B., Hovemann, A., Dumitrescu, R.: Use cases of generative AI in factory planning: potential and challenges. Proc. NordDesign **2024**, 196–205 (2024). https://doi.org/10.35199/NORDDESIGN2024.22

# Digital Engineering Approach for Automation Education: A Case Study in Using Digital Learning Factory

Chuong Chau(✉), Vi Su, Khoa Tran, and Cong Nguyen

Faculty of Engineering, Vietnamese-German University, Thoi Hoa Ward,
Ben Cat City, Binh Duong Province, Vietnam
chuong.ckb@vgu.edu.vn

**Abstract.** In today's fast-evolving technological landscape, automation is crucial for innovation across industries, making education in automation essential for preparing students for modern workplaces. Traditional methods often require significant investment in physical equipment and struggle to keep pace with technological advancements. This study explores the benefits of a Digital Learning Factory (DLF), a virtual environment that offers practical and theoretical automation training without the need for extensive physical resources. The DLF provides scalable, hands-on experience and has shown to improve student learning outcomes, engagement, and satisfaction. The research also examines how incorporating the DLF into academic curricula can enhance education in automation, better prepare students for the workforce, and promote sustainability.

**Keywords:** Automation · Digital Learning Factory · Digital Education

## 1 Introduction

In the rapidly evolving technology landscape, automation is key to innovation across industries [1]. Recent advances in digital engineering, such as Computer-Aided Design (CAD) and Programmable Logic Controller (PLC) programming, have revolutionized education in automation [2]. Conventional methods of teaching automation often rely on costly physical equipment and are limited by space, posing challenges in both scopes of study scenarios and experimentation [3]. DLF has addressed these issues with virtual replicas created using CAD and kinematic models, clearly displayed in [5–10].

This study aims to explore how a DLF can enhance students' understanding of automation principles, practical application, and connection to real-world industrial scenarios.

## 2 Methodology

Education in automation traditionally involves practical training in a physical learning environment where students engage with real instruments as shown in Fig. 1 to understand their processes and control systems. Besides the conventional concept and process,

H. Kohl et al. (Eds.): GCSM 2024, LNME, pp. 772–781, 2025.
https://doi.org/10.1007/978-3-031-93891-7_85

the current generation of engineering students who will become engineers in the future should also be able to cope with new paradigms, concepts, and emergent technologies for workforce readiness [11]. While effective, these setups come with logistical challenges including high costs, limited availability of equipment, and safety considerations [12–14]. Moreover, they may not always align with the latest industry trends and technological advancements. Therefore, it may not guarantee a sustainable method of teaching, particularly in rapidly evolving fields such as automation.

**Fig. 1.** Physical setup of an automation study.

The emergence of digital twins – virtual replicas of physical systems enabled by advanced CAD software and tools like Siemens NX offers a paradigm shift in education in automation [15]. Within a DLF, these replicas allow students to test their PLC program digitally before physical deployment, a practice known as Virtual Commissioning [16–18]. This approach enhances accessibility, scalability, and links theoretical learning with practical industry applications. Fig. 2 shows how the DLF in this study aligns with the established learning factory principle, including the operating model, metric, setting, process, didactic, target, and product according to [19].

In the studied DLF, a static model is first built in a CAD environment, using NX software. Physical and kinematic properties are integrated into the created statics models using the module NX Mechatronic Concept Design (MCD) to transfer them into the mechatronics elements or dynamic virtual objects within a virtual setting of the DLF. In the DLF, virtual models include static representations of uncontrolled elements like work piece and mechanical structures. These static objects allow the simulation of dynamic behaviors, such as collisions, within the virtual environment.

The use of OPC Unified Architecture (UA) enhances the DLF's functionality by enabling seamless data exchange between the virtual model and physical hardware [20]. OPC UA protocols establish communication channels between the digital platform and devices like HMIs or industrial PLCs, allowing students to realistically simulate and validate PLC programs in a controlled environment. Fig. 3 shows the DLF's technical setup from the perspectives of both designers and learners. Designers use CAD and mechatronic tools to create digital replicas of systems, ensuring models correspond

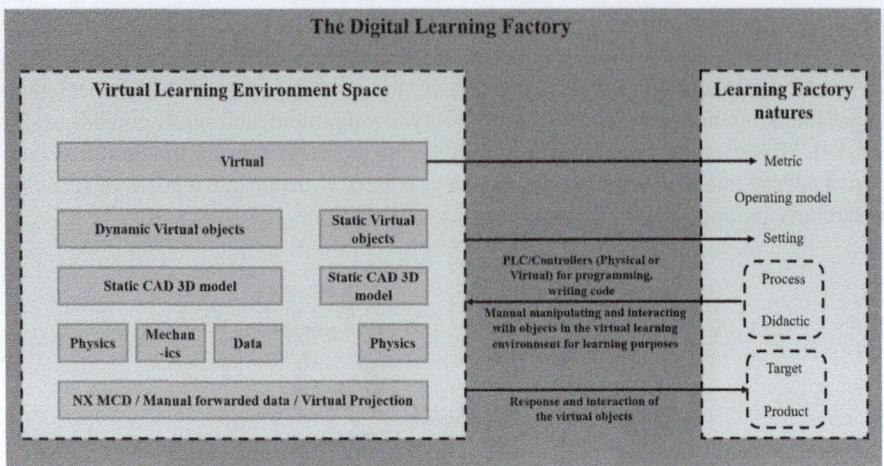

**Fig. 2.** Construction and operational model of the DLF.

accurately with physical behavior. Students then write, test, and debug PLC programs, simulating automation scenarios and observing outcomes.

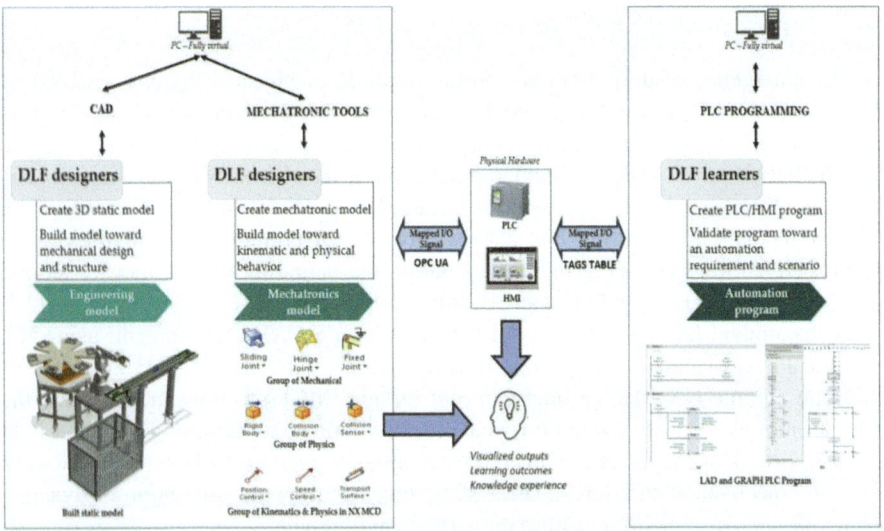

**Fig. 3.** Architecture and the workflow of the DLF.

To further enhance the knowledge and experience gained in using DLF, the modular design of the DLF allows for flexibility and scalability in incorporating different existing industrial models – that can be virtually provided by industries/manufacturers. This scalability ensures that the platform remains relevant and reflective of current industry standards.

# 3 Current State of the Developed DLF

The DLF is constructed based on the concept of digitizing learning factories, incorporating both spontaneously designed virtual systems and virtual replicas of existing physical systems. It is envisioned as a continuously expanding learning environment, unconstrained by physical space or equipment limitations, but rather limited by the educational content that can be effectively delivered by DLF designers. Currently, the DLF operates within segmented virtual subspaces tailored for different educational objectives. The layout of these learning spaces is divided into subspaces, where each virtual subspace offers specific virtual models designed for particular learning activities and outcome.

## 3.1 Operating Model of the DLF

The current configuration of the developed DLF is detailed in Table 1, providing coursework in the following areas:

- Traditional automation topics such as PLC programming, HMI design, and Supervisory Control and Data Acquisition (SCADA) system development.
- Integration of automation solutions and safety considerations for manufacturing shop floors, utilizing virtual replicas of physical systems provided by industrial partners.
- Risk-free test bed for implementing new solutions in automation.

**Table 1.** Current state of the developed DLF

| Feature | Literature | Description |
|---|---|---|
| 1. Operating model | Operation nature: Industrial/Academic/Research | Virtual/digital learning and training for students |
| 2. Target | Learning goals: Competencies development, research, optimization | Develop competencies in sectors: automation programming by PLC, machine shop floor setup/automation integration/safety operation. Machine digital training |
| 3. Process | Learning scheme: Activities, tasks, model of operation | Conduction automation program building. Complete specific requirements with the assigned digital model. Learners can choose to study on-site or remotely |
| 4. Setting | Learning environment: Machine, equipment, tools, other infrastructure, physical or virtual | The setting of DLF is fully virtual and consists of different digital models for distinctive learning purposes |

(*continued*)

**Table 1.** (*continued*)

| Feature | Literature | Description |
|---------|-----------|-------------|
| 5. Product | Learning object: Tangible/Intangible, innovation | Correct solutions for each digital model requirement. Innovation toward digital model improvement of existing physical systems |
| 6. Didactic | Curriculum, methodology: Teaching approach | Single lessons through simple digital models. Assignment or project with more complicated digital models |
| 7. Metric | Operating scale: Number of participants, factory size, equipment quantity | The number of participants can be limited if the study is conducted on-site, but not limited to the remote study. The DLF size is technically not limited, it depends on the digital contents and models that the DLF designers can deliver |

## 3.2   Virtual and Physical Setting of the DLF

The current physical and virtual settings of the DLF are depicted in Fig.4 and Fig.5 accordingly. Utilizing the DLF for work and study necessitates only the essential equipment: automation control hardware and a PC station for demonstrating and writing code.

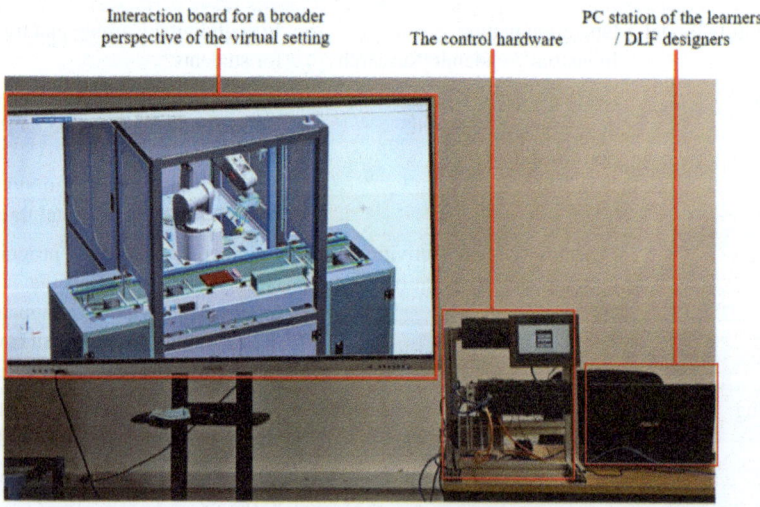

**Fig. 4.**  Physical settings of the DLF

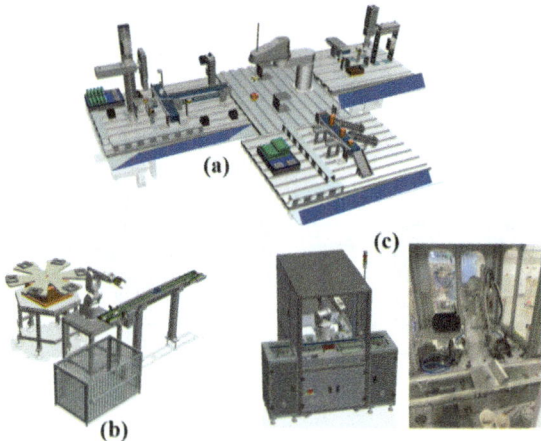

**Fig. 5.** Virtual setting of DLF: (a). Simple didactic automation process; (b). Automation line assembly from single mechatronics components by students; (c). Digital twin of the industrial system.

The DLF's primary virtual space mirrors physical setups like Festo Didactic, enabling students to apply classical automation principles through PLC programming exercises. This platform supports iterative learning and skill development without the limitations of physical equipment or costs.

Another key space focuses on mechatronic integration, where students work with pre-assembled components to design and optimize production lines and automation systems. This hands-on experience in a virtual setting enhances skills in system integration and troubleshooting.

Additionally, the DLF features virtual spaces with digital twins from industry partners. These accurate virtual replicas of industrial systems offer students a realistic learning environment, allowing them to explore complex processes, analyze behaviors, and develop optimization strategies.

## 4    Discussions

The current status of the developed DLF has successfully met several objectives thus far. While the infrastructure, both physical and virtual, requires ongoing development, the operational framework of the DLF has been finalized. This section of the paper reviews aspects of the implementation process, challenges, limitations, and assessments conducted on the developed DLF.

### 4.1   Implementation and Assessment of the DLF

As presented in section 3 of this paper, the operating model of DLF introduces a digital platform where students can enhance their understanding of automation programming and principles through interaction with the DLF virtual setting, as shown in Fig. 6.

It represents a supplement in automation education, offering students a realistic and versatile platform to explore automation concepts in a risk-free virtual environment.

**Fig. 6.** Learners perform commissioning of a virtual setting, projecting to their PLC and HMI program.

To evaluate the DLF's effectiveness, 12 students planning to use it with hardware for an automation module answered a survey with five Likert scale questions (1-strongly disagree to 5-strongly agree). Details of the questions and responses are in Table 2. The recorded average score was above the point of 4 for question number 1 – 4, while question number 5 achieved a result of below 4.

**Table 2.** Evaluation questions and results of the DLF effectiveness

| Number | Question description | Average results |
|---|---|---|
| Question 01 | Working and studying with the DLF proved to be intriguing | 4.7 |
| Question 02 | Virtual settings of the DLF gain practical perspectives of the subject matter | 4 |
| Question 03 | The DLF supports to acquire knowledge in the covered application | 4.5 |
| Question 04 | The DLF is an efficient digital engineering approach to support education | 4.3 |
| Question 05 | The DLF can completely replace the need for physical settings | 3.5 |

An experiment was conducted to measure the effectiveness of the developed DLF. Students from the Automation and Robotics module were divided into four groups to complete automation tasks requiring similar knowledge. They used control hardware (PLC, HMI) in the following configurations: signal and tag simulation without automation process settings, virtual settings in the DLF, and physical settings illustrating the automation process. The experiment recorded the total time taken by each group to reach a sufficient solution, as shown in Fig. 7.

After the experiment, students using signal and tag simulation in the programming software took significantly longer to complete tasks compared to those using the DLF's

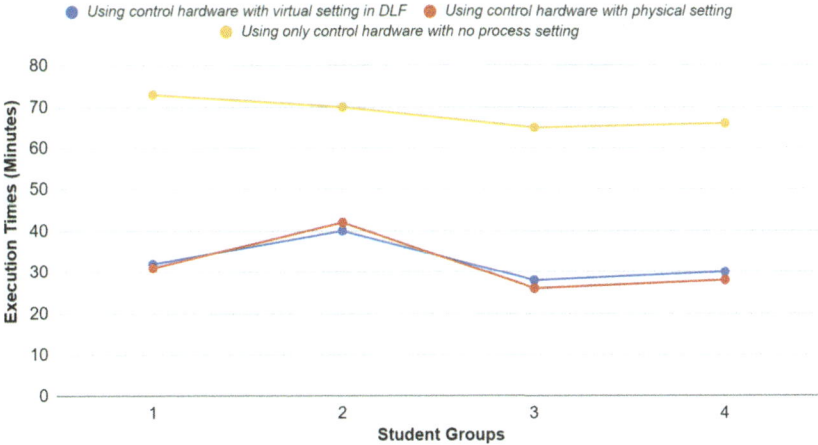

**Fig. 7.** Task execution time from student groups with different setting scenarios.

virtual setting, as recorded. Some students also expressed concerns and hesitated despite achieving sufficient solutions. Similar results were observed for both the physical and virtual settings of the automation process. Questionnaire responses highlighted the DLF's effectiveness in education, though question 5 indicated a preference for working with real physical settings, where students can interact directly with actual automation equipment. Despite the assessment being conducted with a relatively small sample size (12 students), the initial impression of the DLF was still technically achieved

### 4.2   Limitations and Challenges of the DLF

The DLF, built using digital engineering tools, faces challenges due to the extensive IT infrastructure and effort needed to create accurate virtual models. To achieve educational goals, these virtual settings must closely mirror physical systems, increasing the time and effort required. Another challenge is simulating collisions, such as robot grippers handling complex objects or parts colliding with static models. Detailed collision models improve accuracy but require substantial simulation resources, potentially slowing down the process. Fig. 8 shows how collision references impact simulation accuracy and resource consumption within the performance of the used IT infrastructure of the DLF building.

In the current DLF, the inclusion of industry digital twins emphasizes the need for collaboration with industry partners to access advanced knowledge. While these digital twins enhance the realism of learning, not all companies are willing to share proprietary information due to intellectual property concerns. This reluctance underscores the need for innovative simulation methods to provide practical insights despite limited access to industry digital twins. Developing generic digital twins and fostering industry partnerships are technically key to enhancing the DLF's educational impact.

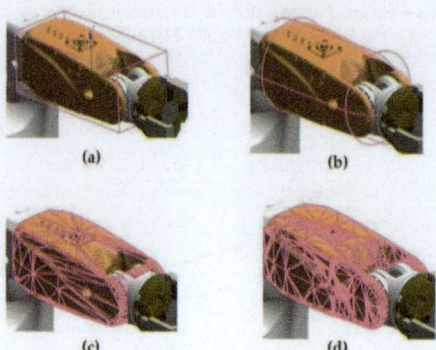

**Fig. 8.** Meshing for collision bodies in mechatronic objects (MO): (a) Box type: Extremely high simulation performance, low accuracy, (b) Capsule type: High simulation performance, moderate accuracy, (c) Convex type: Low simulation performance, high accuracy, (d) Multi Convex type: Extremely low simulation performance, very high accuracy.

## 5    Conclusion

In conclusion, the DLF offers a valuable tool for automation education, providing a virtual platform for students to explore automation concepts without the need for costly physical setups. While still in development, the DLF has already enhanced student engagement and learning. However, challenges remain, including limited industry sharing of digital models and high IT infrastructure demands.

Unlike traditional automation education, which requires significant investment in physical equipment, the DLF promotes sustainability by reducing the need for machinery investment, energy consumption by physical setting and the space for storage. Future improvements will focus on expanding its reach to more students with different focuses of study outcomes, optimizing IT infrastructure, and collaborating with industry partners.

## References

1. Reinhart, G., Wunsch, G.: Economic application of virtual commissioning to mechatronic production systems. Prod. Eng. **1**(4), 371–379 (2007)
2. Zakaria, F.: Revolutionizing industries: the advancement and impact of robotics and automation. https://www.linkedin.com/pulse/revolutionizing-industries-advancements-im-pact-rob otics-zakaria/. Accessed 12 June 2024
3. Mohsin, F.K, Ghulam, S.K, Abdul, K.J, Abdul, H.S.: The role of PLC automation, industry, and education purposes: a review. J. Eng. Technol. Sci. **11**(02), 22–31 (2023)
4. Azadeh, H., Navid, S., Gunilla, S., Thomas, L., Yvonne, E.: Digital learning factories: conceptualization, review and discussion.In: CONFERENCE SPS14, Göteborg (2014)
5. Chi, X., Spedding, A.: A web-based intelligent virtual learning environment for industrial continuous improvement. In: IEEE International Conference, Pp. 1102–1107 (2006)
6. Dessouky, M., Bailey, E., Verma, S., Adiga, S., Bekey, G., Kazlauskas, J.: A virtual factory teaching system in support of manufacturing education. J. Eng. Educ. **87**(4), 459–467 (1998)
7. Goeser, P., Johnson, M., Hamza-Lup, F., Schaefer, D..: VIEW-a virtual interactive web- based learning environment for engineering. Adv. Eng. Educ. (2011)

8. Manesh, F., Schaefer, D.: Virtual learning environment for manufacturing education and training. Comput. Educ. J. **1**(1), 77–89 (2010)
9. Riffelmacher, P., Kluge, S., Kreuzhage, R., Hummel, V., Westkamper, E.: Learning factory for the manufacturing industry: digital learning shell and a physical model factory - ITRAME for production engineering and improvement. In: Proceedings of the 20th International Conference on Computer-Aided Production Engineering, pp. 120–131 (2007)
10. Barney, D., Mark, J.W.L,.: What are the learning affordances of 3-D virtual environments? J. Educ. Technol. **41**(1), 10–32 (2010)
11. Elisa, T., Nicola , C., Emanuele ,M.: Using robotics to train students for Industry 4.0, IFAC-PapersOnline, **52**(9), 153–158 (2019)
12. Soriano, A., Marin, L., Valles, M., Valera, A., Albertos ,P.: Low cost platform for automatic control education based on open hardware. IFAC Proc. **47**(3) (2014)
13. Fisher, D., Gould, P.: Open-source hardware is a low-cost alternative for scientific instrumentation and research. Mod. Instrum. **1**(2) (2012)
14. Gardiner, D.W., Kutsirkova, N., Messer, D.: Robotics and automation in preschool education: opportunities for learning? Front. Psychol. **9**, 1177 (2018)
15. Antti, L., Heikki, P.: Using digital twin technology in engineering education – course concept to explore benefits and barriers. Open Eng. **10**(1), 377–385 (2020)
16. Wünsch, G.: Methoden für die virtuelle Inbetriebnahme automatisierter Produktionssysteme. Utz (2008)
17. Drath, R., Webner, P., Mauser, N.: An evolutionary approach for the industrial introduction of virtual commissioning. In: 2008 IEEE International Conference on Emerging Technologies and Factory Automation, Hamburg, Germany, pp. 5–8 (2008)
18. Hoffman, P., Maksoud, T.M.A., Schuman, R., Premier, G.C.: Virtual commissioning of manufacturing systems – a review and new approaches for simplification. In: European Conference on Modeling and Simulation (ECMS 2010) (2010)
19. Abele, E., et al.: Learning factories for research, education, and training. Proc. CIRP, **32**, 1–6. (2015)
20. Wolfgang, M., Leitner, S.H.: OPC unified architecture: the future standard for communication and information modeling in automation, pp.56–61 (2009)

# An Approach for the Thermal Modelling of Machines and Machine Building Interaction for the Dynamic Assessment of Thermal Loading Conditions

Michael Frank[(✉)], Lukas Theisinger, Fabian Borst, and Matthias Weigold

Technical University of Darmstadt, Institute for Production Management, Technology and Machine Tools (PTW), Otto-Berndt-Straße 2, 64287 Darmstadt, Germany
m.frank@ptw.tu-darmstadt.de

**Abstract.** The development of sustainable manufacturing systems necessitates the knowledge of the thermal loading conditions of the manufacturing equipment in interaction with the manufacturing building. Especially for the early planning phases a dynamic evaluation of different concepts is advantageous and leads to an acceleration of the overall planning procedure. This article introduces an approach for the development of simulation models for the assessment of thermal loading conditions in brownfield manufacturing applications by dynamic simulation. The approach makes use of the novel Modelica Thermal Integration Library, which provides basic models for representative manufacturing equipment in the metal working industry. It enables an efficient simulation of manufacturing equipment with the related manufacturing building to analyse the overall thermal loading conditions. The approach is applied to a machine tool in the manufacturing line at the ETA Research Factory of the Technical University of Darmstadt.

**Keywords:** dynamic simulation · manufacturing equipment modeling

## 1 Introduction

The industrial sector of Germany consumed 723 TWh of final energy in 2021, with 73% attributed solely to thermal energy [1]. This thermal energy consumption predominantly reliant on fossil fuels with a share of 82% in 2022, constitutes a primary source of greenhouse gas emissions, thereby rendering it pivotal for transformative initiatives [2]. This article specifically targets the metal working industry, where less than 1% of total energy consumption is allocated to value-adding processes such as metal part machining, with the remaining 99% transmitted to the surrounding in auxiliary units or utilized for cooling and heating [3]. The disproportionately low share of energy devoted to value addition justifies a focus on transforming industrial supply systems, particularly in addressing thermal loading conditions associated with manufacturing processes. Efforts to

H. Kohl et al. (Eds.): GCSM 2024, LNME, pp. 782–790, 2025.
https://doi.org/10.1007/978-3-031-93891-7_86

enhance energy efficiency in industrial supply systems must include considerations for the manufacturing building and the manufacturing equipment. The current systematic of topology planning, and dimensioning of industrial supply systems follows a sequential process aligned with legislations and norms such as the German HOAI and VDI5200 [4,5]. A lack of information, especially in the early planning stages, significantly drives the tendency to overdimension industrial supply systems. The deficit in understanding the dynamic thermal behavior is also evident in brownfield applications. Consequently, employing dynamic simulation to represent manufacturing equipment offers valuable insights during the early planning phases, facilitating the derivation of transformation possibilities for related industrial supply systems. Augmenting early planning phases with additional information proves especially beneficial for reducing planning uncertainty, particularly in stages where costs for conceptual changes are low and the impact on the overall sustainabiliy of the concept high. To increase the likelihood of application, the proposed modeling approach prioritizes ease of understanding, coupled with swift dynamic simulations. The article advocates a streamlined thermal modeling approach to simplify the modeling and simulation procedures, resulting in reduced development time. The incorporation of this approach within a structured modeling library further enhances applicability, and ensures expandability to accommodate diverse use cases.

## 2    Material and Methods

In accordance with the insights presented by [6], it is imperative to conceptualize a technical system within the framework of techno-economics. This necessitates the analysis of diverse energy and monetary fluxes. The resultant multi-pole system analysis demands detailed knowledge of input variables, system matrices, and output variables. By maintaining mathematical consistency, the fully described system can be addressed employing the multi-pole formalism, rendering it applicable to various systems.[6] This is particularly true in the context of manufacturing equipment and their interaction with buildings. The approach presented in this article, with a focus on thermal energy fluxes, streamlines the multi-pole system analysis by reducing its level of detail, yet effectively delivers detailed insights into the system relevant thermal energy fluxes. In the modeling of physical systems, the object-oriented software development paradigm has proven to be highly suitable. Given that technical systems belong to the subclass of physical systems, which are inherently organized in an object-oriented manner, this paradigm is integrated into the modeling procedure and implemented in a library structure. [7] Modelica, an open-source modeling language, is commonly employed for developing physical models, implementing the object-oriented paradigm, and applying the multi-pole formalism [8,9]. It's suitability is underscored by its user-friendly, fast, and efficient modeling and simulation procedure. The potential to integrate subclass categorization with a library generation process at the outset of the modeling procedure allows for the consistent utilization of the object-oriented paradigm throughout the model development process. Leveraging basic models from the Modelica Standard Library

[10] further expedites and streamlines the development process, particularly in the thermal domain. To address the typically unknown heat transfer coefficients and resistance values, the proposed approach relies on literature-derived values for heat transportation phenomena, taken mainly from [11]. This involves the utilization of standard cases, such as heat transfer at vertically or horizontally oriented walls, as well as combined estimations based on efficiency factors. Thermal connectors and resistors facilitate the connection between components and their respective thermal masses. For instance, machine tools are approximated as cubes with a central thermal mass, parameterized by the overall machine weight, enclosed by four vertically oriented walls and one horizontally oriented top and bottom surface. Energy transformation processes are represented by efficiency factors.

## 3   Results

This section first portrays the general model structure of the manufacturing equipment models in Subsect. 3.1. Based on this, the interaction process between the different models and the relevant base classes is described in Subsect. 3.2. Finally, the Modelica `Thermal Integration Library` is introduced in Subsect. 3.3.

### 3.1   Basic Model and Sequence Control

The basic model framework involves conceptualizing relevant components of the manufacturing system as thermal masses, with the heat flows between these components being characterized by the predominant heat transfers. Convection ($\dot{Q}_{heatconv}$) is considered by the definition of heat transfer coefficients ($\alpha$), surface temperatures ($T_w$) and the sourrounding temperature ($T_{env}$) according to the equation $\dot{Q}_{heatconv} = \alpha \cdot (T_w - T_{env})$. Heat conduction is considered only in one dimension. With the relevant temperature differences ($\Delta T$), thermal resistance values ($R = k \cdot A$), defined by relevant surface extensions ($A$) and the used materials with the thermal conductivity coefficients ($k$), according to equation $\dot{Q}_{heatcond} = k \cdot A(\Delta T)$. [11] Radiation is not considered. Upon differentiation based on heat transfer mechanisms, the thermal behavior of each component can be effectively modeled. At this point it is imperative to account for energy conversion processes, such as the transformation of electrical energy into thermal energy (e.g. in electric motors). This can be achieved either by utilizing nominal values (Fig. 1 (2)) derived from the component data sheet or by employing physical relationships and equations (Fig. 1 (3)). The thermal interactions among the components are modeled in the direction of the temperature gradient. The integration of individual components into an overarching model (Fig. 1 (4)) ultimately encapsulates the representation of the manufacturing system. Figure 1 illustrates a schematic depiction of the fundamental model, with Fig. 1 (1) detailing the implementation of sequence control. In this context, manufacturing programs are instantiated as PLC (programmable logic controller) programs, orchestrating the control of individual components in accordance with the specified manufacturing program through an undirected information flow. This implementation methodology guarantees the independent, dynamic control of components aligned with the manufacturing program. The PLC serves as the pivotal information technology coupling element for these individual components. The accurate execution of the manufacturing program runtime is crucial for enabling full scale simulation capability. [12] This basic model and sequence control implementation procedure is followed for every manufacturing system modeled so far in the `Thermal Integration Library`.

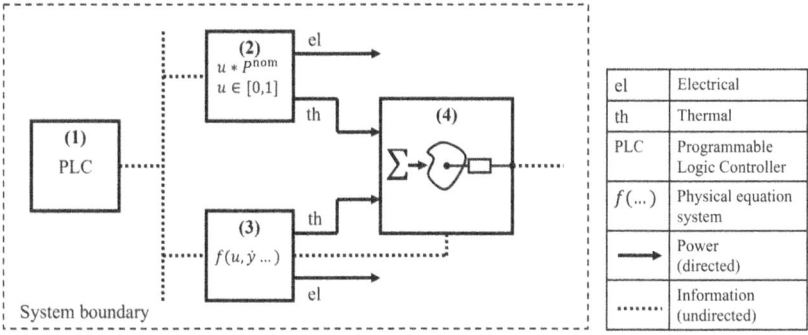

**Fig. 1.** Base model structure with information and power flow [12].

## 3.2    Base Classes and Model Interaction

The modelling approach follows the object oriented concept of inheritance. Therefore, the base classes `BaseProductionEquipment` and `BaseProductionBuilding` are defined, offering besides the possibility of accelerated model development, essential functionalities for the information exchange between the different model instances. By integrating a common interface between the base classes and the `SystemEnergyManager` an information consolidation in a central model instance takes place, offering both evaluation possibilities (e.g. overall energy balances) and further data usage comparable to a global variable. The implementation within the Modelica model class allows a fast implementation of the relevant manufacturing equipment models with the `SystemEnergyManager`.

## 3.3    Introduction of the Modelica Thermal Integration Library

To enable fast and convenient development of simulation models for different manufacturing use cases, a Modelica library was developed consisting of the base classes described in Sect. 3.2 as well as generalized models for manufacturing equipment and buildings. The `Thermal Integration Library` includes concrete manufacturing equipment models for cleaning machines, machine tools and annealing ovens, parameterized by the machinery integrated in the manufacturing line of the ETA Research Factory [13].

## 4    Use Case

For validating the presented approach the machine tool in the manufacturing line at the ETA Research Factory is modeled and simulated in interaction with the manufacturing building. The following Sect. 4.1 portrays the concrete modeling and simulation procedure. In Sect. 4.2 the simulation outcome is evaluated and set into context with the overall goal of identifying energy efficiency measures in brownfield applications to develop sustainable manufacturing systems.

### 4.1    Application

The generalized model of the machine tool is based on the VLC 100 Y vertical lathe manufactured by EMAG GmbH & Co. KG [14]. The core components of this machine tool align with those found in any typical machine tool. Specifically, the primary electrical drives (Fig. 2 (1)), cooling lubricant system (Fig. 2 (2)), hydraulic system (Fig. 2 (3)), cooling system (Fig. 2 (6) and (7)), control cabinet (Fig. 2 (4)), connection to a central cooling unit (Fig. 2 (11)), and central cooling (Fig. 2 (11)) are identified as key components. These components' thermal flows are integrated into a thermal mass known as the thermal system (Fig. 2 (5)). This modeling approach allows the machine tool to be cooled either through a central chilled water or cooling water network or through decentralized compression chillers. Heat transfer to the environment is facilitated by specifying a constant ambient temperature (Fig. 2 (12)). This ensures the capability to simulate the machine tool model independently of a manufacturing building. In cases of coupled simulations involving the machine tool and manufacturing building this temperature specification dynamically adjusts to the temperature within the manufacturing building via the interface to the `SystemEnergyManager`. The manufacturing program is defined through a table input (Fig. 2 (9)) and is communicated to the components by the PLC using a `Modelica` bus connector. The illustrated model in Fig. 2 also permits configuration with the cooling of the control cabinet via an additional cooling system (Fig. 2 (6)). Component parameterization is based on manufacturer specifications, with modeling carried out considering rated outputs and efficiency estimates. It is assumed that 70% of the main electrical power consumption of the machine tool is converted into heat, discharged into the cooling lubricant. The remaining 30% is either released into the surrounding atmosphere via decentralized compression chillers or transferred to a central cooling system, depending on the cooling method for the cooling lubricant [3]. If the cooling lubricant is also cooled decentrally, it is estimated that 100% of the electrical power consumption of the machine tool is converted into heat, which is then discharged into the surrounding atmosphere of the machine. For the machining process a manufacturing program (Fig. 2 (9)) for the control disk of an axial piston pump is implemented. This implementation involves defining electrical power values for each component at specific points in time in the manufacturing program through a table value assignment. The power values are linked to machine states according to [15], distinguishing between off, powering up, standby, operational, working, and powering down. Considering these states results in the power allocations for the manufacturing program for the control disk. Following the implementation of the electrical components, the thermal system of the machine tool (Fig. 2 (5)) is modeled. The machine's mass is determined from data sheets and rating plates and represented as thermal capacity [14]. Heat transfers to the environment and cooling units are modeled based on Sect. 3.1.

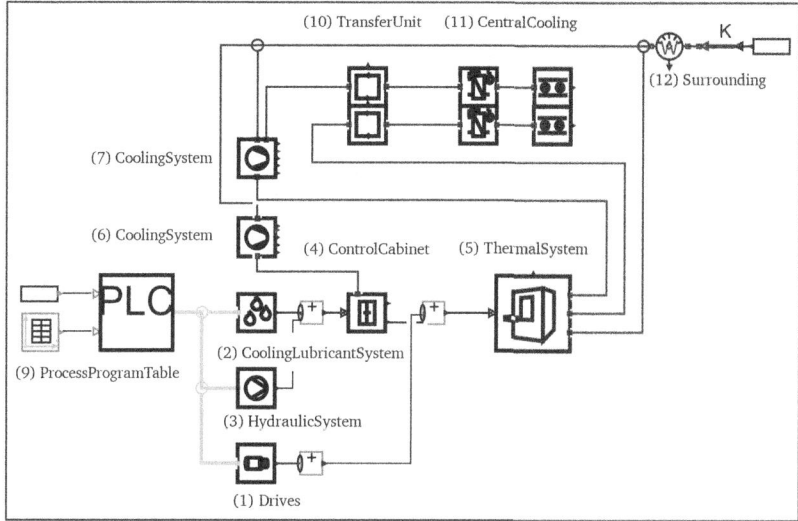

**Fig. 2.** Machine tool model of the **Thermal Integration Library**.

## 4.2    Evaluation

The simulation outcome for the machine tool with the manufacturing program of the control disk in interaction with the manufacturing building for a 6 hour schedule is shown in Fig. 3. Internal heat sources are taken into account via parameter settings which allow the definition of the quantity of machines, workers and lights. The machine states and the related manufacturing program can be identified in the step form of the power consumption of the electrical drives. The correlated dynamic temperature development of the machine tool is depicted in a dotted line, which is correlated with the necessary active cooling demand of the manufacturing building. Based on the cooling demand, energy efficiency measures for the heat supply of the manufacturing building can be developed. E.g. the usage of a heat pump for the heat supply of the manufacturing building could be evaluated as shown in [12].

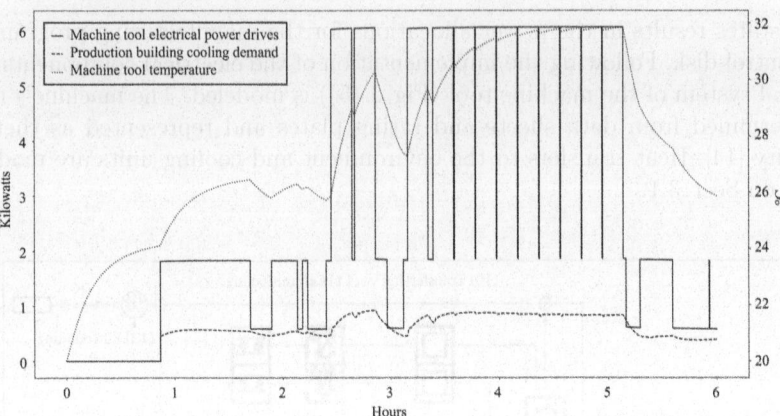

**Fig. 3.** Plot of the combined simulation of the machine tool and the manufacturing building.

# 5   Summary and Conclusion

The object-oriented simulation approach focused on modeling of thermal fluxes proved to be useful for quantitatively evaluating the thermal behavior of manufacturing systems. The incorporation of the modeling approach within the Modelica `Thermal Integration Library` enables a fast model development and allows further enhancement for different manufacturing equipment. Future work should focus on the development of modeling approaches for production schedules which enable the consideration of different start up times of manufacturing equipment and the integration of shift related worker utilization. Furthermore, detailed evaluations of energy efficiency measures require more sophisticated models of the manufacturing building with consideration of the supply systems. Here, innovative approaches for the thermal flux based modeling are necessary to achieve accurate simulation outcomes. Nevertheless, development approaches to increase the sustainabiliy of manufacturing systems can be derived from the simulation outcome of the current library.

**Funding Information.** This research was funded by the German Federal Ministry of Economic Affairs and Climate Action (BMWK) grant number 03EN2053A.

# References

1. "Auswertungstabellen zur Energiebilanz fuer die Bundesrepublik Deutschland 1990 bis 2022 (Datenstand November 2023). https://ag-energiebilanzen.de/daten-und-fakten/auswertungstabellen/. Accessed 16 Feb 2024
2. "Zeitreihen zur Entwicklung der erneuerbaren Energien in Deutschland: unter Verwendung von Daten der Arbeitsgruppe Erneuerbare Energien-Statistik (AGEE-Stat)." https://www.umweltbundesamt.de/dokument/zeitreihen-zur-entwicklung-der-erneuerbaren. Accessed 16 Feb 2024
3. Beck, M., Helfert, M., Burkhardt, M., Abele, E.: Rapid assessment: method to configure energy performant machine tools in linked energy systems. In: 21st CIRP Conference on Life Cycle Engineering, vol. 48, pp. 514–519, 2016
4. VDI Verein Deutscher Ingenieure e.V., "Fabrikplanung: Planungsvorgehen," Berlin, February 2011
5. Bundesgesetzblatt I, "Verordnung über die Honorare für Architekten- und Ingenieurleistungen (Honorarordnung für Architekten und Ingenieure - HOAI): HOAI," 2013
6. Holl, M., Pelz, P.F.: Multi-pole system analysis (MPSA) - a systematic method towards techno-economic optimal system design. Appl. Energy **169**, 937–949 (2016)
7. Clauß, C., Leitner, T., Schneider, A., Schwarz, P.: Object-oriented modelling of physical systems with modelica using design patterns. In: Merker, R., Schwarz, W. (eds.) System Design Automation: Fundamentals, Principles, Methods, Examples, pp. 195–208. Springer US, Boston, MA, 2001
8. Modelica Association, "Modelica Language Specification 3.5," 2000. https://modelica.org/modelicalanguage.html. Accessed 07 Mar 2023
9. Pelz, P.F., Groche, P., Pfetsch, M.E., Schaeffner, M. (eds.): Mastering Uncertainty in Mechanical Engineering, 1st ed., ser. Springer Tracts in Mechanical Engineering. Springer, Cham (2021)
10. Modelica Association, "Modelica Standard Library: Free library to model mechanical (1D/3D), electrical (analog, digital, machines), magnetic, thermal, fluid, control systems and hierarchical state machines." 2020. https://github.com/modelica/ModelicaStandardLibrary. Accessed 12 Apr 2023
11. Stephan, P., Kabelac, S., Kind, M. (eds.): VDI-Waermeatlas: Fachlicher Traeger VDI-Gesellschaft Verfahrenstechnik und Chemieingenieurwesen, 12th ed., ser. VDI Springer Reference. Springer Vieweg, Berlin, Germany (2019)
12. Theisinger, L., Georg Frank, M., Borst, F., Weigold, M.: Energieeffizienzmaßnahmen im Industriebestand. Zeitschrift für wirtschaftlichen Fabrikbetrieb, vol. 118, no. 3, pp. 127–132, 2023
13. Frank, M., Theisinger, L., Borst, F.: Thermal Integration Library: Dymola Simulationsbibliothek," 2024. https://github.com/PTW-TUDa/ThermalIntegrationLibrary.git. Accessed 19 Aug 2024
14. Frank, M., Frieß, T.: TU Darmstadt, "Machine Tool EMAG VLC 100 Y." https://tudatalib.ulb.tu-darmstadt.de/handle/tudatalib/4142.2. Accessed 19 Aug 2024
15. VDMA - Verband Deutscher Maschinen- und Anlagenbau e.V., "Messvorschrift zur bestimmung des energie- und medienbedarfs von werkzeugmaschinen in der serienfertigung," 2019-4. https://www.beuth.de/de/technische-regel/vdma-34179/302624354. Accessed 23 Jan 2023

# Reinventing Multi-Floor Factories: Sustainability-Driven Integrated Planning Process to Mitigate Soil Sealing

Nikolaus Kremslehner[1]([✉]), Maria Antonia Zahlbruckner[2], Julia Reisinger[2], and Sebastian Schlund[1,2]

[1] Fraunhofer Austria Research GmbH, 1040 Vienna, Austria
nikolaus.kremslehner@fraunhofer.at
[2] TU Wien, 1040 Vienna, Austria

**Abstract.** In this paper, we introduce a planning process tailored to address the unique challenges of multi-floor factories, providing practitioners with valuable guidance for their design tasks. Recognizing the critical role of factories in sustainable development, this work addresses their potential to drive economic growth while advancing ecological and social sustainability. Traditional expansion strategies often result in increased soil sealing, adversely affecting agricultural productivity and food security. Our approach seeks to decouple economic growth from land use by incorporating specific considerations for multi-floor factory design into a structured planning process. We analyze the implications of multi-floor buildings on the planning parameters for factories, evaluating the suitability of existing planning methods. Based on our analysis, we propose necessary advancements to these methods and integrate them into a comprehensive new planning framework. The proposed process is a first attempt to encourage practitioners to adopt multi-floor factory designs and lays the groundwork for future research to further refine and validate these methods in real-world applications.

**Keywords:** Sustainable Factory · Integral Planning · Planning Process

## 1 Introduction

The sealing of soil by constructing buildings and roads harms the ecosystem, reducing soil's ability to filter water and prevent flooding, while also causing biodiversity loss and soil fertility decline, which impacts agriculture and food security. Therefore, avoiding soil sealing is crucial for sustainable soil resources, climate, and human health. [1] The EU Soil Strategy aims for net-zero land use by 2050. [2] In Austria, 27.3% of soil sealing is due to commercial and mixed building use. [3] Initially, factories were multi-floor due to limited urban land, but efficient transport systems led to large peripheral factories, ignoring vertical layouts due to cost. [4, 5] Recently, cleaner, space-saving production processes and efficient logistics have made multi-floor factories viable again. [6] Legal sustainability requirements also pressure companies to minimize ecological impacts. [7] Examples include multi-floor production using gravity and mixed-use buildings. [8]

H. Kohl et al. (Eds.): GCSM 2024, LNME, pp. 791–798, 2025.
https://doi.org/10.1007/978-3-031-93891-7_87

As multi-floor production gains traction, planning methods tailored for this purpose are essential, considering the specific possibilities and constraints. Therefore, this paper analyzes how the established factory planning process needs to be adapted to plan new multi-floor factories. To this end, the suitability of existing planning methods and tools for the planning parameters of multi-floor factories is evaluated.

## 2   State-of-the-Art

### 2.1   Factory Planning Process and Methods

**Planning Contents and Process.** Factory planning is defined as "systematic, objective-oriented process for planning a factory, structured into a sequence of phases, each of which is dependent on the preceding phase, and makes use of particular methods and tools". Factory planning projects can investigate different domains, such as production logistics or external logistics and different levels from a work centre to the whole production network [9].

Various formulations of the factory planning process can be found in literature, which can, however, be summarized into a general planning procedure. [10] In the concept planning phase, the factory is designed based on established objectives. Initially, functional units and their relationships are defined according to the business processes. Next, space requirements are determined by selecting production and logistics equipment. Different technological variants for logistics equipment must be compared to develop a material flow concept. The factory layout is created in two steps: arranging functional areas in an ideal layout and then creating real layout variants considering all restrictions. [9] It can be stated that the general procedure for planning factories is well established in both scientific and practical contexts. While this does not exclude the planning of multi-storey factories, it is unclear how the additional planning steps required for this are to be integrated.

**Factory Planning Methods and Tools.** Choosing the right material handling technology involves technical, economic, and organizational considerations and requires holistic systemic thinking. [11] Comparative analysis of suitable technologies is essential, often using methods like utility analysis. [12] Due to the dominance of single-floor factories the possibility of vertical transports is often neglected in selection methods for material handling technologies [13].

Various layout types can be described for flow production, which can serve as a template for the arrangement of the areas in the ideal planning phase. [14] However, these are two-dimensional, so they cannot simply be transferred to a multi-floor building. Layout types for multi-floor factories, can be derived from historical examples as can be found in [15] and in [5]. Such layout types are proposed in [5] and [6]. Industry-specific layout types were discussed for assembly factories [16] as well as so-called micro-factories, which consist of small, standardized production modules [17].

Suitable mathematical methods can facilitate layout optimization, especially in complex production structures like workshop production. [14] Often heuristic methods like the Schmigalla method are mentioned in literature, which can also be carried out without computer support [14, 18–20].

Despite extensive research on the Facility Layout Problem (FLP), multi-level layouts are rarely considered and the examined methods have limitations that are particularly relevant for practical application. [21] The examined methods usually only take into account material handling costs, ignoring other quantitative or qualitative criteria. [22] Moreover, these methods are lacking practical application which is highlighted by the absence of real-life case studies or commercial tools. [21] There are some commercial tools that facilitate the design of real layouts by adapting the ideal layouts according to boundary conditions and expert knowledge. [10] However, to calculate distances between departments on different floors, transports to and from a vertical transporter and the vertical transport itself need to be modelled, hindering the simultaneous visualization and calculation of material flows in these tools. [23] In conclusion, a closer examination of the available methods for the various planning phases reveals that they are not suitable for planning factories with multiple floors.

## 2.2 Integral Planning

The current state-of-the-art in integral planning of industrial buildings and production layout highlights a trend towards integrated and flexible design frameworks. These frameworks focus on combining production layout planning with industrial building design to enhance efficiency and adaptability. A notable contribution in this field is the development of parametric frameworks for multi-objective optimization, which aim to support decision-making by evaluating designs based on life cycle costs, environmental impact and flexibility. [24, 25] Another significant approach involves combining mathematical layout optimization with digital factory tools to handle the complexity of production processes, ensuring comprehensive planning methodologies. [26] Furthermore, innovative parametric evolutionary design methods aim to generate and optimize production layout scenarios that integrate seamlessly with building design, fostering flexibility and coherence [27], but dealing with horizontal expansion. These advancements are crucial for addressing the dynamic nature of production systems and the frequent reconfigurations required by modern manufacturing environments.

# 3   Research Design

This paper focuses on the planning tasks defined in the VDI guideline 5200 [9] with the most impact on the area consumption of the factory, which were outlined in Sect. 2.1. Hence, the concept planning phase is investigated for the planning domain factory and production logistics at the building level. The analysis is further focused on the planning of new factories in contrast to the replanning of existing ones.

Parameters of these planning tasks are derived from literature review to define the relevant planning decisions. Subsequently, the possible values of these parameters in a multi-floor factory are determined through the analysis of literature and case studies, which cover different industries as vehicle manufacturing, food processing and medical equipment manufacturing. Based on this, existing methods and tools for the investigated planning phases, which can be found in literature and practice, are evaluated regarding their ability to consider these parameter values adequately. As a result, necessary adaptations of the planning phases are defined to propose the planning process for multi-floor factories. The application of this process is then demonstrated through a case study to evaluate if a viable multi-floor factory could be designed.

## 4    Design of the Planning Process for Multi-floor Factories

### 4.1   Dimensioning

In the dimensioning phase the parameters personnel resources, required areas, required media supply and operating resources are determined. The operating resources of interest are production equipment and material handling systems. These also determine the required personnel resources, areas and media supply. [10] In the regarded case studies the production equipment was independent from the number of floors of the factory. However, vertical transports required different material handling systems such as elevators of vertical conveyors. A look at historical and contemporary case studies shows that elevators are still the prevailing solution [15, 5], but also potential bottlenecks in the material flow. [19] However, there are more possibilities for material flows between floors, like spiral conveyors [12] or Z-conveyors [13].

Since not only horizontal but also vertical transports are necessary, the selection and dimensioning of material handling technologies is especially complex in the case of multi-floor factories. This is because depending on the layout transports might use different technologies. Consequently, it can be necessary to preselect viable technologies and evaluate their suitability further under consideration of the ideal layout. For preselection a categorization of technologies by their transport function is proposed. For detailed evaluation, it is further necessary to define the cost function for transports that utilize the vertical transport systems.

### 4.2   Ideal Planning

In single-floor factories it is sufficient to find the optimal arrangement of areas on one floor to generate an ideal layout. In multi-floor factories however, it is also necessary to determine the number of floors, the allocation of areas to the floors and the location of elevators as additional parameters [23, 28].

Since neither flow-oriented layout types nor existing tools adequately support the generation of multi-floor layouts with these parameters, new layout types and methods need to be developed for this purpose. Depending on the complexity of the material flow, ideal layouts can be found by transferring best-practice solutions, simple heuristics or more sophisticated algorithms like metaheuristics. To reduce the complexity of the problem, it is proposed to first analyze how many floors could be realized in the specific case. Then ideal layouts for scenarios with different numbers of floors can be generated separately and compared regarding the evaluation criteria.

For multi-floor factories specific interdependencies of economic, ecological and social objectives need to be considered and transferred into evaluation criteria for the layout. Multi-floor factories offer economic benefits by reducing land acquisition costs through higher area productivity, but face higher construction costs and structural complexity, which hinders the changeability of the layout. [29] Moreover, ecological advantages have to be considered, such as reduced soil sealing and higher energy efficiency due to shared walls and centralized systems, compared to single-floor buildings. [30] The social impact includes providing diverse employment opportunities and fostering community development in urban areas. [8] Therefore, it is proposed to evaluate multi-floor layouts regarding area productivity in combination with changeability, construction costs and material handling efforts of the factory to portray its sustainability adequately.

## 4.3  Real Planning

Real planning includes the modeling of the building with its supporting structure and the evaluation of the designed solutions. [9] The supporting structure for multi-floor buildings usually requires additional structural elements such as columns, beams, and slabs to support multiple floors efficiently [29].

Currently no commercial tools are available that support the adaption of layouts with multiple floors adequately. As intermediate solution until such tools are available, we propose an extension of existing tools that aid the remodeling of material flows when areas are moved from one floor to another. Furthermore, the feasibility of the generated layout needs to be examined as additional supporting structures could be necessary. To this end, a parametric modeling approach could provide good solutions. If no feasible solution can be found, it might be necessary to loop back to the ideal planning stage and work with the next best layout option.

Figure 1 depicts how the specific parameters for multi-floor factories can be determined by methods that already exist and methods that need further development.

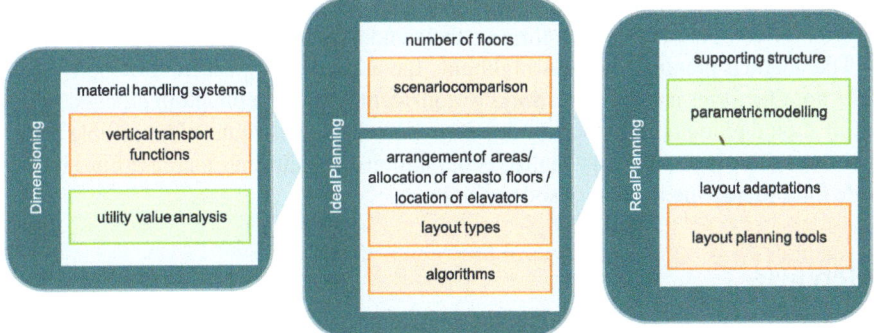

**Fig. 1.** Concept planning process for multi-floor factories with the relevant parameters (grey), existing methods (green) and methods that need further development (orange).

# 5 Demonstration of the Planning Process

The presented case study concerns a European vehicle manufacturer who has integrated assembly, manufacturing, and logistics processes at his site. The company is likely to outgrow its current facilities in the next years. It is demonstrated how the proposed planning process could have been applied to create a viable layout for a new factory on a new plot of land.

In the case study discontinuous transports of piece goods with heavy loads needed to be performed in both directions. Thus, elevators as well as vertical conveyors would have been promising options. An assessment of both technologies according to the initially defined evaluation criteria favored a vertical conveyor, mainly due to the lower personnel costs. Following the possible building heights at the plot, a layout with three floors was aimed at. An ideal layout was created similar to the historical example of the Ford Highland Park factory. [15] By allocating the assembly, preassembly and manufacturing areas on different floors, the flow of production was organized from top to bottom. For efficient evaluation of the material handling efforts and further adaptation of the layout to the building requirements, a script was developed that extends the capabilities of the layout planning tool visTABLE®. The script creates a new material flow table for import into the tool according to the material flow relations and the layers defined in the layout data. Thus, the layout was compared to a single-floor layout, indicating a reduction of material handling costs by 24% with only 35% of the consumed area.

# 6 Conclusion and Future Research

This paper proposes a planning process for multi-floor factories based on their specific planning parameters and the shortcomings of existing planning methods and tools. This motivates future research to develop methods for multi-floor layout optimization, selection of vertical material handling systems and interactive layout planning tools.

The adapted planning process highlights the additional aspects and complexity of multi-floor factories and proposes ways to consider them in the planning process. In this way, it should guide practitioners to develop viable factory layouts with multiple floors. This can enable companies to realize economic growth through increased production while minimizing the necessary soil sealing.

# References

1. Courvoisier, T.J..: Opportunities for soil sustainability in Europe. EASAC Secretariat Deutsche Akademie der Naturforscher Leopoldina, Halle (Saale) (2018)
2. Umweltbundesamt: 13. Umweltkontrollbericht. Umweltsituation in Österreich. Wien (2022)
3. ÖROK: Flächeninanspruchnahme und Versiegelung in Österreich. Kontextinformationen und Beschreibung der Daten für das Referenzjahr 2022. Wien (2023)
4. Lane, R.N., Rappaport, N. (eds.): The Design of Urban Manufacturing. Routlegde/Taylor & Francis Group, New York, London (2020)
5. Haselsteiner, E., et al.: VERTICALurbanFACTORY - Innovative Konzepte der vertikalen Verdichtung von Produktion und Stadt. Wien (2020)

6. Ciaramella, A., Celani, A.: Industry 4.0 and manufacturing in the city: a possible vertical development, pp. 133–142 Pages / TECHNE - Journal of Technology for Architecture and Environment, TECHNE 17 (2019): Horizontality/verticality in architecture, vol. (2019). https://doi.org/10.13128/TECHNE-24009

7. Fang, E.-L., Gassmann, P., O'Connell, K., Picard, N.: The CSRD is resetting the value-creation agenda. https://www.pwc.com/gx/en/issues/esg/csrd-is-resetting-the-value-creation-agenda.html

8. Herrmann, C., Juraschek, M., Burggräf, P., Kara, S.: Urban production: state of the art and future trends for urban factories. CIRP Ann. **69**, 764–787 (2020). https://doi.org/10.1016/j.cirp.2020.05.003

9. Verein Deutscher Ingenieure: Factory Planning. Planning procedures, vol. (2011)

10. Grundig, C.-G.: Fabrikplanung. Planungssystematik - Methoden - Anwendungen. Hanser, München (2018)

11. Martin, H.: Transportlogistik. In: Martin, H. (ed.) Technische Transport- und Lagerlogistik, pp. 61–185. Springer Fachmedien Wiesbaden, Wiesbaden (2021). https://doi.org/10.1007/978-3-658-34037-7_4

12. Hompel, M. ten, Schmidt, T., Dregger, J.: Fördersysteme. In: Hompel, M. ten, Schmidt, T., Dregger, J. (eds.) Materialflusssysteme. VDI-Buch, pp. 125–246. Springer, Berlin, Heidelberg (2018). https://doi.org/10.1007/978-3-662-56181-2_4

13. Wehking, K.-H.: Fördertechnik. In: Wehking, K.-H., et al. (eds.) Technisches Handbuch Logistik, pp. 511–652. Springer Vieweg, Berlin, Heidelberg (2020). https://doi.org/10.1007/978-3-662-60867-8_9

14. Erlach, K.: Wertstromdesign. Der Weg zur schlanken Fabrik. Springer Vieweg, Berlin, Heidelberg (2020)

15. Rappaport, N.: Vertical Urban Factory. Actar Publishers, New York, Barcelona (2019)

16. Schneider, M.: Z-Production – a revolutionary paradigm of production. Zeitschrift für wirtschaftlichen Fabrikbetrieb **117**, 4–8 (2022). https://doi.org/10.1515/zwf-2022-1001

17. Scholz-Reiter, B., Brenner, N., Lütjen, M., Keck, M.: Three-dimensional layout in micro factories. Zeitschrift für wirtschaftlichen Fabrikbetrieb **104**, 791–795 (2009). https://doi.org/10.3139/104.110155

18. Arnold, D., Furmans, K.: Materialfluss in Logistiksystemen. Springer Berlin Heidelberg, Berlin, Heidelberg (2019)

19. Pawellek, G.: Ganzheitliche Fabrikplanung. Grundlagen, Vorgehensweise, EDV-Unterstützung. Springer-Vieweg, Berlin, Heidelberg (2014)

20. Schenk, M., Wirth, S., Müller, E.: Fabrikplanung und Fabrikbetrieb. Springer Berlin Heidelberg, Berlin, Heidelberg (2014)

21. Pérez-Gosende, P., Mula, J., Díaz-Madroñero, M.: Facility layout planning. An extended literature review. Int. J. Prod. Res. **59**, 3777–3816 (2021). https://doi.org/10.1080/00207543.2021.1897176

22. Hosseini-Nasab, H., Fereidouni, S., Fatemi Ghomi, S.M.T., Fakhrzad, M.B.: Classification of facility layout problems: a review study. Int. J. Adv. Manuf. Technol. **94**, 957–977 (2018). https://doi.org/10.1007/s00170-017-0895-8

23. Ahmadi, A., Pishvaee, M.S., Akbari Jokar, M.R.: A survey on multi-floor facility layout problems. Comput. Ind. Eng. **107**, 158–170 (2017). https://doi.org/10.1016/j.cie.2017.03.015

24. Reisinger, J., et al.: Framework for integrated multi-objective optimization of production and industrial building design. In: Proceedings of the 2022 European Conference on Computing in Construction. Computing in Construction. University of Turin (2022). https://doi.org/10.35490/EC3.2022.223

25. Reisinger, J., Kugler, S., Kovacic, I., Knoll, M.: Parametric optimization and decision support model framework for life cycle cost analysis and life cycle assessment of flexible industrial

building structures integrating production planning. Buildings **12**, 162 (2022). https://doi.org/10.3390/buildings12020162

26. Krüger, T.: Entwicklung einer Gesamtmethodik zur Kombination von mathematischer Anordnungsoptimierung und Materialflusssimulation für die Produktionslayoutplanung (2019)

27. Reisinger, J., Zahlbruckner, M.A., Kovacic, I., Kán, P., Wang-Sukalia, X., Kaufmann, H.: Integrated multi-objective evolutionary optimization of production layout scenarios for parametric structural design of flexible industrial buildings. J. Build. Eng. **46**, 103766 (2022). https://doi.org/10.1016/j.jobe.2021.103766

28. Hathhorn, J., Sisikoglu, E., Sir, M.Y.: A multi-objective mixed-integer programming model for a multi-floor facility layout. Int. J. Prod. Res. **51**, 4223–4239 (2013). https://doi.org/10.1080/00207543.2012.753486

29. Vayas, I., Ermopoulos, J., Ioannidis, G.: Multi storey buildings. In: Vayas, I., Ermopoulos, J., Ioannidis, G. (eds.) Design of Steel Structures to Eurocodes. SpringerLink Bücher, pp. 295–336. Springer, Cham (2019). https://doi.org/10.1007/978-3-319-95474-5_7

30. AL-Nassar, F., Ruparathna, R., Chhipi-Shrestha, G., Haider, H., Hewage, K., Sadiq, R.: Sustainability assessment framework for low rise commercial buildings: life cycle impact index-based approach. Clean Tech. Environ. Policy **18**, 2579–2590 (2016). https://doi.org/10.1007/s10098-016-1168-1

# Methodology for Integrative Digital Factory Planning

Jan-Philipp Disselkamp[1]([⊠]) and Roman Dumitrescu[2]

[1] Fraunhofer IEM, Zukunftsmeile 1, 33102 Paderborn, Germany
jan-philipp.disselkamp@iem.fraunhofer.de
[2] University Paderborn - HNI, Fürstenallee 1, 33102 Paderborn, Germany

**Abstract.** The transition to smart factories is a critical challenge within modern manufacturing, characterised by the need to integrate advanced digital technologies with traditional production systems. Existing approaches often fail to address the complexity of such integration, resulting in inefficiencies and missed opportunities for innovation. This paper presents a structured, three-phase model to help companies digitally transform their manufacturing processes. The first phase focuses on the benefits of integrative planning, identifying planning process challenges and assessing readiness for digital planning using aids such as Success Stories and Quick Checks. The second phase looks at planning domain selection, covering the necessary planning processes, interfaces and information flow, supported by detailed process profiles and a reference architecture. The third phase explores digital factory planning methodologies, examining process optimisation opportunities and the application of AI, facilitated by a potential catalogue Together, these phases enable companies to reap the benefits of Smart Factories, overcome planning challenges and improve their digital planning readiness, providing a holistic approach to the digital and integrative development of production systems.

**Keywords:** Design for Manufacturing · Integrative development · Integrated product and production system development

## 1 Introduction

The relentless advance of technology in industry is driving an exponential increase in demand for accelerated innovation cycles [1]. At the same time, small and medium-sized enterprises are faced with the challenge of transforming their factory into a smart factory to produce more economically. Increasing competitive pressures in the global marketplace are further exacerbating these trends, resulting in disparate lead and test times. This represents a further challenge in the development process. To remain competitive, shorter product development times and new factory technologies are necessary to bring innovations faster and more cost efficient to the market [2].

These challenges can be met through intensive collaboration between product and production system development [3]. The optimal and requirement-oriented development

© The Author(s) 2025
H. Kohl et al. (Eds.): GCSM 2024, LNME, pp. 799–806, 2025.
https://doi.org/10.1007/978-3-031-93891-7_88

of products and production systems can be achieved through parallel and closely coordinated development, which is also known as integrative product and production system development [3–5]. The parallelisation of processes has the potential to reduce development times, thereby mitigating the impact of increasingly compressed innovation cycles [2].

Given their inherent complexity, the processes of product and production system development are conducted in a methodical manner [6]. In recent decades, integrative approaches have emerged in response to these considerations. These approaches consider the development of products and production systems in close coordination or as an integrated process [7]. Nevertheless, these approaches have yet to be fully embraced within industry, where product and production system development remains predominantly sequential and isolated from one another [7, 8].

In addition to planning new production facilities in conjunction with the concurrent development of a product, the utilisation of appropriate digital tools to facilitate product development represents a significant challenge. Several integrated digital solutions are already in use in product development, and this paper focuses on production system development. The deployment of software support in production system development offers a multitude of advantages. On the one hand, the occurrence of errors in planning can be avoided. On the other hand, data can be provided that can be used for various applications. These include, for example, the digital factory twin or the virtual workplace design. The current challenge is that it is unclear, especially for beginners, which steps need to be coordinated with product development and how. This challenge is addressed in this paper. To do this, the research methodology is first presented. Subsequently, the issue is subjected to further examination. Thereafter, two methodologies from the contemporary literature are elucidated. This is then followed by an exposition of the developed methodology. Finally, an overview of the findings is provided, along with an outlook on future developments.

## 2 Research Method

This paper employs the research methodology of the Design Research Methodology (DRM) as outlined by Blessing and Chakrabarti [9]. The DRM methodology is structured into four phases: (1) research clarification, (2) descriptive study I, (3) prescriptive study, and (4) descriptive study II. A systematic literature review (SLR) was conducted as part of the **first phase** of the DRM to identify the various challenges of integrative planning [10]. In the **second phase** of DRM, the descriptive study I, the research question *"What are the reasons for the lack of use of integrative methods in industrial product and production development?"* was analysed by a literature review and expert interviews in a previously paper [11]. A brief summary of the study will be presented in the next section. This paper will present the **third phase** of the DRMs, the prescriptive study, which will propose a solution to the issues identified.

# 3    Problem Analysis

Product design, also referred to as **product development**, is the iterative development of a marketable product based on specific requirements [12]. It encompasses activities such as component design, with predefined development goals, including function fulfilment [6, 12].

Production system design (also referred to as **production system development**) encompasses the design or planning of production systems, including the planning of workflows, workplaces, equipment, production logistics and material flow [1, 13]. **Factory planning** encompasses both the construction of the factory itself and the planning of the surrounding site [7, 14].

The intricacy of product and production system development processes is widely acknowledged [15]. To address this, an **integrative approach to product and production system development** is employed, whereby the design of the product and the production system occur simultaneously and in a coordinated manner. This approach aims to optimise both through early consideration of all product lifecycle phases and production requirements, thereby ensuring better alignment across the value chain [16, 17].

In industrial practice, these methods have not yet been implemented across the board. A literature review and expert interviews identified eight reasons for the lack of integrative planning in practice. These can be summarised as follows:

1. **Unclear interfaces:** The unclear connections between product design and production system design result in a high level of uncertainty regarding the process models [3].
2. **Information and organisational management:** Existing methods only partially address the management of information and organisational structures, and there is still a need for research in this area [2, 18].
3. **Insufficient information exchange:** There is a need for effective mechanisms for information exchange between departments [10, 18, 19].
4. **Technical feasibility:** Integrative methods do not sufficiently address technical feasibility, as suitable software support is not considered.
5. **Lack of traceability:** The impact of changes in product design on the production system is not clearly defined, which leads to problems and rework.
6. **Parallelisation instead of interdisciplinary collaboration:** Existing methods tend to parallelise development processes, which means that opportunities for integration are missed [3].
7. **Insufficient consideration of existing knowledge and flexibility:** Current methods do not sufficiently exploit the value of existing knowledge from previous projects [19] and flexibility for new projects [20].
8. **Complexity for beginners:** The integration of different methods is often too complex and difficult to understand for newcomers, which leads to resistance to their introduction.

# 4    State of the Art

This chapter presents two integrative approaches that address the concurrent planning of a production system and product development.

### 4.1  4-Cycle Model of Product Creation

Gausemeier et al. suggest that the product creation process should not be strictly divided into phases and milestones but should be viewed as an interplay of different tasks. They divide the process into four main tasks, which are subdivided into four interlinked cycles. The concept is referred to as the four-cycle model [1]:

- Strategic product planning: This phase is used to define the business and product idea.
- Product development: The aim is to develop a marketable product based on the product idea from strategic planning. This task includes the steps of product conception, design, development, and integration into an overall system.
- Development of the production system: A production system is developed here that is optimally tailored to the product and the specific requirements. The tasks include the design of the production system, work planning and integration into an overall system. Work planning is further subdivided into the following sub-areas: workflow, workplace, material flow and work equipment planning.
- Service development: This phase transforms a service idea into a marketable service offering, considering various aspects and understanding service development as the interaction of various tasks.

### 4.2  Method for Production System Design Based on Early Product Information

Sinnwell (2020) presents a model-based approach to planning production systems. This approach is based on the model developed in the MecPro$^2$ research project and extends it. The concept of integrated development of cyber-physical products and production systems using Model-Based Systems Engineering (MBSE) is applied [21]. The principal objective of the method proposed by Sinnwell is to integrate engineering processes at an early stage to facilitate a more comprehensive mutual understanding through the utilisation of a "common language". The concept is comprised of three principal elements [7]:

- A holistic process model for the development and design of products,
- A systematic methodology for the early design of manufacturing systems that uses a maturity model to evaluate early product information,
- An object-oriented modelling method

The integrated process model is based on the structure of the VDI 2206 V-model [22]. The process model comprises two distinct cycles: the micro cycle and the macrocycle. The macrocycle is employed for coordination, while cooperation and collaboration occur within the micro cycle. While the classic phases of product and production system planning remain, they are integrated into the micro cycle. Each phase commences with at least one iteration of the micro cycle. The conclusion of each phase signifies a significant milestone, at which the intermediate results of product and production system development must be validated [7].

## 5  Methodology for Integrative Digital Factory Planning

This paper presents a three-step procedural model (Fig. 1) and addresses several of the problems in existing methods (Sect. 3). The methodology has two main objectives. Firstly, it enables a **beginner-friendly** integrative development of a new factory or factory hall, which provides orientation in the various phases of factory planning and at the same time considers the coupling to product development. Secondly, the systematics enable the support of this planning by digital and AI-based technologies. The value of this work lies in its ability to enable individuals with no prior knowledge to plan a factory based on the system and to be supported in the selection of digital tools. The method was developed as part of the Datenfabrik.NRW research project in collaboration with two companies.

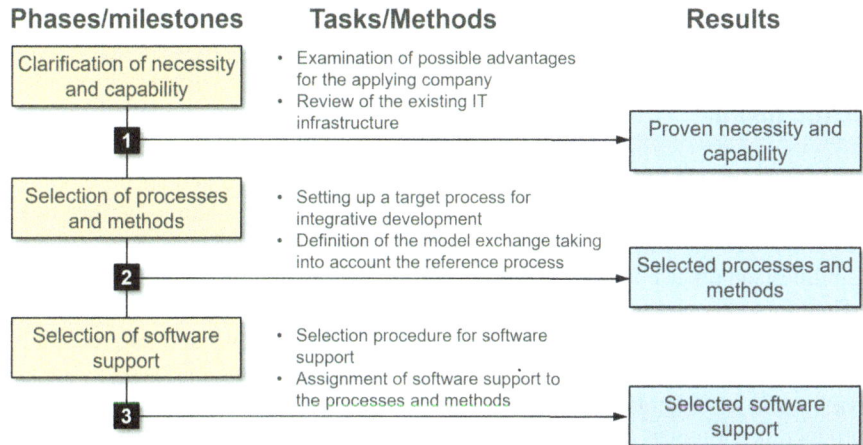

**Fig. 1.** Framework for integrative digital factory planning.

The **first step** of the procedure model entails examining the potential advantages for the respective companies through the examination of the *success stories* of other companies. This involves the assessment of the prerequisites for the utilisation of the system, with particular emphasis on the evaluation of the company's digital requirements. The outcome of the initial phase is the company's verified necessity and capacity to implement the system.

In the **second step**, the selection of appropriate processes and methodologies is undertaken. A *reference process* for integrative planning is considered, and a target process is developed for the respective company [23]. The reference process is characterised by the fact that each factory planning phase is assigned to a corresponding product development phase. This allows less experienced users to identify parallels between the development phases. A *profile* has been created for each phase, describing the individual phases, and suggesting methods for working through the phases. For instance, a material flow analysis is conducted as part of the factory planning process to create a preliminary layout. Various analytical techniques, including an operations sequence diagram, a function diagram, a matrix structure, and a value stream analysis, are proposed as part of the material flow analysis. Furthermore, interfaces between the individual development phases are identified using an *interface matrix,* which is also anchored in the reference process [24]. Subsequently, the selected processes, methods and interfaces between the development departments are documented for future reference.

Once the processes have been defined in the second step, the **third step** involves selecting appropriate software support for these processes [25]. Concurrently, a *catalogue of potential* is employed to illustrate which areas of the company could benefit from further digitisation or the utilisation of technologies such as artificial intelligence. Consequently, company-specific support options for newly implemented processes are proposed. Furthermore, a design guideline for the creation of a 3D factory plan [26] or the development of digital factory twins is integrated into the selection of software support [27, 28]. This leads to the selection of processes, methods, and software support with a definition of the interfaces to product development after the system has been implemented.

# 6  Conclusion

Integrative factory design is required to simultaneously achieve shorter development times and to evolve the factory towards a smart factory. Existing solutions for integrative development are not used in practice, e.g., due to a lack of consideration of technical possibilities and unclear interfaces between product development and factory planning.

Therefore, this paper presents a methodology that enables integrative development. The process model consists of three steps and is aimed at newcomers to integrative digital factory planning. The first step examines the prerequisites and raises awareness of the benefits of integrative planning. In the second step, process steps and interfaces to product development are identified for the applying company. Finally, in the third step, software support for the implementation of integrative digital factory planning is selected.

The methodology was developed in the Datenfabrik.NRW project in cooperation with two companies. However, there is a need for further research to validate the system, as the effectiveness of the approach must be tested comprehensively (e.g., by means of an interview study). In addition, it should be investigated whether the approach can be extended to include technical elements, so that factory planning can be carried out with a digital factory twin and legacy data from the existing factory can also be considered. Further research is needed to analyse the interfaces. It should be investigated whether the exchange between product development and factory planning can be enabled by exchanging and linking some central data artefacts and whether these data artefacts ensure synchronisation.

**Acknowledgments.** This research work is based on "Datenfabrik.NRW", a flagship project by "KI.NRW", funded by the Ministry for Economics, Innovation, Digitalisation and Energy of the State of North Rhine-Westphalia (MWIDE).

**Declaration of Generative AI and AI-assisted Technologies in the Writing Process.** During the preparation of this work the authors used ChatGPT and DeepL to enhance language and readability. After using this tools, the authors reviewed and edited the content as needed and takes full responsibility for the content of the publication.

# References

1. Gausemeier, J., Dumitrescu, R., Echterfeld, J., Pfänder, T., Steffen, D., Thielemann, F.: Innovationen für die Märkte von morgen: Strategische Planung von Produkten, Dienstleistungen und Geschäftsmodellen. Hanser, München, 2019
2. Eversheim, W., Schuh, G., Assmus, D.: Integrierte Produkt- und Prozessgestaltung. In: Eversheim, W., Schuh, G. (eds.), Integrierte Produkt- und Prozessgestaltung (VDI) , pp. 5–20. Springer, Berlin, Heidelberg, 2005
3. Stoffels, P., Vielhaber, M.: Methodical support for concurrent engineering across product and production (system) development. In: Proceedings of the 20th International Conference on Engineering Design (ICED15) Vol 4: Design for X, Design to X, Milan, Italy, 27–30.07.15, C. Weber, S. Husung, M. Cantamessa, G. Cascini, D. Marjanovic, and S. Graziosi, Eds., 2015, pp. 155–162
4. Gausemeier, J., Lanza, G., Lindemann, U.: Produkte und Produktionssysteme integrativ konzipieren. Carl Hanser Verlag, München, 2012
5. Stoffels, P., Kaspar, J., Vielhaber, M.: Product vs. production development ii - integrated product, production, material and joint definition. Proc. Des. Soc. **1**, 2471–2480 (2021). https://doi.org/10.1017/pds.2021.508
6. Gericke, K., Bender, B., Pahl, G., Beitz, W., Feldhusen, J., Grote, K.-H.: Grund-lagen methodischen Vorgehens in der Produktentwicklung. In: Bender, B., Gericke, K. (eds.) Pahl/Beitz Konstruktionslehre: Methoden und Anwendung erfolgreicher Produktentwicklung, 9th ed. , pp. 27–55. Springer, Berlin, Heidelberg, 2021
7. Sinnwell, C.: Methode zur Produktionssystemkonzipierung auf Basis früher Produktinformationen, Dissertation, Universität Kaiserslautern, Kaiserslautern, 2020
8. Humpert, L., Disselkamp, J.-P., Schierbaum, A., Zagatta, K., Dumitrescu, R.: Engineering Autonom Wandelbarer Industrie 4.0-Systeme, 2024. https://doi.org/10.48669/fb40_2024-2
9. Blessing, L.T.M., Chakrabarti, A.: DRM, a design research methodology. Springer, Dordrecht, Heidelberg, 2009
10. Disselkamp, J.-P., Cieply, J., Dyck, F., Grothe, R., Anacker, H., Dumitrescu, R.: Integrated product and production development - a systematic literature review. Procedia CIRP **119**, 716–721 (2023). https://doi.org/10.1016/j.procir.2023.06.198
11. Disselkamp, J.-P., Schütte, B., Dumitrescu, R.: Challenges of the integrative product and production system development. Proc. Des. Soc. **4**, 553–562 (2024). https://doi.org/10.1017/pds.2024.58
12. VDI-Richtlinie 2221 Blatt 1: Entwicklung technischer Produkte und Systeme - Modell der Produktentwicklung, VDI2221-1, Verein Deutscher Ingenieure e.V., November 2019
13. Cochran, D.S., Arinez, J.F., Duda, J.W., Linck, J.: A decomposition approach for manufacturing system design. J. Manuf. Syst. **20**(6), 371–389 (2001). https://doi.org/10.1016/S0278-6125(01)80058-3
14. VDI-Richtlinie 5200 Blatt 1: Fabrikplanung - Planungsvorgehen, VDI5200-1, Verein Deutscher Ingenieure e.V., February 2011
15. Helbing, K., Mund, H., Reichel, M.: Handbuch Fabrikprojektierung, 2nd ed. (SpringerLink Bücher). Springer Vieweg, Berlin, Heidelberg, 2018
16. Bullinger, H.-J., Kugel, R., Ohlhausen, P., Stanke, A.: Integrierte Produktentwicklung: Zehn erfolgreiche Praxisbeispiele (Springer eBook Collection Business and Economics). Gabler Verlag, Wiesbaden, 1995
17. Eigner, M., Stelzer, R.H.: Productlifecycle-Management: Ein Leitfaden für Product Development und Life-cycle-Management, 2nd ed. (VDI). Springer, Berlin, Heidelberg 2009
18. Francalanza, E., Borg, J., Vella, P., Farrugia, P., Constantinescu, C.: An 'Industry 4.0' digital model fostering integrated product development. In: 2018 IEEE 9th International Conference

on Mechanical and Intelligent Manufacturing Technologies (ICMIMT 2018), Cape Town, South Africa, pp. 95–99, 2018. https://doi.org/10.1109/ICMIMT.2018.8340428

19. Albers, A., et al.: Product-Production-CoDesign: an approach on integrated product and production engineering across generations and life cycles. Procedia CIRP **109**, 167–172 (2022). https://doi.org/10.1016/j.procir.2022.05.231

20. Beibl, J., Disselkamp, J.-P., Dumitrescu, R., Krause, D.: Product and production design - rethinking views on flexibility. Procedia CIRP **128**, 204–209 (2024). https://doi.org/10.1016/j.procir.2024.03.010

21. Eigner, M., Koch, W., Muggeo, C.: Modellbasierter Entwicklungsprozess cybertronischer Systeme: Der PLM-unterstützte Referenzentwicklungsprozess für Produkte und Produktionssysteme. Berlin, Heidelberg: Springer Vieweg, 2017

22. VDI-Richtlinie 2206:2021: Entwicklung mechatronischer und cyber-physischer Systeme, VDI2206a, Verein Deutscher Ingenieure e.V., November 2021

23. Disselkamp, J.-P., Seidenberg, T., Lick, J., Ptock, L., Hovemann, A., Dumitrescu, R.: Engineering genetics: towards the double helix model for integrative product and production design. Proc. Des. Soc. 236–245 (2024). https://doi.org/10.35199/NORDDESIGN2024.26

24. Disselkamp, J.-P., Schütte, B., Lick, J., Westphal, S., Hovemann, A., Dumitrescu, R.: Interfaces and models in interdisciplinary integrative product creation. Proceedings CPSL **2024**, 828–837 (2024). https://doi.org/10.15488/17770

25. Disselkamp, J.-P., Kurpick, D., Schutte, B., Hovemann, A., Dumitrescu, R.: Use cases of generative AI in factory planning: potential and challenges. In: Proceedings of NordDesign 2024, pp. 196–205, 12th–14th August 2024. https://doi.org/10.35199/NORDDESIGN2024.22

26. Disselkamp, J.-P., et al.: Towards the digital factory twin – design guide for creating a 3D factory model. Proc. Des. Soc. **4**, 1979–1988 (2024). https://doi.org/10.1017/pds.2024.200

27. Lick, J., et al.: Digital factory twin: a practioner-driven approach for integrated planning of the enterprise architecture. Procedia CIRP **128**, 603–608 (2024). https://doi.org/10.1016/j.procir.2024.03.038

28. Kattenstroth, F., Disselkamp, J.-P., Lick, J., Dumitrescu, R.: Challenges in the implementation of simulation models for the digital factory twin - a systematic literature review. Procedia CIRP **128**, 442–447 (2024). https://doi.org/10.1016/j.procir.2024.07.052

# Student Session

# Metric-Based Sustainability Evaluation of Additive Manufacturing Processes

Solomon Teye Azu[1,2]([✉]), Alan Hensley[1,2], Víctor Shema Ndikumana[1,2], Syed Ibn Mohsin[1,2], Fazleena Badurdeen[1,2], and I. S. Jawahir[1,2]

[1] Department of Mechanical and Aerospace Engineering, University of Kentucky, Lexington, KY 40506, USA
stazz222@uky.edu
[2] Institute for Sustainable Manufacturing, University of Kentucky, Lexington, KY 40506, USA

**Abstract.** As Additive Manufacturing (AM) continues gaining momentum with greater applications, several sustainability assessment methods for AM have emerged in recent times. However, these methods are typically limited in scope and applicability. A review of existing literature on AM process sustainability metrics reveals that a wide variety of the previously proposed AM sustainability evaluation methods consider only environmental and economic aspects, as opposed to the Tripple-Bottom-Line (Environmental, Societal, and Economic). This paper presents a more comprehensive metrics-based methodology for evaluating the sustainability of AM processes to address the shortcomings of the previously established sustainability evaluation methods. To demonstrate the application of the developed method, a numerical case study on Laser Powder Bed Fusion (LPBF) was conducted to evaluate the sustainability of the process, considering the pre-manufacturing and manufacturing stages of the product life cycle. The results enable evaluating the key benefits and tradeoffs with LPBF demonstrating that the metrics-based method proposed in this paper can effectively evaluate the AM process's sustainability performance.

**Keywords:** Metrics · Sustainable Manufacturing · Additive Manufacturing · Laser Powder Bed Fusion

## 1 Introduction

Manufacturing accounts for one-fifth of the global $CO_2$ emissions and 54% of global energy use [1]. The UN Report 2023 [4] emphasizes the importance of reducing greenhouse gas emissions, material footprint, and resource use while improving global living standards, necessitating adoption of more sustainable manufacturing processes. Novel technologies like AM are considered more sustainable than traditional methods due to their capability to reduce material waste and reduced lead time [2]. Over the years, the revenue generated by AM products and services has increased from less than $2 billion in 2008 to over $18 billion in 2022 [3], indicating growing interest in AM technologies worldwide. In addition to those included in ISO/AS 52910-18 [19], two new families

H. Kohl et al. (Eds.): GCSM 2024, LNME, pp. 809–816, 2025.
https://doi.org/10.1007/978-3-031-93891-7_89

of AM which are, solid-state extrusion (e.g., additive friction stir deposition - AFSD) and solid-state Direct Energy Deposition (DED) (e.g., cold spray AM) have emerged in recent years. All these technologies can broadly be classified into three categories as shown in Fig. 1.

Given the growing emphasis on sustainability and the widespread application of AM technologies, a more comprehensive methodology to evaluate different AM processes and understand the trade-offs in Economic, Environmental, and Societal (Triple-Bottom-Line) impacts is vital. However, most of the previous work addressing the sustainability assessment of AM does not comprehensively consider all the required TBL aspects.

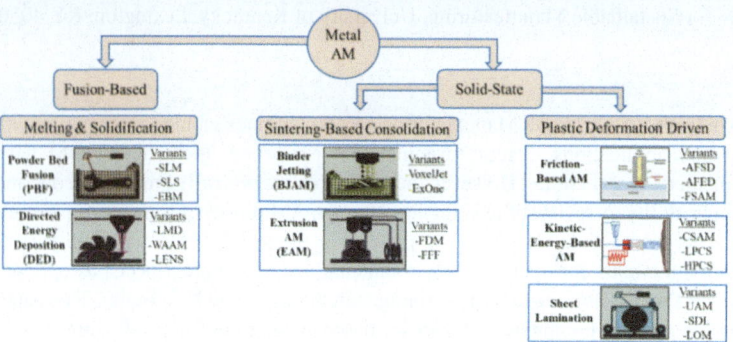

**Fig. 1.** Taxonomy of Metal Additive Manufacturing (with permission from https://hybridman utech.com/)

This paper aims to address this gap through an analysis of sustainability metrics used to assess various AM processes in literature, a comparison with established methods for evaluating TBL sustainability of conventional manufacturing processes, and adapting the latter, as necessary, to evaluate AM processes.

## 2 Literature and Background

Several studies have examined the evaluation of AM process sustainability. Yosofi et al. [7] proposed a comprehensive methodology focused on a general assessment of the environmental impacts of AM processes, considering material, fluid, and electrical inputs. A conceptual model using grey-based methods to assess AM sustainability, highlighting areas like toxicity, carbon footprint, disposal, emissions, energy, cost, and quality for process enhancement is presented in [8]. A multi-criteria assessment for Wire Arc AM (WAAM) is presented in [9], while a comparison of Directed Energy Deposition (DED) and Selective Laser Melting (SLM) in [2] examines sustainability with limited data. These previous works either did not consider postprocessing or did not provide an in-depth analysis of the impacts of postprocessing on overall sustainability. Considering AM's need for various feedstocks and frequent post-processing, it is crucial to evaluate the entire production process for comprehensive sustainability [17]. Several other studies that present sustainability assessment of AM processes are presented in Table 1, indicating which TBL aspects are considered.

**Table 1.** Summary of past work on AM considering TBL

| No | Title & Authors | Year | Scope | TBL aspects considered |
|---|---|---|---|---|
| 1 | Chan et al. [2] | 2017 | SE* | En, Ec, So (Partial) |
| 2 | Kellens et al. [5] | 2021 | SE | En |
| 3 | Yosofi et al. [7] | 2019 | SE | En, Ec |
| 4 | Agrawal et al. [8] | 2019 | SE | En, Ec, So (Partial) |
| 5 | Jung et al. [10] | 2023 | LR | En, Ec, So |
| 6 | Bourhis et al.[11] | 2013 | SE* | En |
| 7 | Huang et al. [12] | 2013 | LR | So |
| 8 | Colorado et al.[13] | 2020 | LR | En |
| 9 | Stefaniak et al.[14] | 2021 | LR* | So |

\* Note: sources for some data used in the case study presented later (SE = Sustainability Evaluation, LR = Literature Review, En = Environmental, Ec = Economical, So = Social)

As indicated, most of the studies focus on environmental factors and often neglect economic and societal aspects. While some studies such as [2, 8], and [10] consider societal aspects, they frequently rely on assumptions. For example, [2] assumes continuous operator monitoring of automated machines and applies injury rates from primary metal manufacturing to additive manufacturing (AM) processes due to data limitations, despite differences in automation and manual labor, and does not consider the material properties or performance of the finished titanium aerospace components. This highlights the need for a thorough evaluation of all TBL aspects.

On the other hand, numerous studies have examined sustainability evaluation of conventional manufacturing processes. One approach identified six key factors for manufacturing sustainability: energy consumption, environmental impact, waste reduction, manufacturing cost, operational safety, and personnel health. [15].

Later, this evolved into the Process Sustainability Index (ProcSI) [6], which has been demonstrated by application to cryogenic machining, and leveraged to develop a method for Product Sustainability Index (ProdSI), incorporating additional aspects [16]. These methods proposed for conventional manufacturing process sustainability evaluation are more comprehensive and include metrics for TBL evaluation. While some characteristics of AM processes differ from those of conventional processes, the methods to evaluate TBL sustainability performance of the latter provide a good foundation to develop sustainability assessment methods for AM.

This paper leverages the ProcSI and ProdSI methods, with findings from AM literature, to develop a comprehensive method for assessing the sustainability of AM processes.

## 3    Methodology

An approach similar to ProcSI is used to organize the metrics and derive three sub-indices of the TBL. The proposed clusters build on the work done in ProcSI by extending its framework. These clusters are further divided into subclusters, which are then broken down into individual, quantifiable metrics. As discussed, the metrics have been selected and grouped based partially on previous literature. However, the inherent differences in AM make some metrics from TM impractical or invalid for this use. New metrics are introduced where relevant to address AM-specific aspects to evaluate sustainability. A bottom-up approach is used, starting with assigning values to the metrics, and then calculating the subcluster and cluster values, followed by the three subindices. These are combined into the AM Process Sustainability Index, with weighting and aggregation at each step.

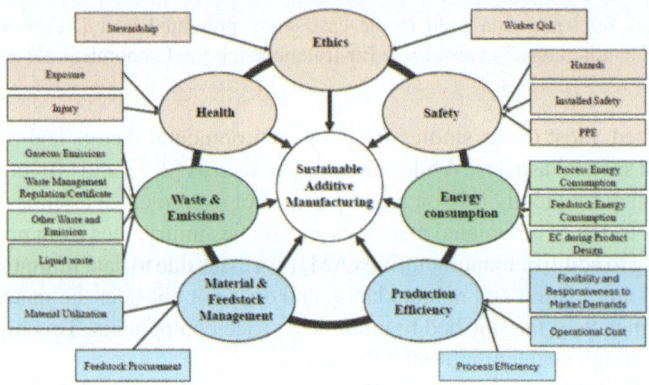

**Fig. 2.**  Proposed Clusters for AM Process Sustainability Assessment

### 3.1    Framework for Metrics

The proposed assessment approach is based on the computational methodology from [16] leveraging the clusters from ProcSI used to evaluate conventional processes and uses new AM-relevant metrics compiled from literature. Seven key clusters, six of which are extensions from previous work on ProcSI and including the newly proposed "Ethics" cluster, are defined to evaluate an AM process. Each cluster is divided into subclusters, as shown in Fig. 2, with specific metrics defined for assessment. Table 2 gives more details of the metrics used in each subcluster. These elements are hierarchically organized with normalization, weighting, and aggregation to create a comprehensive index for assessing AM processes.

### 3.2    Normalization, Weighting, and Aggregation

As the metrics have different units of measurement and cannot be aggregated directly, normalization is used to convert each into a percentage, where a higher value means

**Table 2.** Proposed Hierarchical Framework for AM Process Sustainability Assessment

| Subclusters | Metrics | Subclusters | Metrics |
|---|---|---|---|
| Material Utilization | Material Utilization Rate/Ratio | Solid Waste | Feedstock Waste During AM |
| | Recyclability Rate | | Post-Processing Waste |
| Feedstock Procurement | Cost per Unit of Feedback | Liquid Waste | Unused/Excess Resin |
| | Transportation Cost | | Cleaning Agents/Solvents |
| | Feedstock Price Stability | Other Waste and Emissions | Noise |
| | Storage Cost | | Heat |
| Operational Cost | Energy Cost per Unit Output (KW/h/unit) | | Radioactive emissions |
| | Maintenance Cost | Waste Management Regulation/Certificate | Waste Management Regulation/Compliance |
| | Labor Operation Cost per Unit | | |
| | Machine Cost | PPE | Effectiveness |
| Process Efficiency | Throughput Time | | Availability |
| | Defect Rate | Installed Safety | Effectiveness |
| | Utilization Rate | | Availability |
| Flexibility and Response to Market Demands | Customization Cost Impact | Hazards | Heat |
| | Design Iteration Adaptability Time | | Coolant |
| Feedstock Energy Consumption | Feedstock Production | | Equipment |
| | Feedstock Transportation | Exposure | Particle emission |
| Process Energy Consumption | 3D Printing | | Coolant emissions |
| | Postprocessing/Machining | | Noise |
| EC during Product Design | Electricity | Injury | Severity of Injuries |
| | CO2 at Feedstock Production | | Rates of Injury/Health Incidents |
| Gaseous Emissions | CO2 Emissions during Transportation | Worker QoL | Operator Agency |
| | CO2 Emission during AM | | Workload |
| | CO2 Emission during Postprocessing | Stewardship | Disposals |
| | | | Equipment lifespan |

higher sustainability. The metrics can be either a ratio, scale, or direct value and the normalization procedure will be different depending on the type (see [16] for details). The normalized metrics are weighted (based on data, and assumptions based on trends, from literature). The weighting of subclusters and clusters, as well as the aggregation process used in this paper, also adheres to the equations outlined in [16].

## 3.3 Case Study

A numerical case study for producing a 1 kg Titanium part with LPBF is used to demonstrate the application of the proposed methodology. Most data for the environmental and economic metrics are from the Digital Alloys Inc. Database [18]. Their dataset was synthesized from a diverse array of sources, the main contribution being from industry experts at Reeves Insight Ltd. When data presented conflicting values or ranges, they used midpoints and arithmetic averages to standardize and reconcile the discrepancies. Data for $CO_2$ emissions during feedstock transportation, noise level, and some other metrics are sourced from the US Environmental Protection Agency database and literature sources (some of which are shown in Table 1), as well as insights from other literature (see [20]). Metrics such as injury rates and material utilization were assumed based on trends from existing literature.

Data to evaluate the other metrics in the societal sub-index were not available in the above-mentioned sources. Therefore, values were subjectively determined using a one-to-five scale or assigning binary responses of Yes/No. This necessitates careful attention to ensure reasonableness on the values assigned to ensure validity of the assessment for the societal aspect.

## 4   Results and Discussion

The proposed methodology yields 0.86, 0.78, and 0.67 on a 0 to 1 scale in the economic, environmental, and societal indices, respectively. Unlike previous work, such as that from Yosofi et al. [7] which analyzes only economic and environmental aspects, TBL is considered in this case study. In this case study, the high scores in economic and environmental indices are primarily due to lower material waste, which saves both lost feedstock material and cost of operation. As most AM methods are near-net-shape processes, these indices are likely to be scored consistently high.

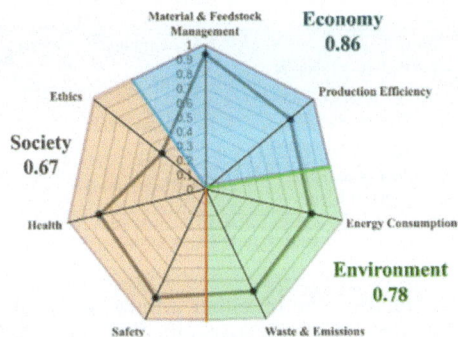

**Fig. 3.** TBL Sustainability Evaluation LPBF Example

Flexibility and responsiveness to market demand are some of the key features of the AM process. This particular aspect is not evaluated in the literature and this case study considers them based on the developed metrics to assess its impact on sustainability.

The societal index falls short due partially to the hazards that are present in powder-based AM processes. These hazards include health hazards associated with contact with metal powder and safety hazards specific to powders such as titanium-like combustibility. Once again, as many of these hazards are typical of AM processes, this is likely to be a similar result to other processes. The usage of low-resolution scales (such as binary or 1 to 5 scales) leads to significant impacts on the index as a whole. The drop in societal index in Fig. 3 can be attributed to the low-resolution scales used in societal aspect assessment, especially, the ethics cluster, which uses mostly low-resolution scales, as being significantly lower than any of the other clusters. This could be resolved by either using higher resolution scales or by further investigating the weight assigned to the metrics within this cluster or the cluster itself. The weighting of clusters about material usage may also warrant investigation as they may be too similar between different systems to derive a meaningful comparison from. Overall, this case study yields that a 1 kg titanium component produced by LPBF receives a score of 0.75 by this methodology.

It's important to emphasize that the score in this case study is not objective and does not provide definitive insights into LPBF on its own. However, when used as a comparative tool, this methodology establishes a quantifiable relative sustainability impact between different AM processes, enabling stakeholders to make informed decisions.

## 5  Conclusions and Future Work

This paper identifies gaps present in the evaluation of the sustainability of AM processes from the TBL perspective and addresses them by proposing a new metric-based AM sustainability assessment tool. This methodology has incorporated comprehensive societal aspects of AM sustainability and also expanded on the work previously done on the economic and environmental aspects. Then, it was used in a numeric case study to demonstrate its application. Future work needs to be done on conducting further case studies involving more AM processes and investigating the weighting and scale resolution of the clusters to ensure that the methodology is viable for creating a robust comparison of the sustainability of AM processes.

## References

1. Zmicier Vaskovich, J.M., Farbstein, E.: The tools and initiatives that enable manufacturers to reduce carbon emissions, 2023
2. Chan, R., Manoharan, S, Haapala, K.R.: Comparing the sustainability performance of metal based additive manufacturing processes. In: Proceedings of the ASME 2017 International Design Engineering Technical Conferences and Computers and Information in Engineering Conference
3. Wohlers Report 2023. Analysis. Trends. Forecasts. 3D Printing and Add. Mfg State of the Industry
4. United Nations., The Sustainable Development Goals Report. 2023
5. Kellens, K., Mertens, R., Paraskevas, D., Dewulf, W., Duflou, J.R.: (Environmental impact of additive manufacturing processes: does AM contribute to a more sustainable way of part manufacturing? Procedia Cirp **61**, 582–587 (2017)
6. Lu, T., Jawahir, I.S.: Metrics-based sustainability evaluation of cryogenic machining. Procedia Cirp **29**, 520–525 (2015). **Cirp**
7. Yosofi, M., Kerbrat, O., Mognol, P.: Additive manufacturing processes from an environmental point of view: a new methodology for combining technical, economic, and environmental predictive models, 2019
8. Agrawal, R.V.S.: Sustainability evaluation of additive manufacturing processes using grey-based approach. Grey Systems: Theory and Application. ahead-of-print. 2019
9. Priarone, P.C., Pagone, E., Martina, F., Catalano, A.R., Settineri, L.: Multi-criteria environmental and economic impact assessment of wire arc additive manufacturing. CIRP Ann. **69**(1) (2020)

10. Jung, S., Kara, L.B., Nie, Z., Simpson, T.W., Whitefoot, K.S.: Is additive manufacturing an environmentally and economically preferred alternative for mass production? Environ. Sci. Technol. (2023)

11. Bourhis, F.L., Kerbrat, O., Hascoet, J.Y., Mognol, P.: Sustainable manufacturing: evaluation and modeling of environmental impacts in additive manufacturing. Int. J. Adv. Manuf. Technol. **69**, 1927–1939 (2013)

12. Huang, S.H., Liu, P., Mokasdar, A., Hou, L.: Additive manufacturing and its societal impact: a literature review. Int. J. Adv. Manuf. Technol. (2013)

13. Colorado, H.A., Velásquez, E.I.G., Monteiro, S.N.: Sustainability of additive manufacturing: the circular economy of materials and environmental perspectives. J. Mat. Res. Technol. (2020)

14. Stefaniak, A.B., Du Preez, S., Du Plessis, J.L.: Additive manufacturing for occupational hygiene: a comprehensive review of processes, emissions, & exposures. J. Toxicol. Environ. Health (2021). (Part B, 24(5), )

15. Wanigarathne, P.C., Liew, J., Wang, X., Dillon Jr, O.W., Jawahir, I.S.: Assessment of process sustainability for product manufacture in machining operations. In: Global Conference on Sustainable Product Development and Life Cycle Engineering (2004)

16. Shuaib, M., Seevers, D., Zhang, X., Badurdeen, F., Rouch, K.E., Jawahir, I.S.: Product sustainability index (ProdSI). J. Ind. Ecol. **18**, 491–507 (2014). https://doi.org/10.1111/jiec.12179

17. Peng, X., Kong, L., Fuh, J.Y.H., Wang, H.: A review of post-processing technologies in additive manufacturing. J. Mfg Mater. Process. **5** (2021)

18. Alloys, D.: Guide to Metal Additive Manufacturing, 2019

19. Additive manufacturing—Design—Requirements, guidelines and recommendations, 2018. https://www.iso.org/standard/67289.html

20. "AM-Sustainability-Metrics/ADDITIONAL REFERENCES." https://github.com/AzuSolomon/AM-Sustainability-Metrics/blob/main/ADDITIONAL%20REFERENCES. Accessed 20 Aug 2024

21. Hybrid Manufacturing Technologies. "About Us." Hybrid Manufacturing Technologies. https://hybridmanutech.com/. Accessed 3 Apr 2024

# A Framework Towards Enhancing Sustainability Performance of Wire Arc Additive Manufacturing

Syed Ibn Mohsin[1,2(✉)], Tamm Omar[1,2], James Caudill[1,2], Fazleena Badurdeen[1,2], and I. S. Jawahir[1,2]

[1] Institute for Sustainable Manufacturing, University of Kentucky, Lexington, KY 40506, USA
im.syed@uky.edu

[2] Department of Mechanical and Aerospace Engineering, University of Kentucky, Lexington, KY 40506, USA

**Abstract.** Manufacturing has undergone tremendous evolution over the past decades with the introduction of Additive Manufacturing (AM) processes. Wire Arc Additive Manufacturing (WAAM), an energy-intensive fusion-based process, is one of the AM technologies that has been at the forefront of the metal AM. Given its widespread application, it is imperative to determine approaches to enhance the sustainability performance of WAAM. Through a detailed critical and comparative structured literature review, this study evaluates the specific contributions of the process towards sustainable manufacturing. The review showed that previous works do not comprehensively incorporate all sustainability aspects in the assessment of the process. Additionally, they lacked consideration of the suitability of WAAM for manufacturing the component and did not include redesigning for WAAM. This paper proposes a framework for sustainable WAAM to address these gaps and enhance the sustainability of the process during both the pre-manufacturing and manufacturing stages. The framework promotes the application of the 6R principles throughout the entire life cycle of the product; it offers guidelines for stakeholders to enhance the potential of WAAM for more sustainable manufacturing solutions.

**Keywords:** Additive Manufacturing · Wire Arc Additive Manufacturing · Sustainability

## 1 Introduction

Sustainable development is a complex term that has gone through several iterations [1]. While early definitions focused on the environmental aspect of sustainability, modern-day sustainability refers to creating a balance between economic benefit, environmental protection, and societal enhancement. These three sustainability lenses of economic, environmental, and societal outcomes are called the triple bottom line (TBL). All 3 components must be balanced and developed in parallel to ensure a prosperous society.

© The Author(s) 2025
H. Kohl et al. (Eds.): GCSM 2024, LNME, pp. 817–825, 2025.
https://doi.org/10.1007/978-3-031-93891-7_90

These sustainable practices must be carried out across the entire life cycle of a product. This begins with Pre-Manufacturing (PM) (extraction and sourcing of materials) followed by Manufacturing (M) (assembly of consumer product), Use (U) (product use), and Post-use (PU) (recovery and disposal of components) stages [2]. The TBL approach to sustainability can be achieved by utilizing the 6R's approach to manufacturing [3]. The 6Rs refer to the practice of Reduce, Reuse, Recycle, Recover, Redesign, and Remanufacture. Reduce emphasizes using less resources, energy, materials, and waste across all life cycle stages. Reuse focuses on reusing the product or its components after its end-of-life (EoL). Recycling converts EoL products/components into new materials. Recover involves salvaging products at EoL and disassembling them into components that can be reused, remanufactured, or recycled. Redesign encompasses changing product designs to improve overall sustainability performance. Lastly, remanufacture involves reprocessing components from products for restoration and improvement [4]. When the 6R's approach is combined with modern technological advancements a much more sustainable economy can be achieved.

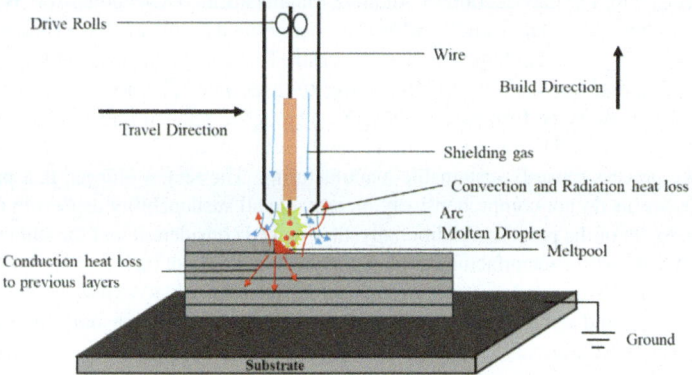

**Fig. 1.** Schematic of a WAAM system

Additive Manufacturing (AM) is one of the technologies that is gaining more prominence in manufacturing. Wire Arc AM (WAAM) is a form of AM in which an electric arc as shown in Fig. 1, melts metal wire feedstock in a layer-by-layer fashion [5]. This process creates products with much less material waste than typical subtractive manufacturing of large- and medium-scale components. It is characterized by high part density, low equipment cost, high energy efficiency, reduced lead times, and high deposition rates. WAAM products, however, often have poor dimensional accuracy and surface finish necessitating post-processing via machining. The freedom of design and superior process performance make WAAM an ideal choice for sectors like aerospace, marine, construction, and defense. Compared to other AM processes, WAAM has a higher potential for wide industrial application due to being relatively cheaper and the workforce already familiar with welding. As the process can be easily adapted and it is one of the most mature large-scale AM technologies [6], making the process sustainable would ensure sustainability across the manufacturing industry [7].

While many prior studies have focused on this topic, enhancing WAAM sustainability requires a structured approach to guide decision-making across all life cycle stages. To facilitate developing a structured approach a literature review was done to analyze if and how aspects such as WAAM suitability, part redesign for WAAM, material selection, etc., that impact sustainability performance has been considered. Based on these insights, the paper presents a 6R principle integrated framework to facilitate better decision-making and enhance WAAM's sustainability performance.

## 2 Background and Methodology

WAAM has been used as early as 1925 to construct 3D decorative [8] used. However, the technique was not widely popular at that time. Technological developments led to more substantial advancements in WAAM in the '90s and early 2000s. Initial work focused on process optimization and exploring the potential of WAAM. The earliest study on the sustainability assessment of WAAM was reported in 2016 [9].

An extensive review of literature focusing on WAAM sustainability was carried out for this work. Web of Science and Scopus were selected as search engines to collect relevant literature. The keywords used were Arc Additive Manufacturing AND [Sustainable, Sustainability, Green, Environment(al), Economic(al)]. In total 660 unique papers were identified. These were filtered down to 40 papers based on the titles. After further examination, 26 papers were selected for detailed review (listed here [10]). The scope of the papers was evaluated using metrics-based methods from [10] and [11] modified to incorporate AM process sustainability considerations. In these methods, the sub-indices represent the TBL, further divided into clusters and sub-clusters. Table 1 contains the summary of the papers that consider the largest number of subclusters.

The earliest study on WAAM sustainability assessment compared it with other AM techniques using life cycle assessment (LCA) to highlight potential gains in material usage compared to Traditional Manufacturing (TM) and superiority in regard to power consumption [9]. Environmental and economic comparison of WAAM and CNC for three different prints, a gear, cylinder, and S-shape demonstrates that WAAM saves between 40–70% of materials and reduces the environmental impact by 12–47% [7]. The comparison of WAAM, Laser Powder Bed Fusion (LPBF), and CNC in steel wall manufacturing showed that the CNC process performed the best followed by WAAM and LPBF [13]. By performing a LCC and LCA it was shown that raw materials and labor were the main cost drivers. Compared to LPBF, the WAAM, and CNC had 45% and 67% environmental impact respectively and CNC milling costs 67% less than WAAM. The interested reader is referred to [10] for further details.

Existing literature uses a WAAM component against a CNC/other process to compare sustainability performance. The suitability of WAAM for that component, however, was not a determining factor. The comparisons often overlook the unique advantages of WAAM, such as its ability to print in a layer-by-layer manner with higher resolution. Also, material sourcing was not considered during the planning stage but merely during post-process analysis to assess the impact of the material source. The lack of comprehensive TBL analysis of WAAM is also evident from the review. There is also a lack of emphasis on redesigning components to fully leverage the capabilities of the

WAAM system. Furthermore, material utilization emerges as a critical criterion driving the sustainability performance of the manufacturing process which needs to be evaluated at an early stage.

**Table 1.** Literature Summary

| Sub-Indices | Clusters | Subclusters | [14] | [13] | [15] | [16] | [17] | [7] | [18] | [9] | [6] | No. of Paper Asses this Subcluster |
|---|---|---|---|---|---|---|---|---|---|---|---|---|
| Economic | Material & Feedstock Management | Material Utilization | ✓ | ✓ | ✓ | ✓ | ✓ | ✓ | ✓ | ✓ | ✓ | 24 |
| | | Feedstock Procurement | ✓ | ✓ | ✓ | ✓ | ✓ | ✓ | | ✓ | | 13 |
| | Machine Cost | AM Machine Cost | ✓ | ✓ | ✓ | ✓ | ✓ | ✓ | | ✓ | ✓ | 16 |
| | | Subtractive Machine Cost | ✓ | ✓ | ✓ | ✓ | | ✓ | | ✓ | ✓ | 13 |
| | Production Efficiency | Operational Cost | ✓ | ✓ | ✓ | ✓ | ✓ | | ✓ | ✓ | ✓ | 15 |
| | | Process Efficiency | ✓ | ✓ | | | | ✓ | ✓ | ✓ | ✓ | 14 |
| | | Flexibility and Responsiveness to Market Demands | | | | | | | | | | 0 |
| Environmental | Energy Consumption | Feedstock Energy Consumption | ✓ | ✓ | ✓ | ✓ | ✓ | | ✓ | | | 13 |
| | | Process Energy Consumption | ✓ | ✓ | ✓ | ✓ | ✓ | ✓ | ✓ | ✓ | ✓ | 24 |
| | | EC during Product Design | | | | | | ✓ | ✓ | | | 9 |
| | Waste & Emissions | Gaseous Emissions | ✓ | ✓ | ✓ | ✓ | ✓ | ✓ | ✓ | ✓ | ✓ | 21 |
| | | Solid Waste | ✓ | ✓ | | ✓ | | | ✓ | | ✓ | 7 |
| | | Liquid waste | ✓ | ✓ | | ✓ | | ✓ | ✓ | | ✓ | 8 |
| | | Other Waste and Emissions | ✓ | ✓ | | | | | | | | 3 |
| | | Waste Management Regulation/Certificate | | | | | | | ✓ | | | 3 |
| Societal | Safety | Personal Protective Equipment | | | | | | | | | | 0 |
| | | Installed Safety | | | | | | | | | | 0 |
| | | Hazards | | | | | | | | | | 0 |
| | Health | Pollution | ✓ | ✓ | | | | ✓ | ✓ | | ✓ | 5 |
| | | Injury | | | | | | | | | | 0 |
| | Ethics | Worker QoL | | | | | | | | ✓ | | 1 |
| | | Stewardship | | | | | | | | | | 0 |

Although various standards exist for AM, they do not incorporate the sustainability aspects of the process. To resolve the gaps that exists in literature a framework to facilitate better decision making for sustainable WAAM is required.

## 3 Framework for Sustainable WAAM

Sustainable WAAM requires considering all the life cycle stages during its design phase. Although the literature predominantly focuses on the PM and M stages when discussing sustainability, the latter significantly impacts both the U and PU phases. As can be seen in Fig. 2, the PM stage consists of material selection and procurement. This phase provides opportunities to select the most sustainable sources for procuring materials. The M stage encompasses design and process optimization. The component must be designed to meet the criteria for the U stage. Additionally, how the component will be disassembled must be considered to recover value in PU.

After PU, the component must be evaluated to determine which parts can be reused and sent to the U stage. Remanufacturable components can be returned to the M stage. Components with recyclable materials will be directed to the PM stage for use in the same or different components and remaining components discarded. Throughout all stages, emphasis should be placed on reducing use of materials and resources as well as minimizing waste and emissions. This enables a 6R-based approach to maximize the sustainability of a WAAM component throughout the life cycle stages.

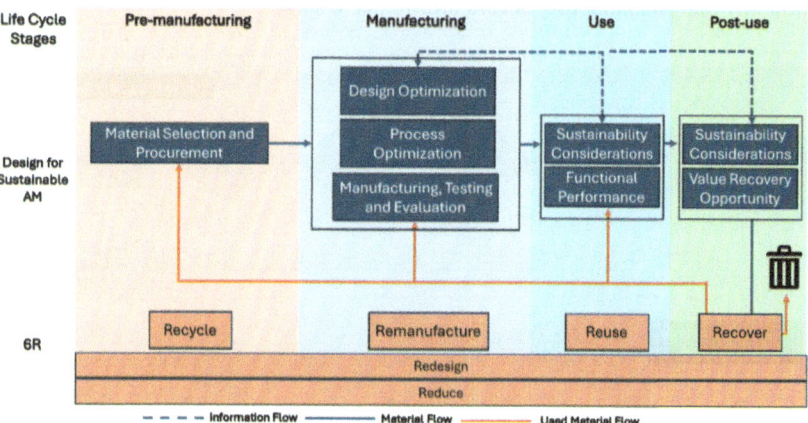

**Fig. 2.** Total Life Cycle Consideration of Sustainable AM Process

To facilitate incorporating these considerations, this paper proposes a framework for sustainable WAAM decision-making adapting ISO/ASTM 52910: 2018 (E), "Additive manufacturing—Design—Requirements, guidelines, and recommendations," [19] (see Fig. 3). The standard provides a general approach for AM without any consideration of sustainability aspects. The proposed approach outlines factors to be considered and the required approach during PM and PM stages for total life cycle WAAM sustainability based on findings from the review and established frameworks [20].

As shown in Fig. 3, the framework consists of four main stages. The first stage involves the 'WAAM Suitability Check' where material-process compatibility is assessed. A key aspect of determining the suitability of an AM process is the amount of material that must be removed from the billet to achieve the desired shape. Literature indicates that TM processes generally require more raw material. If the AM process requires a comparable amount of material to create the final component shape, TM processes may be more economical and environmentally suitable. If WAAM is chosen, it is essential to ensure that the component's functionality can be maintained through the process. Subsequently, "Material Selection and Sourcing" involves various considerations such as if the material meets functional properties, cost of material, transportation cost and emissions, recyclability, and other sustainability aspects.

During "Design and Process Optimization", the component geometry is reevaluated for optimal deposition and redesigned for desired functionality, considering the unique features of WAAM. This will be an iterative process, ensuring optimal design for manufacturing, path planning, and process optimization, all while considering the TBL. In "Manufacturing, Testing, and Evaluation" the first step is part deposition under optimized WAAM conditions. Then, post-treatment and surface or heat treatments are typically applied to impart the desired properties. Various non-destructive tests can then be conducted to check for defects and verify the desired properties. The final step is to ensure the component meets the standards for AM parts and obtain the necessary certification. This proposed framework ensures appropriate decision-making to maximize functionality and sustainability across all life cycle stages.

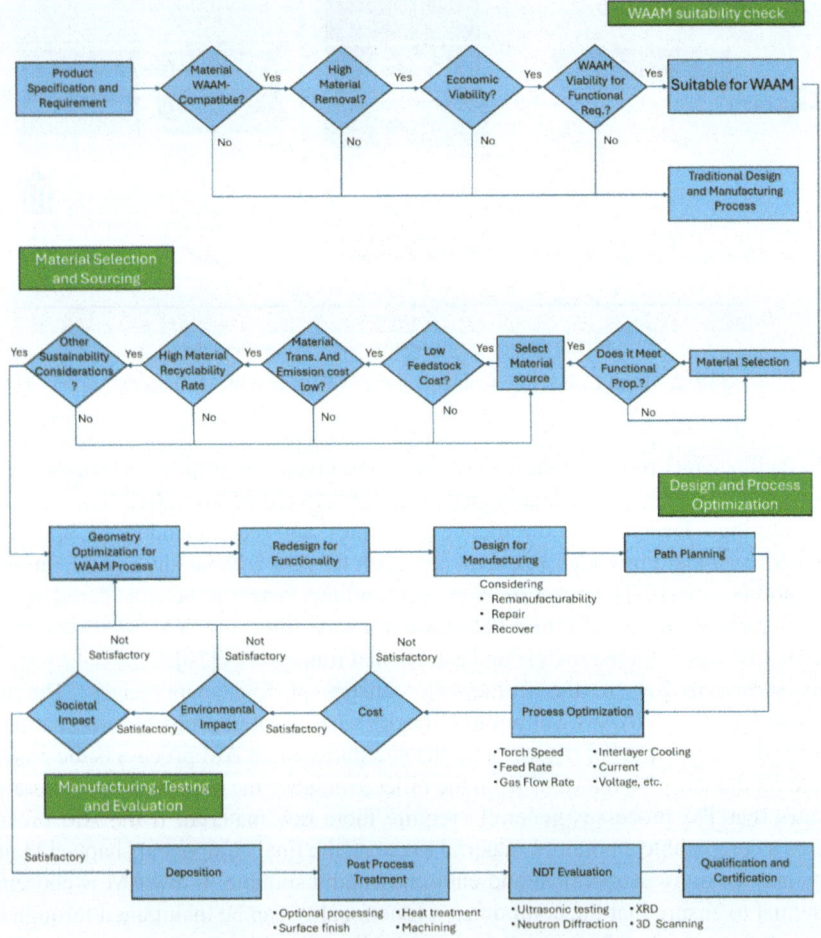

**Fig. 3.** Framework for sustainable WAAM

In Fig. 4, the application of the first step of the framework is demonstrated by considering different components gathered from the literature. The four shapes A, B, C (all from [7]), and D [13], all made with steel, are assessed for the relevant decisions.

The material removal rate (MR) is the ratio between billet to final part weight for TM vs. the weight of wire to final part for WAAM. As the final part is the same, it comes down to the ratio of the billet to wire weight. The higher this value, the more economical and environmental it will be to choose WAAM over TM. As 'D' had a lower MR value TM should be chosen. Next, is an economic assessment to identify the difference in cost for TM vs. WAAM. Suppose they are comparable; then all three components can be assessed for the viability of meeting functional property requirements via WAAM. Given 'A' is a load-bearing part, and the residual stress is higher in WAAM, TM would be more appropriate whereas WAAM would be suitable for the other two components. Thereafter, the following steps in the framework (not discussed due to space) can be applied.

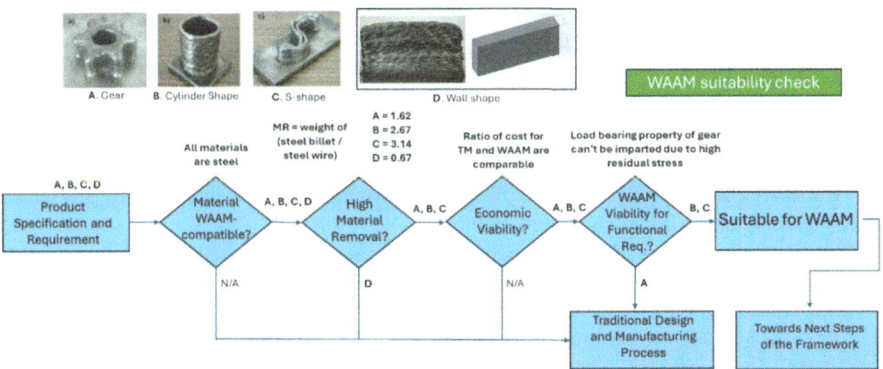

**Fig. 4.** Partial application of framework for sustainable WAAM

## 4   Conclusions and Future Work

A comprehensive review of studies focusing on sustainability in WAAM revealed that most did not consider the TBL or the total life cycle TBL performance when selecting WAAM. To address these gaps, a framework for sustainable WAAM was developed by adapting existing AM standards and integrating 6R principles. Applying the framework for decision-making ensures that economic viability, environmental impact, and social responsibility are all considered for WAAM. Thus, it not only addresses the gaps identified in literature but also offers a practical tool to optimize the sustainability of WAAM operations. Future work will validate this framework through case studies and experimental assessments to demonstrate its effectiveness in real-world applications.

# References

1. Badurdeen, F., Goldsby, T.J.: Design and Management of Sustainable Supply Chains. Cambridge University Press, United Kingdom, Forhtcoming
2. Badurdeen, F., Iyengar, D., Goldsby, T.J., Metta, H., Gupta, S., Jawahir, I.: Extending total life-cycle thinking to sustainable supply chain design. Int. J. Prod. Lifecycle Manag. 4(1–3), 49–67 (2009)
3. Badurdeen, F., Jawahir, I.S., Rouch, K.E.: A metrics-based evaluation of sustainable manufacturing at product and process levels, 2024
4. Hernández, A.E.B., Lu, T., Beno, T., Fredriksson, C., Jawahir, I.S.: Process sustainability evaluation for manufacturing of a component with the 6R application. Proced. Manuf. 33, 546–553 (2019)
5. Xia, C., et al.: A review on wire arc additive manufacturing: monitoring, control and a framework of automated system. J. Manuf. Syst. 57, 31–45 (2020)
6. Bekker, A.C.M., Verlinden, J.C.: Life cycle assessment of wire + arc additive manufacturing compared to green sand casting and CNC milling in stainless steel. J. Clean. Prod. 177, 438–447 (2018). https://doi.org/10.1016/j.jclepro.2017.12.148
7. Reis, R.C., Kokare, S., Oliveira, J., Matias, J.C., Godina, R.: Life cycle assessment of metal products: a comparison between wire arc additive manufacturing and CNC milling. Adv. Ind. Manuf. Eng. 6, 100117 (2023)
8. Ralph, B.: Method of making decorative articles, April 1925
9. Bekker, A., Verlinden, J.C., Galimberti, G.: Challenges in assessing the sustainability of wire+ arc additive manufacturing for large structures, 2016
10. "WAAM_sustainability_ref/list.txt." https://github.com/syedibnmohsin/WAAM_sustainability_ref/blob/main/list.txt. Accessed 24 June 2024
11. Shuaib, M., Seevers, D., Zhang, X., Badurdeen, F., Rouch, K.E., Jawahir, I.: Product sustainability index (ProdSI) a metrics-based framework to evaluate the total life cycle sustainability of manufactured products. J. Ind. Ecol. 18(4), 491–507 (2014)
12. Lu, T., Jawahir, I.S.: Metrics-based sustainability evaluation of cryogenic machining. In: 22nd CIRP Conference Life Cycle Engineering, vol. 29, pp. 520–525, January 2015. https://doi.org/10.1016/j.procir.2015.02.067
13. Kokare, S., Oliveira, J., Santos, T., Godina, R.: Environmental and economic assessment of a steel wall fabricated by wire-based directed energy deposition. Addit. Manuf. 61, 103316 (2023)
14. Dias, M., Pragana, J.P., Ferreira, B., Ribeiro, I., Silva, C.M.: Economic and environmental potential of wire-arc additive manufacturing. Sustainability 14(9), 5197 (2022)
15. Priarone, P.C., Campatelli, G., Montevecchi, F., Venturini, G., Settineri, L.: A modelling framework for comparing the environmental and economic performance of WAAM-based integrated manufacturing and machining. CIRP Ann. 68(1), 37–40 (2019). https://doi.org/10.1016/j.cirp.2019.04.005
16. Kokare, S., Oliveira, J., Godina, R.: Comparison of Wire Arc Additive Manufacturing and Subtractive Manufacturing Approaches from an Environmental and Economic Perspective, presented at the International Conference on Flexible Automation and Intelligent Mfg, pp. 868–878, Springer, 2023
17. Catalano, A.R., Pagone, E., Martina, F., Priarone, P.C., Settineri, L.: Wire Arc Additive Manufacturing of Ti-6Al-4V components: the effects of the deposition rate on the cradle-to-gate economic and environmental performance. Procedia CIRP 116, 269–274 (2023)
18. Maheshwari, P., Khanna, N., Hegab, H., Singh, G., Sarıkaya, M.: Comparative environmental impact assessment of additive-subtractive manufacturing processes for Inconel 625: a life cycle analysis. Sustain. Mater. Technol. 37, e00682 (2023)

19. ISO/ASTM 52910: 2018 (E), "Additive manufacturing—Design—Requirements, guidelines and recommendations", 2018
20. Vaneker, T., Bernard, A., Moroni, G., Gibson, I., Zhang, Y.: Design for additive manufacturing: framework and methodology. CIRP Ann. **69**(2), 578–599 (2020)

# Environmental Impacts of Hydrogen Fuel Cell Systems in Off-Grid Energy Systems

Leonhard Garhammer, Markus Tophoven, Lukas Sturm, and Semih Severengiz[✉]

Bochum University of Applied Sciences, Sustainable Technologies Laboratory, Am Hochschulcampus 1, 44801 Bochum, Germany
semih.severengiz@hs-bochum.de

**Abstract.** The energy transition is challenged by the intermittent nature of renewable energies, unlike conventional sources. This highlights the need for effective and environmentally friendly storage solutions to balance energy generation fluctuations. Green hydrogen emerges as a key sustainable energy carrier and a resource-efficient alternative to batteries, offering clean combustion but requiring infrastructure with significant environmental impact. Hence, assessing the sustainability of green hydrogen systems and finding ways to reduce their environmental impact is crucial. This paper examines the environmental impacts of a fuel cell system used in an off-grid solar-powered hydrogen storage solution in Tema, Ghana, through a life cycle assessment. The global warming potential ($GWP_{100}$) of the system is 342.62 kg $CO_2$-eq per $kW_{el}$. The $GWP_{100}$ of the fuel cell system is composed of four fuel cells, contributing 58%, cabinet (23%), battery (16%), and controller (3%). Significant impacts result from material extraction and processing for membrane electrodes and bipolar plates, despite these making up only 7.5% of the system's weight. This paper is the first to evaluate the environmental impacts of a fuel cell system, ensuring the sustained provision of electricity in off-grid renewable energy systems. Consequently, it is not directly comparable to other LCAs of fuel cell stacks.

**Keywords:** LCA · fuel cell system · PEMFC · PEM fuel cell · life cycle assessment · fuel · electricity grid

## 1 Introduction

To facilitate the energy transition and achieve a sustainable and reliable energy supply without relying on fossil fuels, developing effective and environmentally friendly energy storage solutions is crucial. This is necessary to compensate for the intermittent nature of renewable energy sources and ensure an optimized energy supply. Green hydrogen has emerged as a promising alternative to fossil fuels. Green hydrogen can be utilized not only for national power grids, but also for the sustained provision of electricity to smaller off-grid renewable energy systems. This is of particular importance in regions like sub-Saharan Africa, where 65% of the population lacks access to the national power

L. Garhammer, M. Tophoven and L. Sturm—These authors contributed equally to this work.

© The Author(s) 2025
H. Kohl et al. (Eds.): GCSM 2024, LNME, pp. 826–834, 2025.
https://doi.org/10.1007/978-3-031-93891-7_91

grids [1]. For that reason, diesel generators represent the predominant source for off-grid electricity production in Ghana, which has led to a 36% increase in fossil fuel consumption for power generation alone between 2016 and 2021 [2, 3]. However, the abundant solar irradiance in sub-Saharan countries like Ghana makes them optimal candidates for the production of green hydrogen through the utilization of surplus energy from off-grid energy systems, such as solar mini-grids [4]. Despite their benefits, the necessary infrastructure for those hydrogen systems entails significant environmental costs. Therefore, it is important to examine the sustainability of green hydrogen as an energy source and find ways to improve the environmental footprint of hydrogen systems. A crucial component of this solution is the fuel cell system, which converts hydrogen produced by an electrolyzer during periods of excess electricity generation back into electrical energy. Based on a systematic literature review in Scopus and Web of Science using the keywords "fuel cell", "PEM", "PEMFC", "LCA", and "life cycle assessments" we found that this paper is the first to assess the utilization of fuel cells in off-grid energy systems and their resulting environmental impact. Previous research has primarily focused on fuel cell applications in the transportation sector [5–7]. The research gap is addressed by conducting a cradle-to-gate life cycle assessment (LCA) of a polymer electrolyte fuel cell system deployed in a solar mini-grid in Tema, Ghana [8–11]. The study focuses on a $10\,kW_{el}$ fuel cell system comprising four $2.5\,kW_{el}$ proton exchange membrane fuel cells integrated with a controller into a cabinet.

The paper commences with an introduction to the LCA methodology employed, after which the most significant findings are presented, and the key areas of environmental impact are identified. The results are then discussed, with particular attention paid to the most noteworthy anomalies. The paper concludes by summarizing the study's findings.

## 2 Methodology

To analyze the environmental impact of the fuel cell system, a LCA was carried out in accordance with the ISO standards 14040 and 14044 [12, 13]. The structure of the study thus follows the following scheme: Goal and Scope, Life Cycle Inventory, Impact Assessment, and Interpretation of the results.

### 2.1 Goal and Scope

The system boundaries of the study as shown in Fig. 1, were defined as "cradle to gate". The data sources included the Bill of Materials (BoM) provided by the manufacturing company and existing literature on the subject. The functional unit was defined as $1\,kW_{el}$ of nominal power. The Environmental Footprint 3.1 method was employed to assess the impacts of the manufacturing process on seven environmental indicators, outlined in the results section. The openLCA software was employed to construct the LCA models, using the ecoinvent database version 9.3.

### 2.2 Life Cycle Inventory

All information on the materials and quantities used for the fuel cell system is derived directly from the manufacturer's data provided for this study. The materials used for

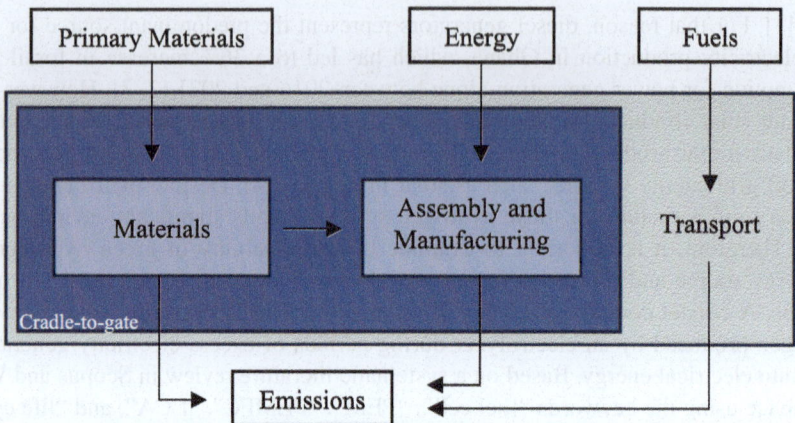

**Fig. 1.** System boundaries of the conducted LCA of the fuel cell system.

manufacturing the system components are outlined in Table 1. For materials not specified in detail in the BoM, assumptions were made based on research into the commonly used material type or the selection of the most suitable material for the application. Due to missing processes in the LCA database, polyamide 12 was approximately modeled using the polyamide 6 process. Additionally, assumptions were made regarding the installed electrical heating as there were no detailed information on its material composition. In the absence of precise information in the BoM on the dimensions of coated metal parts made of galvanized and tin-plated steel, assumptions were made based on the calculation of the surface dimensions of similar products. Further manufacturing steps of the metal parts that were not found in the database were modeled using the "metal working, average" process. If datasets for materials were not available in the database, efforts were made to replicate the material as closely as possible through precursor products and process steps. For Nafion and the production of the gas diffusion layer (GDL) from polytetrafluoroethylene, approximations were derived from literature sources [7, 14].

Materials weighing less than one kilogram and exhibiting minimal environmental impact have been collectively assigned to the category "Other."

## 3  Results

In accordance with the Life Cycle Assessment methodology, this section presents the results of the environmental impact assessment of the fuel cell system. Table 2 displays the results for all impact categories investigated for the systems different components. The global warming potential ($GWP_{100}$) of the entire fuel cell system is 342.62 kg $CO_2$-eq. Per $kW_{el}$. The fuel cells alone contribute 196.85 kg $CO_2$ eq., representing 58% of the system's total $GWP_{100}$.

The fuel cells also have the greatest impact on the acidification potential (4.61 mol $H^+$-eq.), the eutrophication potential for freshwater (0.39 kg P-eq.), marine (0.74 kg N-eq.), and terrestrial (9.23 mol N-eq.) ecosystems. The fuel cells are the primary contributors to these impact categories, with a share of 63% to the systems total acidification

**Table 1.** Material composition of the fuel cell system in kg

| System component | Materials | Weight (kg) |
| --- | --- | --- |
| Fuel Cell | Plastics | 38.31 |
| | Steel | 35.26 |
| | Graphite | 34.00 |
| | Copper | 13.34 |
| | Platinum | 0.01 |
| | Cable | 2.40 |
| | Printed wiring board | 2.39 |
| | Zinc | 1.59 |
| | Other | 2.79 |
| | Total | 130.09 |
| Cabinet | Plastics | 61.10 |
| | Steel | 22.45 |
| | Copper | 92.72 |
| | Aluminum | 5.63 |
| | Iron-nickel-chromium | 2.24 |
| | Printed wiring board | 0.87 |
| | Other | 1.74 |
| | Total | 186.75 |
| Controller | Aluminum | 3.88 |
| | Plastics | 1.10 |
| | Other | 1.26 |
| | Total | 6.24 |
| Packaging & Battery | Carton board | 17.96 |
| | Battery (lead sulfuric acid) | 203.20 |
| | Total | 221.16 |
| | Total fuel cell system | 544.24 |

potential. Regarding the eutrophication potential of ecosystems, the fuel cells accounted for 63% (freshwater), 71% (marine), and 74% (terrestrial) of the systems overall impact in this category.

An exception to this is the material resource consumption of metals and minerals, where the cabinet and system battery have a larger share. This discrepancy can be attributed to the high weight of the battery and cabinet, which account for 37,3% and 34,3% respectively of the system's total weight. In all other categories, these components have the second and third-highest impacts, respectively.

**Table 2.** Contributions to the impact categories of the fuel cell system standardized to 1 kW$_{el}$

| Impact Category | Units | Total | Fuel Cell | Cabinet | Battery | Controller | Other |
|---|---|---|---|---|---|---|---|
| Acidification | mol H + eq | 4.61 | 2.92 | 0.89 | 0.73 | 0.07 | 0.01 |
| Global Warming Potential (GWP$_{100}$) | kg CO$_2$eq | 342.62 | 196.85 | 77.71 | 55.97 | 8.61 | 3.48 |
| Eutrophication, freshwater | kg P eq | 0.39 | 0.25 | 0.10 | 0.04 | 0.01 | 4e-4 |
| Eutrophication, marine | kg N eq | 0.74 | 0.53 | 0.13 | 0.07 | 0.01 | 2.8e-3 |
| Eutrophication, terrestrial | mol N eq | 9.23 | 6.83 | 1.46 | 0.81 | 0.11 | 0.02 |
| Energy Resource, fossil | MJ | 4894.32 | 2737.50 | 1177.46 | 790.54 | 111.38 | 77.45 |
| Material Resource, mineral/metal | kg Sb eq | 0.08 | 0.02 | 0.04 | 0.02 | 1.3e-03 | 2e-05 |

Figure 2 presents the relative contributions to the different impact categories of the fuel cells utilized in the system.

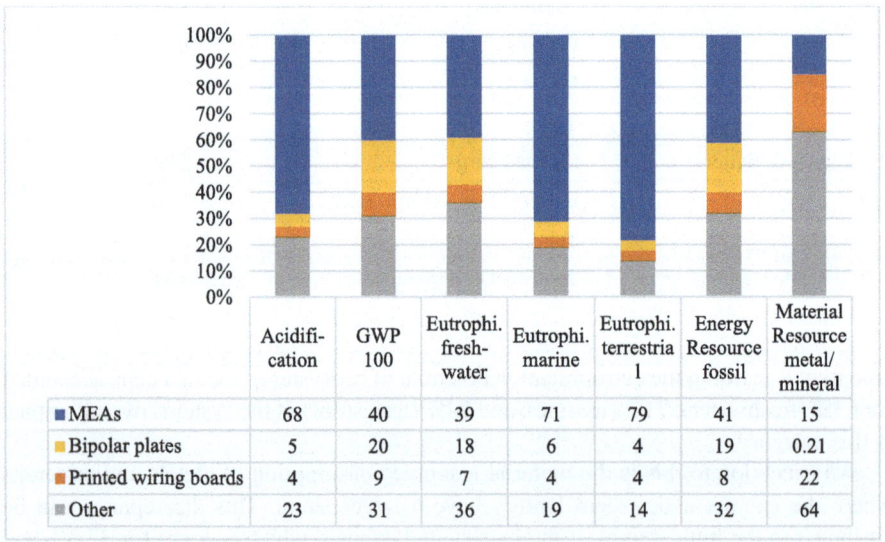

**Fig. 2.** Relative contributions to the impact categories of the fuel cells components.

The MEAs are the largest contributor in all the impact categories except for material resource consumption of minerals and metals. This can be attributed to the platinum used in a MEA resulting in a significant influence of this component on the eutrophication potential of freshwater (32%), marine (68%) and terrestrial (77%) ecosystems. The production of bipolar plates is very energy intensive, therefore their manufacturing contributes 20% to the $GWP_{100}$ of the fuel cells. Furthermore, 19% of the total energy consumption of fossil energy resources can be attributed to the manufacturing of the bipolar plates.

The numerous printed wiring boards are the third-largest contributor to the $GWP_{100}$, accounting for 9% of $CO_2$ eq. Emissions. In the category of material resource consumption of metals and minerals, 22% can be attributed to these.

## 4  Discussion

The results of this LCA deviate from those found in other studies. This can be attributed to several factors. The utilization of fuel cell systems in a solar mini-grid necessitates additional components and materials like cabinet, cables, and printed wiring boards, which are not considered in LCAs of single fuel cell stacks. The significant discrepancies in the quantity of materials required for the entire system in comparison to a fuel cell stack alone can be exemplified by the following illustration. In our study, the quantity of polyamides alone accounts for 1.96 kg, while Mori et al. reported 117.09 g of plastics at the same power output [5]. This discrepancy highlights the importance of considering the entire system, when assessing the environmental impact of fuel cell application in mini-grid energy systems. Furthermore, the limited data available in the Ecoinvent database regarding the composition and manufacturing processes of emission-intensive materials like Nafion, and GDLs posed a challenge. The reliance on approximations for these materials introduces a degree of uncertainty into the results.

As can be seen in Fig. 2, the MEA is identified as the primary contributor to emissions across a range of impact categories within the systems fuel cells. Despite representing only 0.006% of the total mass, the platinum catalyst is responsible for 68% of the acidification potential. The mining of rare earths including platinum and gold is associated with considerable environmental pollution, primarily due to metal contamination resulting from the mining techniques [15]. This pollution of soils and waters is reflected in the high levels of the eutrophication potential of freshwater, marine, and terrestrial ecosystems. Furthermore, the mining and processing of platinum is also a highly energy-intensive operation. With the blasting operations required for mining, this results in the high contribution of the MEAs to the $GWP_{100}$.

The potential for fuel cells that do not rely on precious metals was recently explored, resulting in the technical feasibility of replacing platinum with alternative materials [16]. This would significantly reduce the environmental impact of a fuel cell. The significant contribution of the printed wiring boards to the $GWP_{100}$, can be attributed to the gold content. Both the mining of platinum and gold are associated with high energy demands and environmental disruption [17]. The fossil energy resource category is also significantly influenced by the MEAs and bipolar plates. This is due to the fact that electricity generation is still heavily dependent on fossil fuels, especially in the precious metal mining countries in the global South [18].

The high contributions to different impact categories of the cabinet and battery can be attributed to the considerable material consumption of these components. Additionally, the battery production process is complex, and the lead utilized in the production process causes significant environmental impacts [19].

## 5 Conclusion

The objective of this study was to determine the environmental impacts associated with the production of a fuel cell system for use as an energy-transforming technology in a self-sufficient mini-grid in Ghana. The entire system emits 342.62 kg $CO_2$ eq. Per $kW_{el}$. Of this total, the production of the fuel cells accounts for 196.85 kg $CO_2$ eq., representing more than half of the climate-damaging greenhouse gas emissions. This is particularly associated with the manufacturing of materials such as Nafion, or the mining of precious metals that have significant environmental impacts, as well as energy-intensive manufacturing processes. These findings underscore the necessity to diminish the use of precious metals, such as platinum, in fuel cells and to improve energy-efficiency of manufacturing processes. In particular, platinum, which constitutes only a minor proportion of the total mass, should be replaced or reduced, given its considerable impact on a range of impact categories. Recent research into alternative materials has shown that viable substitutes for platinum are currently difficult to find. While much research has focused on iron-nitrogen as an alternative, it has proven to be unstable in practice and has a relatively short lifespan. However, the reduction and recycling of platinum offer feasible solutions. Moreover, the impact of energy-intensive processes, reflected in the high $GWP_{100}$ contributions of MEAs and bipolar plates, underscores the importance of utilizing renewable energy sources in production and exploring less energy-intensive manufacturing processes.

Future research should focus on the investigation of more sustainable materials for the production of fuel cells. Moreover, an assessment of the entire life cycle of the solar mini-grid, and the hydrogen-based energy system, is necessary. A comparison with other energy transforming technologies like diesel generators has to be done to obtain a comprehensive picture of the sustainability of the application of fuel cells in mini-grids.

**Acknowledgment.** The research work published in this paper was funded by the German Federal Ministry for Environment, Nature Conservation and Nuclear Safety under the on-funding program Export Initative Environmental Protection (EXI). Funding reference: 67EXI6503A.

**Disclosure of Interests.** The authors have no competing interests to declare that are relevant to the content of this article.

## References

1. Agoundedemba, M., Kim, C., Kim, H.-G.: Energy Status in Africa: challenges, progress and sustainable pathways. Energies **16**, 7708 (2023)
2. Asaki, F.A., Amoah, E.K., Abeka, M.J.: The impact of electricity production on environmental quality: the role of institutional quality in Ghana. Oper. Res. Forum **5**(2), 35 (2024)

3. Chaaraoui, S., et al.: Day-ahead electric load forecast for a ghanaian health facility using different algorithms. Energies **14**(2), Art. no. 2 (2021)
4. Kuamoah, C.: Renewable energy deployment in Ghana: the hype, hope and reality. Insight Afr. **12**(1), 45–64 (2020)
5. Mori, M., et al.: Life cycle sustainability assessment of a proton exchange membrane fuel cell technology for ecodesign purposes. Int. J. Hydrog. Energy **48**(99), 39673–39689 (2023)
6. Stropnik, R., Lotrič, A., Bernad Montenegro, A., Sekavčnik, M., Mori, M.: Critical materials in PEMFC systems and a LCA analysis for the potential reduction of environmental impacts with EoL strategies. Energy Sci. Eng. **7**(6), 2519–2539 (2019)
7. Evangelisti, S., Tagliaferri, C., Brett, D.J.L., Lettieri, P.: Life cycle assessment of a polymer electrolyte membrane fuel cell system for passenger vehicles. J. Clean. Prod. **142**, 4339–4355 (2017)
8. Finke, S., Velenderic, M., Severengiz, S., Pankov, O., Baum, C.: Transition towards a full self-sufficiency through PV systems integration for sub-Saharan Africa: a technical approach for a smart blockchain-based mini-grid. Renew. Energy Environ. Sustain. **7**, 8 (2022)
9. Schneider, S., Velenderic, M., Staib, M., Küpper, E., Severengiz, S.: "The Potential of Hydrogen-Based Storage Systems in Sub-saharan Africa," presented at the International Renewable Energy Storage CONFERENCE (IRES 2022), Atlantis Press, pp. 489–499 (2023)
10. Stinder, A.K., Finke, S., Vendeleric, M., Severengiz, S.: A generic GHG-LCA model of a smart mini grid for decision making using the example of the Don Bosco mini grid in Tema, Ghana. Procedia CIRP **105**, 776–781 (2022)
11. Sturm, L., Severengiz, S., Satish Salokhe, D., Bhatia, G.: "Criteria-driven comparison of hydrogen, charcoal and liquefied petroleum gas as cooking fuels based on a systematic literature review," [Unpublished manuscript]. Hochschule Bochum, Bochum
12. International Organization for Standardization, "ISO 14044:2006," https://www.iso.org/standard/38498.html. Accessed 19 June 2024
13. International Organization for Standardization, "ISO 14040:2006," https://www.iso.org/standard/37456.html. Accessed 19 June 2024
14. Simons, A., Bauer, C.: A life-cycle perspective on automotive fuel cells. Appl. Energy **157**, 884–896 (2015)
15. Díaz-Morales, D.M., et al.: Metal contamination and toxicity of soils and river sediments from the world's largest platinum mining area. Environ. Pollut. **286**, 117284 (2021)
16. Gao, Y., et al.: A completely precious metal–free alkaline fuel cell with enhanced performance using a carbon-coated nickel anode, (2022)
17. Aramendia, E., Brockway, P.E., Taylor, P.G., Norman, J.: Global energy consumption of the mineral mining industry: exploring the historical perspective and future pathways to 2060. Glob. Environ. Change **83**, 102745 (2023)
18. Heras, A., Gupta, J.: Fossil fuels, stranded assets, and the energy transition in the Global South: a systematic literature review. WIREs Clim. Change **15**(1), 866 (2024)
19. Raj, K., Das, A.P.: Lead pollution: impact on environment and human health and approach for a sustainable solution. Environ. Chem. Ecotoxicol. **5**, 79–85 (2023)

# Circular Economy Business Model for 3D Printing Filament Production in Vietnam

Vignesh Bhaskaran, Harish Gireesan[✉], Supriya Tumkur Shashishekar,
Avishkar Kanagaraj, Desigan Vincent Selvaraj Virgindevi, and Bernd Muschard

Institute of Machine Tools and Factory Management, Technische Universität Berlin, Berlin,
Germany
{vignesh.bhaskaran,harish.gireesan}@campus.tu-berlin.de

**Abstract.** This paper explores the sustainability potential of low-cost 3D printing in Vietnam by evaluating the impact of a PET recycler called Polyformer, which converts plastic bottles into filament for 3D printing, creating sustainable business models. It addresses the growing challenge of plastic waste by reducing environmental impacts, creating economic opportunities for low-income individuals, and promoting localized manufacturing to strengthen knowledge transfer on circular economy aspects. The paper presents business case scenarios that demonstrate the economic viability of selling recycled filaments and the Polyformer machine itself, thus contributing to Vietnam's transition to a circular economy. It also highlights the importance of stakeholder integration, collaboration between local communities, industry and academia, and emphasizes knowledge transfer to enable broader societal benefits.

**Keywords:** PET Recycler · Polyformer · 3D printing · Sustainable Business Model · Vietnam

## 1 Introduction

In the context of emerging economies such as Vietnam, the challenge of plastic waste management has become increasingly pressing. The country is one of the largest contributors [5] to marine plastic pollution, a result of rapid economic growth and inadequate recycling infrastructure. The Polyformer is in line with Vietnam's national sustainability goals and offers the multiple benefits of reducing plastic waste and promoting local manufacturing. The proposed business model not only addresses the environmental impact, but also explores the economic opportunities of integrating recycled materials into the production process, thereby contributing to a circular economy.

## 2 State of the Art

### 2.1 3D Printing in Emerging Countries

3D printing, also known as additive manufacturing, has the potential to transform manufacturing by reducing waste, improving efficiency, promoting localized production and supporting the circular economy, particularly in emerging economies such as Vietnam

© The Author(s) 2025
H. Kohl et al. (Eds.): GCSM 2024, LNME, pp. 835–846, 2025.
https://doi.org/10.1007/978-3-031-93891-7_92

[1]. Open-source projects such as Polyformer, which recycles plastic waste into fila-ment, illustrate how it can address both environmental and economic challenges in such contexts. In Vietnam, where plastic pollution is a significant problem, this approach not only reduces dependence on imported materials, but also supports the local economy, is in line with sustainable development goals and sets the stage for entrepreneurship [2]. Research supports the technical feasibility and environmental benefits of using recycled plastic for 3D printing. Studies show that this approach can significantly reduce plastic waste and provide economic opportunities, particularly in developing regions [3]. For example, the Polyformer project provides a business model for Vietnam to reduce plastic pollution, create economic opportunities and enable localized production of goods. In addition, the adoption of 3D printing technology could facilitate the transition to a cir-cular economy in Vietnam, especially for small and medium-sized enterprises (SMEs). Local companies such as Minh Chan Ltd. Are already exploring the potential of recycled plastic filaments for 3D printing, in line with national strategies to improve plastic waste management [7].

## 2.2 Sustainability in Vietnam

Vietnam is a major contributor to marine plastic pollution, with an estimated 730,000 tones of plastic waste entering the ocean each year [5]. Rapid economic growth has led to increased waste production, while recycling infrastructure remains underdeveloped, resulting in much of the plastic waste being incinerated or landfilled [2]. The Polyformer could improve recycling rates by making it economically viable for more workers to par-ticipate in the recycling process and by providing a higher value product through upcy-cling. Despite existing recycling efforts, the infrastructure and regulatory framework in Vietnam are insufficient to manage the increasing volumes of waste, as reviewed in [6]. The National Action Plan on ocean plastic waste management [8] demonstrate Vietnam's commitment to addressing these challenges. These policies aim to improve waste man-agement and promote recycling. Economically, Vietnam's landscape is characterized by low-wage workers and a significant informal sector. The Polyformer could improve the economic prospects of waste collectors by enabling them to convert waste (plas-tic bottles) into valuable fiber, thereby providing additional income. In addition, small industries and universities could benefit from affordable filament, fostering knowledge transfer, awareness, innovation and sustainability in manufacturing and research. The economic integration of 3D printing into local production chains could support Vietnam's transition to a circular economy, using the technology's low-cost production capabilities to boost local industries.

## 2.3 Research Gap

The Polyformer project addresses several challenges already mentioned in 2.2. In addi-tion, it has the potential to provide a technical and collaborative basis for exploring further gaps. The infrastructure for collecting and processing plastic waste in Vietnam is still underdeveloped, requiring more robust policies and enforcement mechanisms.

Addressing the variability in the quality of recycled plastic is critical, as inconsistencies can affect the reliability of filament production [10]. Further research is needed to standardize recycling and filament production processes to ensure consistent quality.

The economic feasibility of scaling up recycled filament production is a challenge. While the Polyformer project has demonstrated technical success, its long-term sustainability and scalability remain uncertain. Integrating 3D printing into local economies will require capacity building, including training and access to the necessary technology. Collaboration between academic institutions, industry stakeholders and government agencies are essential to overcome these barriers and realize the full potential of 3D printing in advancing Vietnam's circular economy and environmental sustainability.

## 3 Polyformer Technology

### 3.1 Introduction

The Polyformer is designed to be built largely from 3D printed parts and easily accessible components typically found on 3D printers. The modular architecture allows customers to easily change components and configure the machine to their specifications. The machine consists of three main components: 1. Electrical hub 2. Extrusion chamber 3. Spool & Gearbox as shown in Fig. 1.

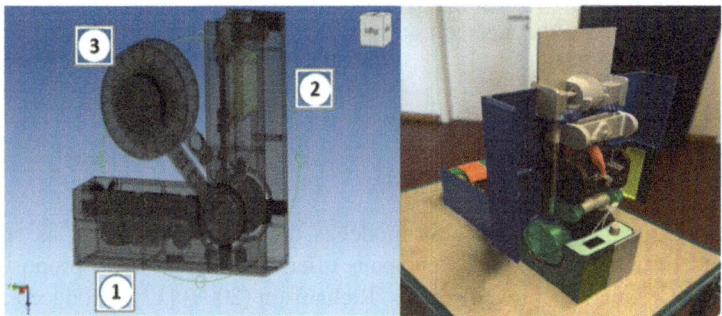

**Fig. 1.** Polyformer (CAD & Final Assembled Model)

### 3.2 Procurement and Assembly

All the mechanical and electronic components available on the market can be ordered online at a reasonable price, or you can buy them yourself. All special components such as housing parts, coil holders or gears must be 3D printed independently. The parts required to build the Polyformer are open sourced on GitHub with links to purchase the components [11] and the STL files for the housing, cutter tool and gear with assembly instructions are also included in the 'STL Files' subsection [11]. The setup costs for the Polyformer are listed below, assuming the user has access to a 3D printing machine to print the Housing, Cutter Tool and Gearbox parts (Table 1).

**Table 1.** Polyformer Setup Cost

| Item | Details | Cost ($) |
|------|---------|----------|
| Electrical Hub and Extrusion Chamber | Heating Block, Motor, PID Controller, Nozzle, Thermostat | $ 90.00 |
| Housing, Spool and Gear | Panels, Gear Parts | $ 30.00 |
| Shipping Cost | Shipping from USA to Vietnam | $ 65.00 |
| Total Initial Investment | Sum of all initial costs | **$ 185.00** |

Shipping costs can be further minimized or eliminated if all electrical and extrusion chamber components can be sourced locally in the Vietnamese market, reducing costs by approximately $65.

### 3.3 Operation Principle

The operation of the Polyformer starts by taking the pre-processed PET bottles and stripping the uniform strips from them, turning on the machine and inserting the collected strip through the brass die into the extrusion chamber as shown in Fig. 1(2). Once the temperature reaches approximately 240° C with the aid of the heating block, as shown in Fig. 1(1). The filament is extruded with the aid of a gear as shown in Fig. 1(3) behind the extrusion chamber and collected as a spool that can be used directly in 3D printers.

## 4  Business Model Development

### 4.1  Sustainable Business Model (SBM)

The central element of any business model is the value proposition, as customers need to understand not only what a company offers, but also its value proposition and how it differs from competitive offerings. Richardson (2008) [12] introduced a widely accepted business model framework that includes the value proposition, value creation and delivery, and value capture.

The literature on SBMs further describes subcategories, subtypes and generic strategies such as product-service systems, which have been explored and synthesized as the so-called 'sustainable business model archetypes' to develop a unified target definition as listed in Table 2.

### 4.2  Stakeholder Integration Model (SIM)

Developing a SIM is important for companies to ensure that they consider the perspectives and interests of all stakeholders in their decision-making processes. Stakeholders can have a significant impact on a company's operations and success, and their opinions and concerns should be taken into account (Fig. 2).

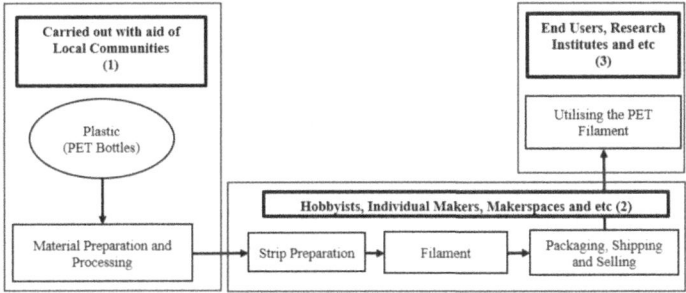

**Fig. 2.** Stakeholders along the Process Flow

## 5 Sustainable Business Model with SIM

This section outlines the value created by the Polyformer across each SBM Archetype as discussed in the Literature review and the Stakeholder Interaction with each other.

**Table 2.** SBM Archetypes with their definition and values

| SBM Archetype | Definition | Value Proposition | Value Creation and Delivery | Value Capture |
|---|---|---|---|---|
| Re-use and Upscaling of Plastic Waste (PET Bottles) **(1)** | Use of non-finite materials and energy sources | Fewer use of virgin resources, generate less waste and emissions than product/services that can deliver the same functionality | Activities and partnerships to maximize PET waste with a focus on product and manufacturing process innovations New partnerships and value network | **Environmental Value:** Used PET Bottles as raw materials influence a positive impact on environment **Economic Value:** Cost of Raw Materials for producing the filament is very minimal |
| Closing resource loops **(2)** | Reuse materials in continuous cycles to create a circular economy, minimizing waste and maximizing the lifecycle of resources | Previously thought waste is eliminated and reused in a new application through life-cycle based approach | Activities and partnerships to eliminate life cycle waste Close material loops New partnerships, potentially across working-class individuals, research institutes and SME's | |

(*continued*)

Table 2.  (*continued*)

| SBM Archetype | Definition | Value Proposition | Value Creation and Delivery | Value Capture |
|---|---|---|---|---|
| Repurpose for society/environment (3) | Create value by redirecting resources or products to serve broader societal or environmental needs Involve various stakeholders in the value creation process | Innovative collaborative cross-sectoral multistakeholder platform Creating income source for waste collectors and working-class individuals and reducing the effect of PET Bottles in their locality | | **Environmental Value:** Used PET Bottles as raw materials influence a positive impact on environment **Economic Value:** Potential source of Income |
| -Deliver functionality, not ownership (4) | Provide access to the benefits of a product or service at free cost | Instructions required to build it will be open sourced with detailed instruction for the public | Delivery through open-source platforms for developing and utilizing it | **Market Expansion:** More enthusiasts to try the business model |
| Encourage recyclability (5) | Promote the use of materials and processes that make recycling easier and more effective | Organizing events for Universities, Schools and other organizations to show them the potential solutions for up-scaling PET Bottles waste | Creating awareness and educating the public and students for ways of recycling and up-cycling the PET Bottles | **Social Value:** Support and Promote sustainability-related behavior among public |
| Develop sustainable scale up solutions (6) | Implement scalable solutions that maximize sustainability benefits across larger contexts and markets | Unlock substantially the performance of PET Filaments through multiple tests and analysis thus maximizing benefits | Partnerships with potential and unusual partners (e.g. government) and other organizations crucial to scale the business | **Economic Value:** From start-up to large scale project ensuring a viable fee is paid for scaling up the solution/venture |

The below Table 3 and Fig. 3 depict the Stakeholder Integration Model for the Polyformer and their Interaction with each other.

**Table 3.** Stakeholders Role and Interest

| Stakeholder Group | Role | Interest |
|---|---|---|
| Local Communities and Waste Collectors (**1**) | Collection of PET bottles for Polyformer process | Economic opportunities, improved environmental conditions |
| Hobbyists and Individual Makers (**2 & 3**) | Individual consumers engaged in DIY and small-scale 3D printing projects | Print and afford cost effective filament for prototyping or for selling the filament as a product to other end users |
| Makerspaces and Fab Labs (**2 & 3**) | Community-driven workshops that provide access to 3D printing technology | |
| 3D Printing Service Providers or Manufacturers (**3**) | Utilize recycled filament for 3D printing | Affordable, sustainable filament for research, prototyping, and educational purposes |
| Educational Institutions and Universities (**3**) | Schools and universities with 3D printing labs | |
| End-Users (Consumers and Businesses) (**3**) | Purchase products made from recycled filament | |

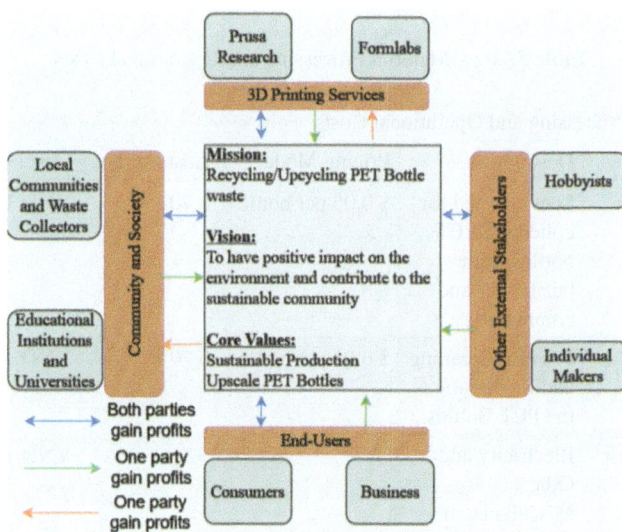

**Fig. 3.** Stakeholder Interaction Model (SIM)

# 6 Business Cases for Circular Economy

The Polyformer offers the user a wide range of business cases that can be implemented to meet all the social, economic and environmental values discussed in Sect. 5. This section provides two business case scenarios of how a fully functional Polyformer can be used in Vietnam.

## 6.1 Business Case for Selling as PET Filament

The Table 4 shows the required plastic bottles, output produced and the machine operational hours for 1 Polyformer per month.

**Table 4.** Key Performance Details of Polymer Process

| Key Performance Details | Value |
| --- | --- |
| Number of Plastic Bottles Required per month (No's) | 1400 Bottles |
| Total Filament production per month (Kg) | 28.6 kg |
| Machine Operational time (Hours) | 588 h |

The Table 5 shows the cost for raw materials processing and other operational costs per month involved in producing the filaments.

**Table 5.** Raw Material Processing and Operational Costs

| Raw Material Processing and Operational Costs | | | | |
| --- | --- | --- | --- | --- |
| Type of Cost | Description | Pricing Model | Cost (USD) | Cost (VND) |
| Collection Cost | Fees charged for collecting PET bottles from businesses and communities | $ 0.05 per bottle | $ 70 | VND 17,39,150.00 |
| Processing Cost | Fees for Cleaning and Processing the PET Bottles | $ 0.05 per bottle | $ 70 | VND 17,39,150.00 |
| Operational Cost | Electricity and Other Miscellaneous Cost | | $ 5 | VND 1,24,225.00 |
| **Total Raw Material Processing and Operational Costs** | | | **$ 145** | **VND 36,02,525.00** |

The Table 6 shows the types of revenue that can be generated when promoting the selling of filaments to the different stakeholders.

## 6.2   Business Case for Selling as Machine/Services

The business case below in Table 7 focuses on selling the Polyformer as a unit or offering parts/kits for maintenance or offering an installation service for initial setup. All costs are approximations based on average costs in Vietnam.

**Table 6.** Revenue Generated from Selling Filaments

| Revenue Generated | | | | |
|---|---|---|---|---|
| Type of Revenue | Description | Pricing Model | Revenue (USD) | Revenue (VND) |
| Filament Sales | Sale of recycled PET filament to end-users, businesses, and 3D printing service providers | $ 12 per kg (Can also be bought in 250 gms spools) | $ 168 | VND 41,73,960.00 |
| Partnership Filament Sales | Selling filament in bulk to 3D printing companies or manufacturers | $ 10 per kg (bulk price minimum 2 kg) | $ 140 | VND 34,78,300.00 |
| Government Incentives/Subsidies | Potential subsidies or incentives for contributing to waste management goals | $ 0.02 per bottle recycled | $ 28 | VND 6,95,660.00 |
| Workshops and Educational Programs | Offering workshops on recycling and sustainability with entry fees | $ 0.5 per participant | $ 13 | VND 3,22,985.00 |
| **Total Sales Generated from Selling Filament** | | | $ 349 | VND 86,70,905.00 |
| **Total Estimated Monthly Revenue (Total Revenue Generated - Total Raw Material Processing and Operational Cost):** | | | **$ 204** | VND 50,68,380.00 |

**Table 7.** Business Case for Selling as Machine/Service

| Revenue Stream | Description | Price per Unit ($) (Approx) | Sales Volume (Approx) | Actual Cost Incurred (Approx) | Monthly Revenue (USD) | Monthly Revenue (VND) |
|---|---|---|---|---|---|---|
| Direct Machine Sales | Sale of the Polyformer machine to educational institutions, SMEs, and hobbyists | $ 300 | 2 Units | $ 400 | $ 200 | VND 49,69,000.00 |
| Maintenance Kits | Sale of periodic maintenance kits (including replacement parts, cleaning tools, etc.) | $ 70 | 3 Kits | $ 24 | $ 138 | VND 34,28,610.00 |
| Training and Installation Services | Offering on-site training and installation services for new customers | $ 15 | 2 Services | $ 7.5 | $ 15 | VND 3,72,675.00 |
| Spare Parts Sales | Sale of individual spare parts (e.g., motors, controllers, etc.) to existing customers | $ 5 | 12 Parts | $ 1.8 | $ 38 | VND 9,44,110.00 |
| **Total Estimated Monthly Revenue** | | | | | **$ 391** | VND 97,14,395.00 |

## 7   Conclusion

In summary, the business cases highlight the significant potential of 3D printing, particularly through Polyformer technology, to address the challenges of plastic waste in Vietnam. By integrating recycled PET bottles into local manufacturing, this sustainable approach not only reduces environmental impact, but also promotes economic growth.

The scalability of this model offers promising prospects for advancing Vietnam's circular economy. Continued collaboration and innovation are essential to realize the full benefits of this sustainable business model.

The next steps will involve further refinement of the model, including proof of concept trials and scalability analysis to ensure long-term viability. It's important to recognize that these calculations are based on current assumptions due to the lack of precise data, which introduces uncertainty. However, with continued research and collaboration, these limitations can be addressed to provide a more comprehensive view of the sustainable potential of 3D printing technologies in emerging markets.

**Acknowledgements.** The authors would like to express our sincere gratitude to all those who supported us in the completion of this research. We are particularly grateful to our academic supervisors Prof. Dr.-Ing. Günther Seliger and Dr.-Ing. Bernd Muschard at the IWF of the Technical University of Berlin for their expert guidance and unwavering encouragement throughout. We would also like to thank the members of the two research teams, whose commitment and cooperation over the past two semesters have contributed significantly to the success of this project. Authors other than those listed above have contributed by name: D. J. Nwaoyibo, N. B. Nguyen, T. A. Phan Tran, S. G. Mane, S. N. Rumale, T. Zametti, Y.-H. Hsu, R. Kalamkar, V. S. Kulkarni, S. Meja Rodriguez, V. A. Rangole, P. Ouentes, C. S. Nwachukwu, P. V. Ghodake, Y. Li.

# References

1. Gebler, M., Uiterkamp, A.J.M.S., Visser, C.: A global sustainability perspective on 3D printing technologies. Energy Policy **74**, 158–167 (2014)
2. Jambeck, J.R., et al.: Plastic waste inputs from land into the ocean. Science **347**(6223), 768–771 (2015)
3. Baechler, C., DeVuono, M., Pearce, J.M.: Distributed recycling of waste polymer into RepRap feedstock. Rapid Prototyp. J. **19**(2), 118–125 (2013)
4. Tymrak, B.M., Kreiger, M., Pearce, J.M.: Mechanical properties of components fabricated with open-source 3-D printers under realistic environmental conditions. Mater. Des. **58**, 242–246 (2014)
5. Lebreton, L.C., Van der Zwet, J., Damsteeg, J.W., Slat, B., Andrady, A., Reisser, J.: River plastic emissions to the world's oceans. Nat. Commun. **8**(1), 15611 (2017)
6. FPTS Plastic Industry Report. (2019). Vietnam's Plastic Industry Overview. Retrieved from FPTS
7. Vietnam Government. (2024). Decision No. 2149/QD-TTG on National Strategy for Integrated Solid Waste Management until 2025, Vision to 2050. Available at: Legal Database
8. Vietnam Government. (2024). Decree No. 59/2007/ND-CP on Solid Waste Management. Available at: Legal Database
9. Vietnam Government. (2020). Directive No. 33/CT-TTg on Strengthening Plastic Waste Management. Available at: Legal Database
10. Zander, N.E., Gillan, M., Lambeth, R.H.: Recycled polyethylene terephthalate as a new FFF feedstock material. Addit. Manuf. **21**, 174–182 (2018)
11. GitHub Homepage. https://github.com/HarishGireesan/Polyformer_gpe24
12. Richardson, J.: The business model: an integrative framework for strategy execution. Strateg. Chang. **17**, 133e144 (2008). https://doi.org/10.1002/jsc.821

# Application of Second Life Electric Vehicles' Lithium-Ion Batteries in Micromobility

Mustafa Anil Safran$^{(\boxtimes)}$, N. Badgujar, and N. Tantharatana

Department for Machine Tools and Factory Management, TU Berlin, Berlin, Germany
m.safran@tu-berlin.de

**Abstract.** Although Lithium Ion Battery (LIB) is an efficient energy storage, lithium extraction from evaporation ponds is time-consuming and often involves large-scale mining operations which can lead to significant sustainability challenges. To combat these challenges, this paper aims to identify opportunities for higher value generation in Second Life (SL) of LIBs. A case study identifies chances for reducing environmental footprint and higher profit by initially applying SL Electric Vehicle (EV) batteries in E-scooters (ESs) and E-bikes (EBs), and later in Stationary Energy Storage Systems (SESS). A Tesla Model 3 battery with SoH 88–75% can conceptually be repurposed to 98 batteries for $5^{th}$ generation ESs/EBs used in Germany by Bolt rental company. Dimensional and electrical capacity calculations of the Repurposed Battery (RB) prove its conceptual applicability in Bolt's ESs and EBs. While providing the required energy output, the RB costs 69% less than a virgin battery based on factors given by 'B2U: Battery Second-Use Repurposing Cost Calculator' provided by National Renewable Energy Laboratory (NREL), U.S. Department of Energy.

**Keyword:** EV Batteries · LIB Repurposing · Micromobility applications

## 1 Introduction

### 1.1 Current Scenario

Most common batteries used in EVs are LIBs due to their comparatively long lifespan, high energy density, longer service life, lightweight and low maintenance costs [1]. Recycling and repurposing of LIBs procured from EVs has gained more attention as it is found that LIB production creates environmental impacts such as scarcity of non-renewable resources e.g. nickel, cobalt, manganese and lithium due to mining process, carbon footprint from the energy demand and intensive utilization of water for lithium extraction [2]. Keeping in mind the sustainability aspects in LIB manufacturing, social impacts such as unhealthy working conditions, child labor and long working hours are also relevant and need to be taken into account [3].

Existing studies currently offer solutions such as extending service life of a battery after the first End-of-Life (EoL) by reusing the retired EV batteries for SESS until they reach the EoL at 40% SoH and then proceed to material recycling [1, 4–6]. The SoH

© The Author(s) 2025
H. Kohl et al. (Eds.): GCSM 2024, LNME, pp. 847–855, 2025.
https://doi.org/10.1007/978-3-031-93891-7_93

at which the SL should start, differs from 88-80% in each of these aforementioned papers. According to research done in 'BATRAW' project by Centre of European Policy Studies (CEPS), by 2027 all the batteries with a capacity of over 2 kilowatt hour (kWh) that are marketed in European Union (EU), including two-wheeled vehicle and industrial batteries are going to be assigned with a Digital Battery Passport (DBP) [7].

## 1.2 Aim and Focus

Taking into account all the regulations and available research which gives us the distribution of First Life (FL), SL and end of life, a scenario is hypothesized where SL applications of EV batteries are redistributed into two phases: SL-A and SL-B according to SoH as shown in Fig. 1.

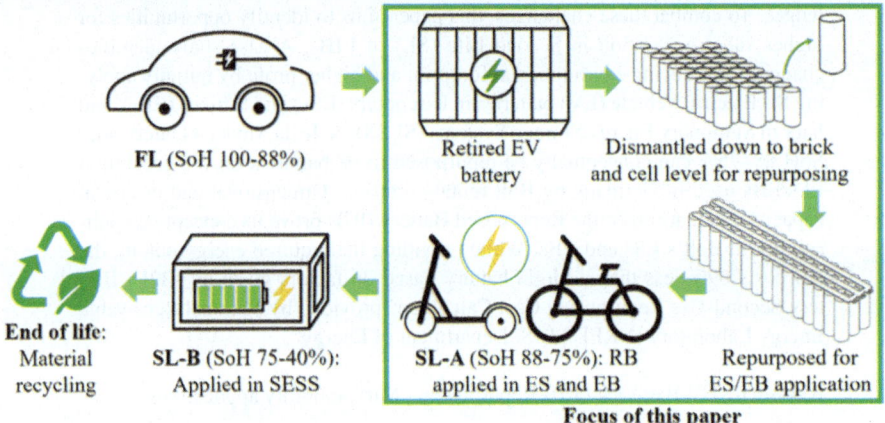

**Fig. 1.** Battery Second Life (SL) modified distribution

Considering that FL of EV batteries ends in 5 years, by 2030, the retired EV batteries available for SL applications will be around 1,100 GWh whereas the forecasted battery demand for SESS and micromobility will be around 300 GWh and 150 GWh respectively [8, 9]. Instead of producing virgin batteries for these applications, using retired EV batteries is environmentally and economically beneficial [8]. Utilizing retired EV batteries in micromobility applications requires careful considerations of dimensional and weight constraints, which are stricter compared to SESS. Though the SL application of EV batteries in micromobility is more complicated than application in SESS due to involvement of repurposing process, SL-A to SL-B approach offers higher profitability than just utilizing in SESS where the profit is primarily generated through energy trading. Therefore, this paper conceptually verifies the use of SL-A batteries (EV batteries with SoH 88-75%) particularly in ESs and EBs by repurposing them. Utilizing retired EV batteries in micromobility as well as SESS applications could also avoid sustainability challenges associated with the production of virgin batteries for both the applications.

## 2   Case Study

### 2.1   Conceptual Applicability

**Selection of Exemplary Battery and Application.** From the data reported by Statista and KBA (Kraftfahrt-Bundesasmt; German Federal Motor Transport Authority), Tesla is one of the leading car manufacturers with global market share (2021 - May 2024) of 18.8% and German market share (2022 - June 2024) of 13.5% [10, 11]. Due to limited technical information disclosures, the 75 kWh Tesla Model 3 battery is selected for this case study. ESs and EBs from Bolt's rental service are selected as exemplary application for RB, as same type of battery is used in both 5$^{th}$ generation ESs and EBs.

There are mainly 6 lithium-ion chemistries in LIBs which include Lithium-iron Phosphate (LFP), Lithium Manganese Oxide (LMO), Lithium Titanate (LTO), Lithium Cobalt Oxide (LCO), Lithium Nickel Cobalt Aluminum (NCA) and Lithium Nickel Manganese Cobalt (NMC) [12]. A Tesla model 3 battery pack of 75 kWh has 4 modules including 2 shorter modules with 23 bricks each and 2 longer modules with 25 bricks each [13]. This battery pack comes with NCA cell model 2170 and has a total of 4,416 cells [13].

**Required Capacity for Micromobility Application.** As per the data given by Bolt, the following specifications are required for ES/EB batteries [14]; Nominal voltage ($N_V$) = 47 V, Nominal capacity ($N_C$) = 14.4 Ah and Capacity in kWh (C) = 0.6768 kWh.

**Remaining Useful Capacity (RUC) of RB.** To calculate the RUC of RB, the capacity of RB with SoH 100% needs to be calculated. By considering the available data of Tesla Model 3 [13]; $N_V$ of a cell = 3.6 V, $N_C$ of a cell = 5 Ah.

As voltage remains constant if the cells are connected in parallel and current remains constant if the cells are connected in series [15], the cells in the RB need to be connected such that 15 cells in each column are connected in series and 3 cells in each row are connected in parallel as shown in Fig. 1 to get the required $N_V$ = 47 V and $N_C$ = 14.4 Ah. Hence, $N_C$ and $N_V$ after repurposing can be calculated as follows:

$$N_C \, of \, RB \, (Ah) = Number \, of \, cells \times N_C \, of \, cell \, (Ah) = 3 \times 5 = \mathbf{15 \, Ah}$$

$$N_V \, of \, RB \, (V) = Number \, of \, cells \times N_V \, of \, cell \, (V) = 15 \times 3.6 = \mathbf{54V}$$

Capacity of the RB in kWh is given as [16]:

$$C(Wh) = \frac{N_C(Ah) \times N_V(V)}{1000} = \frac{15 \times 54}{1000} = \mathbf{0.81 \, kWh}$$

Considering this, RUC of the RB at SoH 88% is given as follows:

$$RUC(kWh) = C(kWh) \times \frac{SoH}{100} = 0.81 \times 0.88 = \mathbf{0.7128 \, kWh}$$

**Dimensional Feasibility.** Same type of battery is used in Bolt's 5th generation ESs and EBs. To repurpose a battery required for Bolt ES/EB from a Tesla Model 3 battery pack, 45 cells are connected as mentioned in 'RUC of RB' above. Considering the dimensions of the cell 2170 that are given as H 70 mm x D 21 mm in [17], approximate dimensions of the RB are L 335.46 mm x B 64 mm x H 71.50 mm. As outer dimensions of Bolt EB/ES batteries were measured to be L 375 mm x B 110 mm x H 95 mm, it is concluded that the RB is within the dimensional limits of Bolt batteries.

## 2.2 Foreseen Economic Advantage

Germany is selected as operation location for precise calculation. 'B2U: Battery Second-Use Repurposing Cost Calculator' provided by National Renewable Energy Laboratory (NREL), U.S. Department of Energy, is used to calculate total cost of RB [18]. Therefore, all the calculations are done in dollars by taking the current exchange rate of $1€ = 1.08\$$ (taken on 2024/08/01) [19].

**Assumptions.** All the assumptions included in B2U calculator for this case study are:

- By simply benchmarking established EV battery repurposing companies, the annual facility throughput is assumed to be 400,000 kWh/year.
- As per German business holidays the module throughput per day is calculated considering 221 working days in a year [20].
- Cost of transport from facility to customer is not considered.
- Facility size is calculated to be 520.88 $/sq. m assuming the factors mentioned in [21] and by considerations of B2U calculator.
- Considering the regulations mentioned in [22] are implemented, the EV OEMs take back the batteries after reaching 88% SoH and forward them for repurposing.

**Location Selection.** Düsseldorf is selected as operation plant location using utility matrix and location selection analysis among viable industrial locations in Germany [21].

**Employment Cost.** The employment costs shown in Table 1 are referred from the industrial wages in Nordrhein-Westfalen region (Düsseldorf) provided by 'IG Metall' [23]. The quantity is scaled by B2U calculator considering annual facility throughput.

**Annual Expenses.** The test equipment and material handling equipment quantities are estimated by scaling formulas in B2U calculator considering the description and number of staff in Table 1. The average property rental cost for industries in Germany is 13.54 $/sq.m according to data given by Statista [24]. The annual percentage share of indirect costs is estimated as given by U.S. NREL [18]. Below in Table 2 the estimated average unit costs and total costs are calculated in each category by B2U calculator.

Based on the above considerations and B2U calculator, the cost of RB with NCA cells is calculated to be 70.79 $/kWh. Whereas the cost of a new 75 kWh NCA battery of Tesla Model 3 is around 231.48 $/kWh [25].

**Table 1.** Estimated annual employment cost

| Description | Quantity | Pay Grade | Annual Wage ($) |
|---|---|---|---|
| Test Technician | 22 | EG7 | 942,612 |
| Forklift/Warehouse Operator | 7 | EG5 | 99,378 |
| MD Truck Driver | 1 | EG3 | 39,023 |
| Supervisor and Operation Manager | 3 | EG11 | 181,491 |
| Chief Executive | 1 | EG12 | 66,290 |
| Electrical Engineer | 4 | EG9 | 210,676 |
| Sales/Logistic Representative | 4 | EG8 | 187,296 |
| Administrative Assistant | 2 | EG3 | 66,252 |
| Employee on site | 17 | EG7 | 728,382 |
| Human Resource Personnel | 1 | EG8 | 46,824 |
| Non-wage compensation | 0.302 | 2,538,108 | 815,619.86 |
| **TOTAL** | | | **3,516,347.86** |

## 3   Results

Considering the data calculated in the economic feasibility and conceptual applicability, the results for cost benefit analysis are presented in the Table 3. The cost to purchase new or repurposed batteries are calculated as:

$$No.\ of\ applications \times Capacity\ req.(kWh) \times cost\ of\ battery(\frac{\$}{kWh})$$

It can be depicted that 98 Bolt ES/EB batteries can be produced from just one 75 kWh battery pack of Tesla Model 3 with SoH of 88%. The conceptual applicability study states that even though an RB with SoH 88% is used, it not only fulfills the capacity and voltage requirements but also exceeds by 0.6 Ah and 7 V which is beneficial.

### 3.1   Profitability Comparison Between SESS and Micromobility (ESs/EBs)

Monthly gross profit-generating potential of SESS (by energy trading) versus RB (by Bolt's rental ESs/EBs) using a SL Tesla Model 3 battery.

**SESS Profit Opportunity.** As per the data in August taken from [26], the average price spread of electricity is approximately 0.25 $/kWh. Amount of energy traded is 66 kWh (RUC of a retired 75 kWh Tesla model 3 battery at SoH 88%).

$$SESS\ Gross\ Profit = 0.25\ \$ \times 66\ kWh \times 31 days = 475.92\ €$$

**RB Profit Opportunity.** Bolt ESs/ EBs rental rate in Germany is 0.32 $/minute [27]. Considering 55 km maximum range and 20 km/h average speed, total range results in 165 min [14].

$$RB\ Gross\ Profit = 0.32\ \$ \times 165\ minutes \times 98\ RB \times 31\ days = 162,411.48\ \$$$

**Table 2.** Estimate of total annual expenses by B2U calculator.

| Category | Description | Quantity | Average Unit cost ($) | Total cost ($) |
|---|---|---|---|---|
| Test Equipment | Battery test channel | 5 | 75,000 | 381,800 |
| | CAN hardware | 5 | 160 | |
| | Computers | 2 | 3,000 | |
| Materials Handling Costs | Estimated upfront battery purchases | 1,777.75 | 710.19 | 1,724,240.27 |
| | Storage racks | 22 | 100 | |
| | Forklifts | 2 | 7,000 | |
| | MD Truck | 1 | 141,000 | |
| | Shipping containers | 597 | 500 | |
| | Workstations and equipment | 12 | 500 | |
| Direct costs | Rent (sq. m) | 482.31 | 13.54 | 115,478.02 |
| | Electricity: Testing (kWh) | 310,583.4 | 0.10 | |
| | Electricity: HVAC & Lightning (kWh) | 5,191.5 | 2.27 | |
| | Transport to facility (miles) | 21,760 | 0.50 | |
| | Other direct costs | 0.02 | 2,761,225 | |
| Indirect costs (%) | Insurance | 0.03 | 18,920,913.19 | 3,241,523.39 |
| | General and administrative costs | 0.05 | 18,920,913.19 | |
| | Warranty | 0.05 | 23,204,459.04 | |
| | Research and development costs | 0.03 | 18,920,913.19 | |
| Others and offices | | | | 100,000 |

## 4   Limitations of this Approach

According to [22], safety parameters e.g. date of manufacture or date of putting into service of a battery, energy throughput, capacity throughput, full equivalent charge-discharge cycles and tracking of harmful events need to be considered to determine expected lifetime of a battery. Furthermore, customer perspective towards the products is a crucial factor and techniques to create positive image and awareness regarding impacts of producing new batteries is vital for increasing attractiveness and market share. This case study is limited to verify the conceptual applicability of repurposing EV batteries with cylindrical type of cells. The experimental analysis is essential for determining real-life applicability of this approach. Also, as the capacity of RB is more

**Table 3.** Cost benefit analysis for the use of RB in ESs/EBs

| SoH | No. of battery pack of 75 kWh | Capacity required for ES/EB (kWh) | No. of applications that can be produced with RUC of 1 T Model 3 battery | Cost required to purchase new batteries at SoH = 100% for given no. of applications | Cost required to purchase repurposed batteries at SoH = 88% for given no. of applications |
|---|---|---|---|---|---|
| 88% | 1 | 0.6768 | ≈ 98 | ≈ 15,353 $ | ≈ 4,695 $ |

than it is originally required, a compatible Battery Management System (BMS) to balance exceeded voltage and capacity must be installed while repurposing which may increase the cost. Cost calculation is based on estimated market value and B2U scaling formulas which gives the approximate cost of RB. A detailed market study may result in higher RB cost, however, results of the calculations conducted in the case study revealed that the cost of RB is less than a new battery.

## 5 Conclusion and Future Scope

As observed in the economic feasibility study, the cost of energy for RB is 69% less than a new battery cost while the required capacity and voltage of the battery are not compromised. Apart from the conceptual and economic aspects, this approach contributes towards 3 of the United Nation Sustainability Development Goals (2015) which are responsible consumption and production by promoting material circularity, affordable and clean energy by offering accessibility to the product in lower price and climate action by reducing Global Warming Potential (GWP) index and carbon footprint caused by LIB production. Given the expected production volume of LIBs for EVs, future development in standardization and knowledge gains from [8], it is predicted that by repurposing just 41% of EV batteries from 2025, 100% of the forecasted SESS and micromobility demand in 2030 can be fulfilled. As the scope of this case study is focused only on ESs and EBs, further market analysis for utilizing the remaining 59% of retired EV batteries in different applications e.g. E-Forklifts, E-Mopeds, etc. is necessary. There is potential scope for developing a simulation of a rental service business model using these RB by assessing the real-time data on SoH in SL-A batteries to predict the degradation for further re-renting in SL-B applications.

## References

1. Guzek, M., Jackowski, J., Jurecki, R.S., Szumska, E.M., Zdanowicz, P., Żmuda, M.: Electric vehicles—an overview of current issues—part 1—environmental impact, source of energy, recycling, and second life of battery. Energies **17**(1), 249 (2024)
2. Halkes, R.T., Hughes, A., Wall, F., Petavratzi, E., Pell, R., Lindsay, J.J.: Life cycle assessment and water use impacts of lithium production from salar deposits: Challenges and opportunities. Resour. Conserv. Recycl. **207**, 107554 (2024)

3. Arvidsson, R., Chordia, M., Nordelöf, A.: Quantifying the life-cycle health impacts of a cobalt-containing lithium-ion battery. Int. J. Life Cycle Assess. **27**(8), 1106–1118 (2022)
4. Braco, E., San Martín, I., Sanchis, P., Ursúa, A., Stroe, D.I.: State of health estimation of second-life lithium-ion batteries under real profile operation. Appl. Energy **326**, 119992 (2022)
5. Wannera, J., et al.: Diagnosis and remaining useful life prediction of end-of-life battery systems for second-life applications (2023)
6. Fischhaber, S., Regett, A., Schuster, S.F., Hesse, H.: Second-Life-Konzepte für Lithium-Ionen-Batterien aus Elektrofahrzeugen. Begleit-und Wirkungsforschung Schaufenster Elektromobilität (BuW) (2016)
7. Rizos, V., Urban, P.: Implementing the EU Digital battery passport – opportunities and challenges for battery circularity. Batraw, CEPS, Brussels (2024)
8. Dr. Thielmann, A, Dr. Neef C. Lithium-ion battery roadmap – industrialization perspectives towards 2030. Frauhnhofer Institute for Systems and Innovations Research ISI, Karlsruhe, Germany (2023)
9. Dong, Q., Liang, S., Li, J., Kim, H.C., Shen, W., Wallington, T.J.: Cost, energy, and carbon footprint benefits of second-life electric vehicle battery use. Iscience **26**(7) (2023)
10. New registrations of passenger cars by brands and model series. https://www.kba.de/DE/Statistik/Produktkatalog/produkte/Fahrzeuge/fz10/fz10_gentab.html?nn=3514348. Accessed 19 Aug 2024
11. Tesla and BYD Claim a Third of the Global BEV Market. https://www.statista.com/chart/27733/battery-electric-vehicles-manufacturers/. Accessed 21 Aug 2024
12. John, T.W.: The Handbook of Lithium-Ion Battery Pack Design: Chemistry, Components, Types, and Terminology. Elsevier, (2024)
13. Munro, S., Ellis, M.: Tesla model 3 teardown: the battery pack. design news webinar by informa markets (2019)
14. Bolt E-Scooter Homepage. https://bolt.eu/de-de/blog/der-bolt-5-unser-e-scooter-der-fuenften-generation/. Accessed 19 Aug 2024
15. Mehta, V.K., Rohit, M.: Basic Electrical Engineering. S. Chand Publishing, (2008)
16. Pop, V., Bergveld, H.J., Danilov, D., Regtien, P.P., Notten, P.H.: Battery management systems: accurate state-of-charge indication for battery-powered applications, vol. 9. Springer Science & Business Media, (2008)
17. Lithium Battery cell models and the industry shift 21700 VS 18650. https://www.epectec.com/articles/lithium-battery-cell-industrial-shifts-21700-vs18650.html#:~:text=The%2021700%20battery%20cell%20has,3.6%20colts%20to%203.7%20volts. Accessed 27 Aug 2024
18. B2U: Battery Second-Use Repurposing Cost Calculator. https://www.nrel.gov/transportation/b2u-calculator.html. Accessed 19 Aug 2024
19. EUR vs USD. https://www.ecb.europa.eu/stats/policy_and_exchange_rates/euro_reference_exchange_rates/html/eurofxref-graph-usd.en.html. Accessed 01 Aug 2024
20. Holidays in Germany 2024. https://www.arbeitstage.org/. Accessed 12 Nov 2024
21. Al-Alawi, M.K., Cugley, J., Hassanin, H.: Techno-economic feasibility of retired electric-vehicle batteries repurpose/reuse in second-life applications: a systematic review. Energy Climate Change **3**, 100086 (2022)
22. Regulation (EU) 2023/1542 of the European Parliament and of the Council of 12 July 2023 on batteries and waste batteries, amending Directive 2008/98/EC and Regulation (EU) 2019/1020, and repealing Directive 2006/66/EC (2023)
23. IG Metall, Elektrohandwerk Löhne und Gehälter (2024). https://www.igmetall.de/download/20240426_Elektrohandwerk_Loehne_Gehaelter_Entgelte_Azubis_b3303481979955cb416f467ae796b0624347f6e0. Accessed 27 Aug 2024

24. Average annual prime industrial rent price per square meter in Europe in 1st quarter 2024, by country. https://www.statista.com/statistics/858110/average-annual-industrial-rent-cost-per-square-meter-by-european-country/. Accessed 19 Aug 2024

25. How much does a new battery actually cost? Are there alternatives? https://tff-forum.de/t/wie viel-kostet-eigentlich-ein-neuer-akku-gibt-es-alternativen/153358. Accessed 19 Aug 2024

26. Electricity production and spot prices in Germany in August 2024. https://www.energy-cha rts.info/charts/price_spot_market/chart.htm?l=en&c=DE&interval=month. Accessed 01 Nov 2024

27. Bolt application version CA.138.0. https://play.google.com/store/search?q=bolt&c=apps& gl=DE. Accessed 01 Nov 2024

# Upcycling for Sustainable Manufacturing: Insights and New Methods

Victor Sodje[1,2], Manpreet Singh[1,2](✉), Jacob Crawford[1], Junwon Ko[1,2], Fazleena Badurdeen[1,2], and I. S. Jawahir[1,2]

[1] Department of Mechanical and Aerospace Engineering, University of Kentucky, Lexington, KY 40506, USA
manpreet.singh@uky.edu
[2] Institute for Sustainable Manufacturing, University of Kentucky, Lexington, KY 40506, USA

**Abstract.** Traditional recycling or downcycling processes, which continually diminish the inherent value of materials, are inadequate to meet the urgent demands of sustainable waste management and resource conservation. Upcycling, transforming end-of-life (EoL) materials into higher value forms, has been gaining traction as a promising alternative. Despite its rising prominence, the varied interpretations across industries and academic studies reveal a lack of consensus on upcycling and a limited understanding of its requirements. This study provides a clear definition of upcycling established through a focused literature review, addressing the current ambiguities associated with the term. In addition, a quantitative criterion, the Sustainability Benefit Factor (SBF), is developed to differentiate upcycling from recycling by evaluating value enhancement across the Triple Bottom Line: economic, environmental, and social. To further clarify upcycling from other EoL management strategies such as reuse, remanufacture, and recycling, the integration of upcycling into the 6R framework is performed based on the established definition and SBF. The findings in this paper contribute to enhancing upcycling practices and facilitating its broader adoption for sustainable waste management.

**Keywords:** Upcycling · downcycling · Sustainability Benefit Factor (SBF) · Triple Bottom Line

## 1 Introduction

With the increase in economic activities and the global population projected to reach approximately 9.7 billion by 2050 [1], natural resource consumption has surged significantly. This increase has placed substantial stress on the environment, leading to critical issues such as resource depletion, waste generation, and pollution. Efficient resource utilization and End-of-Life (EoL) management strategies are essential to mitigate these challenges and promote sustainable development. The 6R framework (reduce, reuse, recycle, recover, redesign and remanufacture) [2], is a comprehensive approach to address related challenges and promote sustainable manufacturing. With the 6Rs,

© The Author(s) 2025
H. Kohl et al. (Eds.): GCSM 2024, LNME, pp. 856–863, 2025.
https://doi.org/10.1007/978-3-031-93891-7_94

**Reduce** relates to minimizing raw material and energy use, and decreasing waste and emissions across all lifecycle stages. **Recover** involves collecting and processing products at EoL to apply other value reclamation strategies. **Reuse** involves extending product lifecycles, and conserving resources. **Remanufacture** restores used products to like-new condition, reducing the need for materials to produce new components. **Redesign** refers to reviewing and changing the design of products, processes, and systems for optimal sustainability performance, taking your time ce. **Recycle**, positioned as the least value recovery strategy in the 6R hierarchy, involves processing components in EoL products to obtain reusable secondary materials, reducing waste going into landfills. Integrating the 6Rs approach enhances manufacturing sustainability and supports a circular economy (CE) [2, 3].

While beneficial in mitigating the environmental impact by reducing the demand for virgin resources, recycling has inherent limitations. Pilz termed the process as 'downcycling' by which used materials are converted into materials of lower quality and diminished functionality over time. Helbig et al.[4] further, elucidate that this gradual degradation in material quality necessitates innovative solutions to enhance material value.

Upcycling has emerged as a viable alternative to address the shortcomings of recycling. However, ambiguities surrounding this concept hinder understanding the requirements for wider application across various sectors. This paper clarifies the upcycling concept by establishing a clear definition through a focused literature review and developing the Sustainability Benefit Factor, a criterion to differentiate upcycling from recycling. Finally, upcycling is incorporated into the 6R framework to improve its understanding, relative to other value recovery strategies.

## 2  Background and Approach

### 2.1  Literature Review

Previous studies have described upcycling in various ways, highlighting its unique benefits over traditional recycling. Upcycling, as defined by Sung et. al. [5] involves the creative reuse of waste materials to produce new items of higher quality or value. This approach contrasts with recycling, which often leads to downcycling, where materials degrade in quality over successive cycles [6]. To distinguish upcycling from other EoL strategies, it is crucial to ensure that value has been added to the material. According to the International Organization for Standardization (ISO), value involves gains or benefits from meeting needs and expectations related to resource use and conservation [7]. However, defining value is inherently subjective and varies across different contexts, making it challenging to establish a universal definition for upcycling. Despite this subjectivity, past studies emphasize that upcycling contributes to sustainability by decreasing the demand for virgin materials and lowering environmental impacts. For instance, upcycling extends the life cycle of materials, thereby conserving resources and reducing waste generation [8]. By creatively reusing waste materials to produce new items of higher quality or value, upcycling reduces the need for new resources and minimizes environmental degradation associated with the extraction and processing of raw materials [5]. This aligns with the principles of the CE, which aims to keep materials in use for as long as possible [2].

The environmental benefits of upcycling are substantial. Upcycling significantly reduces waste by diverting materials from landfills and incineration, thereby reducing greenhouse gas emissions associated with waste decomposition [9]. While recycling shares these benefits, upcycling goes a step further by creating materials of higher intrinsic value and improved functionality (i.e., higher intrinsic value) compared to the original or recycled materials [10]. This process not only extends the lifespan of materials but also provides greater environmental and economic advantages compared to recycling.

Despite all these advantages, upcycling faces several challenges. One significant issue is the discrepancies in the understanding of the concept, leading to inconsistencies in its application and use. Different industries and researchers interpret upcycling differently, which can hinder its widespread adoption and effectiveness [5]. From Table 1 (which shows the most pertinent sources; other relevant literature are compiled and available in [11]), it is evident that various studies interpret upcycling differently. Some literature defines upcycling as converting waste materials directly into finished products, such as turning plastic bottles into fashion accessories or home decor items. Other studies describe upcycling as upgrading or redesigning existing products, focusing more on the product level and creative 'reuse'. While this adds value to the EoL material, it often does not enhance the intrinsic value of the original material itself. Jayasinghe et al. [12] highlight these differing views in their study, emphasizing the need for a clearer understanding and definition of upcycling. However, their study could not establish a clear definition which remains a gap to be addressed.

**Table 1.** An overview of the discrepancies in upcycling processes, highlighting EoL management strategies that are often misidentified as upcycling in the literature.

| Source | Activity Described | EoL Strategy Interpreted as Upcycling |
|---|---|---|
| [10] | Low-nickel polycrystalline cathodes Nickel-rich single-crystal cathodes | Upgrading (i.e., for Reuse) |
| [13] | Waste PVC to value-added organic chlorides | Repurposing (i.e., for Reuse) |
| [14] | Core parts from end-of-life vehicles to reusable components | Refurbishing (i.e., for Remanufacture) |
| [15] | Plastic waste to reusable plastic materials | Recycling |
| [16] | Waste materials to new products | Redesign |
| [17] | Material to product | Redesign |
| [18] | Post-consumer waste to home decor | Remanufacturing |
| [19] | From waste material to valuable products | Chemical Recycling |

## 2.2 Methodology

A clear description of upcycling is essential to clarify the concept, promote research and technology development, and for wider application in industry. To address this, a new definition of upcycling is established. This was achieved by critically assessing previous descriptions of upcycling to examine the underlying value recovery strategies examined in those studies. This enabled differentiating if a study misrepresented any other 'R' as upcycling or, if it involved genuine enhancement of the intrinsic value of the material [5, 12]. The review enabled differentiating instances of value enhanced material processing vs. conventional recycling or application of other Rs. The review also highlighted that upcycling often resembles recycling in terms of processing, input materials, and the nature of the output (although of higher value), creating confusion between the two. Thus, the SBF, a criterion to differentiate between these two concepts, is developed based on the new upcycling definition and the 6R framework proposed by Jawahir. et al. [20]. Integration of upcycling into the 6R framework is carried out to demonstrate how it compares with other strategies.

## 3   Results and Discussion

### 3.1   Upcycling: Establishing a Definition

From the review of literature, it is evident that a variety of practices have been used to describe the term upcycling. Many studies have described upcycling as the transformation of EoL *products* to new, higher-value products, subsequently used for different applications [17]. The new products in these studies have often been utilized for another purpose, complicating the assessment of value, which can vary depending on application and requirements. Other studies referred to upcycling in the context of converting EoL *materials* into a new product [21]. Also, EoL products cannot typically be converted to other new products directly; the transformation process will involve several intermediate steps. This approach which describes upcycling as a product-to-product transformation, therefore, mistakenly leads to grouping all such steps to refer to the process as upcycling.

To address this ambiguity, we posit that upcycling involves:

*"Transformation of material contained in EoL products/components to a higher valued material."*

This definition upholds the notion that upcycling is a transformation process which increases the intrinsic value of the material [5] as opposed to downcycling/recycling [4]. It also limits the description to include only the transformation of resources from one state (EoL products/components) to the immediate next state (materials). Given this description for upcycling, it is vital to establish a method to distinguish it from recycling. The next section proposes a measure to achieve this differentiation.

### 3.2   Sustainability Benefit Factor (SBF)

This section introduces a theoretical measure to clarify the difference between recycling and upcycling. This is done by considering the performance potential (PP) of the resulting material (when used in a product) generated through a process and the negative impact (I) on the environment, economy, and society due to the conversion. The Sustainability Benefit Factor (SBF) is a dimensionless term defined as the ratio of the performance potential (PP) and adverse impacts (I) [Eq. (1)]. Therefore, a higher SBF could be inferred to offer more sustainable value.

$$SBF = PP/I \qquad (1)$$

It must be noted that PP is subjective as it depends on the actual application of the material in an end product and the specific performance requirements. In other words, the PP for the same material could be different based on what product it is used for. Nonetheless, as a concept, it can be a useful tool to evaluate the sustainable value offered by a process.

Similarly, the impact term subsumes all the adverse effects of the process that can affect the environment in terms of resource consumption, use of virgin material, energy use, carbon emissions, etc. The economic impact can include the different costs associated with the material's processing, such as initial investment, direct/indirect costs, labor costs, energy, logistics, etc. The societal aspect covers product quality, durability, and health hazards due to the process.

### 3.3 Differentiating Upcycling from Recycling Using SBF

If a product/component cannot be reused/remanufactured when it reaches EoL, then extraction of the constituent materials must be explored. Also, any by-products, waste and emissions originating from the pre-manufacturing, manufacturing, and use stages can be accumulated for further processing and extraction of value.

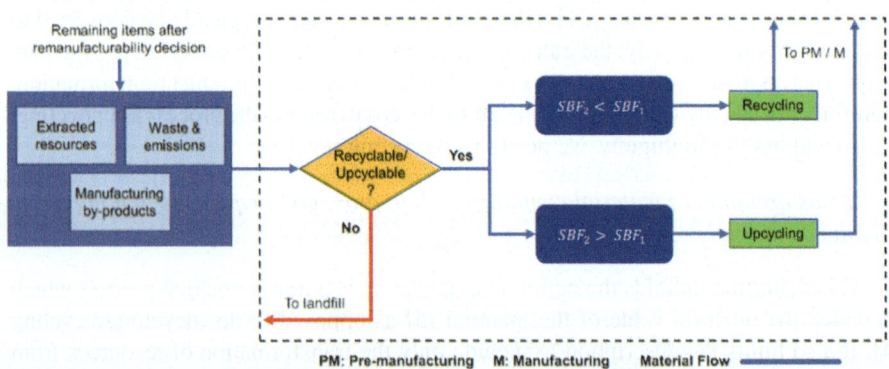

**Fig. 1.** Differentiating upcycling from recycling

As shown in Fig. 1, these items can be assessed for the potential for either upcycling or recycling. Which path of flow occurs will depend on the previously defined SBF. Consider two different SBFs ($SBF_1$ & $SBF_2$) as shown below.

$$SBF_1 = PP_1/I_1 \qquad (2)$$

$$SBF_2 = PP_2/I_2 \tag{3}$$

$SBF_1$ is calculated for the virgin material considering the pre-manufacturing and manufacturing stages. Suppose the impact $(I_1)$ encapsulates the adverse effects resulting from the mining and processing activities for $m_v$ quantity of virgin materials whereas $PP_1$ the derived performance potential by using that material in the end product. Let $SBF_2$ refer to the SBF when the EoL product/components are processed to obtain a secondary material of quantity $m_p$, with $I_2$ being the impact of that transformation and $PP_2$ being the potential performance when that material is used in a product.

If $SBF_2$ is higher than $SBF_1$, then the process shall be classified as "Upcycling"; otherwise, the process can be considered "Recycling." This comparison can be explained by the fact that upcycling is supposed to bring higher value to the material. However, this should not occur at the cost of causing significant damage to the environment, economy, and society.

### 3.4 Upcycling in the 6Rs Framework

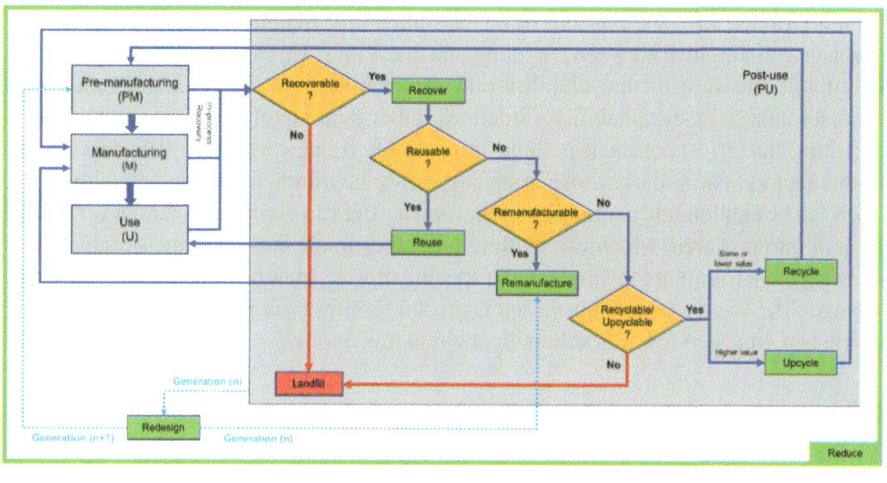

**Fig. 2.** 6Rs for Sustainable Manufacturing updated with upcycling. Adapted from [2]

Figure 2 illustrates the information and material flows in closed loop across the total lifecycle (pre-manufacturing, manufacturing, use and post-use stages) for sustainable manufacturing with 6R application. At the EoL stage, the product, along with by-products and wastes from the pre-manufacturing and manufacturing stages, enters the post-use stage. The first step here is recovery through collection, disassembly, sorting, and cleaning. If recovery is not feasible, the product is discarded in landfills. Recovered resources are preferably reused, reducing the need for virgin materials in new products [2]. If the product or components cannot be reused directly, they are considered for remanufacturing. If remanufacturing is possible, the items are returned to the manufacturing stage for reprocessing and restoration without losing functionality, or else sent for recycling/upcycling, where they can be broken down into material form and processed accordingly.

As illustrated reuse and remanufacturing shall remain preferential over the upcycling/recycling processes as they enable higher value retention. Finally, the output from recycling and upcycling can either go to the pre-manufacturing or manufacturing stage, depending upon the condition and form of the material. For example, if the material is a polymer and is in powder form, then it will go to the pre-manufacturing stage for pelletizing. However, some examples from the literature can also be found in which upcycled material was ready for manufacturing directly, as shown in the study by Yoder et al. [22] demonstrating the conversion of cast aluminum machined chips to the AFSD (Additive Friction Stir Deposition) feedstock by cold pressing. The feedstock in the form of square rods is then utilized for solid-state metal additive deposition.

## 4 Conclusions and Future Work

Upcycling represents a significant advancement in sustainable manufacturing practices. Unlike traditional recycling, upcycling extends the lifespan of materials, contributing to a more robust CE. The concept of the Sustainability Benefit Factor (SBF) offers an approach to distinguish between recycling and upcycling. This approach provides a more quantifiable measure of upcyclability and recyclability, ensuring that manufacturing processes adhere to sustainability guidelines rather than merely assuming compliance. Upcycling has also been incorporated into the 6R framework. The SBF depends on various factors, particularly those associated with environmental and societal aspects, which can be challenging to quantify; therefore, it offers an excellent potential for further research into this area, which can explore how to estimate these into measurable terms and hence help to capture the impact of a specific process in numerical terms, facilitating to gauge SBF. In conclusion, upcycling is pivotal for preserving our planet, and the SBF criteria is a vital step towards achieving a circular economy.

## References

1. Dorling, D.: World population prospects at the UN: our numbers are not our problem? In the struggle for social sustainability, pp. 129–154. Policy Press (2021)
2. Jawahir, I.S., Bradley, R.: Technological elements of circular economy and the principles of 6R-based closed-loop material flow in sustainable manufacturing. Procedia Cirp **40**, 103–108 (2016)
3. Jayal, A., et al.: Sustainable manufacturing: modeling and optimization challenges at the product, process and system levels. CIRP J. Manuf. Sci. Technol. **2**(3), 144–152 (2010)
4. Helbig, C., et al.: A terminology for downcycling. J. Ind. Ecol. **26**(4), 1164–1174 (2022)
5. Sung, K.: A review on upcycling: current body of literature, knowledge gaps and a way forward (2015)
6. Kay, T.: Salvo in Germany-Reiner Pilz. SalvoNEWS (99), 11–14. Hlm (1994)
7. Morris, K., et al.: Standards as Enablers for a Circular Economy. Technology Innovation for the Circular Economy: Recycling, Remanufacturing, Design, Systems Analysis and Logistics, pp. 1–16 (2024)

8. Nallapaneni, M.K., et al.: From Trash to Treasure: Unlocking the Power of Resource Conservation, Recycling, and Waste Management Practices, p. 13863. MDPI (2023)
9. Hossain, M.U., et al.: Environmental and technical feasibility study of upcycling wood waste into cement-bonded particleboard. Constr. Build. Mater. **173**, 474–480 (2018)
10. Qian, G., et al.: Value-creating upcycling of retired electric vehicle battery cathodes. Cell Rep. Phys. Sci. **3**(2) (2022)
11. Additional references (2024). https://github.com/viks996/Addtional-References
12. Jayasinghe, R., Arachchige, P.: Upcycling: A New Perspective on Waste Management in a Circular Economy. In: Waste Technology for Emerging Economies, pp. 237–274. CRC Press (2022)
13. Feng, B., et al.: Waste PVC upcycling: Transferring unmanageable Cl species into value-added Cl-containing chemicals. Appl. Catal. B **331**, 122671 (2023)
14. Wang, R., et al.: A green strategy for upcycling utilization of core parts from end-of-life vehicles (ELVs): Pollution source analysis, technology flowchart, technology upgrade. Sci. Total Environ. **912**, 169609 (2024)
15. Zhao, X., et al.: Upcycling to sustainably reuse plastics. Adv. Mater. **34**(25), 2100843 (2022)
16. Bofylatos, S.: Upcycling systems design, developing a methodology through design. Sustainability **14**(2), 600 (2022)
17. Marques, A.D., et al.: From waste to fashion–a fashion upcycling contest. Procedia CIRP **84**, 1063–1068 (2019)
18. Lisieux, F.: development of home décor from post-consumer waste by implementing upcycling and transformable methods. St. Teresa's college (autonomous), Ernakulam (2022)
19. Kots, P.A., et al.: A two-stage strategy for upcycling chlorine-contaminated plastic waste. Nat. Sustain. **6**(10), 1258–1267 (2023)
20. Zhang, X., et al., On improving the product sustainability of metallic automotive components by using the total life-cycle approach and the 6R methodology (2013). https://doi.org/10.14279/depositonce-3753
21. Zhao, X., et al.: Plastic waste upcycling toward a circular economy. Chem. Eng. J. **428**, 131928 (2022)
22. Yoder, J.K., et al.: Additive friction stir deposition-enabled upcycling of automotive cast aluminum chips. Addit. Manuf. Lett. **4**, 100108 (2023)

# A Standards-Guided Approach to Formalizing Product Circularity Assessment Metrics

Junwon Ko[1]([✉]), Sandwana Sneethan[1], Gisele Bortolaz Guedes[1],
Fazleena Badurdeen[1], I. S. Jawahir[1], Buddhika Hapuwatte[2], and K. C. Morris[2]

[1] Institute for Sustainable Manufacturing (ISM) and Department of Mechanical and Aerospace Engineering, University of Kentucky, Lexington, KY, USA
jko248@uky.edu
[2] National Institute of Standards and Technology (NIST), Gaithersburg, MD, USA

**Abstract.** The industrial transition towards a more sustainable Circular Economy (CE) emphasizes the need for products that align with CE principles and robust methods for assessing how circular they are. However, ambiguities between metrics vs. indicators are not yet addressed in the current discussion of product circularity assessment (PCA). Most PCA methods also lack clarity in the metrics used, including their selection and description, which leads to confusion and limited industry adoption. To address these gaps, this paper proposes clear definitions for indicators and metrics and adapts standards to establish a structured approach for developing PCA metrics. This process involves establishing the objective or purpose of each metric, identifying candidate metrics from literature sources or existing PCA methods, and defining them with detailed context (e.g., description, formula, range) and content information (e.g., notes, references, standards/guidelines). To effectively compile and communicate the developed metrics, an Excel-based database consisting of attributes, indicators, and metrics (AIM) was developed and presented. Over 80 relevant metrics for measuring product circularity were identified, characterized, and organized under the AIM database. The standards-guided approach and database proposed in this paper formalize the development of metrics for PCA and aid data collection processes, enhancing their practical application in the industry. The findings in this paper further contribute to the successful transition toward product-level CE.

**Keywords:** Circular Economy · Product Circularity · Indicators · Metrics

## 1 Introduction

The extraction and processing of resources are continuously increasing to meet the demands of a growing population. These resources are currently extracted, processed, consumed, and disposed of in a manner that leads to the triple planetary crisis: climate change, biodiversity loss, and pollution [1]. Circular Economy (CE) addresses this problem by establishing closed-loop flows of resources through value recovery, retention, and enhancement [2]. In CE, restoration and regeneration practices avoid the loss of value and materials at end-of-life (EoL). Restoration focuses on maximizing the lifespan of

H. Kohl et al. (Eds.): GCSM 2024, LNME, pp. 864–872, 2025.
https://doi.org/10.1007/978-3-031-93891-7_95

products and materials through processes such as repair, reuse, remanufacture, and recycle. On the other hand, regeneration promotes the return of the EoL materials to the biosphere, emulating natural cycles. At its core, CE aims to eliminate waste and pollution through the circulation of products and materials while regenerating nature [3]. CE operates at various scales: micro (individual products and companies), meso (industrial parks or networks), and macro (cities, regions, nations); initiating the transition at the micro level can pave the way for global transformation [4]. Hence, products need to be designed, manufactured, used, and managed based on the CE principles, often referred to as Circular Products (CPs), to support the benefits of circularity across all levels.

Developing a CP and improving its circularity performance necessitates a robust method to assess how well the product aligns with the CE principles [5]. Numerous product circularity assessment (PCA) methods have been proposed by various groups in recent years, including Material Circularity Indicator [6], CE Index [7], C metric [8], and Circularity Potential Indicator [9]. The authors' previous work revealed that existing product circularity assessment methods fail to comprehensively evaluate all relevant CE principles [5]. Additionally, these methods often exhibit ambiguities in the use of the terms *indicator* and *metric*. They also lack clarity in the selection of metrics and provide insufficient detail, such as missing or vague descriptions, undefined scopes, and unclear value ranges. These shortcomings contribute to confusion and difficulty in data collection and implementation, thereby impeding industry adoption. Thus, the main objectives of this paper are to:

1. Clarify and differentiate the terms *indicator* and *metric* in PCA.
2. Establish a structured approach to identify and characterize metrics for PCA.
3. Develop a database to compile PCA metrics with sufficient detail.

This paper is organized as follows: Sect. 2 offers the background and reviews related work. Section 3 presents descriptions of indicators and metrics established and employed in this study, a structured approach for developing PCA metrics adapted from standards, and a database effectively organizing and communicating the metrics. Section 4 provides a discussion of the results, followed by a summary of the main contributions in Sect. 5.

## 2   Background and Related Work

An overview of the framework developed in prior work is presented here to provide a context for the use of attributes, indicators, and metrics in PCA. A review of standards and literature for formalizing Key Performance Indicators (KPIs) is also included in this section, which serves as the basis for formalizing PCA metrics.

### 2.1   Product Circularity Assessment Framework

To develop a measurement system for product circularity, a clear understanding of what makes a product circular is required. Previous studies have explored the necessary conditions for a product to be considered circular [10, 11]. However, most do not address the full range of CP aspects. Some focus on the design side, while others concentrate on resource usage, often without involving industry stakeholders for verification. This limitation reduces the practical applicability of their findings [5]. Previous work by the authors addressed this gap by identifying comprehensive and actionable attributes, which are constitutive characteristics that describe the desired features and properties of CPs. These attributes were then verified for their relevance and applicability by engaging industry stakeholders [5, 12]. Figure 1 presents these attributes, categorized into three groups based on their inherent roles: Drivers and Enablers (D&E), CPs (i.e., the outcome), and Benefits and Implications (B&I). D&E attributes drive the development of CPs, while those in the Outcome group represent the distinctive characteristics of CPs. Collectively, these lead to the attributes in the B&I group, which describe the consequent impacts (more details available in Ko et al. [12] and Guedes et al. [13]).

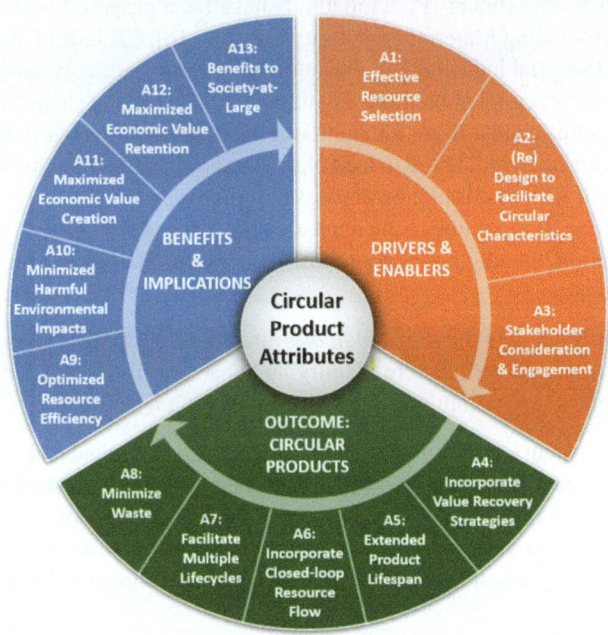

**Fig. 1.** Circular Product (CP) attributes (A1-A13 denote the attributes)

Building on this work, the authors developed AIM-PCA (A: Attributes, I: Indicators, M: Metrics) through a university-industry-government collaborative effort. This hierarchical framework (Fig. 2) was designed to comprehensively measure product circularity while overcoming the limitations of the existing PCAs [13]. At the base of this framework are metrics, which provide quantitative or qualitative measurements of specific indicators. Indicators provide information about the performance of broader attributes, which collectively define CPs. The metrics are then normalized and aggregated into indicators, which are further consolidated into attributes. Sub-indices are established based on the three distinct groups of attributes shown in Fig. 1. Finally, this framework culminates in a Product Circularity Index (PCI). The AIM-PCA framework ensures a detailed and systematic evaluation of product circularity. However, a procedure to identify relevant circularity metrics and details necessary to enable data collection and practical adoption is still largely missing. The work presented in this paper aims to address this limitation.

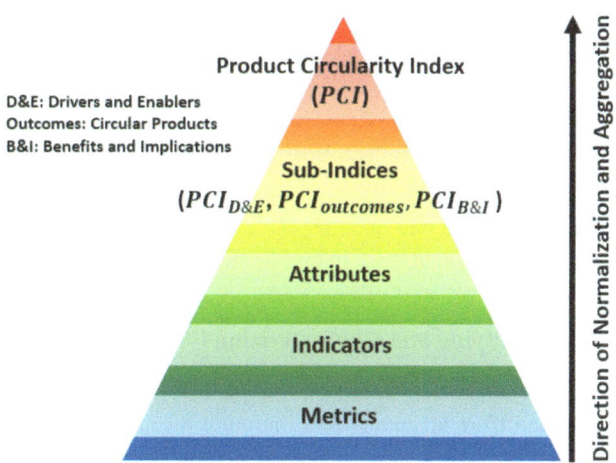

**Fig. 2.** AIM-PCA framework

## 2.2 Standards and Literature for Formalizing Key Performance Indicators

The International Organization for Standardization (ISO) published the ISO 22400 standard to provide a neutral framework for defining KPIs (a measurable value used to assess the achievement of a critical objective) in manufacturing operations management. This standard emphasized the importance of providing content and context information when characterizing a KPI. Content information includes name, ID, description, scope, formula, unit of measure, range, and trend. Context information involves timing, constraints, usage, audience, production methodology, effect model diagram, and notes [14]. The E3096–18 standard by ASTM International offers similar guidelines for defining KPIs but omits some context information items, indicating the information items are adjustable based on specific needs. Additionally, the E3096–18 standard proposes a procedure for developing relevant KPIs, which includes steps such as establishing KPI objectives, identifying candidate KPIs from existing sources or defining new ones, and assessing their effectiveness [15].

Various academic studies have also proposed processes for developing KPIs in the manufacturing industry [16–18]. These processes are similar to the E3096–18 standard in specifying objectives and identifying KPIs from literature sources and include an additional step to engage industry stakeholders to verify the relevance of the identified KPIs. The findings from these standards and academic studies are adapted to establish a structured approach for developing PCA metrics.

## 3   Methodology and Outcomes

This study builds on the previous work described above and defines the method used to select metrics that address PCA indicators. The following definitions of the terms indicator and metric are based on a synthesis of findings from literature sources:

- **Indicator:** *A specific expression that provides information about performance or describes the state of an attribute* [19–21].
- **Metric:** *A means of measuring an indicator, whether absolute or relative, normalized or non-normalized* [22–24].

Each CP attribute can have one or more indicators to describe its performance or state. Similarly, each indicator can contain one or more metrics to measure it.

### 3.1   Approach for Identifying and Characterizing PCA Metrics

This section presents a structured approach for identifying and characterizing PCA metrics (Fig. 3) adapted from standards and literature discussed previously. Once the attributes and indicators are defined, the metrics selection process begins by establishing a metric objective based on its corresponding attribute and indicator. Next, a candidate metric is identified from existing literature and/or PCA methods. If inadequate, a new metric is defined using information categories adapted from the ISO 22400 standard [14]. These categories are also used to further improve metrics that lack sufficient detail or require revision. Stakeholder engagement for defining the metric ensures that it is effective and implementable in real-world applications. Standards and guidelines such as the Global Reporting Initiative (GRI) can also be utilized to support this step.

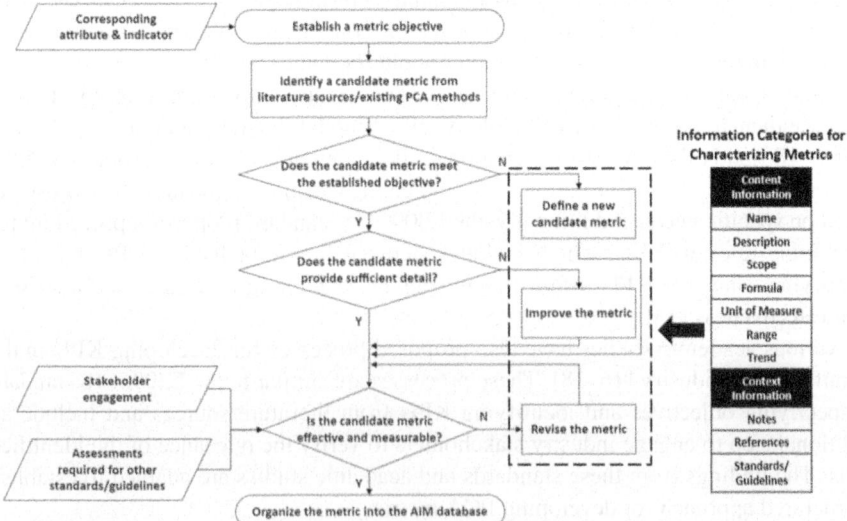

**Fig. 3.** A structured approach for identifying and characterizing PCA metrics

The GRI, for example, provides sustainability reporting standards covering various aspects, from materials usage and emissions to waste generation and occupational health and safety [25]. Due to its extensive adoption across industries, there is typically greater availability of data related to candidate metrics aligning with GRI disclosures. Once the effectiveness and measurability are verified for a metric, it is incorporated into the AIM database. The database was designed based on the standards and the PCA framework described in the background section.

### 3.2 Development of PCA Metrics and AIM Database

The approach described above was used to identify and characterize over 80 relevant PCA metrics which comprehensively assess all the indicators and corresponding attributes. An Excel-based database containing a hierarchical AIM structure was developed to effectively communicate the selected metrics in sufficient detail. This database employs context and content information categories, adapted from both the ISO and ASTM standards [14, 15] as relevant, based on the scope of this paper. Figure 4 shows a quantitative metric in the database. Due to the multifaceted nature of CE, some metrics were identified as qualitative. For example, *Ease of Access to Components*, a metric to assess ease of product disassembly and repair essential for extending its lifecycle, is measured on a categorical scale: low (requires specialized tools), moderate (commonly available tools can be used with minimal guidance), and high (common household tools suffice without any guidance).

| AIM (Attributes, Indicators, and Metrics) Database | |
|---|---|
| Category | Drivers & Enablers |
| Attribute | Effective Resource Selection |
| Indicator | Material Utilization |
| *Metric Characterization* | |
| *Content Information* | |
| Name | Recyclable Materials Usage |
| Description | The ratio of materials that can be recycled with available technology in the product (including its packaging and accessories) relative to the total mass of the product (including its packaging and accessories) |
| Scope | PM |
| Formula | (Total mass of materials that can be recycled with available technology / Total mass of product, packaging, and accessories)*100% |
| Unit of Measure | % (the inputs are measured in grams) |
| Range | 0 to 100% |
| Trend | Higher-better |
| *Context Information* | |
| Notes | - |
| References | (Saidani et al., 2019, Bracquene et al., 2020, Vimal et al., 2021) |
| Standards/Guidelines | GRI Standards, 301-1 |

**Fig. 4.** An example metric in the AIM database

The content information encompasses each metric's name, description, scope, formula, unit of measure, range, and trend. For instance, the name specifies the metric while the description provides a detailed explanation. The scope identifies the relevant lifecycle stage, like pre-manufacturing (PM), and the formula presents the mathematical calculation or categorical scale. The unit of measure defines the basic unit or dimension, the range sets the upper and lower limits, and the trend indicates the direction of improvement, such as higher-better. The context information includes notes for clarity, references if the metric is derived from external sources, and relevant standards or guidelines (useful for locating associated). The systematic and comprehensive approach deployed in the AIM database ensures that each metric is clearly defined and easily understood without ambiguity, facilitating better adoption in the industry.

## 4   Discussion of Results

The proposed distinction between indicators and metrics can resolve the current ambiguities regarding these terms. The AIM database can be an effective tool for clearly communicating how to calculate and interpret each metric; as the related CP attribute and indicator to each metric are identified, it provides insights into enhancing product circularity from various perspectives, such as design, resource selection, manufacturing, and product EoL management. This structure enhances the accuracy and reliability of metrics, facilitating better decision-making in CP development. Moreover, identifying relevant standards and guidelines aids in data collection and usability of the metrics.

Due to the diversity and complexity of products in different sectors, the metrics developed are intended to serve as example metrics. They are customizable based on factors such as product type, the company's specific scope, data availability, and infrastructure capacity. For example, consumer electronics primarily consist of technical materials like metals and plastics, making *Recyclable Materials Usage* or *Critical Materials Usage* more suitable metrics than *Biodegradable Materials Usage* for assessing circularity in the material utilization aspect. Thus, the metrics in the database serve as examples that can be adapted based on the requirements of a product/company.

## 5   Conclusions and Future Work

This paper addresses the ambiguities between indicators and metrics by proposing clear descriptions. Additionally, the development of PCA metrics was formalized through a standards-guided approach and the structured AIM database, improving their practicality and adoption in the industry. This paper further contributes to the development and operationalization of comprehensive PCA methods, which can ultimately enhance Circular Product Design and accelerate the transition toward a more sustainable CE.

Future work will focus on completing ongoing case studies with several global OEMs in the consumer electronics sector. These case studies will verify and validate the AIM framework and database. The PCA metrics developed through this study will also be further improved based on the findings from these case studies.

**Acknowledgements.** The work presented here is supported by grants from NIST (Nos. 70NANB22H104 and 70NANB23H261) and collaborations with industry partners from Amazon, Inc., Ryan Bradley, Ardeshir Raihanian, and Vincenzo Ferrero.

# References

1. UN Environment Programme: Global Resources Outlook 2024: Bend the Trend – Pathways to a liveable planet as resource use spikes. International Resource Panel, Nairobi (2024)
2. International Organization for Standardization: Circular economy - Measuring and assessing circularity performance. Standard No. ISO 59020 (2024)
3. Ellen MacArthur Foundation: Towards the circular economy, vol. 1: An economic and business rationale for an accelerated transition (2013)
4. Kirchherr, J., Yang, N.N., Spuntrup, F.S., Heerink, M.J., Hartley, K.: Conceptualizing the circular economy (revisited): an analysis of 221 definitions. Resour. Conserv. Recycl. **194**, 107001 (2023)
5. Ko, J., et al.: A critical analysis of circular product attributes and limitations of product circularity assessment methods. Resour., Conserv. Recycl. Adv. **23**, 200219 (2024)
6. Ellen MacArthur Foundation and Granta Design: Circularity indicators - An approach to measuring circularity – Methodology (2019)
7. Di Maio, F., Rem, P.C.: A robust indicator for promoting circular economy through recycling. J. Environ. Prot. **6**(10), 1095–1104 (2015)
8. Linder, M., Sarasini, S., van Loon, P.: A metric for quantifying product-level circularity. J. Ind. Ecol. **21**(3), 545–558 (2017)
9. The circularity potential indicator (CPI) tool (beta version). Circularity Indicators. https://www.circulareconomyindicators.com/cpitool.php. Accessed 24 June 2024
10. Suppipat, S., Hu, A.H.: A scoping review of design for circularity in the electrical and electronics industry. Resour., Conserv. Recycl. Adv. **13**, 200064 (2022)
11. Nag, U., Sharma, S., Kumar, V.: Multiple life-cycle products: A review of antecedents, outcomes, challenges, and benefits in a circular economy. J. Eng. Des. **33**(8), 173–206 (2021)
12. Ko, J., Guedes, G.B., Badurdeen, F., Jawahir, I.S., Morris, K.C., Ferrero, V.: Transitioning towards circular consumer electronics products. In: Paper Accepted for Presentation: The 2024 ASME IDETC and CIE Conference, Washington D.C (2024)
13. Guedes, G.B., Ko, J., Badurdeen, F., Jawahir, I.S., Morris, K.C., Ferrero, V.: A systems-based framework for product circularity assessment. In: Proceedings of the 19th Global Con. Sustainable Manufacturing, Buenos Aires, Argentina (2023)
14. International Organization for Standardization: Automation systems and integration - key performance indicators (KPIs) for manufacturing operations management - Part 1: Overview, concepts and terminology. Standard No. ISO 22400–1 (2014)
15. ASTM International: Standard Guide for Definition, Selection, and Organization of Key Performance Indicators for Environmental Aspects of Manufacturing Processes. Standard No. ASTM E3096–18 (2018)
16. Kibira, D., Brundage, M.P., Feng, S., Morris, K.C.: Procedure for selecting key performance indicators for sustainable manufacturing. J. Manuf. Sci. Eng. **140**(1), 011005 (2018)
17. Amrina, E., Vilsi, A.L.: Key performance indicators for sustainable manufacturing evaluation in cement industry. Procedia CIRP **26**, 19–23 (2015)
18. Mohamad, E., Muhamad, M.R., Abdullah, R., Saptari, A.: A study on the development of key performance indicators (KPIs) at an aerospace manufacturing company. J. Adv. Manuf. Technol. **2**(2), 1–17 (2008)
19. ISO: Environmental management – Vocabulary. Int. Organization for Standardization. Standard No. ISO 14050 (2020)
20. OECD: Education at a glance 2014: OECD Indicators. OECD, Paris (2014)
21. Feng, S., Joung, C.: Development overview of sustainable manufacturing metrics. In: Proceedings of the 17th CIRP International Conference on Life Cycle Engineering, Hefei, China (2010)

22. Shuaib, M., Seevers, D., Zhang, X., Badurdeen, F., Rouch, K.E., Jawahir, I.S.: A metrics-based framework to evaluate the total life cycle sustainability of manufactured products. J. Ind. Ecol. **18**(4), 491–507 (2014)
23. Ahi, P., Searcy, C.: An analysis of metrics used to measure performance in green and sustainable supply chains. J. Clean. Prod. **86**(1), 360–377 (2015)
24. Oschlies, A., et al.: Indicators and metrics for the assessment of climate engineering. Earth's Future **5**(1), 49–58 (2017)
25. GRI: Consolidated set of the GRI standards. Global Reporting Initative (2024)

# Exploration of Sustainable Approaches for the Smartphone Industry: A Value Chain Perspective

Gaurav Anant Mahajan[✉], Dhanishtha Jitendra Patole, Maxim Mintchev, and Valentin Eingartner

Institute of Machine Tools and Factory Management, Technische Universität Berlin, Pascalstr. 8 - 9, 10587 Berlin, Germany
gauravmahajan113@gmail.com

**Abstract.** Smartphone manufacturers find themselves in one of the most competitive industries worldwide, while also needing to fulfill their environmental and social responsibilities. These challenges contribute to growing concerns about e-waste, carbon emissions, and ethical sourcing, underscoring the need for sustainable business models. This study evaluates the smartphone industry's progress towards sustainability, amidst growing claims of sustainable practices by assessing the performance of three major companies' smartphones with a Comparative Sustainability Analysis. The cumulative carbon emissions associated with each phase of the Life Cycle Assessment (LCA) for all the smartphones considered, are calculated to understand their environmental impact across these phases which enables the identification and prioritization of areas of potential improvement. The research adopts a value chain approach, also highlighting the gaps in the existing method of reporting. The study assesses the sustainability potential of smartphones and presents approaches for sustainability enhancements. It emphasizes that businesses should have a vision of incorporating sustainability in each primary activity in the value creation process.

**Keywords:** Sustainable Business Models · Smartphone Value Chain · Smartphone Industry · Sustainable Development · Ethical Sourcing · Customer Awareness

## 1 Introduction

The electronics and especially the smartphone industry is reported to affect the environment because of rapid product obsolescence, the large-scale consumption of energy, and electronic waste production. The United Nations University indicates that about 53.6 million metric tons of e-waste are generated annually by the global electronics industry, with only 17.4% being properly recycled [1]. As the demand for smartphones continues to grow, it is essential to develop sustainable business models that minimize environmental concerns. In response to this, companies in the smartphone industry are increasingly adopting sustainable practices [2]. Our study explores sustainability in smartphones and

H. Kohl et al. (Eds.): GCSM 2024, LNME, pp. 873–880, 2025.
https://doi.org/10.1007/978-3-031-93891-7_96

presents approaches through smartphone value chain to aid the sustainable development of smartphones' business model. It aims to provide a foundation for further research and industry initiatives.

## 2  Conceptual Framework and Methodology

### 2.1  Conceptual Framework

Traditionally, Research and Development (R&D) is considered a support activity [3]. However, in the smartphone industry, innovation is a critical driver of competitive advantage. Even global organizations like World Intellectual Property Organization, a specialization agency of United Nations, recognize R&D as a part of smartphone global value chains [4], reflecting the industry's fast-paced, and technology-driven nature, where R&D is a primary driver of value creation.

Extending the functional life of a smartphone is significant in mitigating e-waste generation as it will cut down the volume of the waste streams that need to be recycled and ensure that maximum materials are conserved in use. Furthermore, lower replacement rates mean less energy demand set by the recycling processes themselves. Therefore, extending the life of smartphones should be a priority, as it strongly aligns with the circular economy principles of resource efficiency [5].

Sourcing materials ethically means ensuring that there are no conflict minerals in the supply chain. Mining minerals like tin, tantalum tungsten and gold (3TG) have been linked to human rights abuses, environmental degradation, and conflict financing [6]. Thus, ethical sourcing is an important factor for sustainability in smartphone businesses. Despite several initiatives like Responsible Minerals Initiative (RMI), there are still challenges in necessary implementation of ethical sourcing practices [7].

Sustainable Business Models (SBMs) are "business models that create sustainable value for all stakeholders by integrating economic, social and environmental considerations" [8], and place greater emphasis on long-term sustainability over short-term gains [9]. Despite the growing interest in SBMs, there is a lack of research on their application in the smartphone industry [10]. This research gap highlights the need for further investigation into the development and implementation of SBMs in this sector.

## 3  Methodology

Our research offers an evaluation of the industry's progress towards sustainability in response to the growing trend of smartphone manufacturers claiming to incorporate sustainable practices into their value chains, by analyzing the performance of three major smartphone models- iPhone 15, Samsung Galaxy S24, and Google Pixel 9. This assessment is divided into two parts:

(a) The comparative analysis in the table below highlights the current state of sustainability progress among the selected smartphones (Table 1).

(b) Figure 1 pinpoints which phases of the value chain contribute most to carbon emissions, helping us identify priority areas for intervention. The carbon emissions for each phase of the product life cycle (inbound logistics and operations, product use, outbound logistics, and end-of-life and disposal) were calculated by applying the phase's percentage contribution to the total carbon footprint of each smartphone. These values were then added together across all three smartphones to obtain the total emissions for each phase. Finally, the percentage of total emissions attributed to each phase was calculated and visualized in a pie chart to show the distribution of emissions across the different life cycle stages.

**Table 1.** Comparative Sustainability Analysis

| Sustainability metrics | Apple iPhone 15 | Google Pixel 9 | Samsung Galaxy S24* |
|---|---|---|---|
| Environmental impact - carbon footprint | 56 kg CO2e [11] | 64 kg CO2e [12] | 50.3 kg CO2e [13] |
| Percentage of recycled material | 23% (by weight) [11] | 20% (by weight) [12] | Total percentage by weight not specified [13] |
| Circularity initiatives | • Material recovery lab • Apple Trade-in for takeback • Partnerships for local recycling [14] | Google Recycling Service [15] | • Local Recycling Services and waste product collection system • Circular economy lab [16] |
| Conflict materials and transparency | • List of suppliers • List of smelters, and refiners with 100% third-party auditing of smelters and refiners [17] | List of smelters [18] | • List of suppliers • List of smelters, and refiners with system-based risk management for suppliers [19] |
| Reparability - iFixit score | 4/10 [20] | – | – |

*Our preliminary analysis indicated that the storage capacity greatly affects a smartphone's carbon footprint. To ensure consistency, the models were standardized at 128 GB, except for Samsung Galaxy S24, where storage capacity was not specified in their carbon emission data.

Building on these insights, our contribution explores approaches to enhance sustainability in each value chain element, focusing on addressing both the gaps highlighted in the table Comparative Sustainability Analysis and the high-impact areas shown in the pie chart (as illustrated in Fig. 1).

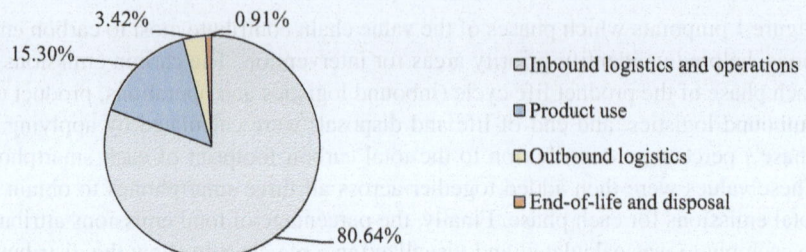

**Fig. 1.** Cumulative carbon footprint shares of each LCA phase of iPhone 15, Pixel9, Galaxy S24

## 4   Approaches for Sustainability in the Smartphone Value Chain

**Value Chain Perspective:**  The current use of Life Cycle Assessment (LCA) by smartphone companies to report environmental impacts, as assessed in this study, presents several challenges which are discussed in detail later in the research. They underscore the importance of adopting a value chain perspective.

The objective of this approach is that businesses should incorporate sustainability into every primary activity in the value creation process. It enables a comprehensive examination of the primary activities in the value chain and can help smartphone companies tackle sustainability issues at each stage thereby enabling targeted actions that collectively build a more sustainable business model for smartphones (Fig. 2).

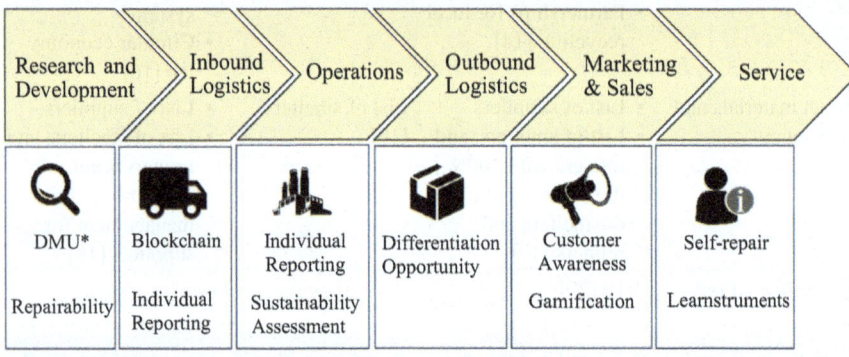

*DMU: Design for Modular Upgradability

**Fig. 2.** Overview of approaches for sustainability in the smartphone value chain

**Research and Development (R&D):**  The focus of the companies in the direction of sustainability, during the research and development phase for the selected smartphones, appears to be on increasing the use of recycled materials and reducing carbon emissions, as indicated in the table. Furthermore, recent smartphone developments, while sophisticated, suggest a phase of refinement rather than groundbreaking innovation, signalling a plateau in technological advancement [21]. Additionally, while recently popular foldable phones introduce form-factor novelty, they do little to extend the product use-phase, address electronic waste and longevity issues.

The research proposes a 'Design for Modular Upgradability' (DMU) framework, which addresses these challenges by contending that key components of a smartphone should be modularized to enable systems that evolve and adapt throughout their lifecycle. This concept shares similarities with the foundational idea of modularity introduced by initiatives like Phonebloks and inspired efforts such as Google's Project Ara which was further discontinued, with no official explanation from the company [22]. Compared to foldable phones, DMU can provide a more sustainable and impactful innovation. It can allow companies to differentiate themselves in a mature market, offering personalization, upgradability, and environmental benefits, meeting the growing demand for long-term, responsible technology. Scalability and feasibility of DMU implementation have to be tested not only in the smartphone industry but beyond.

The DMU framework addresses the performance issues that potentially limited the previous modular projects by involving the following steps: (1) Designing and selling upgraded interchangeable hardware modules, focusing on targeted modularization for components like displays, batteries, cameras, and processors, reducing the bulk and ensuring a user-friendly design; (2) Integrating the modular software updates and compatibility to ensure it keeps pace with hardware upgrades; (3) Designing of modular architecture by adopting open standards to create a broader ecosystem, encouraging third-party manufacturers to contribute in contrast to a closed system to promote interoperability, ensuring seamless integration of new hardware modules.

Repairability of the device contributes to making a smartphone more sustainable by increasing its longevity. The absence of repairability metrics by manufacturers during smartphone releases suggests that device repairability is not a priority in the design and development phases. Additionally, the table of Comparative Sustainability Analysis shows incomplete coverage of repairability scores by independent websites, such as iFixit. This information gap can restrict the market for repairable devices, hindering the growth of environmentally conscious consumerism. Therefore, there is an opportunity for research to formulate a universal Key Performance Indicator (KPI) for smartphone repairability. Such a KPI should be the main emphasis in the design and development phase of a smartphone.

**Inbound Logistics:** Smartphone companies have been publishing detailed list of its smelters, refiners, and suppliers to curb unethical sourcing. However, the complexity and opacity of global supply chains still limit full transparency and traceability, making it challenging for consumers and stakeholders to hold companies accountable. To address that, future research should investigate and implement blockchain technology in their supply chains as it has potential to enhance supply chain transparency through improved traceability, and allow secure recording of each transaction within the supply chain, thereby increasing transparency and accountability [23].

Inbound logistics and operations have a high impact on the overall emissions of the chosen smartphones (see Fig. 1), accounting for over 80% of total carbon footprint shares. Identifying areas for improvement can be challenging because, in LCA, the sourcing and production aspects are calculated and reported together. The value chain approach highlights that, to identify specific areas for targeted environmental action, these elements should also be reported individually, as it emphasizes the need to address sustainability for each element of the value creation process.

**Operations:** As discussed above, the lack of individual data on smartphone operations' environmental impact, due to combined reporting, remains a challenge. Apart from the reporting of this element with the value chain approach, the research about detailed sustainability assessment for the manufacturing processes in the smartphone industry can allow opportunities for optimization measures.

**Outbound Logistics:** Our research found that companies had no clear plan for improving this element of the value chain. Although outbound logistics account for a relatively small portion of the overall carbon footprint at 5.81%, as illustrated in Fig. 1. Optimizing this area could play a pivotal role in achieving long-term sustainability goals. By exploring options such as reducing transportation emissions, smartphone companies can not only minimize their environmental impact but also differentiate themselves in a competitive market through sustainable practices.

**Marketing and Sales:** In today's market situation, customer awareness can play a crucial role in the adoption and hence the success of SBMs in the smartphone industry, since customers' level of understanding is highly dependent on it. One of the emphases of companies in the marketing and sales activities should therefore, be on customer awareness. The viability and profitability of a sustainable smartphone business model might not pay off well without a transformational change made in consumer perception and behaviour related to smartphone usage and its lifecycle.

An innovative approach could include gamification techniques that intend to increase customer engagement and education via interactive and effective learning processes. For example, (a) "Eco-IQ Quiz": a quiz that could challenge people to test what they know about sustainability issues, and practices. (b) "Sustainability Sorting Game": a game that would ask the users to sort phone components into recyclable and non-recyclable. (c) "Smartphone Disassembly Sprints": a timed game which would increase the awareness of smartphone disassembly and repairability. These initiatives would test the knowledge on sustainable manufacturing, circular economy, and topics of ethical supply chain driving demand for sustainable products. In this regard, future study should investigate into new methodologies and tools of customer awareness.

**Service:** In this element of the value chain, take back schemes are employed with respect to circular economy, as shown in the table. Service in sustainability of smartphone business models should go beyond these programs and extend the 'product use' phase, as it has the highest environmental impact after 'inbound logistics and operations' as presented in the Fig. 1. The enterprises should ensure that the original components of the phone are easily available in the market for repair and information on repair is available. Since Augmented Reality-Visual Reality (AR-VR) is gaining increased momentum, companies can incorporate an AR-VR integration of Learnstruments [24], that would allow consumers to self-repair their smartphones, extending its use phase. Advanced exploration should be done on ways such as using AR-VR for self-repair, education, and gamification for engaging consumers.

# 5 Conclusion

This study comprehensively explored the potential for sustainable approaches of smartphone business models from a value chain perspective. It signified the importance of modular design, such as the DMU framework, to extend the longevity of devices and reduce electronic waste. This research has underlined the combined reporting of important metrics and to address that, it presents approaches of how accountable supply chains can be developed with the help of blockchain technology. It also emphasized the need for an in-depth sustainability assessment of manufacturing processes, and highlighted innovative strategies to increase consumer awareness as well as provision of novel ways of facilitating self-repair by integration of AR-VR. It is through such adoption of sustainable practices that smartphone firms can actively contribute to the advancement in sustainability and distinctively position themselves in the competitive market, further fuelling growth in responsible technology demand.

Further innovation and research are needed to lead this industry towards a sustainable and responsible future. This study provides a solid foundation in understanding and developing the sustainable business model in the smartphone industry but also opens doors to a number of avenues for future research.

**Acknowledgement.** We would like to extend our gratitude to the members of our project team- Jay Sanjeev Marwaha, Pramath Prabhakar Nadig, Vivek Kumar Mishra, and Tanisha Deepak Narnaware for their dedication, hard work, and insightful discussions, which greatly contributed to the success of this study.

# References

1. United Nations University, Global E-Waste Surging: Up 21% in 5 Years. https://unu.edu/press-release/global-e-waste-surging-21-5-years. Accessed 24 Aug 2024
2. Global System for Mobile Communications Association GSMA Mobile Industry Impact Report 2023. https://sdgreport2023.gsma.com/. Accessed 24 Aug 2024
3. Porter, M., Book: The Competitive Advantage: Creating and Sustaining Superior Performance, Harvard (1985)
4. World Intellectual Property Report: Intangible Capital in Global Value Chains. https://www.wipo.int/edocs/pubdocs/en/wipo_pub_944_2017-chapter4.pdf. Accessed 24 Aug 2024
5. World Economic Forum: Repairing – not recycling – is the first step to tackling e-waste from smartphones. https://www.weforum.org/agenda/2021/07/repair-not-recycle-tackle-ewaste-circular-economy-smartphones/. Accessed 24 Aug 2024
6. OECD, Costs and Value of Due Diligence in Mineral Supply Chains - OECD Position (2021). https://mneguidelines.oecd.org/costs-and-value-of-due-diligence-in-mineral-supply-chains.pdf. Accessed 24 Aug 2024
7. Barbosa, H., Guido, V., Lezak, S., Natali, P.: Supply chain traceability: looking beyond greenhouse gases, RMI (2022). https://rmi.org/insight/supply-chain-traceability-beyond-greenhouse-gases/. Accessed 24 Aug 2024
8. Stubbs, W., Cocklin, C.: Conceptualizing a "sustainability business model." Organ. Environ. **21**(2), 103–127 (2008)
9. Lozano, R.: Sustainable business models: providing a more holistic perspective. Bus. Strat. Env. **27**, 1159–1166 (2018)

10. Adams, R., Jeanrenaud, S., John Bessant, I., Denyer, D., Overy, P.: Sustainability-oriented innovation: a systematic review. Int. J. Manag. Rev. **18**, 180–205 (2016)
11. Apple, iPhone 15 Impact Report. https://www.apple.com/environment/pdf/products/iphone/iPhone_15_and_iPhone_15_Plus_PER_Sept2023.pdf. Accessed 24 Aug 2024
12. Google, Google Pixel 9 Impact Report. https://www.gstatic.com/gumdrop/sustainability/pixel-9-product-environmental-report.pdf. Accessed 24 Aug 2024
13. Samsung, Samsung Galaxy S24 Impact Report. https://www.samsung.com/global/sustainability/focus/products/mobile/. Accessed 24 Aug 2024
14. Apple, Environmental Progress report 2024. https://www.apple.com/ca/environment/pdf/Apple_Environmental_Progress_Report_2024.pdf. Accessed 24 Aug 2024
15. Google, Circular Economy. https://sustainability.google/operating-sustainably/circular-economy/. Accessed 24 Aug 2024
16. Samsung, Circular Economy. https://www.samsung.com/global/sustainability/planet/circular-economy/. Accessed 24 Aug 2024
17. Apple, Supply Chain Progress Report. https://s203.q4cdn.com/367071867/files/doc_downloads/2024/04/Apple-Supply-Chain-2024-Progress-Report.pdf. Accessed 24 Aug 2024
18. Google, 2023 Appendix I: Smelter List. https://abc.xyz/assets/a2/9e/476d773242d383f4461af1a96178/2023-appendix-i-smelter-list.pdf. Accessed 24 Aug 2024
19. Samsung, Sustainable Supply Chain. https://www.samsung.com/us/sustainability/sustainable-supply-chain/. Accessed 24 Aug 2024
20. Smartphone repairability scores. https://www.ifixit.com/repairability/smartphone-repairability-scores#smartphone-scores. Accessed 24 Aug 2024
21. Moss, S., From "brick" to smartphone: the evolution of the mobile phone. MRS Bull. **46**, 287–288 (2021)
22. Phonebloks: a global campaign to bring to life a modular phone to reduce e-waste at scale. https://www.onearmy.earth/project/phonebloks. Accessed 24 Aug 2024
23. Wannenwetsch, K., Ostermann, I., Priel, R., Gerschner, F., Theissler, A.: Blockchain for supply chain management: a literature review and open challenges. Procedia Comput. Sci. **225**, 1312–1321 (2023)
24. Menn, P., Muschard, B., Schumacher, B., Sieckmann, F., Kohl, H., Seliger, G.: Learnstruments: learning-conducive artefacts to foster learning productivity in production engineering. CIRP Ann. **67**(1), Berlin (2018)

# Transforming Computer Integrated Manufacturing Facility to Use as a Digital Twin Enabled Learning Factory

E. M. H. K. Kumarasinghe[1] ⓘ, N. N. Liyanawaduge[1] ⓘ, A. K. Kulatunga[2](✉) ⓘ,
A. M. N. D. B. Seneviratne[3] ⓘ, and M. Darmawardana[1]

[1] University of Peradeniya, Kandy, Sri Lanka
{e18186,e18200,mahad}@eng.pdn.ac.lk
[2] University of Exeter, Exeter, UK
a.k.kulatunga@exeter.ac.uk
[3] Tecnalia Research and Innovation, Derio, Spain
dammika.seneviratne@tecnalia.com

**Abstract.** Digital Twin is currently getting wider momentum in many engineering disciplines. Manufacturing industry is one of the main sectors which could gain the benefit of DT through taking decisions real time or in advance before physical systems will be tested which will save many resources and production time thereby to promote sustainable manufacturing. Therefore, empowering future manufacturing and industrial engineering professionals with DT enabled manufacturing facilities is greatly beneficial. This paper presents a transformation of CIM facility as a learning factory to educate and to use as a research facility of DT enabled manufacturing system. More than 25 years old Mitsubishi CIM facility (RHM2, PNC300 and RVM2) is retrofitted incorporating the different sensory platforms to monitor real time activities of the Industrial robot arm, product quality inspection facility, mini-CNC center operations and conveyer system integrated through Siemens (S7–1200) PLC and software installations. The Digital platform is developed by Unity and DT is enabled by RS232 and TCP/IP communication protocols and RHM2, Mini-CNC and conveyor belt hardware software integration. By establishing DT facility through retrofitting the aged CIM system, enables DT enabled Learning Factory concept to be used for education and research with very lower investment and without moving into acquire new hardware's (robot arms, conveyers, CNC teaching Units) which itself enhance the sustainability in Manufacturing engineering education.

**Keywords:** Computer Integrated manufacturing · Digital Twin · Sustainable Manufacturing · Learning Factory · Industry 4.0

## 1 Introduction

Digital Twin concept has created many opportunities for wider research communities and knowledge domains capitalize and extend some of the impossible or difficult to implement in the past to be realistic. DT has getting momentum specially with the development

H. Kohl et al. (Eds.): GCSM 2024, LNME, pp. 881–888, 2025.
https://doi.org/10.1007/978-3-031-93891-7_97

of IoT and Industry 4.0. Manufacturing industry is one of the sectors which is reaping benefits in multiple avenues. Since DT enables decisions to be taken real time or in advance using the digital model and some instances before the specific scenario is run through physical system it paves the way to save many resources including production time etc. leading to promote sustainable manufacturing. The benefits from the digital twins in manufacturing could be in-loop planning and validation, production scheduling assurance, enhanced understanding of manufacturing elements, dynamic risk management, part or assembly traceability, and process traceability (Galar & Kumar,2024).

Through the global manufacturing sector has gradually shifted from developed countries to developing countries, there is a gap when it comes to accessibility to latest technologies at the tertiary level manufacturing engineering programmes due to lack of affordability to invest on them. However, industries rapidly adapt latest technologies such as DT enabled systems to move with the trends and to complete in the global business. This creates a gap in knowledge, skill sets of manufacturing engineers when then join the industry. Therefore, empowering future manufacturing and industrial engineering professionals with DT enabled manufacturing facilities is paramount to the sustenance of the manufacturing sector of those countries. This is where learning factory concept could help to bridge the gap.

On the other hand, Learning Factory (LF) concept is not new, where there is a considerable work done in various times with demonstration equipment, simulation software for manufacturing engineering education. A Learning Factory is an environment to support a practice-based engineering curriculum with the possibility of learning the necessary tools and methods, using real life and didactical equipment [6]".

There are many research directions in learning factories around the globe including industry 4.0, machine and system design, and lean manufacturing. Few studies include Gossmann & Nyhuis, (2012) to demonstrate the changeability aspects in manufacturing, Warnecke & Huser, (1995), Bauer et.al., (2018) to teach Lean production, Wagner et.al., (2014) to demonstrate the Product family design for changeability, Rentzos et.al., (2014) on introducing industrial practice using teaching factory paradigm: A construction equipment application Ahamed et.al., (2018) to train reconfigurable assembly process related value stream mapping.

Somaskanthar-Iyer et.al., (2022), has attempted to convert an existing Computer Integrated Manufacturing (CIM) facility of a university to learning factory environment to teach factories of future concept with the adaptation of IoT platform to monitor and control existing CIM System through real time monitoring of hardware via video streaming and sensor network and control the system via web base interface. The main objective of this research work was to demonstrate the controllability of different set of hardware's through web base interface and to experience through a real-time video capturing to realize whether online execution of commands being correctly works etc. Furthermore, Liyanawaduge et.al., (2023) have establish digital Twining of the conveyer system of the same CIM system used by Somaskanthar-Iyer et.al., (2022). In addition to facilitating the DT, a VR platform has been integrated with the digital model of the conveyor system, enabling control of the physical system through a VR setup.

The study proposed in this research basically extension of the previous works of Somaskanthar-Iyer et.al., (2022), and Liyanawaduge et.al., (2023) considered their limitation and to enhance the performance levels with respect to Digital Twining and Learning Factory Concepts. Basically, the present study intends to reestablish the integrated control of different hardware elements of the CIM system initially followed by integration of CIM system with other manufacturing facilities available in the lab environment and establish Digital Twining between Digital model and hardware. Finally, integrated manufacturing lab facility available at Department of Manufacturing & Industrial Engineering (DMIE) will be used as the learning factory with the integration along with establishment of connectivity with digital platform to VR, which will enable learners to get interact both digital platform and hardware platforms through VR enabled environment.

## 2   Methodology

The methodology consists of several phases. In the Phase 1, troubleshooting of Computer Integrated Manufacturing (CIM) system was carried out to decide what type of repair work to be carried out before converting whole facility into work. At that time this system was not fully functional, because this is almost 25 years old system (Fig. 1). For that Initially individual elements (robot arms, mini-CNC, Conveyer, Automated Storage and Retrieval, Quality inspection mechanism) were tested separately to identify the functionality in an isolated manner using the operations and maintenance manuals provided by Original Equipment Manufacturers. In the Phase 2, identified repairs were carried out in each of the elements which includes robot arm, mini-CNC and conveyer system.

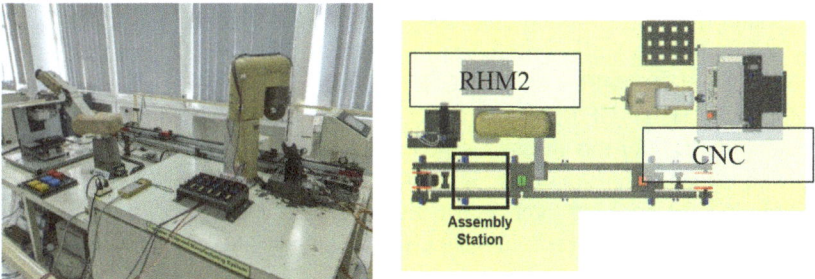

**Fig. 1.** CIM Setup at DMIE (Photo) Plan view

Once these repairs were completed with multiple cycles by replacing requires spare parts, installation of software drivers etc. integrated operations aspects was tested. In the phase 3, reintegration of individual entities (robot arm, CNC, conveyer quality inspection & sorting mechanism) were reestablished through the installation of new SIEMENA S7–1200 PLC along with existing PLC because the existing PLC did not have the necessary communication blocks to communicate with the database (some of the original functionalities were in operational with original PLC of the CIM System).

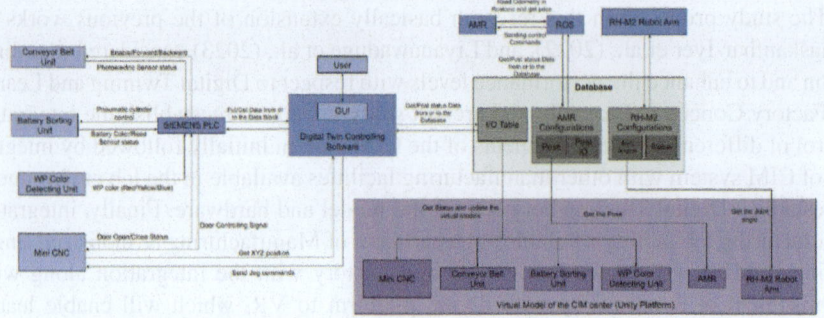

**Fig. 2.** Information flow of the CIM system & DMIE Lab facilities

A SQL database was created to enable the communication between the entities in CIM system and other connected elements. The RHM2 robot was using RS232 protocol to communicate with the CIM system PC and using a python script the ROBOT was controlled through the Digital Model and the IMU sensor readings were taken using the same python script and the database was updated. The mini-CNC's data was accessed using direct memory access method through its software.

In the phase 4, digital platform of the CIM system was developed by connecting sql server database to Unity software to establish the system visualization and Digital Twining capabilities subsequently. Prior to establish the DT capabilities between the hardware and digital platform, the interaction between the entities of CIM were mapped (Fig. 2). This is followed by mapping the information flow between each entity through controllers in the phase 5. The overall DT was established between CIM hardware and digital platform by adding other elements to the previously established BT between the conveyer system of the CIM Liyanawaduge et.al., (2023) by tapping the appropriate status updates through PLCs and other sensors of the CIM hardware and linking those signals to the digital platform to update the status of the physical systems real-time. The Conveyor belt, RHM2 robot, Mini-CNC hardware and supporting accessories were used along with HTTP-GET and HTTP-POST communication methods were used for establishing the DT for this instance. The communication of the different modules of the developed manufacturing system is shown in Fig. 2.

To educate the concepts such as scheduling material handling and supply chain through factory learning concept, automated guided vehicle (AGVs)/autonomous mobile robot (AMR), and other lab facilities and machines of DMIE were integrated into the developed system. A 3D printer was introduced as a process machine as well as a spare part supplier where it prints plastic cylinder pieces. A CNC turning centre was introduced to machine the cylinder workpieces from aluminum. A Coordinate Measuring Machine (CMM) was used to check the measurements of each cylinder pieces as a quality control unit. Figure 4 (bottom right side) shows the plan view of lab facilities and machine tools that is integrated to the CIM system forming enhanced learning factory.

Once the data connectivity and databases were established and tested, the model was expanded to accommodate the manufacturing lab which consist of several industrial scale CNC machine tool and AMR along with digitally twinned CIM system to operate as one single system in phase 6. AMR was introduced to the system for material handling process between the CIM system, CMM machine in metrology lab and 3D printer in Innovation lab. By digital twinning the AMR, can monitor and control the material handling process. Data connectivity was done by accessing to a ROSTOPIC called Odometry that running inside the ROS platform of the AMR. Python script was developed to update the pose inside the database. To get the pose to the UNITY platform C# script was developed. Finally, in the phase 7, the routing aspects were established to operate CIM system, AMR and the CNC turning centre of DMIE. Subsequently DT visualization facilities were customized through user friendly WUI to access the integrated facility in the digital platform and to use it to train undergraduates and self-learning purposes as a learning factory for multiple education goals to teach integrated manufacturing, manufacturing systems, Industry 4.0, DT in manufacturing and to operate as a Flexible Manufacturing System (FMS) for Manufacturing & Industrial Engineering undergraduates of DMIE.

The VR facilities was connected to the DMIE Lab digital model through Vive Cosmos Elite VR hardware. This connectivity enables to explore and interact with the virtual model of the manufacturing system and to control selected hardware facilities too. Steam VR plugin enables to interact with the virtual model that develop inside the UNITY platform and enhance the virtual experience. Figure 3 (left) shows the VR implementation of the DMIE VR lab facility.

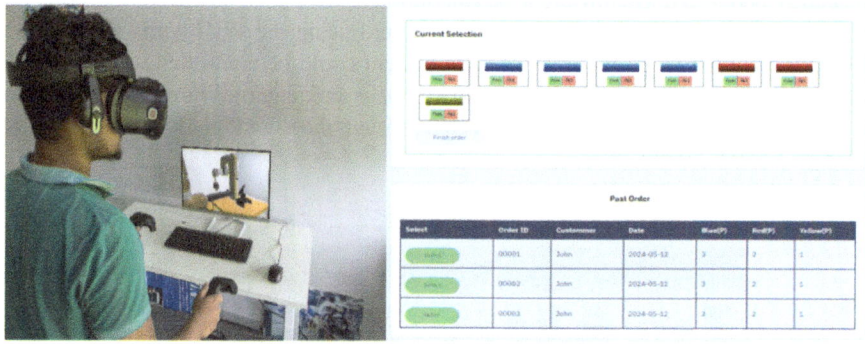

**Fig. 3.** VR implementation of Lab facilities

## 3 Results and Discussion

This section of the paper presents the results obtained in the conversion of CIM system during the digital transformation and digital twin enable learning factory environment. A discussion on obtained results will also be presented in the final paragraph. Integration of other additional hardware to CIM system, development of digital platform, DMIE lab layout enhancement, establishing the digital twining capabilities, and demonstrating routing aspects of AMR are the main outcome of this research project. Furthermore, presentation of GUIs used to control the complete system, different routing within the developed system where operators could use to demonstrate the manufacturing system and its functionalists as a learning factory setup. Finally, communication delays between the digital model and hardware modules have been presented to visualize the performance levels to understand realistic data integration for DT in the retrofitted CIM system.

The Fig. 3 (Right side) show the web user interface (WUI) that has been developed for customers to place orders and operators to see the information about the orders. It also provides the metrology Lab with an interactive interface to mark the workpieces as pass or fail according to their measurements. The WUI was developed using HTML, CSS and JavaScript in front-end while the back end runs by PHP. The WUI has been designed but the full functionality has not yet been implemented. The integrated controlling of different hardware of the DMIE is demonstrated in Fig. 4. The activities planned based on the inputs of WUI is given in the flow chart of Fig. 4 (left side) and the time line of execution of each task (Fig. 4 upper right side) and respective material movement via AMR is shown in Fig. 4 (Right Lower).

In the prevented study following delays were observed after preparing the network communications. The network bandwidth, Latency and packet loss was measured for Local Area Network Arrangement and Wide Area Network Arrangement to check the feasibility of the system through different networks where there were no significant data loss. Since the bandwidth requirement is lower as 5–20 kbps, this can be implemented even with 3G network.

The main interaction with the DT enabled Learning factory is facilitated through VR facility and WUI. The different production scenarios such as different production volumes and process stages etc. can be configured using the WUI while some of the interventions to hardware facilities of CIM system could be controlled using the Vive Cosmos Elite VR hardware. The monitoring of tasks and materials movements between the facilities through AMR could be detected through the digital platform of the DMIE Lab (Fig. 4 bottom-right side).

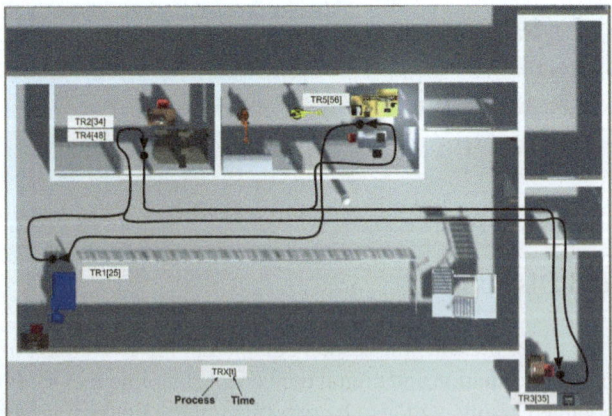

**Fig. 4.** Integrated operation of CIM system and manufacturing facilities

## 4   Conclusion and Future Directions

Learning Factory is a promising method to aware of manufacturing processes and functionalists in undergraduate level. This also helps for training, modeling and research activities, when it comes to industrial level. The developed unite is an invaluable recourse for practice-oriented learning and to understand and experience the advance technology usage in the context of industry 4.0 The cost-effective conversion of old CIM system gave experience to the research students, how an existing facility could transform into a manufacturing facility in digital era of the context of Digital Transformation. Virtual commissioning is one of the industrial used practices in setting up and operating a production plant before it is physically commissioning. There are built in software providers for this purpose such as Siemens virtual commissioning etc. However, there are lack of provisions to convert an old physical manufacturing unit and provide a digital visualization, specifically for the learning factory for engineering education, which was achieved from this research. Further extensions include integration of several other manufacturing facilities such as Injection molding machines, laser cutter machines and introduction of AMRs to the same environment to work as an integrated facility and to develop different scenarios to demonstrate the FMS capabilities within learning factory environment.

Learning factories can significantly enhance sustainability by integrating digital twin (DT) technology into educational environments, allowing real-time decision-making and resource optimization. By transforming existing Computer Integrated Manufacturing (CIM) facilities into DT-enabled learning factories, educational institutions can provide hands-on experience with minimal investment in new hardware, reducing waste and extending the life cycle of existing equipment. This promotes sustainable practices by training future engineers in resource-efficient methods, enhancing their understanding of modern manufacturing technologies, and closing the gap between academia and industry.

**Acknowledgments.** Authors acknowledge the funding support provided by the Autonomous Scholarship Programme of DMIE, technical advice provided by Prof. Gamini Dissanayake and Professor Sanath Ranatunga Memorial Fund for setting up VR facilities at DMIE.

## References

1. Bauer, H., Brandl, F., Lock, C., Reinhart, G.: Integration of Industrie 4.0 in lean manufacturing learning factories. In: Procedia Manuf. **23**, 147–152 (2018). https://doi.org/10.1016/j.promfg. 2018. ISSN 2351–9789
2. Galar, D., Kumar, U., et al.: Digital twins: definition, implementation and applications. In: Varde, P.V., Kumar, M., Agarwal, M. (eds.) Advances in Risk-Informed Technologies. Risk, Reliability and Safety Engineering. Springer, Singapore (2024). https://doi.org/10.1007/978-981-99-9122-8_7
3. Gossmann, D., Nyhuis, H.P.: Learning factory for changeability. Mech. Aerosp., Ind. Mechatron. Manuf. Eng., **6**, 686–692 (2012)
4. Liyanawaduge, N.N., Kumarasinghe, E.M.H.K., Iyer, S.S., Kulatunga, A.K., Lakmal, G.: Digital twin & virtual reality enabled conveyor system to promote learning factory concept. In: 17th IEEE International Conference on Industrial and Information Systems, ICIIS 2023, Peradeniya, Sri Lanka, August 25–26 2023. IEEE (2023). ISBN 979–8–3503–2362–7, 85–90,

5. Ahmad, R., Masse, C., Jituri, S., Doucette, J., Mertiny, P.: Alberta learning factory for training reconfigurable assembly process value stream mapping. In: Procedia Manuf. **23**, 237–242 (2018). https://doi.org/10.1016/j.promfg.2018.04.023. ISSN 2351–9789

6. Rasor, R., Göllner, D., Bernijazov, R., Kaiser, L., Dumitrescu, R.: Towards collaborative life cycle specification of digital twins in manufacturing value chains. In: 28th CIRP Conference on Life Cycle Engineering. Proceedings CIRP, vol. 98, pp. 229–234 (2021)

7. Rentzos, L., Doukas, M., Mavrikios, D., Mourtzis, D., Chryssolouris, G.: Integrating manufacturing education with industrial practice using teaching factory paradigm: a construction equipment application. Procedia CIRP **17**, 189–194 (2014)

8. Somaskantha Iyer, S., Dissanayaka, N., Kulatunga, A.K., Dharmawardhana, M.: Conversion of a manufacturing lab as a learning factory to educate factories of the future concept. In: Proceedings of 18th Global Conference on Sustainable Manufacturing, Oct 5–7 2022, Berlin, Germany (2022)

9. Warnecke, H.J., Huser, M.: Lean production. Int. J. Prod. Econ. **41**, 37–43 (1995)

10. Wagner, U., AlGeddawy, T., ElMaraghy, H., Muller, E.: Product family design for changeable learning factories. Procedia CIRP **17**, 195–200 (2014)

# Author Index

© The Editor(s) (if applicable) and The Author(s) 2025
H. Kohl et al. (Eds.): GCSM 2024, LNME, pp. 889–893, 2025.
https://doi.org/10.1007/978-3-031-93891-7

The manufacturer's authorised representative in the EU is Springer
Nature Customer Service Centre GmbH, Europaplatz 3, 69115 Heidelberg,
Germany. If you have any concerns regarding our products, please
contact ProductSafety@springernature.com

Printed and bound by CPI Group (UK) Ltd, Croydon, CR0 4YY

28/04/2026
02098478-0005